Lecture Notes on Data Engineering and Communications Technologies

Volume 46

Series Editor

Fatos Xhafa, Technical University of Catalonia, Barcelona, Spain

The aim of the book series is to present cutting edge engineering approaches to data technologies and communications. It will publish latest advances on the engineering task of building and deploying distributed, scalable and reliable data infrastructures and communication systems.

The series will have a prominent applied focus on data technologies and communications with aim to promote the bridging from fundamental research on data science and networking to data engineering and communications that lead to industry products, business knowledge and standardisation.

** Indexing: The books of this series are submitted to SCOPUS, ISI Proceedings, MetaPress, Springerlink and DBLP **

More information about this series at http://www.springer.com/series/15362

Jennifer S. Raj · Abul Bashar ·
S. R. Jino Ramson
Editors

Innovative Data Communication Technologies and Application

ICIDCA 2019

 Springer

Editors
Jennifer S. Raj
Department of Electronics
and Communication Engineering
Gnanamani College of Technology
Namakkal, India

Abul Bashar
College of Computer Engineering
and Sciences
Prince Mohammad Bin Fahd University
Al-Khobar, Saudi Arabia

S. R. Jino Ramson
Purdue Polytechnic Institute
Purdue University
West Lafayette, IN, USA

ISSN 2367-4512   ISSN 2367-4520   (electronic)
Lecture Notes on Data Engineering and Communications Technologies
ISBN 978-3-030-38039-7   ISBN 978-3-030-38040-3   (eBook)
https://doi.org/10.1007/978-3-030-38040-3

This Springer imprint is published by the registered company Springer Nature Switzerland AG
The registered company address is: Gewerbestrasse 11, 6330 Cham, Switzerland

We are honored to dedicate the proceedings of ICIDCA 2019 to all the participants and editors of ICIDCA 2019.

Foreword

It is with deep satisfaction that I write this Foreword to the Proceedings of the ICIDCA 2019 held in RVS College of Engineering and Technology, Coimbatore, Tamil Nadu, India, on October 17–18, 2019.

This conference was bringing together researchers, academics and professionals from all over the world, and experts in Innovative Data Communication Technologies.

This conference particularly encouraged the interaction of research students and developing academics with the more established academic community in an informal setting to present and to discuss new and current work. The papers contributed the most recent scientific knowledge known in the field of distributed operating systems, middleware, databases, sensor, mesh, and ad hoc networks, quantum- and optics-based distributed algorithms, Internet applications, social networks, and recommendation systems. Their contributions helped to make the conference as outstanding as it has been. The Local Organizing Committee members and their helpers put much effort into ensuring the success of the day-to-day operation of the meeting.

We hope that this program will further stimulate research in theory, design, analysis, implementation, and application of distributed systems and networks. We feel honored and privileged to serve the best recent developments to you through this exciting program.

We thank all authors and participants for their contributions.

V. Gunaraj
Conference Chair

Preface

This Conference Proceedings volume contains the written versions of most of the contributions presented during the conference of ICIDCA 2019. The conference provided a setting for discussing recent developments in a wide variety of topics including dynamic, adaptive, and machine learning distributed algorithms, game-theoretic approaches to distributed computing, security in distributed computing, cryptographic protocols, fault tolerance, reliability, self-organization, self-stabilization, and so on. The conference has been a good opportunity for participants coming from various destinations to present and discuss topics in their respective research areas.

ICIDCA 2019 Conference tends to collect the latest research results and applications on Innovative Data Communication Technologies. It includes a selection of 95 papers from 274 papers submitted to the conference from universities and industries all over the world. All of the accepted papers were subjected to strict peer reviewing by 2–4 expert referees. The papers have been selected for this volume because of quality and relevance to the conference.

ICIDCA 2019 would like to express our sincere appreciation to all authors for their contributions to this book. We would like to extend our thanks to all the referees for their constructive comments on all papers, and especially, we would like to thank to organizing committee for their hard working. Finally, we would like to thank Springer publications for producing this volume.

V. Gunaraj
Conference Chair

Acknowledgments

ICIDCA 2019 would like to acknowledge the excellent work of our conference organizing the committee, keynote speakers for their presentation on October 17–18, 2019. The organizers also wish to acknowledge publicly the valuable services provided by the reviewers.

On behalf of the editors, organizers, authors, and readers of this conference, we wish to thank the keynote speakers and the reviewers for their time, hard work, and dedication to this conference. The organizers wish to acknowledge Dr. V. Gunaraj for the discussion, suggestion, and cooperation to organize the keynote speakers of this conference. The organizers also wish to acknowledge for speakers and participants who attend this conference. Many thanks are given to all persons who help and support this conference. ICIDCA would like to acknowledge the contribution made to the organization by its many volunteers. Members contribute their time, energy, and knowledge at a local, regional, and international level.

We also thank all the Chair Persons and Conference Committee Members for their support.

Contents

A Secure Identity-Based Three-Party Authenticated Key Agreement Protocol Using Bilinear Pairings 1
Daya Sagar Gupta, S K Hafizul Islam, and Mohammad S. Obaidat

Machine Learning Based Stock Market Analysis: A Short Survey 12
Hrishikesh Vachhani, Mohammad S. Obiadat, Arkesh Thakkar,
Vyom Shah, Raj Sojitra, Jitendra Bhatia, and Sudeep Tanwar

Macroscopic Characterization of Grating Coupled Waveguide Structures for Optical Notch Filtering 27
Aleena Devasia and Manisha Chattopadhyay

Design and Development for Image Transmission Through Low Powered Wireless Networks Using Color Space Conversion Module 37
C. S. Sridhar, G. Mahadevan, S. K. Khadar Basha, B. N. Shobha,
and S. Pavan

Classification and Comparison of Cloud Renderfarm Services for Recommender Systems 45
J. Ruby Annette, W. Aisha Banu, and P. Subash Chandran

Smart Medicine Distributing Tray 57
Rohit Sathye, Sumedh Deshprabhu, Mandar Surve, and Deepak C. Karia

Potential Customer Classification in Customer Relationship Management Using Fuzzy Logic 67
Tanay Kulkarni, Purnima Mokadam, Jnanesh Bhat, and Kailas Devadkar

Classification of Sentiment Analysis Using Machine Learning 76
Satyen M. Parikh and Mitali K. Shah

**Gauging Trowel Shaped Patch Including an Optimized Slot
for 60 GHz WLAN**.. 87
Ribhu Abhusan Panda, Baha Hansdah, Sanjana Misra, Rohit Kumar Singh,
and Debasis Mishra

**Reliable Machine to Machine Communication Using MQTT Protocol
and Mean Stack**.. 94
Nalin Chandeliya, Prashanth Chari, Sameeran Karpe, and Deepak C. Karia

**Lane Keep Assist System for an Autonomous Vehicle Using Support
Vector Machine Learning Algorithm**...................................... 101
M. Karthikeyan, S. Sathiamoorthy, and M. Vasudevan

**An Automated System for Crime Investigation Using Conventional
and Machine Learning Approach**... 109
V. S. Felix Enigo

**Adaptive Neuro Fuzzy Inference System Based Obstacle Avoidance
System for Autonomous Vehicle**... 118
M. Karthikeyan, S. Sathiamoorthy, and M. Vasudevan

**Evaluating the Performance of Machine Learning Techniques
for Cancer Detection and Diagnosis**...................................... 127
Anu Maria Sebastian and David Peter

**Design of Slotted Microstrip Patch Antenna
for 5G Communications**.. 134
C. V. Krishna, H. R. Rohit, and P. Shanthi

**Fourier Descriptors Based Hand Gesture Recognition Using
Neural Networks**.. 140
Rajas Nene, Pranay Narain, M. Mani Roja, and Medha Somalwar

**Improved Feature Based Sentiment Analysis for Online
Customer Reviews**... 148
L. Rasikannan, P. Alli, and E. Ramanujam

**Performance Analysis of Automated Detection
of Diabetic Retinopathy Using Machine Learning
and Deep Learning Techniques**... 156
Nimisha Raichel Varghese and Neethu Radha Gopan

Mobility in Wireless Sensor Networks................................. 165
D. Rajendra Prasad, B. Kiran Kumar, and S. Indraneel

**Distributed Auction Mechanism for Dynamic Spectrum Allocation
in Cognitive Radio Networks**.. 172
D. Sumithra Sofia and A. Shirly Edward

Efficient Abandoned Luggage Detection in Complex Surveillance Videos 181
Divya Raju and K. G. Preetha

Energy Reduction Stratagem in Smart Homes Using Association Rule Mining 188
Karthika Velu, Pramila Arulanthu, and Eswaran Perumal

Analysis of Growth and Planning of Urbanization and Correlated Changes in Natural Resources 194
Supriya Kamoji, Dipali Koshti, and Ruth Peter

Design of Near Optimal Convolutional Neural Network Based Crowd Density Classifier 204
Komal R. Ahuja and Nadir N. Charniya

Face Swapping Using Modified Dlib Image Morphology 213
Anjum Rohra and Ramesh K. Kulkarni

Regular Language Search over Encrypted Cloud Data Using Token Generation: A Review Paper 222
Pranoti Annaso Tamgave, Amrita A. Manjrekar, and Shubhangi Khot

Modeling and Evaluation of Performance Characteristics of Redundant Parallel Planar Manipulator 229
Shaik Himam Saheb and G. Satish Babu

Public Auditing Scheme for Cloud Data Integrity Verification 237
Shivarajkumar Hiremath and R. Sanjeev Kunte

Implementation of Words and Characters Segmentation of Gujarati Script Using MATLAB 247
B. Shah Gargi and S. Sajja Priti

Total Interpretive Structural Modelling: Evolution and Applications 257
Shalini Menon and M. Suresh

Machine Learning Techniques: A Survey 266
Herleen Kour and Naveen Gondhi

Lexicon-Based Text Analysis for Twitter and Quora 276
Potnuru Sai Nishant, Bhaskaruni Gopesh Krishna Mohan,
Balina Surya Chandra, Yangalasetty Lokesh, Gantakora Devaraju,
and Madamala Revanth

Subspace Based Face Recognition: A Literature Survey 284
Bhaskar Belavadi, K. V. Mahendra Prashanth, M. S. Raghu,
and L. R. Poojashree

Medical Image Fusion Using Otsu's Cluster Based
Thresholding Relation   297
C. Karthikeyan, J. Ramkumar, B. Devendar Rao, and J. Manikandan

Weather Prediction Model Using Random Forest Algorithm
and GIS Data Model   306
S. Dhamodaran, Ch. Krishna Chaitanya Varma,
and Chittepu Dwarakanath Reddy

Quantum One Time Password with Biometrics   312
Mohit Kumar Sharma and Manisha J. Nene

Hopfield Network Based Approximation Engine
for NP Complete Problems   319
T. D. Manjunath, S. Samarth, Nesar Prafulla, and Jyothi S. Nayak

The Evolution of Cloud Computing and Its Contribution
with Big Data Analytics   332
D. Nikhil, B. Dhanalaxmi, and K. Srinivasa Reddy

Robust Methods Using Graph and PCA for Detection of Anomalies
in Medical Records   342
K. N. Mohan Kumar, S. Sampath, and Mohammed Imran

Prediction of Cardiac Ailments in Diabetic Patient Using Ensemble
Learning Model   353
Charu Vaibhav Verma and S. M. Ghosh

Emotion Recognition of Facial Expression Using Convolutional
Neural Network   362
Pradip Kumar, Ankit Kishore, and Raksha Pandey

Information Security Through Encrypted Domain Data Hiding   370
Vikas Kumar, Prateek Muchhal, and V. Thanikasiselvan

Person Re-identification from Videos Using Facial Features   380
Ankit Hendre and Nadir N. Charniya

Analysing Timer Based Opportunistic Routing Using Transition
Cost Matrix   388
Chinmay Gharat and Shoba Krishnan

Survey of Load Balancing Algorithms in Cloud Environment
Using Advanced Proficiency   395
Dharavath Champla and Dhandapani Siva Kumar

Comparative Study on SVD, DCT and Fuzzy Logic of NOAA Satellite
Data to Detect Convective Clouds   404
B. Ravi Kumar and B. Anuradha

Question Classification for Health Care Domain Using Rule Based Approach 410
Shubham Agrawal and Nidhi Mishra

A Conjoint Edifice for QOS and QOE Through Video Transmission at Wireless Multimedia Sensor Networks 420
S. Ramesh, C. Yaashuwanth, and Prathibanandhi

Analysis of Temperature Prediction Using Random Forest and Facebook Prophet Algorithms 432
J. Asha, S. Rishidas, S. SanthoshKumar, and P. Reena

Wireless Monitoring and Control of Deep Mining Environment Using Thingspeak and XBee 440
B. Ramesh and K. Panduranga Vittal

Assistive Device for Neurodegenerative Disease Patients Using IoT 447
Saravanan Chandrasekaran and Rajkumar Veeran

Ablation of Artificial Neural Networks 453
Y. Vishnusai, Tejas R. Kulakarni, and K. Sowmya Nag

Precedency with Round Robin Technique for Loadbalancing in Cloud Computing 461
Aditi Nagar, Neetesh Kumar Gupta, and Upendra Singh

Stock Market Prediction Using Hybrid Approach 476
Sakshi Jain, Neeraj Arya, and Shani Pratap Singh

Energy-Efficient Routing Based Distributed Cluster for Wireless Sensor Network 489
R. Sivaranjani and A. V. Senthil Kumar

Autonomous Home-Security System Using Internet of Things and Machine Learning 498
Aditya Chavan, Sagar Ambilpure, Uzair Chhapra, and Varnesh Gawde

Automatic Skin Disease Detection Using Modified Level Set and Dragonfly Based Neural Network 505
K. Melbin and Y. Jacob Vetha Raj

Service-Oriented Middleware for Service Handling in Cloud-Based IoT Services 516
R. Thenmozhi and K. Kulothungan

Intelligent Educational System for Autistic Children Using Augmented Reality and Machine Learning 524
Mohammad Ahmed Asif, Firas Al Wadhahi, Muhammad Hassan Rehman, Ismail Al Kalban, and Geetha Achuthan

Selective Segmentation of Piecewise Homogeneous Regions 535
B. R. Kapuriya, Debasish Pradhan, and Reena Sharma

**Blockchain Technology in Healthcare Domain:
Applications and Challenges** 543
Chavan Madhuri, Patil Deepali, and Shingane Priyanka

**Comparison on Carrier Frequency Offset Estimation in Multi Band
Orthogonal Frequency Division Multiplexing (OFDM) System** 551
C. Rajanandhini and S. P. K. Babu

**A Reliable Method for Detection of Compromised Controller
in Software Defined Networks** 561
Manaswi Parashar, Amarjeet Poonia, and Kandukuru Satish

**Alphabet Classification of Indian Sign Language
with Deep Learning** 569
Kruti J. Dangarwala and Dilendra Hiran

Impact on Security Using Fusion of Algorithms 577
Dugimpudi Abhishek Reddy, Deepak Yadav, Nishi Yadav,
and Devendra Kumar Singh

Smart Stick for Blind 586
B. A. Sujatha Kumari, N. Rachana Shree, C. Radha,
Sharanya Krishnamurthy, and Saaima Sahar

A Neural Network Based Approach for Operating System 594
Gaurav Jariwala and Harshit Agarwal

Randomness Analysis of YUGAM-128 Using Diehard Test Suite 600
Vaishali Sriram, M. Srikamakshi, K. J. Jegadish Kumar,
and K. K. Nagarajan

**Detecting Spam Emails/SMS Using Naive Bayes, Support Vector
Machine and Random Forest** 608
Vasudha Goswami, Vijay Malviya, and Pratyush Sharma

**Feature Based Opinion Mining on Hotel Reviews
Using Deep Learning** 616
Kavita Lal and Nidhi Mishra

**Using Deep Autoencoders to Improve the Accuracy of Automatic
Playlist Generation** 626
Bhumil Jakheliya, Raj Kothari, Sagar Darji, and Abhijit Joshi

**Evolutionary Correlation Triclustering for 3D Gene
Expression Data** 637
N. Narmadha and R. Rathipriya

Contents

A Study on Legal Knowledge Base Creation Using Artificial Intelligence and Ontology 647
Tanaya Das, Abhishek Roy, and A. K. Majumdar

Survey of Progressive Era of Text Summarization for Indian and Foreign Languages Using Natural Language Processing 654
Apurva D. Dhawale, Sonali B. Kulkarni, and Vaishali Kumbhakarna

Zone Safe Traffic Assist System and Automated Vehicle with Real-Time Tracking and Collision Notification 663
K. K. Aishwariya, Sanil K. Daniel, and K. V. Sujeesh

A Survey on Intrusion Detection System Using Machine Learning Algorithms 670
Shital Gulghane, Vishal Shingate, Shivani Bondgulwar, Gaurav Awari, and Parth Sagar

Automatic Greenhouse Parameters Monitoring and Controlling Using Arduino and Internet of Things 676
More Hemlata Shankarrao and V. R. Pawar

withall: A Shorthand for Nested for Loop + If Statement 684
Tomsy Paul and Sheena Mathew

Smart Cane-An Aid for the Visually Challenged 692
Swarnita Venkatraman, Kirtana Subramanian, Chandrima Tolia, and Ruchita Shanbhag

Readmission Prediction Using Hybrid Logistic Regression 702
V. Diviya Prabha and R. Rathipriya

Modified Region Growing for MRI Brain Image Classification System Using Deep Learning Convolutional Neural Networks 710
A. Jayachandran, J. Andrews, and L. Arokia Jesu Prabhu

Effect of Air Substrate on the Performance of Rectangular Microstrip Patch Antenna for the UHF Spaced Antenna Wind Profiler Radar 718
Jayapal Elluru and S. Varadarajan

A System-Level Performance of CS/CB Downlink CoMP in Small Cell LTE-A Heterogeneous Networks 725
Amandeep and Sanjeev Kumar

Machine Learning: An Aid in Detection of Neurodegenerative Disease Parkinson 733
Jignesh Sisodia and Dhananjay Kalbande

Elliptical Curve Cryptography Based Access Control Solution for IoT Based WSN 742
Renuka Pawar and D. R. Kalbande

Novel Exon Predictors Using Variable Step Size
Adaptive Algorithms 750
Srinivasareddy Putluri and Md. Zia Ur Rahman

Tweets Analysis for Disaster Management: Preparedness, Emergency
Response, Impact, and Recovery 760
Archana Gopnarayan and Sachin Deshpande

CFRF: Cloud Forensic Readiness Framework – A Dependable
Framework for Forensic Readiness in Cloud
Computing Environment 765
Sugandh Bhatia and Jyoteesh Malhotra

Face Recognition Based on Interleaved Neighbour Binary Pattern 776
A. Geetha and Y. JacobVetha Raj

Secure Transmission of Human Vital Signs Using Fog Computing 785
A. Sonya, G. Kavitha, and A. Paramasivam

Internet of Things Based Smart Baby Cradle 793
Vedanta Prusty, Abhisek Rath, Pradyut Kumar Biswal,
and Kshirod Kumar Rout

My Friend – A Personal Assistant for Chauffeur 800
P. H. Jayanth, R. Gowda Nishanth, K. Gunashekar, and Kezia Sera Ninan

Prediction of Type 2 Diabetes Using Hybrid Algorithm 809
Aman Deep Singh, B. Valarmathi, and N. Srinivasa Gupta

Discovering Patterns Using Feature Selection Techniques
and Correlation 824
Mausumi Goswami and B. S. Purkayastha

Author Index.. 833

A Secure Identity-Based Three-Party Authenticated Key Agreement Protocol Using Bilinear Pairings

Daya Sagar Gupta[1(✉)], S K Hafizul Islam[2], and Mohammad S. Obaidat[3,4,5,6]

[1] Department of Computer Science and Engineering,
Shershah College of Engineering Sasaram, Sasaram, Bihar, India
dayasagar.ism@gmail.com
[2] Department of Computer Science and Engineering, Indian Institute of Information
Technology Kalyani, Kalyani, West Bengal, India
hafi786@gmail.com
[3] College of Computing and Informatics, University of Sharjah, Sharjah, UAE
msobaidat@gmail.com
[4] KAIST, University of Jordan, Amman, Jordan
[5] University of Science and Technology, Beijing, Beijing, China
[6] Amity University, Noida, India

Abstract. The modern Internet technology era requires secure key agreement protocols, which are playing a significant role in the field of cryptography and network security. These protocols are mainly designed to establish a common session key between different parties. It can be easily found that various key agreement protocols are designed in the literature. However, many of these protocols are either proven insecure or have a burden of communication and computational cost. Therefore, a more secure key agreement protocol is needed. This paper exhibits an identity-based three-party authenticated key agreement (ID-3PAKA) protocol is devised, which securely and efficiently negotiates a common secret session key among three parties over the Internet. This protocol is based on the elliptic curve cryptography (ECC). It uses the idea of identity-based encryption (IBE) with bilinear pairings. The security of the proposed work is based on the hardness assumption of the discrete logarithm problem (DL) and elliptic curve. Further, we show that the proposed protocol ensures the known security properties of the session key.

Keywords: Key agreement · Elliptic curve · Hash function · Bilinear pairing · Forward secrecy

1 Introduction

The twentieth century grew with rapid development in the area of Internet and Mobile Communications Technologies, called Information and Communication Technology (ICT). ICT services havegrown exponentially and have become

© Springer Nature Switzerland AG 2020
J. S. Raj et al. (Eds.): ICIDCA 2019, LNDECT 46, pp. 1–11, 2020.
https://doi.org/10.1007/978-3-030-38040-3_1

essential to the world in different ways. However, these technologies are changing very frequently, and several services with multimedia applications are growing through various real-life applications. In these regards, the security protection to various services becomes essential and challenging as well, and different security mechanisms for data encryption, authentication, integrity and availability are being developed at a rapid pace. In order to design different security mechanisms and meet the challenges, different cryptographic primitives are used in their convenient ways. The key agreement protocol is one of the significant tools of cryptography in which a secret session key is exchanged among a number of parties. In 1976, Diffie and Hellman [4] were the first who gave a new idea of having two separate keys; one for encipherment and other for decipherment. This protocol gave birth to the key agreement protocol, which is named as Diffie and Hellman (DH) key agreement protocol. Their proposal was used to transfer a common key among two authentic entities. Unfortunately, their protocol is vulnerable to a number of attacks including man-in-the-middle attack. To eliminate these difficulties, research has grown in this direction and many researchers have proposed different key agreement protocols [5, 14, 15]. Many of these protocols are not secure and suffer from different attacks. In this paper, we proposed a three-party authenticated key agreement (ID-3PAKA) protocol which is implemented in the framework of identity-based encryption (IBE) and uses bilinear pairings. The proof of correctness and security of our proposal is also presented to ensure the security of our ID-3PAKA protocol.

Cryptography is an art to use security primitives in a way to deal with the security challenges and meet the solutions. Data encryption is divided into two major categories namely, symmetric/private-key and asymmetric/public-key techniques in which the latter one has greater research impact than the former. However, the useful public-key cryptographic techniques like RSA, ElGamal, etc. have some disadvantage as they require extensive public key management overheads. Thus, a new technique called, IBE is introduced recently and is used by researchers to design efficient cryptographic tools for different security applications [6]. In this article, we formulated the idea of this technique to implement our protocol. Shamir [20] has firstly proposed the novel idea of IBE by choosing the known identity of a party which is supposed as his public-key. This known identity may be email id, phone number, physical address, etc. Using the identities, Shamir [20] removed the overhead of certificate management task from public-key cryptography. In addition, an entity called *private key generator* (PKG) is also considered to generate the identity-based private-key of the party. However, the practical IBE was implemented in 2001 by Boneh and Franklin [1]. Further, our proposed ID-3PAKA protocol using bilinear pairings. A bilinear pairing relates two members of a group to a member of another group. To implement our protocol, a bilinear mapping technique takes two members (points) of an elliptic curve group and maps it to a member of another multiplicative group. However, authentication to our proposed ID-3PAKA protocol is provided by means of the elliptic curve cryptography (ECC). The motivation to use the ECC in our protocol is that it provides better security than RSA, *i.e.*, a

160-bit key based on ECC exhibits the equal level of security which is provided by a 1024-bit key based on RSA.

1.1 Related Works

In 2004, Jeong et al. [16] designed many two-party key agreement protocols, which are executed in one round of communication. They claim that their proposed protocols are authenticated and resist known attacks including session key forward secrecy. In 2005, McCullagh and Barreto [18] devised a key agreement (ID-2PAKA) protocol for two parties using IBE. They showed that their protocol is efficient and secure than other existing state-of-the-art protocols. They also presented a comparative analysis for their protocol. Choo [2] reviewed McCullagh and Barreto's ID-2PAKA protocol and showed that their protocols are vulnerable if the attacker has sent the Reveal query. Hölbl et al. [12] devised an ID-2PAKA protocol using bilinear pairing. For their protocol, they used bilinear pairing and also derived a variant of signature scheme, which confirms the security of their protocol. They also showed that their proposal is comparatively secure and cost efficient than the related protocols. Gupta and Biswas [9] proposed two secure bi-partite key agreement protocols using IBE and bilinear pairing. The first protocol is based on the DH protocol; however the later is proposed for elliptic curve. They also proposed secure three-party key agreement protocols which are the extension of their bi-partite scheme. They showed that their protocols are secure against many attacks and claimed that these protocols exhibit better security and efficiency than other similar protocols.

Tseng [21] proposed an ID-2PAKA protocol whose security uses the hard assumption of discrete logarithm problem (DLP). It was also claimed that the performance of their proposal was better than other competing protocols. This protocol is secure and resistance to many possible attacks. Hölbl et al. [11] devised two ID-3PAKA protocols which uses bilinear pairing. They claimed that their protocols are secure on the defined security parameters. Gupta and Biswas [7] proposed a key transfer protocol for secure cellular communication. They also proved that their protocol exhibits strong security. Chou et al. [3] proposed two identity-based key agreement protocols for a mobile environments. Their first protocol was implemented for two-parties and the second protocol was an extension to the two-party key agreement protocol to three-party protocol. They also showed that their protocols achieve strong security notions. Gupta and Biswas [8] proposed an authenticated key agreement protocols in IBE framework. They also analyzed the performance and the security of protocol formally in the random oracle model. He et al. [10] developed a provably secure ID-3PAKA protocol using the elliptic curve for mobile environments. Islam et al. [13] proposed an improved ID-3PAKA protocol using ECC and hash function. They also proved that their protocol exhibits strong security using the AVISPA software.

2   Technical Details

The definitions and notations used in this paper is considered in this section and explained below:

2.1   Elliptic Curve Cryptography (ECC)

Miller [19] firstly introduced the concept of elliptic curve cryptosystem. Later, Koblitz [17] also proposed a cryptosystem based on the elliptic curve.

A cryptographic elliptic curve of prime order p is denoted by E/F_p and is expressed as the following equation:

$$y^2 \bmod p \equiv (x^3 + ax + b) \bmod p$$

In the above equation, a and b are the elements of F_p and satisfy the relation $(4a^3 + 27b^2) \not\equiv 0 \bmod p$. All the points of elliptic curve forms an additive abelian/commutative group. The *"addition operation"* $(+)$ on elliptic curve points is defined as follows:

Let the two point of an additive group E/F_p be L and T. The addition of points L and T can be calculated as per the following conditions:

– Assume that L and T are different point on the curve having different x and y coordinates and the line joining these two points cuts the curve on another point $-Y$, then the reflection of $-Y$ about x-axis gives $L+T$, i.e., $Y = L + T$. Here, the points Y and $-Y$ are known as additive inverse to each other with respect to x-axis.
– Assume that L and T are equal, i.e., $L = T$. In this case, the line joining these same points will be the tangent to the curve and let it cuts the curve on a point $-Y$, then the image of $-Y$ about x-axis gives the addition $L + T$, i.e., $Y = L + L = 2L$ or $2T$.
– Let two points to be added are inverse to each other i.e., the points on the curve are L and $-L$, thus the line joining these points cuts the elliptic curve at infinity called *"point of infinity"* and denoted by O. Thus the addition is $L + (-L) = O$. The *point of infinity* is the identity of elliptic curve.
– The scalar point multiplication on the elliptic curve can be calculated by adding the same point repeatedly i.e., kL can be calculated by adding L to itself k tims using the point addition rules.

2.2   Bilinear Pairing

The technique of bilinear mapping is mainly used in the IBE to pair two groups of same order. In recent years, bilinear pairing has played an essential role in designing many cryptographic primitives. Let there are two groups namely G_1 and G_2 both having the same order q, where q is large prime number. The group G_1 is considered as an additive elliptic curve group, whereas group G_2 is a multiplicative group of same order. A bilinear mapping is denoted by $\hat{e}$: $G_1 \times G_1 \longrightarrow G_2$, which maps two elements of G_1 to an element in G_2 and defined by these three properties:

- **Bilinearity:** $\hat{e}(lL, tT) = \hat{e}(L, T)^{lt}$ for all $L, T \in G_1$ and $l, t \in \mathbb{Z}_q^*$.
- **Non-degeneracy:** $\hat{e}(L, L) = 1_{g_2}$ for all $L \in G_1$ and 1_{g_2} signify the multiplicative identity element of G_2.
- **Computability:** The bilinear mapping $\hat{e}(L, T)$ can be efficiently computed for $L, T \in G_1$ by a polynomial-time-bounded algorithm.

2.3 Identity-Based Encryption (IBE)

A traditional non-identity-based public-key protocol involves lots of computation cost due to the public-key certificate management task. Further, a public-key certificate is also needed to be verified by the party for the validation of the public key. To eliminate the overhead of certificate management tasks in the non-identity-based public-key protocol, as early as 1984, Shamir [20] firstly gave an idea about the identity-based cryptography (IBC) where an identity of the party is considered as the public-key. In IBE, the public-key can be any arbitrary string and the sender uses the identity (say ID) of the receiver as public-key for encryption of a message. For instance, when Alice sends an encrypted message to Bob, she encrypts her message using Bob's email address (public identity) "bob@domainname.com". Then, Bob can decrypt the message using a related identity-based private-key obtained securely from the private key generator (PKG) through proper authentication. This novel concept of Shamir's protocol initiates a paradigm shift in public-key cryptographic techniques. However, until the year 2001, it was an open challenge for the researchers to construct a fully functional and efficient IBE. Boneh and Franklin [1] firstly proposed a fully functional IBE from the Weil Pairing. The IBE scheme depends on a trusted authority, called PKG, who generates the system's parameters *param* and the identity-based private-key of a party. The identity-based private-key is transferred to the party via a secure channel. With the nice functionality that any party can perform the encryption of messages with no prior key distribution among the parties (sender and receiver), IBE gains lots of attention from industry, research and academic community. The four algorithms of IBE are a follows:

- **Setup:** This algorithm is executed by the PKG. The input to *Setup* algorithm is a security parameter t and the outputs are *param* and the master private-key of PKG.
- **Extract:** This algorithm is used to compute the identity-based private key of a party user and it is executed by PKG. The input to *Extract* algorithm is the identity *ID* of a party and the output is the identity-based private key of the party.
- **Encrypt:** This algorithm is executed by the sender of a plaintext m. The input to *Encrypt* algorithm is m and the identity of the receiver, and the output is the encrypted message c, called ciphertext.
- **Decrypt:** This algorithm is executed by the receiver for decryption of the ciphertext c. The input to *Decrypt* algorithm is c and the identity-based private key of the receiver, and the output is plaintext m.

2.4   Computational Problems

- **Elliptic Curve Discrete Logarithm Problem (ECDLP):** This hard problem states that if $L, T \in G_1$ and L and T are known where $T = rL$ for $r \in \mathbb{Z}_q^*$, then it is hard to find r.
- **Bilinear Diffie-Hellman Problem (BDHP):** This hard problem states that if $L \in G_1$ and L, xL, yL and zL are known where $x, y, z \in \mathbb{Z}_q^*$, then it is hard to find $\hat{e}(L, L)^{xyz}$.
- **Computational Diffie-Hellman Problem (CDHP):** This hard problem states that if $L \in G_1$ and L, xL and yL are known where $x, y \in \mathbb{Z}_q^*$, then it is hard to find xyL.

Table 1. Notations

Symbols	Denotation
t	Security parameter
q	A prime modular
$H_1(\cdot)$	A cryptographic hash, where $H_1 : \{0,1\}^* \longrightarrow G_1$
$H_2(\cdot)$	A cryptographic hash, where $H_2 : G_2 \longrightarrow \{0,1\}^q$
G_1	An additive cyclic group on E/F_q
G_2	A multiplicative group of order q
$\hat{e}$	A bilinear map, where $\hat{e} : G_1 \times G_1 \longrightarrow G_2$
$\mathbb{Z}_q$	$\{1, 2, \cdots, q-1\}$
$\mathbb{Z}_q^*$	$\{x : 1 \leq x \leq q-1 \bmod gcd(x,q) = 1\}$
P	Primitive root of G
P_i	Party, where $i = 1, 2, 3$
s	$s \in \mathbb{Z}_q^*$ is the private key of PKG
P_{pk}	Master public key of PKG, where $P_{pk} = s \cdot P$

3   Proposed ID-3PAKA Protocol

The proposed ID-3PAKA protocol is described in this section. Our ID-3PAKA protocol is developed using IBE framework. In this protocol, security of negotiated session key and authentication of the parties are ensured using the hardness assumptions on elliptic curve group and DLP. However, the computational as well as communication complexities are also reasonable. Suppose the three parties, who have participated in key negotiation are P_1, P_2 and P_3 and wants to negotiate a common session key among them. Except P_1, P_2 and P_3, the PKG is also involved in the execution of our ID-3PAKA protocol. For sake of clarity, Table 1 lists the different notations and their descriptions. The phases of our protocol are described as following steps:

3.1 Setup Phase

It takes a security parameter t as input and generates the following system's parameters. PKG chooses:

- a t-bit prime number q and groups G_1 and G_2 of same order q. Here G_1 is considered as an additive elliptic curve group, whereas group G_2 is a multiplicative group.
- a bilinear pairing $\hat{e} : G_1 \times G_1 \longrightarrow G_2$.
- a primitive element $P \in G_1$ and another primitive element $g \in G_2$.
- two hash functions H_1 and H_2 defined as $H_1 : \{0,1\}^* \longrightarrow G_1$ and $H_2 : G_2 \longrightarrow \{0,1\}^q$.
- a master private-key $s \in \mathbb{Z}_q^*$ which is secret to the PKG and master public key $P_{pk} = sP$.

PKG publicly announces system parameters $param = \langle q, H_1, H_2, G_1, G_2, P, \hat{e}, P_{pk} \rangle$ and kept master private-key s as secret to itself.

3.2 Key Extraction Phase

It is also executed by PKG to extract the identity-based private key of the party P_i, $\{i = 1, 2, 3\}$. The PKG executes the steps given below:

- A party P_i requests the PKG with their identities ID_i, $\{i = 1, 2, 3\}$.
- PKG checks the authenticity of P_i and calculates their private keys as $P_{pr_i} = sQ_i$, $\{i = 1, 2, 3\}$. Here, $Q_i = H_1(ID_i)$.
- The public key of P_i is Q_i. It may be noticed that there is no need of public key certificate to verify the public key Q_i of P_i.

3.3 Key Agreement Phase

The parties P_1, P_2 and P_3 are involved in our protocol to negotiate a common session key between them. The steps used to exchange the session key are as follows:

- Party P_1 initiates the protocol and does the following:
    - Chooses an integer $a \in \mathbb{Z}_q^*$ at random and calculates $u = g^a \bmod q$ and $X_1 = aP$.
    - Calculates $Y_1 = aQ_1$ and $\mu_1 = \hat{e}(P_{pr_1}, X_1)$ using his/her private key P_{pr_1}.
    - Computes $\sigma_1 = u \oplus H_2(\mu_1)$.
    - Sends $\langle Y_1, u, \sigma_1 \rangle$ to P_2.
- After receiving $\langle Y_1, u, \sigma_1 \rangle$, P_2 responds in the following manner:
    - Computes $\gamma_2 = H_2(\hat{e}(Y_1, P_{pk}))$ and checks whether $u \stackrel{?}{=} \sigma_1 \oplus \gamma_2$. If it is true, picks an integer $b \in \mathbb{Z}_q^*$ at random and computes $X_2 = bP$, $v = g^b \bmod q$ and $v' = u^b \bmod q$.
    - Calculates $Y_2 = bQ_2$ and $\mu_2 = \hat{e}(P_{pr_2}, X_2)$ using his private key P_{pr_2}.
    - Computes $\sigma_2 = v \oplus H_2(\mu_2)$ and $\sigma_2' = v' \oplus H_2(\mu_2)$.

- Sends $\langle Y_2, v, v', \sigma_2, \sigma_2' \rangle$ to P_3.
- After receiving $\langle Y_2, v, v', \sigma_2, \sigma_2' \rangle$, P_3 does the following:
  - Computes $\gamma_3 = H_2(\hat{e}(Y_2, P_{pk}))$ and checks whether $v \stackrel{?}{=} \sigma_2 \oplus \gamma_3$ and $v' \stackrel{?}{=} \sigma_2' \oplus \gamma_3$. If these verifications are true, picks an integer $c \in \mathbb{Z}_q^*$ at random and computes the session key $K_3 = (v')^c \bmod q = g^{abc} \bmod q$.
  - Computes $X_3 = cP$, $w = g^c \bmod q$ and $w' = v^c \bmod q$.
  - Calculates $Y_3 = cQ_3$ and $\mu_3 = \hat{e}(P_{pr_3}, X_3)$ using his private key P_{pr_3}.
  - Computes $\sigma_3 = w \oplus H_2(\mu_3)$ and $\sigma_3' = w' \oplus H_2(\mu_3)$.
  - Sends $\langle Y_3, w, w', \sigma_3, \sigma_3' \rangle$ to P_1.
- After receiving $\langle Y_3, w, w', \sigma_3, \sigma_3' \rangle$, P_1 does the following:
  - Computes $\gamma_1 = H_2(\hat{e}(Y_3, P_{pk}))$ and checks whether $w \stackrel{?}{=} \sigma_3 \oplus \gamma_1$ and $w' \stackrel{?}{=} \sigma_3' \oplus \gamma_1$. If these verifications are true, then computes the session key $K_1 = (w')^a \bmod q = g^{abc} \bmod q$.
  - Computes $u' = w^a \bmod q$ and $\sigma_1' = u' \oplus H_2(\mu_1)$.
  - Sends $\langle u', \sigma_1' \rangle$ to P_2.
- After receiving $\langle u', \sigma_1' \rangle$, P_2 does the followings:
  - Checks whether $u' \stackrel{?}{=} \sigma_1' \oplus \gamma_2$ and computes the session key $K_2 = (u')^b \bmod q = g^{abc} \bmod q$.

Thus, it may be easily seen that $K = K_1 = K_2 = K_3 = g^{abc} \bmod q$. Note that the proposed protocol executes in only one round of communication and the proof of the verification $u \stackrel{?}{=} \sigma_1 \oplus \gamma_2$ is given below.

$$
\begin{aligned}
\sigma_1 \oplus \gamma_2 &= u \oplus H_2(\mu_1) \oplus \gamma_2 \\
&= u \oplus H_2(\hat{e}(P_{pr_1}, X_1)) \oplus H_2(\hat{e}(Y_1, P_{pk})) \\
&= u \oplus H_2(\hat{e}(sQ_1, aP)) \oplus H_2(\hat{e}(aQ_1, sP)) \\
&= u \oplus H_2(\hat{e}(Q_1, P)^{as}) \oplus H_2(\hat{e}(Q_1, P)^{as}) \\
&= u
\end{aligned}
$$

The key agreement phase of our protocol is given in the Fig. 1.

4   Security Analysis

This section comes with the security analysis of our ID-3PAKA protocol. Our protocol resists all possible security attacks. The different type of security attacks are discussed below and it has been proved that our ID-3PAKA protocol is secure against all these types of attacks.

1. **Man-in-the-middle (MITM) Attack:** Our ID-3PAKA protocol negotiates a common session key and the security of this key is provided using the signature σ_i, $\{i = 1, 2, 3\}$ of the party P_i. In our protocol, P_i generates the signature σ_i using his/her private key. If an adversary $\mathcal{A}$ tries to impersonate any of the party P_i, $\{i = 1, 2, 3\}$, he/she cannot generate a valid signature and thus it is difficult for $\mathcal{A}$ to impersonate P_i. Hence, our ID-3PAKA protocol is not vulnerable to MITM attack.

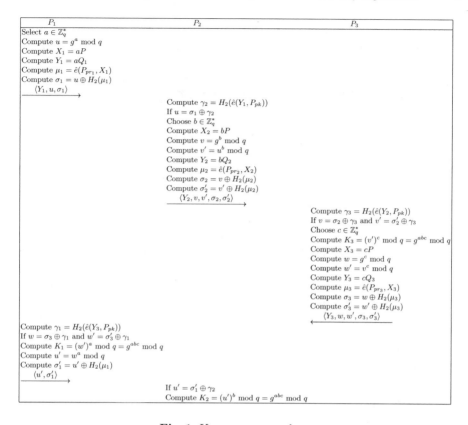

Fig. 1. Key agreement phase

2. **No key control (NKC):** This property ensures that the session key should not be a pre-selected value by any of the party or an adversary. A single party in our protocol cannot derive the actual session key because the ephemeral secrets a, b and c are not known to him/her. It is necessary for our protocol that each party P_i, $\{i = 1, 2, 3\}$ must be involved in the process of generating the session key. An individual party cannot enforce the correct key K to be a pre-selected value. Thus, our ID-3PAKA protocol ensures the NKC property of the session key.

3. **Known key security (K-KS):** This property ensures that an adversary cannot compute a current of any future session key even if any past session key is disclosed to him/her. The negotiated session key K in each session of the proposed protocol is unique because it is generated using the ephemeral secrets a, b and c chosen by P_1, P_2 and P_3 at random. Thus, if an adversary $\mathcal{A}$ has access to previous session keys, he/she might not be able to derive the current or future session keys. Thus, our ID-3PAKA protocol ensures the K-KS property of the session key.

4. **Key-Compromise Impersonation (K-CI) Attack:** This property ensures that if the private-key of a party is disclosed to an adversary, however,

the adversary cannot impersonate other parties to the compromised party to compute the correct session key. Suppose, an adversary $\mathcal{A}$ somehow obtains the private-key, say P_{pr_1} of the party P_1. Now, $\mathcal{A}$ tries to impersonate other parties P_2 and P_3 to P_1, but he/she is not able to do so because he/she does have the knowledge about the private-keys of P_2 and P_3. Thus, the K-CI attack is prevented by out ID-3PAKA protocol.

5. **Perfect Forward Secrecy (PKS):** The PFS property ensures that an adversary cannot compute any past session key even if the private-keys of all parties are disclosed to him/her. An adversary $\mathcal{A}$ cannot recover the previous session keys even after getting the private-keys P_{pr_i} of the party P_i, $\{i = 1, 2, 3\}$ because in the process of key generation, the ephemeral secrets a, b and c are chosen at random. It is difficult for $\mathcal{A}$ to recover a, b and c from g^a, g^b and g^c, respectively because of the DLP. Thus, our ID-3PAKA protocol provides the PKS of the session key.

6. **Unknown Key-Share (UK-S) Attack:** The UK-S attack states that after completion of key agreement phase, the parties, say P_1 and P_2 believe that the session key is established with the party P_3 but, the P_3 feels that the session key is shared with an adversary $\mathcal{A}$. The proposed protocol requires the private keys P_{pr_i} and identities ID_i of parties P_i for $\{i = 1, 2, 3\}$ to generate the signature σ_i. An adversary $\mathcal{A}$ cannot generate the valid signature σ_i which can be verified by the other authentic parties. Hence, our proposed protocol resists UK-S attack.

5 Conclusion

The authors have devised a novel and efficient ID-3AKA protocol using IBE. The proposed protocol is executed for three parties and the negotiated session key is shared among these parties. For the agreement of the session key, we used the concept of ECC and bilinear pairing. The security of the session key, which is generated among three parties in each session, of our protocol is ensured using the hardness assumption of the DLP. Further, different attributes of the session key is ensured in our ID-3PAKA protocol.

References

1. Boneh, D., Franklin, M.: Identity-based encryption from the Weil pairing. In: Annual International Cryptology Conference, pp. 213–229. Springer (2001)
2. Choo, K.K.R.: Revisit of McCullagh-Barreto two-party ID-based authenticated key agreement protocols. Int. J. Netw. Secur. **1**(3), 154–160 (2005)
3. Chou, C.H., Tsai, K.Y., Lu, C.F.: Two ID-based authenticated schemes with key agreement for mobile environments. J. Supercomput. **66**(2), 973–988 (2013)
4. Diffie, W., Hellman, M.: New directions in cryptography. IEEE Trans. Inf. Theory **22**(6), 644–654 (1976)
5. Farash, M.S., Islam, S.H., Obaidat, M.S.: A provably secure and efficient two-party password-based explicit authenticated key exchange protocol resistance to password guessing attacks. Concurr. Comput. Pract. Exp. **27**(17), 4897–4913 (2015)

6. Gupta, D.S., Biswas, G.P.: Identity-based/attribute-based cryptosystem using threshold value without Shamir's secret sharing. In: 2015 International Conference on Signal Processing, Computing and Control (ISPCC), pp. 307–311. IEEE (2015)
7. Gupta, D.S., Biswas, G.P.: Securing voice call transmission over cellular communication. Procedia Comput. Sci. **57**, 752–756 (2015)
8. Gupta, D.S., Biswas, G.P.: An ECC-based authenticated group key exchange protocol in IBE framework. Int. J. Commun. Syst. **30**(18), e3363 (2017)
9. Gupta, D.S., Biswas, G.P.: On securing bi-and tri-partite session key agreement protocol using IBE framework. Wirel. Pers. Commun. **96**(3), 4505–4524 (2017)
10. He, D., Chen, Y., Chen, J.: An ID-based three-party authenticated key exchange protocol using elliptic curve cryptography for mobile-commerce environments. Arab. J. Sci. Eng. **38**(8), 2055–2061 (2013)
11. Hölbl, M., Welzer, T., Brumen, B.: Two proposed identity-based three-party authenticated key agreement protocols from pairings. Comput. Secur. **29**(2), 244–252 (2010)
12. Hölbl, M., Welzer, T., Brumen, B.: An improved two-party identity-based authenticated key agreement protocol using pairings. J. Comput. Syst. Sci. **78**(1), 142–150 (2012)
13. Islam, S.H., Amin, R., Biswas, G.P., Farash, M.S., Li, X., Kumari, S.: An improved three party authenticated key exchange protocol using hash function and elliptic curve cryptography for mobile-commerce environments. J. King Saud Univ. Comput. Inf. Sci. **29**(3), 311–324 (2017)
14. Islam, S.H., Biswas, G.P.: A more efficient and secure ID-based remote mutual authentication with key agreement scheme for mobile devices on elliptic curve cryptosystem. J. Syst. Softw. **84**(11), 1892–1898 (2011)
15. Islam, S.H., Obaidat, M.S., Vijayakumar, P., Abdulhay, E., Li, F., Reddy, M.K.C.: A robust and efficient password-based conditional privacy preserving authentication and group-key agreement protocol for VANETs. Future Gener. Comput. Syst. **84**, 216–227 (2018)
16. Jeong, I.R., Katz, J., Lee, D.H.: One-round protocols for two-party authenticated key exchange. In: International Conference on Applied Cryptography and Network Security, pp. 220–232. Springer (2004)
17. Koblitz, N.: Elliptic curve cryptosystems. Math. Comput. **48**(177), 203–209 (1987)
18. McCullagh, N., Barreto, P.S.: A new two-party identity-based authenticated key agreement. In: Cryptographers' Track at the RSA Conference, pp. 262–274. Springer (2005)
19. Miller, V.S.: Use of elliptic curves in cryptography. In: Conference on the Theory and Application of Cryptographic Techniques, pp. 417–426. Springer (1985)
20. Shamir, A.: Identity-based cryptosystems and signature schemes. In: Workshop on the Theory and Application of Cryptographic Techniques, pp. 47–53. Springer (1984)
21. Tseng, Y.M.: An efficient two-party identity-based key exchange protocol. Informatica **18**(1), 125–136 (2007)

Machine Learning Based Stock Market Analysis: A Short Survey

Hrishikesh Vachhani[1], Mohammad S. Obiadat[2,3,4,5(✉)], Arkesh Thakkar[1],
Vyom Shah[1], Raj Sojitra[1], Jitendra Bhatia[1], and Sudeep Tanwar[6(✉)]

[1] Vishwakarma Government Engineering College, GTU, Ahmedabad, India
hrishikeshvachhani@gmail.com, arkesh.thakar14@gmail.com,
shahvyom18@gmail.com, rajsojitra79@gmail.com,
jitendrabbhatia@gmail.com
[2] Dean of College of Computing and Informatics,
University of Sharjah, Sharjah, UAE
m.s.obaidat@ieee.org
[3] King Abdullah II School of IT, University of Jordan, Amman, Jordan
[4] University of Science and Technology Beijing, Beijing, China
[5] College of Engineering, Al-Balqa Applied University, Al-Salt, Jordan
[6] Institute of Technology, Department of Computer Science and Engineering,
Nirma University, Ahmedabad, India
sudeep.tanwar@nirmauni.ac.in

Abstract. Finance is one of the pioneering industries that started using Machine Learning (ML), a subset of Artificial Intelligence (AI) in the early 80s for market prediction. Since then, major firms and hedge funds have adopted machine learning for stock prediction, portfolio optimization, credit lending, stock betting, etc. In this paper, we survey all the different approaches of machine learning that can be incorporated in applied finance. The major motivation behind ML is to draw out the specifics from the available data from different sources and to forecast from it. Different machine learning algorithms has their abilities for predictions and are heavily depended on the number and quality of parameters as input features. This work attempts to provide an extensive and objective walkthrough in the direction of applicability of the machine learning algorithms for financial or stock market prediction.

Keywords: Stock market prediction · Machine Learning · Stock analysis

1 Introduction

The finance industry, in general, has been an early pioneer of AI technologies. Since the 70s, Wall Street has been using predictive models for predicting the prices of the market. In current era, the prime objective is to predict the market movements and how can these hedge funds be smartly invested at any given point of time based on market movements. Machine Learning techniques can boost the efficiency of an organization's operational accounting, financial reporting, allocations, adjustments, reconciliations, and intercompany transactions. The market for AI in finance is expected to grow to $52.2 billion in 2021 [1]. The main purpose of machine learning in the financial market is to reduce the rate of false positives that a human hasn't anticipated while performing a stock transaction. ML techniques help us to detect the anomaly in a transaction data set.

© Springer Nature Switzerland AG 2020
J. S. Raj et al. (Eds.): ICIDCA 2019, LNDECT 46, pp. 12–26, 2020.
https://doi.org/10.1007/978-3-030-38040-3_2

1.1 Traditional Approach for Stock Market Prediction

There are different conventional approaches available which can help in stock prediction. Among all of these, two important approaches for stock market prediction are Efficient Market Hypothesis (EMH) and Random Walk theory. Efficient Market Hypothesis (EMH) theory states that asset prices fully reflect all available information. The simple explanation is that, it is not possible to perform better than the stock market on a risk-adjusted base as stocks should only respond to new information. There are three different forms in this hypothesis viz., (a) Weak EWH: only the past trading information is considered, (b) Semi-Strong EMH: all unbiased public information is utilized and (c) Strong EMH: all information public and private is used. Random Walk Theory states that stock market prices do not depend on past data and stock prices prediction is not possible due to great fluctuation.

The existence of price fluctuation prices in stock market, undiscovered serial corelations among stock prices, fundamental events and effect of economy are the great motivation behind adapting a better prediction model. This precise prediction can be the backbone for the traders and investors about the possibility of fluctuation in stock market [2]. Also, due to frequent data generation, real time stock prediction is one of the prime challenge. With the recent advancements in cloud computing architecture [3–6], machine learning module can be deployed on this high performance cloud.

1.2 Role of Machine Learning in Stock Market Prediction

Machine learning algorithms help to predict future outcomes based on available or past data. The prediction can be done based on regression and classification. Figure 1 depicts the various data sources, which are fed in to the Machine Learning model. The following subsection describes various machine learning approaches.

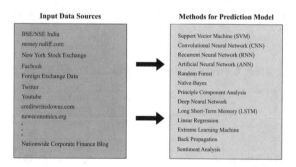

Fig. 1. Data sources and ML techniques for Stock Market Prediction

Multiple Regression Analysis: Regression is an ML technique which describes the relationship between a dependent variable and more than one independent variables. In stock market prediction, many factors are correlated to future stock prices. Regression analysis is one of the non-linear technique used for predicting the stocks. In this method, various market variables are analyzed and then the regression equation is formed using these variables. This equation is then utilized in the prediction model for predicting the dependent variable during the forecast period.

Decision Tree and Random Forest: A decision tree technique performs a recursive action based on a set of instruction and arrives at an end. The simplest way for classification is the decision tree. For the prediction of the class label, the attributes are tested against the decision tree. The quality attribute is used for the best partitions of tuples into particular classes. Popular attributes that are used in the stock market prediction are Information Gain, Gini Index, and Gain Ratio. Multidimensional data handling property made decision tree classifier popular in recent times. In simple terms, Random Forest is the collection of decision trees and classification ranking of variables is done by Random Forest. Importance and accuracy of variables are automatically generated.

Support Vector Machine (SVM): For time series analysis, the viable option is Support Vector Machine. Regression and Classification both can be done in SVM. The two key elements for the implementation of SVM are kernel Function and mathematical programming. The major advantage of SVM is that it scales better to high dimensional data. It is considered to be a better option as a prediction tool for sentiment analysis in stock market prediction.

Neural Networks: Stock Market is considered to be highly non-linear and completely random. The most effective method in Neural Network for prediction is Long-Short Term Memory (LSTM) in general Recurrent Neural Network (RNN) as it has the ability to store certain information for later use. It also extends the capability of analyzing the complex structure between relationships of stock price data. Different models that are generally used for forecasting stock market prices are Artificial Neural Back Propagation (BP), Support Vector Machine Regression (SVMR), General Regression Neural Network (GRNN), and Least Square Support Vector Machine Regression (LS-SVMR). Stock markets generate an enormous amount of data. This opens the door for modern techniques like applying machine learning which makes the system self-functioning and self-Adaptive.

Table 1. Progressive work in stock market prediction

Author	Dataset	Input features	Target output	Number of samples	ML technique used
Patel [7], 2015	India BSE indices	10	Stock Prediction	2400	Support Vector Regression, Random Forest, ANN
Chiang [8], 2016	22 World stock market indices	5	Stock prices	756	ANN, Particle Swarm Optimization
Chourmouziadia [9], 2016	Greece ASE	8	Portfolio Management	3900	Fuzzy Logic System
Arevalo [10], 2016	US Apple Stock	3	Stock Price	19,109	Deep Neural Network
ZHong [11], 2017	US SPDR 500 ETF	60	Market Prediction (Direction up or down)	2518	ANN, Dimension Reduction
Bhattacharya [12], 2018	TAIEX stock index data	–	Stock index prediction	–	Hybrid fuzzy logic
Kohli [13], 2019	India BSE	5	Stock prediction	915	Random Forest, SVM, Gradient Boost

In Table 1, we have summarized the progressive work of last few years related to applicability of machine learning techniques in stock prediction by providing the brief familiarity of ML techniques used. This work focuses on the comprehensive review in the sector of stock prediction, market direction, portfolio management incorporating machine learning techniques like SVM, ANN, Random Forest, Deep Neural Network, Fuzzy logic, among others. Table 2 describes the abbreviations used throughout the paper.

Table 2. Summary of notations

Abbreviation	Description
DL	Deep Learning
NLP	Natural Language Processing
SVM	Support Vector Machine
ANN	Artificial Neural Network
CNN	Convolutional Neural Network
SVR	Support Vector Regression
KNN	K-Nearest Neighbors
RNN	Recurrent Neural Network
ELM	Extreme Machine Learning
LSTM	Long Short-Term Memory
PCA	Principle Component Analysis
BP	Back Propagation
RBFNN	Radial Basis Function Neural Network
ARIMA	Auto-Regressive Integrated Moving Average
ISCA-BPNN	Improved Sin Cosine Algorithm - Back Propagation Neural Network
PSO	Particle Swarm Optimization
MISA-ELM	Mutual Information-based Sentimental Analysis with kernel-based ELM
BPNN	Back Propagation Neural Network

Contribution. The principle objective of this work is to give broad shrewdness of how can Machine Learning be exercised in the field of stock market prediction. We augment the study with key insights into the ML techniques employed, their merits, demerits, and their accuracies. The Summary of our contribution is as the below:

1. We reviewed the existing survey of the past few years by providing the brief familiarity of machine learning techniques applied in stock market, and their shortcomings.
2. We categorize the review work based on different approaches. This categorization shows how machine learning techniques can be incorporated for advancements in stock market prediction.

1.3 Organization

The rest of this paper is organized as follows. We have analyzed the pre-existing methods of stock market prediction. We divided the survey in three different segments that are predictive approach, sentiment analysis, and hybrid approach. Section 2 discusses the related work based on approaches mentioned under each segment. Section 3 describes various trends of machine learning techniques in respective disciplines of stock market prediction along with detailed discussion. Finally, Sects. 4 and 5 concludes the work and highlights the future direction in various disciplines of the stock market prediction proposed, respectively.

2 Related Work

Past studies focus on the statistical techniques which includes different kinds of averaging techniques like Auto-Regressive Moving Average (ARMA). With the recent advancements in stock analysis, we incorporated different approaches of machine learning in stock market prediction as shown in the Fig. 1. The survey focuses on the following three segments: (a) Predictive Approach (b) Sentiment Analysis and (c) Hybrid approaches, as shown in Figure 2.

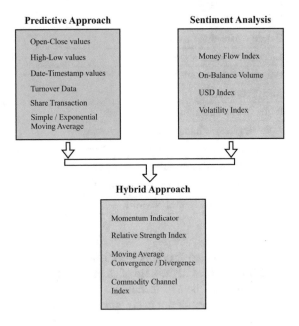

Fig. 2. Taxonomy: Machine Learning techniques for stock prediction

Predictive approach uses indicators that are used in daily basis in stock market like Open-Close value, and High-Low value. Sentiment analysis uses indicators like On-Balance Volume, and Volatility Index which are useful in analyzing the human sentiments that can cause fluctuation in stock price. Hybrid approach is the combination of predictive approach and sentiment analysis in which, the output of sentiment analysis is provided to predictive model as an indicator.

2.1 Predictive Approach

In Predictive approach, models highly make use of quantitative parameters which includes trend indicators, ups, and downs of the stock prices, high and low values of the day, and buy/sell ratio. It also comprises of different types of averages like Simple Moving Average (SMA), Exponential Moving Average (EMA), etc. The progressive work of ML in predictive approach is shown in Table 3.

Yoo *et al.* [14] proposed a model to examine the stock prices and the effect of events on the price of the stock. Authors show that neural networks approach gives better prediction than any other model which uses event-based information. The model also uses web-based event which required web mining technique for prediction in a short time. The authors also highlight the work that may happen in this direction.

Huang *et al.* [15] use Bi-Clustering algorithm along with KNN to discover the trading rules in the financial market along with co-variates. Various technical indicators are also used along with bi-clusters, thus by combination a better accuracy is obtained. It has been applied to 4 data sets and is compared with 4 traditional techniques and it is found to outperform all form of them i.e GP, ANN, etc.

Dixon *et al.* [16] shows a model which uses ANN technique Deep Neural Network for forecasting the stock price. The main objective of this model is to describe an implementation of deep neural networks to financial time series data to classify financial market movement directions. Using an Intel Xeon Phi co-processor with 61 cores model increases the speed enormously.

Patel *et al.* [7], tries to imitate the conventional human approach of investment and decision making in the stock market based on various internal and external factors. The authors proposed various machine learning algorithms like ANN, SVM, Random Forest and Naive-Bayes for predicting the rise and fall in the financial market. Also, authors use deterministic method to segregate the continuous data and classify it in two categories, i.e., −1 or 1 which means fall or rise in the market, respectively. The only consideration of key factors is the prime goal due to which noise is removed and performance is improved to a great extent.

Singh *et al.* [17] proposed a model, which tries to predict the stock market using 2-Directional 2-Dimensional Principal Component Analysis and Neural Networks. The techniques applied for the speech domain can similarly be applied to the stock market as it is also a time series. Here, DNN is used as the trend prediction. DNN provide good performance to nonlinear functions also which can be more useful to predict the stock market.

Chong *et al.* [18] uses some of the deep learning algorithms for stock predictions and also represents seven features with three data representation methods: principal component analysis, auto-encoder, and restricted Boltzmann machine. The advantage of it is that the model analyses the data by its own without prior knowledge or any supervision. Also DNN model gives better accuracy than the auto-regressive model.

Chatzis *et al.* [19] proposed a model about the critical case for the financial crisis of the market. The authors examine the crisis using deep and machine learning techniques

Table 3. Progressive work of ML in predictive approach

Author	ML technique used	Merit	Demerit	Accuracy
Yoo [14], 2005	SVM, Neural Network	–	Model gives the short-term prediction and it may not cover real-time events	Neural Network gives highest Prediction compare to others
Huang [15], 2014	Biclustering Algorithms with KNN (BIC-K-NN)	It outperforms 4 conventional prediction models by using biclustering algorithms which biclusters the data matrix	Only a few technical indicators are selected, And also only a short period of just 10 days is considered for prediction	–
Dixon [16], 2015	Deep Neural Network (DNN)	Efficient implementation of the gradient descent algorithm. Model also reduce the problem of Over-fitting	–	Model gives 73% Accuracy
Patel [7], 2015	ANN, SVM, Random Forest, Naive-Bayes Classifier	Data set is churned into only required 10 technical parameters which give better performance and Data Presentation layer is used to classify continuous value into either −1 or 1	It predicts only for a short term which can hinder the knowledge of long term stoke fluctuations	For continuous values: Naive-Bayes-73.3% and random forest-83.56% For Deterministic Data Representation: ANN-86.69% SVM-89.33% Random Forest-89.98% Naive-Bayes-90.19%
Singh [17], 2017	PCA, Deep Neural Network	This model gives better results than the other models like RBFNN, RNN in terms of the accuracy	It is harder to get desired performance with increasing dimensions and window sizes	4.8% higher than RBFNN and 15.6% higher accurate than RNN. Accuracy for 10 × 10 dimensions is 68%
Chong [18], 2017	DNN, Restricted Boltzmann Machine, ANN	It extracts the features from large data set without relying on prior knowledge	–	–
Chatzis [19], 2018	Deep Neural Networks, SVM, Random Forest	Deep Neural Network gives the strong prediction about Financial crisis	(20 day horizon) maximum hit rate is 46%	
Hu [20], 2018	ANN, SVM, ARIMA, Multiple Regression	–	Trading volumes are been predicted instead of direction of the stock market	Accuracy for the Google data is about 86.81% and for Dow Jones Industry is about 88.98%
Troiano [21], 2018	LSTM	–	Performance decrease on increasing the number of inputs	–
Guo [22], 2018	SVM, Regression	SVM is kept adaptive according to the changes occurred in high frequency data	Historical model's effect are no taken into consideration	–
Naik [23], 2019	ANN, Linear Regression	–	This model is for short term prediction	–
Naik [24], 2019	Recurrent Neural Network, LSTM	Uses recurrent dropout in RNN layers to avoid over-fitting	It does not include other Factor which may have effect on the price of the stock	–
Naik [25], 2019	ANN, SVM	33 technical indicators has been considered and various ML and DL algorithms are applied to have better accuracy on daily stock prices	–	–
Wen [26], 2019	CNN	This Method is more efficient in terms of Computational complexity compare to sequential model	Model only uses past data as a feature for forecasting. They do not consider other factors which will affect the price	Higher than Traditional signal processing method to 6% to 7%

and also give a strong prediction about the stock crisis. The author combines the Neural Network and Random Forest forecasts for prediction of the 1-day horizon and XGBoost and Random Forest forecasts for the 20-day horizon. The model uses DNN to achieve high prediction about financial crisis.

Troiano et al. [21] train robot in their approach to take the decisions for prediction of the stock market with the help of Deep Learning. As different models have their own characteristics, so the hybrid method is build. The model proposed by Hu et al. [20] has ISCA-BPNN which yields the highest accuracy among the different hybrid models. Not only data can be used to predict stock, but it can be used to forecast the trends. It can optimize the weights and the bias to get the prediction.

Adaptive support vector machine technique has been introduced by Guo et al. [22] to analyze the high-frequency data which consists of discrete, non-Linear properties to forecast the stock market. To predict with the constantly changing characteristics of the high-frequency data it has been implemented with the Particle Swarm Optimization (PSO) algorithm.

Naik et al. [24] present a model for stock return prediction using a Recurrent Neural Networks and LSTM. After applying data preprocessing, data is given to the RNN as features and it will pass through many layers using LSTM. The model gives the prediction about stock return prediction.

Naik et al. [25] use Bourta feature selection technique which is useful on selection required technical indicators for predicting the stock movement classification on a daily basis. They focused on two aspects: first the selection of various technical inputs from Bourta algorithm to reduce the noise and second is the application of various machine learning and deep learning techniques on those indicators to give an accurate prediction. Author also mentioned that deep learning gives better performance than machine learning.

Two methods have been proposed by Naik et al. [23]. The methods make a prediction based on the past results where the second based on the different views by the trader's views. This model is for technical analysis which used to take the overall movement happening in the stock market and not the individual company's market.

Wen et al. [26] have proposed a model to predict the stock price using historic data. It uses motif as features to achieve more accuracy. Instead of RNN model, it uses CNN to examine the current trend of the market. The authors basically proposed two algorithms to extract the motif (High Order Characteristic) from the redundant data. The model introduces a novel technique for financial time series trend prediction by reconstructing time series through motif.

2.2 Sentiment Analysis

There are numerous comments from the experts to beginners on a particular company that can affect the prices of the stock. So the sentiment analysis approach analyze those comments which help in analyzing company's perception according to those comments. Comparative analysis of ML-based approaches used in sentiment analysis is shown in Table 4.

Arafat et al. [27] have proposed a model that collects a large data set from various social media platforms like Twitter and various news as input and by performing various operations on it based on timeline and property segmentation for the sentimental

Table 4. Progressive of ML in sentiment analysis

Author	ML technique used	Merit	Demerit	Accuracy
Arafat [27], 2013	Emotion Corpus-Based Method - (vector space model)	Sentiment Analysis based on churned and cleaned data on cloud makes it easy to predict variations	It focuses only on the tweets made by people without taking into consideration any external factors	–
Nguyen [28], 2015	In this paper Joint sentiment/Topic model is used to forecast the stock market	It has an advantage for the large scale evaluation based on the data obtained from the social media	This model does not predicts the drastic variation in the future only shows stock is high or low	Average accuracy for this proposed model is 54.41%
Oliverira [29], 2016	Lexicons: GI, OL, MSOL, MPQA, SWN, FIN	It uses sentiment lexicons which enable an easy and fast unsupervised sentimental analysis	It may results to wrong predictions if proper churning of data is not done on unsupervised SA	Not precisely specified but time-saving approach
Wang [30], 2017	Sentimental analysis with kernel-based Extreme Learning Machine (ELM)	Model has the benefits of both statistical sentimental analysis and ELM which helps in both accuracy and prediction speed	–	Higher Accuracy than Back-propagation neural Network
Ren [31], 2018	SVM	A stop-loss order is used to reduce the loss which can reflect a better performance	Efficiency can be increased by increasing the processing data in a large volume	The accuracy can be as high as 89.93% with a rise of 18.6% after introducing sentiment variables
Shi [32], 2018	CNN, RNN, LSTM	It takes into consideration the trading rules formulated by professional traders along with traditional deep learning to improve the accuracy	CNN and LSTM fails due to noise in the daily feed	Highest accuracy of 93.8% LSTM with highest of 95.3% CNN with highest of 91.4%
Arora [33], 2019	SVM, LSTM, Back Propagation	It can accept the data in any forms like video, text, image	–	LSTM have 66.5% which is more than that of SVM (Support Vector Machine) and Back propagation
Vargas [34], 2019	CNN, RNN	Sentence embedding performs better to word embedding as input, model use a combination of convolutional layer and a recurrent layer	Model uses articles, news from the previous day only to predict for forecasting day	CNN prediction model accuracy 65%

analysis of people's mood swings and its effect on stocks market. It focuses on vector space model which categorizes various mood's and on set space model based on a particular timeline to give a more clear idea and an accurate prediction.

Nguyen *et al.* [28] has introduced the model "Joint sentiment/Topic" (JST) which works on a large scale and predicts the stock market based on the sentiments obtained from the social media. It also bifurcates the data obtained from social media whether it is useful or not. This proposed model is for predicting the stock market based on the mood information.

Oliveira *et al.* [29] use a sentiment lexicon for sentimental analysis of the inputs from various platforms to predict fluctuations in the stock market and stokes of individual companies. Sentiment lexions are faster, cheaper and provide high frequency to both stock market indices and individual stokes. By using this approach, unsupervised SA is possible which is faster and easier than conventional baseline lexicons. Also, some of the baseline lexicons are used for generating an analysis of messages.

Vargas *et al.* [34] has proposed a model to examine and forecast the stock market using financial news and articles as input by performing Deep Learning methods on them. Deep Learning methods help in examining complex patterns and interactions between them to speed up the process. The model focuses on CNN, RNN, and combination of both which can lead to better results in NLP tasks and forecasting.

Wang *et al.* [30] proposed a model named MISA-ELM. It will combine stock price and news related to that stock for forecasting the market price volatility. Using this method, we can achieve high accuracy as well as high prediction speed. An ELM method also uses other network features as input compared to the basic gradient-based learning algorithm. It can help to generate high prediction speed. Here, authors try to examine accuracy using mutual information-based sentimental analysis with kernel-based ELM and compare it with other techniques. Ren *et al.* [31] trained a machine to gain the knowledge from a large number of documents and sentiments. This approach uses SVM to overcome the maximal margin hyperplane problem they have used five-fold cross-validation method. Sentiments are considered as the most precious data.

Shi *et al.* [32] present deepclue system that visually interprets text-based deep learning model. They proposed a three key design from the cutting edge deep learning technologies: hierarchical neural network model, a back propagation-like algorithm, and an interactive visualization interface. It helps to bridge between text-based deep learning and the rules formulated by the professional traders that help to give suggestions to individuals about the fluctuations in the stokes. However, it may face failure because of noise in the social media messages about irrelevant factors.

Arora *et al.* [33] has proposed a model in which data was tested in any form from the social media to predict the stock market. It considers all the factors that can help the model to forecast the model. This model is basically designed for big data analytics so as to increase the ratio of profit.

2.3   Hybrid Approach

Prediction of the stock market includes fundamental analysis of a company which includes the financial status of the company, the quarterly balance, the dividends, audit reports along with the sales data, import/ export volume. The above mentioned is only one portion, the other portion is the sentiments and rumors of the company in the real world as it matters because stock prices may fluctuate because of it. Model-based on hybrid approach has all the parameters that can affect the market. Comparative analysis of various state-of-the art hybrid approaches uses ML-based systems are shown in Table 5.

Dash *et al.* [37] proposed a model which combines technical analysis with machine learning techniques to make decisions about the stock price. Classification using CEFLANN model uses ELM to generate the trend signal. The model analyzes the trend and generates the signal. This model provides a superior profit percentage compared to other classifiers.

Akita *et al.* [36] proposed a model based on LSTM technique which predicts the variations in stock markets by taking into consideration both textual and numerical data. They collect the textual data of the inter-related companies and predicts on the past data in numeric form by applying sentimental analysis on the textual data for a particular time being. Thus, by considering both textual and numeric information accuracy may increase rather than isolated prediction.

Table 5. Progressive work of ML in hybrid approach

Author	ML technique used	Merit	Demerit	Accuracy
Li [35], 2016	ELM, SVM, Back Propagation Neural Network (BPNN)	–	Kernelized ELM uses more CPU resources so CPU scheduling is a big problem	ELM have high accuracy than SVM and BPNN
Akita [36], 2016	Distributed Memory Model of Paragraph Vectors (PV-DM), LSTM	It considers both numerical and textual information to improve the prediction based on sentimental analysis	The model faces the problem of over-fitting	–
Dash [37], 2016	Computational Efficient Functional Link Artificial Neural Network (CEFLANN)	Model give the best time for trading with minimum risk and Technical indicators also give the trend for current market	–	CEFLANN will give higher Accuracy compare to SVM and KNN
Liu [38], 2018	Numerical-Based Attention (NBA)	The method proposed to reduce the noise which can be used to predict the stock market with more accuracy	It is not mentioned that how it is implemented in the industrial area	–
Reddy [39], 2019	SVM Naive Bayes	This model gives the prediction before for 1–10, 15 and 30 days in advance. It will cover sentiment data also to forecasting the model	–	Naive Bayes - 88% SVM - 93%

Li *et al.* [35] have proposed an Extreme Learning Machine that has a tendency to predict fast with high accuracy, as enormous data is generated in the stock market. In this paper, ELM makes prediction based on market price as well as based on the market news. Moreover, in this paper three techniques have been compared; they are BPNN, SVM, and ELM.

According to the work by Liu *et al.* [38], numerical based data and data from the news have been collected as data set to predict the stock market. In this paper, the dataset has been bifurcated into a single data source and double data source. News has been encoded into numerical data and is passed to model for prediction.

Reddy *et al.* [39] proposed a model which uses classification technique i.e., SVM to predict the stock price and it also uses the Term Frequency-In-verse Document Frequency (TF-IDF) technique to find the optimal Features. The model uses Sentiment data like twitter news to predict the price of the stock. The author also gives four classes as an output signal (open, close, high, and low). In TF-IDF algorithm, authors give the priority to the features which will be used as an input in SVM.

3   Discussion

The Fig. 3 depicts the research trends of different types of approach for predicting stock market prices. Predictive analysis approach occurs more often in which the statistical parameters are used to predict the stock prices. Sentiment analysis is the next often used technique which scans particular websites, articles, journals for analyzing the sentiment of the semantics related to the stock market. The next one is the Hybrid approach in which the combination of both predictive and sentiment approach is used for prediction.

Figure 4 shows the machine learning technologies used for developing a predictive model for stock market prediction. SVM is the most frequently used technique. There-

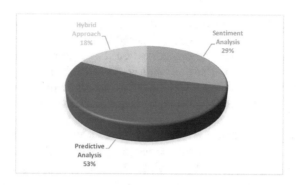

Fig. 3. Trend in various sub-disciplines of stock market prediction

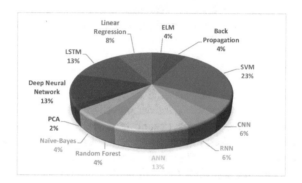

Fig. 4. Usage trend of Machine Learning techniques for stock market prediction

after, ANN, Deep Neural Network and LSTM methods are used often for prediction. Commonly more than one techniques are used in a model to increase the efficiency of prediction.

There are various trading types in financial market like Equity, Derivative or Future and Option. (1) An equity market is where different investors or traders issues Share of equity of a listed company. Several newer ANN approaches such as the LSTM and RNN have given better outcome than others. (2) Future is the extension of a forward contract. These are the 1 month expiry contract and the follow 3 month contract cycle, where you just have to pay a margin amount not actual price of share. SVM, Random Forest, Gradient-boost Decision Trees provide a promising outcome in future. (3) An option gives the buyer the right, but not the bond, to buy (or sell) a certain asset at a specific price at any time during the life of the contract. SVM, SMO, and Bayesian Logistic Regression performs better that the other techniques.

4   Conclusion

In this paper, we provide insights to the readers about how machine learning techniques can be used for efficient prediction in the stock market. We described two traditional theories, Efficient Market Hypothesis (EMH) and Random Walk Theory for the market prediction. Many techniques have been proposed so far for predicting stocks for long and short period segregating in various subsections like sentimental analysis, predictive models and hybrid approach. Many techniques of deep learning and machine learning algorithms like Boruta algorithm, ANN, LSTM (Long Short Term Memory), Biclustering algorithms with KNN (BIC-K-NN) are used for predictive analysis which takes into consideration various factors affecting stock market variation like opening, closing, fall, rise, in the stock prices. Apart from this CNN, SVM, RNN, Lexicons are used for sentimental analysis which is used to predict based on various data collection from news article, social media like Twitter, and Youtube. Moreover, the hybrid approach has also been adapted by various methods to consider both numerical and textual information which leads to better prediction. Furthermore, we represented the merits and demerits of ML techniques like Multiple Regression, Decision Tree and Random Forest, Support Vector Machine and Neural Networks. Amongst them, SVM gives better results with compare to other techniques. Apart from that, ANN, DNN, LSTM also performs well.

5   Future Direction

Future direction in predictive analysis motivates us to incorporate more detailed features like dividend yield, interest rates, insider moves in the currently proposed models which helps to mitigate false positives. In sentiment analysis, unrelated comments, and bluff comments should be taken into consideration because these can affect the accuracy of predicting stock prices based on those sentiments. More robust methods should be proposed in the hybrid approach for maintaining the relevant proportion of input features from both sentiment and predictive approach.

References

1. Ongraph technologies. https://medium.com/datadriveninvestor/development-of-ai-in-financial-services-58e9249e547. Accessed 27 Aug 2019
2. Shah, D., Isah, H., Zulkernine, F.: Stock market analysis: a review and taxonomy of prediction techniques. Int. J. Financ. Stud. **7**(2), 26 (2019)
3. Jaykrushna, A., Patel, P., Trivedi, H., Bhatia, J.: Linear regression assisted prediction based load balancer for cloud computing. In: 2018 IEEE Punecon, pp. 1–3. IEEE (2018)
4. Bhatia, J., Mehta, R., Bhavsar, M.: Variants of software defined network (SDN) based load balancing in cloud computing: a quick review. In: International Conference on Future Internet Technologies and Trends, pp. 164–173. Springer (2017)
5. Bhatia, J.B.: A dynamic model for load balancing in cloud infrastructure. Nirma Univ. J. Eng. Technol. (NUJET) **4**(1), 15 (2015)
6. Bhatia, J., Patel, T., Trivedi, H., Majmudar, V.: HTV dynamic load balancing algorithm for virtual machine instances in cloud. In: 2012 International Symposium on Cloud and Services Computing, pp. 15–20. IEEE (2012)
7. Patel, J., Shah, S., Thakkar, P., Kotecha, K.: Predicting stock and stock price index movement using trend deterministic data preparation and machine learning techniques. Expert Syst. Appl. **42**(1), 259–268 (2015)
8. Chiang, W.-C., Enke, D., Tong, W., Wang, R.: An adaptive stock index trading decision support system. Expert Syst. Appl. **59**, 195–207 (2016)

9. Chourmouziadis, K., Chatzoglou, P.D.: An intelligent short term stock trading fuzzy system for assisting investors in portfolio management. Expert Syst. Appl. **43**, 298–311 (2016)
10. Arévalo, A., Niño, J., Hernández, G., Sandoval, J.: High-frequency trading strategy based on deep neural networks. In: International Conference on Intelligent Computing, pp. 424–436. Springer (2016)
11. Zhong, X., Enke, D.: Forecasting daily stock market return using dimensionality reduction. Expert Syst. Appl. **67**, 126–139 (2017)
12. Bhattacharya, D., Konar, A.: Self-adaptive type-1/type-2 hybrid fuzzy reasoning techniques for two-factored stock index time-series prediction. Soft Comput. **22**(18), 6229–6246 (2018)
13. Kohli, P.P.S., Zargar, S., Arora, S., Gupta, P.: Stock prediction using machine learning algorithms. In: Applications of Artificial Intelligence Techniques in Engineering, pp. 405–414. Springer (2019)
14. Yoo, P.D., Kim, M.H., Jan, T.: Machine learning techniques and use of event information for stock market prediction: a survey and evaluation. In: International Conference on Computational Intelligence for Modelling, Control and Automation and International Conference on Intelligent Agents, Web Technologies and Internet Commerce (CIMCA-IAWTIC 2006), vol. 2, pp. 835–841. IEEE (2005)
15. Huang, Q., Wang, T., Tao, D., Li, X.: Biclustering learning of trading rules. IEEE Trans. Cybern. **45**(10), 2287–2298 (2014)
16. Dixon, M., Klabjan, D., Bang, J.H.: Implementing deep neural networks for financial market prediction on the Intel Xeon Phi. In: Proceedings of the 8th Workshop on High Performance Computational Finance, p. 6. ACM (2015)
17. Singh, R., Srivastava, S.: Stock prediction using deep learning. Multimedia Tools Appl. **76**(18), 18569–18584 (2017)
18. Chong, E., Han, C., Park, F.C.: Deep learning networks for stock market analysis and prediction: methodology, data representations, and case studies. Expert Syst. Appl. **83**, 187–205 (2017)
19. Chatzis, S.P., Siakoulis, V., Petropoulos, A., Stavroulakis, E., Vlachogiannakis, N.: Forecasting stock market crisis events using deep and statistical machine learning techniques. Expert Syst. Appl. **112**, 353–371 (2018)
20. Hongping, H., Tang, L., Zhang, S., Wang, H.: Predicting the direction of stock markets using optimized neural networks with Google Trends. Neurocomputing **285**, 188–195 (2018)
21. Troiano, L., Villa, E.M., Loia, V.: Replicating a trading strategy by means of LSTM for financial industry applications. IEEE Trans. Ind. Inform. **14**(7), 3226–3234 (2018)
22. Guo, Y., Han, S., Shen, C., Li, Y., Yin, X., Bai, Y.: An adaptive SVR for high-frequency stock price forecasting. IEEE Access **6**, 11397–11404 (2018)
23. Naik, N., Mohan, B.R.: Optimal feature selection of technical indicator and stock prediction using machine learning technique. In: International Conference on Emerging Technologies in Computer Engineering, pp. 261–268. Springer (2019)
24. Naik, N., Mohan, B.R.: Study of stock return predictions using recurrent neural networks with LSTM. In: International Conference on Engineering Applications of Neural Networks, pp. 453–459. Springer (2019)
25. Naik, N., Mohan, B.R.: Stock price movements classification using machine and deep learning techniques-the case study of Indian stock market. In: International Conference on Engineering Applications of Neural Networks, pp. 445–452. Springer (2019)
26. Wen, M., Li, P., Zhang, L., Chen, Y.: Stock market trend prediction using high-order information of time series. IEEE Access **7**, 28299–28308 (2019)
27. Arafat, J., Habib, M.A., Hossain, R.: Analyzing public emotion and predicting stock market using social media. Am. J. Eng. Res. **2**(9), 265–275 (2013)
28. Nguyen, T.H., Shirai, K., Velcin, J.: Sentiment analysis on social media for stock movement prediction. Expert Syst. Appl. **42**(24), 9603–9611 (2015)

29. Oliveira, N., Cortez, P., Areal, N.: Stock market sentiment lexicon acquisition using microblogging data and statistical measures. Decis. Support Syst. **85**, 62–73 (2016)
30. Wang, F., Zhang, Y., Rao, Q., Li, K., Zhang, H.: Exploring mutual information-based sentimental analysis with kernel-based extreme learning machine for stock prediction. Soft Comput. **21**(12), 3193–3205 (2017)
31. Ren, R., Wu, D.D., Liu, T.: Forecasting stock market movement direction using sentiment analysis and support vector machine. IEEE Syst. J. **13**(1), 760–770 (2018)
32. Shi, L., Teng, Z., Wang, L., Zhang, Y., Binder, A.: Deepclue: visual interpretation of text-based deep stock prediction. IEEE Trans. Knowl. Data Eng. **31**(6), 1094–1108 (2018)
33. Arora, N., et al.: Financial analysis: stock market prediction using deep learning algorithms (2019)
34. Vargas, M.R., De Lima, B.S.L.P., Evsukoff, A.G.: Deep learning for stock market prediction from financial news articles. In: 2017 IEEE International Conference on Computational Intelligence and Virtual Environments for Measurement Systems and Applications (CIVEMSA), pp. 60–65. IEEE (2017)
35. Li, X., Xie, H., Wang, R., Cai, Y., Cao, J., Wang, F., Min, H., Deng, X.: Empirical analysis: stock market prediction via extreme learning machine. Neural Comput. Appl. **27**(1), 67–78 (2016)
36. Akita, R., Yoshihara, A., Matsubara, T., Uehara, K.: Deep learning for stock prediction using numerical and textual information. In: 2016 IEEE/ACIS 15th International Conference on Computer and Information Science (ICIS), pp. 1–6. IEEE (2016)
37. Dash, R., Dash, P.K.: A hybrid stock trading framework integrating technical analysis with machine learning techniques. J. Financ. Data Sci. **2**(1), 42–57 (2016)
38. Liu, G., Wang, X.: A numerical-based attention method for stock market prediction with dual information. IEEE Access **7**, 7357–7367 (2018)
39. Reddy, R., Shyam, G.K.: Market data analysis by using support vector machine learning technique. In: Proceedings of International Conference on Computational Intelligence and Data Engineering, pp. 19–27. Springer (2019)

Macroscopic Characterization of Grating Coupled Waveguide Structures for Optical Notch Filtering

Aleena Devasia[✉] and Manisha Chattopadhyay

Department of Electronics and Telecommunication,
Vivekanand Education Society's Institute of Technology, Mumbai, India
{2017aleena.devasia,manisha.chattopadhyay}@ves.ac.in

Abstract. The use of optical waveguide gratings that function as spectrally selective loss elements is discussed. The Wavelength Filtering capability of input grating coupled waveguide structure is demonstrated. In addition to the conventional function of such waveguide structures to couple an incident surface beam from an optical source like an optical fiber into a planar waveguide, a modified design to have a predetermined wavelength response of that of a notch filter has been modelled and analysed. The structure is designed for 1310 nm wavelength. An improved waveguide structure with addition of chirped grating section is modelled. Design of planar waveguide structures for optical notch filtering is assessed and its application in optical networking and communication systems is discussed.

Keywords: Optical notch filter · Grating coupler · Optical waveguide · Wavelength filtering · Optical isolator

1 Introduction

Optical grating based devices find various applications due to their broad range of optical effects [4]. The design of grating based waveguide structures for wavelength selective applications has been discussed, particularly for wavelength filtering in optical networking and communication systems. Optical devices such as channel dropping filters, Bragg reflectors and couplers are used in optical communication networks and systems for wavelength division multiplexing applications. Optical notch filters are often employed in optical waveguide and fiber communication systems for removal or attenuation of wavelengths of light to avoid system performance degradation [9]. They are used for gain equalization in erbium-doped amplifiers and in cascaded Raman amplifier/lasers for reduction of unnecessary Stokes' frequency orders [9]. Optical Notch filters are connected between two optical fibers in a system or as devices connected between two waveguide systems [5]. Fiber-based filters are less compact compared to planar waveguide structures and relatively costlier to manufacture [9]. The optical

© Springer Nature Switzerland AG 2020
J. S. Raj et al. (Eds.): ICIDCA 2019, LNDECT 46, pp. 27–36, 2020.
https://doi.org/10.1007/978-3-030-38040-3_3

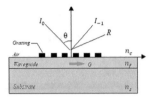

Fig. 1. Structural schematic of linear rectangular grating coupler on planar waveguide.

components in fiber form come with an added challenge of integrating efficiently in photonic integrated circuits [1]. Grating based planar waveguide structures which can be used as notch filters are thus presented. The fundamental concept of grating coupled waveguide can be explained using Fig. 1. The grating provides the necessary coupling when the condition in Eq. (1) is fulfilled [8].

$$n_{eff} = n_c sin\theta + \frac{m\lambda_0}{\Lambda} \tag{1}$$

where n_{eff} is the effective refractive index of the guided mode, n_c is the cover layer (air) refractive index, θ coupling angle, m is the diffraction order, λ_0 free space wavelength, Λ grating period [8] as shown in Fig. 1. This type of interaction is clearly useful for coupling light into or out of an optical waveguide. Such a grating coupled waveguide structure with structural variations which can be used as optical isolator with wavelength filtering capability has been proposed. An improved design using chirped grating section along the waveguide has been modelled and analysed. The structural variations were analysed for macroscopic characterization of the designed structures with the help of Transmission coefficient calculations using FEM based tool [12]. For applications in optical communication systems with $\lambda \sim 1550$ nm, Er-doped $LiNbO_3$ waveguide-lasers are used [6]. The characteristics of the modelled structures suggest their potential for the application in solid-state waveguide lasers. The proposed structures can be used for wavelength filtering in Wavelength Division Multiplexing systems for channel dropping. A narrow bandwidth design can be developed which is highly preferred for filter and switch applications. Such structures can be used in advanced optical signal processing and communication systems as narrowband spectral filters and in arrays of optical switches or modulators [7].

2   Analysis and Structure Model

A linear dielectric grating over planar waveguide structure is used to operate as a beam coupler and convert surface wave to guided beam along the waveguide film as shown in Fig. 1. If the electric field distribution is considered invariant along y-direction, the two-dimensional condition $\frac{\partial}{\partial y} = 0$ can be assumed. For the surface-wave incident over the grating, suppose the field is guided by the waveguide film and progresses longitudinally from the left in the form $exp[i(\beta_{sw} - \omega t)]$, where β_{sw} is real wavenumber in lossless situations. The incoming energy is scattered by the gratings into fields that vary as $exp[i(k_{xn} - \omega t)]$. They are called space-harmonic fields. Here, k_{xn} is complex value given by $k_{xn} = \beta_n + i\alpha$. The wavenumbers of the harmonic components of leaky-wave field produced due to diffraction from grating is given by (2).

$$\beta_n = \beta_o + \frac{2n\pi}{\Lambda} = k_0 \left(N + n\tfrac{\lambda}{\Lambda} \right) \tag{2}$$

where $n = 0, \pm 1, \pm 2...$, N is the effective refractive index of the waveguide and Λ is the grating periodicity. The fundamental term β_0 is approximately equal to the propagation factor β_{sw} of the incident surface wave $\beta_0 \approx \beta_{sw}$ with the decay factor, $\alpha \ll \beta_0$ [10]. The grating scatters the incident energy into diffracted orders and this leakage of energy is given by the factor α called the leakage parameter. Each scattered field forms a leaky-wave beam [10]. The structure in Fig. 1 can thus be considered as one that converts a surface wave into one or more leaky waves [10]. Grating coupled waveguide structure for 1310 nm wavelength transmission is described in Sect. 2.1. The structure is modified for wavelength filtering with different material combinations of gratings. An improved design using chirped gratings is explained with all the simulation results shown in Sect. 4. A Finite Element Method (FEM) based simulation tool [12] is used to obtain transmission coefficients and thus help in macroscopic characterization of these structures.

2.1 Transmission Mode Grating Coupled Waveguide

A grating coupled waveguide transmitting single wavelength along the waveguide can be designed using uniform rectangular gratings with transmission characteristics as shown in Fig. 2(a). The structure consists of linear gratings ($n = 2.87$) with periodicity $\Lambda = 900$ nm, 50% dutycycle and thickness d = 75 nm. The Electric Field distribution for 1310 nm is shown in Fig. 2(b). It can be observed that the electric field is confined within the waveguiding film and wavelength is propagated along the waveguide (shown in red) following the characteristics shown in Fig. 2(a). The structure is modified to obtain wavelength selective transmission characteristics which can be used to reject selective wavelength through the waveguide. This involves complete interference at the specific wavelength

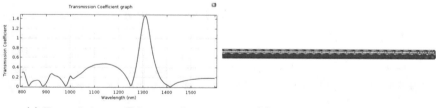

(a) Transmission coefficient graph.   (b) Electric Field confinement.

Fig. 2. Transmission mode grating coupled waveguide output characteristics

incident at a particular angular orientation which results in no transmission of that wavelength. The structural analysis and modelling of a symmetric grating coupled waveguide structure for wavelength filtering is explained. The structure is modified with chirped grating section to improve the notch filtering characteristics of the design model.

2.2   Chirped Grating Design Analysis

Chirped waveguide gratings have a non-uniform grating period (Λ) along the length of the structure [3]. The grating design may alter in period by increasing or decreasing consistently such that different grating constituents cause transmission at disparate wavelengths [3]. The alteration is governed by the chirp parameter (a) which if constant, $a = \pm const$, forms a linearly chirped grating. In the simulated results, the impact of the grating chirp represented by the variable chirp parameter 'a' on the transmission spectra was observed and analysed to design a notch filter.

Formulation of Bloch Wave Problem: Plane waves are the natural modes of free space propagation. Similarly, eigen modes of periodic media are called Bloch waves [11]. The field dispersion and micro-structure properties of aperiodic grating structures can be explained using the Bloch Wave Approach [11]. An eigenvalue system can be built to solve this Bloch Wave Problem. The grating perturbation can be expressed mathematically as [11]:

$$\varepsilon(x, y, z) = \varepsilon_0(x, y) + \Delta\varepsilon(x, y, z) \tag{3}$$

The wave equation reduces to [11]

$$\frac{\partial^2}{\partial z^2}a(z) + k_0^2(z)a(z) = 0 \tag{4}$$

with the wave equation of the TE modes as $E(x, y, z) = a(z)e(x, y)$, where $a(z)$ is the slow varying amplitude of the field and $e(x, y)$ is the electric field in the unperturbed waveguide, $k_0 = \omega\sqrt{\mu_0\varepsilon_0}$ is the wave vector in free space and $N(z) = \sqrt{\varepsilon(z)}$ is the waveguide local refractive index of the unperturbed waveguide [11]. The Eq. (4) is similar to the wave equation describing TE plane wave transmission in a inhomogeneous medium [11]. The modal index for aperiodic grating structure can be represented by a quasi-sinusoidal function [11] as

$$\varepsilon(z) = \overline{\varepsilon}(z) + \Delta\varepsilon(z)cos[Gz + \theta(z)] \tag{5}$$

where G is the mean wave vector of the perturbation and $\theta(z)$ describes the spatial deviations from the mean period [11]. A transformation on the spatial spectrum of $a(z)$ can be performed to express it in terms of wave vector. The unknown functions can be discredited and interpolated to solve the Eigen Value Problem in aperiodic structure, known as a Finite Element Approach, producing a resultant field that consists of a linear combination of the solution eigenvectors [11]. Eigen vector can be split into two block vectors to obtain the reflection/transmission coefficient.

3 Design Methodology

The structure shown in Fig. 1 was modified with combination of different materials (refractive indices) and addition of material defects along the grating section to achieve the transmission characteristics of a filtering element that can be used in optical networks. A grating coupled waveguide structure with following non-uniform refractive index (RI) profile as in Fig. 3(a) is used to design band reject filter. The materials used for grating over the waveguide structure A, B and C are of refractive indices $n_A = 2.222$ (Ta_2O_5), $n_B = 1.65$ (SiO_2) and $n_C = 2.02252$ (Si_3N_4). The waveguide refractive index is 3.37 (n_f) over a substrate of refractive index 1.44 (n_s). The schematic of waveguide filter design model with chirped grating is shown in Fig. 3(b).

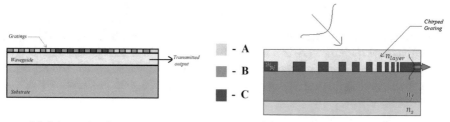

(a) Schematic of grating structure model.   (b) Schematic of waveguide structure with chirped gratings.

Fig. 3. Schematics of design models.

A standard 2 μm thick box with a SOI layer n_f is used along with a top Silicon layer n_g 355 nm thick to model the chirped grating based waveguide design structure. The grating coupler directionality is enhanced by adding a polymer coating with thickness of 1.233 μm as a top cladding layer of refractive index n_{layer} as shown in Fig. 3(b). The geometric aspects of the gratings are such that the use of the upper cladding helps achieve an adequate filling between the gratings. The uniform grating section is followed by a chirped grating section such that the periods of chirped grating section decrease linearly by a parameter 'a' obeying the following equation: $\Lambda_{n+1} = \Lambda_0 - (n \times a)$ where $n = 0, 1, 2..N_c$ and Λ_0 is the grating period of uniform grating section. Here, N_c is the number of chirped gratings. The dutycycle (defined as the ratio of the grating ridge width (w) over the grating period (Λ)) throughout the grating section is constant.

4 Results and Discussion

The structures described in Sect. (3) were modelled and analysed using FEM based simulation tool [12]. A characteristic change in the transmission spectrum from Fig. 2(a) to that of a band-reject filter type of transmission spectrum is observed with the variation in structure. Further, modification of the structure with the help of chirped grating to improve transmission characteristics is done as shown in Sect. 4.1. For the structure schematic in Fig. 3(a), the transmission characteristics with different grating thickness (d) is shown in Fig. 4(a). It is observed that a variation in the grating thickness (d) produces a change in the amount of transmission in the range of 1045 to 1065 nm wavelength. For a specific grating thickness, the transmission output is zero for this wavelength. The transmission characteristics with different grating period (Λ) is shown in Fig. 4(b). Thus, appropriate choice of grating period can be made for wavelength band which is to be rejected. To improve the wavelength selectivity of the structure, a non-uniformity was introduced by varying the width of particular material grating linearly through the structure. The transmission characteristics with variation in width of third grating is shown in Fig. 5(a). Based on the simulation results for transmission coefficient, the structure is optimized to filter 1310 nm wavelength and the transmission characteristics thus obtained is shown in Fig. 5(b). The structural parameters used to obtain the transmission characteristics as in Fig. 5(b) are listed in Table 1. As observed in Fig. 5(b), the transmission characteristics show a broad bandwidth of rejected wavelength around 1310 nm. The characteristics are improved drastically with the addition of chirped gratings along the length of the structure and is described in the following Sect. 4.1.

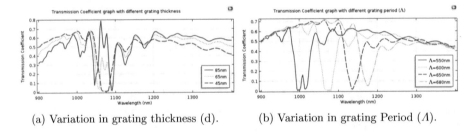

(a) Variation in grating thickness (d).   (b) Variation in grating Period (Λ).

Fig. 4. Transmission characteristics with structural variations

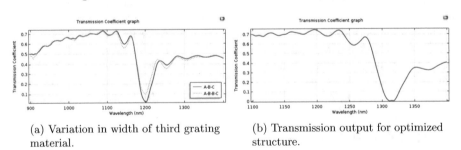

(a) Variation in width of third grating material.   (b) Transmission output for optimized structure.

Fig. 5. Transmission characteristics for 1310 nm.

4.1 Improved Filter Design Using Chirped Grating Waveguide Structure

The transmission output of chirped grating waveguide structure is shown in Fig. 7. The physical elucidation of the chirped grating can be made with parametric changes in the chirped grating section. As the chirp grating section is followed by the uniform grating section, the structure can be seen as a superposition of uniform gratings centered at continuously spaced wavelengths. This is observed in the form of ripples in the transmission output between 1330 nm and 1420 nm wavelength band as shown in Fig. 6, that are shallower as the number of chirped gratings is raised. One can also observe the absence of shallow dips in the 1200 nm to 1300 nm band. The design structural parameters are tabulated in Table 1. The number of chirped gratings were varied to arrive at the optimum number of chirped gratings that would be required to reduce the ripple in the transmission curve and have a single wavelength rejection (dip) in the spectrum. The chirp parameter 'a' is used to model a non-uniform chirp as shown in Fig. 7(a) to improve the transmission characteristics. The first five periods ($\Lambda_1 - \Lambda_5$) follow the relation described by $\Lambda_{n+1} = \Lambda_0 - (n \times 2 \times a)$ where $n = 1, 2..5$ and Λ_0 is the grating period of uniform grating section and last four periods ($\Lambda_6 - \Lambda_9$) follow the relation described by $\Lambda_{n+1} = \Lambda_0 - (n \times a)$ for $n = 6, 7, 8, 9$. Thus, the position of the dropped wavelength and the amount of transmission can be influenced by the value of the grating chirp. The results in Fig. 7 show the effect of different structural variations on the transmission output. A change in the grating period (Λ_0) of the uniform grating section produces a change in the chirp grating section taking the chirp parameter (a) into account. The effect is observed in Fig. 7(c) by a shift in the dip in the transmission spectrum. A variation in the grating dutycycle (w/Λ) shows a clear change in the width of the rejected wavelength bandwidth. For the structure, as the dutycycle is increased, the bandwidth becomes narrower as shown in Fig. 7(d). For a fixed grating period Λ_0, a variation in the grating depth which is constant throughout the structure, shows no change in the shape of the spectrum but a shift in the rejected wavelength (Fig. 7(b)).

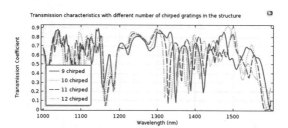

Fig. 6. Variation in number of chirped gratings.

Thus, careful analysis of these structural parameters help in designing and modelling waveguide structures for different wavelength filtering. Figure 8(a) shows the transmission characteristics of an optimized structure for notch filter (1310 nm) with parameters given in Table 1. By varying the grating period ($\Lambda = 681$ nm) of the design structure, a 1550 nm filter is modelled having transmission characteristics as in Fig. 8(b).

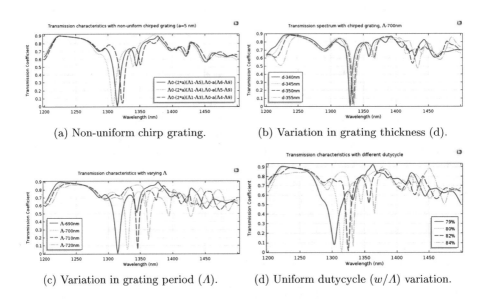

(a) Non-uniform chirp grating.   (b) Variation in grating thickness (d).

(c) Variation in grating period (Λ).   (d) Uniform dutycycle (w/Λ) variation.

Fig. 7. Chirped grating waveguide structure with structural variations.

Table 1. Structural parameters for optimized design models

Parameter	Symbol	Value
Grating Period	Λ	787 nm
RI of film	n_f	3.37
RI of substrate	n_s	1.44
Height of grating	d	56 nm
Width of grating	w	393.5 nm
RI of grating 1	n_A	2.222
RI of grating 2	n_B	1.645
RI of grating 3	n_C	2.02252
Waveguide thickness	t_f	136 nm

a. Parameters for Grating coupled waveguide structure

Parameter	Symbol	Value
Uniform Grating Period	Λ_0	693 nm
Grating dutycycle	dc	0.8
RI of film	n_f	1.43
RI of substrate	n_s	3.48
Height of grating	d	340 nm
Width of grating	w	$\Lambda \times dc$
RI of grating	n_{g1}	3.48
Waveguide thickness	t_f	150 nm
Upper cladding thickness	L	878 nm
Upper cladding RI	n_{layer}	1.38
Number of chirped gratings	N_c	9
Chirp (first 5 gratings)	a	10 nm
Chirp (rest 4 gratings)	a	5 nm

b. Parameters for Chirped grating Waveguide structure

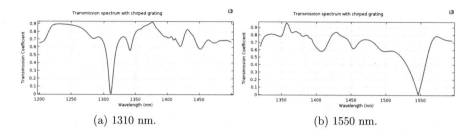

(a) 1310 nm.   (b) 1550 nm.

Fig. 8. Transmission characteristics of optimized Chirped grating filter structure for 1310 and 1550 nm.

5   Conclusion

Optical Notch filters designed using planar waveguide structures with linear rectangular gratings has been presented. The waveguide structures operate as beam couplers and convert surface wave to guided beam along the waveguide, thus acting like optical isolators with filtering capability. The transmission mode grating coupled waveguide structure has been modified with material and structural variations to achieve this functionality. Appropriate combination of grating materials produce zero transmission of particular wavelength at the output. The structural parameters like the grating periodicity and thickness decide the wavelength to be dropped and amount of transmission respectively. The structure is altered with addition of chirped grating section along the waveguide to improve the filter bandwidth significantly. The dropped wavelength and the spectrum bandwidth clearly depend on the chirp parameter along with the other grating parameters. These help in complete characterization of the designed structures and hence attain an optimized structure for the particular wavelength filtering.

References

1. Suresh, L.R.D., Ponmalar, S., Sundaravadivelu, S.: Analysis of optical waveguide grating filter for optical wireless communication. In: IEEE International Conference on Signal Processing and Communications (SPCOM) (2004)
2. Dabos, G., Bolten, J., Prinzen, A., Giesecke, A., Pleros, N., Tsiokos, D.: Perfectly vertical and fully etched SOI grating couplers for TM polarization. Opt. Commun. **350**, 124–127 (2015)
3. Gemzický, E., Müllerová, J.: Apodized and chirped fiber Bragg gratings for optical communication systems: influence of grating profile on spectral reflectance. In: Photonics, Devices, and Systems IV (2008)
4. Quaranta, G., Basset, G., Martin, O.J.F., Gallinet, B.: Recent advances in resonant waveguide gratings. In: Laser & Photonics Reviews, vol. 12, no. 9, p. 1800017 (2018)
5. Vengsarkar, A.M., Lemaire, P.J., Judkins, J.B., Bhatia, V., Erdogan, T., Sipe, J.E.: Long-period fiber gratings as band-rejection filters. J. Lightwave Technol. **14**(1), 58 (1996)
6. Brinkmann, R., Sohler, W., Suche, H.: Continuous-wave erbium-diffused LiNbO3 waveguide laser. Electron. Lett. **27**(5), 415 (1991)

7. Rosenblatt, D., Sharon, A., Friesem, A.A.: Resonant grating waveguide structures. IEEE J. Quantum Electron. **33**(11), 2038–2059 (1997)
8. Waldhausl, R., Schnabel, B., Dannberg, P., Kley, E.-B., Brauer, A., Karthe, W.: Efficient coupling into polymer waveguides by gratings. Opt. Soc. Am. **36**(36), 20 (1997)
9. Narayanan, C., Alpharetta, G., Presby, H.M.: Optical planar waveguide notch filter. U.S. Patent 5,796,906, issued 18 August 1998
10. Tamir, T., Peng, S.T.: Analysis and design of grating couplers. Appl. Phys. **14**, 235–254 (1977)
11. Peral, E., Capmany, J.: Generalized Bloch wave analysis for fiber and waveguide gratings. J. Lightwave Technol. **15**(8), 1295–1302 (1997)
12. Comsol Multiphysics by COMSOL ©, version 5.3a

Design and Development for Image Transmission Through Low Powered Wireless Networks Using Color Space Conversion Module

C. S. Sridhar[1]([⊠]), G. Mahadevan[2], S. K. Khadar Basha[3],
B. N. Shobha[3], and S. Pavan[3]

[1] Bharathiar University, Coimbatore, India
sridhar_cs@yahoo.com
[2] ANNAI College, Kumbakonam, India
g_mahadevan@yahoo.com
[3] SJCIT, Chickballapur, India
basha_skb@rediffmail.com, bnshobha67@gmail.com,
myselfpavans@gmail.com

Abstract. Transmission of images over wireless network requires the utilization of huge Bandwidth, energy, memory space and data rate. Image compression plays a vital role in delivering an efficient processing with the lower end configurations, efficient data transfer, and better utilization of available memory space by significantly reducing the number of bits required for presentation without any appreciable change in its quality and information content. Hence it is widely used for image transmission process.

This paper mainly focuses on the development of hardware module for transmission of color image through low powered wireless network by compressing the color image using Color Space Conversion module. Zigbee being appears as a low powered wireless device, which can offer low Bandwidth and less data rate. It becomes very critical and challenging for transmission of color images without any appreciable delay. Images should be compressed within the given data rate before transmitted. These drawbacks are overcome by using the Color Space Conversion module.

Keywords: Color Space module · Zigbee · MCU · Zigbee TX · Zigbee RX · YUV

1 Introduction

Digital image processing deals with operations on a digital image such as compression, segmentation, resizing etc.

An 2-D image is defined by the function f(x, y). Compression [6] of digital image is possible because color channels are statistically dependent, pixel values are spatially correlated and human vision system does not perceive all details.

© Springer Nature Switzerland AG 2020
J. S. Raj et al. (Eds.): ICIDCA 2019, LNDECT 46, pp. 37–44, 2020.
https://doi.org/10.1007/978-3-030-38040-3_4

Transmission of images over wireless network requires consumption of huge Bandwidth, energy, memory space and data rate. Compression of image plays a very vital role in efficient processing with the low end configurations, [7] efficient data transfer, and better utilization of available memory space by reducing the number of bits required for presentation without any appreciable change in quality and information content.

This research mainly focuses on development of hardware module for transmission of color image through low powered wireless network by compressing the color image using Color Space conversion module. Zigbee being appears as a low powered wireless device with low Bandwidth is used for short range communications. It becomes very critical and challenging for transmission of images without any appreciable delay. Images are to be compressed within the given data rate and transmitted. The drawbacks of this are overcome by using Color Space Conversion module. Hence it is widely used for image transmission in IoTs, Intra-vehicles communication, Wireless Sensor networks, Home-Automation, Surveillance, etc.

2 Methodology

2.1 Design and Implementation of a Color Space Conversion Module Algorithm

Generally, RGB color space format has become an essential part of image processing. Real time captured images and videos are present in Red, Green and Blue colors consisting of three channels and require huge bandwidth [1]. Transmission and storage of high bandwidth images has become very tedious.

To overcome this problem, images are converted into lesser bandwidth color format called YUV or YCbCr or YUV444 format where Y stands for the luminance component (brightness).

Cr and Cb stands for Chrominance namely red and blue. This YUV colors has different formats to describe color images as well as gray images and has three channels. Different YUV formats are YUV444, YUV422, and YUV420 [3].

Thus the high bandwidth YUV 444 is converted into comparatively lesser Bandwidth format of YUV 420. For further reduction in bandwidth of these images again we are converting it into grayscale images by approximating UV samples to zero. Thus finally it has a single channel and as a result of this further reduction in bandwidth is achieved [5]. These less bandwidth images are stored and transmitted efficiently [2]. At the receiving end grayscale or Y images are converted into YUV420 images. This YUV420 images are converted into RGB color images (Fig. 1).

To convert from RGB colors to Gray Initially Convert RGB images and videos into YUV 444 format using OpenCV command The converted image split the image into three different Y, U and V colors using the OpenCV command. After the splitting the image store the separate colors in different buffers This YUV444 color format convert into YUV420 color format In YUV420 format, all the Y values come first, followed by all the U values and finally followed by all the V values. To convert from Gray to RGB First convert Y to YUV420 format Merge all the Y, U and V colors into single YUV

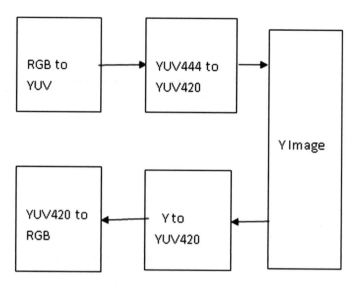

Fig. 1. Shows the framework of the model Color Space Conversion module

format using OpenCV command. Merge all the Y, U and V colors into single YUV format. Finally convert YUV image into RGB color format [1].

2.2   Design and Development of Hardware Module to Transfer Color Image Using Low Powered Wireless Networks

Using low powered wireless network using color space conversion module. The input image is captured by the camera and sent to the SD card. In the SD card the color image to be transmitted is resized using Resizing algorithms. The Resized color image is sent to the microcontroller using com ports. In the microcontroller color image is converted to Gray scale image using Color Space conversion Module algorithm. This algorithm will start generating array of pixels called Minimum Coded Units, The resulting Gray scale image obtained is serially sent to the Zigbee Transmitter module for wireless transmission. At the receiving end Zigbee receiver module receives the incoming data and sends it to the computer using com port communication. On receiving the data, a program converts Gray scale to color and displays the color image [11]. Zigbee, a low powered wireless device offers low Bandwidth is used for short range communications. It becomes very critical and challenging for transmission of images without any appreciable delay. Images are to be compressed within the given data rate and transmitted.

Figure 2 shown below describes the implementation of hardware module to transfer color image using low powered wireless network using color space conversion module for short range communication.

The input image is captured by the camera and sent to the SD card serially. In the SD card the color image to be transmitted is resized using Resizing algorithms. The Resized color image is sent to the microcontroller using com ports. In the microcontroller color

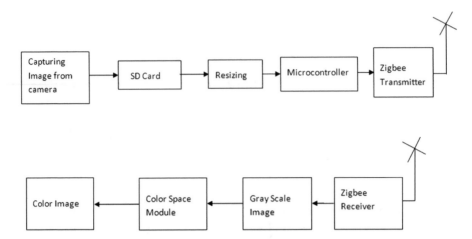

Fig. 2. Shows the block diagram of hardware module for low powered wireless network

image is converted to Gray scale image using Color Space conversion Module algorithm. This algorithm will start generating array of pixels called Minimum Coded Units, The resulting Gray scale image obtained in microcontroller is serially sent to the Zigbee Transmitter module for wireless [9, 10] transmission. Zigbee is a wireless device that operates in 2.4 GHz. Its maximum data rate is 250 Kbits/s. The color image to be transmitted are compressed and then transmitted.

At the receiving end Zigbee receiver module receives the incoming data and send it to computer using com port communication. On receiving the data, decompression algorithm converts Gray scale to color image using color space conversion module [12, 13].

2.3 Parameters for Measurement

A graph plotted taking pixels vs time and distance vs time.

2.4 Hardware Requirements

See Fig. 3.

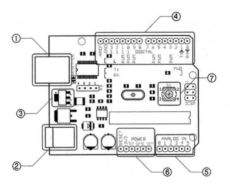

Fig. 3. ATMEL 1080 microcontroller.

ATMEL 1080

1. USB Connecter
2. Power Connecter (7 V–12 V)
3. Automatic power switch
4. Digital pins – 14 Input/Output
5. Analog pins – 6 Input/Output
6. Power pins – 6 pins
7. Reset switch (Figs. 4 and 5).

Fig. 4. Zigbee transmitter module.   **Fig. 5.** Zigbee receiver module.

2.5 Hardware Implementation

See Figs. 6 and 7.

Fig. 6. Hardware module connected to zig-   **Fig. 7.** Hardware module of zigbee
bee transmission module.   receiver.

3 Results

The color image is transmitted using low powered wireless network after compressing the image using Color Space conversion module algorithm. The results of the received image is shown in the form of snapshots.

3.1   Received Colour Image

See Figs. 8, 9, 10, 11, 12 and 13.

Fig. 8. 160×120 Pixels   **Fig. 9.** 320×240 Pixels

Fig. 10. 160×120 Pixels   **Fig. 11.** 320×240 Pixels

Fig. 12. 160×120 Pixels   **Fig. 13.** 320×240 Pixels

3.2   Graphs

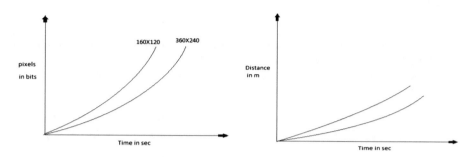

A graph of pixels transmitted vs time taken, distance vs time is plotted.

(i)   As the pixel size increases time taken also increases.

(ii) As the distance increase the time also increases.

4   Conclusion and Future Enhancement

This paper mainly focuses on the compression of images for the transmission of color image using low powered wireless network. The approach involves designing and developing a hardware module for image compression and transmission.

The developed hardware system has the capability of Resizing the incoming color image, color to Gray scale image in 8-bit microcontrollers by using the algorithm of Color Space Conversion module and transmits using Zigbee transmitting module. At the receiving end the zigbee receiver module receives the Gray scale image and using decompression algorithm converts gray scale image into color image. It was found that with an appreciable delay of 40 and 160 s it was able to transmit 160×120 pixels for 10 m and 320×240 pixels for 15 m using low powered wireless network.

In future it is evident that this hardware module can be used to transmit color image in any wireless medium.

Compliance with Ethical Standards

All author states that there is no conflict of interest.

We used our own data.

We used Author's photo only in results and discussion.

References

1. Sridhar, C.S., Mahadevan, G., Khadar Basha, S.K.: Design and implementation of a color space conversion module using open-computer vision. J. Adv. Res. Dyn. Control Syst. Scopus SJR **12**(Special issue), 2321–2329 (2017). ISSN 1943-023X

2. Sridhar, C.S., Mahadevan, G., Khadar Basha, S.K.: Compression of image using hybrid combination of Haar wavelet transform and color space model in noisy environment. Int. J. Eng. Technol. **12**(Special issue), 7(1.5) (2018). 283-285-2018
3. Sridhar, C.S., Mahadevan, G., Khadar Basha, S.K.: Transmission of images using image compressions for low powered wireless networks. J. Adv. Res. Dyn. Control Syst. **12** (Special issue), 2158–2165 (2018). ISSN 1943-023x
4. Pathak, P.H., Feng, X., Hu, P., Mohapatra, P.: Visible light communication, networking, and sensing: a survey, potential and challenges. IEEE Commun. Surv. Tutor. **17**(4), 2047–2077 (2015)
5. Shinde, M.V., Kulkarni, P.W.: Automation of electricity billing process. In: 2014 International Conference on Power, Automation and Communication (INPAC), pp. 18–22. IEEE (2014)
6. Donoho, D.L., Vetterli, M., Devore, R.A., Daubechies, I.: Data compression and harmonic analysis. IEEE Trans. Inf. Theory **44**, 2435–2476 (1998)
7. Haar, A.: Zur theorieder orthogonalen funktionensysteme. Math. Ann. **69**(3), 331–371 (1910)
8. Besslich, Ph.W., Trachtenberg, E.A.: The sign transform: an invertible non-linear transform with quantized coefficients. In: Moraga, C. (ed.) Theory and Application of Spectral Technique, pp. 150–160. University Dortmund Press, Dortmund (1988)
9. Chen, F., Chandrakasan, A.P., Stojanovic, V.M.: Design and analysis of a hardware-efficient compressed sensing architecture for data compression in wireless sensors. IEEE J. Solid-State Circuits **47**(3), 744–756 (2012)
10. Díaz, M., et al.: Compressed sensing of data with a known distribution. Appl. Comput. Harmon. Anal. **45**(3), 486–504 (2018)
11. Duda, K., Turcza, P., Marszałek, Z.: Low complexity wavelet compression of multichannel neural data. In: 2018 International Conference on Signals and Electronic Systems (ICSES). IEEE (2018)
12. Nassar, M.R., Helmers, J.C., Frank, M.J.: Chunking as a rational strategy for lossy data compression in visual working memory. Psychol. Rev. **125**(4), 486 (2018)
13. Sahin, A., Martin, O., O'Hare, J.J.: Intelligent data compression. U.S. Patent Application No. 10/013,170
14. Parvanova, R., Todorova, M.: Compression of images using wavelet functions. In: "Applied Computer Technologies" ACT 2018, p. 90 (2018)
15. Zemliachenko, A.N., Abramov, S.K., Lukin, V.V., Vozel, B., Chehdi, K.: Preliminary filtering and lossy compression of noisy remote sensing images. In: Image and Signal Processing for Remote Sensing XXIV, vol. 10789, p. 107890V. International Society for Optics and Photonics, October 2018
16. Gregor, K., Besse, F., Rezende, D.J., Danihelka, I., Wierstra, D.: Towards conceptual compression. In: Advances in Neural Information Processing Systems, pp. 3549–3557 (2016)

Classification and Comparison of Cloud Renderfarm Services for Recommender Systems

J. Ruby Annette[1], W. Aisha Banu[2(✉)], and P. Subash Chandran[3]

[1] Department of Computer Science, Saveetha School of Engineering, Chennai,
India
rubysubash2010@gmail.com
[2] Department of Computer Science, B.S. Abdur Rahman Institute of Science and
Technology, Chennai, India
aisha@crescent.education
[3] NECAM, Chennai, India
subashchandran.paramasivam@necam.com

Abstract. In India, some of the bollywood movies like the "Lord of the rings"; "Avatar" etc. are exemplifications of the magic that the Visual Effects (VFX) and the 3D animation techniques could create in a movie. In order to create this magic, the animation frames undergo the "rendering" process which takes a long time and contributes to long production time as it is a computationally intensive task. In order to decrease the production time and to satiate the high quality demand; the animation studios have resorted to the cloud renderfarm services. However, the animator's community betrothed to use the services are baffled by the plethora of cloud renderfarm service models and need a classification framework and a recommender system to recommend appropriate services. This work classifies the cloud renderfarm service models by identifying three key characteristics of the services and their components based on which the services could be recommended. The three key characteristics identified are: (a) Functional characteristics, (b) Non Functional characteristics and (c) Cost. The classification helps to analyze and compare the cloud renderfarm services based on these three key characteristics and identify and recommend the suitable, cost effective services for a particular rendering job. The proposed classification framework is explained and verified using the analysis and comparison of 11 popular cloud renderfarm services which could be incorporated in building a recommender system for cloud renderfarm services.

Keywords: Cloud computing · Distributed computing · Data mining · Recommender systems · Recommendation · Cloud renderfarms

1 Introduction

As necessity is the mother of invention, the need of the animation in film industry for rendering high quality within tight time schedules and low cost, was the motivation behind the exploration of the new technologies like Cloud computing for rendering

© Springer Nature Switzerland AG 2020
J. S. Raj et al. (Eds.): ICIDCA 2019, LNDECT 46, pp. 45–56, 2020.
https://doi.org/10.1007/978-3-030-38040-3_5

purposes [1, 2]. Today even a small startup animation studio with less investment potential can compete with large animation studios and produce animation films of high quality at affordable costs using the cloud rendering technology. The animators need to pay only for the resources that they have used and need not invest a huge amount in building the infrastructure or pay for the maintenance cost. Moreover, the Platform-as-a-Service (PaaS) and Software-as-a-Service (SaaS) offerings of cloud rendering services, relieves the animators from buying software licenses for all the software they use. Examples of such Cloud Renderfarm Services FoxRenderfarm [3], Rebusfarm [4], Renderrocket [5]. Several animation studios want to explore the advantages of cloud rendering services. However, they find it hard to compare and analyze the offers to identify the right service provider as the cloud renderfarm services are offered in different models and lack standardization, common protocols and terminology. This work is intended to bridge this gap and provide a classification of the cloud based renderfarm services that could be used in the recommender systems to recommend the cloud renderfarm services. There are many works on the survey and classification of the cloud services on a general basis but classification of cloud renderfarm services specific to the animation rendering is not available according to the best of our knowledge and literature survey. In this work, the cloud renderfarm services have been classified based on three key characteristics namely the: (a) Functional Characteristics, (b) Non Functional Characteristics and (c) Cost which could be incorporated in a recommender system to recommend services that satisfy the three key characteristics identified in this work. A cost analysis tree has also been proposed to perform the cost analysis different cloud renderfarm services quickly. The threefold contributions of this paper are:

- Proposes the classification of the cloud renderfarm service by identifying three key characteristics of cloud renderfarm services available in the current market (Sect. 2).
- Analyses the cost factor of the existing cloud renderfarm services and proposes new cost analysis tree for analysing the cost of the renderfarms (Sect. 3).
- Analyses and compares 11 popular cloud renderfarm services using the proposed classification (Sect. 4). The works related to this work (Sect. 5) and the future ideas (Sect. 6) are also discussed in detail.

2 Classification of Cloud Renderfarm Services

Renderfarms that use the cloud resources to render the frames of the animation files are called the cloud renderfarm services. In cloud renderfarms, the frames are rendered by scheduling them to the render nodes available in the cloud renderfarms. In cloud rendering, the number of render nodes used for rendering can be increased or decreased instantly according to the need and pay only for the resources used. Thus animators can save on the investment and the maintenance cost. However, security measures provided by the cloud renderfarm services should be given higher priority when selecting a service provider as the frames are rendered in the cloud.

The animators who intend to use these cloud renderfarm services get confused on exploring the websites of few popular cloud renderfarm services, about which service to choose, as there are no common protocols followed for describing the cloud renderfarm service models. In an attempt to overcome this problem, this work proposes a new pruned and modified classification of cloud renderfarm services based on the preliminary classification proposed in [20]. The new classification proposed identifies three key characteristics of the cloud renderfarm services and classifies the services based on these three key components namely: (a) Functional characteristics (Fn), (b) Non Functional characteristics (NFn) and (c) Cost (Cn) and their components as given in Fig. 1. The cloud renderfarm service components required to complete the function of animation files rendering like the Compute Unit, Render Engine software supported etc., are classified under the Functional Characteristics (Fn) Level of the classification. Whereas, the service components that are not required to render the files but are important to identify a good renderfarm service like the security and the technical support provided are classified under the Non Functional (NFn) level of classification. The components related to the cost of the cloud renderfarm services are classified under the Cost (Cn) level of classification. The proposed classification is applied to provide a analysis and comparison of 11 popular cloud renderfarm services. The cloud services are compared based on both the main components and the sub components indentified as given in Table 1 and is referred throughout the paper wherever applicable to provide examples for the classification components discussed.

2.1   Functional Characteristics of Cloud Renderfarm Services (Fn)

Cloud Renderfarm Delivery Models (F1)
The cloud computing service models in general are of three types namely the IaaS, SaaS and PaaS [6]. The cloud renderfarm services are usually delivered as IaaS or PaaS models. In most of the PaaS model of cloud renderfarms the required software are included as the SaaS in the services to provide the complete package required for rendering the files. The infrastructure required for computing like the compute unit, and storage space etc. are delivered as a service in the Infrastructure as a Service (IaaS) delivery model. The utilities required can be rented on an hourly basis. The user is charged only for the resources utilized. A leading player in the IaaS services in cloud is the Amazon EC2.

The render farm services also provide the rendering job management tool to control the rendering jobs in the renderfarms. Eg: RevUpRender [7], Rebusfarm [4] etc.

Cloud Renderfarm Deployment Model (F2)
The public Cloud Renderfarm services are advantageous as they allow the animators to scale up or scale down the render nodes and pay based on the pay-per-use model. The main risk to be evaluated before moving to the cloud rendering are the security issues as the frame files are rendered on the render nodes in the Cloud and hence should select a service provider only after verifying the security features. A Private Cloud Renderfarm is created by pooling up the resources of a 3d studio. The advantage of a private cloud renderfarm is that it enables the maximum utilization of the resources. A private renderfarm is more secured compared to a public Cloud renderfarm. However setting up a private renderfarm requires high investment and maintenance charges. The

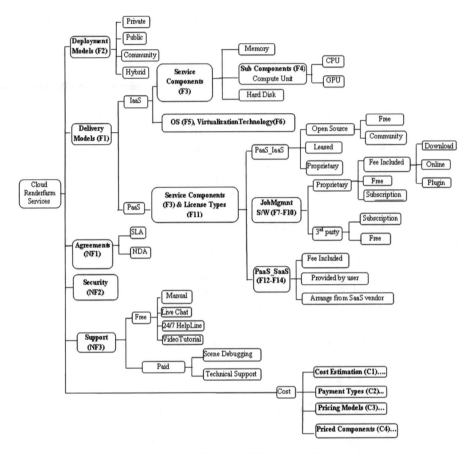

Fig. 1. Overall classification of cloud renderfarm services

combination of public and private clouds is called a hybrid cloud renderfarm services. It minimizes the security issues to be handled by 3d Studios. Whereas, in the Community based Renderfarms the users pool up their computing resources and share it among them for mutual benefits which is usually free of cost. Community renderfarm services come with various challenges and some tradeoffs in terms of privacy, security of data, quality of rendering etc. Eg. VSwarm [8].

Components and Sub Components (F3, F4)
The IaaS delivery model type of cloud renderfarm services provide only the infrastructure required to render the animation files and their components include Compute Unit, Memory, Hard Disk, etc. A compute unit is usually the processor capacity, which differs for each individual provider. The Architecture details of the compute unit comprises of the details like the processor type, speed, no of cores etc. Usually the no of cores may be 8, 12, 32 or 64 bits. Only few of the cloud renderfarm service providers have provided the complete information like the server clock speed, processor architecture etc.

Whereas, the PaaS type of cloud renderfarm services provides the entire infrastructure required like the compute unit, the rendering job management tools, rendering software licenses etc. The PaaS_IaaS is popular than the only-IaaS models among the animators. The Software as a Service model allows the software itself to be offered as a service on the Cloud environment using a web browser. Using the SaaS model, the software license can be obtained for an hour as well. Hence, the animators need not spend a huge amount as the software license fee. As the PaaS_SaaS based model is beneficial for both the software vendors and the animators almost all the software vendors like Autodesk have started to adopt the SaaS model.

The other Functional Components to be considered include the supported Operating System (F5) and Virtualization Technology (F6). The supported operating system and the Virtualization Technology play an important role, when a hybrid cloud is setup using the available resources and the IaaS resources. The Virtualization Technologies popularly used are the Xen and the KVM (Kernel-based Virtual Machine). However, the IaaS service providers usually support all the popular operating systems like the Linux, Windows etc. The PaaS services also support a variety of OS and the details about the supported OS by various IaaS and PaaS service providers is provided in the Table 1.

2.2   Non Functional Components of Classification (NF1–NF4)

The non-functional components of these services include the Formal Agreements signed (NF1), Security (NF2), Free Technical Support (NF3) and Paid Technical Support (NF4). The types of formal agreements popular among the cloud renderfarm service providers are the Service Level Agreements and the Non-Disclosure Agreements (NDA). However from the Table 1, it is clear that most of the cloud render farm companies do not provide SLA related or the QoS (Quality of Service) information expected from them. Though the leading IaaS cloud service providers like the Amazon EC2 have mentioned about the SLA's, only a minimum SLA conditions like the Service uptime are addressed. Most of the cloud render farm services that provide both the PaaS as well as the IaaS type, sign a Non-Disclosure Agreement called the NDA that prohibits the animation model designs from being copied and the works of the animators by other cloud users. In order to implement this, the service provider should be able to provide good security features.

Security is an important issue to be considered when choosing the renderfarm services. There are many security compliance standards [55] that the cloud renderfarm services should adhere to like: (a) FISMA, DIACAP, and FedRAMP (b) ISO 9001/ISO 27001 (c) DOD CSM Levels1–5 (d) SOC 1/SSAE 16/ISAE 3402 (e) SOC 2 (f) SOC 3 (g) MTCS Level 3 (h) ITAR (i) FIPS 140-2. The IaaS type of cloud renderfarm services like the Amazon EC2 adhere to all of these IT security compliance standards given above. However, most of the PaaS Cloud Renderfarm service providers have not published any data about the adherence to these security compliance standards, the information available regarding the security measures are limited to Ordinary FTP, Encrypted SFTP and Encrypted SCP etc.

3 Cost Characteristics of Cloud Renderfarm Services (Cn)

The priced components in IaaS and PaaS render farms are different. Cloud Renderfarm Services Cost Analysis Tree given in Fig. 2 provides an overview of the priced components in PaaS. In IaaS, the priced components are usually the render node unit and the data. The sub components under the render node unit include the compute unit may be a CPU or if required a GPU. Both the activation time and the use time are charged generally. The components related to the data that are priced are the data traffic and the storage space. Both the Incoming and outgoing traffic or the outgoing traffic alone may be charged. In PaaS, apart from the render node unit and the data as in IaaS, other components like the job management application software, software license, technical support; hyper threading may or may not be charged. For example, in Table 1, we see that most of the SP's do not charge for the job management software.

A Cloud Renderfarm Services Cost Analysis Tree has been proposed in this work to enable the animators to identify the hidden priced components, calculate the actual cost and compare the services cost to make informed decisions. In the proposed tree structure the Priced Components (PCn) has been identified for both type of cloud services individually as given in Fig. 2. The actual cost of a cloud renderfarm services is the sum of all the seven priced components identified in the cost analysis tree for each type of renderfarm service. This is important as the cost usually quoted in the websites and other documents is only for the compute unit or the render node and the other hidden costs like the cost of Data traffic and the storage etc. are not explicitly given and the users lack knowledge about these hidden priced components while making comparison of services based on cost.

For example, few Cloud Renderfarm services also charge for the incoming and outgoing data and the data storage. However, the information about the data traffic charges is not explicitly provided by all service providers. As the amount of data processed in animation rendering is very huge, the charges for the incoming and the outgoing data is also one of the important factor to be considered. Amazon EC2 charges more for the outgoing traffic than the incoming data traffic, while others charge equal amount. Similarly, while comparing the PaaS type of services based on cost, one of important factor to be considered is the software licensing cost. For example, comparing the two PaaS type services like Rebus Renderfarm and Rendicity, it can be identified that the render node cost of Rebus renderfarm includes the cost of Data and software license whereas, the Rendicity render node cost does not include the charges of Data and software license. So even if the render node cost of Rendicity is less than that of the Rebus Render farm, it is not the actual case as these hidden costs are not included.

4 Taxonomical Analysis and Comparison of 11 Cloud Renderfarm Services

The comparison table given in Table 1, compares the key characteristics of seven PaaS type of cloud renderfarm services namely the Rebus Renderfarm, RenderNet, Rendicity, RenderingFox, Render Rocket, RenderCore and RenderSolve. The popular IaaS type of cloud service providers namely the Amazon EC2, the Rackspace service characteristics and two other services namely the Vswarm and Sheepit that are of the community type of deployment model are also compared.

From the comparison Table 1, it is clearly evident that some of the cloud renderfarm services offer both PaaS and IaaS type of delivery models. The common component delivered is the CPU though some also offer GPU. The services generally have a proprietary license type and the agreements that are formally signed by the PaaS type of Cloud renderfarm services in the Non-Disclosure Agreement called the NDA. Most PaaS type of Cloud renderfarm services have not provided any details about the Service Level Agreements or about the Quality of Service (QoS) offered by them, except for some IaaS type of cloud services like the Amazon EC2.

The Render nodes of the 'Sheepit' renderfarm that is of the community type are free and are more popular among the students. The components that are generally priced are the render node, Data and the software license fee for the plug-ins and the rendering Engine software used. Though details of the operating system used some of the PaaS services are available, the details about the virtualization technology are not available except for the IaaS type of services.

Most of the PaaS services provide a cost estimation tool to estimate the rendering cost prior to using the actual service and use bench marks like Cine benchmark, Blender benchmark etc. The Rendering Job Manager (RJM) software used is mostly proprietary like Farminiser for Rebus RenderFarm and Wrangler for Rendicty and free of cost. The RJM software is usually delivered as a plug-in or can be downloaded online. Most of the PaaS renderfarm services include the Rendering Engine software license fee with the render node cost. However, some also arrange with the vendor for extra fee or provide limited support. Most of the PaaS services support all popular Render Engine, animation software and plug-ins. They also provide free technical support through various means like the Email, live chat, raising tickets, help centre, forum, user manual etc. Very few offer paid services like scene debugging.

From the comparison tables, we could identify that the IaaS type of cloud services are more standardized and provide good security features. They are more transparent in their operations than PaaS type of services as they provide specific details about the SLA, security features, billing etc. The Community based cloud renderfarm services are free of cost and have limited features compared to other two but are widely used by students as security is not of high priority.

Table 1. Taxonomical analysis and comparison of 11 cloud renderfarm services

Taxonomy element	Rebus renderfarm	Rendernet	Rendicity	Rendering fox	Render rocket	Render core	Render solve	VSwarm	Sheepit	Amazon EC2	Rackspace
F1: DM	IaaS, PaaS	IaaS, PaaS	IaaS	PaaS	PaaS	PaaS	PaaS	PaaS	PaaS	IaaS	IaaS
F2: DPM	Public, Private	Public	Public	Public	Public	Public	Public	Community	Community	Public	Public
F3, F4: SubComp	CPU	CPU	CPU, GPU	CPU	CPU	CPU	CPU	CPU	CPU, GPU	CPU, GPU	CPU
F5: OS	Win 7 (64-bit)	Win 7 (64-bit)	NA	NA	NA	NA	NA	HostMachine Dependent	Host Machine Dependent	Linux, Windows	Linux, Windows
F6: VT	NA	NA	NA	NA	NA	NA	NA	NA	NA	XEN	VMware
F7: RJM	Farminiser	Fxcluster	Rendicity Wrangler	NA	NA	Automated, manual	DropBox	VSwarm web control Center	Web based admin panel	NA	NA
F8: RJM_L	Proprietary	Propriety	Proprietary	NA	NA	Proprietary	Proprietary	Proprietary	Proprietary	NApp	NApp
F9: RJM_P	Free	Free	Free	NA	NA	Free	Free	Free	Free	NApp	Napp
F10:JM_D	Plugin	online	online	NA	NA	Online, download	Online	Downloadable	Online	NApp	NApp
F11:RSL	Preinstalled, fee included	limited	NA	NA	NA	Preinstalled, fee included	Preinstalled, fee included	Provided by User	Provided by user	NApp	NApp
F12: RES	V-ray, MentalRay, Cinema4D	V-ray, MentalRay, Cinema4D	V-ray, MentalRay, Cinema4D	V-ray, MentalRay, Cinema4D	V-ray, MentalRay, Cinema4D	V-ray, MentalRay, Cinema4D	V-ray, MentalRay, Cinema4D	V-ray, MentalRay, Cinema4D	V-ray, MentalRay, Cinema4D	NApp	NApp
F13: S/W	Maya, 3dsMax, Softimage, etc.	Maya, 3dsMax, Softimage, etc.	Maya, 3dsMax, Softimage, etc.	Maya, 3dsMax, Softimage, etc.	Maya, 3dsMax, Softimage, etc.	Maya, 3dsMax, Softimage, etc.	Maya, 3dsMax, Softimage, etc.	Maya, 3dsMax, Softimage, etc.	Maya, 3dsMax, Softimage, etc.	NApp	NApp
F14: Plugin	PhoenixFD, RPC, Hot4Max IvyGrower etc.	PhoenixFD, RPC, Hot4Max IvyGrower etc.	PhoenixFD, RPC, Hot4Max IvyGrower etc.	PhoenixFD, RPC, Hot4Max IvyGrower etc.	PhoenixFD, RPC, Hot4Max IvyGrower etc.	PhoenixFD, RPC, Hot4Max IvyGrower etc.	PhoenixFD, RPC, Hot4Max IvyGrower etc.	PhoenixFD, RPC, Hot4Max IvyGrower etc.	PhoenixFD, RPC, Hot4Max IvyGrower etc.	NApp	Napp
NF1: FA	NDA	NDA	NA	NA	NA	NDA	NA	NA	NA	SLA	SLA
NF2: SM	Ordinary FTP EncryptedSFTP EncryptedSCP		NA	NA	NA	Firewall, SSL, Encryption VPN	NA	NA	NA	Firewall Encrypted storage	NA

(continued)

Table 1. (*continued*)

Taxonomy element	Rebus renderfarm	Rendernet	Rendicity	Rendering fox	Render rocket	Render core	Render solve	VSwarm	Sheepit	Amazon EC2	Rackspace
NF3: FSupport	Live chat, Ticket, Email	Email	NA	Email	Email, Helpcenter Forum, Phone call	Email, Live Chat	Email	Forum, Online, Documents	Forum	Forum Document Paid support	Live chat Forum Email
NF4: P Support	NA	NA	NA	NA	NA	SceneDebugging Trouble Shooting	NA	NA	NA	NA	NA
C1: CEM	Benchmark	Cine benchmark	Blender benchmark	SystemSpec	Test Render	SystemSpec	NA	NA	NA	NApp	NApp
C2: PT	Pay per use	Subscription	Pay per Use	Pay per Use	Pay per use	Pay per Use	Pay per Use	Free Donation Free		Pay per Use	Pay per Use
C3: PM	Priority, Prepaid packages	NA	Config based	Config based	OnDemand, Pre-paid packages	Priority, Prepaid packages, Rental	Hourly Pricing	Free DonationFree		ConfigBased. Traffic (in, out) Storage	ConfigBased. Traffic (in & out) Storage
C4: PC	RenderNode, Data, S/w License	RenderNode, Data, S/w License	RenderNode, Data, S/w License	RenderNode, Data, S/w License	RenderNode, Data, S/w License	RenderNode, Data, S/w License	RenderNode, Data, S/w License	Free	Free	Compute Unit, Data	Compute Unit, Data

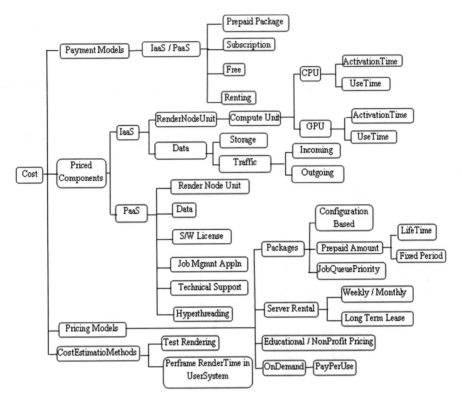

Fig. 2. Cloud renderfarm services cost analysis tree

5 Related Works

Initial research works explored the possibility of rendering files in a network of systems and were successful as the animation frames were independent bag of tasks and the split frames could be processed in different systems. The rendering process evolved along with new technologies like Distributed Rendering Environments popularly known as DRE's [9], rendering using new file distribution systems like the Hadoop [10] and bit torrent [11, 12], Grid based rendering that allows using the systems in an organization when they are free [13]. Cloud computing was also explored for the purpose of rendering with the consideration of the key technical characteristics [14], and the research challenges in cloud computing [15]. Kennedy et al. [1, 16], Carrol et al. [2] and many others have explored cloud technology for 3d animation rendering [17–19]. The evolution of GPU's for 3d animation rendering is also an important contribution in this field. We are working in this direction and have proposed a cloud broker framework that mediates between various cloud renderfarm services [19, 20] to facilitate the selection after comparing the services and also facilitate negotiation, monitoring of the Cloud render farm services. More research works in this direction would help the cloud render farm services to flourish in a better way.

6 Conclusion and Future Work

This work has proposed a framework and has suggested a comprehensive list of cloud renderfarm services characteristics for quick classification and comparison of PaaS cloud renderfarm services services offered in the current market. The classification based comparison of 11 popular cloud render farm services further enhances and clarifies the characteristics used for comparison and enables comparing other cloud render farm services not compared in this work. This work will be further used to build the knowledge base of a cloud render farm services recommender systems and apply machine learning techniques like the decision tree to search for services based on criteria like the functional requirements, non-functional requirements and the cost of the renderfarms. Various machine learning algorithms would also be experimented with to incorporate the user feedback for generating recommendations.

References

1. Kennedy, J., Healy, P.: Method of provisioning a cloud-based render farm. U.S. Patent Application 14/127,705 (2014)
2. Carroll, M.D., Hadzic, I., Katsak, W.A.: 3D rendering in the cloud. Bell Labs Tech. J. **17**(2), 55–66 (2012)
3. FoxRenderfarm. http://www.foxrenderfarm.com/pricing.html
4. Rebusfarm. https://us.rebusfarm.net/en/buy/products
5. Renderrocker. https://www.renderrocket.com/pricing.php
6. Prodan, R., Ostermann, S.: A survey and taxonomy of infrastructure as a service and web hosting cloud providers. In: 2009 10th IEEE/ACM International Conference on Grid Computing, pp. 17–25. IEEE (2009)
7. RevUpRender. https://www.revuprender.com/pages/case- studios
8. Free Community Renderfarm. www.vswarm.com/
9. Misrohim, S.R.: Implementation of Network Rendering Farm (2010)
10. Liu, W., Gong, B., Hu, Y.: A large-scale rendering system based on hadoop. In: 2011 6th International Conference on Pervasive Computing and Applications (ICPCA), pp. 470–475. IEEE (2011)
11. Northam, L., Smits, R.: Hort: Hadoop online ray tracing with mapreduce. In: ACM SIGGRAPH 2011 Posters, p. 22 (2011)
12. Kijsipongse, E., Assawamekin, N.: Improving the communication performance of distributed animation rendering using BitTorrent file system. J. Syst. Softw. **97**, 178–191 (2014)
13. Assawamekin, N., Kijsipongse, E.: Design and implementation of BitTorrent file system for distributed animation rendering. In: 2013 International Computer Science and Engineering Conference (ICSEC), pp. 68–72. IEEE (2013)
14. Gkion, M., et al.: Collaborative 3D digital content creation exploiting a grid network. In: 2009 International Conference on Information and Communication Technologies, ICICT 2009, pp. 35–39. IEEE (2009)
15. Cho, K., et al.: Render verse: hybrid render farm for cluster and cloud environments. In: 2014 7th Conference on Control and Automation (CA), pp. 6–11. IEEE (2014)

16. Weini, Z., et al.: A new software architecture for ultra-large-scale rendering cloud. In: 2012 11th International Symposium on Distributed Computing and Applications to Business, Engineering and Science (DCABES), pp. 196–199. IEEE (2012)
17. Kim, S.S., Kim, K.I., Won, J.: Multi-view rendering approach for cloud-based gaming services. In: The Third International Conference on Advances in Future Internet, AFIN 2011, p. 102107 (2011)
18. Wilkiewicz, J.J., Hermes, D.J.: Cloud-based multi-player gameplay video rendering and encoding. U.S. Patent Application 13/837,493 (2017)
19. Wang, S.H., Li, X.Z., Zhang, L.: The rendering system planning of the 3d fashion design and store display based on cloud computing. In: Applied Mechanics and Materials, vol. 263, pp. 2035–2038 (2013)
20. Ruby Annette, J., Aisha Banu, W., Subash Chandran, P.: Rendering-as-a-Service: classification and comparison. Procedia Comput. Sci. **50**, 276–281 (2015)

Smart Medicine Distributing Tray

Rohit Sathye[✉], Sumedh Deshprabhu[✉], Mandar Surve[✉],
and Deepak C. Karia

Sardar Patel Institute of Technology, Mumbai, Maharashtra, India
{rohit.sathye,sumedh.deshprabhu,
mandar.surve}@spit.ac.in, deepakckaria@gmail.com
https://www.spit.ac.in/

Abstract. This Paper Highlights the development of a Smart Medicine Storing and Delivering Tray system with a unique patient identification by using a RFID technology and a dosage monitoring system enabled by Internet of Things technology. The Smart medicine tray consists of four plastic boxes which are mapped to the individual patients in a particular room. Every box has its own RFID card, which can only facilitate its open and close operations. The Box has a Servo Motor and an IR sensor attached to it. The System is controlled by an ATMega 2560 microprocessor. The opening and closing of the box is done by the servo motor which in turn is operated by the microprocessor signal. This action is possible when the RFID reader correctly identifies the RFID card of a particular box. The presence of medicine in the box is sensed by an Infra-Red (IR) sensor. When the last medicine pill is lifted, the Infrared sensor sends a signal to the microprocessor, which sends it to the Wi-Fi Module. Thus a message is sent to the online Internet of Things platform (named ThingSpeak) thereby signifying that a dosage has been delivered. From this platform the message can be viewed by the doctor in charge, the nurse or even the patient's kin.

Keywords: RFID · Internet of Things · ATMega 2560 · Medicine box · Servo Motor · Infrared Sensor · NodeMCU (Wi-Fi Module) · Medicine delivery

1 Introduction

Healthcare is an important factor in any individual's life. It directly affects the economy and growth of the country. With the advancements in medical technologies, it has become possible for doctors to find better ways to treat patients. The quality of medicines has also improved due to advancements in science which thereby helps the patients to recover effectively and safely.

There is a need for patients to take medication on time which would enhance their recovery. Also patients with chronic diseases need prolonged medication and such patients need to consume medicines at appropriate time. Hence our Smart Medicine Distributing Tray will help patients take medicines at proper time [2].

Our tray will consist of boxes that contain the medication to be taken as well as an RFID card reader. Every patient will be given a specific RFID card which will only open the box containing the medicines to be given to that patient.

© Springer Nature Switzerland AG 2020
J. S. Raj et al. (Eds.): ICIDCA 2019, LNDECT 46, pp. 57–66, 2020.
https://doi.org/10.1007/978-3-030-38040-3_6

The head nurse will place the medicines in the respective boxes and an alarm will notify when the medicines need to be delivered. Every patient will scan their respective card to get their medication as the card will only open the box of the specific patient.

Also another function of this Tray is that it will notify the family members as well as the doctor whether the patient has taken their medication or not. When the patient removes the medicine out of the box a signal will be sent through the Node MCU to the ThingSpeak Internet of Things platform which will help to monitor if the dosage has been taken or not.

2   Objectives

To Develop a Cloud Based Solution for Monitoring Medical Adherence of Patients
Whenever the medicine is picked up by the patient, a signal (message) is sent to the Internet of Things based platform where from it can be seen by the patients kin.

To Enable the Hospitals to Efficiently Utilize Their Manpower
By automating a simple task like medicine delivery, the time of the medical staff can be saved. Consequently the staff can look over to more complex operations.

To Connect the Patient Virtually with His/Her Respective Physician and Monitor the Dosage as and When it is Given
Just as the Medicine delivery can be monitored by the patient's kin, the doctor in charge may also check up virtually on the patient and monitor his dosages. Not only the doctor but the Nurse can also monitor the patients remotely.

3   Hardware Requirements

RFID Reader
The RFID reader is used to read individual patients cards and obtain the UID number of the card.

Servo Motor
The servo motor will be used to open the box when the RFID card is scanned.

IR Sensors
The IR sensor will detect the presence of medicines in the box. When the medicines would be taken out of the box, the IR sensor will detect the absence of medicines and thus will send a signal to the microcontroller which will then close the box [2].

Atmega 2560
The ATMEGA 2560 microcontroller will be used to control the working of the Tray. The RFID reader, Servo motors as well as the IR sensors will be interfaced with this microcontroller. It will send appropriate signals to open and close the boxes based on the input from the RFID reader and the IR sensor.

Node mcu
The NODE MCU will send a signal via the Internet to the ThingSpeak IOT platform once the medicines are delivered to the patient.

4   Software Requirements

Arduino IDE
Arduino IDE is used to program the ATMEGA 2560 microcontroller as well as the NODE MCU.

5   System Design

Smart Pill Box Block Diagram

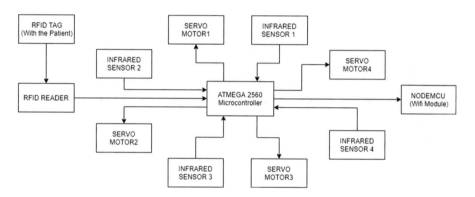

Fig. 1. Block diagram

The ATMega 2560 microcontroller is the heart of the proposed system (See Fig. 1). According to the Block Diagram in Fig. 1, the components connected to the Microcontroller are the 4 Servo Motors, 4 Infrared Sensors and 4 Pins to which the NodeMCU is interfaced. Every Patient will be given a unique RFID Card which will have a unique UID number. The patient has to scan this card on the RFID reader/writer module. The program present in the ATMega will check if the RFID card used has one of the four hardcoded UID numbers or not. Every card has been mapped to an individual box in this system. In this paper we have assumed that the number of patients in a particular Hospital room is 4. Hence we have 4 boxes. The UID of every individual card is also stored in the system. Whenever the card is placed on the Reader module, its UID is recognized by the RFID reader. Pertaining to that specific UID, the particular box is opened.

Let us say that the RFID card for Box number 1 is read. Thus the controller sends a signal to the servo motor which is attached to the box number one. This motor will open the Box-1 so that the patient can take the tablets/pills. Now an Infrared sensor (IR) is also attached to every box. These sensors will act as the input to the microcontroller. So whenever the last tablet from the box is taken by the patient, the IR sensor will give a signal to the microcontroller. Now when this signal is received, the microcontroller will send a signal to the pin connecting the NodeMCU with the controller thereby informing the NodeMCU that the dosage has been delivered. When the NodeMCU will receive this signal, it will update the respective channel and display the delivery of the dosage on the IOT platform. After this process is done, another signal will be sent to the Servo Motor to close the box. Thus in this way a dosage has not only been delivered but monitored as well (See Fig. 4 for the box opening and Fig. 6 for the IOT channel update).

A Similar process follows for all the four boxes. Along with the four unique RFID cards, there also exists a Master Card which will be given to the Medical staff. With this Master card, the person can open all the four boxes (See Fig. 5). So to summarize, the system flow will be as follows:-

RFID Reader - Card detected and Servo Motor (Box opening) - Medicine Taken by patient - Infrared sensor connected to Box sends signal after all medicine in the box is lifted by the patient - Signal sent to Node MCU (Updated on the Internet) - Servo Motor (Box closing). This flow can be understood from the Fig. 1, from the leftmost block RFID Reader to NODE MCU (Wi-Fi Module).

The Respective connections are mentioned in the tables in Sects. 6.1, 6.2 and 6.3.

Algorithm

- 1: Scan RFID card and Obtain UID of the card.
- 2: Compare the Obtained UID with the Stored UIDs.
- 3: If UID of the card matches with any of the Stored UID, go to step 5 else go to step 4.
- 4: Print Invalid card and go to step 3.
- 5: Open the box corresponding to the matched UID and go to step 6: if the master card is scanned go to step 9.
- 6. Read the Output of the IR sensor. If output is zero then repeat step 6 else if output is 1 go to step 7.
- 7: After some delay close the respective box.
- 8: Send a signal to the respective field through NODE MCU to the Thing speak IOT platform that the medication has been delivered. Go to step 11.
- 9: When master card is scanned open all boxes.
- 10: Close all boxes when master card is scanned again.
- 11: Go to step 1.

Flowchart:

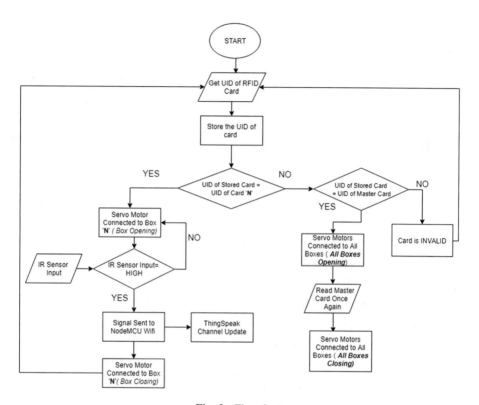

Fig. 2. Flowchart

Here N denotes the number of the Box and the number of the respective RFID card mapped to the boxes and the Servo Motor and Infrared Sensor attached to the Box (Figs. 1 and 2).

6 Figures and Tables

The Tables given below indicate the pin connections of the Servo Motor with the Microcontroller, the Infrared Sensor with the Microcontroller and the Microcontroller with the Wi-Fi Module respectively.

6.1 Servo Motor Connections to ATMega 2560

Motor	Mega pins
1	2
2	3
3	10
4	11

6.2 IR Sensors Connections to ATMega 2560

IR sensor	Mega pins
1	24
2	25
3	26
4	27

6.3 ATMega 2560 to NodeMCU

Mega pins	NodeMCU pins
30	D1
31	D2
32	D3
33	D4

6.4 Smart Medicine Tray

See Fig. 3.

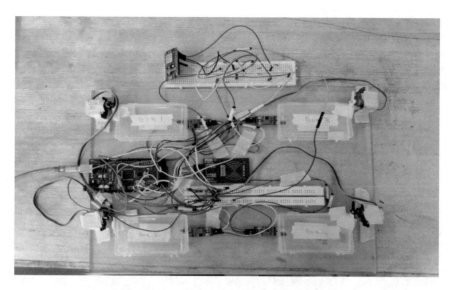

Fig. 3. Smart tray set-up on an acrylic plate

7 Observations and Results

7.1 Individual Card: (In this Case Box 1)

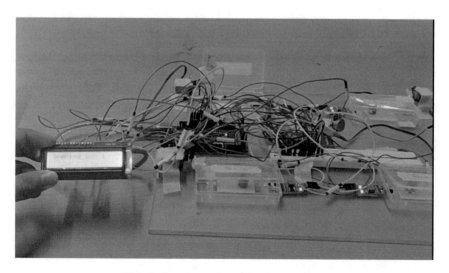

Fig. 4. Box 1 opening using RFID card 1

According to the Fig. 4 the working of the pillbox is depicted when the individual RFID cards are used. Figure 4 shows the working specifically for Box 1. Every Patient will be given an individual RFID card. Each Card will only open the box corresponding to that Individual patient. The box will remain open till the patient removes all the medicines from the box. After the box closes a message will be sent using Node MCU to the ThingSpeak IOT platform indicating that the patient has taken the medicines.

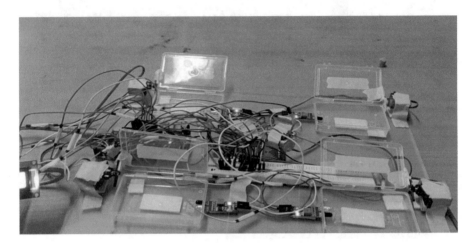

Fig. 5. Master card operation: all boxes opening

7.2   Master Card

According to Fig. 5 the working of the pillbox is depicted when we use the master card. The Head Nurse will have the master card which will be used to open all boxes so as to place medicines required by patients at a specific time inside the boxes. The Head Nurse will scan the card once to open all boxes and then place the medicines inside them after which he/she will scan the card once again to close all boxes.

7.3   Results on Internet of Things Platform-ThingSpeak

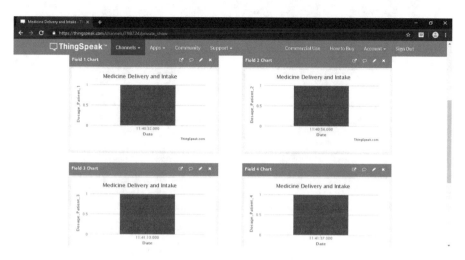

Fig. 6. ThingSpeak IOT platform: dosage 1 update

According to Fig. 6 the result is obtained on the ThingSpeak IOT platform after the first dosage is administered. Each field corresponds to a single patient. For example Field 1 corresponds to patient 1, Field 2 to patient 2 an so on. The chart of each field has the patient name on the Y-axis while the Date and time at which the medication is given on the X-axis. So when the box closes a signal is sent through NODE MCU to the field of the respective patient. The Red bar/column indicates that the dosage has been delivered to the patient. The ThingSpeak channel URL can be shared with the patient's family members and the doctor so that they can understand if the medication is given or not. Every Patient has a specific particular channel which only he can ask. A particular patient cannot see other patient's dosage. However the Medical staff in-charge will be able to see the dosage of all the patients as shown above.

As per Fig. 7, it is indicated that the patient has received proper medication after regular intervals of time. The first dosage was given at 12 pm while the second dosage was given around 7 pm. Thus the family members as well as the doctors can keep a track of the patient's medication.

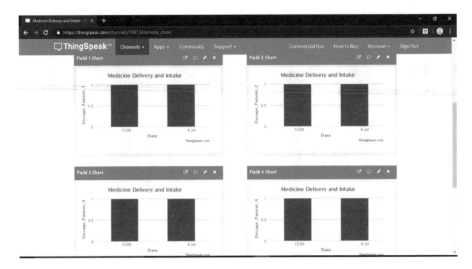

Fig. 7. ThingSpeak IOT platform: dosage 2 update

8 Conclusion

Thus the Smart Medicine Distributing Tray ensures that every patient gets only his respective medicine and not of some other patient. This is made possible by the RFID cards which are given to every individual patient. Figure 4 shows that the medicine delivery is accurate. The event of medicine delivery is marked by an online update thus ensuring that the patient's kin are able to monitor the dosage. This is proved in Figs. 6 and 7.

9 Future Scope

The tray will also be useful if fitted on an autonomous vehicle which can deliver medicines to the patient's bed.

References

1. Hayes, T.L., Hunt, J.M., Adami, A., Kaye, J.A.: An electronic pillbox for continuous monitoring of medication adherence. In: Proceedings of the 28th Annual International Conference of the IEEE Engineering in Medicine and Biology Society, 30 August–3 September 2006, pp. 6400–6403 (2006)
2. Deshpande, S., Choudhari, M., Charles, D., Shaikh, S.: A smart pill box to remind of consumption using IoT. MES College of Engineering, Pune (2018)
3. Zao, J.K., Wang, M.Y., Tsai, P., Liu, J.W.S.: Smart phone based medicine in-take scheduler, reminder and monitor. In: IEEE e-Health Networking Applications and Services (Healthcom), pp. 162–168 (2010)
4. Sharma, U., Chauhan, C., Sharma, H., Sharma, A.: Arduino based medicine reminder. AGU Int. J. Eng. Technol. (AGUIJET) **3**, 37–45 (2016)

Potential Customer Classification in Customer Relationship Management Using Fuzzy Logic

Tanay Kulkarni[✉], Purnima Mokadam, Jnanesh Bhat,
and Kailas Devadkar

Department of Information Technology, Sardar Patel Institute of Technology,
Mumbai, India
tanay.kulkarni@spit.ac.in

Abstract. Customer Relationship Management systems are one of the most significant determinants to maximize sales in any business domain. In the process of CRM, the important step after targeting a lead is to convert these leads into actual customers. In the current CRM system, depending on the lead score, the lead is projected as a potential customer. The paper proposes the application of fuzzy logic in the Customer Relationship Management systems to get the prospects of how potential a lead is to become a customer based on some of the factors that involves the interaction between the lead and the business domain. Fuzzy logic approach is mainly used to identify the important leads, who have the potential to increase the future sales of the business.

Keywords: CRM (Customer Relationship Management) · Fuzzy logic · Leads · Response

1 Introduction

In the real world, the boolean output i.e. true or false is difficult to comprehend and relate and thus its applications in computing systems leads to an inaccurate output. In such cases Fuzzy logic is quite useful where it takes the inaccuracies and uncertainties of any situation into consideration and does not consider any problem in the form of pure 0's and 1's. For example if we talk about a self-driving car and suppose the output to apply the brake is just a 1 or 0, in that case it will suddenly apply brakes or it won't apply breaks at all. But, as we know in real life scenario we need to consider the amount of brakes we apply rather than just applying or not applying brakes, this is the fuzzy logic that we apply in for self-driving cars to handle brakes properly. The output is governed on the basis of rules written in the Rule Base. It has IF-THEN conditions written by experts. In the fuzzy system, the crisp inputs are converted into fuzzy sets on which the rules are applied [8].

After considering the fuzzification method, the rules and their operations, the input is converted into a fuzzified value which is then defuzzified to get a crisp output. Thus a crisp set is converted into a fuzzy set and fuzzy set is then converted into a crisp set. Customer Relationship Management helps to connect companies with leads and customers. CRM is an important tool for interaction with the customers [1]. It is a common platform for everyone to connect at one central place. The fuzzy logic can be used in

© Springer Nature Switzerland AG 2020
J. S. Raj et al. (Eds.): ICIDCA 2019, LNDECT 46, pp. 67–75, 2020.
https://doi.org/10.1007/978-3-030-38040-3_7

CRM to find out the leads to maximize sales. As the number of potential customers increases the business scales i.e. the business grows and expands. Business scales can result in reducing unit cost, increasing the value of a product, the reduced cost for providing services and there is no need for brand awareness.

Therefore fuzzy logic can be implemented in CRM to improve the accuracy of its analysis by considering various real life scenarios [11]. In the current CRM system, various factors like the response to an email by the lead or the buyer alignment towards a particular product is taken into consideration. The traditional CRM systems classifies such factors in terms of 0s and 1s i.e. if we consider the case of the response to an email by the lead, the CRM system of the business considers it to be either responded or not responded which makes it a case of crisp application. When we apply fuzzy logic to it, we mainly consider the time of response that the lead took to reply to the email. It can be an immediate response, delayed response or no response at all. When we also consider such factors, we get an accurate picture of the lead and it also plays a very vital role in identifying the lead behavior for further analysis and insights. The proposed CRM system classifies leads into three broad categories Cold, Warm and Hot. The leads are divided into these categories depending upon their lead scores. The lead score is determined on 4 factors - lead's response to an email sent by the company, how fast the company responds to the lead, for how much time does the lead communicate with the company and lastly buyer's alignment. These four factors when considered together can make the evaluation of a leads very accurate and precise.

2   Methodology

The following are various factors on which the leads score depends.

2.1   Time Taken by the Lead to Respond to the Email

Once some company has sent an email to the lead, then there may arise three scenarios wherein, the lead may not open the mail and there would be "No Response" from his side. Another case that can be considered here is when lead reads the mail and respond to it after some time. So this kind of response would be "with-some-delay" response to the mail. Now in some cases if the lead is too interested to take the offer and responds to them very quickly then in that case it is considered to be a "Quick response".

2.2   Company's Response to Lead's Response

The time frame after which a business responds to its lead's initial queries or enquiry is of paramount importance for them getting converted into actual customers. Let's take an example that the lead wants to buy a fridge, suppose the lead has enquired about the fridge to some company, but in turn the company gets in contact with the lead after 3–4 days, then this kind of response might result into company losing its leads. So, this response to leads enquiries broadly be categorized into 3 categories Quick Response, response with some delay and response given much later. Quick response is where the company responds immediately whereas a delayed response would be a response which is after a few days. Response much later is when the company responds too late to a lead query which is as good as not responding which can make companies lose customers.

2.3   For How Much Time the Lead Communicates with the Company

The time duration for which the lead communicates with the company is essential for getting the extent to which the lead is interested in the company's scheme. For example, if a lead is really interested to buy the product he will communicate for longer time with the salesperson asking him various questions about the product, clearing the doubt he/she has about the product through phone-call or the number of times the lead has enquired using email.

2.4   Buyer's Alignment

If a lead wants to buy a car, he can google it and find all the details about a particular car he likes. He/she will get the list of the features and can compare these features with other cars and can come to a conclusion. But if the lead gets confused about some feature, he/she ends up calling a salesperson. Now the sales person has no idea about what the buyer is interested in and could give some wrong suggestions which will waste buyer's time as well as the salesperson's time. A more fruitful approach is that the salesperson should have known some information about the buyer, know the buyer's mindset, his interest, ask some questions to know exactly what he wants and keep in touch with him throughout the decision making process. This is known as buyer's alignment i.e. how much the company is aligned with its lead or how much are the needs of the buyer aligned to the product of the company.

Therefore this factor also plays an important role in identifying customers from leads. This alignment can be of three types highly aligned, Semi-aligned and Non-aligned. Highly aligned means when the company has full knowledge about his leads requirements. Semi aligned is when a company has only a little knowledge about his lead's requirements and non-aligned is a condition when company has no idea about the lead and his/her interests or requirements.

2.5   Fuzzy System

We propose a fuzzy system using the above mentioned parameters as inputs. The knowledge database consists of the information of the leads and the rules base consists of the rules defined in this particular fuzzy system. The fuzzification unit converts crisp set to fuzzy set and the defuzzification unit converts the fuzzy set to the crisp output [12] (Fig. 1).

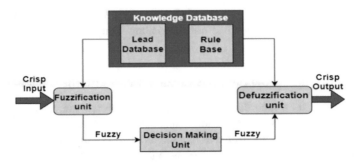

Fig. 1. Flow chart

The following fuzzy control system uses triangular membership function to assign membership values to the crisp input, which states that (Fig. 2).

$$f(x, a, b, c) = \begin{cases} 0, & x \le a \\ \dfrac{x-a}{b-a}, & a \le x \le b \\ \dfrac{c-x}{c-b}, & b \le x \le c \\ 0, & c \le x \end{cases}$$

Fig. 2. Membership function formula

Here the membership value is calculated for each input value x for the parameters a, b and c. The defuzzification method used for the following case is the centroid method, which gives the center of area under the curve [3].

3 Implementation

The implementation consists of applying the fuzzy control system to get a lead score and classify a lead as cold, warm or hot. The lead score generated using the fuzzy control system would be from 0 to 100 and the fuzzy set will help it to be classified as cold, warm or hot [4].

The problem can be formulated as follows:

3.1 Antecedents (Inputs to the Problem)

3.1.1 Response to Email

- Universe (Crisp value range/Crisp set): It tells us how fast the customer responded to the email in the value range from 0 to 10.
- Fuzzy set (fuzzy value range): It is divided into three fuzzy value ranges-quickly, with some delay and no response (Fig. 3, Table 1).

Table 1. Fuzzy set for response to email.

Range	Fuzzy set
0–5	Quickly
0–10	With-some-delay
5–10	No response

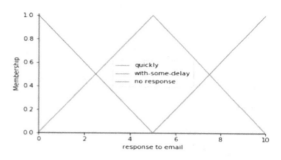

Fig. 3. Membership function for response to email

3.1.2 Company Response

- Universe (Crisp value range/Crisp set): It tells us how fast the company responded back to the customer after his/her response to the email in the value range from 0 to 10.
- Fuzzy set (fuzzy value range): It is divided into three fuzzy value ranges-immediate, with some delay and much later (Fig. 4, Table 2).

Table 2. Fuzzy set for company response.

Range	Fuzzy set
0–5	Immediate
0–10	With-some-delay
5–10	Much later

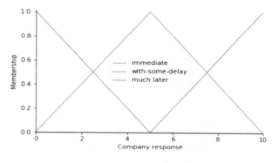

Fig. 4. Membership function for company response.

3.1.3 Contact Time Spent

- Universe (Crisp value range/Crisp set): It tells us the contact time spent between the company and the customer via email or phone-call in the value range from 0 to 10.
- Fuzzy set (fuzzy value range): It is divided into three fuzzy value ranges-less, medium and more (Fig. 5, Table 3).

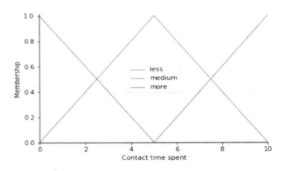

Table 3. Fuzzy set for contact time

Range	Fuzzy set
0–5	Less
0–10	Medium
5–10	More

Fig. 5. Membership function for contact time

3.1.4   Buyer Alignment

- Universe (Crisp value range/Crisp set): It tells us how the needs of the buyer aligns with the product of the company in the value range from 0 to 10.
- Fuzzy set (fuzzy value range): It is divided into three fuzzy value ranges-aligned, semi-aligned and not aligned (Fig. 6, Table 4).

Table 4. Fuzzy set for buyer alignment.

Range	Fuzzy set
0–5	Aligned
0–10	Semi-aligned
5–10	Not aligned

Fig. 6. Membership function for Buyer alignment.

3.2   Consequent (Output)

3.2.1   Lead

- Universe: What exactly is the lead score between 0 to 100
- Fuzzy set: cold, warm, hot.

3.3   Rules

The rules were defined in consensus with CRM experts and therefore taken into consideration for further implementation. There are a total of 81 rules that were generated because of 4 Antecedents. The Table 5 below represents some of those 5 randomly selected rules.

Table 5. Few rules from rule base

Sr. No.	Response to email by lead	Company's response	Communication time	Lead's alignment	Output
1	Quickly	Immediate	More	Aligned	Hot
2	Quickly	With-some-delay	Medium	Aligned	Warm
3	With-some-delay	With-some-delay	Medium	Aligned	Warm
4	Quickly	Much later	More	Not aligned	Cold
5	No response	With-some-delay	More	Semi-aligned	Cold

4 Output and Result

To test the fuzzy control system, we gave the system an input of the four terms namely, response to email, company response time, contact time spent and buyer alignment for 2 test-cases as (Table 6).

Table 6. Output for random inputs

Sr. No.	Response to email	Company's response	Contact time spent	Buyer's alignment	Lead score	Output
1	3	2	5	7	55.99676	Warm
2	1	2	9	2	90.01111	Hot

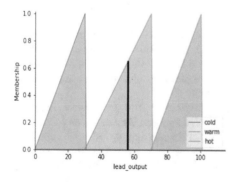

Fig. 7. Output Graph I

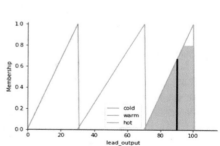

Fig. 8. Output Graph II

The crisp output represented in Fig. 7 is in the warm range thus the lead is warm and needs to be nurtured to make him/her a customer, while crisp output represented in Fig. 8 is in the hot range thus the lead here is hot and can be a customer soon and thus the company can generate sales from that customer.

5 Conclusion

The above result shows that using fuzzy logic we can classify leads into three distinct categories i.e. Cold, Warm and Hot. These categories give us information about the customers' needs and help company project its future customers. In the case where the lead is found to be in the hot category, just by using simple marketing and targeting techniques he or she can be converted into an actual customer. Since, they are most probable customers they can be targeted with Customer Retention Techniques of the company in order to make them a loyal customer to the business. The ones who fall in the category of warm leads, can be nurtured with further recommendations on the interests they have shown and also we can use interesting and attractive schemes to convert them into customers. Therefore by using fuzzy logic, businesses can derive meaningful insights on the lead's behavior especially their requirements and eventually know the areas they need to target upon.

References

1. Ullah, Z., Al-Mudimigh, A.S.: CRM scorecard measurement: the case for the banking sector in Saudi Arabia. In: 2009 Third UKSim European Symposium on Computer Modeling and Simulation, pp. 269–273 (2009)
2. Daif, A., Eljamiy, F., Azzouazi, M., Marzak, A.: Review current CRM architectures and introducing new adapted architecture to Big Data. In: 2015 International Conference on Computing, Communication and Security (ICCCS), pp. 1–7 (2015)
3. Yu, T., Zhou, J., Zhang, Y., Dong, S., Wang, W.: Research on CRM performance evaluation based on fuzzy comprehensive algorithm. In: 2008 International Conference on Information Management, Innovation Management and Industrial Engineering, pp. 329–334 (2008)
4. Khan, N., Khan, F.: Fuzzy based decision making for promotional marketing campaigns. Int. J. Fuzzy Logic Syst. (IJFLS) 3(1), 1148–1158 (2006)
5. Zhou, J., Yu, T., Zhang, Y., Dong, S., Wang, W.: System of CRM performance evaluation based on fuzzy comprehensive algorithm. In: 2008 International Conference on Information Management, Innovation Management and Industrial Engineering, pp. 382–385 (2008)
6. Hu, G.L.: Performance evaluation of enterprise CRM system based on grey-fuzzy theory. In: 2010 International Conference on Networking and Digital Society, pp. 32–35 (2010)
7. Bibiano, L.H., Marco-Simó, J.M., Pastor, J.A.: An initial approach for improving CRM systems implementation projects. In: 2014 9th Iberian Conference on Information Systems and Technologies (CISTI), pp. 1–6 (2014)
8. Fathi, M., Kianfar, K., Hasanzadeh, A., Sadeghi, A.: Customers fuzzy clustering and catalog segmentation in customer relationship management. In: 2009 IEEE International Conference on Industrial Engineering and Engineering Management, pp. 1234–1238 (2009)

9. Beyadar, H., Gardali, K.: The study of customer relationship management method. In: 2011 5th International Conference on Application of Information and Communication Technologies (AICT), pp. 1–4 (2011)
10. Prabha, D., Subramanian, R.S.: A survey on customer relationship management. In: 2017 4th International Conference on Advanced Computing and Communication Systems (ICACCS), pp. 1–5 (2017)
11. Jiang, H., Cui, Z.: Study on tourism CRM based on fuzzy evaluation. In: 2009 Sixth International Conference on Fuzzy Systems and Knowledge Discovery, pp. 428–431 (2009)
12. Alang-Rashid, N.K., Heger, A.S: A general purpose fuzzy logic code. In: Proceedings of IEEE International Conference on Fuzzy Systems, pp. 733–742 (1992)

Classification of Sentiment Analysis Using Machine Learning

Satyen M. Parikh[1] and Mitali K. Shah[2(✉)]

[1] Ganpat University, Ahmedabad, India
satyen.parikh@ganpatuniversity.ac.in
[2] Ganpat University, Kherva, India
shahmitali07@gmail.com

Abstract. An application of the computational linguistics is identified as Natural Language Processing or NLP. With the help of NLP, the text can be analyzed. Any person's opinion through which the emotions, attitude and thoughts can be communicated is known as sentiment. The surveys of human beings towards specific occasions, brands, items or organization can be known through sentiment analysis. Every one of the assumptions can be ordered into three unique categories, they are positive, negative and neutral. Twitter, being the utmost mainstream microblogging webpage, is utilized to gather the information for perform analysis. Tweepy is utilized to extract the source of information from Twitter. Python language is utilized in this exploration to execute the classification algorithm on the gathered information. In sentiment analysis, two steps namely feature extraction and classifications are implemented. The features are extracted using N-gram modeling technique. The opinion is classified among positive, negative and neutral by utilizing a supervised machine learning algorithm. In this research work, SVM (Support Vector Machine) and KNN (K-Nearest Neighbor) classification models are utilized. Also, we have shown both comparison for sentiment analysis.

Keywords: Twitter · Sentiment analysis · Machine learning · SVM · KNN

1 Introduction

1.1 Sentiment Analysis

The word opinion mining is used for sentiment analysis in various applications. Its aim is to examine the people's opinion of the item's services and attributes. In basic terms, sentiment analysis is a sort of common approach utilized for recognizing whether the people are interested or not in the product. The feelings or sentiments of user are communicated textually. These views are gathered from various websites or mobile applications [1].

Sentiment analysis is utilized to communicate or express the feelings of people. Positive and negative opinions are given by people for the services. These sentiments are very useful for client in setting on some buying choices [2]. In the context of vast textual dataset and unstructured datasets, the exact recognition of right view is very

© Springer Nature Switzerland AG 2020
J. S. Raj et al. (Eds.): ICIDCA 2019, LNDECT 46, pp. 76–86, 2020.
https://doi.org/10.1007/978-3-030-38040-3_8

important. In order to identify the features of an unstructured dataset, a productive technique should be structured imperatively [3].

Sentiment analysis has three principle part and analysis methods for the most part preprocessing, feature extraction and classification [4]. Few groups skip the neutral text acceptance. These texts utilize the limit of the binary classification. A few scientists propose polarity issues for implementing three classes [5]. The neutral classes, presentation etc. are very important for the classifiers entropy and SVM. It increases the overall classification accuracy. Certain standards should be considered by the neutral classes for execution. The first algorithm recognizes the neutral language and filters the residual opinions of people. This algorithm finishes the three level classifications in just single phase [6].

In every class, the probability and distribution is computed in the next methodology [7]. At the point when the information is the most neutral with the deviation to positive and negative outcomes, then the implementation of this methodology becomes more difficult [8]. Sentiment analysis is utilized in a few different manners. It is useful for the advertisers estimating the reputation and achievement of some novel item dispatch and sees which version of an item are in demand and feature the famous highlights of the new item [9]. For example, assume the surveys on a site are for the most part positive audits regarding a computerized camera attributes yet explicitly negative with reference to its more density. An extremely clear image related to the view of a person is provided to the marketer by identifying this sort of information in required format. There are certain issues involved in sentiment analysis. There are some words that give a positive meaning in one context while negative in other context. As an example, word "long" can be used both in positive and negative contexts. Consider a sentence "the battery life of PC is long". In this sentence, word "long" gives positive impression. When this word is used in some other sentence like "PC takes a long time to start", then, this word gives negative impression. Taking care of these changes is extremely important in sentiment analysis as it needs framework prepared datasets for getting views feelings on item as well as its attributes. The next test isn't communicating the sentiment somehow or another.

1.2 Various Faces of Sentiment Analysis

1.2.1 Data Extraction

The original tweets are included in the input data set. A tool named "Tweepy" is utilized to extract these tweets. API is utilized to keep the question running after it has been created. The yield of this question will be all the important twitter source data. In each record that is produced however a tweet, data, for example, tweet id, content, customer name, etc. can be separated. In the website that any client makes his location open (public), the information applicable to the location from where tweet is posted is produced as latitude and longitude from the Twitter API. Since 2012, people stopped to share their location because of the security issues and customer protections. The location is utilized as a sifting parameter in the standard question later Twitter. Along these lines, contingent on the settled arrangement of locations, the tweets are extracted (Fig. 1).

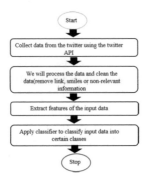

Fig. 1. Flow chart for sentiment analysis (Source: ICACSE-2019, p. 92)

1.2.2   Data Pre-processing

In the pre-processing stage, the input data set is cleaned and transformed into a form suitable for feature extraction. There is sure measure of irrelevant information accessible inside the data that is extracted from Twitter. Any type of arbitrary characters or pointless data should be filtered from the tweet data [11]. The Natural Language Processing device is connected for filtering through this pointless information. Any types of syntactic relations that exist in the middle of the words of sentences are given as output by this NLP tools. The data examiners consider the main word relations as significant even though semantics characterizes a couple of other word relations inside a sentence. The tweets that contain important data are recognized using these relations. The results are not helped at any rate through the encouraging filtering alongside more relations. The relations among nouns and adjectives words or action words are found utilizing nsubj connection inside any noun sentence. Regardless to complementing a noun in a sentence or not, this is viewed as of high significance.

1.2.3   Feature Extraction

This stage extracts features from the tweets. In the stage, hash tags and emojis are extracted. Based on polarity of feelings they depict; particular loads are allocated to the emojis on the basis of their division. The positive emojis are appointed with weight "1" and negative emojis with "−1" weight. It also works for hash tag to be certain and negative. Inside the vector of features, they are incorporated as individual features. The pre-processed data is applied as input to the feature extraction algorithm. The distribution of weights to the keywords is done in this stage. Therefore, this stage prepares them for classification. In this paper, N-Gram modeling approach is utilized to design the feature extraction algorithm.

1.2.4   Sentiment Classification

Standard classifiers, for example, SVM, k-nearest neighbor, ensemble classifiers, Naïve Bayes classifiers, etc. inside the arrangement step are designed in feature vector. In based on their polarity. A classification model named k-nearest neighbor is used in this work for classification purpose. This classifier classifies tweets into positive, negative and neutral.

2 Methodology

The sentiment analysis of twitter data is the main objective of this study. A classification model named SVM is utilized in the existing framework for classifying input data into seven classes using the SANTA Tool. The proposed study replaces SVM classification model with the KNN classification model. This classification model classifies input data into seven classes. The performance analysis of both approaches is performed on the basis of accuracy.

2.1 SVM Classifier for Sentiment Analysis

The major objective of introducing SVMs was to utilize them for classification. Further, the applications also extended to regression and preference learning with the advancements of SVMs. The original form of SVM algorithm is known as binary classifier. This classifier achieves either a positive or negative output of learned function [12]. A multiclass classification is implemented by integrating multiple binary classifiers using a pair-wise coupling technique. The mapping is made from input space to feature space by SVM to support the constraints related to non-linear classification.

The kernel trick is utilized by preventing accurate formulation of the mapping function. This generates the curse of dimensionality. This approach makes the linear classification in new space equal to the non-linear classification in original space. To the higher dimensional space in which a maximal separating hyperplane is generated, the mapping is done for input vectors by the SVM classifier.

For instance, there are N training data points $\{(x_1, y_1), (x_2, y_2), \ldots (x_N, y_N)\}$ Here, $x_i \in R^d$ and $y_i \in \{+1, -1\}$. Equation (1) shows the issue of identifying a maximum margin that separates the hyperplane as:

$$\min_{w,b} \frac{1}{2} w^T w \text{ subject to } y_i \left(w^T x_i - b \right) \geq i = 1, \ldots N$$

In general, this is a convex quadratic programming problem. Lagrange multipliers α is introduced to achieve Wolfe dual in the given equation:

$$\text{maximize}_\alpha \mathcal{L}_D \equiv \sum_{i=1}^{N} \alpha_i - \frac{1}{2} \sum_{i,j} \alpha_i \alpha_j y_i y_j x_i . x_j \tag{1}$$

Subject to

$$\alpha \geq 0, \sum_i \alpha_i y_i = 0.$$

Following is the Eq. (2) of providing primary solution:

$$w = \sum_{i=1}^{N} \alpha_i y_i x_i \tag{2}$$

2.2   KNN Classifier for Sentiment Analysis

K Nearest Neighbors (KNN) is known as a non-parametric classifier. In various cases, this algorithm is quite simple and effectual. In pattern recognition, KNN is one of the most common neighborhood classifiers because of its effectiveness and competent outcomes. The feature of this classifier can be used easily. There are certain applications in which this algorithm is utilized. These applications include pattern recognition, machine learning, text categorization, data mining, object recognition etc. [13]. Memory requirement and time complexity are the limitations of this classifier due to its dependency on each instance in the training set. This classifier resolves the issue of clustering. As compared to other classification algorithms, this algorithm is the simplest unsupervised learning algorithm. A simple process is used by the fixed amount of clusters easily and efficiently to classify the given data earlier.

These algorithms are implemented during the non-existence of labeled data. Therefore, this technique is based on the distance function as well. The minimal average distance is computed using Euclidean distance. This technique normalizes the overall features in the same range. K-nearest neighbor classification model computes the optimum performance of the best values of k. This is a conventional non-parametric classification model. K-nearest neighbor algorithm is used for the classification of pattern x. A class label is allotted to the pattern. This pattern is characterized most regularly in the picture. This technique assigns this pattern within the k nearest patterns [14]. The class having minimal average distance to the test pattern is allotted in case of tie between two samples or patterns. A global distance function can be computed on the basis of individual attribute by combining local distance functions. The easiest method is to add the values as described by the given equation:

$$\text{dist}(x, q) = \text{def} \sum_{i=1}^{N} \text{dist}_{A_i}(x.A_i, q.A_i) \tag{3}$$

The global distance is identified as the weighted sum of local distances. For calculating the overall distance, there are different levels of importance provided by weight w_i. The values amongst zero and one are sometimes the weight of values. A completely irrelevant attribute might be generated by the weight of zero. Therefore, the modified form of Eq. (3) can be inscribed as (Fig. 2):

$$\text{dist}(x, q) = \text{def} \sum_{i=1}^{N} w_i \times \text{dist}_{A_i}(x.A_i, q.A_i) \tag{4}$$

There is a common weighted average which is given as in Eq. (5):

$$\text{dist}(x, q) = \text{def} \frac{\sum_{i=1}^{N} w_i \times \text{dist}_{A_i}(x.A_i, q.A_i)}{\sum_{i=1}^{n} w_i} \tag{5}$$

Fig. 2. Proposed system

3 Literature Review

Korovkinas et al. [15] presented that rise of social communities and spread of Internet-associated smart gadgets was followed by explosion in information accessible for collection and preparing, which offered genuine mechanical and computational difficulties together with new attractive outcomes in research, appropriation and use of new and existing information science and machine learning techniques. Author concluded, by propose a hybrid technique to improve SVM classification precision using training data sample and hyper parameter tuning. The proposed system applies clustering to choose training data and parameter tuning to improve classifier viability. The paper reports that better outcomes were gotten utilizing our proposed strategy in all investigations, compared with past results displayed technique's work [15].

Tyagi et al. [16] presented that any sentiment of a people through which the emotions, attitude and thoughts can be communicated is known as opinion. The sorts of information examination which is accomplished from the news reports, client surveys, social media updates or micro blogging sites are called sentiment analysis or opinion mining. The surveys of people towards specific occasions, brands, item or organization can be known through sentiment analysis. The reactions of overall population are gathered and improvised by specialists to perform evaluation. The fame of sentiment analysis is developing today since the quantities of perspectives being exchanged by public on the micro blogging sites are additionally expanding. Every one of the opinions can be ordered into three distinct classifications called positive, negative and neutral. Twitter, being the most well-known micro blogging web page, is utilized to gather the information to perform examination. Tweepy is utilized to extract the source information from Twitter. Python language is utilized in this exploration to execute the algorithm on the gathered information. Author concluded using N-gram modeling method the features are to be extracted. The opinion is ordered among positive, negative and neutral utilizing a supervised learning algorithm known as K-Nearest Neighbor [16].

Iqbal et al. [17] presented a hybrid system to assist the lexicon-based and machine learning techniques. The main objective of proposed system was to attain improved precision and scalability. A new genetic algorithm (GA)-based feature lessening method was proposed to provide solution of scalability problem. This issue rose along with the growth of feature set. The size of feature set could be reduced up to 42% with the help of proposed integrated algorithm without affecting the accuracy. The proposed feature lessening method was compared with extensively utilized feature lessening methodologies called principal component analysis (PCA) and latent semantic analysis (LSA). The proposed approach surpassed PCA algorithm and LSA algorithm by 15.4% and 40.2% respectively in terms of accuracy. In addition, several other performances parameters such as precision, recall, F-measure, and feature size were considered for evaluating the sentiment analysis system. A novel cross-disciplinary region of geopolitics was proposed as well for representing the efficiency of GA-based design. The novel approach was proposed as a case study application for the sentiment analysis system. The tested results depicted that people sentiments and views were measured precisely by the proposed approach on various subjects. These subjects include violence, global clashes, and social problems etc. The relevance of presented approach was envisaged in different domains. These fields included security and observation, law-and-order, and public administration [17].

Kulkarni et al. [18] presented sentiment analysis is broadly utilized in a large portion of real time applications. The exact recognizable proof of text features collected from the unorganized textual information is significant study dispute. A few strategies that present the retrieval of sentiment-based component in dataset mining samples are utilized by particular survey accumulation and avoid by the non-trivial uniqueness distributed attributes of sentiment trait. These techniques cannot forecast the feedback of public effectively. Sentiment analysis is an alternate strategy directly from mining classification to advance sentiment analysis. To start with, traditional procedures were examined for taking care of the issue of sentiment analysis. At that point they examined the ongoing technique for sentiment features, classification and information recovery. They likewise examined the near investigation among every one of these methods.At last, they watched and saw the present research issues of interest based on expanded study of ongoingtechniques. From their investigation, they can reason that sentiment analysis is an attraction in numerous researchers [18].

Shahnawaz et al. (2017) stated that a process utilized to recognize the opinion or views uttered in the opinioned data for finding the approach of an author regarding a particular subject is called sentiment analysis. The views are classified into positive, negative or neutral. This approach gives idea to the purchaser to identify the satisfactory level of the product or service prior to its purchase. For gathering and extracting public opinions, the scientific groups and business world were mainly concerned about the views of people on different types of social media platforms. The existing methods have several major problems. These problems include lack of accuracy and incapability to give good performance in various areas. These issues can be resolved by utilizing semi-supervised and unsupervised learning-based models. With the help of these models, the inadequacy of labeled data can be minimized easily even during the availability of sufficient unlabeled data [19].

4   Result and Discussion

Python is a high-level programming language. This tool comprises dynamic semantics. This language is generated within the data structures. This tool can easily be used by The Rapid Application Development due to the integration of data structures with dynamic typing and binding. In this tool, the previously accessible components get interrelated using scripting. As this language is extremely simple and easy to learn, therefore, it can be read easily. This also minimizes the maintenance cost of program.

Fig. 3. Classification report plotting

As shown in Fig. 3, the classification parameters like precision, recall and f-measure is calculated. The precision, recall and f-measure is calculated for each class like positive, negative. The values of each parameter are plotted in the form of figure.

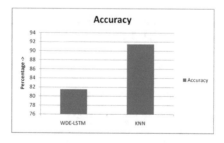

Fig. 4. Accuracy analysis

The Fig. 4 shows the comparison of WDE-LSTM and KNN in terms of accuracy. KNN classification model shows better accuracy than WDE-LSTM for sentiment analysis.

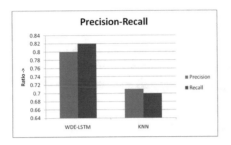

Fig. 5. Precision-recall analysis

The Fig. 5 shows the comparison of WDE-LSTM and KNN in terms of precision recall. The values of the precision and recall is shown in the above figure.

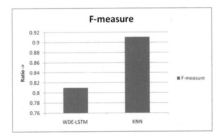

Fig. 6. F-measure analysis

The Fig. 6 shows the comparison of WDE-LSTM and KNN in terms of f measure. KNN classification model has higher f measure than WDE-LSTM for the sentiment analysis (Table 1).

Table 1. Performance analysis

Parameter	WDE-LSTM	KNN
Accuracy	81.51	91.45
Precision	0.80	0.71
Recall	0.82	0.70
F-measure	0.81	0.91

5   Conclusion

The sentiments of the users were studied in this work by analyzing the twitter data. For feature extraction, this work used N-Gram method. The tweets are classified into positive, negative and neutral classes by employing the k-nearest neighbor classification model. The proposed approach gives performance results in terms of accuracy, precision, recall and F1 score correspondingly. In order to analyze sentiments, various techniques have been designed in the last few years. But, absolute competence is still not achieved. Object identification, anaphora resolutions, negation expressions, sarcasms, abbreviations, misspellings, etc. are the major issues faced during sentiment analysis.

References

1. Duygulu, P., et al.: Object recognition as machine translation: learning a lexicon for a fixed image vocabulary. In: European Conference on Computer Vision. Springer, Heidelberg (2002)
2. Zhang, X., Cui, L., Wang, Y.: Commtrust: computing multi-dimensional trust by mining e-commerce feedback comments. IEEE Trans. Knowl. Data Eng. **26**(7), 1631–1643 (2014)
3. Yu, X., et al.: Mining online reviews for predicting sales performance: a case study in the movie domain. IEEE Trans. Knowl. Data Eng. **24**(4), 720–734 (2012)
4. Xu, X., et al.: Aspect-level opinion mining of online customer reviews. China Commun. **10**(3), 25–41 (2013)
5. Ren, F., Ye, W.: Predicting user-topic opinions in twitter with social and topical context. IEEE Trans. Affect. Comput. **4**(4), 412–424 (2013)
6. Poria, S., et al.: Enhanced SenticNet with affective labels for concept-based opinion mining. IEEE Intell. Syst. **28**(2), 31–38 (2013)
7. Garcia-Moya, L., Anaya-Sánchez, H., Berlanga-Llavori, R.: Retrieving product features and opinions from customer reviews. IEEE Intell. Syst. **28**(3), 19–27 (2013)
8. Cheng, V.C., et al.: Probabilistic aspect mining model for drug reviews. IEEE Trans. Knowl. Data Eng. **26**(8), 2002–2013 (2014)
9. Hai, Z., et al.: Identifying features in opinion mining via intrinsic and extrinsic domain relevance. IEEE Trans. Knowl. Data Eng. **26**(3), 623–634 (2014)
10. Liu, K., Liheng, X., Zhao, J.: Co-extracting opinion targets and opinion words from online reviews based on the word alignment model. IEEE Trans. Knowl. Data Eng. **27**(3), 636–650 (2015)
11. Clavel, C., Callejas, Z.: Sentiment analysis: from opinion mining to human-agent interaction. IEEE Trans. Affect. Comput. **7**(1), 74–93 (2016)
12. Juneja, P., Ojha, U.: Casting online votes: to predict offline results using sentiment analysis by machine learning classifiers. In: 2017 8th International Conference on Computing, Communication and Networking Technologies (ICCCNT). IEEE (2017)
13. Hassan, A.U., et al.: Sentiment analysis of social networking sites (SNS) data using machine learning approach for the measurement of depression. In: 2017 International Conference on Information and Communication Technology Convergence (ICTC). IEEE (2017)
14. Rezaei, Z., Jalali, M.: Sentiment analysis on twitter using McDiarmid tree algorithm. In: 2017 7th International Conference on Computer and Knowledge Engineering (ICCKE). IEEE (2017)
15. Korovkinas, K., Danėnas, P., Garšva, G.: SVM and k-means hybrid method for textual data sentiment analysis. Baltic J. Mod. Comput. **7**(1), 47–60 (2019)
16. Tyagi, P., Tripathi, R.C.: A review towards the sentiment analysis techniques for the analysis of twitter data. Available at SSRN 3368718 (2019)
17. Iqbal, F., et al.: A hybrid framework for sentiment analysis using genetic algorithm based feature reduction. IEEE Access **7**, 14637–14652 (2019)

18. Kulkarni, D.S., Rodd, S.F.: Extensive study of text based methods for opinion mining. In: 2018 2nd International Conference on Inventive Systems and Control (ICISC). IEEE (2018)
19. Astya, P.: Sentiment analysis: approaches and open issues. In: 2017 International Conference on Computing, Communication and Automation (ICCCA), pp. 154–158. IEEE (2017)
20. Chauhan, C., Smriti, S.: Sentiment analysis on product reviews. In: 2017 International Conference on Computing, Communication and Automation (ICCCA). IEEE (2017)
21. Al Hamoud, A., et al.: Classifying political tweets using Naïve Bayes and support vector machines. In: International Conference on Industrial, Engineering and Other Applications of Applied Intelligent Systems. Springer, Cham (2018)
22. Rathor, A.S., Agarwal, A., Dimri, P.: Comparative study of machine learning approaches for amazon reviews. Procedia Comput. Sci. **132**, 1552–1561 (2018)
23. Haque, T.U., Saber, N.N., Shah, F.M.: Sentiment analysis on large scale Amazon product reviews. In: 2018 IEEE International Conference on Innovative Research and Development (ICIRD). IEEE (2018)

Gauging Trowel Shaped Patch Including an Optimized Slot for 60 GHz WLAN

Ribhu Abhusan Panda[1(✉)], Baha Hansdah[2], Sanjana Misra[2],
Rohit Kumar Singh[2], and Debasis Mishra[1]

[1] Department of El&TC Engineering,
V.S.S. University of Technology, Burla, Odisha, India
ribhupanda@gmail.com, debasisuce@gmail.com
[2] Department of ECE, GIET University, Gunupur, Odisha, India
bahahansdah15@gmail.com, sanjanamisra840@gmail.com,
rs9761731@gmail.com

Abstract. The proposed research work provides a novel design which includes both semi-circular and triangular patch leading to a shape like gauging trowel with specific dimensions. To shift the resonant frequency, the antenna can be operated exactly at 60 GHz with a desired value of return loss (S_{11} < 10 dB), a circular slot has been inserted at the middle of the modified patch. To enrich antenna gain a metal boundary has been implemented on the substrate. Dimension of the substrate has been taken as 30 mm × 30 mm × 1.6 mm. Ground plane has same dimension as that of the substrate but the height is 0.01 mm. The circular slot has been optimized to get the resonant frequency at 60 GHz so that the designed structure can be operable for 60 GHz based WLAN application. A metal strip of 8 mm acts like an antenna gain booster, which has been implemented around the patch on the substrate. HFSS software is the software by which the simulation has been done.

Keywords: Gauging trowel · Semi-circular patch · Triangular patch · Optimized circular slot · Metal boundary · 60 GHz WLAN

1 Introduction

The past decade provided many novel designs that include unique perturbations for different applications [1–7]. In the past few year different gain enhancement techniques that are used with the modified patch provided a new way of research in the field of planar antennas [8–13]. A slot of defined shape can play a very vital role of shifting the resonant frequency and bandwidth increment can also be done with the implementation of the slot. Different slots have been inserted in a patch in recent years [14–17]. In this paper the semi-circular and triangular patch shape have been combined with a resulting gauging trowel shape that is used for plastering with cement on the walls. Proposed structure is designed for 60 GHz WLAN. Copper is for both ground plane and patch, for substrate the dielectric FR4 epoxy is used which has dielectric constant ε_r = 4.4. Heights of the ground and dielectric substrate are 0.01 mm and 1.6 mm respectively.

© Springer Nature Switzerland AG 2020
J. S. Raj et al. (Eds.): ICIDCA 2019, LNDECT 46, pp. 87–93, 2020.
https://doi.org/10.1007/978-3-030-38040-3_9

2   Antenna Design with Specific Design Parameters

The diameter of the semicircular patch is having same value as that of twice of the corresponding wavelength (λ) that has been calculated from the operating frequency 60 GHz. Base of the triangular patch in this design is having same value as that of the diameter of the semi-circular patch. The other two sides have been designed with the length equals to the corresponding wavelength (λ). Optimized value of the diameter of circular slot has been found out to be D = 5.4 mm. The width of the metal boundary is 8 mm. Figure 1 illustrates whole design structure of the projected antenna (Table 1).

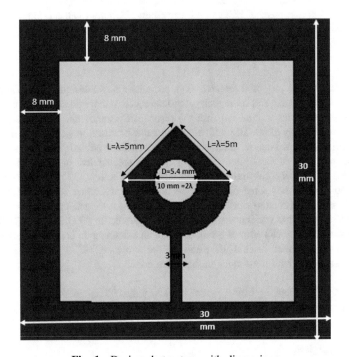

Fig. 1. Designed structure with dimensions

Table 1. Specific design parameters for the antenna design

Parameters	Symbol	Value (mm)
Width of ground	WG	30
Length of ground	LG	30
Width of substrate	WS	30
Length of substrate	LS	30
Height of ground	HG	0.01

(continued)

Table 1. (*continued*)

Parameters	Symbol	Value (mm)
Height of substrate	HS	1.6
Radius of optimized circular slot	R	2.7
Diameter of optimized circular slot	D	5.4
Diameter of semicircular patch	DS	10
Length of sides of triangular patch	LT	5
Width of metal boundary	WB	8

3   Outcomes from the Simulation

3.1   Resonant Frequency, Return Loss and Standing Wave Ratio

The emphasis has been given to the S_{11} Vs Frequency plot to find out the resonant frequency. From the optimized valued of the radius of the circular slot different plots have been determined where it can be clearly observed that although at some values the value of the return loss is appreciable but only at 2.7 mm radius (Diameter = D = 5.4 mm) the resonance is at 60.1 GHz with S_{11} (return loss) of −39.1 dB. Considering the projected structure as the transmission line model, standing wave ratio measured in terms of voltage has been found to be 1.0224 at 60.1 GHz which is nearer to the desired value 1. Figures 2 and 3 represents the distinction of the return loss with frequency for diverse values of the slot and for optimized valued slot respectively. Figure 4 represents the VSWR plot.

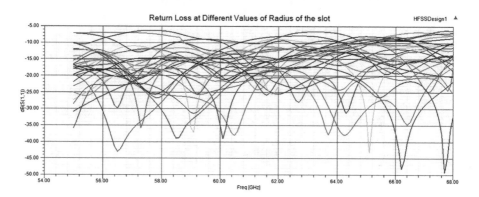

Fig. 2. Different S_{11} plots for different values of the circular slot in optimization process

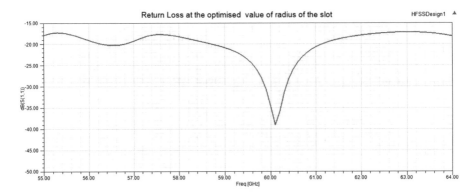

Fig. 3. Return loss (S_{11}) at the optimized value of circular slot (R = 2.7 mm, D = 5.4 mm)

3.2 Antenna Gain and Directivity

To measure the performance of an antenna the antenna gain and directivity are considered as vibrant parameters. For this novel design the antenna gain at resonant frequency has been found to be 3.915 and the directivity has been found to be 10.73 dB. Figures 5 and 6 provides the variation of antenna gain and frequency in the frequency range.

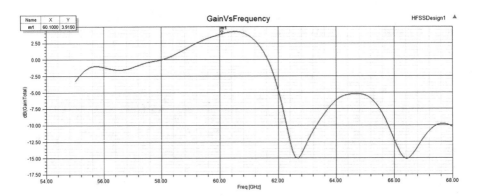

Fig. 4. Antenna gain vs frequency

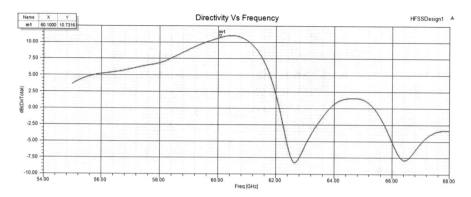

Fig. 5. Variation of directivity with frequency

3.3 Surface Charge Distribution on the Patch with Metal Boundary and 3D Gain Pattern

3D gain pattern is shown in Fig. 6 which provides the observation that the designed antenna is radiation uniformly in all the directions. In Fig. 7, the charge distribution clearly provides a reflection that the current passing through the metallic structure that includes both the metal boundary and the perturbed patch with optimized circular slot (Table 2).

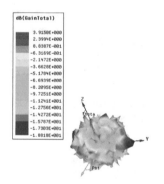

Fig. 6. Antenna gain in 3D

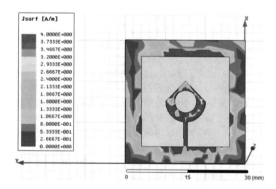

Fig. 7. Current density on the projected structure

Table 2. Simulated parameters

Parameters	Value
Antenna gain (dB)	3.915
Directivity (dB)	10.732
Resonant frequency (GHz)	60.1
Return loss (dB)	−39.1
VSWR	1.0224

4 Conclusion

The gauging trowel patch with optimized circular slot provides a sharp resonance at 60.1 GHz with broadband characteristic having a return loss of −39.1 dB. So it can be used efficiently for 60 GHz WLAN. The values of antenna gain and directivity for this proposed assembly also gives an indication that the antenna is a suitable alternative for 60 GHz WLAN.

References

1. Panda, R.A., Mishra, D., Panda, H.: Biconcave lens structured patch antenna with circular slot for Ku Band application. In: Lecture notes in Electrical Engineering, vol. 434, pp. 73–83. Springer (2018)
2. Lee, W., Kim, J., Cho, C.S., Yoon, Y.J.: Beam forming lens antenna on a high resistivity silicon wafer for 60 GHz WPAN. IEEE Trans. Antennas Propag. **58**, 706–713 (2010)
3. Panda, R.A., Mishra, D.: Reshaped patch antenna design for 60 GHz WPAN application with end-fire radiation. Int. J. Mod. Electron. Commun. Eng. **5**(6) (2017)
4. Panda, R.A., Mishra, D., Panda, H.: Biconvex patch antenna with circular slot for 10 GHz application. In: SCOPES 2016, pp. 1927–1930. IEEE (2016)
5. Panda, R.A., Panda, M., Nayak, P.K., Mishra, D.: Log periodic implementation of butterfly shaped patch antenna with gain enhancement technique for X-band application. In: ICICCT-2019, System Reliability, Quality Control, Safety, Maintenance and Management, pp. 20–28. Springer (2020)
6. Panda, R.A., Mishra, D.: Log periodic implementation of star shaped patch antenna for multiband application using HFSS. Int. J. Eng. Tech. **3**(6) (2017)
7. Panda, R.A., Mishra, D.: Modified circular patch and its log periodic implementation for Ku Band application. Int. J. Innov. Technol. Explor. Eng. **8**(8), 1474–1477 (2019)
8. Panda, R.A., Panda, M., Nayak, S., Das, N., Mishra, D.: Gain enhancement using complimentary split ring resonator on biconcave patch for 5G application. In: International Conference on Sustainable Computing in Science, Technology & Management (SUSCOM-2019), pp. 994–1000 (2019)
9. Attia, H., Yousefi, L.: High-gain patch antennas loaded with high characteristic impedance superstrates. IEEE Antennas Wirel. Propag. Lett. **10**, 858–861 (2011)
10. Rivera-Albino, A., Balanis, C.A.: Gain enhancement in microstrip patch antennas using hybrid substrates. IEEE Antennas Wirel. Propag. Lett. **12**, 476–479 (2013)

11. Kumar, A., Kumar, M.: Gain enhancement in a novel square microstrip patch antenna using metallic rings. In: International Conference on Recent Advances and Innovations in Engineering (ICRAIE-2014), Jaipur, pp. 1–4 (2014)
12. Ghosh, A., Kumar, V., Sen, G., Das, S.: Gain enhancement of triple-band patch antenna by using triple-band artificial magnetic conductor. IET Microw. Antennas Propag. **12**(8), 1400–1406 (2018)
13. Panda, R.A., Dash, P., Mandi, K., Mishra, D.: Gain enhancement of a biconvex patch antenna using metallic rings for 5G application. In: 6th International Conference on Signal Processing and Integrated Networks (SPIN), pp. 840–844 (2019)
14. Sun, X., Cao, M., Hao, J., Guo, Y.: A rectangular slot antenna with improved bandwidth. Int. J. Electron. Commun. (AEU) **66**, 465–466 (2012)
15. Rafi, Gh., Shafai, L.: Broadband microstrip patch antenna with V-slot. IEE Proc. Microw. Antennas Propag. **151**(5), 435–440 (2004)
16. Lu, J.-H.: Bandwidth enhancement design of single-layer slotted circular microstrip antennas. IEEE Trans. Antennas Propag. **51**(5), 1126–1129 (2003)
17. Bao, X.L., Ammann, M.J.: Compact annular-ring embedded circular patch antenna with cross-slotground plane for circular polarisation. Electron. Lett. **42**(4), 192–193 (2006)

Reliable Machine to Machine Communication Using MQTT Protocol and Mean Stack

Nalin Chandeliya[(⊠)], Prashanth Chari, Sameeran Karpe,
and Deepak C. Karia

Sardar Patel Institute of Technology,
Munshi Nagar, Andheri West, Mumbai 400058, India
{nalin.chandeliya, prashanth.chari,
deepak_karia}@spit.ac.in

Abstract. IoT (Internet of Things) and Industrial Automation is emerging as a revolutionizing technology across the globe. Presently, Machine to Machine (M2M) communication is implemented by using Pushing protocol. Push protocol is lightweight and is highly productive. Out of all push protocols used MQTT is considered as the best of all and more favorable than others in terms of speed, security and bandwidth efficiency which can also framed as salient features of MQTT when used to compare with other protocols. The proposed work aims to deliver the importance of MQTT in sensor networks by making a reliable Machine to Machine communication (M2M) system using MQTT and MEAN stack.

Keywords: Industrial Automation · MQTT · MEAN stack · Machine to Machine

1 Introduction

The MQTT protocol has been declared as more reliable and efficient protocol for Internet of Things by many industry experts [1, 2]. It has been researched that MQTT is more power efficient than HTTPs and other Internet of Things protocols, its main purpose that separates MQTT protocol from others is its speed, reliability and power efficiency to meet different test environments. Presently Facebook's Messenger uses MQTT protocol for message transfer between different devices and has been successful in implementing it. Companies like CISCO is planning to introduce Beyond MQTT [3], a version which will be having more upgradation as compared to present MQTT which will further increase the scope of the protocol used in the industry. MQTT was invented back in the 20th Century end, this protocol when invented was never even thought of contributing to upcoming technology Internet of Things in the coming years [4]. MQTT when invented had a view to create a protocol that is more concerned with less power, i.e. power efficient and least bandwidth for a reliable Machine to Machine communication.

J. S. Raj et al. (Eds.): ICIDCA 2019, LNDECT 46, pp. 94–100, 2020.
https://doi.org/10.1007/978-3-030-38040-3_10

2 Methodology

Using advanced MQTT protocol as transmission protocol, we connect the server and the various clients. MQTT (Message Queuing Telemetry Transport) uses 3 stage network for communication which includes publisher, subscriber and broker which works on TCP/IP protocol. Here, our server is the message broker, the machines are the publishers and the database is the subscriber. To facilitate the reception of data from all the connected machines, each database is subscribed to a topic known only to a particular set of publishers. The message broker in turn, receives the messages, segregates them as per topics and sends them to various sections of the database for precise recording. A security layer over the data has been added (SSL) to encrypt the data while sending and decrypt the data only when received by the intended recipient. The server runs on Node.js, a leading technology based on Javascript for backend development as it provides robust and faster data handling as well as caters to many client requests at the same time. In terms of reliability of the system, MQTT has the additional features like Quality of Service (QoS) and Last Will and Testament over other protocols, which ensures a guarantee that the data sent is being delivered to the receiver.

3 System Layout

System design is a vital part of the project as it completely determines the workings of the system. The system layout shows the two level communication using MQTT protocol having client, subscriber and a broker connecting them. Using advanced MQTT protocol as transmission protocol, we connect the server and various clients. Here, our server is the message broker, the machines are the publishers and the database is the subscriber. To facilitate the reception of data from all the connected machines, each database is subscribed to a topic known only to a particular set of publishers. The message broker in turn, receives the messages, segregates them as per topics and sends them to the various sections of the database for precise recording. The server runs on Node.js, a leading technology based on JavaScript for backend development as it provides robust and faster data handling as well as ability to cater to many client requests at the same time (Fig. 1).

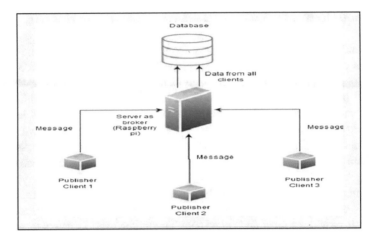

Fig. 1. Block diagram

4 Simulation and Results

We carried out the following steps to build the experimental setup (Figs. 2 and 3):

1 Installed Raspbian OS in raspberry pi to setup the PC
2 Setup in headless mode and connected to the common network
3 Installed Mosquitto client in raspberry pi PC
4 Started Mosquitto as a background process (daemon)
5 Initialized NodeMCU, uploaded the code which connected it to the common network (Wi-Fi hotspot).

Fig. 2. Headless mode raspberry pi PC

A code is established to take/read data from DHT11, where DHT11 mounted on NodeMCU is acting as a publisher and above IDE is the server, i.e. the broker where all readings will be displayed. These readings are stored in a database as real time data and are used for further processes to examine time delay and other parameters (Fig. 4).

The readings as captured from DHT11 on NodeMCU are displayed, which is then sent to the database, which is the subscriber.

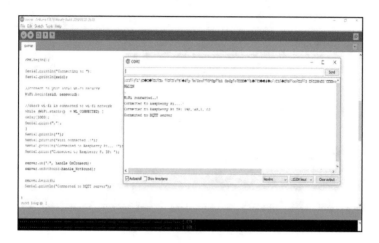

Fig. 3. Connected to the server

Fig. 4. Data received

5   Hardware Requirements

5.1   Flowchart

See Fig. 5.

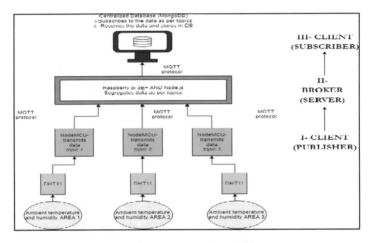

Fig. 5.   Complete flowchart

5.2   Requirements and Their Usage

5.2.1   DHT-11

The DHT11 is a basic low cost humidity and temperature measuring sensor. The sensor consists of a thermistor and a capacitor having certain dielectric which work together and evaluate the surrounding environment. It does not require any analog pins since it has an inbuilt ADC converter which yields a digital signal and can be connected directly to the digital pin of NodeMCU. It is easy to use but care has to be taken regarding its accuracy [6].

5.2.2   Node MCU

NodeMCU is a microcontroller which is mainly used in IoT projects and applications. It is based on ESP8266 Wi-Fi SoC and primarily utilizes Lua scripting. It has 17 general purpose I/O pins out of which six of them can be used; the rest are used by the flash memory of the board. It also has an inbuilt ADC pin for specific types of sensors which cannot transmit digital signal [7]. It is easy to use NodeMCU for the IoT device applications as the libraries necessary for running codes on the board are easily available.

5.2.3   Raspberry Pi 3b+

The Raspberry Pi is a debit card sized board which can be used as a computer as well as a microcontroller, as per the application. Not only it has all the required ports to connect I/O devices but also extensive connectivity options like Bluetooth, Ethernet and WiFi connectivity are supported. It is a revolutionary invention bringing people of all ages together to learn the marvels of computing [8].

6 Software Requirements

6.1 Raspbian OS

Raspbian is Raspberry Pi's Debian-based operating system. Out of the two different versions of Raspbian, it has been lustrum since when the Raspbian Os is officially provided for Raspberry Pi computers [9]. There is still active development of the operating system. Raspbian is extremely optimized for the low-performance ARM CPUs of the Raspberry Pi line. It consists of a modified LXDE desktop environment and a fresh theme and few other modifications to the Openbox stacking window manager.

6.2 Eclipse Mosquitto

Eclipse Mosquitto is a free platform message broker of the MQTT protocol [10]. Mosquitto is part of the Foundation for Eclipse and is a project for iot.eclipse.org. Eclipse Mosquitto offers a lightweight MQTT protocol implementation server appropriate for all circumstances which includes full power to embedded computers and low power machines. Sensors and actuators can be very tiny, consume very less power and this also applies to the integrated computers they are attached to, which is precisely where one can operate Mosquitto client. Typically, Mosquitto's current implementation has a 120 kB executable that consumes about 3 MB of RAM linked to 1000 customers. There were reports of effective test cases at small message rates with 100,000 linked customers. In addition to accepting MQTT client links, Mosquitto has a bridge which establishes connectivity with all the MQTT servers. This enables MQTT server networks being built, sending messages using MQTT protocol from one network place to another, depending on the bridge setup.

6.3 MongoDB

MongoDB is a platform which specializes in storing real time data and acts as a cloud to store data in unstructured stacks by using NoSQL [11] which makes it faster than SQL [12]. We have used a computer having MongoDB installed as the primary subscriber, which receives all data from the nodes and stores it. The thought process behind keeping a database as a receiver was to check timestamps of data from all the nodes and infer whether they are the same for data sent at the same time. In an ideal case, the data sent from all the nodes should periodically be stored at the same instant of time, which would show that using Node.js as the server, we can address many clients at once.

7 Conclusion

This project aims to demonstrate a Machine to Machine Communication system based on MQTT and MEAN stack. For attaining MQTT protocol mechanism, we have developed a three stage network between publisher, broker and subscriber to ensure end-to-end transmission of information with secured flow of messages; we take DHT11 mounted on NodeMCU as publisher, broker is Raspberry Pi and the subscriber is the MongoDB-enabled machine. The advantage of the protocol is its lightweight and the server's ability to cater multiple requests at a time will make this system more reliable than any others.

Acknowledgement. We would like to convey our gratitude to Dr. Surendra Rathod, Head of Department, Electronics Engineering, SPIT Mumbai, for his endearing support and words of encouragement.

References

1. Jackson, J.: OASIS: MQTT to be the protocol for the Internet of Things (2013). http://www.pcworld.com/article/2036500/oasis-mqtt-to-be-the-protocol-for-the-internet-of-things.html
2. Gupta, R.: 5 Things to know about MQTT - the protocol for Internet of Things (2014). https://www.ibm.com/developerworks/community/blogs/5things/entry/5_things_to_know_about_mqtt_the_protocol_for_internet_of_things?lang=en
3. Duffy, P.: Beyond MQTT: a Cisco view on IoT protocols (2013). https://blogs.cisco.com/digital/beyond-mqtt-a-cisco-view-on-iot-protocols
4. HiveMQ (2015). http://www.hivemq.com/blog/mqtt-essentials-part-1-introducing-mqtt
5. Eslava, H., Rojas, L.A., Pereira, R.: Implementation of machine-to-machine solutions using MQTT protocol in Internet of Things (IoT) environment to improve automation process for electrical distribution substations in Colombia. J. Power Energy Eng. **3**, 92–96 (2015). http://www.scirp.org/pdf/JPEE_2015041409413629.pdf
6. Adafruit sensors product description. https://www.adafruit.com/product/386
7. NodeMCU GitHub repository. https://github.com/NodeMCU
8. Raspberry pi description. https://www.raspberrypi.org/help/what-%20is-a-raspberry-pi/
9. Raspbian description. https://www.raspberrypi.org/downloads/raspbian/
10. Official Eclipse Mosquitto website. https://mosquitto.org/
11. Techopedia. https://www.techopedia.com/definition/27689/nosql-database
12. Official MongoDB website. https://www.mongodb.com

Lane Keep Assist System for an Autonomous Vehicle Using Support Vector Machine Learning Algorithm

M. Karthikeyan[1], S. Sathiamoorthy[2(✉)], and M. Vasudevan[3]

[1] Division of Computer and Information Science, Annamalai University,
Chidambaram, India
mkarthi82@gmail.com
[2] Tamil Virtual Academy, Chennai, India
sathiamoorthy2019@gmail.com
[3] HCL Technologies Limited, Chennai, India
vasudevan.m@hcl.com

Abstract. Autonomous self driving vehicles are getting greater attention and this would be the future requirement in Automotive domain. However, Fail proof driving is the only solution to reduce the rate of accidents and that makes the driverless vehicles as a possible one. In man handled vehicles, by using Advance Driver Authorization System (ADAS), accident free driving can be ensured. This paper focuses on one of the ways to contribute towards accident free driving of autonomous vehicles by deploying a novel Lane Keep Assist (LKA) system. A Machine Learning algorithm has been used in proposed LKA system for tracking the lane of the autonomous vehicles by providing the required inputs. Proposed LKA system has been demonstrated in Matlab/Simulink platform and the results have been presented in this paper.

Keywords: Autonomous vehicle · Support Vector Machine · Lane Keep Assist · Machine learning algorithm · Classification and regression techniques

1 Introduction

In recent years, Research and Development of autonomous vehicle has reached several milestones and commercial makers have implemented various features in existing vehicles in the form of advanced driver authorization system [1]. Self driving cars may have many challenges but they are being resolved by proper control system algorithms and artificial intelligence (AI). Autonomous vehicles have the unique features such as lane keep assist, traffic jam assist, automatic park assist, Highway autopilot, etc. [2]. In order to accomplish these features, four main control steps should be applied and they are steering control, auto pedaling, auto braking and lighting control. One of the interesting and important features of autonomous vehicle is Lane Keep Assist and if we achieve this feature then safety of driving is almost ensured. In order to achieve the efficient LKA, proper control of Steering angle & auto pedaling & braking procedures are required. Figure 1 shows the block diagram of autonomous vehicle. Input images of Lane &road map, reference vehicle speed and reference steering angle are given as

© Springer Nature Switzerland AG 2020
J. S. Raj et al. (Eds.): ICIDCA 2019, LNDECT 46, pp. 101–108, 2020.
https://doi.org/10.1007/978-3-030-38040-3_11

inputs to the AI based control system which in turn activate the power train and transmission system of vehicle that resultant into staggering angle variation with speed control and braking. This Fig. 1 clearly gives the overview of operation of autonomous vehicle [9–12].

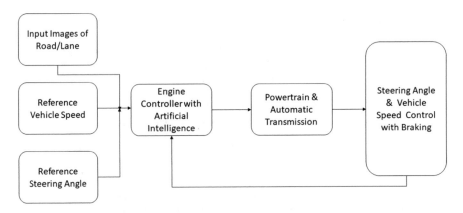

Fig. 1. Block diagram of autonomous vehicle

In this paper, authors focused to demonstrate the autonomous vehicle with Lane Keep Assist mode by applying the proposed Support Vector machine learning algorithm.

2 Principle of Lane Keep Assist (LKA) System

Lane keep assist system is an important feature for safe driving and to prevent the accidents [3]. This system controls the vehicle within the lane by providing required speed and torque information. Figure 2 shows the schematic of LKA with Machine learning algorithm based controller. Among various algorithms of Machine Learning, Support Vector Machine has been selected for this problem as it is a straight forward method and inputs and outputs are well determined. Scope of the problem statement is to keep the vehicle in the middle of the lane at static condition and during the dynamic condition (moving condition), distance between the centre point of vehicle and the lane at both left and right side has to be monitored and this information needs to be given as input to algorithm. Apart from this information, if any obstacles on the road & destination information also to be given as inputs in order to determine the steering angle. With the determined steering angle, vehicle will follow either acceleration and deceleration based on the input information. Authors have found that advanced SVM would be right algorithm for this identified LKA problem. Proposed solution has been designed and implemented in Matlab/Simulink platform. Authors have presented the details of proposed SVM based LKA in upcoming sections.

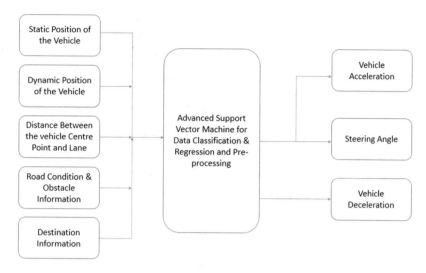

Fig. 2. Lane Keep Assist algorithm

3   Proposed Support Vector Machine (SVM) for Images Classification

SVMs are quite popular since last decade to analyses the known data and sensing the remote data [4]. In general, support vectors are used to find the optimal solution which is very close to required decision. Support vectors are lying on the plane which is very close to the decision point [5]. If a set of data has multiple classifications and the required output to be separated out, then either linear or nonlinear support vectors are useful to this job. Linear Kernel support vector machines [6, 7] are being considered in this work. As LKA problem is almost a straight forward one, authors have decided to proceed with linear kernel. In this work, linear kernel is used based on the Eq. (1) [8].

$$f(x) = G(0) + sum(ai * (x, xi)) \tag{1}$$

Where $x = input\ vector$
$xi = supporting\ vector$
$G0\ \&\ ai = Coefficient\ for\ each\ input.$

In the view of regularization of parameters during tuning the algorithm, SVM optimization could be done by kernel, Gamma and Margin. Figure 3 shows the selection of SVM considered for this work. Steering angle control is the main objective this work and by controlling the steering angle, turning of vehicle's wheel would be possible either in left side or right side. Steering angle values are clearly defined in this work as positive & negative angles and lies between $-360^0 < 0 < +360$. Positive angles represent the right turning of steering angle and negative angle represents the left turning of steering angle. Positive and negative classes are split across the hyperplane. Image of the lane has been modelled and the same is used for modelling of the entire LKA system.

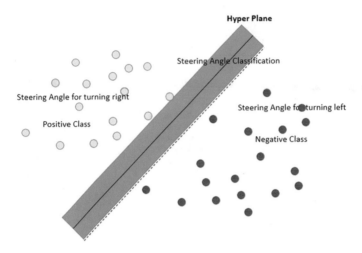

Fig. 3. Proposed Support Vector Machine for lane and path tracking

Figure 4 shows the model lane taken for LKA & SVM modeling and simulation work considered in this paper. Figure 4(a) shows the centre point of the road wherein the vehicle needs to be started with the maximum distance between its centre point and the lanes. At this condition, vehicle's steering angle is 0° and no further changes in the steering angle is required. Figure 4(b) shows the Vehicle tracking within the lane and left and right marking of lane as well. In both the figures, road path is marked with A & B. A represents the right direction and B represents the left directions which are being

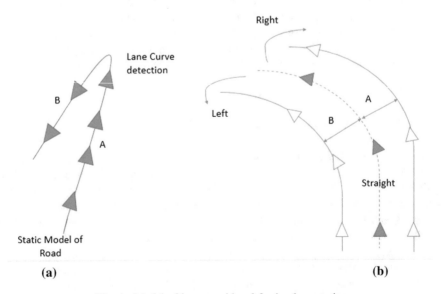

Fig. 4. Model of lane considered for implementation

marked for modeling convenience. Following parameters are considered for Modeling and Simulation: d1, d2, d3, …. are the instant distance between the centre point of the vehicle and the lane

- x1, x2, x3, …. are the steering angles and are the support vectors
- y1, y2, y3, ….. are the vehicles' wheel turning angle.

4   Implementation of SVM in Matlab/Simulink Platform

Algorithm of SVM based LKA has been designed and implemented in Matlab/Simulink platform. Statistics and Machine Learning algorithm toolbox has also been considered for modeling and simulation. Steering angle to wheel turn angle ratio has been considered as 360:24 and this is very common in any passenger vehicles. That is, for one full rotation of steering (360°), wheel will turn for 24°. Sensor & actuator models have been considered to capture the position of vehicle in order to calculate the distance between the center point and the lane. This data is an input to Machine learning algorithm and support vectors have been identified and placed. Table 1 shows the portion of support vectors (Steering Angle Right & Left). Figure 5 shows the Simulink model of LKA algorithm and which is integrated with SVM algorithm presented in this paper. To calculate the steering angle various position of vehicle within the lane are required from which the distance between these positions and respective lane curvature point needs to be calculation, This function has been accomplished by developing the state flow chart as shown in Fig. 5. Support vectors output from this model are taken as input to SVM algorithm designed within Machine learning toolbox.

Table 1. Support vectors

Steering angle right	Wheel turn angle right	Steering angle left	Wheel turn angle right
0	0	0	0
10	0.67	−10	−0.67
20	1.33	−20	−1.33
30	2.00	−30	−2.00
40	2.67	−40	−2.67
50	3.33	−50	−3.33
60	4.00	−60	−4.00
70	4.67	−70	−4.67
80	5.33	−80	−5.33
90	6.00	−90	−6.00

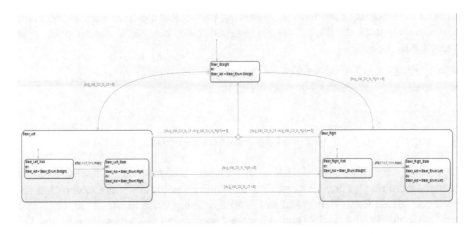

Fig. 5. Steering angle calculation

5   Results and Discussion

Different SVMs have been analyzed and compared as below

(a) Linear SVM
(b) Quadratic SVM
(c) Cubic SVM
(d) Fine Gaussian SVM
(e) Medium Gaussian SVM
(f) Coarse Gaussian SVM.

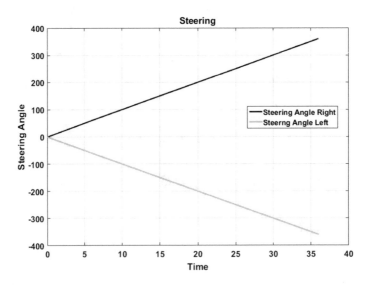

Fig. 6. Derived steering angle

Figure 6 shows captured steering angle graph simulated in Matlab/Simulink platform. It shows both right and left turning steering angles and proportionately wheel turn angles were also varied. Figure 7 shows the wheel turn angle with reference to the estimated steering angle.

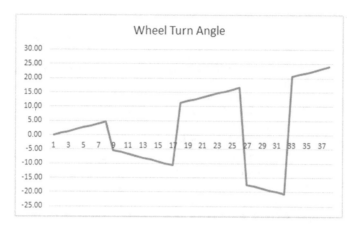

Fig. 7. Wheel turn angle

Table 2 shows the results of various types of SVMs used for Lane keep assist system. It is evident that medium Gaussian SVMs are more accurate than the other types though it took more training time. So, based on the application, type of SVM could be selected.

Table 2. Accuracy Levels of Various Types of SVM

Types of SVM	Accuracy	Training time	Prediction period
Linear SVM	87.3%	2.53 s	8500 obs/sec
Quadratic SVM	91.2%	2.53 s	8000 obs/sec
Cubic SVM	90.4%	2.53 s	8500 obs/sec
Fine Gaussian SVM	92.3%	2.53 s	9000 obs/sec
Medium Gaussian SVM	94.2%	2.53 s	9000 obs/sec
Coarse Gaussian SVM	86.3%	2.53 s	7500 obs/sec

6   Conclusion

This paper is mainly focused on the Machine Learning based Lane Keep Assist system of Autonomous Vehicle. Authors have identified the solution called Support Vector Machine algorithm as the right solution to resolve this problem. This solution has not been published earlier as per author's literature survey. In this paper, work has been split into two stages such as Steering angle calculation module and SVM Modules.

Both of these modules were integrated to achieve the goal of Lane Keep Assist system. Contributions of the work have been articulated and it has summarized the merits and efficiency of the identified algorithm.

References

1. Blundell, H., Thornton, S.M.: Object classification for autonomous vehicle navigation of Stanford campus, Technical report, CS 229 (2015)
2. Chen, Z.: Computer vision and machine learning for autonomous vehicles, dissertation, Worchester Polytechnic Institute (2017)
3. Mountrakis, G., Im, J., Ogole, C.: Support vector machines in remote sensing. A review. ISPRS J. Photogramm. Remote. Sens. **66**(3), 247–259 (2011)
4. Albousef, A.A.: Vehicle lane departure prediction based on support vector machines, Dissertation, Wayne State University (2014)
5. Apostolidis-afentoulis, V.: SVM classification with linear and RBF kernels, 0–7 July 2015
6. Mathworks Inc.: http://www.mathworks.com
7. Kavzoglu, T., Colkesen, I.: A kernel functions analysis for support vector machines for land cover classification. Int. J. Appl. Earth Obs. Geoinformation **11**(5), 352–359 (2009)
8. Paterl, S.: Chapter 2: SVM (Support Vector Machine), Theory (2017)
9. Hernández, D.C., Seo, D., Jo, K.-H.: Robust lane marking detection based on multi-feature fusion. In: 2016 9th International Conference on Human System Interactions (HSI), pp. 423–428(2016)
10. Yeniaydin, Y., Schmidt, K.W.: A lane detection algorithm based on reliable lane markings. In: 2018 26th Signal Processing and Communications Applications Conference (SIU), pp. 1–4 (2018)
11. Zhou, B., Wang, Y., Yu, G., Wu, X.: A lane-change trajectory model from drivers' vision view. Transp. Res. Part C Emerg. Technol. **85**, 609 (2017)
12. Sharma, S., Tewolde, G., Kwon, J.: Behavioral cloning for lateral motion control of autonomous vehicles using deep learning. In: 2018 IEEE International Conference on Electro/Information Technology (EIT), pp. 0228–0233 (2018)

An Automated System for Crime Investigation Using Conventional and Machine Learning Approach

V. S. Felix Enigo$^{(\boxtimes)}$

Department of Computer Science and Engineering, SSN College of Engineering,
Chennai, India
felixvs@ssn.edu.in

Abstract. Crime causes significant damage to the society and property. Different kinds of physical or direct methods are devised by the law and order department to spot out the criminals involved in the crime. This techniques will explore the evidences at crime site. For instance if it finds a fingerprint then the system will capture and send it to forensic department for fingerprint matching, which can be later used for identifying the suspects or criminals by investigations etc. Yet, it is a huge challenge for them to find the criminal due to less or no evidence and incorrect information, which can change the direction of investigation to the end. This paper proposes a data analysis approach to help the police department by giving them first-hand information about the suspects. It automates the manual process for finding criminal and future crime spot by using various techniques such as pattern matching, biometric and crime analytics. Based on the availability of information, the system is able to produce the expected accuracy.

Keywords: Crime · Criminal · Fingerprint matching · Pattern matching · Biometric · Crime analytics

1 Introduction

Crime analysis has increased to high magnitude in today's world than before due to poverty, unemployment and technological advancement. Law enforcement department uses various strategies physically to spot and uncover criminals, which are based on the analysis of physical identities obtained from witness testimony, personality traits analysis and fingerprints analysis or crime scene photographs in the crime scene.

Several researches have been done to automate these processes to help the law and order officials. Simple statistical methods of mean and standard deviation [1] are used to find the personality traits of normal and criminal persons after collecting the socio-demographic variables from a personality factor questionnaire. This knowledge helps to identify whether the suspect is a criminal or normal. Text mining techniques are used on criminal dataset [2] to understand the unstructured data, perform concept extraction based on frequent words, pattern analysis and detailed reporting using visualization.

© Springer Nature Switzerland AG 2020
J. S. Raj et al. (Eds.): ICIDCA 2019, LNDECT 46, pp. 109–117, 2020.
https://doi.org/10.1007/978-3-030-38040-3_12

Law and enforcement department maintains huge criminal database that it is infeasible to analyse such data manually without errors to predict crimes. Data science [3] approach is popularly used to analyse such huge datasets to explore hidden patterns. Some of the real-world crimes problems have been successfully predicted using this approach. One such application is drug-related criminal activities in Taiwan [4], which used machine learning algorithms using spatio-temporal analysis to predict the crime hotspots. Similarly, machine learning prediction models such as K-nearest neighbour and boosted decision tree is applied to Vancouver crime data [5] that contains last 15 years of data and able to get crime prediction accuracy in between 39% to 44%. Using the crime statistics in India obtained for past 14 years (2001–2014), different machine learning methods such as supervised, semi-supervised and unsupervised have been applied based on the requirements in predicting crimes with the notion to help the local police stations in spotting crimes [6].

The challenge with respect to machine learning techniques is the need for large datasets to train the machine learning algorithms. Since, criminal dataset contains individual criminal information each of different data; it renders useless in identifying a criminal based on patterns. Existing criminal databases allow performing analysis based on overall statistics such as trends in crime [7], crime classification [8] and hotspot detection [9].

Many researches were done in automating the manual fingerprint analysis process obtained at the crime site using image processing based fingerprint analysis. This technique tends to match the fingerprint images obtained at the crime-spot with the fingerprints in the criminal biometric database to identify the suspects. Most of the fingerprints found at crime site are partial and so partial fingerprint matching techniques were used. Majorly, it uses minutiae based partial fingerprint matching technique [10–12] than other techniques as it relies on local features and so invariant to translation, rotation and displacement.

All the above existing researches focus on crime analysis using any one of the approach such as pattern matching, machine learning and fingerprint analysis. In general, a crime site may have very few evidences and so based on the available proofs, any one or combination of techniques is required to find the suspects. Hence, in this paper we propose a combined approach to automate the process of identification of suspects. Based on the available evidence either pattern matching or fingerprint analysis or both can be applied. Additionally, machine learning algorithms are used to find the crime hotspot region.

2 Proposed System

The proposed system automates the process of predicting the suspects by considering two evidences such as the physical identification of the criminal based on the witness and the fingerprints of the persons in the crime spot. If the fingerprint of the criminal is available, then fingerprint analysis using image processing is performed. Otherwise, based on the data collected at the site, pattern matching techniques are used. Additionally, the system employs crime analytics to predict the future crime hotspots. The workflow of the system is shown in Fig. 1. To accomplish the aforementioned tasks, the system is divided into three components namely, pursuit based on physical features, pursuit based on finger print analysis and crime hotspot identification.

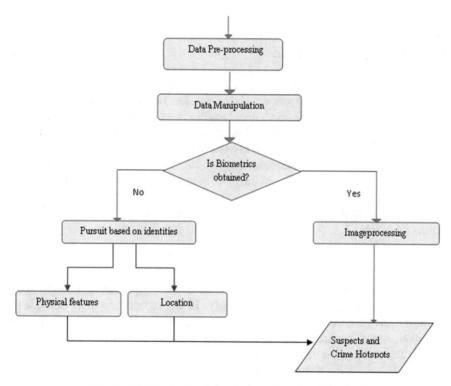

Fig. 1. Workflow of automated crime investigation system

2.1 System Components

Pursuit Based on Identities and Physical Features. This component performs the search for the suspects in the criminal database based on physical identities and physical identities with location information. In first approach, the physical identities of the criminal obtained at the crime site are matched with the patterns in existing criminal records and the percentage of similarity is computed. In the location based approach, it additionally takes into account the latitude and longitude of the crime spot with the specified range and searches the records for pattern matching.

Pursuit Based on Finger Print Analysis. With the finger print data obtained from the crime site, this module compares it with the biometric dataset of the criminals and computes the percentage of match. Persons whose finger print matches 50% and above are considered as suspects to undergo further investigations. Since, most of the fingerprints obtained are partial; partial fingerprint matching algorithms are applied to detect the suspects.

Crime Hotspot Detection. To prevent crimes before onset, it is essential to find out the areas of frequent occurrence of crime. To accomplish this, machine learning technique such as clustering algorithm is applied to identify the crime hotspot.

3   System Implementation

Various algorithms are implemented in the process of finding the criminal and crime hotspot detection. For finding suspects using physical feature matching Aho Corasick pattern matching algorithm is used. For location based crime suspect identification, midpoint circle algorithm is applied to set the boundary area for the crime analysis. Within the set range, Aho Corasick algorithm is applied to find the suspects that matches the maximum physical features in the dataset. To find the fingerprint matching of the suspects, SourceAFIS, an image processing based partial fingerprint matching algorithm is used. Crime hotspot detection is done using the Density based spatial Clustering of applications with Noise (DBSCAN) Algorithm to locate the area of high density of crime.

3.1   Aho Cosick Algorithm

Aho Corasick is a string searching algorithm which finds the match of the input string from the set of strings found in the dictionary. In this context, the input string is the physical evidence collected at the crime site such as height, hair type, unique features such as scar etc. This input string is compared with the criminals features present in the dataset. The striking feature of the algorithm is the ability to find the matches for all the substring present in the string simultaneously which makes it useful algorithm in data science area. To perform this, it constructs a finite state machine that mimics the trie structure.

Aho-corasick algorithm works as follows: It basically performs three functions: (1) Goto (2) Failure and (3) Output. The Goto function, maps the state with the input character in the string and produces the output. The failure function notifies the Goto function to which state it should make the transition if the character read did not match. The output function maps the state with the outputs. At the first stage, the algorithm builds all the functions and in the next stage it output all the string matches by iterating over the input text.

In our case, the input is the physical identity available at crime site as in Fig. 2 and the output of the Aho-corasick algorithm is the string pattern that matches the data shown in Fig. 3

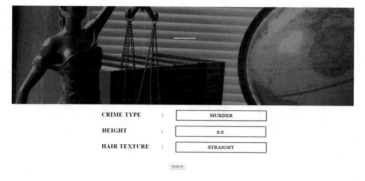

CRIME TYPE	:	MURDER
HEIGHT	:	5-5
HAIR TEXTURE	:	STRAIGHT

Search

Fig. 2. Input to Aho-corasick algorithm

12 KARAN 5 3 ERODE 11 3410 77 7172 RIGHT THUMB MOLE MURDER STRAIGHT 18 KATHIR 5 4 UDUPI 13 3409 74 7421 RIGHT KNEE MOLE MURDER STRAIGHT 19 ZEVA 5 2 EGMORE 13 003168 80 28018 SCARE IN RIGHT ELBOW MURDER STRAIGHT 20 IAIN 5 2 TIRUNELVELI 13 201411 77 75868 SCARE ON CHIN MURDER STRAIGHT 25 REKA 5 5 ARAKKONAM 13 05 7 02 NO RIGHT HAND MURDER STRAIGHT 26 REMA 5 4 ARCOT 12 56 79 24 LEFT THUMB MOLE MURDER STRAIGHT 29 RAMESH 5 8 DHARMAPURI 12 0933 78 2020 RIGHT EAR MOLE MURDER STRAIGHT 34 IAGA 5 5 NAMAKKAL 11 13 78 13 SCAR ON FACE MURDER STRAIGHT

Fig. 3. Pattern matched output of Aho-corasick algorithm

3.2   Midpoint Circle Algorithm

The midpoint circle algorithm works as follows: Given any point, here it refers to the latitude and longitude of the crime site, the algorithm accepts two inputs, the center (x, y) and the radius 'r'. Here center refers to the location of the crime and the radius is the range of distance around the crime site to be investigated. It creates a circle with the given radius and computes all perimeter the points in the first octant and it mirrors these in the other octants.

For a given coordinate (x, y), the next point is either (x, y + 1) or (x − 1, y + 1), where 'x' refers to latitude and y refers to longitude. This is chosen based on two conditions. Find the midpoint p of latitude and longitude. If p lies inside or on the circle boundary, (x, y + 1) is plotted, else it plots (x − 1, y + 1). The boundary condition, to check p lies inside or outside the circle will be decided upon the formula given below.

$$F(p) = x^2 + y^2 - r^2 \tag{1}$$

If $F(p) < 0$, the point is inside the circle.
IF $F(p) = 0$, the point is on the perimeter
IF $F(p) > 0$, the point is outside the circle

Using this algorithm, the crime site boundary is fixed and the suspect identification process is carried out within this perimeter.

3.3   Source AFIS Algorithm

This is a minutiae based algorithm. A minutiae is a point of interest in a fingerprint that captures smaller details which is important for fingerprint matching as shown in Fig. 4. Minutiae has three features: ridge endings, bifurcations and dot. The line connecting two minutiae is called as Edge. Edge possesses two relative angles and a length and these three properties do not change on moving or rotating the edge. These properties are utilized to analyse the fingerprint matching.

SourceAFIS algorithm employs nearest neighborhood algorithm to the finds the matches in edge in two fingerprints. The first pair that is matched is taken as root pair and taking it as reference, it moves outwards to build several such pairs. Finally based on the features matched, it gives the matching scores. The partial scores are summed and compared with a threshold value to decide whether the two fingerprints are matched or not.

In our case, we have taken the threshold for percentage of match as 50% and above. If the fingerprint of the suspects matches the above threshold, it is taken into account as evidence. The output of the algorithm is shown in Fig. 4.

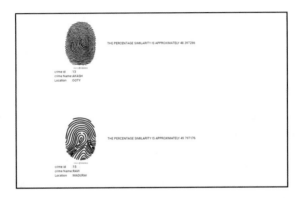

Fig. 4. Percentage of match from fingerprint matching

3.4 DBSCAN Algorithm

DBSCAN is applied to find the crime hotspot. It is a non-parametric density based clustering algorithm. It groups the closely related points based on Euclidian distance as a cluster neglecting the distance points in the low density region as outliers. DBSCAN works on the principle of density reachability and density connectivity.

Density reachability is termed as within a radius 'r' if a point 'a' and 'b' has sufficient neighbours to reach each other. Density connectivity is a chaining process where 'b reaches c', 'c reaches d', 'd reaches a' implies that 'b reaches a' as it got sufficient neighbours in its neighbourhood.

DBSCAN Clustering Algorithm

1. Start arbitrarily from any point unvisited point
2. Find out the neighbourhood of this point within ε
3. If it has sufficient neighbourhood, perform clustering around this point
4. Mark the point as visited, otherwise as noise

5. If the point is a member of the cluster then its ε neighbourhood is also a part of the cluster. Do step 2 for all ε neighbourhood points till all points in the cluster is determined
6. Do step1 to step6 for new unvisited point to discover further clusters
7. Repeat step 7 until all points are visited.

The crime hotspot region detected using DBSCAN algorithm is shown in Fig. 5.

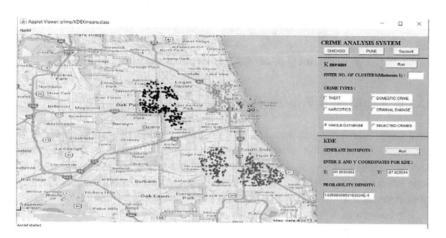

Fig. 5. Hotspot region for Chicago dataset

4   Performance Evaluation

The performance of the system for pattern matching is evaluated by computing the accuracy between the actual value and the observed value as given below:

$$\% \ of \ accuracy = (Va - Vo) \ / \ Va \ x \ 100 \tag{2}$$

where Va and Vo are actual value and observed value respectively. We have found that the accuracy of the system in retrieving the suspects using pattern matching Aho-Corasick algorithm varies, depending on the number of physical identities given as input and the number of matches. For example, if the crime type, height and hair texture is given as input. The search accuracy based on pattern matching algorithm is given below in Fig. 6

Fig. 6. Accuracy of the system for pattern matching algorithm

5   Conclusion and Future Work

In this paper, we have automated the crime investigation process of law and order department using ensemble of techniques that includes machine learning algorithms, pattern matching and fingerprint analysis to find the suspects based on the evidences obtained at the crime site. It was found that the accuracy of system differs depending of the evidence match and availability. Thus we have shown the feasibility of realizing a complete automated system that finds the suspects purely based on available historic data and current data without physical investigation.

In future, we have planned to improve the performance of the system using better and fast data retrieval and matching mechanisms. We further intent to explore the possibilities of using crime data analytics in deeper sense to predict future crimes and criminal links.

References

1. Sinha, S.: Personality correlates of criminals: a comparative study between normal controls and criminals. Indian J. Psychiatry **25**(1), 41–46 (2016)
2. Ananyan, S.: Crime pattern analysis through text mining. In: 10th Americas Conference on Information Systems, New York, pp. 1969–1981 (2004)
3. Wang, T., Rudin, C., Wagner, D., Sevieri, R.: Learning to detect patterns of crime. In: Joint European Conference on Machine Learning and Knowledge Discovery in Databases, Machine Learning and Knowledge Discovery in Databases. LNCS, vol. 8190, pp. 515–530. Springer, Heidelberg (2016)
4. Lin, Y.-L., Chen, T.-Y., Yu, L.-C.: Using machine learning to assist crime prevention. In: 6th IIAI International Congress on Advanced Applied Informatics, vol. 1, pp. 1029–1030 (2017)
5. Kim, S., Joshi, P., Kalsi, P.S., Taheri, P.: Crime analysis through machine learning. In: 9th Annual Information Technology, Electronics and Mobile Communication Conference (IEMCON), Vancouver, BC, Canada, pp. 415–420 (2018)
6. Yadav, S., Timbadia, M., Yadav, A., Vishwakarma, R., Yadav, N.: Crime pattern detection, analysis & prediction: In: International conference of Electronics, Communication and Aerospace Technology (ICECA), Coimbatore, India, pp. 20–22 (2017)

7. Brown, D.E., Oxford, R.E.: Data mining time series with applications to crime analysis. In: International Conference on Systems, Man and Cybernetics, Tucson, AZ, USA, pp. 1453–5148. IEEE Press (2001)
8. Babakura, A., Sulaiman, Md.N., Yusuf, M.A.: Improved method of classification algorithms for crime prediction. In: International Symposium on Biometrics and Security Technologies (ISBAST), Kuala Lumpur, Malaysia, pp. 250–255 (2014)
9. Das, S., Choudhury, M.R.: Geo-statistical approach for crime hot spot prediction. Int. J. Criminol. Sociol. **9**(1), 1–11 (2016)
10. Ahmed, N., Varol, A.: Minutiae based partial fingerprint registration and matching method. In: 6th International Symposium on Digital Forensic and Security (ISDFS), Antalya, Turkey, (2018)
11. Jea, T.-Y., Govindaraju, V.: A minutia-based partial fingerprint recognition system. J. Pattern Recogn. **38**(10), 1672–1684 (2005)
12. Chaudhary, U., Bhardwaj, S., Sabharwal, H.: Fingerprint recognition using orientation features. Int. J. Adv. Res. Comput. Sci. Softw. Eng. Res. **4**(5), 1403–1413 (2014)

Adaptive Neuro Fuzzy Inference System Based Obstacle Avoidance System for Autonomous Vehicle

M. Karthikeyan[1], S. Sathiamoorthy[2(✉)], and M. Vasudevan[3]

[1] Division of Computer and Information Science, Annamalai University,
Chidambaram, India
mkarthi82@gmail.com
[2] Tamil Virtual Academy, Chennai, India
sathiamoorthy2019@gmail.com
[3] HCL Technologies Limited, Chennai, India
vasudevan.m@hcl.com

Abstract. Adaptive Neuro Fuzzy Inference System (ANFIS) is a well proven technology for predicting the output based on the set of inputs. ANFIS is predominantly used to track the set of inputs and output in order to achieve the target. In this paper, authors have proposed a Nonlinear ANFIS algorithm to track the distance between the autonomous vehicle and the obstacle while vehicle is moving and the brake force required. By tuning neuro fuzzy algorithm, accurate brake force requirement has been achieved and the results are captured in this paper. Back propagation algorithm based neural network & Sugeno model based Fuzzy inference system have been used in the proposed technique. Matlab/Simulink software platform is used to implement the proposed algorithm and proven the expected results.

Keywords: ANFIS · Autonomous vehicle · Neural network · Fuzzy Inference System · Obstacle identification and Matlab

1 Introduction

In recent years, autonomous vehicles have reached several milestones and commercial makers are testing these features on their vehicles and the test models are running on the roads [1]. Autonomous vehicles have the unique features such as obstacle avoidance, lane keep assist, traffic jam assist, automatic park assist, Highway autopilot, etc [2] and moreover, autonomous tractors are already in the field for farming & cultivation [3]. Among these features, obstacle avoidance is a key feature to make this vehicle successful on the roads and this is achieved by AI techniques only [4]. Although Various control algorithms & Image processing algorithms are being used over the years in various format of the vehicles including Autonomous vehicles, Artificial Intelligence (AI) is one of the successful algorithm in autonomous vehicle domain and it has been used in control systems, image processing, data analytics, output estimation, etc., Artificial Neural Network, Genetic Algorithm, Ant colony algorithm, Fuzzy Logic, swarm intelligence, and reinforcement learning are mainly used AI techniques in

© Springer Nature Switzerland AG 2020
J. S. Raj et al. (Eds.): ICIDCA 2019, LNDECT 46, pp. 118–126, 2020.
https://doi.org/10.1007/978-3-030-38040-3_13

industrial applications [5]. Most popular and proven AI techniques are Artificial neural networks, Fuzzy logic and Genetic algorithm which are being used in many industrial applications. In this paper, authors are using advanced non-linear neuro fuzzy algorithm to estimate the brake force in order to stop the vehicle by sensing the obstacles. ANFIS is being used in many applications such as Heating Ventilating and Air Conditioning Systems (HVAC), Water Management Systems, Actuator and Motor control, Automotive systems, Bio Chemical systems, etc., [6–12]. ANFIS has been a proven technology in prediction which gives the accurate results and it has been proved by authors also in this paper that ANFIS could be useful to get the accurate results from obstacle avoidance system as well.

2 Principle of Obstacle Avoidance System with ANFIS Controller

Autonomous vehicle mainly runs based on the inputs such as target place information, obstacle presence in front of the vehicle, obstacle behind the vehicle, brake force requirement, steering angle, Lane distance, etc., [14]. Figure 1 clearly gives the overview of operation of autonomous vehicle in the view of obstacle identification, steering control and automatic braking. In order to avoid the obstacle while self-driving car is on the way, distance between an obstacle and the vehicle should be sensed accurately. In this paper, obstacle identification during the forward motion of the car is considered and the forward motion data has been processed using ANFIS algorithm. There are three set of data have been given as input to ANFIS controller as below:

(i)   Distance between obstacle and front side of vehicle on the straight line
(ii) Distance between obstacle and vehicle on the right side
(iii) Distance between obstacle and vehicle on the left side

Output of ANFIS controller is braking force and this will influence the angular velocity, steering angle and brake pedal position. In Fig. 1, all these three inputs are given to ANFIS controller and brake force output of ANFIS controller is given to powertrain of the vehicle. Powertrain model reflects the mathematical expressions of vehicle dynamics and estimated brake force is substituted in the vehicle dynamic equations. Based on this processing, steering control, angular velocity and brake pedal positions are identified as the end output. Authors have focused in this paper that use of ANFIS controller in the estimation of brake force with sensory information of distance between the vehicle and obstacle.

Figure 2 shows the architecture of ANFIS network with multiple layers. ANFIS is the combination of neural networks and fuzzy logic and it has the hybrid nature of these two intelligence systems [13]. Self-tuning nature of this hybrid technique used the learning algorithm contributed by neural networks [3, 13]. Figure 2 is known as Takagi-Sugeno-fuzzy model as given in [13].

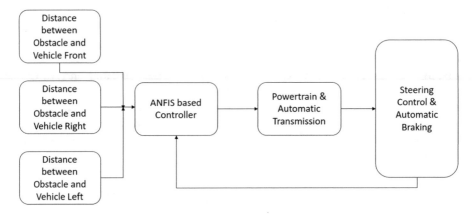

Fig. 1. Block diagram of autonomous vehicle with Artificial Intelligence

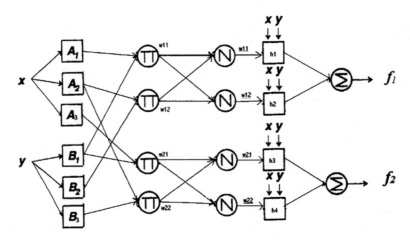

Fig. 2. Architecture of ANFIS

ANFIS network has 5 inbuilt layers and the function of each layer is depicted in Table 1. This table shows the nodes, functions and input/output parameters of every layer clearly. First and fourth layers of ANFIS architecture are playing vital role as both are having adaptive nodes and modifiable parameters. Both premise and consequent parameters are modifiable parameters. These parameters are adjusted to make the ANFIS output to match with training data. ANFIS has both feed forward and back propagation process while training the data. However, back propagation algorithm contributes to adapt the premise parameters with output error.

Table 1. Different Layers of ANFIS Network

Layers	Node	Function	Input/output parameters
Layer: 1	Adaptive	Membership function variation for given Fuzzy set	Premise parameters
Layer: 2	Fixed	Fuzzy AND & node function	Product of all incoming signals
Layer: 3	Fixed	Firing strength ratio calculation	Normalized firing strength
Layer: 4	Adaptive	Firing strength normalization	Consequent parameters
Layer: 5	Fixed	Summation of incoming signals	Overall output (f1.f2)

3   Proposed ANFIS Algorithm with Obstacle Avoidance System and Results

Proposed ANFIS controller is mainly designed to avoid the obstacle in forward motion with three directions as described above in section. Sensory signals have been captured and given as inputs to ANFIS controller (Sugeno). Figure 3 shows the implemented ANFIS controller with three inputs and one output. These sensory signals are classified as 0, −1 and +1 where 0 indicates the front side direction and −1 indicates the left direction and +1 indicates the right direction.

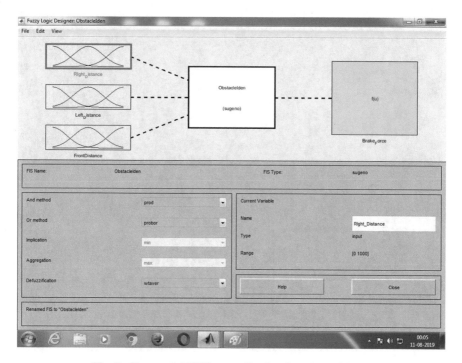

Fig. 3. Proposed ANFIS controller for obstacle avoidance

Input data is generated based on the direction and the distance between the obstacle and vehicle. Input data given to ANFIS model is tabulated in Table 2.

Table 2. ANFIS Inputs

Sl. No.	Distance_Front (meters)	Distance_Right (meters)	Distance_Left (meters)
1	0.9	0.9	0.9
2	0.7	0.5	0.7
3	0.5	0.9	0.9
4	0.9	0.9	0.5
5	0.9	0.7	0.9
6	0.6	0.6	0.4
7	0.4	0.6	0.6
8	0.6	0.4	0.6
9	0.5	0.5	0.5
10	0.4	0.4	0.4

Proposed Obstacle avoidance system with ANFIS has been implemented in Matlab/Simulink platform. Figure 4 shows the membership function selected by Layer 1 of ANFIS and each input is represented by the respective membership functions such as mf1, mf2 and mf3.

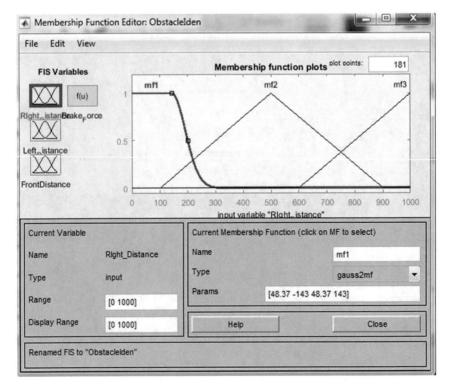

Fig. 4. FIS variables with membership function format

ANFIS learning technique predicting the braking force and it is shown in Fig. 5. Figure 5 is ANFIS rules viewer and inputs can be viewed and varied manually in order to minimize the predicting error. This figure clearly defines that inputs can be varied simultaneously or any one variation at a time. This method helps to predict the obstacle either in one direction or all the three directions.

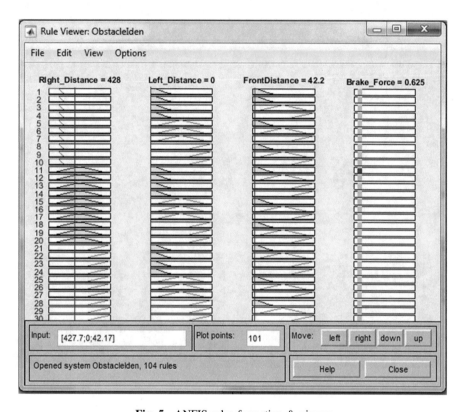

Fig. 5. ANFIS rules formation & viewer

Proposed ANFIS learning technique predicts the braking force for both right and left wheels as follows:

- Number of nodes - 215
- Number of linear parameters - 115
- Number of non linear parameters - 35
- Number of total parameters - 150
- Number of Fuzzy rules – 104
- Error in predicting brake force – 5%
- Epoch number - 75

Figure 6 shows the ANFIS rule editor wherein rules can either be added or removed. Number of rules can be defined and modified in this editor. In order to achieve the better results, number of rules are optimized with total number of 104.

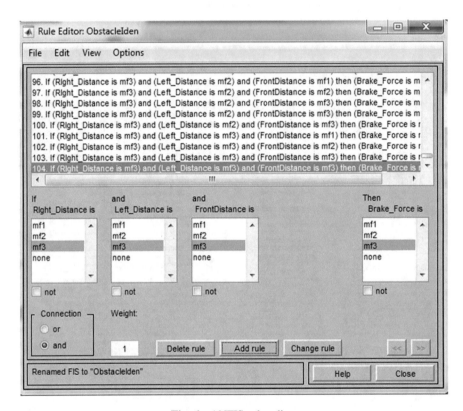

Fig. 6. ANFIS rule editor

Figure 7 shows the surface plot of proposed ANFIS controller. It defines the distribution of inputs and output values. Surface plots shown in Fig. 7 exhibits the distribution between Left and right sides obstacles with braking force output. Similarly, other two combinations like front and left obstacles with braking force output and front and left with braking force output are also possible. Predicted braking force is varied from 600 N to 5000 N as shown in Fig. 5. Brake force shown in Fig. 7 is in the unit of kN.

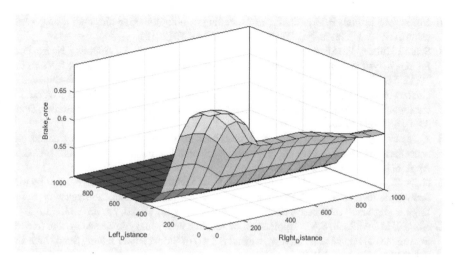

Fig. 7. Distribution of inputs (LD, RD & brake force)

4 Conclusion

This paper mainly focused on the ANFIS controller based obstacle avoidance system for an autonomous vehicle. This is achieved mainly by predicting the braking force from the input values of the distance between the vehicle and obstacle from three different directions. Authors have introduced ANFIS algorithm in autonomous vehicle for obstacle avoidance system as no similar work exists in the past with reference to passenger vehicle. Many researchers introduced ANFIS to Robots for this problem but authors have introduced the same newly to autonomous passenger vehicle and brought the results successfully. Future work would be the addition of multiple ANFIS controllers in autonomous vehicle to make the entire system robust and accurate.

References

1. Blundell, H., Thornton, S.M.: Object classification for autonomous vehicle navigation of Stanford campus. Technical report, CS 229 (2015)
2. Stentz, A., Dima, C., Wellington, C., et al.: Auton. Robots **13**, 87 (2002)
3. Al-Mayyahi, A., Wang, W., Birch, P.: Adaptive neuro-fuzzy technique for autonomous ground vehicle navigation. Robotics **3**, 349–370 (2014)
4. Li, X., Choi, B.: Design of obstacle avoidance system for mobile robot using fuzzy logic systems. Int. J. Smart Home **7**, 321–328 (2013)
5. Chiou, J.S., Liu, M.T.: Using fuzzy logic controller and evolutionary genetic algorithm for automotive active suspension system. Int. J. Automot. Technol. **10**(6), 703 (2009)
6. Termeh, S.V.R., Khosravi, K., Sartaj, M., Keesstra, S.D., Tsai, F.T.-C., Dijksma, R., Pham, B.: Optimization of an adaptive neuro-fuzzy inference system for groundwater potential mapping. Hydrogeol. J. **27**, 2511–2534 (2019). https://doi.org/10.1007/s10040-019-02017-9

7. Hosoz, M., Ertunc, H.M., Alkan, A.: Modelling of an automotive air conditioning system using ANFIS. J. Therm. Sci. Technol. **33**(1), 127–137 (2013)
8. Senthil Kumar, P., Sivakumar, K., Kanagarajan, R., Kuberan, S.: Adaptive Neuro Fuzzy Inference System control of active suspension system with actuator dynamics. J. Vibro Eng. **20**(1), 541–549 (2018)
9. Ghafari-Nazari, A., Reddy, S., Fard, M.: A new approach for optimizing automotive crashworthiness: concurrent usage of ANFIS and Taguchi method. Struct. Multidisc. Optim. **49**(3), 485–499 (2014)
10. Kusagur, A., Kokad, S.F., Ram, B.V.S.: Modeling, design & simulation of an adaptive neurofuzzy inference system (ANFIS) for speed control of induction motor. Int. J. Comput. Appl. **6**(10), 29–44 (2010)
11. Hajar, S., Mohammad, A., Jeffril, M.A., Sariff, N.: Mobile robot obstacle avoidance by using fuzzy logic technique. In: Proceedings of the 3rd IEEE International Conference on System Engineering and Technology, Shah Alam, Kuala Lumpur, 19–20, pp. 331–335 (2013)
12. Ahmed, M., AliShah, S.M.: Application of adaptive neuro-fuzzy inference system (ANFIS) to estimate the biochemical oxygen demand (BOD) of Surma River. J. King Saud Univ. Eng. Sci. **29**(3), 237–243 (2017)
13. Jang, J.R.: ANFIS: adaptive network based fuzzy inference system. IEEE Trans. Syst. Man Cybern. **23**, 665–685 (1993)
14. Driankov, D., Saffiotti, A. (eds.): Fuzzy Logic Techniques for Autonomous Vehicle Navigation, vol. 61. Physica, Heidelberg (2013)
15. www.mathworks.com

Evaluating the Performance of Machine Learning Techniques for Cancer Detection and Diagnosis

Anu Maria Sebastian[✉] and David Peter

Department of Computer Science, CUSAT, Kochi 682022, India
anumseb@gmail.com, davidpeter123@gmail.com

Abstract. Machine Learning (ML) techniques find value in healthcare due to its ability to process huge data sets and convert them into clinical insights. These insights help physicians and healthcare providers in planning, quick decision making and providing timely care to patients with higher accuracy, lower costs and increased customer satisfaction. This paper evaluates the performance of ML algorithms for cancer detection and diagnosis. It compares the performance of different cancer detection algorithms for different types of datasets. It also investigates how the performance of the ML algorithms improve with the use of feature extraction methods, specifically for lung cancer. Developing generic feature extraction methods still remain as a challenge. Healthcare providers need to reorient their treatment approach for diseases like cancer with more focus on detecting them at a very early stage so as to maximize the chances of recovery for the patients.

Keywords: Machine learning · Cancer · Feature extraction · Disease detection and diagnosis · Performance comparison

1 Introduction

Machine learning (ML) uses algorithms or models with data to train and develop the capacity for a computing device with the ability to learn and perform specific tasks. ML algorithms can learn the pattern in the supplied healthcare data and can perform detection, prediction, and classification of diseases. The recent advancements in ML have triggered a paradigm shift in healthcare from interventional state to predictive, preventive and personalized medicine [2]. ML applications in healthcare include disease identification and diagnosis, diagnosis in medical imaging, disease prediction, epidemic outbreak prediction, drug discovery, robotic surgery, personalized medicine, clinical trial research, radiology, radiotherapy, etc. Another potential application of ML is in predictive modeling of biological processes and drugs, based on molecular and clinical big data, where new potential-candidate molecules can be developed into drugs [3]. To offer better health services for patients, the different sectors of healthcare such as preventive, diagnostic, remedial, and therapeutic medicine should work together in a holistic manner [4]. Late and erroneous diagnosis of the diseases, delayed and inappropriate treatment received, etc. can be fatal for patients. In India, it is estimated that

© Springer Nature Switzerland AG 2020
J. S. Raj et al. (Eds.): ICIDCA 2019, LNDECT 46, pp. 127–133, 2020.
https://doi.org/10.1007/978-3-030-38040-3_14

nearly 5,000,000 people die every year due to medical negligence. It is suggested that with the help of proper technology and training, these medical negligence and death rates can be brought down substantially.

2   Literature Review

Schmidhuber et al. [5] performed a study on the different ML algorithms such as supervised, unsupervised, reinforcement and evolutionary algorithms. Shivade et al. [6] performed a review on systems developed for automated techniques to identify cohorts of patients with specific phenotypes using electronic health records (EHRs). They claimed that ML systems together with natural language processing (NLP) techniques are becoming more popular than rule-based systems. Atmaja et al. [7] have given a brief description of several intelligent ML techniques used in big data predictive analytics solutions. Nithya et al. [8] studied various prediction techniques and tools for ML in healthcare. Their main objective was to reduce the healthcare costs and they have studied how EHR management and disease prediction and diagnosis influence the costs.

Srivastava et al. [9] have presented their research works in disease prediction, data visualization, assistive technology development and healthcare data security and privacy, based on deep learning (DL). For technology to improve care for the patients, the EHRs provided to the doctors need to be enhanced by the power of analytics and ML which can accurately analyze the medical data for disease prediction, early disease detection, and treatments.

Kourou et al. [25] performed a review on ML approaches deployed for developing predictive models of cancer progression. Hu et al. [26] had given an overview of the deep learning application's progress in cancer detection and diagnosis using images. They focus on aspects such as objectives of ML applications, domain chosen, and the type of data used for the application.

3   Comparing the ML Algorithms for Oncology Applications

Manogaran et al. [11] developed a scalable ML approach to deal with the problems in detecting DNA copy number change. Sun et al. [13] studied how deep structured algorithms can be used for extracting and diagnosing the features in lung nodule CT images.

Syed et al. [14] developed a system for the early detection of breast cancer using ML techniques. Their experimental results showed that the multi-layer perceptron (MLP) is the optimal classifier, followed by Random Forest and KNN. Singh et al. [15] demonstrated a supervised algorithm-based technique for the detection and classification of lung cancer using CT scan images. The proposed approach processes the images first and then classifies them using seven different classifiers such as KNN, SVM, MLP, etc. The model with high accuracy and precision was found to be MLP with 3 hidden layers and 100 neurons in each layer.

Gareau et al. [16] a method for generating quantitative image analysis metrics for melanoma that are analyzed using ML algorithms to generate a risk score, with higher scores indicate increased probability of lesion being a melanoma. This method yielded good sensitivity and specificity for melanoma detection. The performance of the method was close to that of expert lesion evaluation. Hornbrook et al. [17] developed the ColonFlag algorithm which is an ML-based colorectal cancer detection model and its prediction performance was evaluated. The model was found to have high accuracy in identifying right-sided (than left-sided) colorectal cancers. The model could flag colorectal tumors 6 months to 1 year ahead in comparison with the usual clinical diagnosis.

Table 1. Performance comparison of cancer detection algorithms for different types of data.

Type of data	ML algorithm	Performance metric(s)	Reference
Genomic data	Logistic regression	AUC 0.77	[10]
	Bayesian HMM with gaussian mixture clustering	Accuracy 0.80	[11]
Images	SVM-RFE+SMOTE	AUC 0.955	[12]
	CNN	AUC 0.899	[13]
	MLP	Accuracy 0.969, RMSE 0.1387	[14]
	MLP	Accuracy 0.886	[15]
	Q-score	Sensitivity 0.98 and specificity 0.36	[16]
Statistical data	Colon Flag	AUC 0.80	[17]
Textual data	CRCP	Accuracy 0.872	[18]

In order to assist cancer registrars in identifying reportable cancer cases accurately, Osborne et al. [18] developed a Cancer Registry Control Panel (CRCP). The CRCP could identify clinical records with cancer symptoms and add this information with the codes from the Clinical Electronic Data Warehouse (CEDW). Using NLP and supervised learning, they could accurately detect patients' cancer status from the clinical data. This showed good accuracy, precision, and recall rate.

Hadavi et al. [19] developed a technique for lung cancer detection using cellular learning automata (CLA). With the help of image enhancement, segmentation, and feature extraction, they could achieve 0.90 accuracy in detecting lung cancer. Zhang et al. [20] discovered a set of biomarkers for detecting lung cancer from urine samples. Pearson correlation was used for selecting the biomarkers. The RF algorithm used five biomarkers which could identify the lung cancer patients from the control group and also from other cancer patients. Zhou et al. [21] proposed a method to predict the distant metastasis in lung cancer patients. Use of SVM and stochastic gradient descent for classification have achieved an AUC of 0.728 with radiomic features. The AUC increased to 0.890 when combined with clinical features.

Singh et al. [15] experimented with many ML methods for lung cancer detection with CT scans and found that MLP was the best performer with gray level co-occurrence matrix feature extraction. Tajbhaksh et al. [22] developed a massive-training artificial neural network (MTANN) for lung cancer detection which had an AUC of 0.8806. Ma et al. [23] identified a set of four serum proteins for lung cancer detection through logistic regression which could further improve its performance by adding three demographic features. Than et al. [24] developed a lung cancer risk stratification method using SVM with directional and textural features.

4 Discussion

The AUC obtained from plotting the ROC graph indicates better performance when the value approaches 1, but it does not explain the individual parameters such as sensitivity and specificity. However, it is good for comparing two or more tests results. Sensitivity and specificity are not generally influenced by disease prevalence. Therefore, the results from one study setting could easily be mapped to another setting with a different prevalence of disease in the population but is largely dependent on the spectrum of diseases in the studied group. Table 1 compares the performance of different ML algorithms for cancer detection, based on the types of data. It can be observed that better AUC was possible with SVM-RFE+SMOTE algorithm with limited image data. Similarly, the MLP algorithm yielded better accuracy with image data. In general, the performance of the ML algorithms was better with image data.

Table 2 compares the performance of different ML algorithms for lung cancer detection. It can be observed that the highest accuracy was obtained for the SVM algorithm with CT images dataset. This improved detection accuracy of the lung cancer was made possible by the use of a feature extraction method along with the ML algorithm. Table 3 depicts the feature extraction methods used along with the ML methods in Table 2 for lung cancer detection. From the Tables 2 and 3, it can be inferred that regional and morphological feature extraction, followed by tissue characterization with Reiz and Gabor transforms is a preferable feature extraction method when using images. The [21] justifies the argument that adding clinical features along with image features improves the detection accuracy.

Table 2. Performance comparison of lung cancer detection algorithms.

Dataset	ML algorithm	Performance metric(s)	Reference
CT images	CLA	Accuracy 0.90	[19]
Proteomic analysis of urine	RF	AUC 0.9853	[20]
CT images and clinical data	SVM	AUC 0.728 without clinical data	[21]
		AUC 0.890	
CT images	MLP	Accuracy 0.885	[15]
Low-dose CT images (LDCT)	MTANN	AUC 0.8806	[22]
Clinical data	Multivariate logistic regression	AUC 0.89	[23]
CT images	SVM	Accuracy 0.995	[24]

Table 3. Feature extraction methods associated with the ML methods.

Feature extraction	Reference
Image enhancement and segmentation	[19]
Pearson correlation	[20]
Concave minimization	[21]
Gray-level co-occurrence matrix	[15]
Mann Whitney U test	[23]
Regional and morphological feature extraction, and tissue characterization with Reisz and Gabor transforms	[24]

5   Conclusion

One of the major ML applications in healthcare is to detect hard to diagnose diseases like cancers at an early stage. ML plays a key role in cancer detection which involves the development of new medical procedures and handling of patient data. This paper studied and compared the performance of some popular ML algorithms used for cancer detection and diagnosis for different datasets. It also investigated how the performance of these ML algorithms improve with the use of feature extraction methods, specifically for lung cancer. The performance of any ML algorithm depends on the data set and the feature extraction methods used. However, the feature extraction is subjective in nature and there is no generic feature extraction method which works in all cases. Exploring the possibilities of using regression algorithms and score values for cancer diagnosis is suggested as a topic for further research. Developing more generic feature extraction methods is also a challenge which could be another topic for further research.

References

1. Murphy, K.: Machine Learning. MIT Press, Cambridge (2012)
2. Nikolaiev, S., Timoshenko, Y.: Reinvention of the cardiovascular diseases prevention and prediction due to ubiquitous convergence of mobile apps and machine learning. In: 2015 Information Technologies in Innovation Business Conference (ITIB) (2015)
3. Cattell, J., Chilikuri, S.: How big data can revolutionize pharmaceutical R&D. https://www.mckinsey.com/industries/pharmaceuticals-and-medical-products/our-insights/how-big-data-can-revolutionize-pharmaceutical-r-and-d
4. Bhardwaj, R., Nambiar, A., Dutta, D.: A study of machine learning in healthcare. In: 2017 IEEE 41st Annual Computer Software and Applications Conference (COMPSAC) (2017)
5. Schmidhuber, J.: Deep learning in neural networks: an overview. Neural Netw. **61**, 85–117 (2015)
6. Shivade, C., Raghavan, P., Fosler-Lussier, E., Embi, P., Elhadad, N., Johnson, S., Lai, A.: A review of approaches to identifying patient phenotype cohorts using electronic health records. J. Am. Med. Inform. Assoc. **21**, 221–230 (2014)
7. Athmaja, S., Hanumanthappa, M., Kavitha, V.: A survey of machine learning algorithms for big data analytics. In: 2017 International Conference on Innovations in Information, Embedded and Communication Systems (ICIIECS) (2017)
8. Nithya, B., Ilango, V.: Predictive analytics in health care using machine learning tools and techniques. In: 2017 International Conference on Intelligent Computing and Control Systems (ICICCS) (2017)
9. Srivastava, S., Soman, S., Rai, A., Srivastava, P.: Deep learning for health informatics: recent trends and future directions. In: 2017 International Conference on Advances in Computing, Communications and Informatics (ICACCI) (2017)
10. Saha, A., Harowicz, M., Wang, W., Mazurowski, M.: A study of association of Oncotype DX recurrence score with DCE-MRI characteristics using multivariate machine learning models. J. Cancer Res. Clin. Oncol. **144**, 799–807 (2018)
11. Manogaran, G., Vijayakumar, V., Varatharajan, R., Malarvizhi Kumar, P., Sundarasekar, R., Hsu, C.: Machine learning based big data processing framework for cancer diagnosis using hidden Markov model and GM clustering. Wireless Pers. Commun. **102**, 2099–2116 (2017)
12. Feng, Z., Rong, P., Cao, P., Zhou, Q., Zhu, W., Yan, Z., Liu, Q., Wang, W.: Machine learning-based quantitative texture analysis of CT images of small renal masses: differentiation of angiomyolipoma without visible fat from renal cell carcinoma. Eur. Radiol. **28**, 1625–1633 (2017)
13. Sun, W., Zheng, B., Qian, W.: Automatic feature learning using multichannel ROI based on deep structured algorithms for computerized lung cancer diagnosis. Comput. Biol. Med. **89**, 530–539 (2017)
14. Syed, L., Jabeen, S., Manimala, S.: Telemammography: a novel approach for early detection of breast cancer through wavelets based image processing and machine learning techniques. In: Advances in Soft Computing and Machine Learning in Image Processing, pp. 149–183 (2017)
15. Singh, G., Gupta, P.: Performance analysis of various machine learning-based approaches for detection and classification of lung cancer in humans. Neural Comput. Appl. **31**, 6863–6877 (2018)
16. Gareau, D., Correa da Rosa, J., Yagerman, S., Carucci, J., Gulati, N., Hueto, F., DeFazio, J., Suárez-Fariñas, M., Marghoob, A., Krueger, J.: Digital imaging biomarkers feed machine learning for melanoma screening. Exp. Dermatol. **26**, 615–618 (2016)

17. Hornbrook, M., Goshen, R., Choman, E., O'Keeffe-Rosetti, M., Kinar, Y., Liles, E., Rust, K.: Early colorectal cancer detected by machine learning model using gender, age, and complete blood count data. Dig. Dis. Sci. **62**, 2719–2727 (2017)
18. Osborne, J., Wyatt, M., Westfall, A., Willig, J., Bethard, S., Gordon, G.: Efficient identification of nationally mandated reportable cancer cases using natural language processing and machine learning. J. Am. Med. Inform. Assoc. **23**, 1077–1084 (2016)
19. Hadavi, N., Nordin, M., Shojaeipour, A.: Lung cancer diagnosis using CT-scan images based on cellular learning automata. In: 2014 International Conference on Computer and Information Sciences (ICCOINS) (2014)
20. Zhang, C., Leng, W., Sun, C., Lu, T., Chen, Z., Men, X., Wang, Y., Wang, G., Zhen, B., Qin, J.: Urine proteome profiling predicts lung cancer from control cases and other tumors. EBioMedicine **30**, 120–128 (2018)
21. Zhou, H., Dong, D., Chen, B., Fang, M., Cheng, Y., Gan, Y., Zhang, R., Zhang, L., Zang, Y., Liu, Z., Zheng, H., Li, W., Tian, J.: Diagnosis of distant metastasis of lung cancer: based on clinical and radiomic features. Transl. Oncol. **11**, 31–36 (2018)
22. Tajbakhsh, N., Suzuki, K.: Comparing two classes of end-to-end machine-learning models in lung nodule detection and classification: MTANNs vs. CNNs. Pattern Recogn. **63**, 476–486 (2017)
23. Ma, S., Wang, W., Xia, B., Zhang, S., Yuan, H., Jiang, H., Meng, W., Zheng, X., Wang, X.: Multiplexed serum biomarkers for the detection of lung cancer. EBioMedicine **11**, 210–218 (2016)
24. Than, J., Saba, L., Noor, N., Rijal, O., Kassim, R., Yunus, A., Suri, H., Porcu, M., Suri, J.: Lung disease stratification using amalgamation of Riesz and Gabor transforms in machine learning framework. Comput. Biol. Med. **89**, 197–211 (2017)
25. Kourou, K., Exarchos, T.P., Exarchos, K.P., Karamouzis, M.V., Fotiadis, D.I.: Machine learning applications in cancer prognosis and prediction. Comput. Struct. Biotechnol. J. **13**, 8–17 (2015)
26. Hu, Z., Tang, J., Wang, Z., Zhang, K., Zhang, L., Sun, Q.: Deep learning for image-based cancer detection and diagnosis—a survey. Pattern Recogn. **83**, 134–149 (2018)

Design of Slotted Microstrip Patch Antenna for 5G Communications

C. V. Krishna, H. R. Rohit[(⊠)], and P. Shanthi

Department of Telecommunication Engineering, R.V. College of Engineering,
Bangalore, India
krishnacv23@gmail.com, hrrohit98@gmail.com,
shanthip@rvce.edu.in

Abstract. 5G is emerging as a significant research topic and it has been a difficult task to design antennas with small sizes which are required for the design of high frequencies that are capable of supporting 5G technology. This paper consists of the design and comparisons between patch and slotted antenna parameters with a symmetrical U shaped slots about the centre of the patch. The various parameters were observed for the designed frequency of 25 GHz with some practical limits. The tool used for the design and simulation is CST STUDIO SUITE 2018. RT duroid 5880 was used as dielectric substrate. Low cost, low profile, and ease of fabrication are the major highlights for using patch antenna for the design.

Keywords: 5G communications · Antennas · Microstrip · Patch · RT duroid

1 Introduction

5th generation cellular technology is called 5G technology. The usage of millimeter waves in 5G have made the cell sizes to be small. On comparison with 4G, 5G technology have higher speed of data transfer. The major applications of 5G are defined by ITU-R are Enhanced Mobile Broadband (eMBB), Ultra Reliable Low Latency Communications (URLLC), and Massive Machine Type Communications (mMTC).

Antenna of low profile has become an important criteria and requirement in today's communication industry. With the decrease in size of the portable mobile devices, it has been a great requirement for the need in the design of antennas with small size and with higher efficiencies. To achieve this micro strip patch antenna may be used because it can be easily integrable in the circuits.

In the 5th generation technology which uses a wider bandwidth, to support this need the bandwidth can be increased by introducing slots inside the patch antenna. Most of the new features are included in the 5G technology and it uses data rates which are greater than 100 Mbps and greater the 1 Gbps with full and low mobility respectively. As there is continuous increase in the demand 5G technology will become more powerful.

The dielectric constant and the height of substrate determines the performance offered by the micro strip antennas. For substrates with lower height there is a disadvantage of having a low impedance bandwidth. but on the other hand the hand held

© Springer Nature Switzerland AG 2020
J. S. Raj et al. (Eds.): ICIDCA 2019, LNDECT 46, pp. 134–139, 2020.
https://doi.org/10.1007/978-3-030-38040-3_15

devices needs to have small dimensions as possible, so there is a requirement of trade off in the size and impedance bandwidth [7].

2   Literature Survey

Microstrip or patch antennas are ending up progressively valuable since it is very easy to legitimately print it onto a circuit board. These antennas can be easily fabricated, have low profile and are inexpensive. Microstrip patch antenna are generally reasonable to produce and design as a result of its basic planar geometry. A maximum of 6–9 dBi directive gain is provided by a single patch.

Due to the unique advantages of microstrip, it has become a popular material choice for individual antenna or antenna array implementations. The ease of fabrication, compatibility and low cost, especially after the development of various enhancement techniques, which are focusing in neutralizing the shortcomings of the microstrip technology such as bandwidth limitations as well as spurious emissions from the feeding lines are the most important advantages of utilizing microstrip [2].

3   Design

The Microstrip patch antenna is designed for the centre frequency of 25 GHz with a microstripline feeding. The dimensions of the patch, feedline are calculated using the standard formulas [1–4] as shown below,

The design equations [3] are as follows:

$$W = \frac{c}{2f_r} x \sqrt{\frac{2}{\varepsilon_r + 1}} \tag{1}$$

$$\varepsilon_{reff} = \frac{\varepsilon_r + 1}{2} + \frac{\varepsilon_r - 1}{2} \left[1 + \frac{12h}{W} \right]^{-1/2} \tag{2}$$

$$\frac{\Delta L}{H} = 0.412x \frac{\left(\varepsilon_{reff} + 0.3\right) x \left(\frac{W}{h} + 0.264\right)}{\left(\varepsilon_{reff} - 0.258\right) x \left(\frac{W}{h} + 0.8\right)} \tag{3}$$

$$L = \frac{c}{2f_r \sqrt{\varepsilon_{reff}}} - 2\Delta L \tag{4}$$

Taking $\varepsilon_r = 2.2$

Using the above equations, the following values are obtained as shown in Table 1.

Where L_g, W_g, h_g represent length, width and height of ground plane, L_d, W_d, h_d represent length, width and height of dielectric respectively and L_p, W_p represent length, width of the patch respectively [4].

Table 1. Theoretical values calculated using formulas.

Parameters @ f_r = 25 GHz	Length (mm)	Width (mm)	Height (mm)
Ground plane	$L_g = 6$	$W_g = 6$	$H_g = 0.03$
Dielectric substrate	$L_d = 6$	$W_d = 6$	$H_d = 1$
Microstrip patch	$L_p = 2.89$	$W_p = 4.23$	

4 Working Principle

There is a presence of ground plane on which the dielectric substrate of the required thickness is placed and the material of the dielectric provides the dielectric constant of the material required for the design as shown in Fig. 1 The patch is on the top of the dielectric which needs a feeding element which can be feed with different feeding techniques whereas here the designed slotted antenna uses the centre feed technique [5]. The patch antenna can be of different shapes such as rectangle circle or square depending on the design parameters and its applications of use the shape needs to be chosen [6].

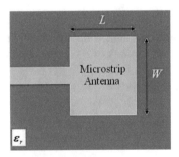

Fig. 1. Microstrip patch top view

There is presence of fringing fields which causes the spreading of the waves [8]. To have an astounding antenna performance the lesser value of dielectric constant is chosen with small thickness. Using this will offer higher productivity along with better radiation. Hence there must be trade-off between the antenna physical dimensions and the performance offered by it.

5 Schematics

The below Figs. 2, 3, 4, 5, 6 and 7 shows the simulation results of micro strip antenna.

Fig. 2. Slotted top view

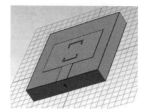

Fig. 3. Slotted bottom view

6 Results

The directivity near the slots are poor but it gradually increases as we move towards the end of the antenna [9]. The VSWR is minimum at the designed frequency. Although the radiation efficiency is not the maximum at the designed frequency of 25 GHz but the power radiated is highest at that frequency and it reduces on both the sides of the graph power radiated vs frequency.

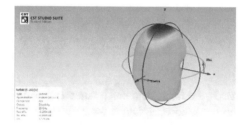

Fig. 4. Directivity

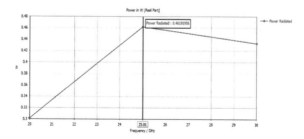

Fig. 5. Power in watts

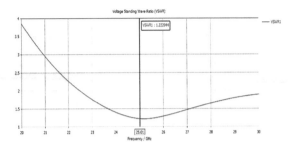

Fig. 6. VSWR

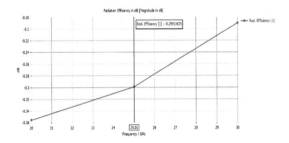

Fig. 7. Radiation efficiency

7 Conclusions

Also as a part of the work, the microstrip patch antenna was designed for both slotted and unslotted patch and the following values were obtained (Table 2).

Table 2. Comparison table

Parameters	Slotted	Un-slotted
S11	−20.153 dB	−20.24 dB
Power radiated	0.4619 W	0.4624 W
Total efficiency	92.38%	92.41%
Radiation efficiency	93.34%	93.35%
VSWR	1.222	1.215
Y11	6.97×10^{-3}S	7.25×10^{-3}S
Z11	143.3618 Ω	137.85 Ω
Directivity	3.129 dB	3.080 dB

As seen from the above values there is not much distinction between the two design parameters, slotted antenna can be used to get higher bandwidth with the reduction in q factor might be a consideration [10]. The slotted antenna is more directive then normal patch but by a very small amount.

There may be design considerations while practically designing the slotted antenna which may require a precision work and might be time consuming. comparing the values of the basic parameters if the bandwidth is not the major concern then normal patch antenna can be used but in 5G communication which require a large bandwidth the slotted antenna can be used [11]. The slots of different shapes and at different positions can also be used and the results may vary.

References

1. Tarpara, N.M., Rathwa, R.R., Kotak, N.A.: Design of slotted microstrip patch antenna for 5G application. Int. Res. J. Eng. Technol. **5**, 2827–2832 (2018)
2. Sam, C.M., Mokayef, M.: A wide band slotted microstrip patch antenna for future 5G. EPH Int. J. Sci. Eng. **2**, 19–23 (2016)
3. Balanis, C.A.: Antenna Theory, Analysis and Design, 3rd edn. Wiley, New York (1997)
4. Kumar, G., Ray, K.P.: Broadband Microstrip Antennas. Artech House, Boston (2003)
5. Yadav, D.: L-slotted rectangular microstrip patch antenna. In: 2011 International Conference on Communication Systems and Network Technologies, pp. 220–223. IEEE, June 2011
6. Singh, N., Singh, S., Kumar, A., Sarin, R.K.: A planar multiband antenna with enhanced bandwidth and reduced size. IJEER Int. J. Electron. Eng. Res. **2**(3), 341–347 (2010). ISSN No 0975-6450
7. Ali, M.M.M., Azmy, A.M., Haraz, O.M.: Design and implementation of reconfigurable quad-band microstrip antenna for MIMO wireless communication applications. In: 31st National Radio Science Conference (NRSC), pp. 27–34. IEEE (2014). https://doi.org/10.1109/NRSC.2014.6835057
8. Singh, I., Tripathi, V.S.: Microstrip patch antenna and its applications: a survey. Int. J. Comput. Technol. Appl. **2**(5), 1595–1599 (2011)
9. James, S.E., Jusoh, M.A., Mazwir, M.H., Mahmud, S.N.S.: Finding the best feeding point location of patch antenna using HFSS. ARPN J. Eng. Appl. Sci. **10**(23), 17444–17449 (2015)
10. Velip, J., Virani, H.G.: Design of slot patch antenna and comparative study of feeds for C-band applications. IJIRST Int. J. Innov. Res. Sci. Technol. **1**(12) (2015). ISSN (online) 2349-6010
11. Krishna Jyothi, N., Anitha, V.: Design of multiple U slotted microstrip antenna for Wimax and wideband applications. Int. J. Innov. Technol. Explor. Eng. **8**(4), 152–155 (2019). ISSN 2278-3075

Fourier Descriptors Based Hand Gesture Recognition Using Neural Networks

Rajas Nene[(✉)], Pranay Narain, M. Mani Roja, and Medha Somalwar

EXTC, Thadomal Shahani Engineering College, Mumbai University,
Mumbai, India
rajas.nene16@gmail.com, narain.pranay@gmail.com,
maniroja@thadomal.org

Abstract. With the advanced computing and efficient memory utilization, vision-based learning models are being developed on a large scale. Sign language recognition is one such application where the use of Artificial Neural Networks (ANN) is being explored. In this article, the use of Fourier Descriptors for hand gesture recognition is discussed. The system model was set up as follows: Images of 24 hand gestures of American Sign Language (ASL) were subjected to edge detection algorithms, the contours were extracted and Complex Fourier descriptors were obtained as features for classification using a 2 layer feed-forward neural network The effects of subsampling of the contour, number of hidden neurons and training functions on the performance of the network were observed. Maximum accuracy of 87.5% was achieved.

Keywords: American Sign Language (ASL) · Artificial Neural Networks (ANN) · Contours · Feature extraction · Fourier descriptors · Gesture Recognition

1 Introduction

One of the major problems faced by people with speaking and hearing disabilities is communicating with people who do not know sign language. Therefore, to facilitate effective communication, this article presents the effects of changes to the inputs and network parameters for hand gesture recognition. For vision-based Hand Gesture Recognition, handling the Degrees of Freedom (DOF) play a key role. Spatial resolutions (details to be resolved), temporal resolutions (details resolved during movement) and camera point of view can change this. When analyzing any vision-based solution, it is important to understand how the performance of the system is affected by its ease of use, scalability. For high accuracy systems, the processed data size could be large, thereby increasing the memory requirements. For a robust system, a tradeoff between these parameters should be considered.

The research work incorporates image processing, feature extraction, and a Feed-Forward Neural Network model to accurately determine the sign language gesture(s) shown in the image(s) and translate it into the corresponding English word(s). Image datasets were subjected to edge detection algorithms and subsequently, features in the form of Fourier descriptors were extracted and fed to the neural network. The language

© Springer Nature Switzerland AG 2020
J. S. Raj et al. (Eds.): ICIDCA 2019, LNDECT 46, pp. 140–147, 2020.
https://doi.org/10.1007/978-3-030-38040-3_16

of choice is American Sign Language (ASL) since it is widely used by the deaf and mute communities in North and South America, West Africa and East Asia with an estimate of up to 500,000 users. For the scope of our article, we have considered 24 symbols corresponding to English alphabets that use a single gesture frame and exempted the letters J and Z as they are multi-frame hand gesture symbols (Fig. 1).

Fig. 1. ASL (excluding 'J' and 'Z')

2 Literature Review

Two primary approaches involved in hand gesture recognition are glove-based and vision-based systems. Parvini and Shahabi [1] used a glove based approach to recognize ASL signs and achieved a 75% recognition rate. In their study, Dong [2] presented a vision-based approach for real-time human-vehicle interaction. Image segmentation and feature extraction were carried out along with template matching for the classification of gestures. Ionescu [3] discussed static and dynamic hand gesture recognition by skeleton extraction and comparison using Baddeley distance. Swapna [4] proposed another system consisting of three important steps, i.e. image capturing, thresholding application, and number recognition and achieved a recognition rate of 89%. Various approaches have been tried for image-based recognition including the use of Artificial Neural Networks (ANN) [5–7], SVM [8], Hidden Markov models [9]. Feature extraction refers to the extraction of useful information from an image to be given as inputs to the neural network by avoiding redundant information in the bigger size images. Discrete Cosine Transform (DCT) [10, 11], Fourier descriptors [12] Point of Interest (POI) and track point [5].

3 Process and Approach

3.1 Structure of Work

The research work is divided into 3 parts:

1. Pre-processing and Feature Extraction
2. Building and Training the Neural Network
3. Testing the Neural Network.

3.2 Preprocessing and Feature Extraction

A dataset of 24 hand gestures of ASL was clicked with 20 images for each gesture. The preprocessing stage involves resizing the RGB images, converting them into grayscale images, applying Sobel Filter for edge detection of the hand gesture and finally extracting Fourier descriptors for the input to Neural Network. The original images in the datasets are RGB JPEG images with a $3456 \times 3456 \times 3$ size. The images were resized to $64 \times 64 \times 3$. Then converted from RGB to Gray, to obtain two-dimensional images with 0–255 pixel values representing different levels from Black to Gray to White in order. Next edge detection filters are applied to the images. Canny [13] and Sobel filters were both experimented with however it was observed that the Canny filters included unnecessary details such as palmar flexion creases while a Sobel filter with a threshold of 0.1, based on subjective fidelity, neglected such features.

After this stage, the images are in binary format with 0s representing black pixels and 1s representing white pixels. The boundary was traced from the edge of the image and the values for rows and columns corresponding to the coordinates of the white pixels were stored. Because the length of the edge was different for each image, a constant length was needed. The approach followed was to study the effects of sampling the varying contour sizes of each gesture over 128, 64 and 32 values. These were obtained by incrementing the average length of each sample over the length of the contour resulting in a consistent size of contours for each image irrespective of the size of the traced edge. The contour was divided into 128 parts initially. Even with 16/32/64 parts we still get a distinguishable shape for the dataset (Fig. 2).

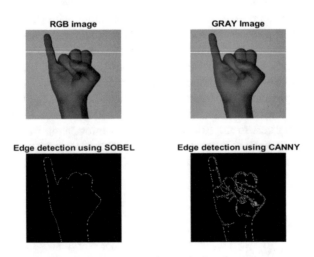

Fig. 2. Image processing and edge detection

This process was followed for the feature extraction of the images. Fourier descriptors were used for this stage. One of the important reasons for using Fourier descriptors is the 'Affine Fourier Invariants' property [14], which render them invariant to translations, rotations and scalings. As seen in Fig. 3, the Fourier descriptors can be used to retrieve the images by taking the Inverse Fourier transform. The coordinates of the row and column in which the white pixels were detected, were arranged in the form of a complex number with the columns as Real part and rows as the Imaginary part. The frequency-domain equivalents of these points were obtained by applying Fast Fourier Transforms. These real and imaginary parts of the Fourier descriptors are then separated and arranged as row vectors with all the real parts first followed by imaginary parts. In this way, features were extracted from the image(s) (Figs. 4 and 5).

Fig. 3. Image retrieval using only 8, 16, 32 and 64 points of the whole contour and grayscale image

$$s(k) = x(k) + j * y(k) \tag{1}$$

where,

x(k) = the column coordinate of the contour
y(k) = the row coordinate of the contour
k ∈ [0, N]

$$a(u) = \sum_{k=0}^{K-1} s(k)e^{\frac{-j2\Pi uk}{K}} \tag{2}$$

where,

s(k) = complex format of the contour.

3.3 Building and Training the Neural Network

For the training dataset, 15 images per hand gesture were used. The preprocessing and feature extraction was implemented on each image and row vectors were obtained. The input matrix to the neural network was formed by arranging the rows in a matrix. For targets of the neural network, the presence of an image is represented by 1 and absence by 0. A 2 layer feed-forward Neural Network was designed with 1 hidden layer and 1 output layer. The Hyperbolic tangent sigmoid transfer function [15] was used as the activation function for the hidden layer. At the hidden layer, the number of neurons was varied. The output layer had 24 neurons representing 24 hand gestures. Values closer to

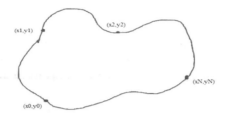

Fig. 4. Points represented on contour

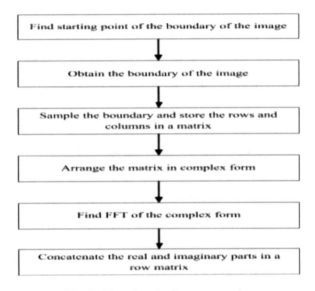

Fig. 5. Flowchart for feature extraction

1 would represent correct identification and otherwise a misidentification. Training, validation and testing set with ratios of 70%, 15% and 15% respectively were obtained. For the training functions, the results of two popular algorithms, Scaled Conjugate Gradient backpropagation [16] and Levenberg–Marquardt back propagation [17, 18] algorithms were observed (Figs. 6 and 7).

3.4   Testing the Neural Network

A set of 5 new images of each hand gesture was used for testing the neural network. The images were processed similar to the training dataset for edge detection and feature extraction. Further, this data was arranged as rows into a matrix and used for testing the network. With each change of training parameters, the network accuracies varied. The neural network was trained so that the error is minimized by calculating MSE. The network weights are re-adjusted to adapt to the calculated error and continue for achieving better performances. The learning rate was initially kept constant at 0.1 but was later decremented on a schedule of every 10 epochs by a factor of 0.2.

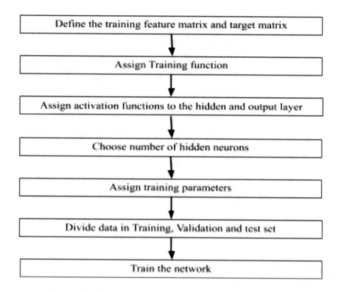

Fig. 6. Flowchart for training of neural network

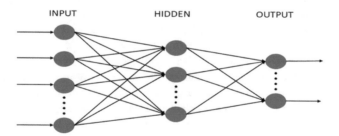

Fig. 7. Two layer feed forward neural network

4   Results

As observed in Table 1, using the Levenberg-Marquardt backpropagation algorithm, an accuracy of 87.5% was achieved for 128 points on the contour and 40 hidden layer neurons. When the number of neurons was decreased for the hidden layer, the accuracy dropped. When the parts of contour were decreased, thereby directly decreasing the data size, the performance again increased. As with the Scaled Conjugate Gradient algorithm, it was similarly found that better accuracies were achieved with a balance of data size and number of neurons in the hidden layer. All the tests were performed using MATLAB 2018a software. The system configuration was Intel(R) Core(TM) i5-8250U CPU @ 1.60 GHz 1.80 GHz Processor, 8 GB of internal RAM and 64-bit WINDOWS 10 HOME Operating System.

Table 1. Percentage accuracies for different training functions

Samples on the contour	Neurons in the hidden layer	Accuracy using Levenberg - Marquardt backpropagation	Accuracy using Scaled Conjugate Gradient backpropagation
128	N = 40	87.5%	75%
	N = 30	72.22%	84.72%
	N = 20	77.77%	70.83%
64	N = 40	72.22%	79.16%
	N = 30	81.94%	62.5%
	N = 20	84.72%	87.5%
32	N = 40	83.33%	70.83%
	N = 30	75%	80.55%
	N = 20	75%	83.33%

5 Conclusion and Future Scope

The article showcased the implementation of hand gesture recognition using ANN. Feature extraction using Fourier descriptors was showcased and an accuracy of 87.5% was achieved. The effects of the number of neurons in the hidden layer, changes in the input data size were also demonstrated. Better results could be expected with larger data sizes and conditions for clicking images. For gesture recognition, image processing plays an important role. This research work was limited to the recognition of static images of the ASL. Multiframe gestures recognition involves tracking the movement of the hand across a number of frames. Along with this, the background of the dataset images was also kept constant and without any obstructions. Image processing with background subtraction can be further explored to simulate real-world conditions.

Fourier descriptors are ideal for feature extraction when utilized as Affine Fourier Invariants. Only a part of the descriptors changes subject to translation, rotation and scaling of the images. For neural networks, Scaled Conjugate Gradient and Levenberg-Marquardt backpropagation algorithms give good results. The learning rate decay, hidden layer size and the number of hidden layers can be changed for further testing. Early stopping algorithms can be used for improving network generalization and to avoid overfitting.

References

1. Parvini, F., Shahabi, C.: An algorithmic approach for static and dynamic gesture recognition utilising mechanical and biomechanical characteristics. Int. J. Bioinform. Res. Appl. 3(1), 4–23 (2007)
2. Dong, G., Yan, Y., Xie, M.: Vision-based hand gesture recognition for human-vehicle interaction. In: Paper Presented at the Proceedings of the International Conference on Control, Automation and Computer Vision (1998)

3. Ionescu, B., Coquin, D., Lambert, P., Buzuloiu, V.: Dynamic hand gesture recognition using the skeleton of the hand. EURASIP J. Appl. Signal Process. **2005**, 2101–2109 (2005)
4. Swapna, B., Pravin, F., Rajiv, V.D.: Hand gesture recognition system for numbers using thresholding. In: Das, V.V., Thankachan, N. (eds.) Computational Intelligence and Information Technology, vol. 250. Springer, Heidelberg (2011)
5. Mekala, P., Gao, Y., Fan, J., Davari, A.: Real-time sign language recognition based on neural network architecture. In: 2011 IEEE 43rd Southeastern Symposium on System Theory (2011)
6. Murakami, K., Taguchi, H.: Gesture recognition using recurrent neural networks. In: Proceedings of the SIGCHI Conference on Human Factors in Computing Systems Reaching Through Technology - CHI 1991 (1991)
7. Hasan, H., Abdul-Kareem, S.: Static hand gesture recognition using neural networks. Artif. Intell. Rev. **41**(2), 147–181 (2012)
8. Ren, Y., Zhang, F.: Hand gesture recognition based on MEB-SVM. In: 2009 International Conference on Embedded Software and Systems (2009)
9. Chen, F., Fu, C., Huang, C.: Hand gesture recognition using a real-time tracking method and hidden Markov models. Image Vis. Comput. **21**(8), 745–758 (2003)
10. Tolba, A., Elsoud, M.A., Elnaser, O.A.: LVQ for hand gesture recognition based on DCT and projection features. J. Electr. Eng. **60**, 204–208 (2009)
11. Paulraj, M., Yaacob, S., Desa, H., Hema, C., Ridzuan, W., Majid, W.: Extraction of head and hand gesture features for recognition of sign language. In: 2008 International Conference on Electronic Design (2008)
12. Bourennane, S., Fossati, C.: Comparison of shape descriptors for hand posture recognition in video. Signal Image Video Process. **6**(1), 147–157 (2010)
13. Canny, J.: A computational approach to edge detection. IEEE Trans. Pattern Anal. Mach. Intell. **8**(6), 679–698 (1986)
14. Arbter, K., Snyder, W., Burkhardt, H., Hirzinger, G.: Application of affine-invariant Fourier descriptors to recognition of 3-D objects. IEEE Trans. Pattern Anal. Mach. Intell. **12**(7), 640–647 (1990)
15. Baughman, D., Liu, Y.: Fundamental and practical aspects of neural computing. In: Neural Networks in Bioprocessing and Chemical Engineering, pp. 21–109 (1995)
16. Møller, M.: A scaled conjugate gradient algorithm for fast supervised learning. Neural Netw. **6**(4), 525–533 (1993)
17. Yu, H., Wilamowski, B.: Levenberg–Marquardt training. In: Electrical Engineering Handbook, pp. 1–16 (2011)
18. Levenberg, K.: A method for the solution of certain non-linear problems in least squares. Q. Appl. Math. **2**(2), 164–168 (1944)

Improved Feature Based Sentiment Analysis for Online Customer Reviews

L. Rasikannan[1], P. Alli[2], and E. Ramanujam[3(✉)]

[1] Department of Computer Science and Engineering,
Alagappa Chettiar College of Engineering and Technology,
Karaikudi, Tamil Nadu, India
laxmanrasivalli@gmail.com
[2] Department of Computer Science and Engineering,
Velammal College of Engineering and Technology, Madurai, Tamil Nadu, India
alli_rajus@vcet.ac.in
[3] Department of Information Technology, Thiagarajar College of Engineering,
Madurai, Tamil Nadu, India
erit@tce.edu

Abstract. The evolution of E-commerce site tends to produce a huge amount of data nowadays. These data consist of very novel knowledge to compete with the other E-commerce sites in terms of business perspective. Customers often use these E-commerce sites to manage decision on the purchase of products based on comments or reviews given by the existing customer who bought the same product. The concept of Opinion Mining enables these processes of selection and decision easier. Several techniques have been proposed for the opinion mining and provided their own advantages. However, those techniques contain certain drawbacks in the selection of features and opinions with respect to the priority of product feature given by the individual user. This paper proposes a novel idea of incorporating weight which is automatically calculated according to the attributes evolved. The reason to do certain weight calculation is an assumption of weight and weight based on priority may differ from person to person. Experimental results show the performance of the proposed with various techniques for the online review collected from different sites.

Keywords: Opinion mining · Sentiment analysis · Reviews · Comments · Weights · Priority · Opinion feature · Product feature

1 Introduction

Nowadays, E-commerce sites are gaining more importance and popular as it is used by people to purchase products and to express their views about that product. These reviews can be of huge importance for the seller to improve their business. The user review comprises of emotions, texts, sentiments, and patterns. Opinion mining or sentiment analysis processes these reviews to aid better decision to the customers and business people [1, 2]. Earlier days, the reviews are collected through polls or in the form of relational table or text. Mostly, Machine Learning (ML) algorithms were used to analyze those relational data. However, due to the increase of e-commerce sites such

© Springer Nature Switzerland AG 2020
J. S. Raj et al. (Eds.): ICIDCA 2019, LNDECT 46, pp. 148–155, 2020.
https://doi.org/10.1007/978-3-030-38040-3_17

as amazon, e-bay, flipkart, shopclues etc., and advancement in technology the user express their reviews and comments in the form of text, audio or emojis i.e. unstructured in nature. It makes difficult for a company to analyze manually. Also, the review analyzed by persons through different ML algorithms yields a different opinions. So, automated analysis is required for a company to extract the users' opinions. In addition, certain reviews posted are more specific to technical issues, but some may be general. For example, "Sony is Good" is a general comment, however, "Processor of Nokia A310 is slow" is a review that describes the processor is slow, but not the quality of the mobile phone. In particular, some reviews may be positive, others may be negative and some may be neutral. To analyze these polarity level and sentiment of the reviews, enhanced sentimental analysis technique feature-based sentiment analysis evolved. In this approach, the user reviews are parsed to find out the sentiment of the users, to recommend the product to other user even based on the feature of the product. Various papers has been proposed for automated feature based sentiment analysis. However, these techniques have certain drawbacks in identifying the opinion with respect to the priority of feature of the product given by individual user. In the existing system, there is no provision for the prioritization of feature of the product. To resolve this issue, this paper proposes an improved feature based sentiment analysis through automated weight. In the proposed work, mutidimensional user reviews are dimensionally reduced into fact table with entity and features of the product. The fact table is mapped with the predetermined ontology using automated weights to recommend the product based on the priority of feature of the product. The rest of the proposed work has been organized as follows. Related feature-based sentimental analysis is discussed in Sect. 2. Section 3 describes the proposed methodology of the work. Section 4 describes the experimental results and Sect. 5 provides the conclusion.

2 Literature Review

With the emerging usage of e-commerce sites, users' purchase more products and shares their views about the product purchased through the site. Sentiment analysis emerges due to these sites to analyze the user review/comments. The sentiment analysis process determines the sentiment and its polarity termed to be positive, other may be negative and neutral given in the user review based on the scope of the review. Sentiment polarity is categorized into three levels, document/sentence level, entity level and aspect level [3]. Document level determines the polarity of sentiments in the entire document. Sentence level deals with categorization of sentiments in each sentence. For categorization of sentiments in sentence level, the research work [4] has proposed a lexicon based and ML based approach. In a lexicon based approach, the sentiment lexicons are represented by either positive or negative sentiment words. These sentiment words are already pre-compiled and pre-determined into dictionaries/word corpus/Ontology for further processing [5, 6]. ML based approach uses syntactic or linguistic features along with text mining algorithms for categorization [7, 8]. For example, Parts Of Speech (POS), Sentiment words, negation are pre-processed and the reviews are classified using ML algorithms. The major complication in ML-based approach is the extraction of opinion features of the product from the user reviews. Recently, various papers have proposed the concept of opinion feature extraction. Agarwal in [9] has proposed a method to analyze the twitter data using POS-tagging specific to linguistic features. To extract opinion feature of the product, Hu et al. [10] has proposed a concept of text summarization technique. Popescu et al. [11] has proposed a system named OPINE to mine the customer reviews based on the features of a particular product. Similarly, Siqueira et al. [12] has presented a system termed "WhatMatter" to extract the opinion features of the product. To recognize the semantic orientation of feature in addition to opinion feature, Eirinaki et al. [13] has proposed an algorithm by integrating with an opinion search engine to determine the sentiment of feature of the product. Yaakub et al. in [14] has proposed a multidimensional model which integrates the users' characteristics, comments and features of the product. The system initially identifies the entities and sentiments from the table. Further, the table has been reduced to fact table using the polarity of sentiments. The research work [15] has proposed a system "RedOpal", which identifies the best products with respect to set of automatically extracted and evaluated features. This work considers each word of the review as features by finding its adjacent opinions on probability. Htay et al. [16] have proposed an efficient idea to discover the sentiment words of each feature from the customers review. The drawbacks of the existing systems are, there is no provision for the importance/priority of the feature of the product considered by each user. To resolve this, the proposed work uses an automatic weight with feature-based sentiment analysis to recommend the other product based on the feature of the product considered by other user.

3   Proposed Methodology

Customer reviews are the only important aspects that e-commerce sites need to emphasize especially to recommend the product with the loyalty to be retained to the customer. The proposed work extends the opportunities produced by the research work in [14]. The proposed work uses pre-determined ontology [17] just to maintain the sentimental lexicon of set of features given by the users' in their reviews. Generally, the features of the product are categorized into specific and technical terms. The proposed work uses the review comments given by the user only in technical terms of the product. Sample pre-determined ontology has been shown in Table 1 with the set of technical term and its relevant sub levels of mobile phone as given in [14]. The product is Camera and its technical term mentioned by the user is body, lens, flash etc. Further its sub levels may be mentioned as accessories, glass, color etc.

Initially, the user reviews of any online sites are fetched using Extraction Transformation and Loading (ETL) as given in [16] and stored as source file with an extension .txt. The ETL process itself preprocesses the source file using POS tagging, negation etc. After preprocessing of source file, the user review has been transformed

Table 1. Sample Ontology for mobile phones.

Technical term	Level 1	Level 2
Camera	Body	Accessories
	Lens	Glass
		Shutter
		Magnify
	Flash	–
	Picture	Light
		Resolution
		Color
		Compression
	Delay	Time
	Video	Capture
		Image
		Resolution

into customer information along with data and pairs of entity. Then, the nouns (sentiment words) are extracted from the source file to match with *Feature Ontology(FO):* $<A, R>$ for the better understanding of customer reviews. In this case, A is a set of attributes/features of the product purchased by the customer C with their relationships R. If necessary, the Feature Ontology is further generalized into more general concepts from lower to higher order concept hierarchy or vice versa. An Opinion Sentence *(OS)* are created by the pairs of sentiment feature pairs after generalization, defined as $OS = \{(f_1, s_1), \ldots, (f_m, s_m)\}$ where, f and s are sentiment-feature pairs. Then to mine the customer opinion about the specific feature of the product, each sentiment feature pairs are clustered with closest sentiment words in the ontology. To cluster this feature pairs, the polarities of each sentiments word are used for processing. Then the closest feature pair of sentiment words is considered to calculate the opinion of other customer in the group to categorize the products using Eq. (1).

$$ogc(g, c) = \sum_{z=-3}^{3} (z * polarity(c, g, z)) \qquad (1)$$

Where, z is sentiment value, g is group of customers, c is the feature of the product and polarity is the total support. The orientation of the customer $o(g, c)$ is termed to be positive if $ogc(g, c)$ is greater than zero, negative if $ogc(g, c)$ is lesser than zero, else neutral. This user orientation doesn't promptly recommend you the product based on other users. As, some user may have priority on one attribute than the other attribute while purchasing a product. For example, some user has higher priority on camera pixels and some may have battery life while purchasing a mobile phone. To prioritize the product feature, the research paper [14] assigned weights to the attribute for the process and is calculated using Eq. (2).

$$Mogc(g,c) = ogc(g,c) * W_i \forall W_i = \sum_{i>0}^{1} W_i \qquad (2)$$

However, the priority of each customer varies, and weight assignment for each of the priority makes a process very tedious. To resolve this issue, the proposed work follows a novel idea of weighting an item using non-negative real number, to reflect the importance of each subcategory of the ontology. Based on ogc, weight w for each subcategory is calculated as given in Eq. (3).

$$w = (\Sigma z_i * \max(polarity(c,g,z))) / (\Sigma polarity(c,g,z) * 10) \qquad (3)$$

Where Σz_i, $\forall z_i > 0$ and $\Sigma polarity(c,g,z)$, $\forall polarity(c,g,z) > 0$.

4 Experimental Results and Discussions

The proposed concentrates on extracting the sentiment features of the customer reviews based on the features of the product. The proposed uses pre-determined ontology to define the features of the products as in [17]. To evaluate the proposed work, all the users' reviews are read and evaluated manually. Similarly, the opinions of the customers are mined to be positive or negative opinion based on the polarity representation. The polarity representation is framed in the scale of seven point polarity from -3, -2, -1, 0, 1, 2 3 with the representation of Very Poor, Poor, Weak, Neutral, Excellent, Very Good and Good. The effectiveness of the proposed work is evaluated using Precision (p), Recall (r) as given in Eqs. (4) and (5) which is taken from [17].

$$p = \frac{TP}{(TP + FP)} \qquad (4)$$

$$r = \frac{TP}{(TP + FN)} \qquad (5)$$

Further, to evaluate the proposed work, the sample of 100 reviews for the product 'fitbit' has been considered from www.amazon.in as also been used in [18]. Table 2 shows the detailed review comments with its polarity and attribute values.

The number of attributes based on sub categorization levels along with the Polarity scale value for each and every comment is cumulated and grouped on their attributes is shown in Table 3. The Column W_i represents the weight assumed and the modified OGC shows the result of opinion as per paper [18]. The proposed OGC shows the value of OGC as per the proposed technique with novel weight value given in the equation. The efficiency of the proposed when compared with the paper [18], it clearly shows the positive, negative and neutral opinions. However, the research work [18] has shown only positive and negative opinions. This provides a better efficiency of the proposed paper. To provide the efficiency of the proposed in a robust manner, the comparison has been made with the paper [14, 18] and it is shown in Table 4. The project Hu and Liu's [19] used a dataset which contains 1000 reviews of Nokia 6610

Table 2. Sample 30 reviews with attributes and polarity

Comment	Attributes and polarity						
	Battery	Display	Waterproof	Synch.	General	Durable	Connec.
1.	2	2	−3				
2.				1			
3.	2						
4.	2				2	2	
5.	2						2
6.				1	2		
7.		2	−1			3	
8.	3			0	1		2
9.		2	1	3			
10.	1		1			1	
11.	3						
12.		3					
13.	2		2		−2		0
14.		2					
15.	2		3		−2	−1	
16.	-3						
17.			0	1		3	
18.	−1	1	2		3	2	
19.	−3		1	2	1		
20.		2					
21.	1		3			−3	
22.		0					
23.	−3						
24.	−1		−2		0		
25.							2
26.						3	
27.			2				
28.		1					
29.	2		0		−3		
30.	−1		2		2		

customer, provided their comments on Amazon. The same dataset is used for evaluation by the proposed work. Minqin Hu and Bing Liu has compared their efficiency with 5 different techniques along with the performance of the paper in [14, 18] which are considered here for performance comparison of the proposed work. Minqin and Bing's technique emphasized the frequently appeared nouns and ignored the other nouns. This makes hard change in the extraction of feature of the product. On comparing the performance of the proposed with papers [14, 18, 19], the proposed perfectly recommends the product with positive, negative and neutral opinions and also outperforms in terms of precision and Recall.

Table 3. Calculation of OGC, modified OGC and proposed OGC

Attributes	Polarity							OGC	W_i	Modified OGC	$\sum Z$	Max (Polarity)	$\sum polarity * 10$	Proposed OGC	Orientation
	−3	−2	−1	0	1	2	3								
Battery	0	0	0	0	0	4	1	11	0.2	2.2	5	4	20	1	Positive
Display	0	0	0	0	0	3	0	6	0.1	0.6	2	3	10	0.6	Positive
Accuracy	0	0	0	0	4	4	1	15	0.1	1.5	6	4	30	0.8	Positive
Water proof	5	0	0	0	0	0	0	−15	0.2	−3	−3	5	10	−1.5	Negative
Synchronization	0	0	3	0	10	1	0	9	0.1	0.9	0	10	30	0	Neutral
General	0	0	0	0	3	4	0	11	0.01	0.11	3	4	20	0.6	Positive
Durable	0	0	0	0	0	3	0	6	0.09	0.54	2	3	10	0.6	Positive
Alarm	0	0	0	2	0	0	0	0	0.01	0	0	2	10	0	Neutral
Connectivity	0	0	0	0	1	5	0	11	0.09	0.99	3	5	20	0.75	Positive
Reliable	0	0	0	0	0	1	0	2	0.01	0.01	1	1	10	0.1	Positive
Features	0	0	0	0	0	1	0	2	0.01	0.09	1	1	10	0.1	Positive

Table 4. Performance of proposed with feature-based sentiment analysis models

Methodology	Precision	Recall
Integration model	0.855	0.923
Opinion sentence extraction	0.675	0.815
Frequent feature	0.731	0.563
Compactness pruning	0.716	0.676
P-support pruning	0.716	0.828
Infrequent feature	0.761	0.718
Proposed model	**0.913**	**0.945**

5   Conclusion

The proposed work evaluates the user reviews given in the e-commerce sites using feature-based sentiment analysis through automated weight. Several techniques have also proposed the concept of sentimental analysis for user reviews but only a few techniques are proven to be significant by providing additional importance to user priority for feature of the product. The proposed work uses automated weight to provide prioritization for the specific features of the product. The recommendation given to the new customer greatly depends on the type of reviews. Several reviews were collected from Amazon for the experiment purpose and evaluated. By assigning the automated weight, the proposed outperforms the existing technique in terms of precision and recall value of 0.913 and 0.945, perfectly recommends the product to new customer with its polarity level termed to be positive, negative or neutral.

References

1. Pang, B., Lee, L.: Opinion mining and sentiment analysis. Found. Trends® Inf. Retrieval **2** (1–2), 1–135 (2008)
2. Das, S., Chen, M.: Yahoo! for Amazon: extracting market sentiment from stock message boards. In: Proceedings of the Asia Pacific Finance Association Annual Conference, vol. 35 (2001)

3. Liu, B.: Sentiment analysis and opinion mining. Synth. Lect. Hum. Lang. Technol. **5**(1), 1–167 (2012)
4. Ravi, K., Ravi, V.: A survey on opinion mining and sentiment analysis: tasks, approaches and applications. Knowl.-Based Syst. **89**, 14–46 (2015)
5. Ding, X., Liu, B., Yu, P.S.: A holistic lexicon-based approach to opinion mining. In: Proceedings of the 2008 International Conference on Web Search and Data Mining, pp. 231–240. ACM, February 2008
6. Turney, P.D.: Thumbs up or thumbs down?: semantic orientation applied to unsupervised classification of reviews. In: Proceedings of the 40th Annual Meeting on Association for Computational Linguistics, pp. 417–424. Association for Computational Linguistics, July 2002
7. Meena, A., Prabhakar, T.V.: Sentence level sentiment analysis in the presence of conjuncts using linguistic analysis. In: European Conference on Information Retrieval, pp. 573–580. Springer, Heidelberg (2007)
8. Pang, B., Lee, L., Vaithyanathan, S.: Thumbs up?: sentiment classification using machine learning techniques. In: Proceedings of the ACL-2002 Conference on Empirical Methods in Natural Language Processing-Volume 10, pp. 79–86. Association for Computational Linguistics (2002)
9. Agarwal, A., Xie, B., Vovsha, I., Rambow, O., Passonneau, R.: Sentiment analysis of Twitter data. In: Proceedings of the Workshop on Language in Social Media (LSM 2011), pp. 30–38, June 2011
10. Hu, M., Liu, B.: Mining opinion features in customer reviews. In: AAAI, vol. 4, no. 4, pp. 755–760, July 2004
11. Popescu, A.M., Etzioni, O.: Extracting product features and opinions from reviews. In: Natural Language Processing and Text Mining, pp. 9–28. Springer, London (2007)
12. Siqueira, H., Barros, F.: A feature extraction process for sentiment analysis of opinions on services. In: Proceedings of International Workshop on Web and Text Intelligence, pp. 404–413, October 2010
13. Eirinaki, M., Pisal, S., Singh, J.: Feature-based opinion mining and ranking. J. Comput. Syst. Sci. **78**(4), 1175–1184 (2012)
14. Yaakub, M.R., Li, Y., Algarni, A., Peng, B.: Integration of opinion into customer analysis model. In: Proceedings of the 2012 IEEE/WIC/ACM International Joint Conferences on Web Intelligence and Intelligent Agent Technology, vol. 03, pp. 164–168. IEEE Computer Society (2012)
15. Scaffidi, C., Bierhoff, K., Chang, E., Felker, M., Ng, H., Jin, C.: Red Opal: product-feature scoring from reviews. In: Proceedings of the 8th ACM Conference on Electronic Commerce, pp. 182–191. ACM, June 2007
16. Htay, S.S., Lynn, K.T.: Extracting product features and opinion words using pattern knowledge in customer reviews. Sci. World J. **2013**, 5 pages (2013)
17. Balachander, J., Ramanujam, E.: Rule based medical content classification for secure remote health monitoring. Int. J. Comput. Appl. **165**(4), 21–26 (2017)
18. Malik, M., Habib, S., Agarwal, P.: A novel approach to web-based review analysis using opinion mining. Procedia Comput. Sci. **132**, 1202–1209 (2018)
19. Hu, M., Liu, B.: Mining and summarizing customer reviews. In: Proceedings of the Tenth ACM SIGKDD International Conference on Knowledge Discovery and Data Mining, pp. 168–177. ACM, August 2004

Performance Analysis of Automated Detection of Diabetic Retinopathy Using Machine Learning and Deep Learning Techniques

Nimisha Raichel Varghese$^{(\boxtimes)}$ and Neethu Radha Gopan

Rajagiri School of Engineering and Technology, Kakkanad, Kochi, India
nimishakovelil@gmail.com, neethurg@rajagiritech.edu.in

Abstract. Diabetic Retinopathy is related to a combination of eye disorder due to difficulty of mellitus. This disorder leads to complete blindness or vision loss. Automated methods for detecting and classifying the type of disease into normal or abnormal have important medical application. Here, deep learning and machine learning techniques are used to classify a given set of images into normal or abnormal classes. In machine learning section, local binary pattern (LBP) technique is used for feature extraction. Random forest (RF) and Support Vector Machine (SVM) are the two best machine learning algorithms taken for classification purpose. AlexNet, VGG16 and Long Short Term Memory (LSTM) are used as the deep learning techniques. Single algorithms are optimise with respect to their parameters, and are compare the parameters in terms of their accuracy, sensitivity, specificity, precision and F1-score. The accuracy of SVM, RF, AlexNet, VGG16 and LSTM were found to be 88.33%, 94.16%, 98.35%, 99.17% and 97.5%. Also, the performance evaluation table of machine learning and deep learning algorithms were tabulated using these parameters.

Keywords: Diabetic retinopathy · Convolutional neural network · Support Vector Machine · Long Short Term Memory · Random Forest

1 Introduction

Diabetic retinopathy is an eye problem in which, damages happens to the retina mainly due to diabetes mellitus. Nowadays, diabetic retinopathy has become a curious problem to human eye. The rate of DR is increasing day by day. In 2010, 126.6 million people were having this disease, and the DR predictions growth reaches upto 191 million population in 2030. The images of retina are realized to examine the texture of blood vessels, microaneurysms, hemorrhages, neovascularization, and exudates. These are certain symptoms of diabetic retinopathy. Based on these symptoms it is decided whether the fundus images are normal or abnormal. Therefore, the common methods for diagnosis this disease is become

J. S. Raj et al. (Eds.): ICIDCA 2019, LNDECT 46, pp. 156–164, 2020.
https://doi.org/10.1007/978-3-030-38040-3_18

inefficient to DR patients and to identify the symptoms of diabetic retinopathy as soon as possible.

de la Calleja [1] proposed a technique to identify the detection of diabetic retinopathy using image analysis method. The hard and soft exudates were segmented by using morphological operations. The proposed algorithm for the exudates detection were tested over the retinal images of dataset. Neera and Tripathi [2] proposed a technique based on Local binary pattern and machine learning for the detection of DR. This technique mainly have two steps: step one is Texture operator to extract features of diabetic retinopathy. The second step are random forest, artificial neural network and support vector machines were introduced for the purpose of detection. At first level of results shows 94.46% accuracy of Random forest. Andonov, Pavloviov, Kajan [3], contributed on image recognition methods that are suitable for diagnostic purposes in ophthalmology. To classify the images according to whether or not they have some types of anomalies, a convolutional neural network (CNN) with 4 convolutional layers was proposed. Li, Pang, Xiong, Liu [4] introduced a neural network related to transfer learning for the classification of DR. The proposed CNN model consists of seven convolutional layers, two fully connected and a softmax layer. Results demonstrate that CNN systems related to transfer learning can reach maximum accuracy with less number of datasets.

2   Methodology

In this paper, deep learning and machine learning methods are used for the detection of retina images. In machine learning section, the classification is done by Support Vector Machine and Random Forest algorithms. In deep learning, AlexNet, VGG16 and LSTM are the techniques used to detect the retina images. The most important difference between machine and deep learning model is based on feature extraction. Features to be extracted are determined manually in machine learning whereas in deep learning models, it figures out the feature by itself. Input image is taken from a database that consists of 402 fundus retina images of diabetic retinopathy patients collected from publicly available dataset like DIARETDB0, STARE and DRIVE. Both machine learning and deep learning techniques are tested on the same set of data.

3   Machine Learning

The significant stages in automated detection of diabetic retinopathy are (a) Image Preprocessing (b) Feature Extraction (c) Classification and (d) Disease Identification. Classification of retinal images as normal or diabetic, has two stages in case of machine learning. The first step is feature extraction and second step is image classification. First step is done by the characterization of retina images using local binary pattern (LBP). LBP is a texture descriptor mainly used for texture identification of retina images. Shape and blood vessels are the two features extracted for classification. SVM and RF are the two algorithm used for the classification purpose. The flow chart of machine learning is shown in Fig. 1 [7].

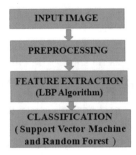

Fig. 1. Machine learning flowchart.

3.1   Preprocessing

The shape and blood vessels are the two features taken for the LBP texture descriptor. The following preprocessing steps are used for shape detection [6]: The first step is convert color image (Irgb) to grayscale image (Ig). The second step is to apply a filter to the grayscale image, the filter is median filter. After that adaptive histogram equalization (CLAHE) is given to median filtered image. Again another filter i.e., gaussian filter is given to the CLAHE image for further noise removal. Shade correction is applied to contrast enhanced image. During the shade correction, the back portion of the image is viewed by apply a 68 by 68 large median filter. Then the image is normalized and the vessels of blood can be scene by morphological method i.e., black top-hat operation. The shape detection of diabetic and normal image is shown in Fig. 2. The second feature required for LBP texture descriptor are blood vessels. The following steps are used for blood vessel detection [6]. First step is to change a color to grayscale image. Last step is apply median filter to the grayscale image. After that CLAHE technique is applied to the image. Then the difference between gray and the median filtered image is taken, then converted into binary image. Morphological dilation operation is applied and again binary is changed to grayscale image. The blood vessel detection of DR is shown in Fig. 3.

3.2   Feature Extraction

The basic idea of extraction is to extract the important characters or features such as color, shape, boundaries etc. Local Binary Pattern technique [5] is used for feature extraction in both SVM and Random Forest. LBP is the most recent texture operator which replaces the pixel estimation of an image with decimal values. Each central value of pixel is contrasted and its eight neighbors pixel values; the neighbors pixel value having smaller value than that of the central value of pixel then the value will become bit 0 value, and neighbors pixel having value equivalent to or more prominent than that of the central pixel value will have the bit value 1 [5].

Fig. 2. Shape detection of DR and normal image.

Fig. 3. Blood vessel detection of DR and normal image.

3.3 Classification Algorithms: SVM and RF

SVM and Random Forest are the two machine algorithms used for the classification. SVM [8] is a unequal classifier mainly characterized by an isolating hyperplane. As a basic model, first step of SVM is to find the hyperplane (i.e., decision boundary) linearly separating the classes. The boundary equation is $w^T x + b = 1$. The support vector points above the decision boundary should have label 1 i.e., $w^T x_i + b > 0$ will have corresponding $y_i = 1$. Similarly, the points below the decision boundary should have label -1 i.e., $w^T x_i + b < 0$ will have corresponding $y_i = -1$. Now, the data points is rescaled such that the points on or above boundary level $w^T x + b = 1$ is of class one and the points on or below the boundary point $w^T x + b = -1$ is of the class second. The data points from the two classes has to lie far away from each other as possible. The distance between the two classes is called the margin. The second algorithm used here for classification is Random Forest [9]. In Random Forest classification algorithm, different trees are taken. To classify a test data, the new data is add on to another one of trees. Every trees estimates a class of trees have a test data, known as voting. The voting is taken for the each class. The more voted class tree is taken as the final class of the test data by the classifier. RF classification is the important classifier model. This classifier runs on huge amount of database it will become a time consuming method. The following steps are used for RF Algorithms.

4 Deep Learning: AlexNet, VGG16 and LSTM

4.1 AlexNet

AlexNet is a CNN framework is mainly for the computer vision tasks [10]. The framework has 8 deep layers - 5 Convolutional Layers and 3 fully Connected Layers. In a fully connected layer, every neuron gets contribution from the components of the last layer. The three layers of convolutions are followed by an overlapping Max Pooling layer. The final output of Max Pooling layer is given to the two fully connected layers whose output is given to the Softmax layer. In total, 402 images are taken from publically available datasets. The framework has an input image size of 227-by-227. The first layer of convolution, AlexNet contains 96 kernels or filters of size $11 \times 11 \times 3$. Each layer having the width and height is the same and the depth is the same as the number of channels. Overlapping Max Pooling contains first and second layers of convolution. Overlapping Max Pooling layers are common to the Max Pooling layers. The size of pooling windoe 3×3 with a stride of 2 between the adjacent windows. The second layer consists of 256 filters of size $5 \times 5 \times 48$. 384 filters of size $3 \times 3 \times 256$ is used in third convolutional layer. 256 filters of size is $3 \times 3 \times 192$ is used for fourth convolutional layer. The fifth layer of convolution is followed by an Overlapping Max Pooling layer (3×3 with stride 2). Then the output of fourth layer goes into a series of two fully connected layers. 4096 neurons are used in fully connected layer. The second fully connected layer gives into a Softmax classifier with 2 class labels. ReLU non linearity is used after all the convolution and fully connected layers. ReLU (Rectified Linear Unit) is used for the non-linear part. ReLU is $f(x) = \max(0, x)$. The architecture of AlexNet is shown in Fig. 4.

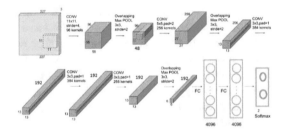

Fig. 4. Architecture of AlexNet.

4.2   VGG16

VGG16 is a kind of multi-layer CNN, is used to recognize visual patterns from image pixel. The main feature of this architecture was the increased depth of the network. VGG16 [10] have more development over AlexNetmainly due to large kernel- filters with 3×3 multiple kernel- filters. VGG16 contain 16 layers - 13 Convolutional layer and 3 fully connected layer. During training period, VGG16 input is fixed size 224×224 RGB image. The image is given through into the convolutional layers, where the filters used with a very less field size: 3×3. For the first layer, the network have 64 filters with size $3 \times 3 \times 3$ pixel one is set as stride; convolutional layer has spatial padding as input is such that the spatial resolution is preserved for convolution. Spatial pooling leads by 5 max-pooling layers, which have some of the layers of convolutional. 2×2 pixel window operation done in max pooling, with stride 2. Three layers of fully connected follow a stack of convolutional layers. The units in the last layer of convolution is $3 \times 3 \times 512$. The last layer is the softmax layer. The VGG16 architecture is given in Fig. 5.

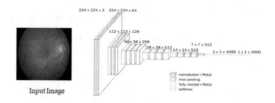

Input Image

Fig. 5. Architecture of VGG16.

4.3 LSTM

A long short-term memory network is a special kind of RNN. The preprocessing steps are used for the LSTM techniques. The first step is convert color image (Irgb) to grayscale image (Ig). The second step is to apply a filter i.e., median to the grayscale image. After that CLAHE is applied to median filtered image. Again another filter i.e., gaussian is applied to the CLAHE image for further noise removal. Shade correction is given to contrast the enhanced image. During shade correction the back portion of image is enlarged by apply a 68 by 68 large median filter. Then the image is resize into 250-by-250. It has two hidden layers in each layer has 100 nodes. The important kinds of an LSTM network are a LSTM and a sequence input layer. A sequence input mainly have the input sequence data given to the network. The network begins mainly with sequence input layer and followed with LSTM layer [11]. The class predict labels, the network contains fully connected layer, a classification output and a softmax layer. The block diagram is shown in Fig. 6 [12].

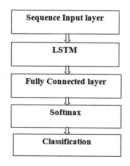

Fig. 6. Block diagram architecture of LSTM.

5 Performance Evaluation of Deep and Machine Learning Algorithms

In machine algorithms the retina image is extracted to gray scale image and it undergoes certain preprocessing steps to extract the features. The feature extraction is done by Local binary pattern texture descriptor. The classification is done by Random Forest (RF) and Support Vector Machine (SVM). In total 402 images are taken, out of images 120 are taken for the purpose of testing and remaining taken for training purpose. The accuracy is calculated by using confusion matrix. The confusion matrix of SVM and Random forest is shown in Fig. 7. In deep learning section, the Algorithms is done by using AlexNet, VGG16 and LSTM. The training section takes more time than the testing phase. The learning rate is taken as 0.0001 for both networks. In case of LSTM, the learning rate is 0.001. The VGG16 takes more time compared to other techniques. The accuracy of the deep learning techniques is calculated by using Confusion matrix. The simulation results were carried out on MATLAB 2018b. The confusion matrix of VGG16, AlexNet and LSTM is shown in Figs. 8 and 9. The confusion matrix is used to calculate different parameters such as sensitivity, accuracy, specificity, precision and f1-score. These parameters were used for the performance analysis of machine learning and deep learning techniques.

6 Result and Discussion

The performance evaluation of machine learning and deep learning algorithms are calculated based on accuracy, specificity, sensitivity/recall, precision and F1-score. The calculations are based on the confusion matrix. Confusion matrix consists of false positive (FP) false negative (FN) true positive (TP) and true negative (TN). Comparison of machine learning and deep learning algorithms based on the above parameters are shown below in the Table 1. The equation of the parameters is shown below:

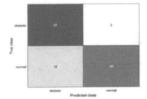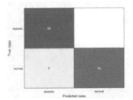

Fig. 7. Confusion matrix of Support Vector Machine and Random Forest.

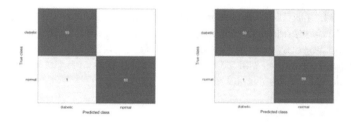

Fig. 8. Confusion matrix of VGG16 and AlexNet.

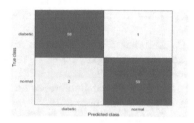

Fig. 9. Confusion matrix of LSTM.

Table 1. Comparison table of machine learning and deep learning

Technique name (%)	Sensitivity (%)	Accuracy (%)	Specificity (%)	Precision (%)	F1-score (%)
SVM	96.61	88.33	80.33	82.6	89.04
Random Forest	100	94.16	88.52	89.39	94.39
AlexNet	98.47	98.33	98.33	98.33	98.53
VGG16	100	99.17	98.36	98.33	99.15
LSTM	98.30	97.5	96.72	96.66	97.53

7   Conclusion and Future Work

The performance analysis of deep learning algorithms are better compared to machine learning algorithms. In deep learning section, VGG16 shows 99.17% accuracy compared to AlexNet and LSTM. Sensitivity, specificity, precision and F1-score value are 100%, 98.36%, 98.33% and 99.15% for VGG16. AlexNet has second better accuracy of 98.35% and it takes less time for training and testing compared to VGG16. Specificity, precision, sensitivity and F1-score values of AlexNet are 98.33%, 98.33%, 98.33% and 98.53%. In machine learning section, SVM and RF shows less accuracy of 88.33% and 94.14% specificity, sensitivity, precision and F1-score value are 80.33%, 96.61%, 82.6% and 89.04% for SVM. Sensitivity, precision, specificity and F1-score value of RF are 100%, 89.39%,

88.52% and 94.39%. The future work in this field is to find out the different stages levels of DR. The number of features and dataset required to identify the stages of diabetic retinopathy can also be increased. The hybrid combination of different deep learning techniques can be used in this field and also the performance of these techniques can be evaluated.

References

1. De la Calleja, J., Tecuapetla, L., Medina, M.A., Bárcenas, E., Nájera, A.B.U.: LBP and machine learning for diabetic retinopathy detection. In: International Conference on Intelligent Data Engineering and Automated Learning (IDEAL), September 2014
2. Neera, S., Tripathi, R.C.: Automated early detection of diabetic retinopathy using image analysis techniques. Int. J. Comput. Appl. **8**, 18–23 (2010)
3. Andonová, M., Pavloviová, J., Kajan, S., Oravec, M.: Diabetic retinopathy screening based on CNN. In: 7th International Conference on Computer and Knowledge Engineering (ICCKE 2017), 26–27 October 2017
4. Li, X., Pang, T., Xiong, B., Liu, W., Liang, P.: Convolutional neural networks based transfer learning for diabetic retinopathy fundus image classification. In: 10th International Congress on Image and Signal Processing, BioMedical Engineering and Informatics, Shanghai (2017)
5. Du, N., Li, Y.: LBP and machine learning for diabetic retinopathy detection. In: Proceedings of the 32nd Chinese Control Conference, 26–28 July 2013
6. Habib, M.M., Welikala, R.A., Hoppe, A., Owen, C.G., Rudnicka, A.R., Barman, S.A.: Microaneurysm detection in retinal images using an ensemble classifier. In: Sixth International Conference on Image Processing Theory, Tools and Applications (IPTA), Oulu (2016)
7. Bhatia, K., Arora, S., Tomar, R.: Diagnosis of diabetic retinopathy using machine learning classification algorithm. In: 2nd International Conference on Next Generation Computing Technologies (NGCT), Dehradun (2016)
8. Shetty, S., Kari, K.B., Rathod, J.A.: Detection of diabetic retinopathy using support vector machine (SVM). Int. J. Emerg. Technol. Comput. Sci. Electron. (IJETCSE) **23**(6), 207–211 (2016)
9. Roychowdhury, A., Banerjee, S.: Random forests in the classification of diabetic retinopathy retinal images. In: Advanced Computational and Communication Paradigms, vol. 475. Springer (2018)
10. Nikhil, M., Rose, A.: Diabetic retinopathy stage classification using CNN. Int. Res. J. Eng. Technol. (IRJET) **6**, 5969 (2019)
11. Tang, T.B., Meriadeau, F., Quyyum, A.: CNN-LSTM: cascaded framework for brain tumour classification. In: 2018 IEEE-EMBS Conference on Biomedical Engineering and Sciences (IECBES), Sarawak, Malaysia (2018)
12. Pulver, A., Lyu, S.: LSTM with working memory. In: International Joint Conference on Neural Networks (IJCNN), Anchorage, AK (2017)

Mobility in Wireless Sensor Networks

D. Rajendra Prasad[1]([⊠]), B. Kiran Kumar[1], and S. Indraneel[2]

[1] Department of ECE, St. Ann's College of Engineering and Technology,
Chirala, India
rp.devathoti@gmail.com, bandikiran17@gmail.com
[2] Department of CSE, St. Ann's College of Engineering and Technology,
Chirala, India
sreeram.indraneel@gmail.com

Abstract. A cluster of sensors are locally scattered and offered to monitor, track as well as to record physical circumstances of the surrounding and systematizing the data collection at the central position defines Wireless Sensor Network (WSN). The sensor nodes substantially move about from their location due to the motion medium or through particular motion hardware in the mobile sensor network. In this research, we have surveyed the related works done for mobility in WSN.

Keywords: Wireless Sensor Network (WSN) · Cluster · Motion · Mobility

1 Introduction

To monitor the physical world, Sensor networks is recognized as a capable tool to utilize self-organizing networks with powered battery to sense, practice and communicate with wireless sensors. In this network, a critical energy resource for the applications to exhibit restricted characteristics sets is analyzed [1]. Among the users and the network, a sink or base station can be active as an edge. The necessary information can be recovered through the network by introducing queries and collecting outcomes using sink. Usually wireless sensor network includes numerous sensor nodes. These nodes can communicate between themselves by means of radio signals. The operational modes for these nodes are permanent or incident driven [2]. Global Positioning System (GPS) and local positioning algorithms are used for obtaining location and to position the information.

The sensor nodes mobility has been developed to improve and enable sensing and communication together [3]. This Mobility node can expand WSN applications. In addition, mobility can delay the lifetime of the nodes because of the data transfer among the nodes and it does not typically employ the similar transmitted nodes within the path route.

Additionally, it has been served for increasing the connectivity among nodes as well as it helps the communication amid isolated nodes to expand the coverage area of interest. Though, mobility causes some issues of disconnection nodes in tender process to cause data failure and have an unenthusiastic crash on application performance [4].

© Springer Nature Switzerland AG 2020
J. S. Raj et al. (Eds.): ICIDCA 2019, LNDECT 46, pp. 165–171, 2020.
https://doi.org/10.1007/978-3-030-38040-3_19

A numerous approaches make use of the mobility nodes for collecting data. These approaches are classified as [5]: sink and node motilities. In sink mobility, in WSNs the sensed data within the vital destination be in motions and courses itself from static nodes for gathering data. In mobility nodes entity sensor nodes are in motion away as of position to position in group to hold an end-to-end communication relation.

Sink mobility can classified in accordance with pattern movement of mobile sinks and its data accumulation manner.

Random mobility: The mobile data gathers casually and the bumpered models are collected. From the single hop sensors the received data moved into a wireless access point. As the mobile data collector route is casual, then the transmission interruption may be higher.

Predictable mobility: In this prototype, the moving route is known to the static nodes of mobile data collector and to expect the data transfer time this information is being used. According to the expected time and location, sensors program its sleep stages. By this way, the energy consumption is optimized in the network.

Controlled mobility: In some situations at multiple rates data are transmitted because of the events change or event interval. It may lead to a data loss when the transmission is not completed before buffer overflows to the mobile data collector.

Node mobility have further advantages like improving coverage in circumstances where nodes are disconnected because of the initial irregular or random arrangements and unexpected breakdowns; or if the battery about to die. When the mobile nodes subjected on the attached human beings, mobility can concerned whether a macroscopic or microscopic feature. Since a **macroscopic** feature, it imitates the mobility custom on day to-day behaviors [6]. When a **microscopic** feature, it imitates the humans interrelate way to their nearby surroundings. Hence the observed movements at the radio choice of wireless interfaces are significance; they are extra pressured by the microscopic mobility.

In this paper we have conducted a survey for the mobility in wireless sensor network.

2   Literature Survey

Liang et al. [7] discovered the WSN node mobility to expand the network lifespan. Through petrol and electricity the mobile sinks mechanical progress was constrained and the entire travel distance was bounded. Initially the authors had devised the issue as a mixed integer linear programming (MILP) because of their NP-hardness. Widespread experiments had been carried out through models for evaluating this projected algorithms performance in network lifespan. The trial outcomes had demonstrated that this proposed heuristic was optimum when compared to the MILP formulation however with a great shorter running time.

Akkaya et al. [8] offered PADRA for detecting probable separations to renovate the network connectivity by organized transfer of movable nodes. The major purpose was to confine the recovery scope and to reduce the overhead inflicted on the nodes. PADRA promote expanded to hold different node breakdowns. This approach, MDAPRA strived in relocating the nodes to reinstate connectivity to a common exclusion system. The efficient of this proposed approach was validated by simulation experiments.

A novel location management system was proposed by Melodia et al. [9, 10] to monitor the actors mobility with minimum energy expenditure based on hybrid strategy including updating and predicting locations for the sensors. This location management system had allowed effective geographical routing on optimum energy-selecting rule had been obtained for sensor communication. Lastly, a system was suggested to allocate chores to actors in a synchronized way and manage their motion to achieve on the attributes events. Performance assessment had showed the efficiency of this projected solution.

Sugihara and Gupta [11, 12] devised latency minimizing problem data mule system. The data mule system (DMS) was a system issue with location and time restrictions together. This heuristic algorithm ran faster and yielded good solutions within.

In SLMP, localization had been carried out in a hierarchical manner, and this entire procedure was partitioned. Widespread simulations were carried out and the consequences had showed that the SLMP can diminish localization communication to a great extent cost as maintaining higher localization coverage and exactness.

Luo and Hubaux [13] constructed a frame to investigate the joint sink mobility and routing issue through sink constraint to a fixed location number. Formally the issue proved the NP-hardness and examined the stimulated sub-problems. Especially, an effective primal-dual algorithm had been expanded to resolve the sub-problem, this algorithm had been generalized. Finally, this algorithm had been applied to typical topological graphs sets; the consequences were demonstrated the advantages of sink mobility and recommended the popular moving traces of a sink.

Yang and Liu [14] investigated the previous works restriction and suggested a new node localizability concept. It was possible o examine a nude number to situate thin or reasonable connected networks by means of essential possible conditions for node localizability. To authenticate this design, this proposed solution executed on true-globe system and the trial consequences had showed the localizability of the node had provided necessary instructions to deploy the network and some other locality services.

A cooperative localization algorithm was proposed by Chen et al. [15] for considering the obstacles existence in mobility-assistance. To achieve higher localization of accurateness and exposure, a new curved pose evaluation algorithm had been offered to resolve the problem for the occurrence of infeasible points due to radio abnormality and obstruction effects. Simulation consequences were showed the efficiency of this projected algorithm.

Caldeira et al. [16] overviewed handover systems employed in wireless sensors mobility in healthcare and projected a novel ever-present mobility intended in Body sensor networks. A case study had been developed with this novel handover system meant for a hospital sanatorium and highlighted the performance gain of this projected solution.

Tunca et al. [17] presented a survey on currently disseminated mobile sink routing protocols. To offer an imminent balanced mobile sink routing protocol, devise conditions and disputes related to the issue of mobile sink routing were decided and explicated. An ultimate and featured categorization had been made and the protocols' merits and demerits were decided regarding its target applications.

Wang et al. [18] analyzed the mobile heterogeneous coverage in WSNs. Authors had examined the asymptotic coverage in consistent organization system. From this, 1-dimensional random walk mobility was able to amplify coverage in particular delay patience, and hence diminishes the sensing energy utilization. In addition, the heterogeneity characterized role in coverage and energy presentation by means of two mobility forms, and had presented the inconsistency of heterogeneity impact in multiple forms. Beneath the Poisson employment system, the lively k-coverage had examined with 2-dimensional random walk mobility form. Both k-coverage were explored and obtained the expected fraction instantaneously and over an occasion in the entire functional area with k-covered and it recognized the coverage advancement brought through the mobility.

Mahboubi et al. [19] studied the needed sensing and communication sensor radii and their location at every time instant to meet the recommended specifications set. These specifications included end-to-end connectivity conservation to a definite destination from the target at the same time the sensors stability was maximized and the on the whole energy utilization was reduced. This issue had been devised as a forced optimization, and to resolve it, a process was offered. Simulation consequences were demonstrated the efficiency of this proposed techniques.

Tan et al. [20] developed reactive mobility for improving the goal discovery act of WSN. In this system, the mobile sensors were collaborated with stationary sensors to attain the essential discovery performance. The precision was improved in final detection outcome as the mobile sensors measurements contain high Signal-to-Noise Ratios following the involvement. Authors had expanded a SMS rule that had attained near-optimal performance in a known detection interruption place. This effective proposed approach had been validated by means of widespread models by means of the original data outlines gathered via 23 sensor nodes.

He et al. [21] measured an on-demand data compilation circumstances to transmit service demands if its bumper was about to complete. All the mobile elements progress towards to collect data on accepting such requests, and upload the data into the sensor nodes sink if probable. An M/G/1 queue-based logical form had been presented, and logical consequences were derived on some significant system performance metrics. In addition, an advanced service system was proposed to combine the requests each time they were in closeness. This proposed work was assessed by means of extensive simulations to validate the precision of the presented model. This efficient proposed service system was verified in improving the system.

Park et al. [22] projected a new geo casting, called M-Geo casting (Mobile Geocasting). M-Geocasting offered the delegate location information to the sources of a sink cluster. The location information including facts regarding a limited area in all members sinks in the existing cluster. A source distributes data to the nearest node in the area and within the area the node limitedly deluges the data. Also, for supporting local movement in sink member sinks beyond scope region, the data were managed and offered through nodes on boundary region to member sinks.

The mobility node manipulated on the junction time of regular rumor algorithms in networks was studied by Sarwate and Dimakis [23]. It was revealed that a few of completely mobile nodes be able to produce a major reduce in meeting time.

Akbaş and Turgut [24] proposed an aerial wireless sensor and actor network (WSAN) for actor positioning approach with the deadly plume observation circumstances following a volcanic outbreak was the promising appliances of aerial UAV networks. This proposed algorithm had made relationship among molecular geometry and the atom figure in that. The functioning of this projected realistic algorithm had been presented by means of an extensive simulation.

The well-organized rendezvous intend algorithm proposed for mobile base stations with this verifiable performance bounds with changeable and predetermined tracks, respectively. The efficiency of this proposed approach had been authenticated by means of theoretical analysis and widespread simulations.

Li et al. [26] proposed the scheme it means of zero inter-sensor group effort overhead stand upon sparse random projections. Authentic urban surroundings data sets were employed in the experiments under diverse vehicular mobility models for testing the renovation precision and energy competence. The results had showed that this proposed approach was better to the usual sampling and interpolation approach of propagated data in an uncompressed figure, by means of 4–5 dB gain in reform excellence and 21–55% economy cost for the similar sampling times in communication.

3 Conclusion

The network topology had been very much affected by such node joins or breakdowns. This usual network topology alters and entity node connects and collapses were defined as weak mobility. The concurrent node joins/failures and corporal node mobility defined as strong mobility. Mobility leads to worsening in value of a recognized link and, hence, data transmission was prone to breakdown to amplify the packet retransmission rate. Mobility directs to recurrent route alters and effects in a significant packet delivery delay. A mobile node cannot instantly begin its data transmission on connecting a network; its occurrence was discovered by their neighbor and decided about their collaboration with it.

References

1. Biradar, R.V., Patil, V.C., Sawant, S.R., Mudholkar, R.R.: Classification and comparison of routing protocols in wireless sensor networks. Spec. Issue Ubiquit. Comput. Secur. Syst. **4** (2), 704–711 (2009)
2. Matin, M.A., Islam, M.M.: Overview of wireless sensor network. INTECH Open Access Publisher (2012)
3. Basagni, S., Carosi, A., Petrioli, C.: Mobility in Wireless Sensor Networks, pp. 1–33
4. Bouaziz, M., Rachedi, A.: A survey on mobility management protocols in Wireless Sensor Networks based on 6LoWPAN technology. Comput. Commun. **74**, 3–15 (2014)
5. Dong, Q., Dargie, W.: A survey on mobility and mobility-aware MAC protocols in wireless sensor networks. IEEE Commun. Surv. Tutor. **15**(1), 88–100 (2013)

6. Dong, Q., Dargie, W.: Analysis of the cost of handover in a mobile wireless sensor network. In: International Conference on Wired/Wireless Internet Communications, pp. 315–322. Springer, Heidelberg, June 2012

7. Liang, W., Luo, J., Xu, X.: Prolonging network lifetime via a controlled mobile sink in wireless sensor networks. In: 2010 IEEE Global Telecommunications Conference (GLOBECOM 2010), pp. 1–6. IEEE, December 2010

8. Akkaya, K., Senel, F., Thimmapuram, A., Uludag, S.: Distributed recovery from network partitioning in movable sensor/actor networks via controlled mobility. IEEE Trans. Comput. **59**(2), 258–271 (2010)

9. Gao, S., Zhang, H., Das, S.K.: Efficient data collection in wireless sensor networks with path-constrained mobile sinks. IEEE Trans. Mob. Comput. **10**(4), 592–608 (2011)

10. Melodia, T., Pompili, D., Akyildiz, I.F.: Handling mobility in wireless sensor and actor networks. IEEE Trans. Mob. Comput. **9**(2), 160 (2010)

11. Sugihara, R., Gupta, R.K.: Optimal speed control of mobile node for data collection in sensor networks. IEEE Trans. Mob. Comput. **9**(1), 127–139 (2010)

12. Zhou, Z., Peng, Z., Cui, J.H., Shi, Z., Bagtzoglou, A.: Scalable localization with mobility prediction for underwater sensor networks. IEEE Trans. Mob. Comput. **10**(3), 335–348 (2011)

13. Luo, J., Hubaux, J.P.: Joint sink mobility and routing to maximize the lifetime of wireless sensor networks: the case of constrained mobility. IEEE/ACM Trans. Netw. (TON) **18**(3), 871–884 (2010)

14. Yang, Z., Liu, Y.: Understanding node localizability of wireless ad hoc and sensor networks. IEEE Trans. Mob. Comput. **11**(8), 1249–1260 (2012)

15. Chen, H., Shi, Q., Tan, R., Poor, H.V., Sezaki, K.: Mobile element assisted cooperative localization for wireless sensor networks with obstacles. IEEE Trans. Wireless Commun. **9**(3), 956–963 (2010)

16. Caldeira, J.M., Rodrigues, J.J., Lorenz, P.: Toward ubiquitous mobility solutions for body sensor networks on healthcare. IEEE Commun. Mag. **50**(5), 108–115 (2012)

17. Tunca, C., Isik, S., Donmez, M.Y., Ersoy, C.: Distributed mobile sink routing for wireless sensor networks: a survey. IEEE Commun. Surv. Tutor. **16**(2), 877–897 (2014)

18. Wang, X., Wang, X., Zhao, J.: Impact of mobility and heterogeneity on coverage and energy consumption in wireless sensor networks. In: 2011 31st International Conference on Distributed Computing Systems (ICDCS), pp. 477–487. IEEE, June 2011

19. Mahboubi, H., Momeni, A., Aghdam, A.G., Sayrafian-Pour, K., Marbukh, V.: An efficient target monitoring scheme with controlled node mobility for sensor networks. IEEE Trans. Control Syst. Technol. **20**(6), 1522–1532 (2012)

20. Tan, R., Xing, G., Wang, J., So, H.C.: Exploiting reactive mobility for collaborative target detection in wireless sensor networks. IEEE Trans. Mob. Comput. **9**(3), 317–332 (2010)

21. He, L., Zhuang, Y., Pan, J., Xu, J.: Evaluating on-demand data collection with mobile elements in wireless sensor networks. In: 2010 IEEE 72nd Vehicular Technology Conference Fall (VTC 2010-Fall), pp. 1–5. IEEE, September 2010

22. Park, S., Lee, E., Park, H., Lee, H., Kim, S.H.: Mobile geocasting to support mobile sink groups in wireless sensor networks. IEEE Commun. Lett. **14**(10), 939–941 (2010)

23. Sarwate, A.D., Dimakis, A.G.: The impact of mobility on gossip algorithms. IEEE Trans. Inf. Theory **58**(3), 1731–1742 (2012)

24. Akbaş, M.İ., Turgut, D.: APAWSAN: actor positioning for aerial wireless sensor and actor networks. In: 2011 IEEE 36th Conference on Local Computer Networks (LCN), pp. 563–570. IEEE, October 2011

25. Xing, G., Li, M., Wang, T., Jia, W., Huang, J.: Efficient rendezvous algorithms for mobility-enabled wireless sensor networks. IEEE Trans. Mob. Comput. **11**(1), 47–60 (2012)
26. Yu, X., Zhao, H., Zhang, L., Wu, S., Krishnamachari, B., Li, V.O.: Cooperative sensing and compression in vehicular sensor networks for urban monitoring. In: 2010 IEEE International Conference on Communications (ICC), pp. 1–5. IEEE, May 2010

Distributed Auction Mechanism for Dynamic Spectrum Allocation in Cognitive Radio Networks

D. Sumithra Sofia$^{(\boxtimes)}$ and A. Shirly Edward

SRM İnstıtute of Science and Technology, Vadapalani, Chennai, India
sumithrasoof@gmail.com, edwards@srmist.edu.in

Abstract. Dynamic spectrum allocation is an efficient approach to enhance the efficiency of spectral utilization by cognitive users (unlicensed users). Auction mechanism is one approach by which the cognitive users get a portion of the unused licensed band for lease from the primary users (licensed users). The multi-winner auction is a new challenge to the existing auction mechanisms like (Vickrey–Clarke-Groves) VCG and Second price auction. In this work, we have manipulated the pricing based auction mechanism among cognitive users. Specifically, the proposed mechanism enhances the spectral efficiency of the cognitive users and also motivates the primary users to lease the bands to the cognitive users with increased primary user revenue. The proposed mechanism also ensures fairness among cognitive users. Simulation results show that our mechanism increases the spectral efficiency of cognitive users and the revenue of primary users.

Keywords: Cognitive radio · Dynamic spectrum access · VCG · Auction mechanism · Second price auction · Multi-winner auction

1 Introduction

The fifth generation of wireless technology, or "5G", offers new possibilities in various applications with increased bandwidth usage and more scarcity of spectrum resources. The improvement of cognitive radio technology [1] resolves the bandwidth and energy scarcity problem by dynamically allocating the spectrum to the users in wireless networks. The primary licensed user who pays to access the spectrum leases the unused portion of the spectrum to the cognitive users based on network architecture and access behavior of cognitive users (SU) [2]. In dynamic spectrum sharing, cognitive users compete for limited primary user spectrum resources and maximize their self-benefit of utilizing the primary user spectrum resources [3]. The cognitive users only aim to utilize the available spectrum by avoiding interference to the primary user based on the graph-theoretical model [4]. For dynamic spectrum sharing a primary prioritized Markovian method is proposed [5] to reduce the tradeoff between spectrum efficiency and fairness of the cognitive users. In [10], the authors proposed the multi-winner framework using the Nash bargaining solution. The interference based papers all uncover that the intellectual CUs offer the careful cost for bidding [6–8]. Every

© Springer Nature Switzerland AG 2020
J. S. Raj et al. (Eds.): ICIDCA 2019, LNDECT 46, pp. 172–180, 2020.
https://doi.org/10.1007/978-3-030-38040-3_20

essential PU gives a bit of radio assets for renting and furthermore offers its range band with a specific level of nature of administration (QoS) [15]. A various leveled game hypothesis system is structured as two level game where PU are pioneers and SU are supporters [16]. In first-value bidding game recurrence and schedule vacancy sets are considered as parameters for SU to act as bidders [17]. The (Vickrey–Clarke-Groves) VCG game [11] guarantees reality telling as an overwhelming system for subjective SUs. In any case, even in this component, there are not many disadvantages like diminished vender income and progressively powerless against intrigue assaults. In our changed system, the multi-victor closeout is planned as an advancement issue and the champs are chosen dependent on the exhibition criteria [18].

2   System Model

We consider a cognitive Adhoc network in which there are L cognitive users with U primary users, and the primary users lease the bands to cognitive users for improving their revenue. We model this scenario as an auction mechanism in which leasers are the primary users and bidders are cognitive users. For reduced complexity, we assume that the primary user leases only one band (G = 1) to the cognitive users, and also the demand of each cognitive users is only one band from the primary user there is a common channel to trade the data among the cognitive users. In this paper, the auction framework is implemented in a single band (G = 1) and extend it to multi-band. The leasing period of T is fixed considering the channel and overhead scenarios in the account by the cognitive users. At the start of each leasing period, the primary user notifies to the best cognitive user regarding the intention to sell the spectrum. This information is forwarded to the remaining cognitive users, all efficient users bids different values q = [q_1, q_2, ..., q_L] to the auctioneer where qi is the user i bid. Based on bidding values, the auctioneer determines the efficient allocation y = [y_1, y_2, ..., y_L] and pricing s = [s_1, s_2, ..., s_L]. When y_i = 1 states that user i wins the band similarly when y_i = 0 states that user i lose the band. When s_i = 5 states that the user i needs to pay an amount of 5 for occupying the band. The set of possible winners for which if and only if is given by W = [1, 2, 3, ..., L]. The system utility $U_a(y)$ for cognitive users with different valuations a = [a_1, a_2, ..., a_L] is given by (1).

$$U_a(y) = \sum_{i=1}^{L} a_i y_i \tag{1}$$

For example in Fig. 1(a) there are six cognitive users, each cognitive user competes with the remaining users to occupy the available band from the primary user. User one interferes with the user three, four and two. Similarly, user five interferes with user six. Figure 1(b) shows the constructed L × L matrix denoted by C. With the condition constraint if, it means cognitive users i and j cant share the same band because of interference between them. Similarly when the users i and j can share the same band because there is no interference between them. The matrix C should be updated by the auctioneer every time, as the cognitive users is continuously moving. In case of multi-winner auction in cognitive adhoc network, there are two important collusion attacks namely sublease collusion [12] and loser collusion [13] which reduces the efficiency of the VCG mechanism and second price auction mechanism.

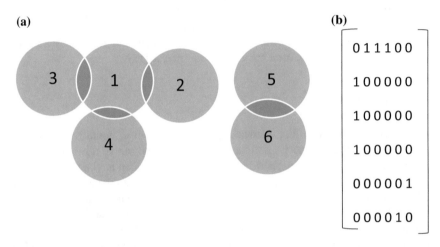

Fig. 1. Interference structure of the system model. (a) System model; (b) Adjacency matrix

3 VCG Mechanism

Auction mechanism is a game theory in which one analyses the interactions among auctioneer (primary user) and bidders (secondary users) and decides the auction outcome on the basis of bids and auction rules. When more than one spectrum band is leased to the cognitive users each cognitive user who bids to each spectrum band individually. This type of bidding by cognitive users is a combinatorial auction [11]. The cheating behavior of some cognitive users reduces cooperation among all the cognitive users and decreases the spectrum usage [14]. But the VCG mechanism got several drawbacks when it is applied to multi-winner auction makes it less efficient (Fig. 2).

Algorithm for VCG Mechanism: Step1: Among all the cognitive users who compete for occupying the spectrum, the winner set of the cognitive users is obtained by solving binary integer programming (2):

$$\max_{y} U_a(y) = \sum_{i=1}^{L} a_i y_i$$

$$\text{s.t. } a_i + a_j \leq 1, \forall i, j \text{ if } C_{ij} = 1$$

$$a_i = 0 \text{ or } 1 \quad i = 1, 2, \ldots, L$$

(2)

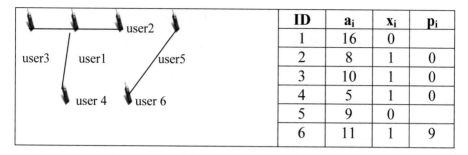

ID	a_i	x_i	p_i
1	16	0	
2	8	1	0
3	10	1	0
4	5	1	0
5	9	0	
6	11	1	9

Fig. 2. Case study for VCG mechanism

The best of all the compactable combinations is formulated using binary integer programming problem for L variables in (2) to obtain the maximum utility obtained is given by $U_a^* = U_a(y^*)$. In order to calculate the prices for the VCG, and to determine the seller revenue value of VCG, assume if any user i of winner set is absent and remaining users except user i is present in the system, the system utility for this case is given by U_{a-i}^*. The system utility can be computed by solving (2) with a_i replaced by a_{-i}. Where $a_{-i} = [a_1, a_2 \ldots a_{i-1}, a_{i+1} \ldots a_L]$ with the i-th entry of the user i is excluded. Similarly when a user i exist in the system utility is given by U_i^*.

3.1 The Second Price Auction Mechanism

In the second price auction mechanism in which the winner among the secondary users is the one bidding the highest value compared to all the remaining users. And the winner need not pay his/her bided value, instead he/she pays second highest bid to the system. Further this mechanism motivates the players to bid higher values to occupy the spectrum and also enforces the player's truth telling behavior.

4 Proposed Auction Based Game Theory

We have formulated the VCG mechanism to reduce the collusion among the cognitive users with added interference and power constraints in the selection of the auctioneer. In our proposed method, selection of the auctioneer is done to reduce the possibility of the auctioneer to cheat the buyers(cognitive users). These additional constraints included in the selection of the auctioneer not only reduces the cheating behavior of the auctioneer, but also it increases the fairness among the buyers to take turns in being an auctioneer.

4.1   Algorithmic Evaluation of the Proposed Mechanism

In the proposed algorithm first, we select auctioneer based upon the power constraint.

$$(p_1, p_2, \ldots, p_N) = \max(p_i) \tag{3}$$

A. Steps for proposed mechanism:

1. Each node sets the power levels, when a node satisfies Eq. (3) it is selected as auctioneer for that period of time.
2. Interference matrix is updated for the nodes based upon the distance vector.
3. Each node bids value to the auctioneer for occupying the spectrum hole leased from the primary user.
4. With the steps (2) and (3) as a constraint in selecting the auctioneer, winner set is formulated using binary integer programming.
5. Loser set pricing strategy is formulated by the winner set.
6. The winner set pays the loser set pricing to the auctioneer for reducing the collusion from the loser set.

We group all the secondary users of negligible interference together and form set of virtual bidders. From the case study the virtual bidders are $v(\{1\}) = 16$, $v(\{2\}) = 8$, $v(\{3\}) = 10$, $v(\{4\}) = 5$, $v(\{5\}) = 9$, $v(\{6\}) = 11$, $v(\{2,4\}) = 13$, $v(\{3,4\}) = 15$, $v(\{5,6\}) = 20$, $v(\{2,3,4\}) = 23$, $v(\{2,3,4,5,6\}) = 43$. These eleven virtual bidders are hidden competition to occupy the spectrum. As in second price mechanism the virtual bidder with highest bidding value is allowed to access the spectrum. When any two bidders quote the same bidding value in that case, randomly any one bidder among them gets chance to occupy the spectrum. Total payment by the winner is the highest bid value made by the virtual bidders. This can be solved by formulating two binary integer programming in steps, without showing the virtual bidders explicitly. In step (i) the solution for optimized allocation is obtained for the winner set W using Eq. (2). In step (ii) the winner of the step (i) is removed from the system and taking the remaining players into consideration the best allocation strategy is again formulated to calculate the system utility U_{a-W}^*. The winners of the proposed mechanism needs to pay U_{a-W}^* in total to occupy the spectrum. This problem could be resolved by using the Nash bargaining solution [14]. In the proposed auction mechanism, bargaining for an individual is not necessary because the prices are set for the bands available and this value is fixed. Therefore the solution to this proposed mechanism is given in (4).

$$\max_{s_i, i \in W} \prod_{i \in W} (a_i - p_i) \tag{4}$$

such that $\sum_{i \in W} p_i = U_{a-W}^*$ for $0 \leq p_i \leq a_i$ colluders raise their bid values more than U_{a-W}^*.

The winner pays more for even the band worth amount. The losers or colluders can't bid that much value more for the band worth for. Therefore this mechanism eliminates the collusion in VCG mechanism. For example the winners set pays a price of $U^*_{a-w} = 10$ to occupy the band. Even if losers have the motivation to collude the winner set they need to pay more than U^*_{a-w}. However the price of winner set is already very high than the worth of the band is. So the colluders can't afford that much amount to produce the collusion.

5 Simulation Results

In this section, performance of the proposed mechanism is compared with other generalized mechanisms like VCG, Second Price Auction. Consider the area of 1000×1000 m², in which L secondary users are deployed randomly based upon uniform distribution. The power values for the nodes are generated in the range of [50 100]. Figure 3 shows the randomly generated nodes in the specified area. Interference matrix is formulated based upon the location of all the nodes in the restricted area. Each of the nodes provides the information about its location and the neighboring nodes location to auctioneer for updating the L × L adjacency matrix C.

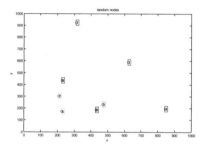

Fig. 3. Nodes deployment in the specified area.

Assuming each cognitive user is deployed in center of a coverage radius of R_i-meter. If any two secondary users are away twice the radius they exchange the spectrum band without any interference. The bidding values of each cognitive users are generated randomly based upon uniform distribution in between the interval [20 30]. We consider that the primary user leases one band for auction, i.e. G = 1 means to the cognitive user. Figure 4 shows the seller's revenue of the primary user for different auction mechanisms for R_i. When nodes are grouped based upon the minimum distance. Each player gains maximum payoff and reduced power utilization.

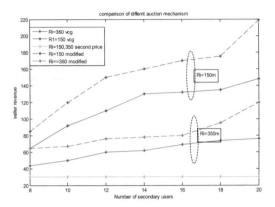

Fig. 4. Primary user revenue for different auction mechanisms

From Fig. 5, the performance based upon system utility for the proposed mechanism and the existing VCG mechanism is compared.

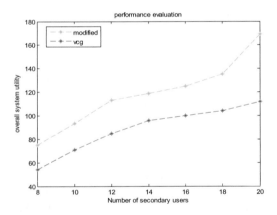

Fig. 5. Performance of system utility for the proposed mechanism and VCG mechanism

From Fig. 6, when the players are more in number the probability of an auctioneer to be an malicious auctioneer is drastically reduced. When more players enter the game (N = 10, 12, 14, 16, 18, 20 …) the percentage of a node to be malicious is drastically reduced. The graph is simulated for both coverage radius of r_i = 150 and 350. From Fig. 7, in our proposed mechanism, processing time is comparatively lesser than the existing VCG mechanism. For example when number of users (N = 8) are eight the processing time is comparatively reduced to about 5% from the existing VCG mechanism.

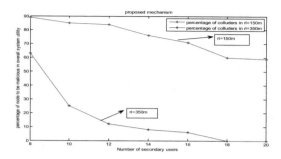

Fig. 6. Percentage of malicious node in overall system utility

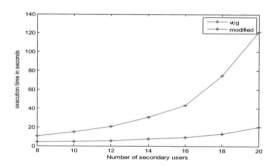

Fig. 7. Comparison plot between proposed and VCG mechanisms based on execution time.

6 Conclusion and Future Work

In this paper, we have proposed pricing based auction mechanism in dynamic spectrum allocation for cognitive adhoc networks to resolve the collusion problems among the cognitive users to occupy the primary user spectrum. This proposed mechanism achieves higher seller revenue which motivates the primary users to lease their bands to cognitive users. In this work we have reduced the collusion among the cognitive users by formulating a pricing based auction mechanism. We have also introduced the additional constraints in selection of the auctioneer to reduce the malicious behavior of the auctioneer. The future extension can be done based on different strategies like power etc.

References

1. Mitola III, J.: Cognitive radio: an integrated agent architecture for software defined radio. Ph. D. thesis, KTH Royal Institute of Technology, Stockholm, Sweden (2000)
2. Akyildiz, I.F., Lee, W.Y., Vuran, M.C., Mohanty, S.: NeXt generation/dynamic spectrum access/cognitive radio wireless networks: a survey. Comput. Netw. **50**, 2127–2159 (2006)

3. Ji, Z., Liu, K.J.R.: Dynamic spectrum sharing: a game theoretical overview. IEEE Commun. Mag. **45**(5), 88–94 (2007)
4. Zheng, H., Peng, C.: Collaboration and fairness in opportunistic spectrum access. In: IEEE International Conference on Communications, ICC 2005, Seoul, pp. 3132–3136, May 2005
5. Wang, B., Ji, Z., Liu, K.J.R.: Primary-prioritized Markov approach for dynamic spectrum access. In: IEEE Symposium on New Frontiers in Dynamic Spectrum Access Networks (DySPAN 2007), Dublin, pp. 507–515, April 2007
6. Ileri, O., Samardzjia, D., Mandayam, N.B.: Demand responsive pricing and competition spectrum allocation via a spectrum server. In: IEEE Symposium on New Frontiers in Dynamic Spectrum Access Networks (DySPAN 2005), Baltimore, pp. 194–202, November 2005
7. Huang, J., Berry, R., Honig, M.L.: Auction-based spectrum sharing. Mob. Netw. Appl. **11**(3), 405–418 (2006)
8. Kloeck, C., Jaekel, H., Jondral, F.K.: Dynamic and local combined pricing, allocation and billing system with cognitive radios. In: IEEE Symposium on New Frontiers in Dynamic Spectrum Access Networks (DySPAN 2005), Baltimore, pp. 73–81, November 2005
9. Gandhi, S., Buragohain, C., Cao, L., Zheng, H., Suri, S.: A general framework for wireless spectrum auctions. In: IEEE Symposium on New Frontiers in Dynamic Spectrum Access Networks (DySPAN 2007), Dublin, pp. 22–33, April 2007
10. Wu, Y., Wang, B., Liu, K.J., Charles Clancy, T.: A multi-winner cognitive spectrum auction framework with collusion-resistant mechanisms. In: IEEE Symposium on New Frontiers in Dynamic Spectrum Access Networks (DySPAN 2007), pp. 1–9, October 2008
11. Cramton, P., Shoham, Y., Steinberg, R.: Combinatorial Auctions. MIT Press, Cambridge (2006)
12. Sodagari, S., Attar, A., Bilén, S.G.: On a truthful mechanism for expiring spectrum sharing in cognitive radio networks. IEEE J. Sel. Areas Commun. **29**(4), 856–865 (2011)
13. Niyato, D., Hossain, E.: Competitive pricing for spectrum sharing in cognitive radio networks: dynamic game, inefficiency of Nash equilibrium, and collusion. IEEE J. Sel. Areas Commun. **26**(1), 192–202 (2008)
14. Osborne, M.J.: An Introduction to Game Theory. Oxford University Press, Oxford (2004)
15. Yi, C., Cai, J.: Two-stage spectrum sharing with combinatorial auction and stackelberg game in recall-based cognitive radio networks. IEEE Trans. Commun. **62**(11), 3740–3752 (2014)
16. Xu, L.: Joint spectrum allocation and pricing for cognitive multi-homing networks. IEEE Trans. Cogn. Commun. Netw. **4**(3), 597–606 (2018)
17. Eraslan, B., Gozupek, D., Alagoz, F.: An auction theory based algorithm for throughput maximizing scheduling in centralized cognitive radio networks. IEEE Commun. Lett. **15**(7), 734–736 (2011)
18. Smys, S., Thara Prakash, J., Raj, J.S.: Conducted emission reduction by frequency hopping spread spectrum techniques. Natl. Acad. Sci. Lett. **38**(3), 197–201 (2015)

Efficient Abandoned Luggage Detection in Complex Surveillance Videos

Divya Raju[✉] and K. G. Preetha

Department of Information Technology,
Rajagiri School of Engineering and Technology,
Rajagiri Valley, Kochi 682037, India
divyarajul996@yahoo.com,
preetha_kg@rajagiritech.edu.in

Abstract. Abandoned luggage is seen as a major threat as they may carry explosives and smuggled goods and it affects the lives of normal people. It is necessary to have an automated surveillance system that provides alerts about potential threats in the environment. Abandoned luggage detection suggested in this paper automatically detects abandoned luggage items using camera installed in a particular scene and produces a real time alarm. Image processing techniques are applied on multiple video frames to detect objects, which are recognized as luggage using convolutional neural networks (CNN). If a luggage is detected as unattended for a huge amount of time an alarm sound will be raised.

Keywords: Static object detection · Luggage recognition · Convolutional neural network · Background subtraction · Image processing

1 Introduction

The importance of abandoned luggage detection arises in highly crowded public areas like railway stations, airports, malls etc. Most of the terrorist attacks are carried out using different types of bombs. Luggage bombs can easily be detected when compared to the remaining types of bombs like vehicle bombs and suicide bombs. Baggage bombs can be prevented as there is enough time to evacuate from the area. Suspicious items can hence be removed from the scene with the help of abandoned luggage detection system. Abandoned luggage can also carry prohibited items, smuggled goods, ancient artifacts or items of biological warfare. The basic aim of abandoned luggage detection system is to identify possible threats in the environment. Negligence of humans can cause errors in video data analysis as huge amounts of video data are generated. These problems can be avoided by using efficient automated detection systems that produces alarm whenever an abandoned luggage is detected.

The rest of the paper is organized as follows. Section 2 reviews the literature in the abandoned object detection. Proposed method is presented in Sect. 3. Results and discussion is given in Sect. 4 and a conclusion in Sect. 5.

© Springer Nature Switzerland AG 2020
J. S. Raj et al. (Eds.): ICIDCA 2019, LNDECT 46, pp. 181–187, 2020.
https://doi.org/10.1007/978-3-030-38040-3_21

2   Related Research

Automatic luggage detection is an interesting research area for the last decade in the research community. Manual detection of abandoned luggage is not accurate due to the weariness of humans. Existing methods employed for automatic abandoned luggage detection is presented below.

The method proposed in [2] detects static objects in the video using background subtraction and motion estimation. İnitially, background subtraction is done by subtracting a roughly calculated background from the video frames. The foreground mask will contain objects in motion as well as the static objects. To detect moving objects, motion is calculated by subtracting frames that are five frames apart from each other. İn the final stage luggage recognition is done by using cascade of convolutional neural networks (CCNN).

The authors in [3] describes five stages for abandoned luggage detection. İnitially blurring is done to eliminate image noise. İntended objects are then retrieved from the background using thresholding. Blob detection is done to recognize images with dissimilar image features like brightness, contrast etc. Next, motion is estimated by identifying the position of the object in the current and previous frames. Finally for accurate results human beings are distinguished from abandoned objects by employing RGB to HSV conversion.

A deep learning technique called R-CNN to detect unattended objects is described in [4]. It is a region proposal method that identifies different regions of interest and it uses CNN to find the presence of object within that region. In R-CNN selected frames are fed into a CNN which acts as a feature extractor. The extracted features are then fed into a SVM to classify the presence of objects.

A comparison of the above methods is presented below. Object detection is the first step in abandoned luggage detection. The list of objects detected shouldn't contain items that are already in the background of the scene. The technique presented in [2] is very efficient in object detection as it subtracts the roughly calculated background from each video frame. Motion detection in [3] is less effective than the other methods as it considers only the current and previous frames whereas other two methods take into consideration multiple frames. Human detection is highly important as a person who hasn't moved in multiple frames is also detected as an abandoned object. Human detection used in [3] that uses skin colour thresholding provides less accurate results as it may identify any object with skin colour as human. The methods described in [3] and [4] that uses convolutional neural networks are more accurate.

3   Proposed Method

The proposed system consists of two stages for detecting abandoned luggage items. Initially, all the steady objects in the scene are detected.

3.1 Static Object Detection

Video from the installed camera is divided into image frames. Static object detection (SOD) is done using background subtraction and foreign object detection. The detailed architecture of SOD is given in Fig. 1. Background subtraction is used to subtract the calculated background from each video frame. Foreign object detection identifies contours in the frame and the centroids of the different contours are tracked. Centroid is calculated using moments as shown in Eqs. 1 and 2.

$$Cx = M10/M00 \tag{1}$$

$$Cy = M01/M00 \tag{2}$$

Objects are tracked with respect to the sum of the centroids $Cx + Cy$. If the object appears in more than 300 consecutive frames then it is identified as a static object. A contour dictionary is maintained for storing the static object along with the last frame in which the object was static. If a static object is moved after a while then the entry is removed from the contour dictionary. In the next stage, the static objects tracked are further processed to identify if it is luggage or not.

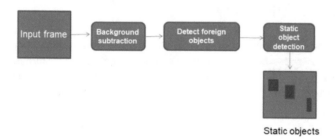

Static objects

Fig. 1. Static object detection

3.2 Luggage Recognition

A second stage is mainly used to recognize the static objects detected. In order to identify the static luggage objects a convolutional neural network (CNN) is used. Bounding box is used to crop out the image of the static item to detect if it is a luggage or not. The CNN is trained with positive samples of luggage items, e.g. hand bags, backpacks, suitcases. If the object is recognized as a luggage then an alarm is raised. The process discussed is represented in Fig. 2.

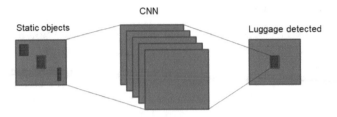

Fig. 2. Luggage recognition

Convolutional neural network (CNN) is used for image recognition. It takes an input image which is the static object detected. It then processes it and classifies it either as luggage or non-luggage item. The neural network consists of input, output and different hidden layers. It is a sequential model that stacks sequential layers of the network in order from input to output. A 2D convolutional layer processes the input images. A max pooling layer is used to down-sample the input representation. Dropout layer prevents over fitting and flatten layer flattens the matrix into vector. Finally it outputs the class using an activation function and classifies images.

4   Results and Discussions

The abandoned luggage detection system was trained with various images and tested with sample test cases. Initially the estimated background frame is subtracted from each video frame. Both the video frame and the background image is converted into grayscale images before subtracting it. The background subtracted frame is depicted below in Fig. 3.

Fig. 3. Frame after background subtraction [1]

From a grayscale image, thresholdinig is used to create binary images. Binary image is created by converting all the pixels under a certain value to 0 and all the others to 1 (Fig. 4).

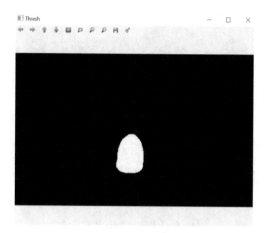

Fig. 4. Threshold image

Foreign object detection (Fig. 5) identifies contours in the frame and the centroids of the different contours are tracked. Using the contours bounding boxes are drawn around the foreign objects.

Fig. 5. Example foreign objects in the frame

The object is identified as static when it retains for 300 frames in a row. The static object detected is shown in Fig. 6.

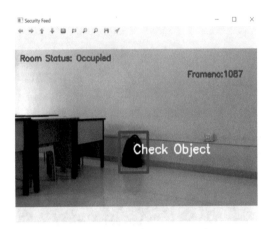

Fig. 6. Static object detected

These static objects are recognized as unattended luggage items by using a CNN. A classification report is used to examine the quality of classification by CNN. The alarm is produced once the object is recognized as luggage. A classification report represented in Table 1 shows the main classification metrics on a per-class basis.

Table 1. Classification Score Report

	Precision	Recall	f1-score	Support
Class 0 (BAGS)	0.90	0.98	0.93	132
Class 1 (NOT BAG)	0.93	0.73	0.82	55
Micro avg	0.90	0.90	0.90	187
Macro avg	0.91	0.85	0.88	187

5 Conclusion

Abandoned luggage detection plays an important role in places with high security alerts. The system is used to detect abandoned luggage in public areas using surveillance cameras. Image processing is applied on multiple video frames to detect static objects. Then humans from arbitrary objects are distinguished and security alarm is raised. The method discussed can be applied in crowded areas for abandoned luggage detection as it reduces false alarms.

Compliance with Ethical Standards

All author states that there is no conflict of interest.
We used our own data.

References

1. Chang, J.-Y., Liao, H.-H., Chen, L.-G.: Localized detection of abandoned luggage. EURASIP J. Adv. Signal Process. **2010**, 9 (2010). Article ID 675784
2. Smeureanu, S., Ionescu, R.T.: Real-Time Deep Learning Method for Abandoned Luggage Detection in Video, University of Bucharest, 14 Academiei, Bucharest, Romania (2018)
3. Bangare, P.S., Yawle, R.U., Bangare, S.L.: Detection of human feature in abandoned object with modern security alert system using Android Application. In: 2017 International Conference on Emerging Trends & Innovation in ICT (ICEI) Pune Institute of Computer Technology, Pune, India, 3–5 February 2017 (2017)
4. Pham, N.T., Leman, K., Zhang, J., Pek, I.: Two-stage unattended object detection method with proposals. In: 2017 IEEE 2nd International Conference on Signal and Image Processing (2017)
5. Lin, K., Chen, S., Chen, C., Lin, D., Hung, Y.: Left-luggage detection from finite-state-machine analysis in static-camera videos. In: Proceedings of ICPR (2014)
6. Ferryman, J., Hogg, D., Sochman, J., Behera, A., Rodriguez-Serrano, J.A., Worgan, S., Li, L., Leung, V., Evans, M., Cornic, P., et al.: Robust abandoned object detection integrating wide area visual surveillance and social context. Pattern Recogn. Lett. **34**(7), 789–798 (2013)
7. Wen, J., Gong, H., Zhang, X., Hu, W.: Generative model for abandoned object detection. In: Proceedings of ICIP, pp. 853–856 (2009)
8. Auvinet, E., Grossmann, E., Rougier, C., Dahmane, M., Meunier, J.: Left-luggage detection using homographies and simple heuristics. In: Proceedings of the 9th IEEE International Workshop PETS, pp. 51–58 (2006)
9. Luo, W., Xing, J., Milan, A.: Multiple object tracking: a literature review. arXiv (2017)
10. Lin, K., Chen, S.C., Chen, C.S., Lin, D.T., Hung, Y.P.: Abandoned object detection via temporal consistency modeling and back-tracing verification for visual surveillance. IEEE Trans. Inf. Forensics Secur. **10**(7), 1359–1370 (2017)
11. Bhargava, M., Chen, C.C., Ryoo, M.S., Aggarwal, J.K.: Detection of object abandonment using temporal logic. Mach. Vis. Appl. **20**(5), 271–281 (2016)
12. Ren, S., He, K., Girshick, R., Sun, J.: Faster R-CNN: towards realtime object detction with region proposal networks. In: NIPS (2015)

Energy Reduction Stratagem in Smart Homes Using Association Rule Mining

Karthika Velu[✉], Pramila Arulanthu, and Eswaran Perumal

Computer Applications, Alagappa University, Karaikudi, India
mkmoshi2016@gmail.com, pramimark@gmail.com,
eswaran@alagappauniversity.ac.in

Abstract. Electricity price forecasting is performing a challenging role. The basic purpose of this study is energy diminution which saves forcefulness in smart homes without threatening the comfort of living soul. Due to high demand of energy resources, pivotal challenges like increasing cost, decreasing resources, lack of stability in electricity consumption has to be perceived. An increase or diminution problem can be solved with efficiency using evolutionary algorithmic rule. It presents an algorithmic program that mines user activeness and usage data then applies association regulation mining to reduce the energy absorption of their domestic. The system looks for predominant structure in the event data furnished by the home equipment system. The dataset used in this work is one apartment's one year electricity power consumption average details. The proposed **Association Rule Mining (ARM)** discusses how usage structure and predilection of inhabitants can be learned expeditiously. And also this mechanism specifies the way to allow the smart homes to self-governing and to achieve energy savings in future. Association between power consumption and temperature of the circumstances assists to obtain the maximum accuracy in this research. This system can be deployed to any other set of test locations like hospital, organizational buildings, etc., and it gives a methodological analysis for reducing the ingestion of electrical energy for house hold area.

Keywords: Energy reduction · Consumer behavior · Association Rule Mining · Energy consumption · Smart home

1   Introduction

Each and every home has plenty of home appliances for day to day work. Increasing of world's population energy efficiency buildings are the importunate and efficient factor. Energy efficiency and renewable energy are the twin towers of the sustainable energy resources. In many countries energy consumption is the most pivotal aspect. Solution to this kind of problems has been implemented here. The electricity producing plants are really unidirectional due to the centralized approach. The smart grids are the bi-directional and it is an opportunity to optimize consumption cost with Peak to Average Ratio (PAR) reduction. The main aim of this study is over all energy savings and to increase energy sustainability. It's better to meet artificial intelligence to make a smart thing. Smart home product anticipates a user's needs and preferences, it become more precious to a user's life and comfort.

© Springer Nature Switzerland AG 2020
J. S. Raj et al. (Eds.): ICIDCA 2019, LNDECT 46, pp. 188–193, 2020.
https://doi.org/10.1007/978-3-030-38040-3_22

Association rule mining plays an identical, essential part in employ such as catastrophe management, water management, assemblage analysis, land usage observation, weather condition monitoring, power usage observation, solid waste administration and many others. It describes how the energy ingestion of a convenience is related to the state that the user's performance, temperature of that location and the time utilization of an appliances. Furthermore, the knowledge to automatically derive substance about the normal usage of an instrumentality can also enable more brilliant systems that alert users of anomalous activity.

The atmosphere also takes responsibility to the consumption of the energy usage by the living soul. Appliances used in the home based on the factors of that particular circumstance. Energy reduction is possible when we associate the home appliances based on the temperature.

2   Background Investigation

In [1] Baig et al. expressed the well-organized capacity management system to protect energy. It consists two parts namely Energy Management Center (**EMC**) which is a User interaction medium and Capacity Scheduling which is used Single Knapsack Problem. EMC software developed in LABVIEW and load scheduling is completed by MATLAB. Hardware model is specified as a Human Machine Interface (**HMI**). The approach read the data by HMI and sends it to EMC. Zigbee transceiver is acted as an interface.

In [2] Blumandori characterized the styles and experiments in sustainable smart homes strategy. Smart home technology plays a vital role in the society. A sustainable home should give the comfort in all the ways such as water and energy. Sustainability and its technology have been explained.

In [3] Roda, explained the earlier and upcoming technologies related to Smart Home. It used to afford voltage permanency. It reduced electricity usage cost and GHG (Green House Gas) secretions. The future of this skill is Smart Grid for communication. Finally the customer definitely has the chance to take smart home energy management dynamism to reduce electricity cost.

In [4] Dickmann proposed a digital STROM structure used to communicate with the sectors. The Meter which is a concentrator in the board acts as a communication medium. The smart meter measures the electricity of a household. When the power consumption of a particular device has been monitored, it is easy to control with the certain conditions. Based on the load profile concept this methodology helped to optimize the energy consumption.

In [5] Rashidi et al. summarized an approach for finding the patterns of a household applications by using mining methods and cluster those patterns for compact representation. Sensor ID has been initiated. Visualization components implemented in order to help the enhanced apprehend the activity patterns.

In [6] Cook presented the approach for sequential prediction based on the density techniques. It is fully an algorithm based approach. This approach used for smart environment which helps the living soul to predict and make it as an automated system. Consecutive prediction obtained based on the consecutive events.

2.1   Proposed Work

This structure specifies a working model of this prediction approach (Fig. 1).

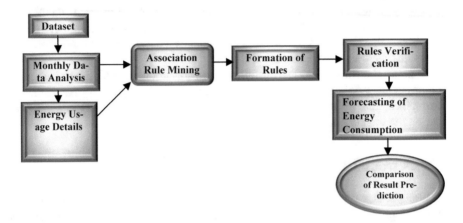

Fig. 1. Proposed system architecture.

The preliminary task is to load the dataset (which contains the details of average electricity usage). The dataset used in this work is apartment's one year electricity average consumption details. In that monthly usage data, applies the association rule mining with certain criteria. Energy consumption details used to predict the usage details based on the temperature. Finally the obtained result is compared with the prevailing methodology for the performance comparison.

2.2   Methodology

2.2.1   Pattern Mining Methodology

The system projected excavation of patterns from a historical event collection from an energy consumption of a home. A pattern may be a set of items, subsequences and any other criteria which is related to the dataset. The quality of the pattern excavation algorithm and the pattern identification is elucidated here.

2.2.2   Data Analysis

Event data is remarkableness in determination of regular and intermittent patterns. And this is referred to as natural covering frequent and natural event in sequences of excavation action acknowledgment. In this formulation, the algorithmic rule starts by excavation patterns with a minimal length of two active data, which is prolonged until the algorithmic rule is not able to find frequent pattern any longer. Data from the dataset analyzed based on the temperature of the circumstance.

2.2.3   Definition and Identification of Patterns

Temperature and Hour are the leading attributes of this work. When the average of temperature upraised then the usage of power consumption also increased. Analysis of the information discovered that the power measuring data is not transcribed as incisively. And this data is an indispensable for obtaining the power ingestion of a single device.

2.2.4 Identification in Relevant Types of Behavior Patterns

All predominant or intermittent patterns result in energy savings is not possible to take for prediction approach. Only some relevant patterns are constrained some characteristics to determine the under consideration behavior patterns. Once the lowest consumption of energy pattern has been obtained for prediction, then the pattern takes as relevant patterns for further consideration. To guarantee that a relevant structure container differs utilized to suggest a human activity, it must be imperturbable of two main constituents:

(i) An applicable pattern must incorporate at least one action to lower physical phenomenon usage.
(ii) The decoration must consist of mean events (not actions), which serve as a precondition to intimate the activeness at the right period of time.

2.3 Algorithm Description of Proposed Work

2.3.1 Association Rule Mining

Association rules are useful for predicting user behavior. Therefore this methodology proposes the user to execute the lower physical phenomenon usage. Because one-event precondition is insufficient to suggest an action, the general length of an applicable structure must be at least three psychological features (one activeness and two normal psychological features). Consequently, a relevant structure can be defined as a patterns that is thirsty than two psychological feature and incorporates at least one action to lower physical phenomenon usage. An affiliation rule is an entailment of the form

$P \rightarrow Q$, where P, $Q \subset R$,
And $P \cap Q = \emptyset$, where
P is a chronological sequence of average events,
Q is individual activeness,

And R is the set of all achievable events. The affiliation rule states that when P occurs, Q passes off with certain quantity. To reiterate, the types of activeness patterns that are necessary to command some activity must fulfill the shadowing criteria:

The structure must occur oftentimes and/or sporadically in the information.

The structure must be applicable for action anticipation and incorporate at least one activeness to lower berth energy usage.

The structure must have a minimum length of three events, exclude normal psychological feature and actions.

2.4 Result and Discussion

The anticipated result obtained by using Association Rule mining. The implementation section mentioned below. Figure 2 specifies the energy consumption of a particular month on a particular day based on the temperature. Figure 3 illustrates that the power consumption during summer has been increased. Obviously during summer days power consumption should increase based on the temperature increase and the usage.

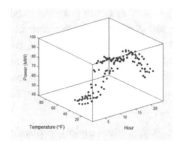

Fig. 2. Plots represent energy consumption based on the dataset in ARM.

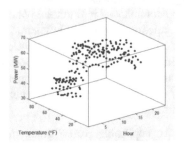

Fig. 3. Plots represent energy consumption based on the dataset in ARM during summer.

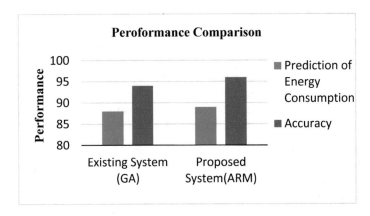

Fig. 4. Performance comparison between the existing and the proposed system

To obtain the desired result in energy saving, the implementation task has been done using MATLAB tool. Association among home appliances helps to achieve a preferred outcome. Performance analysis has been noticed in Fig. 4 with the prediction of energy consumption and accuracy. Based on this prediction methodology we can get some idea to save energy in the household.

2.5   Conclusion and Future Trends

It presents a case study of a system that brings forth to prevention forcefulness in a smart location without reduction the comfort of the inhabitant. This approach used to reduce energy consumption in a home. The usage of power definitely changed based on the temperature in the house. The period of time between two physical phenomenal (or the action) is thoughtful either by the excavation algorithmic rule or by the grouping technique. Time of day, Unit of power consumption and the season based on this either or three only the prediction approach met the desired result.

We can implement this methodology to any other industrial areas, educational institutions, hospital and other public buildings to protect energy. In future there is a scarcity in a resources certainly. Due to this problem the energy forecasting in building need the improvement. For that this article helps us to protect energy.

Acknowledgment. This article has been written with financial support of RUSA–Phase 2.0 grant sanctioned vide Letter No. F. 24-51/2014-U, Policy (TNMulti-Gen), Dept. of Edn. Govt. of India, Dt. 09.10.2018.

References

1. Baig, F., Mahmood, A., Javaid, N., Rassaq, S., Khan, N., Saleem, Z.: Smart home energy management system for monitoring and scheduling of home appliances using zigbee. J. Basic Appl. Sci. Res. **3**, 880–891 (2013)
2. Blumendorf, M.: Building sustainable smart homes. In: Proceeding of the First International Conference on Information and Communication Technologies for Sustainability, February 2013
3. Roda, C.: Home energy management system. A Technical report, May 2014
4. Dickmann, G.: DigitalSTROM: a centralized PLC topology for home automation and energy management (2011)
5. Rashidi, P., Cook, D.J.: Mining and monitoring patterns of daily routines for assisted living in real world settings (2010)
6. Cook, D.J.: Online sequential prediction via incremental parsing: the active Lezi algorithm, February 2017
7. Koch, S., Meier, D., Zima, M., Wiederkehr, M., Andersson, G.: An active coordination approach for Thermal household appliances-local communication and calculation tasks in the household, Bucharest, Romania, 28th June–2nd July (2009)
8. Sagi, M., Mijic, D., Milinkov, D., Bogovac, B.: Smart home automation. In: 20th Telecommunication Forum TELEFOR 2012, 20–22 November 2012
9. Khalid, M.U., Javaid, N.: An optimal scheduling of smart home appliances using heuristic techniques with real time co-ordination, Institute of Information Technology, Islamabad
10. Teng, R., Yamazaki, T.: Load profile-based coordination of appliances in a smart home. IEEE Trans. Consu. Electron. **65**(1) (2019)
11. Zhou, B., Li, W., Chan, K.W., Cao, Y., Kuang, Y., Liu, X.: Smart home energy management system concepts, configurations, and scheduling strategies. Renew. Sustain. Energy Rev. **61**, 30–40 (2016)
12. Hussain, B.: An innovative heuristic algorithm for IOT enabled smart homes for developing. CIIT/FA/ISB, Comsats University Islamabad (2018)

Analysis of Growth and Planning of Urbanization and Correlated Changes in Natural Resources

Supriya Kamoji, Dipali Koshti$^{(\boxtimes)}$, and Ruth Peter

Fr. Conceicao Rodrigues COE, Mumbai, India
{supriyas,dipalis}@fragnel.edu.in,
ruthpeter2490@gmail.com

Abstract. Ever since the earth was inhabited, all the living beings are dependent on air, water and land that exist freely in nature to survive. However these natural resources are getting rapidly exhausted due to a number of factors like over population, excess land usage, climate change, environmental pollution etc. To have an environmentally acceptable secure future, where we can still enjoy natural resources, we need to transform the way we use natural resources. The proposed system uses the deep learning model U-Net to classify satellite tiles in land categories like forest, water bodies and urban area. The system is also capable of analysing satellite images to study the decline of various natural resources through the years and provide analysis of decline in natural resources. Based on analysis, recommendations regarding alternate solutions to urban planning can be provided.

Keywords: Natural resources · Semantic segmentation · U-Net · Deep learning · Computer vision

1 Introduction

Natural resources are important for the development of any country. However, with increasing population, they are getting scarce and hence it is very essential to conserve them. The current system for analysis and development of a city is dependent on manual analysis and surveys are done by sending people to particular regions. This system, however, is not only time consuming but also very expensive. Hence the proposed system uses satellite tiles to identify natural resources near cities for effective utilization. Since Landsat data is readily available [1], the proposed system provides a cost effective solution to the existing problem by analysing these data. By analysing the data, we aim to study the impact on the growth of a city causes on its surrounding natural resources.

Given a satellite tile of a particular region, the system classifies the tile into three categories: Forest, water bodies and urban. If we provide a satellite tile of the same area (or region) for past few years, then the system also generates graph showing changes occurred in respective categories. For example, decline in water bodies or forest area in

© Springer Nature Switzerland AG 2020
J. S. Raj et al. (Eds.): ICIDCA 2019, LNDECT 46, pp. 194–203, 2020.
https://doi.org/10.1007/978-3-030-38040-3_23

the given region can be easily identified from the graph. This data can be used for planning cities, making sure that these areas are kept under consideration.

2   Literature Review

Albert, Kaur, Gonzalez [2], have applied deep convolutional neural network on large scale satellite imagery data obtained from Google Map Static APIs and analyzed land usage patterns in urban environment. The authors have done analysis of 6 example cities in Europe. They have used VGG-16 and ResNet with transfer learning to classify cities into 10 urban environmental classes.

In [3] authors have envision the application of Convolution Neural network on high resolution satellite imagery of a city to explore different socio economic regions, especially slum areas. They have used Convolutional neural network to extract features from high resolution satellite tiles of a city. Based on these features the images were organized into visually similar groups. The grouping was done by using k-means clustering. The validation of model was done using demographics data of Mumbai. And finally, they applied this model to 6 different cities across the world to test the system.

Chhor, Aramburu, Bougdal-Lambert [4], focus on identifying buildings using satellite images. This application can be used to assess building damages in case of a natural disaster. It can also be used to formulate an appropriate plan for solar manu-facturers who want to evaluate the available roof surface in a particular region. The authors examined an end-to-end approach for semantic segmentation of satellite images for building detection using few data of relatively low resolution. They implemented a CNN based on U-Net architecture and used the Mapbox API in OpenStreet to collect dataset. They used high-level API Keras to implement the model and to facilitate data augmentation.

3   Proposed System

The proposed system uses a deep learning model U-net for semantic segmentation of the various landforms. Given a satellite tile, it categories the tile into three regions: Forest, water bodies and urban areas. The system also analyses the data over the years and provides a means for effectively planning urban areas taking into consideration theses natural resources. The outline of the proposed system:
Create Dataset
The dataset used for training is obtained from Google maps. It includes the satellite tiles attached along with its respective masked image which is used as labels for our neural network.

Once the dataset is ready, it needs to be transformed in a usable format. Hence, we resize our images to 256*256 pixels and then convert it to a tensor object to feed it to the neural network model.

Splitting of Data into Training and Testing Data
Next step is to create a train set and test set. We split our dataset into two parts with ratio 80:20. That is, 80% of the data is used for training the model, 10% is used for validation and remaining 10% is for testing the model.

Fitting the Training Data into our Deep learning algorithm
Next step is to train the Deep Learning model using the training data.

Testing the trained model on test data
We apply the stated algorithms and test the algorithm for accuracy, efficiency, and performance.

To Choose the best accuracy and efficiency model
Various parameters of the model such as number of hidden layers, nodes in a hidden layer, number of epochs are varied and the accuracy and efficiency of the model is calculated. Here we come to know which model is better and use that one in the final application.

To Develop the server-based application
We developed an application which calls the Deep Learning model at the back end and is used for semantic segmentation of the map tiles and further analysis is done based on the results. Figure 1 depicts the high level architecture of the proposed system.

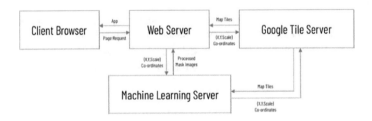

Fig. 1. Overall model of the system

Architecture of the proposed system focuses on building web application which provides us with information about the various geographic features of an area, provide segmentation of the various landforms and further analysis based on the results. The web server requests for processed masked images from Machine Learning server by providing Cartesian coordinates. The Google tile server returns a map tile to the Machine Learning server based on these coordinates.

The final output is presented to the client in the form of overlays on the original map tiles.

3.1 Dataset

The dataset we used is from Google Maps which is readily available at no additional cost. Figure 2 shows a few examples of dataset used.

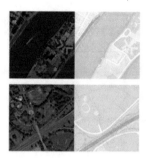

Fig. 2. Google maps dataset

3.2   Exploring Landsat Bands

We combine various Landsat bands from Landsat 8 for analysis of certain type [5]. Figure 3 shows some of the common band combinations applied to Landsat 8.

3.3   Models and Algorithms Used

Deep Learning using Artificial Neural Network(ANN). Deep learning is a hierarchical neural network structure that involves cascading of many layers of processing units. Output from the previous layer is fed as an input to the next consecutive layer. This deep structure learns by itself and extracts features from the given input image [6]. Compared to traditional neural network, deep networks have shown more promising results in almost all the areas ranging from image classification, pattern recognition, segmentation, prediction etc.

Semantic Segmentation. Semantic segmentation involves understanding and recognizing the image at a deeper level i.e. at a pixel level. Each pixel in an image is assigned a class label of the enclosing object [7, 9]. As an example, consider an input image shown in Fig. 4. If we apply a semantic segmentation on this image, the resulting output is shown in Fig. 4. Unlike classification, Semantic segmentation involves not only recognizing different objects but also recognizes objects boundaries. As can be seen in Fig. 4, two objects - the bike and the person riding on it along with their boundaries have been recognized.

Analysis Type	Band 1	Band 2	Band 3
Natural Color	4	3	2
False Color (urban)	7	6	4
Color Infrared (vegetation)	5	4	3
Agriculture	6	5	2
Atmospheric Penetration	7	6	5
Healthy Vegetation	5	6	2
Land/Water	5	6	4
Natural With Atmospheric Removal	7	5	3
Shortwave Infrared	7	5	4
Vegetation Analysis	6	5	4

Fig. 3. Band combinations

 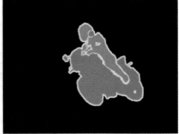

Fig. 4. Input and output image for semantic segmentation

In 2014, Long et al. from Berkeley [8] presented Fully Convolution Networks (FCN), a CNN architecture for dense prediction with no fully connected layers. Using FCN, it was now possible to generate segmentation maps from any arbitrary sized input. Almost all the subsequent approaches on semantic segmentation have adopted FCN model.

The problem of fixed size input was solved by FCN. Another major problem in CNN architecture when used for segmentation was pooling layers. By applying pooling, we down sample the image, thus reducing the width and the height (size) of the image. This helps the filters in the deeper layers to aggregate the larger context, but at the same time the where information is lost. In other words, Pooling helps us to understand what is present in the image but loses the where information. For semantic segmentation we need to retain both what and where information.

Two different solutions evolved in the literature to tackle this problem. The first one is the Encoder-Decoder network and the second one is dilated convolutions.

In the former method, the encoder gradually increases the dimensionality of the input by using pooling layers and decoder gradually reduces the dimensionality by adding pooling layers. Decoding is done to gain back the features that might have been lost during dimensionality reduction. U-Net is a very popular example of this class. The second class of architecture is called dilated convolutions that helps in aggregating knowledge of larger context and do away with pooling layers [9].

UNET for Semantic segmentation as discussed earlier UNET uses fully convolution network model with no dense layer in it. As shown in Fig. 5, the architecture has two paths: Contraction path also called the encoder and the symmetric expansion path also called the decoder. The contraction path captures the context in the given image. The encoder is made up of many contraction blocks with each block having convolution layers followed by max pooling layers. The decoder uses transposed convolution [10] to enable precise localization. Each contraction block applies consecutive two times 3×3 convolution followed by a 2×2 max pooling on an input image. After each contraction block, the number of filters doubles; this helps in extracting more complex features in the given image. During the contraction process, the image size reduces so the feature maps [11]. Expansion path exactly does opposite of contraction. Each expansion block applies consecutive of 2×2 Up-convolution and two times 3×3 Convolution to recover the size of segmentation map. After each expansion block, the number of feature maps gets half, thus maintaining symmetry. This means that we lose the localization information. To regain the localization information (where information), after each up-convolution, the feature map of corresponding contraction layer gets concatenated (shown by Gray arrow). This allows the contraction path to pass the localization information to the corresponding expansion path.

Since the output feature map have only 2 classes, cell and membrane, finally, 1×1 convolution is applied to map the feature map size from 64 to 2. Here we have used unpadded convolution so final output image size is smaller than input size.

4 Results and Conclusions

Figure 6 shows the original images and in Fig. 7 corresponding masked images are shown. The satellite tiles are trained with masked images which is used as labels for our UNET model. Figure 8 shows the model output for the input images shown in Fig. 6. The output is a processed image which is classified as Urban areas, water bodies, forests and roads (Fig. 9).

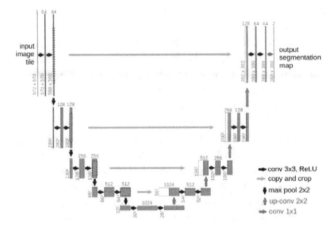

Fig. 5. U-Net architecture

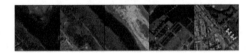

Fig. 6. Original images

Fig. 7. Image masks

Fig. 8. Output images

Fig. 9. Error in prediction

Fig. 10. Distribution of land forms

In order to analyse the different classified regions, we calculate the percentage of each segmented region in a given tile and display it in the form of a pie chart. The pie charts formed over the years for a particular region is shown in Fig. 10 can be compared. In Fig. 10, the green color shows forests areas, blue color shows water bodies and grey shows urban areas.

The user is provided with the UI shown in Fig. 11 with a view of the map so that he can pan in or pan out to any desired tile. This can also be achieved by specifying coordinates in the latitude and longitude input space. Zoom levels from 6 to 18 are also provided to users in order to zoom into a particular tile for better view. When the user selects a particular area, the analysis of that area is displayed along with the current segmented image and a graph depicting the change in the land forms over the years.

Figure 13 depicts the graph for the area of Aarey milk colony in Mumbai from the year 1988 to 2018. As we can see a few errors in the segmentation of the tiles shown in Fig. 12, it has occurred due to unavailability of a good dataset. We have considered another scenario in Fig. 14 from 1984 to 2017 where we have considered the mining scenario in Alberta, Canada where the vegetation in the area has declined drastically. Hence on using our software efficiently we can calculate if urbanization is a feasible option for this region using the past satellite tiles available to us. The inconsistencies in the segmentation is due to the lack of a good dataset.

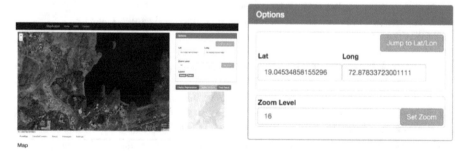

Fig. 11. User interface

Fig. 12. Segmentation over the years for Aarey milk colony

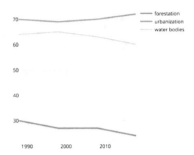

Fig. 13. Graph showing change over the years for Aarey milk colony

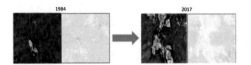

Fig. 14. Deforestation in Alberta

5 Conclusions

In this paper we proposed the use of deep learning U-Net model for statistical monitoring of natural resources with an automated analysis process. Further, the tweaking
parameters such as the learning rate of the U-Net model, the number of epochs, the size
of the dataset and the batch size, we arrived at the nal model with a learning rate of
0.001, number of epochs 100, and the batch size as 5, along with training dataset nearly
as much as 2000 images. This gave us a loss (mean squared error) of 0.0028 on the
training data and a higher loss of 0.023 on testing data. In future we will further try to
segment various kinds of buildings, such as hospitals and police stations.

6 Future Enhancements

Due to the modular structure of all components of the proposed system, new algorithms
can be implemented to achieve better accuracy. A certain neural network with a future
scope which we have looked at is PSPNet [12], also known as pyramid scheming
parsing network which is used for semantic segmentation. It uses FCNs and pyramid
pooling module for pixel level scene parsing. The global pyramid mid pooling feature
provides additional contextual information.

The refineNet [13] also seems to be a promising candidate for semantic segmentation. It iteratively pools features that increase resolutions using special RefineNet
blocks for several ranges of resolutions and finally produces a high resolution segmentation map for semantic segmentation. As compared to our U-net model, it uses
input at multiple resolutions, fuses extracted features and passes them to the next stage
in the network.

The proposed system can also be modified to categorize various buildings such as hospitals and police stations, Slum areas, different kinds of water bodies like sea, lakes or rivers. This will be possible if we have a better training and testing data.

References

1. Patel, R., Vadher, B.M., Waikhom, S., Yadav, V.G.: Change detection of Land Use/Land Cover (LULC) using remote sensing and GIS in Surat City. Glob. Res. Dev. J. Eng. (2019). eISSN 2455-5703
2. Albert, A., Kaur, J., Gonzalez, M.C.: Using convolutional networks and satellite imagery to identify patterns in urban environments at a large scale. In: Proceeding of 23rd ACM SIGKDD International Conference on Knowledge Discovery and Data Mining, pp. 1356–1366, 13 September 2017 (2017)
3. Block, J., Yazdani, M., Nguyen, M., Crawl, D., Jankowska, M., Graham, J., DeFanti, T., Altintas, I.: An unsupervised deep learning approach for satellite image analysis with applications in demographic analysis. In: 2017 IEEE 13th International Conference on e-Science (e-Science), Auckland, pp. 9–18 (2017). https://doi.org/10.1109/escience.2017.13
4. Chhor, G., Aramburu, C.B., Bougdal-Lambert, L.: Satellite Image Segmentation for Building Detection using U-Net (2017)
5. Acharya, T.D., Yang, I.: Exploring landsat 8. Int. J. IT Eng. Appl. Sci. Res. (IJIEASR) 4(4) (2015). ISSN 2319-4413
6. van Dijk, S.G., Scheunemann, M.M.: Deep learning for semantic segmentation on minimal hardware, 15 July 2018. arXiv:1807.05597v1[cs.LG]
7. Guo, Y., Liu, Y., Georgiou, T., Lew, M.S.: A review of semantic segmentation using deep neural networks. Int. J. Multimed. Inf. Retrieval 7(2), 87–93 (2018). https://doi.org/10.1007/s13735-017-0141-z
8. Long, J., Shelhamer, E., Darrell, T.: Fully convolutional networks for semantic segmentation. In: 2015 IEEE Conference on Computer Vision and Pattern Recognition (CVPR), Boston, MA, pp. 3431–3440 (2015). https://doi.org/10.1109/cvpr.2015.7298965
9. Ronneberger, O., Fischer, P., Brox, T.: U-Net: convolutional networks for biomedical image segmentation. MICCAI (2015). arXiv:1505.04597
10. Ibtehaz, N., Rahman, M.: MultiResUNet: rethinking the U-Net architecture for multimodal biomedical image segmentation, February 2019. arXiv:1902.04049v1
11. Xie, C.-W., Zhou, H.-Y., Wu, J.: Vortex pooling: improving context representation in semantic segmentation. arXiv, abs/1804.06242 (2018)
12. Zhao, H., Shi, J., Qi, X., Wang, X., Jia, J.: Pyramid scene parsing network. In: 2017 IEEE Conference on Computer Vision and Pattern Recognition (CVPR), Honolulu, HI, 2017, pp. 6230–6239 (2017). https://doi.org/10.1109/cvpr.2017.660
13. Lin, G., Milan, A., Shen, C., Reid, I.: RefineNet: multi-path refinement networks for high-resolution semantic segmentation. In: 2017 IEEE Conference on Computer Vision and Pattern Recognition (CVPR), Honolulu, HI, pp. 5168–5177 (2017). https://doi.org/10.1109/cvpr.2017.549

Design of Near Optimal Convolutional Neural Network Based Crowd Density Classifier

Komal R. Ahuja$^{(\boxtimes)}$ and Nadir N. Charniya

Vivekanand Education Society Institute of Technology,
Chembur, Mumbai, India
{2017komal.ahuja, nadir.charniya}@ves.ac.in

Abstract. Crowd density estimation and crowd counting has acquired importance towards Machine learning and Deep learning industry due to the improvement in performance, when compared to traditional computer vision techniques. This paper presents deep learning based optimal dimension convolutional neural network (CNN) for estimating crowd density, which is used to classify images of crowd into various density levels such as low crowd, very low crowd, moderate crowd, high crowd, very high crowd. This approach is experimented on existing datasets and gives the better accuracy with optimum network dimensions.

Keywords: Crowd density estimation · Convolutional neural network architecture · Deep learning

1 Introduction

A crowd is a complete countless accumulation of people together in a compact place. With the increase in human population, crowd gathering takes place at many public places such as railway stations, metro stations, platforms, stadium, marathon etc., so there is a risk of stampede. There were various incidents took place because of over-crowd and many people died. The incidents are in 2005, more than 640 people died in Religious procession at Baghdad [1], In 2013, 115 people died in Hindu Festival at Datia District, in 2014, during Dussehra festival in India around 32 people died and in 2017, 23 people died at Elphinston Road Mumbai. Therefore, to overcome this critical situation some efficient and an effective crowd management techniques are required. Sometimes in such overcrowded sciences there are faces or heads occlusions and therefore density estimation by analyzing individuals count is a difficult task. To estimate the crowd density, crowd features are extracted in the first step and in the next step classifier is designed to classify the images into various density levels, such as, low crowd, very low crowd, moderate crowd, high crowd, and very high crowd. There were various methods developed to build optimal classifier for classification task. Recently various researchers adopted computer vision based techniques for crowd density estimation. Computer vision based techniques with deep learning on different existing datasets gives the better accuracy and proves to be beneficial for the analysis of overcrowd incidents. Nowadays various machine learning and deep learning techniques have attracted researchers as it provides best results in crowd counting and

© Springer Nature Switzerland AG 2020
J. S. Raj et al. (Eds.): ICIDCA 2019, LNDECT 46, pp. 204–212, 2020.
https://doi.org/10.1007/978-3-030-38040-3_24

density estimation. This paper proposes to build optimal Convolutional Neural Network (CNN) to estimate crowd density. In this paper, author have experimented on near optimal CNN design using single layer and two layers to determine optimal network architecture without compromising the classification accuracy.

This paper is organized as Sect. 2, which reviews about the related work, Sect. 3 which proposes optimal CNN design methodology, in Sect. 4 experiments with various graphs, results and finally Sect. 5 concludes the paper with future scope.

2 Related Work

Initially researchers main focus was to count the number of people from video sequence or from still images instead of density estimating. Traditionally, they used Closed Circuit Television cameras to capture the crowd information. For this purpose, human operators were needed for continuously monitoring the camera to detect any unusual and suspicious action in the crowd. For this task large number of staff were required and hence the effective cost of the system was higher [2]. In this field of crowd analysis lots of work has been done. Crowd analysis is divided into four categories, such as, Detection based approach, Regression based approach, Density estimation based approach, CNN based approach.

2.1 Detection Based Approach

In this method, crowd analysis is carried out by identifying each individual from the dataset by using scanning window which helps to detect individuals from video scenes from left to right and from top to bottom. After detecting each individual, counting was carried out. This kind of methodology requires Haar cascade classifier. As the whole body detection is a difficult task, part based detection was implemented by researchers. This method is not suitable for highly dense flow [3].

2.2 Regression Based Approach

To overcome the limitation of part based detection, regression based approach is adopted. This method focuses on local and global image patches which are essential for crowd counting. There are various features for estimating the crowd, they are, foreground features, edge features, and texture features. Foreground is subtracted from background by using some background subtraction techniques. Then number of people are counted from the foreground images. The background subtraction techniques are based on Fourier analysis and Scale-invariant feature transform technique. Background subtraction technique is a crucial step in a clogged scene is the limitation of this method [4, 5].

2.3 Density Estimation Based Approach

In this approach, linear density map concept is implemented. That is extracted features are linearly mapped with the corresponding density level. Linear mapping is difficult task for this approach. [10] proposed non-linear mapping between the image features with the corresponding density levels.

2.4   CNN Based Approach

In recent years, crowd density estimation and crowd counting proves to be an effective research area topic under artificial intelligence for public safety and security [6, 7] In this approach, Deep learning CNN is used for crowd density estimation by extracting deep features in the crowd image samples. From experimentations done by researchers it is evident that CNN based approach gives the accurate results over all traditional approach discussed above.

3   Implementation of Crowd Density Classifier

CNN directly extract crowd features and classify the images into low crowd, very low crowd, moderate crowd, high crowd, and very high crowd. With the help of convolution operation, features are acquired by taking images as input directly. Typically CNN consist of Convolution layer, Pooling layer, neurons layer, Activation layer, Flatten, Fully-connected layer as shown in the Fig. 1. Convolution layer maps the input image with the linear filter and the output represents the response of the filter. Pooling layer reduces the dimensions of the feature vector by taking the mean of all the pixels or by taking maximum value of all the pixels. Activation layer introduces non-linearity to the input neurons. Common activations functions are sigmoid function, softmax function, tangent hyperbolic function, rectified linear unit etc. The last layer is fully-connected layer which computes the output [10, 12].

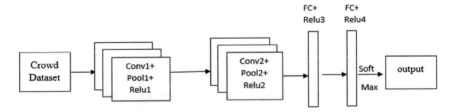

Fig. 1. Structure of deep crowd classification method

Where, Conv: Convolution layer
Pool: pooling layer
Relu: ReLU
FC: Fully Connected

3.1   The Designed CNN Architecture

In this work, the fixed convolution layer of 3×3 kernel size is implemented with different number of neurons. The pooling size is taken as 2×2 with activation function as ReLu. The designed architecture of CNN is given in Table 1.

Table 1. Designed configuration of CNN architecture

Layer	Layer name	Layer properties
01	Image input size	$64 \times 64 \times 3$
02	Convolution	3×3 kernel with varying neurons
03	Activation function	ReLu
04	Max pooling	2×2
05	Flatten	–
06	Fully-connected	64
07	Activation function	ReLu
08	Fully-connected	05
09	Activation function	Softmax
10	Classification output	Cross entropy with 5 classes

3.2   System Block Diagram

(See Fig. 2).

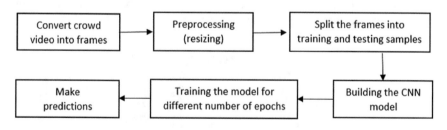

Fig. 2. System Block Diagram

3.3   Preprocessing

Initially all the crowd images are of different size so by using preprocessing step the dataset is converted into a fixed size. The experimentation is carried out for different image size and the best result is obtained for image size of 64×64.

3.4   Data Augmentation

The main aim of data augmentation is to produce more training data from the existing datasets by augmenting the existing data with some random transformation processes such as zooming the images, flipping horizontally or vertically, viewing from different orientations [8, 9]. Some examples of data augmentation is given in Fig. 3.

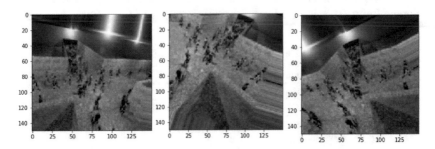

Fig. 3. Data augmentation example

4   Experiments and Results

4.1   Dataset

The existing dataset 2015 UNC is used for crowd classification task which is a video sample. This video is converted into frames for estimating the crowd. There are several frames of different density levels, such as, low crowd, very low crowd, moderate crowd, high crowd, very high crowd. The dataset is categorized into training set and validation set. Training set consist of 75% of total frames and validation set consist of 25% of total frames. CNN has been designed using single layer and two layers cases. The dataset distribution for five classes is shown below in Fig. 4.

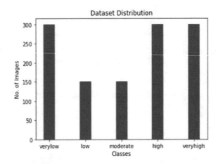

Fig. 4. Dataset Distribution for five classes [11]

4.2   Experimental Analysis for Single Layer and Two Layer Architecture

4.2.1   Single Layer Architecture

Table 1 represents the training and validation accuracy for single layer CNN architecture in which convolution layer size is fixed taken 3×3 and pooling size is taken as 2×2 for activation function ReLu and fully-connected layer consist of 64 neurons.

The experimentation is performed for different neurons and the best accuracy is observed for 08 neurons (Table 2).

Table 2. Performance of single-layer CNN for training and validation accuracy with variable number of neurons

Neurons	Training accuracy	Validation accuracy	Training loss	Validation loss
04	0.9814	0.9912	0.0320	0.0176
08	0.9912	0.9900	0.0265	0.0501
16	0.9875	0.9825	0.028	0.0423
32	0.9894	0.9825	0.0238	0.0623
64	0.9898	0.9750	0.0268	0.1186

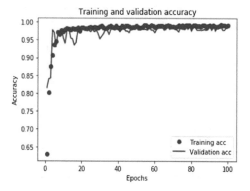

Fig. 5. Training and Validation Accuracy verses Epochs for Single Layer Architecture with 08 neurons

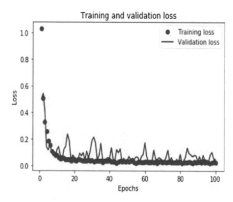

Fig. 6. Training and Validation Losses verses Epochs for Single Layer Architecture with 08 neurons

Accuracy curves and loss curves are illustrated in the above Figs. 5 and 6. From the accuracy curves it has been observed that the training and validation accuracy is almost the same. CNN models have been trained for 100 epochs and training and validation accuracy of 99 percent with 08 neurons for convolution layer have been obtained. From the loss curve it is evident that training and validation loss follow nearly the same trend.

4.2.2 Two-Layer Architecture

Two-layer CNN architecture has been implemented for different number of neurons for both the convolution layers. Different combinations are analyzed and the best results obtained for 32 neurons for both the layers. The training and validation accuracy of 99 percent is obtained with 100 epochs.

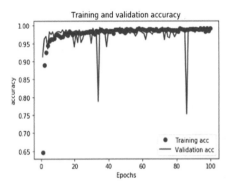

Fig. 7. Training and Validation Accuracy verses Epochs for Two Layer Architecture with 32 neurons

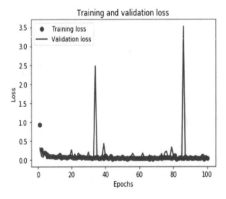

Fig. 8. Training and Validation Losses verses Epochs for Two Layer Architecture with 32 neurons

Accuracy and loss curves for two-layer architecture is depicted in the above Figs. 7 and 8. From above graphs it is observed that the best accuracy is obtained for 32 neurons for both the layers. On increasing the number of neurons further there was no further improvement in validation accuracy. Optimal number of epochs were found to be 25. CNN gives the best results as compared to conventional computer vision methods for crowd density estimation. Single layer CNN architecture and two-layer architecture are compared in the below Table 3.

Table 3. Performance of Single Layer and Two Layer architecture

Accuracy	Single layer CNN architecture	Two-layer CNN architecture
Training accuracy	99.12%	99.0%
Validation accuracy	99.2%	99.1%

5 Conclusion and Future Scope

Various experiments are performed for estimating crowd density by using CNN architecture for single layer and for two layers. CNN with single layer of kernel size 3×3 and with 08 neurons and 100 epochs gives the training accuracy of 99.12% and validation accuracy of 99%. CNN with two layers of kernel size 3×3 and with 32 neurons for both the layers and 25 epochs gives the training accuracy of 99.2% and validation accuracy of 99.1%. The experiment is performed on the existing dataset UCN 2015 crowd sample video in which the frames are divided into training and validation data. Data augmentation was performed to increase the training samples.

From experimentation results it is evident that the designed CNN model gives best accuracy in real world applications.

The optimal CNN network designed with single layer and two layer will helps in future work for estimating the crowd size in the railway compartment so that the people will come to know about the vacant spaces in compartment before the train arrives.

References

1. Saleh, S.A.M., Suandi, S.A., Ibrahim, H.: Recent survey on crowd density estimation and counting for visual surveillance. Eng. Appl. Artif. Intell. **41**, 103–114 (2015)
2. Saqib, M., Khan, S.D., Blumenstein, M.: Texture-based feature mining for crowd density estimation: a study. In: International Conference on Image and Vision Computing New Zealand (IVCNZ), pp. 1–6. IEEE, November 2016
3. Sindagi, V.A., Patel, V.M.: A survey of recent advances in CNN-based single image crowd counting and density estimation. Pattern Recogn. Lett. **107**, 3–16 (2018)
4. Saqib, M., Khan, S.D., Sharma, N., Blumenstein, M.: Crowd counting in low-resolution crowded scenes using region-based deep convolutional neural networks. IEEE Access **7**, 35317–35329 (2019)
5. Tagore, N.K., Singh, S.K.: Crowd counting in a highly congested scene using deep augmentation based convolutional network. SSRN 3392307 (2019)

6. Deep Learning Toolbox. http://cs231n.github.io/convolutional-networks/. Accessed Mar 2019

7. Deep Learning Book: Deep Learning with Python, François Chollet. Accssed Jan 2019

8. Cao, L., Zhang, X., Ren, W., Huang, K.: Large scale crowd analysis based on convolutional neural network. Pattern Recogn. **48**, 3016–3024 (2016)

9. Al-Hadhrami, S., Altuwaijri, S., Alkharashi, N., Ouni, R.: Deep classification technique for density counting. In: 2nd International Conference on Computer Applications & Information Security (ICCAIS), pp. 1–6. IEEE, May 2019

10. Mayur, C., Archana, G.: A study of crowd detection and density analysis for safety control. In: International Conference of Computer Science and Engineering (IJSE), pp. 424–428, April 2018

11. Dataset Distribution. https://towardsdatascience.com/boost-your-cnn-image-classifier-perfor mance-with-progressive-resizing-in-keras-a7d96da06e20. Accessed June 2019

12. CNN Layer Architecture. https://www.ncbi.nlm.nih.gov/pmc/articles/PMC6458916/. Accessed July 2019

Face Swapping Using Modified Dlib Image Morphology

Anjum Rohra[✉] and Ramesh K. Kulkarni

Vivekanand Education Society Institute of Technology,
Chembur, Mumbai, India
{anjum.rohra, ramesh.kulkarni}@ves.ac.in

Abstract. Morphing is an image processing technique used for the metamorphosis from one image to another. Apart from its application in entertainment industry, image morphing is also used in computer based trainings, electronic book illustrations, presentations, education purposes etc. The idea is to get the transition from source image to target image with maximum matching. To accomplish this, Image Morphing has gained attention from multimedia users and entertainment seekers in order to obtain fancier transitions and animations. The proposed Face Swapping technique is used to transform the source image to target image and vice-versa. The results are compared to the available pre-trained Dlib model for landmarks and the results are most encouraging. The landmarks highlight the important facial attributes.

Keywords: Morphing · Face Swapping · Dlib · Landmarks

1 Introduction

An image is a two dimensional representation of an array of picture elements (pixels). Digital Image Processing is the most popular field, which works on various image transformations so that the output image is of a better quality, as per the user requirements. Different types of processing are done on the images. Image Morphing is one of them. It is highly used in media and entertainment industries, especially for animation applications. Image Morphing is a best substitute to the traditional method, where one image was merely replaced by another image. The method is termed as Cross Dissolving, resulting in poor quality, misaligned output, represented in Fig. 1.

In case of Image Morphing, there is a smooth transition from source image to target image by generating various intermediate images in between as illustrated in Fig. 2. The amount of morphing in intermediate images is determined by Morphing Index. Morphing Index is the most important parameter in the morphing process, denoted by α. It determines how much the intermediate images matches to the source or the target image. The least value of α is 0 and the maximum value is 1. When α is 0, the resultant image is the source image and when α is 1, the resultant image is the target image. The intermediate images contain the combined characteristics of both the source and the target images.

© Springer Nature Switzerland AG 2020
J. S. Raj et al. (Eds.): ICIDCA 2019, LNDECT 46, pp. 213–221, 2020.
https://doi.org/10.1007/978-3-030-38040-3_25

Fig. 1. Cross dissolving process [10]

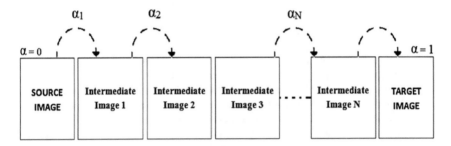

Fig. 2. The morph sequence

The morphing process is explained in three steps:

- Extraction of facial landmarks: Landmarks highlight the important facial features. Landmarks can be extracted either manually or by using the available pre-trained Dlib model for landmarks.
- Warping using Delaunay Triangulation: Delaunay Triangulation splits the image surface into a mesh of small triangles. Each triangle in source image is warped separately into corresponding triangle in the target image.
- Colour transition: The quality of resultant morphs is based on colour transition. It is done in order to avoid colour mismatch between the two images in the generated morph.

This paper is organized as Sect. 2, which reviews about the related work, Sect. 3, which proposes the implementation of proposed technique and Sect. 4, which comprises of experiments and results and finally Sect. 5 concludes the paper, along with future work.

2   Related Work

Various methods are used by researchers to generate artificial faces for many experiments. There are various uses of morphing technique. For example, generation of composites of other faces to compare the original faces with their composites. By this, we can know that which face is more attractive [1]. Composites of faces can also be used to investigate that whether there is a categorical perception of faces [2]. Morphing also finds its application in generating caricatures of faces [3, 4]. In a caricature of a face, the distinctive facial features are highlighted. Caricatures simplify the task of face recognition [5].

Etcoff and Magee were first to investigate the categorical perception of face expressions [6]. They converted various images using the Ekman and Friesen (1976) series [7] into line-drawings. These line drawings were obtained by using the positions of 169 control points by merging them into 37 different lines.

Face Alignment is significant in case of Image Morphing to obtain better results. This can be done by using the concept of Ensemble of Regression trees. In this method, the position of landmarks can be identified based on the pixel intensities, with high quality predictions [8]. Dlib library uses the concept of regression trees, which is a Machine Learning Supervised algorithm, to assign the landmarks. With the help of this algorithm, it is possible to accurately detect the shape of any input face, using the ensemble of regression trees [8].

3   Implementation of Proposed Technique

3.1   Block Diagram

Figure 3 illustrates complete morphing process. The source and the target images are warped using Delaunay Triangulation, where the image surface is split into number of tiny triangles, thereby forming a triangular mesh, on both the source and the target images. Each triangle in the source image is warped to corresponding triangle in the target image. The triangular mesh is formed based on the landmarks or the control points extracted, seen in Fig. 7. After warping, the images are colour blended so as to avoid any colour mismatch in the morphed image. The landmarks are extracted for both the images (i.e.) the source and the target images.

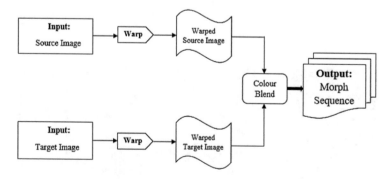

Fig. 3. Block diagram of image morphing process

3.2   Methodology

Image Morphing can be used for swapping two faces. Block Diagram of Fig. 4 represents the detailed process of Face Swapping using Face Morphing.

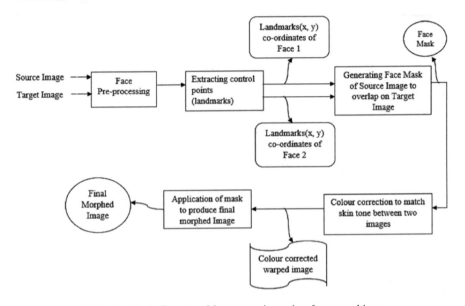

Fig. 4. Block diagram of face swapping using face morphing

The source and the target images are first pre-processed to obtain size equalization and the landmarks or control points. The landmarks can either be obtained using the pre-trained Dlib model of landmarks or can also be marked manually. Based on the landmarks selected, triangular mesh is formed on the source and target images, represented in Fig. 7. A face mask is formed corresponding to the shape of the triangular mesh of target image shown in Fig. 5. The colour correction matches the skin tone to the source image. Finally, this colour corrected mask is overlapped on the target image to generate the final morphed image.

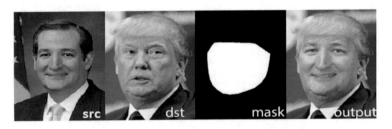

Fig. 5. Generation of mask from source and target images [9]

3.3   Delaunay Triangulation on Images

The Landmarks on source and destination images are obtained by manually marking the control points. Based on the positions of these points, a triangular mesh is formed by joining these points. This process is known as 'Delaunay Triangulation'. The triangles are formed based on an important property that Delaunay Triangulation 'maximizes the minimum angles of the triangles' in the image plane. The circumcircle of the triangle obtained by Delaunay Triangulation passes through all the three vertices of the triangle. The circumcircle so formed contains no other control points in its interior plane as seen in Fig. 6. The results obtained are seen in Fig. 7.

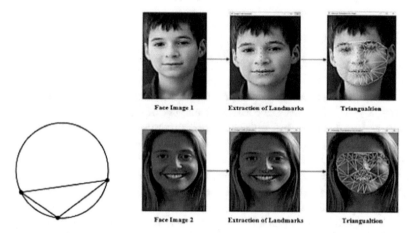

Fig. 6. Circumcircle of a Delaunay Triangle

Fig. 7. Results of Delaunay Triangulation

3.4   Morphing Sequence

The Morphing Sequence comprises of Source Image, Target Image and various intermediate images in between. The intermediate images determine how the source image slowly transforms to target image. The amount of morphing in intermediate images is determined by the morphing index α. The value of α is between 0 and 1. The Morphing Sequence obtained is represented in Fig. 8.

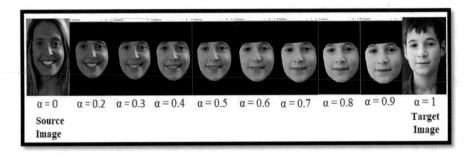

α = 0   α = 0.2   α = 0.3   α = 0.4   α = 0.5   α = 0.6   α = 0.7   α = 0.8   α = 0.9   α = 1

Source   **Target**
Image   **Image**

Fig. 8. Result of image morphing sequence for different morphing index

4   Experiments and Results

4.1   Dataset

The existing dataset used for implementation of Image Morphing is the HELEN Dataset [11, 12]. This dataset consists of number of face images with different orientations and each image has different face expressions. The dataset is created using Flickr images which are annotated. The dataset is split into 2000 training images and 330 test images. The HELEN dataset is also trained for detecting facial landmark positions based on the pixel intensities using the concept of ensemble of regression trees. Dlib library in Python uses the concept of regression trees, which is a Machine Learning Supervised algorithm, to assign the landmarks. With the help of this algorithm, it is possible to accurately detect the shape of any input face, using the ensemble of regression trees.

4.2   Face Swapping Using Image Morphing

Face Swapping is an application of Image Morphing where the source image face is replaced by the target image face in accordance to the landmarks selected and the face mask generated for each face. The shape of the face mask depends upon the position of the landmarks. Table 1 below represents different Face Swapping morphed outputs for different input (Source and Target) images. Row 3 represents morphed outputs using Dlib model for landmarks whereas Row 4 represents morphed outputs where in the landmarks are manually marked. After marking the landmarks, their co-ordinate values are noted and saved as a text file for each image. Face Swapping is tested on the HELEN dataset and the results are tabulated in Table 1.

Table 1. Face swapping outputs using image morphing

4.3 Structural Similarity Index (SSI)

Face Swapping (using Image Morphing) outputs for six examples from HELEN Dataset are obtained, which are represented in Table 1. Outputs are obtained for two methods. First is by using pre-trained Dlib landmarks model for extracting landmarks of input images. Second method is by marking the control points manually on the image and saving their co-ordinates in a text file, to use them for morphing purpose. Then, each morphed output for each method is compared with the corresponding Target Image for its SSI. SSI determines the perceptual difference between the two images. The value of SSI lies in between 0 and 1. The method which gives the greater SSI value is better [13].

Table 2 is the SSI table which represents SSI outputs obtained for each example in Table 1. Each Target image in Table 1 is compared with both the Morphed outputs.

Table 2. SSI outputs for six examples in Table 1

Sr. No.	Structural similarity index between the target image and the morphed output (using Dlib)	Structural similarity index between the target image and the morphed output without Dlib (proposed method)
1.	0.549395	0.559974
2.	0.468333	0.445579
3.	0.541889	0.474677
4.	0.577885	0.606967
5.	0.862394	0.737669
6.	0.756254	0.702059

5 Conclusion and Futurescope

The demand for high quality image transitions has highly increased in multimedia industries. The traditional method (i.e.) Cross Dissolving is rarely used nowadays as the resulting morph obtained by this method is quiet disturbing. Morphed image can be obtained either by using the pre-trained Dlib landmarks model or by marking the control points manually on the image. The SSI output for proposed method depends upon the manner in which the control points are selected by the user. More the control points for each facial feature, higher accurate is the SSI.

As a future work, instead of using just two input images, multiple input images can be applied to obtain a morph sequence. Image Morphing can also be used for secure money transactions as internet banking has highly increased for online transactions. Morphing can be used to improve the security standards.

References

1. Rhodes, G.: The evolutionary psychology of facial beauty. Annu. Rev. Psychol. **57**, 199–226 (2006)
2. Beale, J.M., Keil, F.C.: Categorical effects in the perception of faces. Cognition **57**(3), 217–239 (1995)
3. Benson, P.J., Perrett, D.I.: Perception and recognition of photographic quality facial caricatures: implications for the recognition of natural images. Eur. J. Cogn. Psychol. **3**(1), 105–135 (1991)
4. Czajkowski, K., Fitzgerald, S., Foster, I., Kesselman, C.: Grid information services for distributed resource sharing. In: 10th IEEE International Symposium on High Performance Distributed Computing, pp. 181–184. IEEE Press, New York (2001)
5. Mauro, R., Kubovy, M.: Caricature and face recognition. Mem. Cogn. **20**(4), 433–440 (1992)
6. Etcoff, N.L., Magee, J.J.: Categorical perception of facial expressions. Cognition **44**(3), 227–240 (1992)
7. Ekman, P., Friesen, W.V.: Felt, false, and miserable smiles. J. Nonverbal Behav. **6**(4), 238–252 (1982)

8. Kazemi, V., Sullivan, J.: One millisecond face alignment with an ensemble of regression trees. In: Proceedings of the Conference on Computer Vision and Pattern Recognition, pp. 1867–1874. IEEE (2014)
9. Face swap using OpenCV. https://www.learnopencv.com/face-swap-using-opencv-c-python/. Accessed Aug 2019
10. Face morph using OpenCV. https://www.learnopencv.com/face-morph-using-opencv-cpp-python/. Accessed Feb 2019
11. Helen dataset. http://www.ifp.illinois.edu/~vuongle2/helen/. Accessed Feb 2019
12. Helen dataset for facial landmark localization. http://www.f-zhou.com/fa_code.html. Accessed Feb 2019
13. Structural similarity index. http://www.imatest.com/docs/ssim/. Accessed Apr 2019

Regular Language Search over Encrypted Cloud Data Using Token Generation: A Review Paper

Pranoti Annaso Tamgave[(⊠)], Amrita A. Manjrekar[(⊠)], and Shubhangi Khot

Computer Science and Technology, Department of Technology,
Shivaji University, Kolhapur, Maharashtra, India
pranotitamgave84@gmail.com,
shubhangikhot521@gmail.com, aam_tech@unishivaji.ac.in

Abstract. Cloud computing is a continuous and rapidly propelling model, with new points and capacities being declared consistently. Cloud is the best solution for sharing, maintaining, handling large amount of data. Using the cloud computing the user can access the data anywhere and anytime through any mobile devices. Cloud server provides the storage services but this storage service is not fully trusted by users, hence many times users will upload encrypted data on cloud. In this paper different techniques related to regular language search on encrypted data are studied. These techniques are compared based on attributes like privacy, efficiency, accuracy and overhead. Here we propose regular language search over encrypted data store on public cloud using DFA based token generation method.

Keywords: Encryption algorithm · Regular Language · Cloud storage · Searching Techniques · Token generation · Encrypted data

1 Introduction

Cloud computing is an emerging technology which provides ubiquitous data access, flexible data management and pay as you use service to customers. Cloud computing is fast growing technology which is based on various requirements (different resources like storage, hardware, or applications) of the users. Users can remotely manage data using cloud services. Cloud offers infrastructure, storage and different application as a service. Cloud computing is the most recent and fast increasing technology which is founded on the intake of resources. These resources are distributed via the internet. The idea of cloud computing turns out to be significantly more justifiable when one starts to consider what present-day IT situations consistently require. The association needs to build limit or abilities of their framework progressively 0% putting cash in the acquisition of new foundation, new programming or increment stockpiling limit [6]. Cloud model that incorporates a membership-based or pay per use worldview gives assistance that can be utilized over the web and broaden the association's current ability one of the significant administrations given by the cloud is capacity as a help. Here cloud specialist organization

© Springer Nature Switzerland AG 2020
J. S. Raj et al. (Eds.): ICIDCA 2019, LNDECT 46, pp. 222–228, 2020.
https://doi.org/10.1007/978-3-030-38040-3_26

gives a consistent pool to store the computerized information. Clients can undoubtedly store and impart their information to one another utilizing the distributed storage innovation (e.g. Google Drive, Dropbox). Many users are not assured about integrity and privacy of stored data [7, 8].

Numerous clients store their significant information on the cloud. Usually important data is stored in the encrypted form by the customers. As data is important, customer stores it in the encrypted format. The searching operation on data usually needs decryption followed by searching. If the amount of data is very large then it becomes difficult to decrypt and search. Especially the cloud users have to download huge amount of data and apply decryption and then search. The searchable encryption technology [4, 5] here protects the data and supports effective search data by maintaining data secrecy. Users generate the token using his own private key to search any content from stored data. With the assistance of this produced token cloud server look on the encoded information without decrypting the information. This scheme will support the search pattern as a single keyword search. This system will use DFA to search encrypted data. The data is stored on public cloud in encrypted format. The consumer will pass search token to cloud server, then this token will be used by system to find the encrypted ciphertext using DFA.

2 Literature Survey

There are different subjects defined as Data security on cloud, Searchable encryption Techniques and keyword Searching. Few important parameters are described below.

2.1 Data Security on Cloud

The introduction of cloud services generates various challenges for business organization. When an organization migrates to cloud services especially public cloud services, then much of its computing infrastructure will now be under the control of third party Cloud Service Provider (CSP) [11]. Data security has been major issue in digital world. In cloud environment it becomes more serious as data is migrated and located in various places all over the world. Organizations are mainly concerned with security and privacy of the data.

This framework gives probabilistic confirmation that the outsider store a document. Here framework creates the ownership confirmation by testing arbitrary arrangements of squares from the server. The customer keeps up the confirmation metadata to check the evidence.

"Ensuring Data Security in Cloud based social Networks." Praveena and Smys [9] Social network generates huge amount of data. Secure storage of this data is one of the important aspect. Here author proposed a framework for storing data securely on the cloud based social networks. Data is stored in encrypted format using private key encryption and proxy re-encryption technique is used to give more security to data.

"A hybrid multilevel authentication scheme for private cloud environment" Sridhar and Smys [13] here author proposed multilevel authentication for private cloud network. Here new authentication mechanism is proposed which provides security against virtualization and insider attacks. The system depends upon symmetry based AES encryption algorithm.

Various mechanisms have been proposed to check data integrity on cloud. Usually in these systems a mark is appended to each square in information and honesty of information depends on the rightness of the considerable number of marks. Wang et al. [12] here proposed Panda a novel open inspecting instrument for the honesty of shared information is proposed. It additionally underpins the renouncement of the client in the cloud. This is an adaptable system that proficiently underpins countless clients to share information and furthermore ready to deal with numerous evaluating assignments all the while.

2.2 Searchable Encryption Techniques

Accessible encryption gives a successful component that accomplishes secure inquiry over encoded information has become essential and important techniques for cloud computing as it search data without leaking any information.

In enable search over encrypted cloud data with concealed search patter [10]. The secure search token for plaintext keywords are generated randomly. In this plan, it is infeasible to tell whether the basic plaintext catchphrases are a similar given to two secure inquiry tokens, which abstains from utilizing explicit secure hunt token which is the fundamental driver of pursuit design spillage.

In this paper authors implements [4] a novel privacy preserving useful encryption-based inquiry component over encoded cloud information. The benefit of this plan is, advance plan contrasted and existing looking through open key-based frameworks is that it bolsters an extraordinary expressive inquiry mode, customary language search. This framework is secure and more proficient than some accessible frameworks with high expressiveness.

Here authors introduced [2] a searchable attribute-based proxy re-encryption system. The contrasted with existing frameworks just a single capacity underpins accessible property based intermediary re-encryption and quality based usefulness, this framework bolsters the two capacities and gives adaptable catchphrase update administration. In particular, the framework gives the capacity to an information proprietor to productively share his information to a predefined gathering of clients coordinating a sharing arrangement and in the interim, the information will keep up its accessible property yet additionally the closest search keyword(s) can be refreshed after the information sharing. This framework component is applied to some continuous applications, for example, electronic wellbeing record frameworks.

2.3 Keyword Searching Techniques

The following Fig. 1 graph shows development trends of different keyword searching methods over the period of time. Graph shows that Boolean and ranked searching methods were develop in early 20's. Recently multi-ranked and Regular Language Keyword Searching techniques are proposed.

Boneh and Waters [4] this system proposed Boolean Keyword search technique is proposed which supports with conjunctive keyword over encrypted data. Boolean

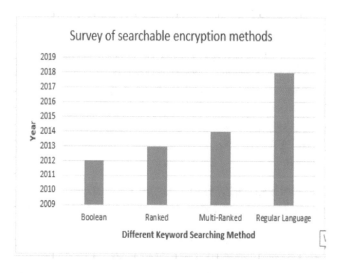

Fig. 1. Development of searching methods

keyword search problem is solved by this scheme. In this paper, it defines one of the technique which describes the communication last over number of documents whose security is provided by using Discretional Diffie-Hellman (DDH) assumption.

Wang et al. [2] this system proposed Ranked Searchable Symmetric Encryption (RSSE) scheme with support secure ranked keyword search over encrypted data on cloud. This plan is enormously improved framework convenience by restoring the coordinating records in a positioned request saving mapping method. This method secures delicate score data. Positioned Search gives the positioning rather than undifferentiated outcomes. The subsequent structure can encourage effective server-side positioning without losing catchphrase security.

Liu et al. [3] here this system proposed Efficient information retrieval for ranked Query (ERIQ) schemes. This plan diminishes question overhead acquired on the cloud. In this plan, inquiries are ordered into numerous positions, where a higher-positioned question can recover a higher level of coordinated documents. A client can recovers records on interest by picking inquiries of various positions.

Yang and Zheng [1] it provides techniques to searching regular language with the help of DFA. It supports the regular language encryption and deterministic finite automata (DFA) based data retrieval.

The following table shows comparison between various keyword based searching methods based on properties like privacy, efficiency, scalability, accuracy and over-head. In the paper [1] authors compare Boolean method with ranked keyword search and in [2] paper author compares ranked method with multi ranked searchable encryption method and with the same scenario we compare the regular language search method (Table 1).

Table 1. Comparison between searching methods

Sr. No.	Properties	Methods			
		Boolean	Ranked	Multi-ranked	Regular language
1	Privacy	Low	Medium	Medium	High
2	Efficiency	Low	Medium	Medium	High
3	Scalability	No	Yes	Yes	Yes
4	Accuracy	Low	Medium	Medium	High
5	Overhead	Low	Medium	High	Low

3 Proposed Method: Regular Language Search over Encrypted Cloud Data Using Token Generation

Existing accessible encryption plans don't bolster ordinary language search. Liang's plan depends on standard language accessible encryption characterized as accessible deterministic limited automata-based practical encryption. Still Liang's scheme [4] demonstrates that a few restrictions recorded as pursues:

- Small Universal Construction, The size of the image is predefined when the framework is set up. The size of the ace open key increment with the image set, which expends a major extra room when the predefined image set is huge. To add another image to the framework, the KGC needs to again assemble the whole framework.
- The encryption and search token age calculations are low efficiencies due to the information clients need to execute a great deal of vitality expending exponentiation figurings, which acquires immense calculation overhead.
- It is a ward trapdoor age. The information client needs to interface with the key generation center (KGC) to produce an inquiry question. The information client doesn't freely give a DFA question.
- In existing system does not resist off-line keyword guessing attack (KGA).

This proposed will search the regular language using token generation DFA based. The following Fig. 2 shows basic system block diagram. The system will be consists of 5 entities which are listed follows:

1. Data User
2. Searching token algorithm
3. Cloud Server
4. Data Owner
5. Key Generation Center (KGC)

In this system User can register and login using Mobile devices if the user is a new user then it's important to enroll the new client and get a username and secret key for login credentials. In the wake of logging the client android application give an alternate arrangement of administrations. The User can access the data, download data and search particular encrypted data with help of token based DFA search on a cloud

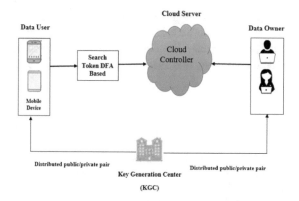

Fig. 2. Basic system architecture

server. Data owner and the user can access cloud stored data from anywhere. The data owner stores data in encrypted format. Data user can access the data and search on encrypted data.

In email service normally the users have access email in plaintext so that lot of security problems arises. If we provide security to sensitive email message and encrypt the E-mails then searching of data in E-mail will be one of the major concern. In this case one can use this proposed system. The following Fig. 3 shows that the user can ask to search the encrypted email and retrieve mail as urgent through smart phone or send request to the Mail Service Provider to her encrypted email via internet.

Fig. 3. An example of encrypted data for email

4   Conclusion

Cloud computing enables information proprietors to redistribute information stock-piling to cloud (storage as a service - Dropbox, Google Drive). However this facility of data outsourcing generates information security and privacy challenges on cloud. Encrypted storage raised new issue like searching on encrypted information. Search-able encryption is one of new and useful technique that could preserve privacy of data and enable keyword query over encrypted document. Here we have discussed Data Security on Cloud, different Searchable Encryption Techniques with their pros and cons and specific Keyword Searching Techniques with their development trend and their comparison based on five properties is illustrated. Here we would like to propose a system based on universal regular language searchable encryption scheme for the cloud. It will provide data privacy and security against the off-line key-word guessing attack (KGA).

References

1. Yang, Y., Zheng, C., Rong, C., Guo, W.: Efficient regular language search for secure cloud storage. IEEE Trans. Cloud Comput. (2018). https://doi.org/10.1109/tcc.2018.2814594
2. Wang, C., Cao, N., Ren, K., et al.: Enabling secure and efficient ranked keyword search over outsourced cloud data. IEEE Trans. Parallel Distrib. Syst. 23(8), 1467–1479 (2012)
3. Liu, Q., Tan, C.C., Wu, J., et al.: Towards differential query services in cost-efficient clouds. IEEE Trans. Parallel Distrib. Syst. 25(6), 1648–1658 (2014)
4. Boneh, D., Waters, B.: Conjunctive, subset, and range queries on encrypted data. In: Vadhan, S.P. (ed.) Theory of Cryptography, vol. 4392, pp. 535–554. Springer, Heidelberg (2007)
5. Liang, K., Huang, X., Guo, F., et al.: Privacy-preserving and regular language search over encrypted cloud data. IEEE Trans. Inf. Forensics Secur. 11(10), 2365–2376 (2016)
6. Mather, T., Kumaraswamy, S., Latif, S.: Cloud Security and Privacy an Enterprise Perspective on Risks and Complience. O'Reilly, Sebastopol (2009)
7. Dattatray Kankhare, D., Manjrekar, A.A.: A cloud based system to sense security vulnerabilities of web application in open-source private cloud IAAS. In: 2016 International Conference on Electrical, Electronics, Communication, Computer and Optimization Techniques (ICEECCOT), Mysuru, pp. 252–255 (2016)
8. Kasunde, D.S., Manjrekar, A.A.: Verification of multi-owner shared data with collusion resistant user revocation in cloud. In: 2016 International Conference on Computational Techniques in Information and Communication Technologies (ICCTICT), New Delhi, pp. 182–185 (2016). https://doi.org/10.1109/icctict.2016.7514575
9. Praveena, A., Smys, S.: Ensuring data security in cloud based social networks. In: 2017 International Conference of Electronics, Communication and Aerospace Technology (ICECA), 20 April 2017, vol. 2, pp. 289–295. IEEE (2017)
10. Yao, J., Zheng, Y., Wang, C., Gui, X.: Enabling search over encrypted cloud data with concealed search pattern. IEEE Trans. (2019). https://doi.org/10.1109/2018.2810297
11. Krutz, R.L., Vines, R.D.: Cloud Security a Comprehensive Guide to Secure Cloud Computing. Wiley, New York (2010)
12. Wang, B., Li, B., Li, H.: Panda: public auditing for shared data with efficient user revocation in the cloud. IEEE Trans. Serv. Comput. 8(1) (2015)
13. Sridhar, S., Smys, S.: A hybrid multilevel authentication scheme for private cloud environment. In: 2016 10th International Conference on Intelligent Systems and Control (ISCO), 7 January 2016, pp. 1–5. IEEE (2016)

Modeling and Evaluation of Performance Characteristics of Redundant Parallel Planar Manipulator

Shaik Himam Saheb[1]([⊠]) and G. Satish Babu[2]

[1] Faculty of Science and Technology, IFHE, Hyderabad, India
himam.mech@gmail.com
[2] Department of Mechanical Engineering, JNTUHCEH, Hyderabad, India

Abstract. This paper presents the mathematical modeling of 3RRR and 4 RRR Planar parallel manipulator and the performance analysis (Dexterity and manipulability indices) by using direct kinematic solutions, A parallel mechanism consists of the base of mechanism and the moving platform where the end-effectors are located. In this work the moving platform is replaced with revolute joint connected with end effectors, the base and the platform are connected by a series of linkages, first the direct kinematic relations are developed subsequently performance indices are compared for 4RRR and 3RRR parallel planar manipulators, with the addition of redundancy the 4RRR manipulators shows better performance indices than 3RRR parallel manipulators.

Keywords: Planar parallel manipulator · 3-RRR · 4-RRR · Redundant PPM · Degree of freedom · Dexterity index · Manipulability indices

1 Introduction

Generally, the robot manipulators are classified into two types, serial and parallel robots which have open, closed loop respectively. Last two decades, robot manipulators were designed analysed and operated using the open loop serial chains, most of the industrial manipulators are belongs to serial type, which has less pay load and less stiffness. In general most of the robots used in industries for pick and place, manufacturing applications, medical applications most of them are serial cantilevered structure where pay load capacity is less due to this problem parallel type robots are developed where the load carrying capacity is more the load carrying capacity is poor but besides it has the advantage of reaching large area of workspaces and dexterous maneuverability.

Joao et al. and Andres et al. compared 4RRR and 3RRR concluded that the 4RRR PPM has better dynamic performance [8, 9] The links tend to bend at high load on one hand, while on the other, to satisfy the strength requirements, the links become bulky, which leads to vibration at high speed. Though possessing a large work space, their precision positioning capability is rather poor due to their cantilever structure after a decade, parallel manipulators became a popular research topic and in general it was addressed as Stewart-Gough Platform. In the next section, motivations and need of

© Springer Nature Switzerland AG 2020
J. S. Raj et al. (Eds.): ICIDCA 2019, LNDECT 46, pp. 229–236, 2020.
https://doi.org/10.1007/978-3-030-38040-3_27

parallel manipulator is discussed. The following methodology is also can be applied to 3-RPR, 3-PRR PPMs. In this analysis geometric parameter variations are not considered even these parameters are effects the output of manipulator. First architecture of this paper is described. Then the dexterity indices of the 3RRR and 4RRR PPMs are calculated which is the measure of a manipulator to get different pose (position and orientation) for each point in the reachable work space.

About 3-RRR Parallel Planar Manipulator: Figure shows the parallel manipulator with three rotational joints P_i Q_i R_i and two links R_i, L_i are connected to each other, where i represents the 1, 2, 3 for respective joints (Fig. 1).

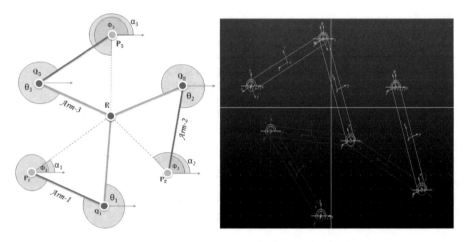

Fig. 1. Showing 3RRR planar parallel manipulator

Mathematical modeling of 3-RRR Planar parallel manipulator

$$(R_x - Q_{1x})^2 + (R_y - Q_{1y})^2 = L_1^2 \tag{1}$$

$$(R_x - Q_{2x})^2 + (R_y - Q_{2y})^2 = L_2^2 \tag{2}$$

$$(R_x - Q_{3x})^2 + (R_y - Q_{3y})^2 = L_3^2 \tag{3}$$

$$M_1 R_x + M_2 R_y + M_3 = L_1^2 - L_2^2 \tag{4}$$

$$M_4 R_x + M_5 R_y + M_6 = L_3^2 - L_2^2 \tag{5}$$

$$\begin{bmatrix} R_x \\ R_y \end{bmatrix} = \begin{bmatrix} M_1 & M_2 \\ M_4 & M_5 \end{bmatrix} \begin{bmatrix} L_2^2 - L_1^2 - M_3 \\ L_3^2 - L_1^2 - M_6 \end{bmatrix} \tag{6}$$

$$M_1 = 2Q_{1x} - 2Q_{2x}$$

$$M_2 = 2Q_{1y} - 2Q_{2y}$$

$$M_3 = \left(Q_{2x}^2 + Q_{2y}^2 \right) - \left(Q_{1x}^2 + Q_{1y}^2 \right)$$

$$M_4 = 2Q_{1x} - 2Q_{3x}$$

$$M_5 = 2Q_{1y} - 2Q_{3y}$$

$$M_6 = \left(Q_{3x}^2 + Q_{3y}^2 \right) - \left(Q_{1x}^2 + Q_{1y}^2 \right)$$

About 4-RRR Parallel Planar Manipulator: The 4 RRR planar parallel manipulator which is redundant type parallel manipulator the manipulator is consist of a moving platform and a fixed platform the moving platform is connected with four links in the proposed mechanism all the links of same length all four links have an actuated joint which are kept at equal distance like a square. For this mechanism the dexterity, manipulability and stiffness indices are calculated and compared with 3RRR Parallel Planar Manipulator (Fig. 2).

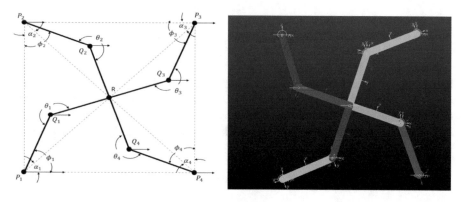

Fig. 2. Showing the 4RRR planar parallel manipulator

$$(R_x - Q_{1x})^2 + (R_y - Q_{1y})^2 = L_1^2$$

$$(R_x - Q_{2x})^2 + (R_y - Q_{2y})^2 = L_2^2$$

$$(R_x - Q_{3x})^2 + (R_y - Q_{3y})^2 = L_3^2$$

$$(R_x - Q_{4x})^2 + (R_y - Q_{4y})^2 = L_4^2$$

$$M_1 R_x + M_2 R_y + M_3 = L_1^2 - L_2^2$$

$$M_4 R_x + M_5 R_y + M_6 = L_3^2 - L_1^2$$

$$M_7 R_x + M_8 R_y + M_9 = L_4^2 - L_1^2$$

$$M_1 = 2Q_{1x} - 2Q_{2x}$$

$$M_2 = 2Q_{1y} - 2Q_{2y}$$

$$M_3 = \left(Q_{2x}^2 + Q_{2y}^2 \right) - \left(Q_{1x}^2 + Q_{1y}^2 \right)$$

$$M_4 = 2Q_{1x} - 2Q_{3x}$$

$$M_5 = 2Q_{1y} - 2Q_{3y}$$

$$M_6 = \left(Q_{3x}^2 + Q_{3y}^2 \right) - \left(Q_{1x}^2 + Q_{1y}^2 \right)$$

$$M_7 = 2Q_{1y} - 2Q_{4y}$$

$$M_8 = 2Q_{1y} - 2Q_{4y}$$

$$M_9 = \left(Q_{4x}^2 + Q_{4y}^2 \right) - \left(Q_{1x}^2 + Q_{1y}^2 \right)$$

Direct Kinematics: To solve any kinematic problem we have two ways i.e. direct and inverse kinematics, in direct kinematics the link length and link angle known to us and we will find end effectors position with position and orientation (pose) where as in inverse kinematics with end effectors pose the link lengths and angles are determined.

$$\left[Q_{ix} Q_{iy} \right] T = \left[P_{ix} + R_i \cos \varphi_1 P_{iy} + R_i \sin \varphi_i \right] T$$

By Applying Trigonometry i.e. law of cosines the following are expected

$$(Q_1 R)^2 = (P_1 Q_1)^2 + (P_1 R)^2 - 2(P_1 Q_1)(P_1 R) \cos(\alpha_1 - \varphi_2)$$

$$\alpha_1 = \text{Arccos} \frac{(P_1 Q_1)^{\wedge} 2 + (P_1 R)^{\wedge} 2 - (Q_1 R)^{\wedge} 2}{2(P_1 Q_1)(P_1 R)} + \theta_1$$

$$\varphi_2 = \text{Arctan} \frac{R_y - P_{1y}}{R_x - P_{1x}}$$

$$P_1R = \sqrt{(R_y - P_{iy})^2 + (C_x - P_{ix})^2}$$

$$R_x = x_{ai} + r_i \cos(\varphi_i) + l_i \cos(\theta_i)$$

$$R_y = y_{ai} + r_i \sin(\varphi_i) + l_i \sin(\theta_i)$$

$$R_x = -r_i \sin(\varphi_i)\varphi_i + l_i \sin(\theta_i)\theta_i$$

$$R_y = +r_i \cos(\varphi_i)\varphi_i + l_i \cos(\theta_i)\theta_i$$

$$R_x \cos(\theta) = -r_i \sin(\varphi_i)\varphi_i \cos(\theta) - l_i \sin(\theta_i)\theta_i \cos(\theta)$$

$$R_y \sin(\theta) = r_i \cos(\varphi_i)\varphi_i \sin(\theta) + l_i \cos(\theta_i)\theta_i \sin(\theta)$$

$$R_x \cos(\theta) + R_y \sin(\theta) = -r_i \sin(\varphi_i)\varphi_i \cos(\theta) - l_i \sin(\theta_i)\theta_i \cos(\theta)$$
$$+ r_i \cos(\varphi_i)\varphi_i \sin(\theta) + l_i \cos(\theta_i)\theta_i \sin(\theta)$$

$$R_x \cos(\theta) + R_y \sin(\theta) = r_i(\sin(\theta_i - \varphi_i))$$

The jacobian of 4-R$\underline{R}$R planar parallel manipulator is

$$J_x = \begin{bmatrix} \cos(\theta_1) & \sin(\theta_1) \\ \cos(\theta_2) & \sin(\theta_2) \\ \cos(\theta_3) & \sin(\theta_3) \\ \cos(\theta_4) & \sin(\theta_4) \end{bmatrix}$$

$$J_q = \begin{bmatrix} r_1 \sin(\theta_1 - \varphi_1) & 0 & 0 & 0 \\ 0 & r_2 \sin(\theta_2 - \varphi_2) & 0 & 0 \\ 0 & 0 & r_3 \sin(\theta_3 - \varphi_3) & 0 \\ 0 & 0 & 0 & r_4 \sin(\theta_4 - \varphi_4) \end{bmatrix}$$

And the Jacobian for 3-R$\underline{R}$R PPM

$$J_x = \begin{bmatrix} \cos(\theta_1) & \sin(\theta_1) \\ \cos(\theta_2) & \sin(\theta_2) \\ \cos(\theta_3) & \sin(\theta_3) \end{bmatrix}$$

$$J_q = \begin{bmatrix} r_1 \sin(\theta_1 - \varphi_1) & 0 & 0 \\ 0 & r_2 \sin(\theta_2 - \varphi_2) & 0 \\ 0 & 0 & r_3 \sin(\theta_3 - \varphi_3) \end{bmatrix}$$

Manipulability Indices: In kinematic evaluation of robot manipulators the manipulability index is a good measure for identifying the end effectors manipulating capacity in the workspace. It is clearly observed from literature that the usage of this index has high values and clearly identified that maximum dexterity position is the best place to place the object and end effectors position. These indices are introduced by Yoshikawa

[7] in 1985 further so many researchers are used these indices for performance evaluation. This indices are defined as the second root of the determinant product of the parallel manipulator Jacobian and its transpose, The normalized manipulability indice is ranges between 0 to 1.

$$\mu = \sqrt{\det(JJ^T)} = \sqrt{\lambda 1, \lambda 2, \ldots, \lambda m} = \sigma_1, \sigma_2, \ldots \sigma_m$$

where $\lambda 1, \lambda 2, \ldots, \lambda m$ are the eigen values of JJ^T and the $\sigma_1, \sigma_2, \ldots \sigma_m$ are the singular values of J [6].

The manipulability indices are calculated from the above Yoshikawa formulation for 3 RRR (maximum and minimum 5 nos) planar parallel manipulator which are tabulated below.

S. no.	θ_1	θ_2	θ_3	φ_1	φ_2	φ_3	Manipulability indices
1	10	10	70	70	70	100	0.063730095
2	130	70	70	40	10	130	0.069788301
3	10	10	70	130	70	40	0.077949008
4	70	70	130	10	160	160	0.08257052
5	130	70	70	160	100	10	0.083244601
6	10	40	130	130	160	40	0.917695
7	10	100	130	160	40	70	0.92467
8	10	40	130	70	100	10	0.938468
9	10	40	70	160	160	40	0.948954
10	10	40	70	160	160	160	0.962994

Similarly the manipulability indices are calculated from the above Yoshikawa formulation for 4RRR PPMS (maximum and minimum 5 nos) which are tabulated below.

S. no.	θ_1	θ_2	θ_3	θ_4	φ_1	φ_2	φ_3	φ_4	Manipulability indices
1	40	10	70	70	70	100	10	10	0.06393
2	100	70	70	40	10	130	130	70	0.069988
3	40	10	70	130	70	40	10	10	0.078149
4	70	70	130	10	160	160	70	70	0.082771
5	130	70	70	160	100	10	130	70	0.083445
6	10	40	130	130	160	40	10	40	0.917895
7	10	100	130	160	40	70	10	100	0.92487
8	10	40	130	70	100	10	10	40	0.938668
9	10	40	70	160	160	40	10	40	0.949154
10	10	40	70	160	160	160	10	40	0.963194

Dexterity indices are the indication of a manipulator to achieve different pose (position and orientation) for each point in the possible workspace of manipulator. The use of condition number which is the ratio of maximum singular value to minimum singular value it represents the amount of change in end effectors position and orientation. The dexterity indices ranges from zero to one, the dexterity indices explains the best position and accuracy point with this we can place the object to be manipulated with less energy. The following table showing the dexterity indices of 3RRR parallel planar manipulator (maximum and minimum 5 nos).

S. no.	θ_1	θ_2	θ_3	φ_1	φ_2	φ_3	Dexterity indices
1	10	10	70	70	130	160	0.08494
2	10	10	70	70	160	160	0.087707
3	10	10	70	70	100	160	0.089806
4	10	10	70	40	70	160	0.090514
5	70	10	70	40	100	10	0.090514
6	70	10	70	10	130	10	0.800451
7	40	10	70	160	100	160	0.898504
8	40	10	130	100	70	40	0.90402
9	40	10	130	160	160	40	0.916261
10	100	10	130	40	160	70	0.92467

The following table showing the dexterity indices of 4RRR parallel planar manipulator (maximum and minimum 5 nos).

S. no.	θ_1	θ_2	θ_3	θ_4	φ_1	φ_2	φ_3	φ_4	Dexterity indices
1	10	70	70	130	160	10	70	130	0.08394
2	10	70	70	160	160	10	70	160	0.086707
3	10	70	70	100	160	10	70	100	0.088806
4	10	70	40	70	160	10	40	70	0.089514
5	10	70	40	100	10	70	40	100	0.089514
6	10	70	10	130	10	70	10	130	0.899451
7	10	70	160	100	160	40	160	100	0.897504
8	10	130	100	70	40	40	100	70	0.90302
9	10	130	160	160	40	40	160	160	0.915261
10	10	130	40	160	70	100	40	160	0.93367

2 Conclusion

In this article the robot modelling and analysis is done, the 3RRR and 4RRR i.e. redundant and hyper redundant parallel manipulators are compared in terms of dexterity and manipulability the maximum and minimum dexterity and manipulability is obtained at different positions of robot that all values are tabulated with joint angles the results are obtained from Matlab and modelling is done in the msc Adams software, the other effects of robot like stiffness sensitivity are not taken into consideration while doing this analysis, though the degree of freedom of joint is less the extra actuators are taken for analysis as the degree of freedom is 2 but taken 3RRR, 4RRR analysed the performance of the robot manipulator the 4RRR manipulator shown good performance at orientation angles from 10 to 70°. With this can conclude that the addition of redundancy to the parallel planar manipulators achieves good performance results.

References

1. Saadatzi, M.H.: Multi-objective scale independent optimization of 3-RPR parallel mechanisms. In: 13th World Congress in Mechanism and Machine Science, Guanajuato, México, 19–25 June 2011 (2011)
2. Selvakumar, A.A.: Simulation and workspace analysis of a tripod parallel manipulator. World Acad. Sci. Eng. Technol. **33**, 583–588 (2009)
3. Merlet, J.P.: Direct kinematics of planar parallel manipulators. In: Proceedings of IEEE International Conference on Robotics and Automation, vol. 4, pp. 3744–3749. IEEE (1996)
4. Rezania, V., Ebrahimi, S.: A comparative study on the manipulability index of RRR planar manipulators. Modares J. Mech. Eng. **14**, 299–308 (2015)
5. Relative kinematic analysis of serial and parallel manipulators. https://www.researchgate.net/publication/329788391_Relative_Kinematic_Analysis_of_Serial_and_Parallel_Manipulators
6. Salem, A.A., Khedr, T., El Ghazaly, G., Mahmoud, M.: Modeling and performance analysis of planar parallel manipulators, pp. 13–23 (2018). https://doi.org/10.1007/978-3-319-64861-3_2
7. Yoshikawa, T.: Manipulability of robotic mechanisms. Int. J. Robot. Res. **4**(2), 3–9 (1985). https://doi.org/10.1177/027836498500400201
8. Fontes, J., Venter, G.S., Silva, M.: Assessing the actuation redundancy trade-off effects on the dynamic performance of planar parallel kinematic machines (2015). https://doi.org/10.6567/iftomm.14th.wc.os13.010
9. Ruiz, A.G., Fontes, J., da Silva, M.M: The impact of kinematic and actuation redundancy on the energy efficiency of planar parallel kinematic machines. In: Proceedings of the XVII International Symposium on Dynamic Problems of Mechanics, DINAME 2015, February 2015 (2015)
10. Saheb, S.H., Babu, G.S.: Sensitivity analysis of 3-PRRPlanar manipulator. Int. J. Eng. Res. Manag. (2016)
11. Zarkandi, S., Vafadar, A., Esmaili, M.: PRRRRRP redundant planar parallel manipulator: kinematics, workspace and singularity analysis. In: Proceedings of the IEEE Conference on Robotics, Automation and Mechatronics, RAM, pp. 61–66 (2011)

Public Auditing Scheme for Cloud Data Integrity Verification

Shivarajkumar Hiremath[1(✉)] and R. Sanjeev Kunte[2]

[1] Department of ISE, SKSVMACET, Lakshmeshwar, India
`shivaraj2323@gmail.com`
[2] Department of CSE, JNNCE, Shivamogga, India
`sanjeevkunte@gmail.com`

Abstract. Cloud computing is one that provides an on-demand accessibility for computer resources, particularly data storage and computing power by the user. It is generally used to describe data hubs that remain accessible to users over the internet. The users place their data in cloud to save their local memory. The cloud service providers will control the data that is placed in the cloud. As users upload their data in cloud that is controlled by service providers, they lose control over their data. Hence, we must have few safety measures to retain the user's data securely. Here, we described an auditing system that makes use of external auditor. We utilized Advanced Encryption Standard (AES) and Merkle Hash Tree (MHT) for ensuring integrity of cloud data. This work is accomplished in Amazon EC2 ubuntu16.04 server instance. From the results gained, we determine that our scheme is intact and considers a firm time to process files.

Keywords: Third Party Auditor (TPA) · Merkle Hash Tree · Cloud storage · Public auditing

1 Introduction

In simple terms, cloud computing is the distributed on demand accessibility of computer resources like data storage and virtual computing power. It changed the world around us and now it is everywhere. Dropbox, Google drive, Amazon S3 and many others are extensively deployed for storing data. By 2020, it is estimated by international data corporation (IDC) that about 40% of data or information might be included in cloud computing [1]. The cloud storage generally involves two entities, cloud servers and cloud service providers. There are many advantages of cloud storage like increased flexibility, minimization of software/hardware requirements, pay as per use etc. [2].

In spite of many advantages of cloud technology, it raises a security concern specifically on confidentiality, availability and data integrity. There was a leak of google mails user data in February 2009 and March 2011. Corruption of AmazonS3, because of its internal failure triggered by mismatching of files with client's hashes is another security breach incident [2].

It is obvious that cloud service provider cannot be trusted completely. Usually, CSP will execute operations properly and doesn't change or deletes user's data. But in few cases CSP may act negatively to the safety of user's data. They might remove data that

© Springer Nature Switzerland AG 2020
J. S. Raj et al. (Eds.): ICIDCA 2019, LNDECT 46, pp. 237–246, 2020.
https://doi.org/10.1007/978-3-030-38040-3_28

is not frequently retrieved to save their storage space or they might hide data corruptions triggered by internal errors and hackers to maintain the reputation [3]. Security issues like data integrity, availability and confidentiality are the key hurdles in cloud technology to adopt completely.

When the CSP cannot be trusted completely, an important aspect is how to achieve outsourced data guarantee. One simple solution is that, the entire data can be downloaded at regular intervals to check for its integrity. But the problem is, it will consume a high communication cost. Hence for delicate and private data, we must have few security procedures to deliver safety of cloud data.

In the proposed system, importance is given to attain data confidentiality as well as data integrity by developing a public or external auditing system that enhances cloud data security.

The remaining part of the paper is prepared as follows. A literature survey based on data integrity techniques is presented in Sect. 2. A proposed public auditing system is enlightened in Sect. 3. The Sect. 4 describes the analysis of our results and discussions. Lastly, an conclusion is specified in Sect. 5.

2 Literature Survey on Data Integrity Techniques

In order to resolve integrity investigation in cloud storage, many works have been proposed based on different models. In the earlier period, two important solutions had been proposed. Firstly, provable data possession (PDP) and the later one is proof of retrievability (POR). The method of generating proof of clients file by not retrieving the complete file is considered as proofs of data possession. If the data of client is extracted, then the system is considered as proof of retrievability [4].

2.1 Survey on PDP Schemes

The scheme based on PDP is primarily explained by Ateniese et al. [5]. They proposed public auditability titled "Provable Data Possession (PDP) at Untrusted Stores". For auditing remote data, they used homomorphic linear authenticators (HLA) based on RSA. The structure generates proofs of ownership by sampling unspecified sets of blocks that significantly reduces I/O costs. The data owner uses metadata (hash function) to validate the proof. The challenge/response procedure communicates with small volume of data, which decreases network communication. Conversely, as the auditing in this scheme uses the linear grouping of sampled blocks uncovered to peripheral auditor, the method is not completely privacy preserving and might disclose user's data to the auditor.

Dynamic data operations are supported in PDP by Ateniese et al. [6]. They developed a dynamic PDP procedure entitled "Scalable and Efficient Provable Data Possession". They constructed an effective provable safe PDP technique built exclusively on symmetric key cryptography. Proposed PDP technique allows operations like data modification, deletion and append. It exceeds the prior work in many ways containing storage, bandwidth etc. and specially the provision for dynamic operations. Yet, as it is built on symmetric key, it is inappropriate for third-party (public/external) verification.

However, due to less randomness in the challenge, the servers can cheat clients by using previous metadata. The number of challenges are restricted and fixed.

Erway et al. [7] explored construction of dynamic PDP. The PDP scheme of [5] is enhanced by Erway et al. to allow updates to stored file by adopting rank based authenticated skip lists. The scheme is completely dynamic form of the PDP model. In particular, they tried to eliminate index formation in the tag computation in [5] to support updates like block insertion. But this method imposes computational complexity at client side losing efficiency.

To assure the data integrity of the file placed in cloud storage, authors Shah et al. [8] suggested message authentication code (MAC) based PDP. By using a set of secret keys, the owner calculates a MAC of the file. Later he saves that MAC of the file locally before storing it in cloud storage. The local system stores only the MAC value and drives the file to CSP. To confirm data integrity, client generates a request to CSP to retrieve the file from cloud. Once CSP acknowledges with the MAC of the file, client compares MAC with the previously stored MAC. The proposed system cannot be adopted for large files.

A successful PDP scheme is described by Zhu [9] for distributed cloud storage. Their proposed system supports the scalability of service and data migration. They considered several CSP's to store and preserve the client's data. The cooperative PDP (CPDP) uses hash index hierarchy and homomorphic verifiable response. This system presents lesser computation cost and less communication overheads. But due to its high complexity, the large sized files are influenced by the bilinear mapping operations.

2.2   Survey on POR Schemes

The POR scheme primarily was developed by Jules et al. in 2007 [10]. The proposed system uses sentinel based scheme and error correcting codes for achieving both integrity and retrivability. The insertion of sentinals and error corrected codes leads to storage overhead at the server side. The challenge algorithm offers restricted number of implementations. This is because the number of sentinels embedded in the file are precomputed. Also, public or external auditability concept is not explained in their scheme.

Shacham and Waters [11] in the year 2008 developed an enhanced version of POR termed as Compact POR. Their work consists of two factors, the first, based on the concept of pseudorandom functions (PRFs) and second, based on BLS signatures. Pseudo random functions makes a POR scheme safe in the ideal model. BLS signatures supports POR scheme to allow for public verifiability. While [10] offers restricted number of challenge implementations, compact POR permits an unlimited quantity of queries that will result in less communication overhead.

The first dynamic POR based on fairness concept was introduced by Zheng and Xu [12]. They presented a fairness dynamic POR (FDPOR) by employing two tools namely range based 23 trees and the later is incremental signature system considered as hash compress and signature (HCS). Unfortunately, the concept of how to update the redundant encoded data is not mentioned. Also the concept of public verifiability is not considered in their scheme.

In 2012, authors Mo, Zhou and Chen [13] proposed a dynamic POR with O (log n) complexity. They proposed dynamic POR by using merkle hash tree (MHT) and B+ tree named merkle B+ tree (CMBT) integrated through BLS signature and accomplished the low communication complexity. This model assumes a semi trusted CSP and will detect data corruptions even if the CSP attempts to mask them. This scheme also supports dynamic update.

A scheme that validates cloud storage retrievability based on third party auditing was developed by Qin et al. [1]. This proposed model meets the issue of data integrity, confidentiality and data recovery. It has 3 phases like setup, verification and data recovery and extraction. The assurance of TPA is achieved through mutual authentication protocol. This method has computational overhead cost.

3   Proposed System

This section presents the proposed public auditing system which delivers confidentiality and integrity of cloud data storage. The main intention of proposed work is to store the cloud data in a secure manner. To provide confidentiality to cloud data, AES is used intended for encryption and Merkle Hash Tree (MHT) is built using SHA-2 to generate hash function.

We used merkle tree because it is a construction that provides a secure verification of data blocks in a file. The Merkle hash trees are formed by considering pairs of nodes and repetitively hashing those nodes till we get only single hash called the merkle root. These are built on the concept of bottom up approach from the hashes of discrete transactions. Every leaf nodes indicates the hashes of transactional data whereas intermediate non-leaf nodes indicates the hashes of its previous hashes as presented in Fig. 1.

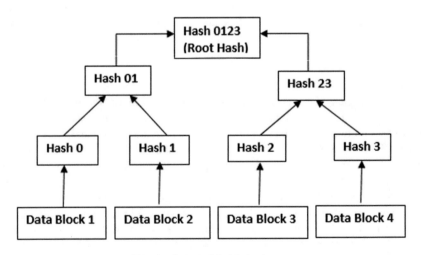

Fig. 1. Sample Merkle hash tree

The proposed scheme consists of 3 entities, cloud user, cloud server and third party auditor. The cloud user is one who wishes to store file in cloud storage, cloud server provides services to the users to reside data and it is controlled by CSP's and third party auditor is an external auditor who reviews the data files for its integrity. The following Fig. 2 demonstrates the model of our proposed scheme.

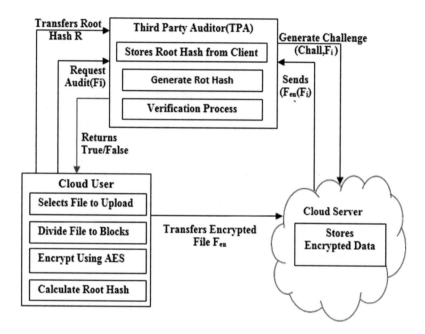

Fig. 2. Architecture design of proposed scheme

The user selects and splits file in to blocks that is to be uploaded in cloud. AES encryption algorithm is applied on each block and hash is generated for the file blocks by using SHA-2 algorithm. The strategy is to encrypt the data first to achieve confidentiality and then merkle hash tree is constructed by calculating hash for each blocks of data till we get single root hash. Encrypted data or file is placed in cloud.

If client's wishes to test the integrity of file kept in the cloud then he will directs a request to TPA to audit a file. Upon request from client, TPA challenges cloud to send a file for audit. Once cloud acknowledges to TPA by sending required file, TPA performs verification process and later intimates client regarding the status of a file stored in cloud.

3.1   Providing Data Confidentiality

The process of keeping users data secret from the cloud service providers is regarded as data confidentiality. The system should keep users data safe from CSP's if they try to get users delicate data. The users can place their data in the cloud after successfully validating with proper username and password. After proper authentication, the user will select the file that is to be placed in cloud. The file is divided in to number of small data blocks and AES encryption is applied on all data blocks. Later SHA-2 is applied to get root hash of the file. Merkle hash tree is developed by repeatedly hashing file blocks till we get single merkle root. The root hash R is forwarded to TPA and the file that is in encrypted form is uploaded to cloud. Since the cloud contains the encrypted form of a file, the CSP cannot access user's delicate data. Thus the proposed system achieves confidentiality.

3.2 Achieving Data Integrity

In the proposed work, the auditing is done by TPA to evaluate the integrity of a file F_i. The user posts or directs a request to TPA for performing data audit of a file Fi that is placed in cloud. After getting a request for an audit from user, TPA produces a challenge response (chall, Fi) protocol to the cloud to obtain the data file Fi that is to be audited. Cloud responds to TPA's request by forwarding the requested file Fi. After cloud acknowledges with file Fi, TPA starts the process of auditing a file. TPA determines the root hash R of the file Fi employing the same SHA-2 algorithm.

During the evaluation process, the TPA examines the new root hash value of file Fi with the root hash of file Fi stored earlier. If the new root hash value of file Fi calculated by TPA (Fi, Rnew) is same as TPA's R that was stored earlier (Fi, Rstored), then integrity is achieved. Later, the TPA acknowledges the status of file to the user. The entire working process is explained with the following steps.

(1) Client selects and splits the file Fi in to number of data chunks or blocks that is to be uploaded in cloud.

$$D = \{b1, b2, b3............bn\}$$

(2) The data chunks are then encrypted using AES and hash is created for each and every data block using SHA-2 algorithm.

$$Hi = H(Esk(bi)) \text{where bi is ith data block}$$

$$S = \{Hi\} \text{where Hi is hash of ith data block and S is set of hashes.}$$

(3) Merkle Hash Trees are built by recurrently hashing pairs of nodes until we get single merkle root

$$R = H(\sum Hi) \text{or} R = H(S) \text{where R represents Root Hash}$$

(4) Client uploads encrypted file Fen to cloud and transfers root hash R (H(S)) to TPA
(5) Client requests to TPA to audit file Fi where Fi represents ith file.
(6) TPA accepts request from client and generate challenge (chall, Fi) to the cloud server to send requested file.
(7) Cloud responds to TPA's request by forwarding (Fen (Fi)) for the requested file Fi

(8) For the file (Fen (Fi)) received from the cloud, TPA calculates the root hash value by using the same steps as of client. If the new root hash value of file Fi calculated by TPA (Fi, Rnew) is same as TPA's R that was stored earlier (Fi, Rstored), then integrity is achieved.

$$\text{i.e. if}\left((Fi, Rstored) = (Fi, Rnew)\right)$$
$$\text{res} \leftarrow \text{true}$$
$$\text{else}$$
$$\text{res} \leftarrow \text{false}$$

(9) Returns status of the file (True/False) to the client.

The proposed model delivers a technique to securely keep the data in cloud with the help of MHT and AES to achieve its integrity.

4 Results and Discussions

Here, within this segment, we provide an experimental setup and assessment of the proposed system. Python programming is used to implement the proposed scheme and is executed in amazon ubuntu EC2 instance by launching a 64-bit Ubuntu Server 16.04 instance. An experimental construction of the proposed system is as shown below.

- Programming Language: Python
- Virtual Instance: Amazon EC2's Ubuntu Server 16.04(Intel Xeon Family 1 vCPUs, 2.5 GHz, 1 GB memory)
- Input - Files ranging from 100 KB to 1000 KB
- Output - Status of file (file is intact/corrupted)

Merkle Hash Tree is constructed by repeatedly hashing each blocks of file till root hash (merkle root) is obtained. The size of merkle root is 32 bytes and represents a summary of all transactions data.

To represent the hashes of merkle tree nodes, we considered a data file of size 4 KB. The file is separated or divided in to data blocks of 1 KB such that we get four data chunks 1, 2, 3 and 4. Hash is applied for each data chunk using SHA2 algorithm to get leaf nodes h (1), h (2), h (3) and h (4). Once we get hashes of each data chunks, SHA2 is again applied for both h (1) and h (2) to obtain the hash of parent node 1. Similarly, hash is calculated for h (3) and h (4) to get hash of parent node2. Finally, root hash (merkle root) is obtained by applying SHA2 algorithm to parent node 1 and parent node 2. The following Table 1 represents the 32 byte hash values that we obtained for leaf nodes, internal nodes and root hash during the analysis of our proposed scheme.

Characteristics like, time for encryption, and time to audit files are considered for files of varying sizes. To attain continuous bandwidth, we measured the data files of the range 100 KB to 1000 KB. In Table 2, the various results obtained by executing our proposed model are presented.

Following the results obtained, we determine that the proposed model will take finite time to encrypt and audit files. Note that, the time for encryption increases proportionally as the volume of the file increases. The encryption time increases because, the increase in file size will result in number of blocks and obviously more time to encrypt.

Table 1. Sample hash values obtained for proposed scheme

Data blocks	Hash of leaf nodes	Hash of internal nodes	Root hash (Merkle root)
1	H(1): 5feceb66ffc86f38d95278 6c6d696c79c2dbc239dd4e914b 6279d73a27fb57e9	H(parent 1): fa13bb36c022a69 43f37c638126ac288fc8d008eb5 a9fe8fcde17026807feae4	Root Hash: 862 532e6a3c9aafc2016810598ed0cc3025a f5640db73224f586b6f1138385f4
2	H(2):6b86b273ff34fce19d6b 8044eff5a3f5747ada4eaa22f1d49c0 1e52ddb7875b4b		
3	H(2):6b86b273ff34fce19d6b80 44eff5a3f5747ada4eaa22f1d49 c01e52ddb7875b4b	H (parent 2): 70311d9d203b 2d7e4ff70d7fce219f82a4fcf73 a110dc80187dfefb7c6e4bb87	
4	H(2):6b86b273ff34fce19d6b80 44eff5a3f5747ada4eaa22f1d49 c01e52ddb7875b4b		

Table 2. Results of proposed system

Sl. no	File size (in KB)	Time to encrypt (in Seconds)	Time taken to audit by TPA (in Seconds)
1	100	0.22	1.02
2	200	0.48	1.03
3	300	0.71	1.05
4	400	1.02	1.07
5	500	1.31	1.11
6	600	1.56	1.15
7	700	1.73	1.18
8	800	1.85	1.22
9	900	1.97	1.26
10	1000	2.10	1.33

We characterized the comparison results of our proposed scheme with Sudarsan Rao et al. [14]. We retained the same file size as that of [14] and compared the execution of time taken for encrypting a file. We observed that, in terms of encrypting files, our system will takes less time for encryption over existing scheme. The comparison of results are shown in Fig. 3.

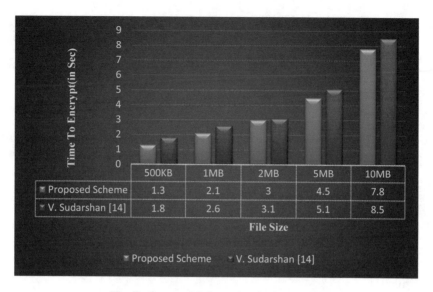

Fig. 3. Proposed Scheme Vs. Sudarsan [14]

5   Conclusion

Cloud Computing offers cloud users and numerous organizations or administrations to deposit enormous quantity of data in cloud. The key issue is to evaluate or recognize the integrity of file that is kept in cloud. We described a TPA auditing scheme that achieves data integrity and data confidentiality with the concept of AES+MHT. TPA performs auditing without recovering the data copy of a cloud user. The integrity of a file or data is assessed by TPA with the help of root hash or merkle root. In the proposed model, TPA examines that the file in the cloud is reformed or damaged and returns the status of file to the user. The system is developed and tested in Ubuntu 16.04 Amazon EC2 instance. The results obtained are promising. In future, data recovery technique and data dynamics are to be accomplished.

References

1. Qin, Z., Song, Y., Zhang, Q., Huang, J.: Cloud storage retrievability based on third party audit. In: International Workshop on Cloud Computing and Information Security (CCIS 2013), pp. 303–307, October 2013
2. Albeshri, A., Boyd, C., Nieto, J.: A security architecture for cloud storage combining PoR and fairness. In: CLOUD COMPUTING 2012: The Third International Conference on Cloud Computing, GRIDs, and Virtualization, pp. 30–35 (2012). ISBN 978-1-61208-216-5
3. Ren, Z., Wang, L., Wang, Q., Xu, M.: Dynamic proofs of retrievability for coded cloud storage systems. IEEE Trans. Serv. Comput. **PP**(99) (2015)
4. Kumar, D.: Review on task scheduling in ubiquitous clouds. J. ISMAC **1**(01), 72–80 (2019)

5. Ateniese, G., Burns, R.C., Curtmola, R., Herring, J., Kissner, L., Peterson, Z.N.J., Song, D. X.: Provable data possession at untrusted stores. In: Proceedings of ACM Conference on Computer and Communications Security, CCS 2007, pp. 598–609 (2007)
6. Ateniese, G., Pietro, R.D., Mancini, L.V., Tsudik, G.: Scalable and efficient provable data possession. In: Proceedings of the 4th International Conference on Security and Privacy in Communication Networks, Secure Communication, pp. 1–10 (2011)
7. Erway, C., Küpçü, A., Papamanthou, C., Tamassia, R.: Dynamic provable data possession. ACM Trans. Inf. Syst. Secur. (TISSEC) **17**(4), 15 (2015)
8. Shah, M.A., Baker, M., Mogul, J.C., Swaminathan, R.: Auditing to keep online storage services honest. In: Proceedings of 11th Workshop on Hot Topics in Operating Systems (HotOS 2007). Berkeley, CA, USA, pp. 1–6. USENIX Association (2007)
9. Zhu, Y., Ahn, G., Yu, M., Hu, H.: Cooperative provable data possession for integrity verification in multicloud storage. IEEE Trans. Parallel Distrib. Syst. **23**(12), 2231–2244 (2012)
10. Juels Jr., A., Kaliski, B.S.: POR'S: proofs of retrievability for large files. In: Proceedings of the ACM Conference on Computer and Communications Security, CCS 2007, pp. 584–597 (2007)
11. Shacham, H., Waters, B.: Compact proofs of retrievability. J. Adv. Cryptogr. **26**(3), 442–483 (2013)
12. Zheng, Q., Xu, S.: Fair and dynamic proofs of retrievability. In: First ACM Conference on Data and Application Security and Privacy, CODASPY 2011, San Antonio, TX, USA, 21–23 February 2011
13. Mo, Z., Zhou, Y., Chen, S.: A dynamic proof of retrievability (PoR) scheme with O(log n)complexity. In: ICC 2012, pp. 912–916 (2012)
14. Sudarsan Rao, V., Satyanarayana, N.: Experimental analysis and comparative study of secure data outsourcing schemes in cloud. Int. J. Cloud Comput. **8**(1), 83–101 (2019)

Implementation of Words and Characters Segmentation of Gujarati Script Using MATLAB

B. Shah Gargi[1]([⊠]) and S. Sajja Priti[2]

[1] Department of Computer Science, VNSGU, Surat, India
gbshah@vnsgu.ac.in
[2] G. H. Patel Post Graduate Department of Computer Science and Technology,
SPU, VVN, Vallabh Vidyanagar, India
priti@pritisajja.info

Abstract. From past decades, many research works have been carried out for identifying characters from images in Gujarati language. However, a few solutions are available for segmentation of words and characters in Gujarati language from a text file. In this paper, we propose a novel algorithm which considers input from a text file containing Gujarati script and segments words and characters. The proposed algorithm has been validated with 10 numbers written in words, each with varying length, implemented using MATLAB. The screenshots are attached with this paper, which shows segmentation of words and characters. This system is considered as a subsystem of a much bigger system, where Gujarati text is to be converted into to its equivalent speech. Without proper segmentation of words and characters, it is not possible to achieve the intended task. Prior to this, existing solutions and tools are analysed and the summary is presented in the literature survey. Based on the limitations found through the survey, the proposed research work is designed. An experiment is also carried out and discussed in detail in this paper along with the results achieved. At end, conclusion and possible future enhancements are also presented.

Keywords: Character encoding · Consonants · Vowel modifier · Gujarati language · Text to speech · Synthesizer

1 Introduction

The Indo-European language family has Indo-Aryan languages and Gujarati language is one of them. It is adapted from the Devanagari script [1]. Total of 50 million people from India are speaking Gujarati language [2]. This mass also includes physically impaired and illiterate people. The objective behind this research work is to develop text to speech system which will take in from a text document, segments individual characters of a string and which then be used to record in a sound (wave) file to develop a synthesizer. In current scenario, many open source Text to Speech systems are available, which reads data from an image, but none of them takes input from the text file. So, our aim is to develop an application which takes input from the text file.

J. S. Raj et al. (Eds.): ICIDCA 2019, LNDECT 46, pp. 247–256, 2020.
https://doi.org/10.1007/978-3-030-38040-3_29

The syllable (Phonets) of Gujarati language are briefly divided as follows, which includes 33 consonants (Fig. 1), 12 vowels, 12 Matras (Fig. 2), 4 diacritics (Fig. 3), 1 halanta (Fig. 4) and 1 nukta (Fig. 5). Each individual are individual syllable (phonet) in Gujarati language [3].

Gujarati Consonant											
ક	ખ	ગ	ઘ	ઙ	ચ	છ	જ	ઝ	ટ	ઠ	ડ
ka	kha	ga	gha	cha	chha	ja	za	ta	tha	da	dha
ણ	ત	થ	દ	ધ	ન	પ	ફ	બ	ભ	મ	ય
aNa	ta	tha	da	dha	na	pa	fa	ba	bha	ma	Ya
ર	લ	વ	સ	શ	ષ	હ	ળ	ક્ષ	જ્ઞ		
ra	la	va	sa	tha	shha	ha	ala	ksha	gna		

Fig. 1. Consonants

Fig. 2. Vowels and Matras/Vowel Modifier

ં	Anushwar
ઁ	Chandrabindu
ઃ	Visarga
ઽ	Avagraha

Fig. 3. Diacritics

Fig. 4. Halanta

Fig. 5. Nukta

2 Literature Survey

Mohammed Javed, P. Nagabhushan, B.B. Chaudhari proposed an algorithm for segmentation of compressed text documents in to lines, words and characters for Bengali, Kannad and English script [4].

Character recognition as an aid to the visually handicapped was the first attempt by the Russian Scientist Cyurin in 1900 [5].

According to Shailesh A. Chaudhari and Ravi M. Gulati, They have classified pattern of bilingual English – Gujarati digit. Hybrid statistical and structural feature and template matching classifier are used for classification purpose. The achieved classification rate of 98.30% and 98.88% for Gujarati and English digit respectively [6].

According to R.C. Tripathi and Vijay Kumar, They have suggested Neural Network for character recognition, but unable achieve success rate at an impressive level, it results at worse level [7].

Amit Choudhary, Rahul Rishi and Savita Ahlawat have focused on mining features from image of handwritten English character using binarization technique. Multi-layered feed forward artificial neural network was used as a classifier. The combine efforts of binarization features and the neural network classifier employing back-propagation algorithm, which achieves accuracy of 85.62% [8].

Manthan Khopkar has proposed a new approach to zone based segmentation and numeral recognition for Gujarati numbers. The algorithm proposed by him operate on input image and efficiently recognize the individual characters. The network has been trained and tested for Arial font type in the Gujarati alphabet set. The system identifies individual character with an accuracy of 81.5% [9].

Nikisha Jariwala, Bankim Patel have surveyed different TTS systems. They have observed that work is found on Text-to-Speech conversion system. But very less amount of work is done on Gujarati Text-to-Speech system. There are several methodology and techniques to develop TTS system. If full-fledge Gujarati Text-to-Speech conversion system is developed, then visually impaired and illiterate mass will get advantage of it [10].

Mitul Modi, Fedrik Macwan, Ravindra Prajapati, presented different techniques through which GOCR (Gujarati Optical Character Recognition) is possible. In this technique, character is identified first, segmented thereafter, then noise reduction takes place by feeding resultant character to pre-processor. After classification, original symbol string is reconstructed again by grouping identified characters, and context may then be applied to detect and correct errors [11].

Milindkumar Audichya, Jitendrakumar R. Saini, used an open-source Tesseract OCR for different font, font style and font size and they have found that, it gives some ambiguity for complex and similar looking characters [12].

Payel Rakshit, Chayan Halder, Subhankar Ghosh and Kaushik Roy presented tri-level segmentation of Bangla handwritten text, which includes line, word and character. Input data are taken with different writing instruments along with people of different age, gender and education qualification group. They have achieved 90.46% accuracy for line segmentation, 90.06% accuracy for word segmentation and 75.97% of average accuracy achieved for character segmentation [13].

Parul Sahare and Sanjay B. Dhok have worked for multilingual printed and handwritten documents for character recognition and proposed two algorithms under their research work. Heuristic based and SVM algorithms are integrated for segmenting characters. Maximum SR of 98.66% is achieved for Latin script database. Three new structural geometry based features: FCDF, FCCF and NCF are proposed for character recognition. The accuracy achieved on numeral database is more higher than the database containing alphabets with proposed algorithms [14].

Vishal Naik and Apurva A. Desai have used hybrid features for handwritten Gujarati character recognition and proposed an algorithm for the same. The proposed algorithm was compared with SVM (linear) and SVM (polynomial) classifiers. 94.13% of accuracy and 0.103 s of average accuracy execution time per stroke is achieved with proposed algorithm. They have concluded that the accuracy of second layer classifier depends upon the result clarity of first layer [15].

Ajay P. Singh and Ashwin kumar Kushwaha have focused more on Brahmi script. They have reviewed for various segmentation approaches with respect to different languages and determined the best method for segmentation of Brahmi script-based 'Rumandei inscription'. But this method may not produce a similar satisfactory result of other inscription of Brahmi [16].

We tried to observe the limitations of research carried out from literature survey. Our observation says that all work have been done for image processing. Data are retrieved from images, processed and then words and characters are identified. Very less amount of work has been done for text inputted by text document. Apart from this, more works are carried out on languages other than Gujarati language. Taking all this in to consideration, our research work focuses on Gujarati language as well as input will be taken from text document.

The presented work proposes, word and character segmentation inputted by a text file in Gujarati language. The word and character segmentation algorithm is proposed in this paper, which is implemented by MATLAB programming in windows environment.

This paper is organized in following sections. Section 3 is the proposed algorithm, Sect. 4 presents experimental results along with discussion and Sect. 5 concludes the research work.

3 Proposed Algorithm

We have developed following algorithm for segmentation of word and character (in terms of phonet) of Gujarati language. The algorithm was implemented in "MATLAB" programming environment on Windows 10 platform. Numbers in words are taken as an input test data and algorithm works successfully and gives an accurate output. The input data are provided by the text file.

The default character encoding style of MATLAB is 'Windows – 1252', which we have changed to 'UTF-8' and also the text file was saved with 'UTF-8' encoding style.

Algorithm : Segmentation of words and characters

Input : Text file containing Gujarati numbers written in words

Output : Segmented words and characters

1   feature('DefaultCharacterSet') ;
     // feature function displays default character set of MATLAB (It is 'US-ASCII' by default)

2   feature('locale') ;
     //Displays current character encoding , it is Windows – 1252 by default

3   **if** CharacterEncoding $\neq$ 'UTF – 8'

4   CharacterEncoding $\leftarrow$ 'UTF-8' ;
                    // using slCharacterEncoding() function

5   **End**

6   filename $\leftarrow$ 'C:\Matlab Programs\TextFile1.txt' ;

7   fid $\leftarrow$ fopen(filename) ;

8   str $\leftarrow$ READ data from file pointed by fid ;

     // DETERMINE number of words in a string.

9   NoOfWords $\leftarrow$ length(str{1}) ;

     // DISPLAY each word individually

10   **for** I $\leftarrow$ 1 **to** NoOfWords **do**

11   DISPLAY str{1}{I} ;

12   **End**

     // DISPLAY character of each word individually

13   **for** I $\leftarrow$ 1 **to** NoOfWords **do**

14   L $\leftarrow$ length(str{1}{I} ;

15   **for** J $\leftarrow$ 1 **to** L **do**

16   DISPLAY s(j) ;

17   **end**

18   **end**

Note: The feature('locale') reads the user locale and system locale of windows platform. The user locale and system locale must be the same value. If not, it might display garbled text or incorrectly displayed characters.

In this proposed algorithm, the first step displays character set supported by MATLAB using feature('DefaultCharacterSet') function, which is 'US-ASCII' by default. In the second step character encoding pattern is displayed using feature('locale') function, which is 'Windows-1252'. That needs to be changed to 'UTF-8' as text file is saved with 'UTF-8' character encoding pattern, which is done in third step. Sixth and seventh steps opens the file to be read and file id is returned by fopen() function, which then stored to variable 'fid' for further use. Initially fid points to the first character of the string in file, using which entire string from the file is read in 'str' variable. Then after first words are segmented and then characters are segmented of each word individually. The length() function gives number of words in a sting, which is stored in 'NoOfWords' variable. The 'str' ultimately becomes two dimensional array, the row dimension the sentence number and column dimension is words in each sentence. Steps from 10 to 12, displays each word of a sentence. Same way from steps 13 to 18, length of each word is identified and then characters are displayed. By this way, the algorithm is implemented and successfully segmented words and characters from Gujarati text inputted by text file.

4 Experimental Result

To analyse an algorithm, the input test data were taken from the text file saved with 'UTF-8' character encoding style. Number written in words of Gujarati language are taken as an input each with varying length. The experimental result shows Number of words in a string, individual word, length of each word, and each individual characters of a word (Table 1).

The experimental result shows that, vowels are considered as a separate character while consonants and vowel modifiers are considered as individual characters. Following screen shots shows results of an algorithm implemented using MATLAB. Which includes display of character set and character encoding pattern (Fig. 6), word segmentation (Fig. 7), length of words and character segmentation (Figs. 8 and 9).

Table 1. Experimental results

Sr. No	Inputted String (from text file)	Array Size	No. of Words	Individual Words	Length of Each word	Individual Characters of Word	Status / Remark
1	બે	1 x 1	1	બે	2	બ , ે	Successfully Separated
2	સાત	1 x 1	1	સાત	3	સ , ા , ત	Successfully Separated
3	અગિયાર	1 x 1	1	અગિયાર	6	અ , ગ , િ , ય , ા , ર	Successfully Separated
4	ત્રણસો નવ	2 x 1	2	ત્રણસો	6	ત , ્ , ર , ણ , સ , ો	Successfully Separated ('ત્ર' is not an individual consonant, it is made up of 'ત' and 'ર'
				નવ	2	ન , વ	
5	આઠ હજાર છ	3 x 1	3	આઠ	2	આ , ઠ	Successfully Separated (Vowels are considered as 1 single character like 'આ')
				હજાર	4	હ , જ , ા , ર	
				છ	1	છ	
6	નવ હજાર સાતસો પચાસ	4 x 1	4	નવ	2	ન , વ	Successfully Separated
				હજાર	4	હ , જ , ા , ર	
				સાતસો	5	સ , ા , ત , સ , ો	
				પચાસ	4	પ , ચ , ા , સ	
7	સાત હજાર બસો સત્તાવીસ	4 x 1	4	સાત	3	સ , ા , ત	Successfully Separated
				હજાર	4	હ , જ , ા , ર	
				બસો	3	બ , સ , ો	
				સત્તાવીસ	8	સ , ત , ્ , ત , ા , વ , ી , સ	
8	એક લાખ નવ હજાર નવસો નેવું	6 x 1	6	એક	2	એ , ક	Successfully Separated (Vowels are considered as 1 single character, here 'એ')
				લાખ	3	લ , ા , ખ	
				નવ	2	ન , વ	
				હજાર	4	હ , જ , ા , ર	
				નવસો	4	ન , વ , સ , ો	
				નેવું	5	ન , ે , વ , ુ , ં	
9	દસ લાખ બે હજાર એકસો અગિયાર	6 x 1	6	દસ	2	દ , સ	Successfully Separated
				લાખ	3	લ , ા , ખ	
				બે	2	બ , ે	
				હજાર	4	હ , જ , ા , ર	
				એકસો	4	એ , ક , સ , ો	
				અગિયાર	6	અ , ગ , િ , ય , ા , ર	
10	નવ હજાર પાંચસો ઓગણત્રીસ	4 x 1	4	નવ	2	ન , વ	Successfully Separated
				હજાર	4	હ , જ , ા , ર	
				પાંચસો	6	પ , ા , ં , ચ , સ , ો	
				ઓગણત્રીસ	8	ઓ , ગ , ણ , ત , ્ , ર , ી , સ	

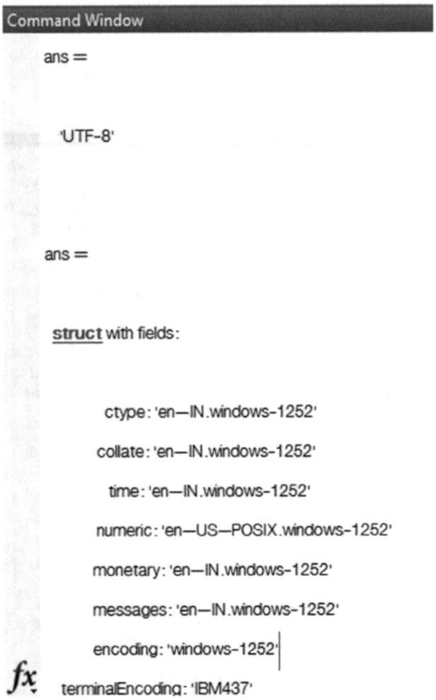

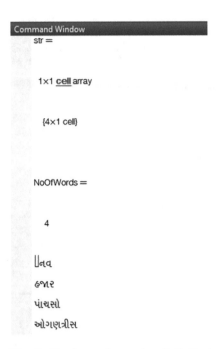

Fig. 6. Output of Default Character Encoding and locale values

Fig. 7. Number of words, Individual words of a string

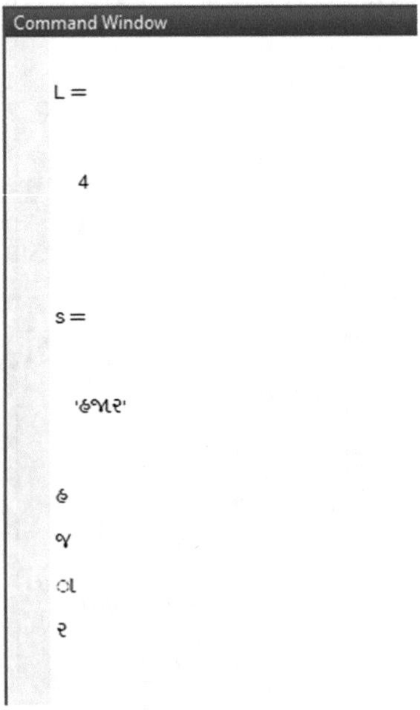

Fig. 8. Length of each word and Individual characters Ex. હજાર

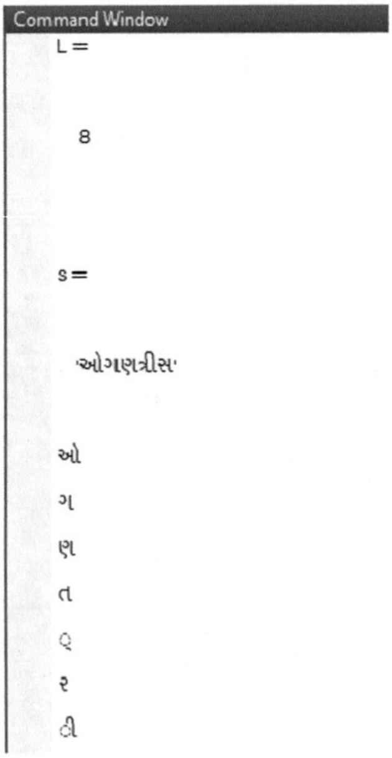

Fig. 9. Length of each word and Individual characters Ex. ઓગણત્રીસ

5 Conclusion

As a result of this algorithm, consonants and vowel modifiers are considered as different characters. We can combine connected consonants and vowel modifiers as single character. For example., in word 'ૈજ઼ાર', individual characters are 'ૈ, જ, ◌ા, ર', which can be separated as 'ૈ, જા, ર'. This paper proposes an algorithm for segmentation of words and characters of Gujarati language inputted from text file saved as 'UTF-8' encoding style. The input data are numbers in Gujarati, written as words of varying length. The algorithm was implemented by MATLAB programming environment on windows 10 platform. The algorithm successfully separates each words and characters of an inputted string. The work documented in this paper, will be useful to develop a text-to-speech system for Gujarati language.

Compliance with Ethical Standards

All author states that there is no conflict of interest.

Humans/Animals are not involved in this work

We used our own data.

References

1. Rakholia, R.M., Saini, J.R.: A study and comparative analysis of different stemmer and character recognition algorithms for Indian Gujarati script. Int. J. Comput. Appl. **106**, 45–50 (2014)
2. Tailor, J.H., Shah, D.B.: Speech recognition system architecture for Gujarati language. Int. J. Comput. Appl. **138**(12), 28–31 (2016)
3. Prasad, J.R.: Image normalization and preprocessing for Gujarati character recognition. Int. J. Comput. Sci. Netw. **3**(5), 334–339 (2014)
4. Javed, M., Nagabhushan, P., Chaudhari, B.B.: Extraction of line word character segments directly from run length compressed printed text documents
5. Nallapareddy, P., Mandal, R.: Line and word segmentation approach for printed documents. IJCA Special Issue on "Recent Trends in Image Processing and Pattern Recognition, pp. 30–36 (2010)
6. Chaudhari, S.A., Gulati, R.M.: Character level separation and identification of English and Gujarati digits from bilingual (English-Gujarati) printed documents. Int. J. Comput. Appl. (2011)
7. Tripathi, R.C., Kumar, V.: Character recognition: a neural network approach. In: National Conference on Advancement of Technologies – Information Systems & Computer Networks (ISCON 2012) (2012). Proceedings published in International Journal of Computer Applications
8. Choudhary, A., Rishi, R., Ahlawat, S.: Off-line handwritten character recognition using features extracted from binarization technique, pp. 306–3012. ScienceDirect (2013)
9. Khopkar, M.: OCR for Gujarati numeral using neural network. Int. J. Sci. Res. Dev. **1**(3), 424–427 (2013)
10. Jariwala, N.B., Patel, B.: Gujarati text to speech conversion: a review (2014)
11. Modi, M., Macwan, F., Prajapati, R.: Gujarati character identification: a survey. Int. J. Innov. Res. Electr. Electron. Instrum. Control Eng. **2**(2), 939–943 (2014)

12. Audichya, M.K., Saini, J.R.: A study to recognize printed Gujarati characters using tesseract OCR. Int. J. Res. Appl. Sci. Eng. Technol. (IJRASET) **5**, 1505–1510 (2017)
13. Rakshit, P., Halder, C., Ghosh, S., Roy, K.: Line, word, and character segmentation from Bangla handwritten text—a precursor toward Bangla HOCR. In: Advanced Computing and Systems for Security. Advances in Intelligent Systems and Computing (2017)
14. Sahare, P., Dhok, S.B.: Multilingual character segmentation and recognition schemes for Indian document images. IEEE Access **6**, 10603–10617 (2018)
15. Naik, V., Desai, A.A.: Multi-layer classification approach for online handwritten Gujarati character recognition. Advances in Intelligent Systems and Computing, vol. 2, pp. 595–606 (2018)
16. Singh, A.P., Kushwaha, A.K.: Analysis of segmentation methods for Brahmi script. J. Libr. Inf. Technol. **39**(2), 109–116 (2019)

Total Interpretive Structural Modelling: Evolution and Applications

Shalini Menon and M. Suresh[✉]

Amrita School of Business, Amrita Vishwa Vidyapeetham, Coimbatore, India
binushalini74@gmail.com, drsureshcontact@gmail.com

Abstract. Total Interpretive Structural Modelling (TISM) is an approach that is used for theory building as it helps the researchers to answer the fundamental research questions of what, how and why. The method helps identify and define the variables, the relationship between them and the reason for causality between variables. This study gives an insight in to TISM methodology is fast growing modelling tools adopted by many researchers in organizational research. The study provides an overview of the TISM tool, its evolution into a sound theoretical model and its application across manufacturing and service sectors. This study will help researchers and practitioners to understand the importance of TISM applications in theory building.

Keywords: Total Interpretive Structural Modelling · Theory building · Organizational research · Manufacturing and service sector

1 Introduction

[1] Called the relationship between theory and research dialectic as for development of theory, research is required and for any research, theory is important. According to [2] any research should address the six fundamental questions what, how, why, where and when. TISM method answers the three fundamental questions- what, how and why. A theory should first define the dimensions, constructs and characteristics of the individual, group, situation or an event [1]. TISM helps identifying and defining the dimensions and hence answers the question "what". The next question "how" is answered by establishing relationships between the variables identified and the "why" would explain the causality between the variables [3].

In organizational research, the researcher may rely on the existing literature or it may use opinion of experts and professionals where very little is known about the subject. Existing literature and interviews with experts may help identify the variables and the backing of the past theories may help with the how and why question but a grounded theory may lack explaining the relatedness between the variables and the logic behind the relationship. [3] Recommended a system theory approach which would help the researcher understand the structure of the organization within the system and facilitate decision making process.

© Springer Nature Switzerland AG 2020
J. S. Raj et al. (Eds.): ICIDCA 2019, LNDECT 46, pp. 257–265, 2020.
https://doi.org/10.1007/978-3-030-38040-3_30

2 TISM and Its Evolution

[4] Introduced the Interpretive Structural Modelling (ISM). ISM helps in formation of well-defined mental models. The modelling makes use of graph theory simplifying the complex system exhibiting the interrelationship between the various elements in the system. The ISM was used by a number of researchers to identify the elements in a system and the linkages between them but the model miserably failed in interpreting these linkages. The ISM diagram used nodes to explain the elements and the links depicted the relationship and the direction of relationship but lacked explaining why there is a relationship between the variables and what causes the relationship. To overcome the shortages of ISM, [3] modified and refined it by enhancing the interpretation of structural modelling thereby making it more logical and transparent leaving no room for multiple interpretations by the users. This author called Total Interpretive Structural Modelling abbreviated as TISM. TISM model is same as the ISM with a major difference that it interprets the relation between the elements at every stage.

TISM has been applied for theory building in varied fields. The method has found its application predominantly in the manufacturing sector: Flexible manufacturing, sustainable manufacturing, lean performance, agile manufacturing, legible manufacturing, and flexible procurement. Lately TISM application has found grounding in supply chain management where issues like sustainable and green supply chain, technological capabilities, re-configurability and structural changes in supply chain, integrated logistics, big data initiatives in supply chain and handling risk and uncertainty in supply chain have been explored. Other applications have been in healthcare, higher education, food management, waste management, ICT, telecommunication and other areas. The Table 1 below summarizes the area of application of TISM modelling.

Table 1. Classification of TISM applications

Sl. no.	Area of application	References
1	Manufacturing	[5–20]
2	Supply chain management	[21–34]
3	Food management	[35, 36]
4	Waste Management	[37–39]
5	Higher education	[40–43]
6	Healthcare	[44–50]
7	Telecommunication	[51–53]
8	Airline	[54]
9	Construction	[55]
10	Pharmaceutical	[56]
11	IT/ICT/IOT	[57–63]
12	Information and organization management	[64]
13	Fly ash handling	[65]
14	Inland water transport	[66]
15	Social commerce in emerging markets	[67]
16	Group decision making	[68]

(*continued*)

Table 1. (*continued*)

Sl. no.	Area of application	References
17	Aligning technology and business strategy	[69]
18	Alignment for effective strategy execution	[70]
19	Product innovation management	[71]
20	Organizational excellence	[72]
21	Asymmetrical motives of cross border joint ventures	[73]
22	Organizational citizenship behaviour	[74]
23	Organization vitalization processes	[75]
24	Trade of halal commodities	[76]

3   Steps Involved in TISM

- **Identification and defining of elements:** TISM model begins with the identification and defining of elements in particular study. The identification is carried out by reviewing the literature available and by contacting experts with the help of a structured questionnaire. The questionnaire consists of each element in paired comparison with all other elements.
- **Establishing contextual relationship between elements:** The paired comparison is carried out to derive the contextual relationships between the factors. The relationship between the elements is explained on basis of intent, priority, dependence and enhancement of attributes (e.g. A influences/enhances/changes/alters B) supported by a logical reasoning explaining how an element influences/enhances the other element.
- **Reachability Matrix and Transitivity Check:** After the questions are answered as YES or NO with explanation for the given answer, all the answers with Yes are represented by binary digit 1 and No are represented as 0. The so obtained matrix is then checked for transitivity. The entries in the initial reachability matrix that are denoted by 0 become eligible for checking possibility of transitivity between the pair compared. In case of transitivity established: the first level, where say factor C=D and D=E, then C=E is denoted by 1*, second level, where F=G, G=H, H=I, then H=I is denoted by 1** and the third level, where K=L, L=M, M=N and N=O, then K=O is denoted by 1*** respectively in the final reachability matrix. The pair that shows no transitivity remains as 0.
- **Partition of the final reachability matrix:** Partitioning of the variables into different iteration levels takes place at this level. The partition matrix consists of reachability set, antecedent set and intersection set. From the final reachability matrix for each factor the rows and the columns corresponding to that particular factor are considered. The row elements that are represented by number 1 are entered in the reachability set and likewise the Columns that carry number 1 are entered in the antecedent set. The common numbers between these two sets enter the intersection set. The factors that have the same elements in reachability and intersection set are assigned to the first level and are not considered in the next level. The process goes on till all the factors are assigned to different levels.

- **Developing Diagraph:** The elements and the relationship between elements are presented in a graphical form showing the nodes and links. Nodes representing the elements and the links are represented by arrows (unidirectional/bidirectional) with the interpretation of the relationship shown on the arrow. By this stage only the crucial transitive links are retained. Diagraph contains factors assigned to first level at the top followed by the second level and so on.

TISM approach involves both qualitative and quantitative logic. Qualitative when the variables are compared in pairs by taking expert opinion and quantitative when comes to transitivity and reachability matrix check [77]. The model has evolved over the years. Changes were made to overcome the shortcomings of the ISM model and to reduce errors related to relationships between elements, transitivity, partitioning of the elements and other such inconsistencies [78].

Another challenge that is faced by the researchers is when the elements identified are more and this leads to increasing number of paired comparisons. [78] suggested checking of pair comparisons and transitivity between elements simultaneously to filter out the redundant transitive relationships and obtain only those that are distinctive and require expert intervention. The author also recommended easy visualization of relationships between elements by using user friendly software with graphical interface.

TISM model was further refined by [64] including the polarity of relationship between elements and also for transitive links. The polarity would help understand whether one variable influences the other variable in a positive way or in a negative manner. One of the shortcomings of the TISM model was that the respondents could answer only in "yes" or "no". To provide respondents with varied degree of response authors [3, 68] recommended assigning linguistic terms such that it clearly spells out the strength of relationship between elements in terms of influence of one element on another. On the scale of 0–1, No influence is assigned value 0, very low–0.25, low-0.5, high-0.75, and very high-1 respectively. This was called Fuzzy TISM that provided wide flexibility to express the level of influence [6, 9, 23, 58, 68, 76].

TISM model explains the relationship between the variables but to know the strength of relationship between the variables many researchers have used MICMAC [14, 42] along with TISM enabling researchers to rank the enablers and segregate them on the basis of driving and dependency power as Autonomous enablers(weak driving and dependence power), Linkage factors (strong driving and dependence power), Dependence enablers (weak driving power but strong dependence) and Driving enablers (strong driving but weak dependence). Researchers have also used TISM along with other methods like SAP–LAP linkages, Flowing Stream Strategy (FSS) [77], TISM and Analytic Hierarchy Process [10, 71].

4 Conclusion

The study revealed TISM approach as a fast growing methodology among researchers for theory building. The methodology has evolved over the years and developed into a sound theoretical model. The past decade has seen a number of researchers adopting it in both manufacturing and service sector. TISM approach was predominantly used in manufacturing sector but lately its acceptance in service sectors has grown. Limitation of this methodology of not being statistically valid is arguable as its growing application in varied areas of research establishes it as a supportive analytical tool that is very much capable of developing an initial model.

Compliance with Ethical Standards

All author states that there is no conflict of interest.

Humans/Animals are not involved in this work

We used our own data.

References

1. Downs, F.S., Fawcett, J.: The Relationship of Theory and Research. McGraw-Hill/Appleton & Lange, London (1986)
2. Whetten, D.A.: What constitutes a theoretical contribution? Acad. Manag. Rev. **14**(4), 490–495 (1989)
3. Sushil, S.: Interpreting the interpretive structural model. Glob. J. Flex. Syst. Manag. **13**(2), 87–106 (2012)
4. Warfield, J.N.: Intent structures. IEEE Trans. Syst. Man Cybern. **3**(2), 133–140 (1973)
5. Bamel, N., Dhir, S., Sushil, S.: Inter-partner dynamics and joint venture competitiveness: a fuzzy TISM approach. Benchmarking Int. J. **26**(1), 97–116 (2019)
6. Jain, V., Soni, V.K.: Modeling and analysis of FMS performance variables by fuzzy TISM. J. Model. Manag. **14**(1), 2–30 (2019)
7. Dwivedi, A., Agrawal, D., Madaan, J.: Sustainable manufacturing evaluation model focusing leather industries in India: a TISM approach. J. Sci. Technol. Policy Manag. **10**(2), 319–359 (2019)
8. Agrawal, R., Vinodh, S.: Application of total interpretive structural modelling (TISM) for analysis of factors influencing sustainable additive manufacturing: a case study. Rapid Prototyp. J. (2019). https://doi.org/10.1108/RPJ-06-2018-0152. ahead-of-print
9. Virmani, N., Saha, R., Sahai, R.: Evaluating key performance indicators of leagile manufacturing using fuzzy TISM approach. Int. J. Syst. Assur. Eng. Manag. **9**(2), 427–439 (2018)
10. Singh, M.K., Kumar, H., Gupta, M.P., Madaan, J.: Analyzing the determinants affecting the industrial competitiveness of electronics manufacturing in India by using TISM and AHP. Glob. J. Flex. Syst. Manag. **19**(3), 191–207 (2018)
11. Solke, N.S., Singh, T.: Application of total interpretive structural modeling for lean performance-a case study. Int. J. Mech. Eng. Technol. (IJMET) **9**(1), 1086–1095 (2018)
12. Betaraya, D.M., Nasim, S., Mukhopadhyay, J.: Modelling subsidiary innovation factors for semiconductor design industry in India. In: Connell, J., Agarwal, R., Sushil, D.S. (eds.) Global Value Chains, Flexibility and Sustainability. Flexible Systems Management. Springer, Singapore (2018)
13. Chaple, A.P., Narkhede, B.E., Akarte, M.M., Raut, R.: Modeling the lean barriers for successful lean implementation: TISM approach. Int. J. Lean Six Sigma (2018). https://doi.org/10.1108/IJLSS-10-2016-0063
14. Sindhwani, R., Malhotra, V.: A framework to enhance agile manufacturing system: a total interpretive structural modelling (TISM) approach. Benchmarking Int. J. **24**(2), 467–487 (2017)

15. Nagpal, S., Kumar, A., Khatri, S.K.: Modeling interrelationships between CSF in ERP implementations: total ISM and MICMAC approach. Int. J. Syst. Assur. Eng. Manag. 8(4), 782–798 (2017)
16. Bag, S.: Flexible procurement systems is key to supply chain sustainability. J. Transp. Supply Chain Manag. 10(1), 1–9 (2016)
17. Jayalakshmi, B., Pramod, V.R.: Total interpretive structural modeling (TISM) of the enablers of a flexible control system for industry. Glob. J. Flex. Syst. Manag. 16(1), 63–85 (2015)
18. Jain, V., Raj, T.: Modeling and analysis of FMS flexibility factors by TISM and fuzzy MICMAC. Int. J. Syst. Assur. Eng. Manag. 6(3), 350–371 (2015)
19. Dubey, R., Gunasekaran, A., Sushil, Singh, T.: Building theory of sustainable manufacturing using total interpretive structural modelling. Int. J. Syst. Sci. Oper. Logist. 2(4), 231–247 (2015)
20. Dubey, R., Ali, S.S.: Identification of flexible manufacturing system dimensions and their interrelationship using total interpretive structural modelling and fuzzy MICMAC analysis. Glob. J. Flex. Syst. Manag. 15(2), 131–143 (2014)
21. Biswal, J.N., Muduli, K., Satapathy, S., Yadav, D.K.: A TISM based study of SSCM enablers: an Indian coal-fired thermal power plant perspective. Int. J. Syst. Assur. Eng. Manag. 10(1), 126–141 (2019)
22. Mohanty, M.: Assessing sustainable supply chain enablers using total interpretive structural modeling approach and fuzzy-MICMAC analysis. Manag. Environ. Qual. Int. J. 29(2), 216–239 (2018)
23. Lamba, K., Singh, S.P.: Modeling big data enablers for operations and supply chain management. Int. J. Logist. Manag. 29(2), 629–658 (2018)
24. Soda, S., Sachdeva, A., Garg, R.K.: Green supply chain management drivers analysis using TISM. In: Sushil, S.T., Kulkarni, A. (eds.) Flexibility in Resource Management. Flexible Systems Management. Springer, Singapore (2018). https://doi.org/10.1007/978-981-10-4888-3_8
25. Mohanty, M., Shankar, R.: Modelling uncertainty in sustainable integrated logistics using Fuzzy-TISM. Transp. Res. Part D Transp. Environ. 53, 471–491 (2017)
26. Sandeepa, S., Chand, M.: Analysis of flexibility factors in sustainable supply chain using total interpretive structural modeling (T-ISM) technique. Uncertain Supply Chain. Manag. 6(1), 1–12 (2018)
27. Luo, Z., Dubey, R., Papadopoulos, T., Hazen, B., Roubaud, D.: Explaining environmental sustainability in supply chains using graph theory. Comput. Econ. 52(4), 1257–1275 (2018)
28. Rajesh, R.: Technological capabilities and supply chain resilience of firms: a relational analysis using Total Interpretive Structural Modeling (TISM). Technol. Forecast. Soc. Chang. 118, 161–169 (2017)
29. Shibin, K.T., Gunasekaran, A., Dubey, R.: Explaining sustainable supply chain performance using a total interpretive structural modeling approach. Sustain. Prod. Consum. 12, 104–118 (2017)
30. Biswas, P.: Modeling reconfigurability in supply chains using total interpretive structural modeling. J. Adv. Manag. Res. 14(2), 194–221 (2017)
31. Deshmukh, A.K., Mohan, A.: Analysis of Indian retail demand chain using total interpretive modeling. J. Model. Manag. 12(3), 322–348 (2017)
32. Yadav, D.K., Barve, A.: Modeling post-disaster challenges of humanitarian supply chains: a TISM approach. Glob. J. Flex. Syst. Manag. 17(3), 321–340 (2016)
33. Shibin, K.T., Gunasekaran, A., Papadopoulos, T., Dubey, R., Singh, M., Wamba, S.F.: Enablers and barriers of flexible green supply chain management: a total interpretive structural modeling approach. Glob. J. Flex. Syst. Manag. 17(2), 171–188 (2016)

34. Mangla, S.K., Kumar, P., Barua, M.K.: Flexible decision approach for analysing performance of sustainable supply chains under risks/uncertainty. Glob. J. Flex. Syst. Manag. **15**(2), 113–130 (2014)

35. Zhao, G., Liu, S., Lu, H., Lopez, C., Elgueta, S.: Building theory of agri-food supply chain resilience using total interpretive structural modelling and MICMAC analysis. Int. J. Sustain. Agric. Manag. Inform. **4**(3–4), 235–257 (2018)

36. Balaji, M., Arshinder, K.: Modeling the causes of food wastage in Indian perishable food supply chain. Resour. Conserv. Recycl. **114**, 153–167 (2016)

37. Gopal, G.C., Patil, Y.B., Shibin, K.T., Prakash, A.: Conceptual frameworks for the drivers and barriers of integrated sustainable solid waste management: a TISM approach. Manag. Environ. Qual. **29**(3), 516–546 (2018)

38. Singh, A., Sushil: Flexible waste management practices in service sector: a case study. In: Connell, J., Agarwal, R., Sushil, Dhir, S. (eds.) Global Value Chains, Flexibility and Sustainability. Flexible Systems Management. Springer, Singapore (2018). https://doi.org/10.1007/978-981-10-8929-9_20

39. Singh, A.: Developing a conceptual framework of waste management in the organizational context. Manag. Environ. Qual. Int. J. **28**(6), 786–806 (2017)

40. Kashiramka, S., Sagar, M., Dubey, A., Mehndiratta, A., Sushil, S.: Critical success factors for next generation technical education institutions. Benchmarking Int. J. (2019). https://doi.org/10.1108/BIJ-06-2018-0176

41. Jain, A., Sharma, R., Ilavarasan, P.V.: Total interpretive structural modelling of innovation measurement for Indian universities and higher academic technical institutions. In: Flexibility in Resource Management, pp. 29–53. Springer, Singapore (2018). https://doi.org/10.1007/978-981-10-4888-3_3

42. Yeravdekar, S., Behl, A.: Benchmarking model for management education in India: a total interpretive structural modeling approach. Benchmarking Int. J. **24**(3), 666–693 (2017)

43. Prasad, U.C., Suri, R.K.: Modeling of continuity and change forces in private higher technical education using total interpretive structural modeling (TISM). Glob. J. Flex. Syst. Manag. **12**(3–4), 31–39 (2011)

44. Pradhan, R., Sagar, M., Pandey, T., Prasad, I.: Consumer health risk awareness model of RF-EMF exposure from mobile phones and base stations: an exploratory study. Int. Rev. Public Nonprofit Mark. **16**(1), 125–145 (2019)

45. Vaishnavi, V., Suresh, M., Dutta, P.: A study on the influence of factors associated with organizational readiness for change in healthcare organizations using TISM. Benchmarking Int. J. **26**(4), 1290–1313 (2019)

46. Vaishnavi, V., Suresh, M., Dutta, P.: Modelling the readiness factors for agility in healthcare organization: an TISM approach. Benchmarking Int. J. (2019). https://doi.org/10.1108/BIJ-06-2018-0172

47. Ajmera, P., Jain, V.: Modelling the barriers of Health 4.0–the fourth healthcare industrial revolution in India by TISM. Oper. Manag. Res. (2019). https://doi.org/10.1007/s12063-019-00143-x

48. Ajmera, P., Jain, V.: Modeling the factors affecting the quality of life in diabetic patients in India using total interpretive structural modeling. Benchmarking Int. J. **26**(3), 951–970 (2019). https://doi.org/10.1108/BIJ-07-2018-0180

49. Mishra, D., Satapathy, S.: An assessment and analysis of musculoskeletal disorders (MSDs) of Odisha farmers in India. Int. J. Syst. Assur. Eng. Manag., 1–17 (2019). https://doi.org/10.1007/s13198-019-00793-x

50. Patri, R., Suresh, M.: Modelling the enablers of agile performance in healthcare organization: A TISM approach. Glob. J. Flex. Syst. Manag. **18**(3), 251–272 (2017)

51. Singh, B.P., Grover, P., Kar, A.K.: Quality in mobile payment service in India. In: Conference on e-Business, e-Services and e-Society, pp. 183–193. Springer, Cham (2017)
52. Jena, J., Fulzele, V., Gupta, R., Sherwani, F., Shankar, R., Sidharth, S.: A TISM modeling of critical success factors of smartphone manufacturing ecosystem in India. J. Adv. Manag. Res. **13**(2), 203–224 (2016)
53. Yadav, N.: Total interpretive structural modelling (TISM) of strategic performance management for Indian telecom service providers. Int. J. Prod. Perform. Manag. **63**(4), 421–445 (2014)
54. Singh, A.K., Sushil: Modeling enablers of TQM to improve airline performance. Int. J. Product. Perform. Manag. **62**(3), 250–275 (2013)
55. Sandbhor, S.S., Botre, R.P.: Applying total interpretive structural modeling to study factors affecting construction labour productivity. Constr. Econ. Build. **14**(1), 20–31 (2014)
56. Wasuja, S., Sagar, M., Sushil: Cognitive bias in salespersons in specialty drug selling of pharmaceutical industry. Int. J. Pharm. Healthc. Mark. **6**(4), 310–335 (2012)
57. Patil, M., Suresh, M.: Modelling the enablers of workforce agility in IoT projects: a TISM approach. Glob. J. Flex. Syst. Manag. **20**(2), 157–175 (2019)
58. Kaur, H.: Modelling Internet of Things driven sustainable food security system. Benchmarking Int. J. (2019). https://doi.org/10.1108/BIJ-12-2018-0431. ahead-of-print
59. Sehgal, N., Nasim, S.: Total Interpretive Structural Modelling of predictors for graduate employability for the information technology sector. High. Educ. Ski. Work Learn. **8**(4), 495–510 (2018)
60. Chawla, N., Kumar, D.: Structural modeling of implementation enablers of cloud computing. In: Advances in Computer and Computational Sciences, pp. 273–286. Springer, Singapore (2018)
61. Sharma, A., Sagar, M.: New product selling challenges (key insights in the ICT sector). J. Indian Bus. Res. **10**(3), 291–319 (2018)
62. Prasad, S., Shankar, R., Gupta, R., Roy, S.: A TISM modeling of critical success factors of blockchain based cloud services. J. Adv. Manag. Res. **15**(4), 434–456 (2018)
63. Manjunatheshwara, K.J., Vinodh, S.: Application of TISM and MICMAC for analysis of influential factors of sustainable development of tablet devices: a case study. Int. J. Sustain. Eng. **11**(5), 353–364 (2018)
64. Sushil: Incorporating polarity of relationships in ISM and TISM for theory building in information and organization management. Int. J. Inf. Manag. **43**, 38–51 (2018)
65. Singh, S.P., Singh, A.: Deriving the hierarchical relationship of factors of fly ash handling. Manag. Environ. Qual. Int. J. **29**(3), 444–455 (2018)
66. Kumar, P., Haleem, A., Qamar, F., Khan, U.: Modelling inland waterborne transport for supply chain policy planning: an Indian perspective. Glob. J. Flex. Syst. Manag. **18**(4), 353–366 (2017)
67. Kumar, H., Singh, M.K., Gupta, M.P.: Socio-influences of user generated content in emerging markets. Mark. Intell. Plan. **36**(7), 737–749 (2018)
68. Khatwani, G., Singh, S.P., Trivedi, A., Chauhan, A.: Fuzzy-TISM: a fuzzy extension of TISM for group decision making. Glob. J. Flex. Syst. Manag. **16**(1), 97–112 (2015)
69. Kedia, P.K.: Strategy alignment of critical continuity forces wrt technology strategy and business strategy and their hierarchical relationship using TISM. In: Global Value Chains, Flexibility and Sustainability, pp. 145–159. Springer, Singapore (2018)
70. Srivastava, A.K.: Alignment: the foundation of effective strategy execution. Int. J. Prod. Perform. Manag. **66**(8), 1043–1063 (2017)
71. Haleem, A., Kumar, S., Luthra, S.: Flexible system approach for understanding requisites of product innovation management. Glob. J. Flex. Syst. Manag. **19**(1), 19–37 (2018)

72. Agarwal, A., Vrat, P.: A TISM based bionic model of organizational excellence. Glob. J. Flex. Syst. Manag. **16**(4), 361–376 (2015)
73. Hasan, Z., Dhir, S., Dhir, S.: Modified total interpretive structural modelling (TISM) of asymmetric motives and its drivers in Indian bilateral CBJV. Benchmarking Int. J. **26**(2), 614–637 (2019)
74. Yadav, M., Rangnekar, S., Bamel, U.: Workplace flexibility dimensions as enablers of organizational citizenship behavior. Glob. J. Flex. Syst. Manag. **17**(1), 41–56 (2016)
75. Bishwas, S.K., Sushil: LIFE: an integrated view of meta organizational process for vitality. J. Manag. Dev. **35**(6), 747–764 (2016)
76. Khan, M.I., Khan, S., Haleem, A.: Using integrated weighted IRP-Fuzzy TISM approach towards evaluation of initiatives to harmonise Halal standards. Benchmarking Int. J. **26**(2), 434–451 (2019)
77. Sushil: How to check correctness of total interpretive structural models? Ann. Oper. Res. **270**(1–2), 473–487 (2018)
78. Sushil, A.: Modified ISM/TISM process with simultaneous transitivity checks for reduced direct pair comparisons. Glob. J. Flex. Syst. Manag. **18**(4), 331–351 (2017)

Machine Learning Techniques: A Survey

Herleen Kour[(✉)] and Naveen Gondhi

Sri Mata Vaishmo Devi University, Katra, India
{18mms005,Naveen.Gondhi}@smvdu.ac.in

Abstract. Artificial intelligence (AI) is a technique, which makes machines to mimic the human behavior. Machine learning is an AI technique to train complex models, which can make the system or computer to work independently without human intervention. This paper is a survey on Machine learning approaches in terms of classification, regression, and clustering. The paper concludes with a comparative analysis between different classification techniques based on its applications, advantages, and disadvantages.

Keywords: Classification · Regression · Clustering

1 Introduction

The main objective of Machine learning is to design a program for accessing data and make the system learn by itself. In order to obtain the better decision the procedure is started by inspecting the data and the pattern is searched [3]. The aim of machine learning is to make system work independently, without the human intervention and the patterns present in data are identified (Fig. 1).

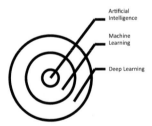

Fig. 1. Basic architecture

However, in diverse environment machine learning algorithm are found to be difficult. The machine learning techniques are classified into four categories viz. Supervised learning, Unsupervised learning, Semi-Supervised learning and Reinforcement learning (Fig. 2).

© Springer Nature Switzerland AG 2020
J. S. Raj et al. (Eds.): ICIDCA 2019, LNDECT 46, pp. 266–275, 2020.
https://doi.org/10.1007/978-3-030-38040-3_31

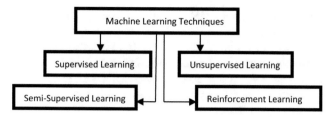

Fig. 2. Machine learning techniques

2 Supervised Learning Technique

It is a kind of machine learning technique that makes use of the previous data and knowledge to forecast events with the help of labels [3]. It requires human to provide input & feedback to predict the accuracy in the training process (Fig. 3).

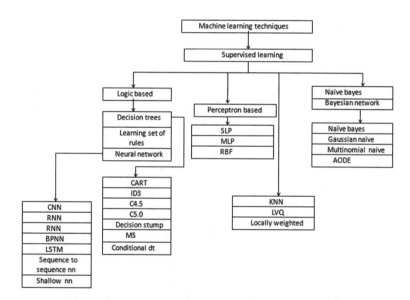

Fig. 3. Categories of machine learning techniques

2.1 Logic Based Supervised Learning

Techniques which involves the automated construction of logic-based programs from examples and existing domain knowledge.

Logic based learning includes the following techniques. **Decision trees** is a widely used technique to split the data continuously according to the certain parameter thus capable of handling large datasets. It is the widely used algorithm [27]. The decision trees are further classified into four categories viz, **CART (classification and regression tree)**, predicts next questions depending upon the current structural sequence of question and answers. **ID3 (iteration dichotonise)** is generated in the form of tables. In order to classify the data ID3 uses top down, greedy search to find and select the best attribute in classification of data. **Decision stump** is a form of decision tree in which the root (internal) node is immediately connected to the leaf (terminal) node. It consists of one level of decision tree. **MS (Microsoft decision tree)** is a hybrid decision tree algorithm used for modeling and predictiveness of models for both continuous and discrete attributes.

CNN (convolution neural network) is kind of neural network that uses machine learning unit named perceptron for learning and analyzing data. It is used in natural language processing, image processing. **RNN (recurrent neural network)** is neural network that uses patterns to predict the next likely sequence character of data. RNN (recursive neural network) network is created when same set of weights is recursively applied over the structure, in order to produce structural prediction over variable length input. **LSTM is a modified version of RNN** network, and the RNN when build with LSTM helps in determining which data is important, should be remembered, needs to be looped back to the network and what data to be forgotten. It has a feedback connection.

2.2   Perceptron Based Supervised Learning

It is an algorithm used for supervised learning of binary classification that enables the function to decide the class of the input. It is divided into following techniques. **SLP (single layer perceptron)** is a network that consists of only one layer of input nodes and send the weighted input to the receiving nodes using given activation function. **MLP (multi layer perceptron)** consists of three layers: input layer, hidden layer and output layer. It uses back propagation machine learning technique for training. Each node has its own non activation function. The properties like non-linear activation and makes it different from single layer perceptron. The radial base function is a function defined whose value depends upon distance from the origin and the neural network that uses RBF as activation function is called Radial based function neural network [3]. The output **of RBFN** is the combination of RBF of input and neuron parameters.

Instance based supervised learning are further classified as: **KNN (k nearest neighbour)** is a classification technique that stores all the available cases and helps to group the nearest data points. It is robust to noise [12]. **LVQ (learning vector quantization)** uses have labeled input data. A supervised version learning of vector quantization called LVQ can be used. **Locally weighted** method are non-parametric where the prediction is done by a function called local function, using the subset of data. It is used by the data analyst to manage the missing data.

Statistical based supervised learning are further classified into techniques including naive bayes classification techniques, **Naïve bayes (NB)** classifier is an algorithm that classifies object using bayesian theorem NB classifier specifies the strong independence between the attribute of data pattern. When features are continuos **Gaussian NB** is used whereas, **Multinomial NB** is used when features have discrete values.

3 Unsupervised Learning

The unsupervised learning is contrary supervised learning technique with the absence of Training process. The unsupervised learning explores the hidden patterns in the unlabeled data (Fig. 4).

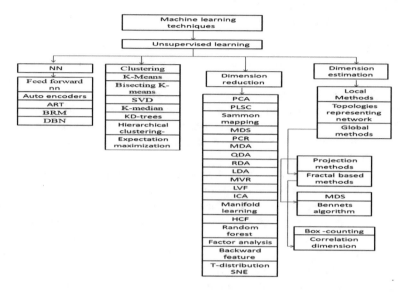

Fig. 4. Unsupervised learning techniques

Neural Network Based Unsupervised Learning

Feed forward neural network is simple kind of neural network in which there is no cyclic formation of connections between the nodes and thus no loops are formed. The information moves only in forward direction that is from the input layer to hidden layer and then to the output layer. In single layer feed-forward network layer is formed by taking processing elements and combining it with other elements. Input and output are linked to each other. **Adaptive resonance theory** is a philosophical driving unsupervised neural network model that keeps the existing model in place and helps some neural network to build the capacity for new learning. The number of neural networks which uses supervised & unsupervised methods are described by this theory. **Deep belief neural network** is a generative neural network that uses an unsupervised learning model to get the results [1]. It consists of multiple layer of graphical model having both directed and undirected graph. They are used to recognise, cluster, generate images. **Auto encoders** is an algorithm that is used by neural network to learn compressed and encoded representation of data, mostly for dimension reduction and pre-training of feed-forward neural network [1].

K-means clustering is the method of dividing the observations into clusters such that each observation belonging to a particular cluster with the nearest mean. **K-Median** is the variation of K-means clustering and it works to calculate the median for cluster. **Hierarchical Clustering** is an alternative to KNN approach for labelling dataset and is used to identify groups in datasets without specifying the number of clusters. **Expectation maximization** is used by the data analyst to deal with the missing data

3.1 Dimension Reduction Is the Process of Reducing the Number of Random Variables Under Consideration

These techniques are of following types

PCA is used to explore, visualize, discover patterns in the dataset of high dimension. It is a method that is highly used in learning statistical approaches. Similar to PCA, **PLSC** has a different approach in which the linear model by projecting the objective variable and predicting variable to a new space. **Sammon Mapping** helps in mapping the high dimension space to a low dimension space and applies the non linear methods to determine points that have similar distance from each other in low dimension or the high dimension.

MDS (multidimensional scaling) is an attempts to construct the coordinates of n pairs in Euclidian space using information about distance between n patterns while preserving the pair wise distance. **PCR (principal component regression)** is based on PCA, PCR works on a high dimension dataset in which we can perform dimension reduction and then compared to a smaller set of variables. **LDA (linear discriminant analysis)** is useful for pattern recognition and machine learning that helps in finding the combination of features that separates two or more classes. It is a classical statistical method [6]. **QDA (quadratic discriminant analysis)** is used to avoid over fitting by projecting the dataset into low dimension space with efficient class separability in order to and to reduce the computation. It is used in preprocess step. It is the variant of LDA that helps in non linear separation of data. **RDA (regularized discriminant analysis)** is an intermediate technique between LDA and QDA to model the non linear separation of data and to deal with large number of features. **FDA (flexible discriminant analysis)** is an extension of LDA that uses the flexible extension to LDA that uses the splines: non-linear combination of predictors. Within each group it is useful to model the non-linear relation and multivariate avariable. **LVF (low variance filter)** is helpful in removing the column with low variance. This technique helps in filtering the column whose value is below the threshold defined by the user. **ICA (independent component analysis)** is used to separate the multivariate signals into additive subcomponents. It assumes that the subcomponents that comprises the signal source are independent from each other. The maximal variation in dataset can be explained by manifold learning. It is a non linear dimension reduction approach. **Random forest** technique is useful for feature extraction and classification [12]. The most informative subset of features is found by generating, then constructing the large set of trees against target attribute and then using each attribute statistics. **High correlation filter** is used as the information is carried out by the data column with similar trends but only one will be allowed to feed the machine learning model. By adapting non-linear and unsupervised learning this algorithm achieves high quality visualization.

3.2   Dimension Estimation

The various techniques of dimension estimation are:

Local methods are also called as topological methods. The topological dimension of data is estimated by topological methods [11], it is considered to be the basic dimension of local linear approximation of hyper surface in which the data resides that is the tangent space. The best sub-space for projection of data is provided by projection methods by minimizing the projection error. **Projection methods** are of two types: **Multidimensional** scaling used to preserve the distance between any datasets [11] and the image needs to be projected in a way that the projections are similar and matches the given dataset [9]. **Bennett's algorithm** is based on assumptions that data are uniformly distributed inside the sphere of radius r in an L-dimension space.

4   Semi-supervised Learning

This technique uses the labeled data and the unlabeled data. The supervised learning and unsupervised learning has their own pros and cons [4], therefore it lies in between the supervised learning and unsupervised learning (Fig. 5).

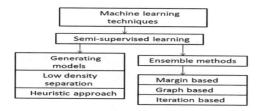

Fig. 5. Semi-supervised learning

It considers the small quantity ratio of labeled data and large quantity ratio of unlabeled data. These techniques are capable of adjusting in order to attain higher accuracy. These are further classified into viz, **Generative models** is either an extension of supervised learning or an unsupervised learning. It determines whether the unlabeled data is decreasing or increasing the performance. If we have a distribution parameterized by θ and assume to take the form $p(x/y, \theta)$. The unlabeled data will increase the performance once the assumptions are correct, otherwise it will decrease the performance.

Low density separation techniques are used in the region where few data points are located and we want to place the boundaries in these regions. Its application can be found in SVM, TSVM. Heuristic based, **Graph based** method deals with graphical representation of data with each node for labeled and unlabeled data. These methods aim to construct the graphs connecting to the similar observations [9]. In order to control the interplay between labeled and unlabelled data, a methodology uses the grouping information from labeled data together with th concept of margins called as **margin based methods**.

5   Reinforcement Learning

Reinforcement learning technique helps the machine to maximize its performance by allowing machines to determine the ideal behavior within the specific context. It comes from the animal learning theory that contains no prior knowledge, it is capable of getting optimal policy from knowledge that is acquired from trial and error method with continuous interaction with the non static environment.

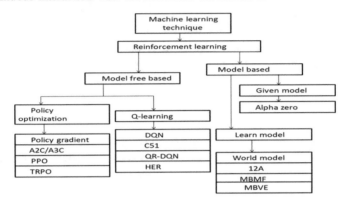

The reinforcement learning has been further divided into following categories

The **Model based** reinforcement learning is used in order to know the different non static state of the environment and to understand the logic how these states are leading to reward, a model is designed. It tends to emphasize planning. The learning where no model is designed and the main emphasize is on learning, is called **Model free based** reinforcement learning, this learning can be done either from past experience or some other ways.

Model based learning are more efficient than model free based learning as instead of using directly the information from environment. These techniques are further classified into different categories. World model (**12A**) is an hybrid approach that combines both model and model free based methods. It conducts implicit plans by learning to interpret the predictions from the learned environment. Learn model is further divided into many Predictiveness is a deep policy network. **MBMF (model based model free)** allows to bridge the gap between model based and model free based reinforcement learning. By allowing imagination of fix depth **MBVE (model based value expansion)** method allows to control uncertainty in a model. It helps in improving the value estimation and reducing the learning complexity. **Given model (Alpha zero)** is a kind of a model based reinforcement learning that is the biggest hype hype in the recent games, in which it defeats the best players in words.

Tables 1 and 2 describes the advantages, disadvantages & application of classification technique:

Table 1. Classification techniques with advantages and disadvantages

Name	Applications	Advantages	Disadvantages
ID3	Fraud detection	Easily understandable prediction rules	If a small sample is tested data can be over fitted
C4.5	Soil quality prediction	Handle issues of incomplete data	Cannot handle variations
C5.0	Application on individual credential evaluation of bank	Robust in presence of problems with missing data, can have two or more outcomes	Does not deal with missing data
CART	Blood donor classification	Flexible, develop interactions among variables	Has unstable decision tree splits by only one variable

Table 2. Classification techniques with advantages and disadvantages

Name	Type	Applications	Advantages	Disadvantages
CNN	Supervised	Face recognition, image recognition	Most accurate	High computation cost
RNN	Supervised	Language modeling and prediction, image recognition	Offers memory that can be used in any application	Vanish gradient problem
LSTM	supervised	Text generation, image recognition	Handle noise	Unlimited states number
BPNN	Supervised	Classification, image recognition, time prediction	Computing gradient	Paralysis occurs when weight is adjusted to a large value
AE	Unsupervised	Dimension reduction	Study coding, compression of data	Captures irrelevant information
DBN	Unsupervised	Feature detection, numerical data classification	Helps in handling the problems that cannot be solved through structured techniques	Requires too much hard work time & memory
RBN	Unsupervised	Dimension reduction	Easy design, increase in computational capacity	Slow in comparison to MLP
NB	Supervised	Text classification, realtime predictions	Fastest to train, easy to implement	Interactions between variables is low
SVM	Supervised	Scene classification, predict financial distress	Produce robust and accurate classification result	Multilabel classification
BAYSIAN	Supervised	Examine dental pain, sig verification	Handles missing data	Multidimensional data

6 Conclusion

The review of all the four types Machine learning techniques has been discussed in this paper. These techniques are different from each other in every aspect, either in terms of applications, advantages or disadvantages. There are variety of applications of machine learning techniques: classification, clustering, pattern recognition etc. Taking classification as the parameter Tables 1 and 2 are constructed containing the techniques used for classification with applications, advantages and disadvantages. Different type of data require different classification technique, depending on the user problem domain selection of the classification technique is done. Every technique has its own merits and demerits. A lot of work has been done in the field of classification domain but there is still difficulty when we deal with big data.

References

1. Dike, H.U., Zhou, Y., Deveerasetty, K.K., Wu, Q.: Unsupervised learning based on artificial neural network: a review. In: The Proceedings of the IEEE International Conference on Cyborg and Bionic Systems, Shenzhen, China, p. 2 (2018)
2. Sun, A., Yen, G.G., Yi, Z.: Evolving unsupervised deep neural networks for learning meaning representations. IEEE Trans. Evol. Comput. 1089–1778
3. Saravanan, R.: A state of art techniques on machine learning algorithms: a perspective of supervised learning approaches in data classification. In: Proceedings of IEEE Second International Conference on Intelligent Computing and Control Systems, p. 261 (2018)
4. Dayan, P.: Helmholtz machines and wake-sleep learning. In: Arbib, M. (ed.) Handbook of Brain Theory Neural Network. MIT Press, Cambridge (2000)
5. Ghahramani, Z.: Unsupervised learning. In: Advanced Lectures on Machine Learning, p. 612. Springer (2004)
6. Zou, K., Sun, W., Yu, H., Liu, F.: ID3 decision tree in fraud detection application. In: IEEE International Conference on Computer Science and Electronics Engineering, pp. 8–9 (2012). https://ieeexplore.ieee.org/xpl/conhome/6187453/proceeding
7. Saravanan, K., Sasithra, S.: Review on classification based on artificial neural networks. Int. J. Ambient Syst. Appl. 2, 4–7 (2014)
8. Priyadarshini, R.: Functional analysis of artificial neural network for dataset classification. Spec. Issue IJCCT 1(2), 106 (2010). For International Conference
9. Kégl, B.: Intrinsic dimension estimation using packing numbers. In: Advances in Neural Information Processing Systems, p. 69 (2003)
10. Hehui, Q., Zhiwei, Q.: Feature selection using C4.5 algorithm for electricity price prediction. In: International Conference on Machine Learning and Cybernetics, pp. 175–180 (2014)
11. Kirby, M.: Geometric Data Analysis: An Empirical Approach to Dimensionality Reduction and the Study of Patterns. Wiley, Hoboken (2001). p. 201
12. Mannila, H.: Data mining: machine learning, statistics, and databases. In: Proceedings of the Eighth International Conference on Scientific and Statistical Database Systems, p. 6. IEEE (2004)

13. Zhou, Z.H.: Ensemble methods: foundations and algorithms. In: Indurkhya, N., Damerau, F. J. (eds.) Handbook of Natural Language Processing, 2nd edn. Chapman & Hall/CRC, Boca Raton (2012)
14. Ongsulee, P.: Artificial intelligence, machine learning and deep learning. In: IEEE 15th International Conference on ICT and Knowledge Engineering (ICT&KE), p. 45 (2017)
15. Mannila, H.: Data mining: machine learning, statistics, and databases. In: Proceedings of Eighth International Conference on Scientific and Statistical Database Systems, p. 4. IEEE (2008)

Lexicon-Based Text Analysis for Twitter and Quora

Potnuru Sai Nishant[1(✉)], Bhaskaruni Gopesh Krishna Mohan[1],
Balina Surya Chandra[1], Yangalasetty Lokesh[1], Gantakora Devaraju[1],
and Madamala Revanth[2]

[1] Department of Computer Science and Engineering,
Koneru Lakshmaiah Education Foundation, Vaddeswaram, A.P., India
`sai.nishant8@gmail.com, gopeshbhaskaruni@gmail.com,`
`balinasuryachandra2015@gmail.com,`
`lokesh19191@gmail.com, g.devaraju0143@gmail.com`
[2] Department of Computer Science and Engineering, IIIT Kurnool,
Kurnool, A.P., India
`revanthrex1001@gmail.com`

Abstract. Nowadays Social Media is a trending platform for freedom of speech. So, this became a cakewalk to know the opinion of people. It has demonstrated, apart from social media uses, that it plays a crucial role in analyzing the trends in elections on the contrary to the biased predictions belong to the same region, community, class, and religion with the help of sentimental Analysis. Sentimental Analysis refers to identifying and categorizing opinions, especially in terms of positive, negative, neutral. In this paper, we study the trends of Andhra Pradesh Election 2019 using websites like Quora and Twitter by using Lexicon based approach and calculating the polarity score. The experimental results infer that Quora can also be used to obtain the behavior of different political parties.

Keywords: Social Media · Sentiment Analysis · Opinion mining · Twitter · Quora · Election

1 Introduction

The scientific community is hopefully more accurate in examining web information, such as blog entries or the activity of users of social networks, as an alternative way of predicting election results [1]. Furthermore, traditional polls are too expensive, while online information is easily obtainable and freely available, which is a suggestive approach for going into Sentiment Analysis. It can be implemented into two ways one is the Machine Learning approach, and the other is the Lexicon based approach. We use Lexicon Based Approach to understand the feelings of the people in Social Media. We study the possibility of using query websites like Quora besides micro-blogging sites like Twitter information as a data source to forecast the trend in Andhra Pradesh about elections, leaders and parties.

© Springer Nature Switzerland AG 2020
J. S. Raj et al. (Eds.): ICIDCA 2019, LNDECT 46, pp. 276–283, 2020.
https://doi.org/10.1007/978-3-030-38040-3_32

Quora is a site for queries and answers, where questions are asked, replied, altered, and sorted out in the form of opinions by its user community. Twitter is one of the social-media where cataloged clients can post tweets, like the posts of others, and re-tweet them, yet unregistered clients can only go through them. Now, on a monthly premise, it has over two hundred million visitors and 500 million daily messages [2]. Tweets were initially intended to post user status updates, but tweets can be about every possible topic these days.

Our work primarily is to compare and understand the feelings of netizens but not for any final prediction stating that this party is going to win. This work focuses on three parties, namely Telugu Desam Party (TDP), Yuvajana Shramika Rythu Congress Party (YSRCP) and Janasena Party (JP).

2 Related Work

We are going to discuss related works about anticipating the consequence of decision-making using Twitter. We saw that researchers use Twitter and Google patterns for information concerning the election prediction issue. Zhang et al. suggested a technique of polarity-based sentiment analysis in which modifiers and words are combined into sentences to help discover text polarity [3]. The classification algorithm based on the lexicon is used to evaluate and predict the polarity of the user [4]. It explicitly, positively, comparatively, and superlatively used degrees of comparison on words. Besides that, during Pakistani General Elections 2013, the reputation of twitter and other social media apps was also disclosed, which trigger twitter as an essential aspect of the election campaign. They focused on five political parties and carried out sentiment analysis while they did not discover a relationship between the proportion of votes and tweets they got [5].

These days working on Twitter data, especially for predicting election results are predominantly deciding the lead in elections. The predicted results are very closer to the actual results in recent UP elections [6]. Wang suggested a unique technique for twitter projections of candidate popularity that depends indirectly on the election [9]. This algorithm considered neutral tweets related to definite candidates. But it did not clearly explain how they classified into positive neutral and negative tweets. Burnap et al. proposed a model for the prediction of the United Kingdom's general elections in 2015 [7]. They gave a threshold score ranging from −5 to +5 to each word like for good 5 and for bad - 4 and thus calculated Sentiment. A score ranging from −5 to +5 was ascribed to each word as excellent as 5 and bad as 5 and therefore calculated sentiment. Nausheen used text blob and NLTK packages in python to estimate polarity and subjectivity of US Presidential Elections [8]. Ramteke performed data set development first by gathering information using twitter API of 60,000 tweets for each user then pre-processing is done to remove unique characters, and data labeling is attained first manually using hashtag labeling and a lexicon-based tool called VADER was used [9]. Ikoro et al. combined two sentiment lexica for better analysis where the first lexicon is used for finding out essential terms for analysis which carry high sentiments and the other lexicon is used for classification of data [10]. Pollacci et al. proposed an epidemic model for sentiment analysis of sentences which are longer which is a challenging task and requires a lot natural language processing and understanding and this is method is also based on lexicons [11].

3 Proposed Model

The proposed model consists of three stages where we perform sentiment analysis on Andhra Pradesh elections.

3.1 Data Extraction

The information extraction step is the underlying stage in this examination, where information is gathered from Twitter and Quora shown in Fig. 1.

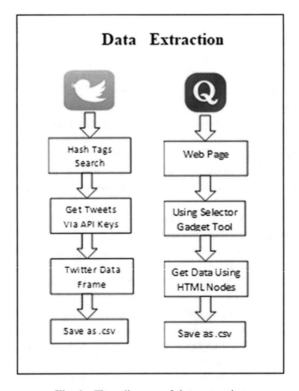

Fig. 1. Flow diagram of data extraction

From twitter, we use the Twitter API to retrieve the tweets pertaining to the hashtags consider for our research. We use these API keys in the R language to retrieve the tweets. We use these hashtags shown in Table 1 respective to the parties to get the tweets. A total of 45,000 tweets were collected and 20,000 comments from Quora. All the webpages on Andhra Pradesh Election 2019 were scraped.

Table 1. Twitter hashtags

Party	Hashtags
TDP	#ncbn, #jaiTDP, #vote4TDP
YSRCP	#ysjagan, #YSRCParty, #Vote4YSRCP
JP	#JanaSenaParty, #PawanKalyan #JSPForAPSpecialStatus

In Quora, we select articles based on queries accessible with respect to that party or those specific contestants challenging in the election. The general people give their opinion about the queries available in Quora. For a particular gathering, there are articles, and we are choosing them for our analysis. Later we go to every page and scrap the information using the web scraping tools. We used selector gadget tool to catch the HTML nodes where the comment lies on. In this way, the whole page is scraped and stored into a CSV file. Every site page is an article. At the end of scraping, we bind the data pertaining to the respective party. For every party, there are 50 articles scrapped.

3.2 Pre-processing

Pre-processing is one of the text mining techniques, where raw data is transformed into a structured format. Due to the unnecessary characters, numerical and symbols in the text, the model doesn't process this data for further analysis. Several pre-processing stages improve the quality of text like Redundant Tweets, Case Transformation, URL Removal, stripped white spaces Removal, Non-ASCII characters Removal, Stop Words Removal, and Stemming. We can observe the transformation of text given in Table 2.

Table 2. Before and after pre-processing

Before	After
@JaiTDP, "NCBN is doing good for the AP." Great!!!! https://t.c	Jaitdp ncbn doing good ap great

3.3 Proposed Algorithm

After pre-processing, we divide the entire text corpus into a stream of sentences. For every sentence, we calculate the polarity score using the proposed algorithm. This new algorithm proposed and shown below classifies each sentence as positive, negative and neutral based on the polarity score. It maintains a positive count, negative count, and neutral count. The willingness of any party is calculated based on the positive percentage. In this Algorithm, Polarity is calculated based on the sentiment based on the proposed algorithm where sentiment is our defined function which gives score according to positive, negative lexicons in each sentence. The standard dictionary of positive and negative words is determined by natural language processing team of MIT, which is given here as input.

In this score function, we use sentiment jockers hash value for sentiment words which is a lexicon-based dictionary for all positive and negative words along with their sentiment score. POS words are adverbs, adjectives, conjunctions, etc. For finding the sentiment of the POS word, we use hash value which contains all POS words with their valency. By calculating the count of POS words before a word, we are finding the polarity score using the below function [12].

The emotions are based on the words categorization where the dictionary is extracted from [12] and the classified into respective categories like anger, joy, trust, and disgust. For instance, if a sentence has more words related to anger or any synonyms pertaining to it, then it falls under Anger category.

```
Initialize positive, negative and neutral counts to zero
for each sentence in data:
    Sentence is preprocessed
    Calculate Polarity as sco=sentiment(sentence)

If sco>0 Pos++;
If sco<0 Neg++;
If sco=0 Neu++;
Total = Pos + Neg + Neu
Pos_percent = (Pos/Total)/100
Neg_percent = (Neg/Total)/100
Neu_percent = (Neu/Total)/100

Procedure Score (for each sentence):
    Initially score = 0;
    For word in sentence:
        Count occurrence of POS words before words
        If word in positive dictionary:
                Score = count*hash(word)+hash(POS)
        Else in negative dictionary:
                Score = count*hash(word)+hash(POS)
Return score
```

4 Experimental Results

Quora Results. We observed that in sentiment analysis, positive comments were more in TDP data than that of YSRCP and JP. Neutral tweets were more for JP as this party is contesting for the first time, and people are not much aware of his regime. We can see the comparison of positive, negative, and neutral in Fig. 2 after we take four different variables of emotion analysis, such as joy, trust, anger, and disgust. We observe that TDP has more joy and trust, followed by YSRCP and JP, respectively. The percentages of Anger and disgust are also less for TDP when compared to the other two parties. We can compare the result in Fig. 3.

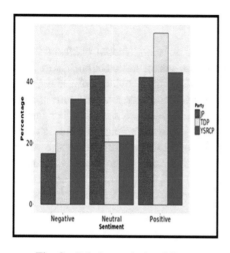

Fig. 2. Polarity analysis of Quora

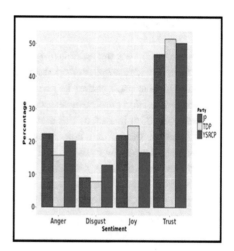

Fig. 3. Emotion analysis of Quora

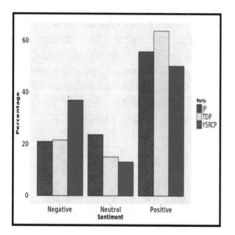

Fig. 4. Polarity analysis of Twitter

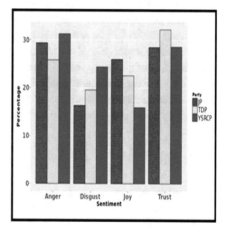

Fig. 5. Emotion analysis of Twitter

Twitter Results. We observe that in this polarity classification also TDP gets the highest actual percentage followed by JP and YSRCP. After TDP, JP has a more favorable rate on Twitter due to the professional background of the party chief as he is a reputed film actor in the state. We can compare the result in Fig. 4. From Fig. 5, we observe that TDP has more trust, and when coming to joy JP is leading, as said earlier the party chief of JP is an actor and has a great following on Twitter. Besides, there is a minority of anger and disgust for TDP when compared to the other two parties.

5   Conclusion

In this research, we are concluding only the trends in social media platforms. This paper is in no way a prediction for who is going to win in the election. Apart from Twitter which is used earlier in predicting elections in other provinces of India Quora was never used for predicting elections which makes this research work more unique. Our research is unbiased to the election result because there are a lot of other factors that impact the outcome. Majority of the demographics in the country are rural voters. Meanwhile, people who tweet and comment on social media are mostly the urban society and cannot have projected them for prediction, and in the same way, many other factors impact the result of the analysis. To conclude, our paper entails the feelings and emotions of netizens towards the election of Andhra Pradesh state election 2019.

6   Future Scope

The sentiment analysis is always helpful to understand the people's inner voice through social media platforms in many ways. The election result could be used by a party to understand the feelings even if it is positive or negative and improve their strategies for winning the campaign and do good for society. In the same way, when a company launches a new product, people tweet or comment on that product and these reviews could be used for sentiment analysis and perceive the negatives and positives of that product and care could be taken when a new product is going to be released. The model presented in this paper could be used for any data for sentiment analysis.

References

1. Salunkhe, P., Deshmukh, S.: Twitter based election prediction and analysis (2017)
2. GSR_Social_Media_Research_Guidance_Using_social_media_for_social_research (n.d)
3. Zhang, C.G., Liu, P.Y., Zhu, Z.F., et al.: A sentiment analysis method based on a polarity lexicon. J. Shandong Univ. **47**(3), 47–50 (2012)
4. Razzaq, M.A., Qamar, A.M., Bilal, H.S.M.: Prediction and analysis of Pakistan election 2013 based on sentiment analysis. In: 2014 IEEE/ACM International Conference on Advances in Social Networks Analysis and Mining (ASONAM 2014), Beijing, pp. 700–703 (2014)
5. Ramzan, M., Mehta, S., Annapoorna, E.: Are tweets the real estimators of election results? In: 2017 Tenth International Conference on Contemporary Computing (IC3), Noida, pp. 1–4 (2017)
6. Wang, L., Gan, J.Q.: Prediction of the 2017 French election based on Twitter data analysis. In: 2017 9th Computer Science and Electronic Engineering (CEEC), Colchester, pp. 89–93 (2017)
7. Burnap, P., Gibson, R., Sloan, L., Southern, R., Williams, M.: 140 characters to victory? Using Twitter to predict the UK 2015 General Election. J. Elect. Stud. **41**, 230–233 (2016)

8. Nausheen, F., Begum, S.H.: Sentiment analysis to predict election results using Python. In: 2018 2nd International Conference on Inventive Systems and Control (ICISC), Coimbatore, pp. 1259–1262 (2018)

9. Ramteke, J., et al.: Election result prediction using Twitter sentiment analysis. In: International Conference on Inventive Computation Technologies (ICICT), vol. 1. IEEE (2016)

10. Ikoro, V., Sharmina, M., Malik, K., Batista-Navarro, R.: Analyzing sentiments expressed on Twitter by UK energy company consumers. In: 2018 Fifth International Conference on Social Networks Analysis, Management, and Security (SNAMS), pp. 95–98. IEEE, October 2018

11. Pollacci, L., Sîrbu, A., Giannotti, F., Pedreschi, D., Lucchese, C., Muntean, C.I.: Sentiment spreading: an epidemic model for lexicon-based sentiment analysis on Twitter. In: Conference of the Italian Association for Artificial Intelligence, pp. 114–127. Springer, Cham, November 2017

12. Jockers, M.: Package 'syuzhet' (2017). https://cran.r-project.org/web/packages/syuzhet

Subspace Based Face Recognition:
A Literature Survey

Bhaskar Belavadi[✉], K. V. Mahendra Prashanth, M. S. Raghu,
and L. R. Poojashree

Department of Electronics and Communication Engineering,
SJB Institute of Technology, Kengeri, Bengaluru, India
bhaskar.brv@gmail.com, kvmprashanth@yahoo.co.in,
raghusrinivasaniyengar@gmail.com,
l.r.poojashree@gmail.com

Abstract. It has become a necessity than a mere requirement in computer vision and image analysis technology over years, for image database identifications and identity verifications through iris scan, color, occlusions, texture of the face images. Also we could see security cameras installed under surveillance in sensitive areas such as Airports, Security on defense system, ATMs, in schools and colleges, banks, offices etc. In order to recognize the image of a person and identify, there are plethora of approaches. However, there are many problems which are still unsolved in the face recognition. This paper provides a till-date survey of the image face recognition showcasing the complete review of the research underwent on linear face recognition approaches and throw light upon certain non-linear approaches for the future case of study. The survey is made only with principal component analysis as the key on algorithms which are analyzed for its pros and cons in the purview of databases and classifiers. The survey shows the linear phase with least complex to the most complex issues with solutions and challenges faced by many authors.

Keywords: Principal Component Analysis (PCA) · Databases · Classifiers · Algorithms

1 Introduction

The biometric fingerprint requires the information of the thumb impression; the ATM machine requires the PIN to withdraw the cash, an internet for retrieving the data, the account number to deposit a cash and passbook entry, the onetime password (OTP) for online transactions through internet, Facebook applications requires the name to be tagged by the user and many more. However, all these have a constraint that they require the information in one or the other form from the user end. For the recognition of personnel without the cooperation from the personnel is more effective. Hence there is a more scope in the face recognition research [15].

Face recognition provides a communication interface among the researchers. Here, numerous approaches (algorithms) are studied for the face recognition development.

© Springer Nature Switzerland AG 2020
J. S. Raj et al. (Eds.): ICIDCA 2019, LNDECT 46, pp. 284–296, 2020.
https://doi.org/10.1007/978-3-030-38040-3_33

In the framework of face recognition, the concept of reduction in dimensionality of images is known as subspace analysis. The face recognition is a hot topic among the researchers due to this theory. Subspace is the subclass of the loftier space with the properties of the loftier space. When an image is considered for recognition, it is of high dimensional. Considering this image is a burden for the processor which encounters computational problems and other miscellaneous threats for processing. By adopting the theory of subspace, the image is reduced to low dimensions. Hence, all or most of the features of the image are represented in the subspace image than the original image. Further by using these subspace images, face recognition is performed efficiently. Many researchers had inherited all these topics in investigations from past many years but still a single almighty approach is to be explored. Thus, face recognition is the hot cake for the researchers [16].

With the increasing day to day challenges and demand in digital world, there are enormous algorithms being proposed in the computer vision and image analysis. The face manifolds in subspace can be either linear or nonlinear. The classification is as shown in Fig. 1.

Hybrid approach involves the idea of how the human vision reception of the individual feature and the entire face i.e., the combination of holistic and feature based approaches [17].

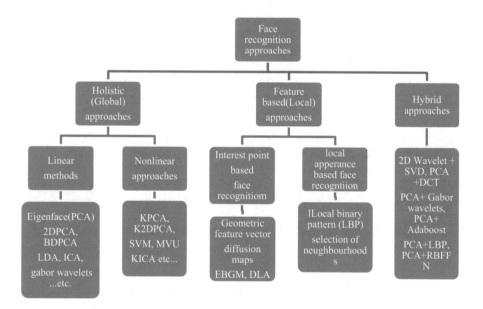

Fig. 1. Recognition approaches

In this paper, we keenly concentrate on the subspace part with principal component analysis as the topic on the main stage.

2 Principal Component Analysis

The PCA also known as Karhunen-Loeve (KL) transformation or Eigen space approach is a very simple yet effective approach of face recognition. It was by Kirby and Sirovich [15] in 1987; who made an attempt to treat images with reduced dimensions through subspace methods for image identification. Then in 1991, Turk and Pentland [18] extended the concept to eigenface projection. Further many developments introduced the same concept with solutions for extensive variety of images with various distur-bances [2, 6, 7]. Principal component analysis is linear combination of weighted vectors called Eigen faces. The images from the dataset are resized to low dimension. The database is divided into test and train images. The train images covariance matrix is calculated from which, Eigen vectors are determined. These Eigen vectors are the numerical measures that represent an entire image. The image to be tested is projected on the vector space for similarity measure between the image to be tested and the trained images. The classification of the image to which class it belongs is done by different classifiers. The most popular and prevalent is the Euclidean distance classifier.

2.1 Mathematical Formulation of PCA

- We obtain N training images I1, I2, I3 ……. IN each of dimension w × h. These images are converted into a vector space through concatenation. The concatenation produces a 1D Eigen vector transformed from the 2D matrix. Represent each image Ii with its corresponding vector.

$$\begin{bmatrix} a11 & a12 & \cdots & a1h \\ \vdots & & \ddots & \vdots \\ aw1 & aw2 & \cdots & awh \end{bmatrix} \xrightarrow{\text{concatenation}} \begin{pmatrix} a11 \\ \vdots \\ a1h \\ \vdots \\ awh \end{pmatrix} = \alpha i$$

- Calculate the mean vector using the following the equation:

$$\text{Mean}(x) = \frac{1}{N} \sum_{i=1}^{N} \alpha i$$

- Subtract the mean face vector from each image vector to obtain a set M i.e.,

$$M = \alpha i - x$$

The motto of subtracting is to keep only the distinguished features by eradicating the common information.

- Find the covariance of the matrix C using the following equation:

$$C = K^T K \qquad \text{Where, } K = [M1, M2, M3, \ldots .MN]$$

- Now find the eigenvalues of each vector for C and sort them in descending order for the first m highest values. Now each Eigen vector will have $N \times 1$ dimension.

3 Algorithms

– The 2-D facial images of dimension $N \times N$ from training set are represented as a column vector of dimension N^2. **Principal component analysis** [1], a face recognition and facial expression identification technique is used, both of which are based on PCA. It is used for identifying whether the facial expression is anger, sad, happy, neutral, disgust, fear etc. PCA creates a feature space where the faces are projected with reduced dimensions and the feature vectors (principle components) are arranged in decreasing order in correspondence with eigenvalues. In this method the images are converted to 1D vector and with the single row or column vector are further processed. Significant difference in the recognition accuracy was observed when the number of images under test are constant, but for two different databases. For effective performance evaluation, the recognition accuracy is calculated which is defined as the ratio of the number of test images correctly recognized to the total number of test images. What if the face recognition is for different and clumsy environments in background for image face recognition?
Ch. Satyanarayana et al. evaluates the performance of **Incremental PCA (IPCA)** for face recognition. The key advantage of this algorithm lies in the ease of including every new addition to the facial images which is done with the support of predictor. With the predictor, the accumulated data is used for the reshaping of the fresh arrived image. The problem with face recognition for heavy dataset is studied, and a model that uses PCA approach is evaluated.
In the above two papers there is difficulty in computation for accuracy and they follow a batch processing of images which is a tedious and a time consuming process. Hence, an optimum method called **Sequential Principal component analysis (SPCA)** which is the sequential way of calculating the principal components for corresponding Eigen values of covariance matrix, has been studied. Hsu et al. proposed this procedure in USA 2015 [20]. Here, two methods of PCA are adopted for given P number of data vectors and Q set of samples for each data vectors. On the first kind, for $P \leq Q$, i.e., covariance matrix, the eigenvectors are generated by employing Direct cosine IIT method whereas for $P > Q$ i.e., cross correlation matrix, the eigenvectors with IIT method are processed. Thus cross correlation matrix method with PCA is an effective sequential update procedure and, the processing can be done due to the fixed size of the cross correlation matrix for $Q \times Q$. On the other hand, the PCA processing with covariance matrix is more effective for $p < Q$ (p is the number of measurements). This algorithm updates the

database dynamically in a minimal amount of time. But with the increase in the value of p, the sequential updating of the corresponding PCA values will be difficult with excess memory for storage and process.

The Eigen values are selected as 9 and 10. Further, recognition and reconstruction are processed. A productive attempt is made for dynamic optimization for real time applications with less memory which outperforms the existing techniques. But, how to improve the performance of the proposed method in terms of accuracy and capability in order to prove the human authentication under some real wide circumstances would be the question left to think over.

- The problem of human verification is elucidated to an extent in 2016 by Matin et al. [4]. Face recognition process for a single person uses two stages. In the first stage, the test image is compared with all the trained images and the best matched images are selected and grouped as another set of trained images for the second stage. In both the stages, PCA is applied for feature extraction of images with Euclidean distance as the classifier criteria. Standard **ORL** (92.5%) face database and **Face94** (92.10%) database is used for simulations. ORL database extended a better result of 1.5%, whereas Face94 showed a better result of 0.52%. Thus, it can be inferred that with PCA double fold (**PCA + PCA**), the algorithm provides best result than with traditional PCA with considering different features in a database. Instead of dynamic optimization [3], double fold PCA is employed for better performance. But, the quest for the method to apply in a generalized manner with analysis of time as complex variable needs to be answered.
- In 2013, [5] gave an answer to the question raised in [4] using **Localized PCA** instead of double fold. The problem that the images of other individuals may be affected with Dynamical Data Updates (DDU) i.e., reduced dimensionality increase with increase in the database is solved in this paper. Euclidean distance is used as classifier. The dataset is divided into subsets of one image of a person and each subset contains the respective Eigen vectors. Each person has a different system for recognition with their Eigen vectors which are autonomous of each other. Thus if the dataset needs to update an image with the latest one, the algorithm allows this feature. This update is included in a separate space constructed and the Eigen vectors are calculated. Then this Eigen vector is place in the old Eigen vectors of the feature space. Due to the autonomous nature of Eigen vectors, other values are unaffected. Another significant feature of this algorithm is that, irrespective of the size of the images in the database, localized PCA method would retain the image size with the efficient features. Hence, the images can be dynamically updated. Localized PCA employed standard database which had a superior quality of images. What is the face recognition is for inferior quality of images?

The LPQ provides the histograms for each class of image. Use of LBQ helps in distortion invariant and provides a productive frame for features. This makes this method more effective. The major limitation was found out in the event of face occlusion due to spectacles.

- In the trending research, it is found that 2DPCA is more efficient than traditional PCA. A thought of improving the robustness in PCA where the images are treated as 1D vectors and thus the **Two dimensional principal component analyses**

(2DPCA) was proposed by Yang et al. [8]. 2DPCA is based on image matrices i.e., the images are treated as matrices rather than 1D vector as in PCA. Thus, holds only the important features (highest Eigen vectors) of the images as the Eigen values unlike PCA, where all the data values are converted into single vector. Here the direct computation of covariance is carried out and the eigenvectors are derived for feature extraction. 2DPCA is mainly based on 2D eigenvectors (principal components). Experiments are performed on this algorithm using ORL, YALE and AR database. With the simulations, it was found that 2DPCA is more simple and straight forward for image feature extraction, since it treats images as matrices. Also it is computationally more efficient and improves the feature extraction speed drastically than traditional PCA. Also it evaluates the covariance matrix more accurately than PCA algorithm. But, it requires more number of coefficients required for image representation (more than PCA) when small number of PC's is taken i.e., over fitting and time feasting is more even for small databases.

- Though 2DPCA is robust enough it encounters an over fitting problem as described in [8]. If any human face image is given to identify a person, the system should be able to recognize the person with relevant information stored in the database or it needs to acknowledge that the person is new to the database, which is quite a challenging task because, there exists a greater similarity between human facial images. [9], solves the over fitting problem of 2DPCA by proposing the **Random subspace Two dimensional PCA (RS-2DPCA)** algorithm for face recognition system. It is the combination of 2DPCA and the random subspace approach in which it defines a set of classifiers. As the first step, it computes the covariance matrices P of trained image like 2DPCA. Then the random subspaces R are defined by selecting randomly from the P. By the application of each random space, classifiers are defined. The classifiers are used to label the trained images sequentially, after which the test image is identified based on the labels of pre classified groups. In this way the test image is recognized and labeled with the respective class. The number of random sub spaces (R) affects the performance of RS-2DPCA on all the datasets. Experiments are carried out on ORL, YALE face database and Extended YALE face database. When the K value is very small, it cannot retain the discriminant information when face image is projected onto K random subspaces. But, when K is above a certain value (threshold), algorithm can retain the information and there is no improvement further even with increase in K. Also, the number of fixed Eigen vectors (M) and random Eigen vectors (N) also has a great impact on recognition accuracy. Thus, to get a high accuracy for the RS-2DPCA, author suggests selecting small M, large N and K. Thus RS2DPCA is more reliable than 2DPCA for Euclidean distance classifier. But, what if other classifiers other than Euclidean distance classifiers are used? [9].

The Gabor-face representation is nothing but convolution of the images of size $m \times n$ with a filter in the tier of Gabor face images. The required coefficients are reduced by combining the 2D Gabor-face representation with the 2D feature extraction algorithms (i.e., 2D PCA and (2D) ^{2} PCA). Thus the proposed method which is opposite to the conventional Gabor-face representation with adopting

EFGR and MGFR method achieved the best recognition performance in comparison with PCA, 2DPCA, (2D) ^2PCA.

- The efficiency in identification can be enhanced by combining the advantages of E2DPCA (more information than 2DPCA) and the main objective of 2D2DPCA (2 side projections). Nedevschi et al. [12] made a constructive approach on **Extended 2D2D (2 dimensional 2 directional) PCA**, which retains more information than previous algorithms. Euclidean distance is used as the classification pattern. In the 2D2DPCA the dimensionality reduction takes place in both directions for a given matrix. Similar computation is followed for the extended dimension case. This algorithm adds on two additional features-this extended version is better-off than 2D2DPCA and it is also better than E2DPCA, since it involves the comparison of both rows and columns in a matrix. With this model, more efficiency in recognition, especially for small number of Eigen vectors which equals with the 2DPCA is obtained. There is point to think over, if other distance measures like Mohalnobis… are used?

- In [8–11] and [12], 2DPCA and its improved methods with the base algorithm as 2DPCA is discussed. With 2DPCA, it is not possible to estimate the low rank components while the training process and if we want any excess data to be added or updated for the training matrix, all the trained items has to be retrained. The maximum of the data variance can be captured from the very initial Eigen vectors hence proved that, 2DPCA is better than GLRAM and 2DIPCA is better than IGLRAM in image construction. Thus an incremental method is more efficient than a batch method as it shows the most optimal execution time.

- However, 2DPCA and its incremental methods [13] are direct techniques of human face representation and recognition. Diagonal PCA is the diagonal representation of the face images whereas **BDPCA** is the bilateral representation of images i.e., simultaneous projections in both the directions (right and left). The dimensionality of the images is reduced in the later method (proposed) in both rows and column directions. A new projection subspace is obtained by computing the two matrices obtained by DPCA. This is based on the diagonal representation of images developed for corresponding eigenvectors and eigenvalues of the images. Then in both right and left directions of the original images, projection is done simultaneously. K-nearest neighbor (KNN) classifier is employed for classification pattern. To test and evaluate the performance of the BDPCA, the experiments are simulated on subset of FERET, YALEB and PF01 database. Also the results are compared for 4 different distance matrix such as Yang, Frobenius, AMD and distance metric classifiers [14].

The PCA, 2DPCA, BDPCA al belongs to the linear classification of face recognition. Now we shall ponder over the nonlinear aspects of face recognition such as KPCA, KLDA …

- [20] adopted **Kernel principal component analysis (K- PCA)** for face recognition. **Kernel- PCA** is a non-linear extension of PCA. Author used Kernel- PCA for extracting the facial information where the input (A) is mapped into feature space (S) and a fundamental principal component analysis (PCA) is applied in feature space. To incorporate the proposed method, the author used the standard **ORL**

(Olivetti Research Laboratory) database. It is observed that higher the degree (d) of polynomial and Eigen vectors (v), smaller is the error rates. Hence, by using SVM as a recognizer he got a least error rate of 2.5% which overtook the 4% error rate with linear PCA and other systems. Thus the author by using SVM as recognizer with ORL database, presented the value of the Kernel-PCA based feature extraction.

- [21] introduced a **Compound Kernel PCA** algorithm for face recognition. The problems of nonlinearity with the pixels in traditional PCA are solved. Compound kernel PCA is based on PCA with inheriting the benefits of Kernel PCA. CKPCA is used by combination of two individual kernels. Each kind of kernel has its own features and range of application in various representations and high dimensional space, which produces different classification results. Depending on kernel functions, they are divided into global kernels (polynomial kernel) and local kernel (Gauss kernel). Hence, according to the characteristics of gauss and polynomial kernel, CKPCA is developed. With the non-linear mapping, KPCA can use better sample information than traditional PCA. But with the diversification in characteristics with each Kernel PCA, two kernels with two different characteristics are combined in CKPCA. Thus the CKPCA has a better recognition rate than traditional PCA and KPCA algorithms, but the former requires much time for processing. Much concentration must be on time consumption for analysis to be cost effective for large samples.

3.1   Table Summary

Sl. No	Algorithms	Classifiers	Databases	Dimensions (pixels)	Training samples	Pc's	ARR (%)	C_Time (sec)
1	PCA	Euclidean	ORL CSU	–	3	–	85.5 81.3	–
2	IPCA	Euclidean (0.25) (1.25)	FERET ORL JNTU	–	–	–	100 70.5	–
3	PCA + DCT	–	ORL	20 × 20	20	9	100	–
4	PCA + PCA PCA	Euclidean	FACE94	–		8	92.10 91.58	
5	Localized PCA PCALDA DLDA	Euclidean	OWN	–	–	–	96 86 92	0.411 0.436 0.134
6	LGPQ DSR S2R2	–		–	–	–	85.22 82.46 81.21	–
8	2DPCA PCA KPCA ICA LDA	–	YALE	–	–	–	84.24 71.53 72.33 71.52 –	–

(*continued*)

(continued)

Sl. No	Algorithms	Classifiers	Databases	Dimensions (pixels)	Training samples	Pc's	ARR (%)	C_Time (sec)
			ORL	–	–	–	96 – 94 85 94.5	
			ORL (leave one out technique)				98.3 97.5 98 93.8 98.5	
9	RS2DPCA	–	YALEB	–	11 2 3 4 5	–	91 96 91 92 94.5	–
			YALE	–	1 2 3 4 5	–	67 89 88 92 93	
10	2D^2PCA EFGR based MFGR based	Frobenius	ORL YALE	– 32 × 32	– 5	–	93.5 91.11 96.67 98.87	–
	2D^2PCA EFGR based MFGR based 2DPCA PCA		ORL				98 98 100	
12	E2D2D--PCA E2DPCA 2D^2PCA 2DPCA PCA	Euclidean	ORL	–	10	–	95 94.5 94 95 93	
13	2DIPCA 2DPCA	SVD	FERET AR YALE	–	–	1 5 10 15 20	94.2 91.2 83.8 76.1 68.4	66.17 309.9

(continued)

(continued)

Sl. No	Algorithms	Classifiers	Databases	Dimensions (pixels)	Training samples	Pc's	ARR (%)	C_Time (sec)
14	BDPCA	Yang Frobenius AMD Weighted	FERET	10×15 14×8 8×9 –	–	–	87 86.50 86.00 –	–
	DPCA	yang		90×12 90×8 90×12 90×15			83.50 82.50 84.00 83.00	
	BDPCA	Yang Frobenius AMD Weighted	YALEB	11×20 8×19 10×12 –			88.88 84.44 87.77 –	–
	DPCA	yang		90×13 90×12 90×16 90×14			87.77 84.44 86.66 87.77	

4 Result

The table summary shown above has been depicted in the graphical model with the accuracy on the y–axis and different algorithms on the x-axis as shown in Fig. 2. Three different databases are chosen for analysis of the algorithms. i.e., FERET, YALE and ORL. The main intention is to explain the algorithm which is efficient in simple ORL database is proven to be efficient at the average level YALE database and, also at the complex FERET database (as per FRGC). With PCA being trailed, it produces an 85.5% accuracy while it to lowered to 81.33% at Yale and Feret database. Inference is that PCA is consistent in an average accuracy of 82.72% in recognition. Hence, with the same algorithm, when trailed successively (PCA + PCA) and, when incremented (IPCA) its efficacy is 92.5% and centum at ORL. This shows that the PCA is the most simple but a very efficient algorithm for 2D images. The nature of 2DPCA in the consideration of 2D matrices instead of 1D vector in PCA makes it more suitable for efficient computation of eigenvectors which is a straight forward than PCA. When we see the variation in 2DPCA, it is much clearer that the accuracy is better than PCA with the mean accuracy of 87.44%. This algorithm is quite efficient at Feret than Yale, which is noteworthy. 2DPCA is 2 dimensional, which can be analyzed even with 2 dimensional, 2 directional PCA. 2DPCA encountered with an over fitting problem which is further improvised in RS2DPCA and (2D) ^2PCA. It could be seen that (2D) ^2PCA is getting a result of 100% at ORL, while it is 94.2% and 96.67% for the other two. It is 96.95% on an average accuracy for all the databases which is an auxiliary than PCA and 2DPCA. We could see the plots for recognition accuracy in case of ORL, FERET & YALE B database.

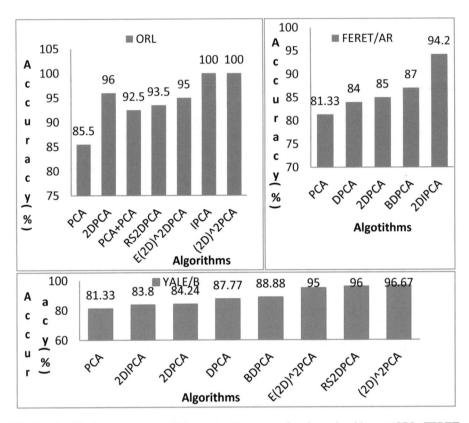

Fig. 2. Graphical representation of the recognition rates of various algorithms at ORL, FERET and YALE/B databases.

Extended version and random subspace of this algorithm provides greater accuracy in the FERET as well. From the above discussions it is clear that if we consider a 2D images, it is preferred to use a simple PCA algorithm while for 2D image with both the directions taking into account, it's always a greater and the best choice to go with (2D) ^2PCA.

5   Summary and Conclusions

This literature survey is all about the till-date assessment of the linear and nonlinear face recognition in two dimensional space using principal component analysis and its extended versions. The basic PCA is used for subspace analysis. But the major problems are with illumination, pose and outside environmental variations in the background. To solve these problems various subspace methods like, treating images a 2D vectors (matrix form) and its extension, bidirectional methods, kernel methods (in case of nonlinear format) were adopted, withstanding PCA as the basic idea behind size reduction. With application of various classifiers, the image identification and recognition is made more robust and accurate. Various strategies were used like; leave one out strategy and majority voting to improve the efficiency of the system. Significant work has been done over segmentation, dimensionality reduction, efficiency and time complexity of 2D face recognition for the standard databases, even, highly complicated FERET database (as per the FRGC).

It is evident from all these analysis that, the principal component analysis is the simplest and the most fundamental approach in the field of face recognition systems. However there exist some conditions or scenarios where the solutions are still in the research mode. Thus, all the researchers are in their best to provide a unique approach that can overcome all these circumstances for robust and user friendly face recognition systems for the future world.

References

1. Meher, S.S., Meben, P.: Face recognition and facial expression identification using PCA. IEEE (2014)
2. Satyanarayana, Ch., Potukuchi, D.M., Pratap Reddy, L.: Performance evaluation of incremental training method for face recognition using PCA. Springer (2007)
3. Zhao, Z., Li, B.: Dynamically optimizing face recognition using PCA. In: 2015 International Conference on Control, Automation and Information Sciences (ICCAIS), Changshu, China, 29–31 October 2015
4. Matin, A., Mahmud, F., Shawkat, M.T.B.: Recognition of an individual using the unique features of human face. In: 2016 IEEE International WIE Conference on Electrical and Computer Engineering (WIECON-ECE), Aissms, Pune, India, 19–21 December 2016
5. Palghamol, T.N., Metkar, S.P.: Constant dimensionality reduction for large databases using localized PCA with an application to face recognition. In: Proceedings of the 2013 IEEE Second International Conference on Image Information Processing (ICIIP-2013) (2013)
6. Bhattacharya, S., Dasgupta, A., Routray, A.: Robust face recognition of inferior quality images using Local Gabor Phase Quantization. In: 2016 IEEE Students' Technology Symposium (2016)
7. Sajid, I., Ahmed, M.M., Taj, I.: Design and implementation of a face recognition system using fast PCA. In: International Symposium on Computer Science and Its Applications. IEEE (2008)
8. Yang, J., Zhang, D., Frangi, A.F., Yang, J.-y.: Two-dimensional PCA new approach to appearance-based face representation and recognition. IEEE Trans. Pattern Anal. Mach. Intell. **26**(1), 131–137 (2004)
9. Nguyen, N., Liu, W., Venkatesh, S.: Random subspace two-dimensional PCA for face recognition. In: Ip, H.H.-S., Au, O.C., Leung, H., Sun, M.T., Ma, W.Y., Hu, S.M. (eds.) Advances in Multimedia Information Processing – PCM 2007. Springer, Heidelberg (2007)
10. Wang, L., Li, Y., Wang, C., Zhang, H.: Face recognition using Gaborface-based 2DPCA and (2D)2PCA classification with ensemble and multichannel model. In: 2007 IEEE Symposium on Computational Intelligence in Security and Defense Applications (CISDA 2007) (2007)
11. Vijay Kumar, B.G., Aravind, R.: Computationally efficient algorithm for face super-resolution using (2D) 2-PCA based prior. IET Image Proc. **4**(2), 61–69 (2009)
12. Nedevschi, S., et al.: An improved PCA type algorithm applied in face recognition. IEEE (2010)

13. Nakouri, H., Limam, M.: An incremental two-dimensional principal component analysis for image compression and recognition. In: 2016 12th International Conference on Signal-Image Technology & Internet-Based Systems (2016)
14. Rouabhia, C., Tebbikh, H.: Bilateral diagonal principal component analysis based on matrix distance metrics: a projection technique for face identification. J. Electron. Imaging **22**(2), 023020 (2013)
15. Sirvoch, L., Kirby, M.: Low dimensional procedure for the characterization of human faces. J. Opt. Soc. Am. A **4**, 519 (1987)
16. Rao, A., Noushath, S.: Survey "Subspace methods for face recognition". Elsevier (2009)
17. Mulla, M.R., Patil, R.P., Shah, S.K.: Facial image based security system using PCA. In: 2015 International Conference on Information Processing (ICIP), Vishwakarma Institute of Technology, 16–19 December 2015
18. Turk, M., Pentland, A.: Eigenfaces for recognition. J. Cogn. Neurosci. **3**(1), 71–86 (1991)
19. Hsu, C., Szu, H.: Sequential principal component analysis. SPIE (2015)
20. Kim, K.I., Jung, K., Kim, H.J.: Face recognition using kernel principal component analysis. IEEE (2002)
21. Liu, C., Zhang, T., Ding, D., Lv, C.: Design and application of compound kernel-PCA algorithm in face recognition. In: 35th Chinese Control Conference, Chengdu, China, 27–29 July 2016
22. Eftekhari, A., Forouzanfar, M., Moghaddam, H.A., Alirezaie, J.: Block-wise 2D kernel PCA/LDA for face recognition. Elsevier (2010)

Medical Image Fusion Using Otsu's Cluster Based Thresholding Relation

C. Karthikeyan[(⊠)], J. Ramkumar, B. Devendar Rao,
and J. Manikandan

Department of Computer Science and Engineering, Koneru Lakshmaiah
Education Foundation (K L University), Vaddeswaram, Andhra Pradesh, India
ckarthik2k@gmail.com

Abstract. Medical imaging is the method of making pictorial illustrations of the inner parts of a human body. Multimodality medical images are required to help more precise clinical information for specialists to make do with medical diagnosis, for example, Computer Tomography (CT), Magnetic Resonance Imaging (MRI), and X-ray images. The fused images can regularly prompt extra clinical data not evident in the different pictures. Another preferred standpoint is that it can diminish the capacity cost by putting away simply the single combined image rather than multisource images. The necessities of MIF are that the resultant fused image ought to pass on more data than the distinct images and should not present any artifacts or mutilation. In this paper, the existing DTCWT-RSOFM method is modified by using the Otsu's cluster based thresholding relation with fuzzy rules are carried out on each sub band independently and merges the approximation and detailed coefficients. The inverse DTCWT is used to obtain the fused image. The results of simulation indicate that our method has attained maximum PSNR of 54.58 whereas the existing one with maximum PSNR of 34.71. Thus, there is an improvement of 35% in PSNR. It's quite evident from visual quality of the proposed method output that, the edge based similarity measures are preserved fine. The simulation results indicate that modified technique has attained maximum EBSM of 0.9234 whereas the existing one with maximum EBSM of 0.3879. Thus there is an improvement of 42% in EBSM. The fused image offers improved diagnosis without artifacts.

Keywords: Edge based similarity measure (EBSM) · Dual Tree Complex Wavelet Transform · Fuzzy rule · Peak Signal to Noise Ratio (PSNR) · Robust Self Organizing Feature Map

1 Introduction

The existing methods like Principal Component Analysis (PCA), Hue Intensity Saturation (HIS), Wavelets, Discrete Wavelet Transform (DWT) [1], Fast Discrete Curvelet Transforms (FDCT), Non sub sampled Contorlet Transform (NSCT), Dual Tree Complex Wavelet Transform (DTCWT) etc., for decomposing the images and different fusion rules for fusing the resultant coefficients. These approaches had the drawback of poor PSNR and EBSM. FDCT method [2] is less complex, less redundant and faster than DWT. But, it suffers from multi directional decomposition [3]. Image fusion

© Springer Nature Switzerland AG 2020
J. S. Raj et al. (Eds.): ICIDCA 2019, LNDECT 46, pp. 297–305, 2020.
https://doi.org/10.1007/978-3-030-38040-3_34

(IF) method based on NSCT and robustness analysis can achieve better fusion performance than other fusion methods [4]. But this method is lack of multi resolution feature. An efficient IF technique using wavelet combined transformation for multi sensor lunar image information is implemented in [5]. In this approach, Wavelet combined Hue Intensity Saturation (HIS) and PCA transformations are applied and it provides better fused image. But it suffers from artifacts and distortions, therefore the original features are not preserved in this method. DTCWT based multimodal MIF is implemented in [6, 7] has the advantages of limited redundancy, perfect reconstruction, good selectivity and directionality, and approximate shift invariance.

Fast Discrete Curvelet Transform method cannot perform multi scale and multi directional decomposition. It also leads to reduce contrast which is undesirable. This method is also sensitive to noise sources from different images and also affected by the movement of the image. In DTCWT-Fuzzy based multimodal MIF method, move invariance can be acquired by utilizing two parallel wavelet trees but, they are subsampled differently, which leads to the lack of frequency localization. The IF technique based on the DTCWT and robust self organised feature map method suffers from artifacts and distortions, original features are not preserved. Otsu's cluster based thresholding relation with fuzzy rules is introduced in our modified fusion method which is based on DTCWT and SOFM. In this paper, Sect. 2 presents the modified method with Otsu's Cluster based thresholding relation and fuzzy rules. Section 3 presents the experimental results of the modified technique. Section 4 discusses the results of our work and Sect. 5 concludes the paper.

2 Otsu' Cluster Based Thresholding Relation with Fuzzy Rules (OCBT)

The existing work uses intensity based registration [11, 12] followed by DTCWT for decomposing the input images and Robust Self Organised Feature Map (RSOFM) is utilized to extract and recognize the features for the approximation and detailed coefficients. The fusion rule specifies the method of fusing the approximate and detail coefficients. Fusion rule can be classified as pixel based and region based. Different rules can also be applied for approximate and detail coefficients. The rules like min, mean, max and fuzzy rule, Neural Networks and Robust Self Organizing Feature Map can be applied [8]. The existing DTCWT-RSOFM method is modified by using the Otsu's cluster based thresholding relation; with fuzzy rules are carried out on each sub band independently and merges the approximation and detailed coefficients. The inverse DTCWT is used to obtain the fused image (Fig. 1).

That is, the pixels that either falls in below threshold (background) or above threshold (foreground). To estimate the weighted variance

With in class variance $\sigma_w^2(T) = \omega_b \sigma_b^2 + \omega_f \sigma_f^2$

Where ω_b is weight of the pixels in the background

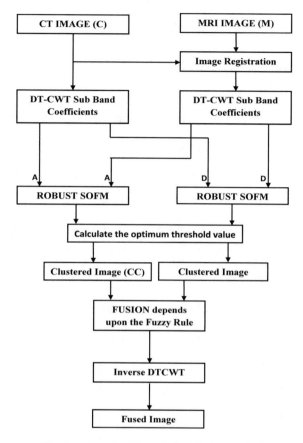

Fig. 1. Flow chart for modified fusion method

$$\omega_b(T) = \sum_{i=0}^{i=T-1} \omega_i$$

σ_b^2 is the Variance of the pixels in the background

$$\sigma_b^2(T) = \frac{\sum_{i=0}^{T-1}(i - m_b)^2 (\text{No. of Pixels})}{\text{Total number of pixels}}$$

m_b is the mean of the pixels in the background

$$m_b = \frac{\sum_{i=0}^{T-1}(i)(\text{No. of Pixels})}{\text{Total number of pixels}}$$

Where ω_f is weight of the pixels in the foreground

$$\omega_f(T) = \sum_{i=T}^{i=N-1} \omega_i$$

σ_f^2 is the Variance of the pixels in the foreground

$$\sigma_f^2(T) = \frac{\sum_{i=T}^{N-1}(i - m_f)^2(\text{No. of Pixels})}{\text{Total number of pixels}}$$

m_f is the mean of the pixels in the foreground

$$m_f = \frac{\sum_{i=T}^{N-1}(i)(\text{No. of Pixels})}{\text{Total number of pixels}}$$

[0, N − 1] is the range of intensity levels. All pixels with a level less than the final selected threshold are background, all those with a level equal to or greater than the final selected threshold are foreground. From the above Otsu's cluster based thresholding relation, the corresponding optimum threshold value is calculated (Fig. 2).

(CT) Registered image (C)					Clustered mage (CC)				
172	146	140	133	131	0	1	1	0	0
133	123	126	129	127	0	1	1	1	0
125	125	123	126	128	0	0	1	1	0
126	131	129	121	127	0	1	0	0	0
140	145	133	128	133	0	0	0	0	0

Fused Image (F)

136	162	171	133	131
124	145	149	142	127
115	128	133	133	125
126	131	140	132	115
140	145	138	123	104

(MRI) Image to be Fused(M)					Clustered image (MM)				
136	162	171	161	141	0	1	1	0	0
124	145	149	142	130	0	1	1	1	0
115	128	133	133	125	0	1	1	1	0
127	136	140	132	115	0	0	1	1	0
140	146	138	123	104	0	0	1	0	0

Fuzzy rule - if $CC_{i,j} = 1$ and $MM_{i,j} = 0$ then $F_{i,j} = C_{i,j}$
if $CC_{i,j} = 0$ and $MM_{i,j} = 1$ then $F_{i,j} = M_{i,j}$
if $CC_{i,j} = 1$ and $MM_{i,j} = 1$ then $F_{i,j} = \max(C_{i,j}, M_{i,j})$
if $CC_{i,j} = 0$ and $MM_{i,j} = 0$ then $F_{i,j} = \min(C_{i,j}, M_{i,j})$

Fig. 2. Mathematically analyses the Otsu's cluster based threshold relation with fusion rules

The specific pixel value of resultant image is minimum value of ($C_{i,j}$, $M_{i,j}$). At that point, apply inverse DTCWT, and then get the final resultant fused image. This is the reverse process of synthesis of Filter Bank for DTCWT. Low and High pass filter bank will combine to one and 2N coefficients combined to N point signal.

3   Results

This section discusses the various parameters to assess the efficiency of the resultant fused image with relevant charts and tables. With various MRI and CT images, the visual results of the modified IF method are analyzed and discussed.

3.1   Performance Metrics

Resultant image performance is assessed by using the following metrics;

Peak Signal to Noise Ratio: In image enhancement process, PSNR measure the quality of reconstruction.
Edge Based Similarity Measure: In the fusion process, EBSM provides the resemblance between the transferred edges.

3.2   Validation of Modified IF Method Using Otsu's Cluster Based Thresholding with Fuzzy Rules

The modified method for multimodal MIF using Otsu's Cluster based Thresholding with fuzzy rules is quantitatively evaluated with PSNR and EBSM. The performance of this proposed work is assessed more than 4 sets of MR and CT images, in comparison with prevailing state of the art approaches such as DTCWT-RSOFM, DTCWT-FUZZY, and FDCT.

3.2.1   Visual Results
Figure 3 shows the visual quality comparison of fused image results with Modified fusion method and other existing methods. These images are in grey scale mode with JPEG format of size 512×512 resolution.

3.2.2   Performance Evaluation
The performance of the modified IF method is quantitatively evaluated with PSNR, and EBSM. Table 1 shows the efficiency of the modified IF method in compare the PSNR values with existing techniques. The results indicate that our method has achieved maximum PSNR of 54.58 whereas the existing one with maximum PSNR of 34.71. Thus there is an improvement of 35% in PSNR.

From the values of Table 1 shows, in all 4 image sets, proposed (OCBT Fuzzy) work only gives the better PSNR values than the other methods. Figure 4 shows the Comparison chart on PSNR value of the modified IF method and other existing methods.

MRI Image	CT Image	FDCT	DTCWT-FUZZY	DTCWT-RSOFM	OCBT-FUZZY

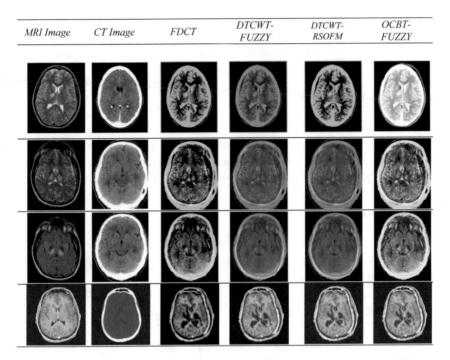

Fig. 3. Visual quality performance of modified method with existing methods [7]

Table 1. PSNR performance of the modified method (OCBT Fuzzy)

Image set	Peak Signal to Noise Ratio			
	FDCT	DTCWT FUZZY	DTCWT RSOFM	OCBT FUZZY
Image 1	23.62	27.29	45.33	54.58
Image 2	22.55	27.82	48.28	54.43
Image 3	25.62	28.72	48.79	55.33
Image 4	21.18	24.25	43.89	52.75

Table 2 demonstrates the operational efficiency of the modified IF method in comparison with available methods based on edge based similarity measure. It's quite evident from visual quality of the proposed method output that, the EBSM values are preserved fine. The results of simulation indicate that modified technique has attained maximum EBSM of 0.9234 whereas the existing one with maximum EBSM of 0.3879. Thus there is an improvement of 42% in EBSM. From the values of Table 2 shows, in all 4 image sets, proposed (OCBT Fuzzy) work only gives the better EBSM values than the other methods.

Figure 5 shows the Comparison chart on EBSM value of the modified IF method and other existing methods. Conclude that the proposed (OCBT Fuzzy) work only gives the better EBSM values than the other methods.

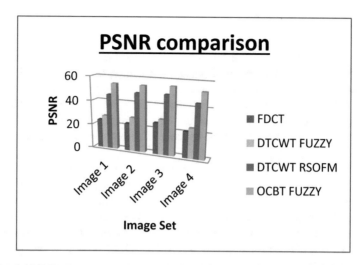

Fig. 4. PSNR Comparison chart between modified method and existing methods

Table 2. EBSM Performance of the Modified technique (OCBT Fuzzy)

Image sets	Edge Based Similarity Measure-(EBSM)			
	FDCT	DTCWT FUZZY	DTCWT RSOFM	OCBT FUZZY
Image 1	0.4248	0.5948	0.6102	0.9234
Image 2	0.4178	0.5843	0.6345	0.8974
Image 3	0.4012	0.5745	0.5865	0.8753
Image 4	0.3879	0.5354	0.5534	0.8013

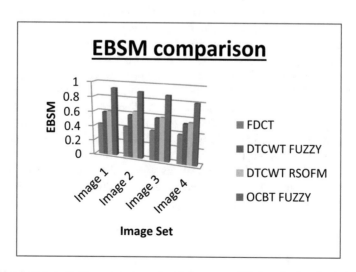

Fig. 5. Edge based similarity comparison chart between modified method and existing methods

4   Discussion

It is observed from the Fig. 3 that the obtained resultant images using modified fusion method using Otsu's Cluster based Thresholding with fuzzy rules (OCBT-Fuzzy) are more accurate and suitable for human visual. The edge preserving is an important feature of interest in complementary details of input images, such as MR and CT images. The Visual results shows that the edge preserving feature of the obtained resultant fused image is better than the existing rules such as Min rule, Mean rule and Max rule.

It is observed from the Tables 1 and 2 that the quantitative evaluation results demonstrate that the modified fusion method using Otsu's Cluster based Thresholding with fuzzy rules (OCBT-Fuzzy) gives better performance in comparison to the existing state of the art methods for multiclass object classification. Experiment results show that the PSNR and EBSM of the modified fusion method are better than the fusion methods of FDCT, DTCWT-Fuzzy and DTCWT-RSOFM. The modified method when compared to existing methods, displays better performance in terms of PSNR values. The results indicate that our method has achieved maximum PSNR of 54.58 whereas the existing one with maximum PSNR of 34.71. Thus there is an improvement of 35% in PSNR. It's quite evident from visual quality of the proposed work output that, the EBSM values are preserved fine. The results of simulation indicate that modified technique has attained maximum EBSM of 0.9234 whereas the existing one with maximum EBSM of 0.3879. Thus there is an improvement of 42% in EBSM. Hence the fused image without artifacts provides better diagnosis.

5   Conclusion

In Medical imaging applications, particularly in CT and MRI images, the edge preserve is vital information in the input images' complementary details. This work investigates and modified as an efficient MIF using Otsu's Cluster based Thresholding with fuzzy rules (OSBT-Fuzzy). This enables us to combine two modalities, MRI and CT images for the visual assessment of information on a single image. The simulation results shows that the performance of the multimodal MIF using Otsu's Cluster based Thresholding with fuzzy rules method is more effective in preserving contrast and complementary details than the existing methods. From the simulation results, as associated with the FDCT, DTCWT-Fuzzy and DTCWT-RSOFM methods, it can be determined that the modified IF method using Otsu's cluster based thresholding relation with fuzzy rules for medical applications gives better PSNR and improved edge based similarity measure. There are numerous research directions possible in future to further enhance the performance of IF methods and for the specific applications.

References

1. Qu, Z., Zhang, J., Feng, Z.: Image fusion algorithm based on two-dimensional discrete wavelet transform and spatial frequency. In: IEEE Fifth International Conference on Frontier of Computer Science and Technology (2010)
2. Candes, E., Demanet, L., Donoho, D., Ying, L.: Fast discrete curvelet transforms. Multiscale Model. Simul. **5**, 1–44 (2006)
3. Sapkal, R.J., Kulkarni, S.M.: Innovative multilevel image fusion algorithm using combination of transform domain and spatial domain methods with comparative analysis of wavelet and curvelet transform. Int. J. Comput. Appl. (2013)
4. Xing, S., Lian, X., Chen, T.: Image fusion method based on NSCT and robustness analysis. In: IEEE International Conference on Computer Distributed Control and Intelligent Environmental Monitoring (2011)
5. Lavanya, A., Vani, K., Sanjeevi, S.: Image fusion of the multi sensor lunar image data using wavelet combined transformation. In: IEEE International Conference on Recent Trends in Information Technology (2011)
6. Karthikeyan, C., Ramadoss, B.: Non linear fusion technique based on dual tree complex wavelet transform. Int. J. Appl. Eng. Res. **9**(22), 13375–13385 (2014). ISSN 0973-4562
7. Singh, R., Srivastava, R., Praksh, O., Khare, A.: DTCWT based multimodal medical image fusion. In: International Conference on Signal, Image and Video Processing (2012)
8. Karthikeyan, C., Ramadoss, B.: Comparative analysis of similarity measure performance for multimodality image fusion using DTCWT and SOFM with various medical image fusion techniques. Indian J. Sci. Technol. **9**(22), 1–6 (2016). https://doi.org/10.17485/ijst/2016/v9i22/95298
9. Otsu, N.: A threshold selection method from gray-level histograms. IEEE Trans. Syst. Man Cybern. **SMC-9**(1), 62–66 (1979)
10. Zhang, J., Hu, J.: Image segmentation based on 2D Otsu method with histogram analysis. In: International Conference on Computer Science and Software Engineering (2008). https://doi.org/10.1109/csse.2008.206
11. Karthikeyan, C., Ramadoss, B.: Fusion of medical images using mutual information and intensity based image registration schemes. ARPN J. Eng. Appl. Sci. **10**(8) (2015). ISSN 1819-6608
12. Karthikeyan, C., Ramadoss, B.: Image registration and segmentation for the quantitative evaluation of complex medical imaging and diagnosis over head-a systematic review. Int. J. Comput. Technol. Appl. **2**(4) (2011)

Weather Prediction Model Using Random Forest Algorithm and GIS Data Model

S. Dhamodaran$^{(\boxtimes)}$, Ch. Krishna Chaitanya Varma,
and Chittepu Dwarakanath Reddy

Computer Science and Engineering,
Sathyabama Institute of Science and Technology, Chennai, India
s.dhamodaran07@gmail.com,
chaitanya.carbon8989@gmail.com,
c.dwarakanathreddy97@gmail.com

Abstract. Weather forecasting has significant impacts on the society and on our prosaic life from betterment to disaster measures. Foregoing weather forecasting and prediction models utilized the intricate combination of mathematical instruments which was inadequate to get a superior classification rate. In this project, we propose new novel methods for predicting monthly rainfall using machine learning algorithms. By accumulating the quantitative data regarding the contemporary state of atmosphere weather forecasts are being made. Exclusively through samples and require limited the intricate mappings from input to output are learnt by machine learning algorithms. Due to the presence of dynamic behaviour in atmosphere, achieving meticulous prediction of weather conditions is a strenuous task. The past year's weather condition variations should be used for prediction of the future weather conditions. There is a very high chance of probability that it will match within the range of an adjacent fortnight of the preceding year. By considering parameters such as wind, temperature and humidity for weather forecasting system, Random forest algorithm and linear regressions are utilized. The weather forecasting of the proposed model is based on the previous record. Hence, this prediction will prove to be much reliable. The performance of the model is more accurate when compared with traditional medical analysis as it uses a fused image having higher quality.

Keywords: Weather · Linear regression · Classification · Accuracy · Random forest · Prediction

1 Introduction

Deep Learning and Machine Learning Techniques are some of the latest technologies that have developed due to the rapid growth in Artificial Intelligence. In applications like land cover classification [4], flood detection in a particular city or area [3], robot path planning [2], and medical image analysis [1], Machine Learning practices are used extensively. Machine learning enables mastering of tasks by continuous learning process and learning to perform tasks without the interaction of humans. Supervised [5] and unsupervised [6] learning are the two learning types. In supervised learning, the training dataset features are labelled and in unsupervised learning, labels are not available and are to be provided by the system to the dataset features. Feature

© Springer Nature Switzerland AG 2020
J. S. Raj et al. (Eds.): ICIDCA 2019, LNDECT 46, pp. 306–311, 2020.
https://doi.org/10.1007/978-3-030-38040-3_35

extraction is an important process in machine learning techniques. Regression, classification and several such operations can be performed by using these extracted features. The commonly used machine learning classifiers are Random Forest Trees [11], Naive Bayes [9], Decision Trees [8], Logistic Regression [10] and SVM classifiers [7]. Artificial Neural Networks (ANNs) are the best choice for feature extraction in applications that involve huge datasets as they provide more accurate results [12]. Application of ANN is widely termed as Deep Learning. In this method, every layer is learned thoroughly by the neural network. The resultant layer is used as a cascading input to the successive layer. Weights are used to connect the layers that resembles biological neurons. These are classified by ANNs [13]. Based on the input image, the number of pixels or pixel elements are decided in image processing. An image is a replica of reality and provides the maximum visual information that is available with respect to an object. Network structures represents the arrangement and connection patterns of neurons and the layers it forms [14].

Herein this weather prediction model for predicting the weather beforehand to decrease several impacts on society and to make predictive measures for disaster reliefs. The Logistic regression and Random Forest algorithms are used for classification of images and prediction of weather forecast.

2 Related Literature

Based on several machine learning methodologies, several research works are being undertaken. Lu et al. [15] proposed a technique for estimating the tumour cell volume in the liver by means of a semi-automatic segmentation method. The CT image is used for localization of the boundaries of the tumour cells. The process segments the tumour cells into slices and estimated the volume efficiently despite the large computation time required for the process. Haris et al. [16] used region-based and combined edge techniques for watersheds based on morphological algorithm for hybrid image segmentation. This technique largely reduced the false edge detection count and hence proved to be efficient. The computational processing time is increased indirectly by the emission of noise. Meng [17], proposed an active contour model called colour reward for segmentation of supervised images. The system is robust and efficient and produced minimum error and performed efficient pairing when tested using several images from certain standard databases. Weather forecasting is also a major field of research in the recent days. Sharma et al. [18] predict the generation of solar power through weather forecasts by means of machine learning algorithms. The amount of rainfall in specific locations are predicted using several classifiers [19].

3 Proposed Technique

The system proposed in the current research makes use of logistic regression and random forest trees to automatically predict the weather though training using previous weather forecasts. Given a particular time or date the classifier is able to predict the weather of that given time using the trained data. A dataset is used to train the classifier that consists of various information about the previous weather conditions such as the amount of rainfall, the temperature of the day, the intensity of moisture and lots more Logistic regression technique is used for binary classification. Here the classifier needs to classify based on only two conditions and does not give an output apart from the given labels.

The dataset issued to train the classifier. The random Forest Tree is used on multiple decision trees. The information stored in the parent node entire depends on the information stored in the child node. There are numerous child nodes present in the classifier and the root node is always one. The root node is responsible to determine the output of the classification which depends on the other nodes of the tree. Figure 1. Shows the comprehensive composition of the proposed model. The model comprises of the weather dataset that first enter the initial stage of pre-processing where it is converted into grayscale to get the intensity of the data obtained. Then the various approaches discussed like Random Forest Trees and Logistic Regression are performed to predict the weather. Forests or random call forests are associate composite learning technique for regression and different tasks, by composing a mess of call trees on coaching time and output the category that's mode of categories (classification) or regression of the individual trees. Forest builds multiple call trees and joins them along to induce a lot of correct and stable prediction. One of the most massive advantages of using random forest tree is that, it may be used separately for each classification of the data and also for many of the regression related issues, which is achieved by the use of various machine learning techniques. We tend to train our system with dataset and make the model for future prediction.

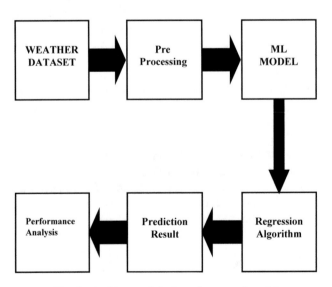

Fig. 1. Architectural design of proposed model

4 Analysis of Experimental Outputs

The experimental outputs were generated on numerous datasets. The dataset consisted of several weather-related data of a particular city for a specific period of time. The experiment was conducted in MAT Lab R2017b where a neural network was built to predict the weather of the city on a specific date. Using the GIS web application and server based maps are using here for data collection and pre-processed for prediction. Using this methods we have developed the good efficiency as shown in Fig. 3. We achieved it for prediction values of weather.

The network made use of Logistic regression and Random Forest trees to make the decision and prediction about the weather. The Computer Vision toolbox available in MAT Lab is used for generation of the network. We use the data to train the machine learning model (Random forest) to predict the weather. Once the model is built the live data is from API as a data source and updated real time in a website with a map interface (Which is also known as GIS). A website that consists of a map on which different data points (different locations) are located when clicked on a data point the predicted weather of that particular area is displayed [20]. Figure 2 displays the experimental outcomes of the classifier trained on previous weather forecasts. Whereas, Fig. 3 depicts a sample weather prediction made by the classifier. The network could efficiently detect the weather prediction for the specified time duration.

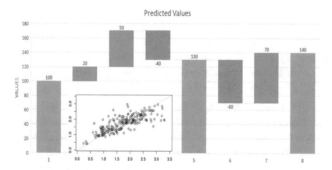

Fig. 2. Predicted values of the weather forecast

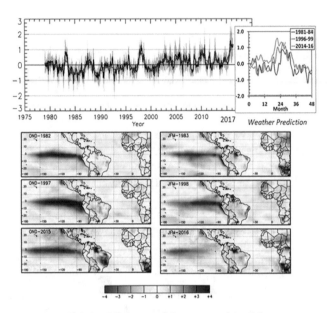

Fig. 3. Efficiency of the proposed model

5   Conclusion

Weather plays an important part in all traits. It is wise to forecast the future weather and take if any necessary steps need to be taken to reduce life and resource losses. The proposed work in this paper that predicts the weather of a specified location with based on the outcomes of the classifier that is trained through several iterations of the pervious weather data. For performing this classification, we use Neural Networks based on Logistic Regression and Random Decision Trees as the algorithms. The proposed model offers improved accuracy. Advanced prediction of the weather is made possible by extraction of features from the training datasets. Several other parameters, namely computational time and efficiency were also improved in comparison to the traditional methods. Future scope involves application of other deep learning methods and improving the security of the neural networks.

References

1. Rahman, M.M., Bhattacharya, P., Desai, B.C.: A framework for medical image retrieval using machine learning and statistical similarity matching techniques with relevance feedback. IEEE Trans. Inf. Technol. Biomed. **11**(1), 58–69 (2007)
2. Morales, M., Tapia, L., Pearce, R., Rodriguez, S., Amato, N.M.: A machine learning approach for feature-sensitive motion planning. In: Algorithmic Foundations of Robotics VI, pp. 361–376. Springer, Berlin, Heidelberg (2004)

3. Ireland, G., Volpi, M., Petropoulos, G.P.: Examining the capability of supervised machine learning classifiers in extracting flooded areas from Landsat TM imagery: a case study from a mediterranean flood. Remote Sens. **7**(3), 3372–3399 (2015)
4. Huang, C., Davis, L.S., Townshend, J.R.G.: An assessment of support vector machines for land cover classification. Int. J. Remote Sens. **23**(4), 725–749 (2002)
5. Dhamodaran, S., Shrthi, A., Thomas,A.: High-resolution flood hazard mapping using remote sensing data. In: International Conference on Computation of Power, Energy Information and Communication (ICCPEIC). IEEE. ISBN: 978-1-5090-0901-5
6. Cheriyadat, A.M.: Unsupervised feature learning for aerial scene classification. IEEE Trans. Geosci. Remote Sens. **52**(1), 439–451 (2014)
7. Bauer, S., Nolte, L.P., Reyes, M.: Fully automatic segmentation of brain tumor images using support vector machine classification in combination with hierarchical conditional random field regularization. In: International Conference on Medical Image Computing and Computer-Assisted Intervention, pp. 354–361. Springer, Berlin, Heidelberg, September 2011
8. Dhamodaran, S., Lakshmi, M.: Design and analysis of spatial–temporal model using hydrological techniques. In; IEEE International Conference on Computing of Power, Energy and Communication 22–23 March 2017. ISBN 978-1-5090-4324-8/17
9. Dumitru, D.: Prediction of recurrent events in breast cancer using the Naive Bayesian classification. Ann. Univ. Craiova-Math. Comput. Sci. Ser. **36**(2), 92–96 (2009)
10. Chou, Y.H., Tiu, C.M., Hung, G.S., Wu, S.C., Chang, T.Y., Chiang, H.K.: Stepwise logistic regression analysis of tumor contour features for breast ultrasound diagnosis. Ultrasound Med. Biol. **27**(11), 1493–1498 (2001)
11. Liaw, A., Wiener, M.: Classification and regression by randomForest. R News **2**(3), 18–22 (2002)
12. Sarhan, A.M.: Cancer classification based on microarray gene expression data using DCT and ANN. J. Theor. Appl. Inf. Technol. **6**(2), 208–216 (2009)
13. Rogers, S.K., Ruck, D.W., Kabrisky, M.: Artificial Neural Networks for Early Detection and Diagnosis of Cancer, pp. 79–83. Elsevier Scientific Publishers, North-Holland (1994)
14. Badrul Alam Miah, Md., Tousuf, M.A.: Detection of lung cancer from ct image using image processing and neural network. In: Proceedings of 2nd International Conference on Electrical Engineering and Information and Communication Technology. IEEE (2015)
15. Lu, R., Marziliano, P., Thng, C.H.: Liver tumor volume estimation by semi-automatic segmentation method. In: 2005 27th Annual International Conference of the Engineering in Medicine and Biology Society, IEEE-EMBS 2005, pp. 3296–3299. IEEE (2006)
16. Haris, K., Efstratiadis, S.N., Maglaveras, N., Katsaggelos, A.K.: Hybrid image segmentation using watersheds and fast region merging. IEEE Trans. Image Process. **7**(12), 1684–1699 (1998)
17. Meng, F., Li, H., Liu, G., Ngan, K.N.: Image cosegmentation by incorporating color reward strategy and active contour model. IEEE Trans. Cybern. **43**(2), 725–737 (2013)
18. Sharma, N., Sharma, P., Irwin, D., Shenoy, P.: Predicting solar generation from weather forecasts using machine learning. In: 2011 IEEE International Conference on Smart Grid Communications (SmartGridComm), pp. 528–533. IEEE, October 2011
19. Revathy, B., Parvatha Varthini, S.: Decision theoretic evaluation of rough fuzzy clustering. Arab Gulf J. Sci. Res. **32**(2/3), 161–167 (2014)
20. Geetha, K., Kannan, A.: Efficient spatial query processing for KNN queries using well organised net-grid partition indexing approach. IJDMMM **10**(4), 331–352 (2018)

Quantum One Time Password with Biometrics

Mohit Kumar Sharma(✉) and Manisha J. Nene

Department of Computer Science and Engineering, Defense Institute of Advance Technology, Pune, India
majesticbrat@gmail.com, mjnene@diat.ac.in

Abstract. Use of One Time Passwords (OTP) have proved to be more secure method, than one factor authentication and is therefore currently utilised in online transactions. However, it authenticates the device rather than the user. Also, number of fraud cases is being highlighted owing to the inherent insecurity in classical methods of OTP generation and communication. The study in this paper utilises mathematical concepts of the *Quantum Entanglement* property to generate *Quantum OTP (QOTP),* for authenticating the user based on its biometrics. The proposed methodology enhances security in two factor authentication methods, as per mathematically proven properties of *Quantum Cryptography.*

Keywords: One Time Password · Biometrics · Quantum computing · Quantum entanglement · Quantum cryptography · Quantum OTP

1 Introduction

The rise of digital wallets like *Paytm, Phone Pe, Google Pay* and payment have further reduced the complexities of operating by traditional online banking methods. The provision of security is the biggest challenge faced in the roll out plan for online transaction methods which brings in the use of multifactor authentication [1] of user with One Time Password (OTP) using phone based Short Message Service (SMS) or email services [2, 3]. However, technological advancement has also led to increase in online frauds [4, 5]. The work in this paper focuses towards improving OTP security by proposing use of mathematically proven facts of *Quantum Computing* to generate *Quantum OTP (QOTP)* and integrate it with user biometrics for user based operations.

The rest of the paper is organised as follows: Sect. 2 describes background and related work in the field of security of current OTP methods leading to motivation for the proposed work. Section 3 discusses the preliminaries and notations. Section 4 presents the proposed three phase RIQ model and its security analysis and Sect. 5 summarizes the paper with future work envisaged in this field.

© Springer Nature Switzerland AG 2020
J. S. Raj et al. (Eds.): ICIDCA 2019, LNDECT 46, pp. 312–318, 2020.
https://doi.org/10.1007/978-3-030-38040-3_36

2 Background and Related Work

2.1 Background

OTP. The multifactor authentication by using OTPs [6, 7] is considered very safe wherein the OTP is provided to the user by the authorizing device or service provider. It is generated based on time of transaction as well as type of transaction, with a limited lifetime for use [8, 9]. OTP is shared as per mutually decided method of delivery of OTP to the user, which may be an SMS [10] or email or both.

Biometrics. Role of biometrics, as a tool of information security, is important since it can uniquely identify a user for who it is and also, it is difficult to copy the same [11, 12]. But because captured data is noisy [13], errors in exact generation of digital biometric code at both ends are handled [14] with proposed architectures like *Fuzzy Extractors* [15], *Fuzzy Commitment* and *Secure Sketch* for fingerprint based biometrics.

2.2 Related Work

Security of OTP. SMS based OTP sharing is carried out in clear mode thus, making it prone to Man in the Middle Attack (MITM) [16]. The randomness in OTP generation also depends only on timestamp/seed and OTP generating function. And if the seed or function gets disclosed to the attacker, it can clone or reconstruct the OTPs [17, 18]. To improve the randomness use of genetic algorithms and hardware based seed generation [19–22] have been proposed. An effective solution has been proposed where user biometrics are utilised to thwart MITM attacks [23–25]. A method of creating random one time pads has been proposed using quantum superposition states [26].

2.3 Motivation

The researchers have made considerable efforts to enhance secure and reliable authentication of user with OTP, as discussed in Sect. 2.2. The work in this paper is motivated with following facts to further develop and improve OTP security:

- OTP generated should have high entropy or high degree of randomness.
- OTP, generated at server end is more secure than generating OTP at user end.
- A user must be authenticated rather than the device.

The proposed approach endorses the above mentioned facts by use of quantum communication properties to utilise inbuilt randomness of photon generation and use of biometrics for authentication of the user rather than the device.

3 Preliminaries and Notations

3.1 Quantum Computing

A *qubit* is a bit which is in state 0 and 1 at the same time in some fixed probability. Qubit is a state given by $|\psi\rangle = \alpha|0\rangle + \beta|1\rangle$ [24]. Similarly, when two qubits are considered together, they stay in a probabilistic superposition of four possible states i.e. 00, 01, 10 and 11. We represent '0' and '1' in computational basis as $\begin{bmatrix} 1 \\ 0 \end{bmatrix}$ and $\begin{bmatrix} 0 \\ 1 \end{bmatrix}$ respectively, so the two qubit state will be the tensor product of the respective qubits [27]. Thus, a state of a *n*-qubit system can be decomposed into its individual qubit states which implies measurement of one of the qubit state doesn't affect the state of other qubits. However, there are certain system states which can't be factored into individual qubits and are called as *Entangled States* [28, 29]. States in which sum of observables of individual entangled states is always zero are called as *singlet states*. In two qubit system, *Bell States* (also called as Bell Basis or EPR pairs) are maximally entangled and are given by [27]:

$$|\psi_+\rangle = \frac{|01\rangle + |10\rangle}{\sqrt{2}}, \; |\psi_-\rangle = \frac{|01\rangle - |10\rangle}{\sqrt{2}}, \; |\varphi_+\rangle = \frac{|00\rangle + |11\rangle}{\sqrt{2}}, \; |\varphi_-\rangle = \frac{|00\rangle - |11\rangle}{\sqrt{2}}$$

3.2 Quantum Entanglement

Quantum entanglement is a property by which two states, on a brief interaction, get entangled in such a manner that, when these entangled states are kept so far from each other that no concept of *locality of reference* is involved, then on measuring one state in a particular basis, its outcome can predict the result of measurement of the other entangled state, at a far off place, in the same basis with 100% probability without carrying out the measurement physically. The generation of entangled pair and their distribution [30] holds the key for the proposed work in this paper.

3.3 Notations

Let the variables used in explanation of proposed model be as given in Table 1.

4 Proposed RIQ Model

4.1 RIQ Model

A three phase RIQ model is proposed for two factor authentication by using *QOTP*, using quantum entanglement, and user biometric based operations. The proposed RIQ model has following three phases as shown in Fig. 1:

Table 1. Notations.

Notation	Symbol	Notation	Symbol
User	U	Server	S
Attacker	A	Random seed	R_g
Username	U_n	Transaction code	T_r
Password	P_w	Timestamp	T_s
Classical code	C	Hash function	H
Qubit	q	XOR operation	O_p
Biometric code	B_c	Lifetime of OTP	L_t

Phase I: Registration

- U registers himself to the service provider and they agree upon the two authentication schemes.
- Biometrics can be provided either as a digital code or a biometric image depending on the infrastructure available.

Phase II: Initialisation

- U when wants to avail any service carries out its authentication by using first scheme with S of service provider.
- U enters U_n and P_w, which is verified by S to activate third phase.

Phase III: Quantum

- Figure 1 illustrates the conceptual model for $QOTP$ based two factor authentication with U but can perform classical operations only. Hence, all quantum actions will be carried out by S.

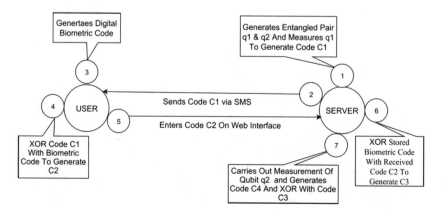

Fig. 1. RIQ: conceptual model of quantum phase

- On successful completion of this phase only, U is able to avail the services provided by the service provider.

4.2 Steps for Proposed RIQ Model

Event flow diagram of RIQ model is depicted in Fig. 2. The major steps involving quantum concepts are described in details as given below:

1. S randomly generates entangled pair of qubits q_1 and q_2, using photons as *QOTP*, which in this case is assumed to be two entangled pairs of qubits with each pair in one of the bell basis, about which even the S is not aware of.
2. One of the entangled qubit, q_1, will be measured by S to generate code C_1 and by property of quantum entanglement of singlet states the measurement value of other qubit q_2 which is C_4 will be exact opposite to the measured value of q_1 i.e. the C_1 and C_4 will give all zeros. The same has been highlighted with example in Event flow diagram of RIQ model in Fig. 2.

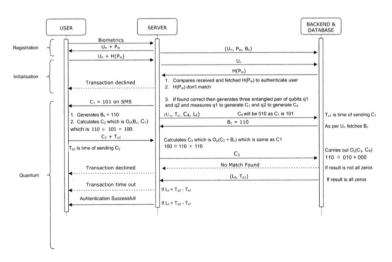

Fig. 2. RIQ: Event FLow diagram

Security Analysis. The current OTP system involves use of biometrics with classical communication and computation techniques. As a result the challenges of MITM attack due to malware infected devices connected on network and vulnerable classical media, faced by existing models and proposed model, stay the same.

However, the proposed model improves upon the security of OTP based transactions by gradual replacement of classical computation methods with quantum computation methods. Due to inherent and mathematically proven quantum communication security principles of *no cloning* and *wave function collapse*, the drawback of utilising numerous pairs of the codes C_1 and C_2 to discover function used for seed generation as well as OTP generating function is eliminated. The generation of entangled bell states are created at server, but cannot be predicted even by server, as their generation is completely random in nature due to use of photons. At the same time the user is getting authenticated rather than device. Thus meeting all the concerns of our motivation.

There are no quantum capabilities available with U, thus restricting to employment of only classical information and operations at user end. Hence, the user based vulnerabilities existing in current OTP execution will still persist. However quantum operation at S to generate $QOTP$ brings the randomness in OTP generation which still provides one step better security than existing methods.

5   Conclusion and Future Work

Classical computing methods bring inherent security challenges for OTP, especially with regards to randomness in OTP generation, which can be overcome using quantum computing. User authentication should also limit itself to authenticating a user rather than a device for which biometrics play a major role. The work in this paper proposes a method to implement two factor authentication using quantum computing to generate QOTP, which is completely random in nature and user biometrics for authenticating the user. This paper presents an effective method of utilising properties of quantum entanglement communication and cryptography for making OTP transactions more secure. Future research work in this domain needs to address the challenges of practical implementation of the proposed model in real world environment, along with addressing server end vulnerabilities once all quantum capabilities of server are available with attacker.

Acknowledgements. All the views and opinions expressed in this paper are those of the authors only. The authors would like to extend gratitude to the colleagues at Defence Institute of Advanced Technology, Pune for their valuable assistance and genuine feedbacks. However, any errors remaining in this clause are solely the author's responsibility.

References

1. Potter, E.R.: Multi-factor authentication using a one time password. US Patent App. 11/697,881 (2008)
2. Varadarajan, R., Malpani, A.: One-time use password systems and methods. US Patent 9,665,868 (2017)
3. Cox, C.T.: Onetime passwords for mobile wallets. US Patent 8,565,723 (2013)
4. Agarwal, S., Khapra, M., Menezes, B., Uchat, N.: Security issues in mobile payment systems, pp. 142–152 (2007)
5. Wang, Y., Hahn, C., Sutrave, K.: Mobile payment security, threats, and challenges, pp. 1–5. IEEE (2016)
6. Haller, N., Metz, C., Nesser, P., Straw, M.: A one-time password system. Network Working Group Request for Comments 2289 (1998)
7. Rockwell, P.: Two factor authentication using a one-time password. US Patent 9,378,356 (2016)

8. Ahn, T.H.: Transaction-based one time password (otp) payment system. US Patent App. 13/555,442 (2013)
9. M'Raihi, D., Machani, S., Pei, M., Rydell, J.: Time-Based One-Time Password Algorithm. Internet Engineering Task Force RFC2011; 6238
10. Eldefrawy, M.H., Khan, M.K., Alghathbar, K., Kim, T.H., Elkamchouchi, H.: Mobile one-time passwords: two-factor authentication using mobile phones. Secur. Commun. Netw. **5**(5), 508–516 (2012)
11. Liu, C.H., Wang, J.S., Peng, C.C., Shyu, J.Z.: Evaluating and selecting the biometrics in network security. Secur. Commun. Netw. **8**(5), 727–739 (2015)
12. Schultz, P.T.: Multifactor multimedia biometric authentication. US Patent 8,189,878 (2012)
13. Jain, A.K., Ross, A., Pankanti, S.: Biometrics: a tool for information security. IEEE Trans. Inf. Forensics Secur. **1**(2), 125–143 (2006)
14. Hao, F., Anderson, R., Daugman, J.: Combining crypto with biometrics effectively. IEEE Trans. Comput. **55**(9), 1081–1088 (2006)
15. Dodis, Y., Reyzin, L., Smith, A.: Fuzzy extractors: How to generate strong keys from biometrics and other noisy data, pp. 523–540. Springer (2004)
16. Peotta, L., Holtz, M.D., David, B.M., Deus, F.G., De Sousa, R.: A formal classification of internet banking attacks and vulnerabilities
17. Kelsey, J., Schneier, B., Wagner, D., Hall, C.: Cryptanalytic attacks on pseudorandom number generators, pp. 168–188. Springer (1998)
18. Ambainis, A., Rosmanis, A., Unruh, D.: Quantum attacks on classical proof systems: The hardness of quantum rewinding, pp. 474–483. IEEE (2014)
19. Jain, A., Chaudhari, N.S.: An improved genetic algorithm for developing deterministic OTP key generator. Complexity **2017**, 1–17 (2017)
20. Popp, N., M'raihi, D., Hart, L.: One time password. US Patent 8,434,138 (2013)
21. Juels, A., Triandopoulos, N., Van Dijk, M., Brainard, J., Rivest, R., Bowers, K.: Configurable one-time authentication tokens with improved resilience to attacks. Int. J. Comput. Sci. Inf. Technol. **3**(1), 186–197 (2011). 2016. US Patent 9,270,655
22. Hamdare, S., Nagpurkar, V., Mittal, J.: Securing SMS based one time password technique from man in the middle attack. arXivpreprint arXiv:1405.4828 (2014)
23. Plateaux, A., Lacharme, P., Jøsang, A., Rosenberger, C.: One-time biometrics for online banking and electronic payment authentication, pp. 179–193. Springer (2014)
24. Hosseini, Z.Z., Barkhordari, E.: Enhancement of security with the help of real time authentication and one time password in e-commerce transactions, pp. 268–273. IEEE (2013)
25. Zhu, H.: One-time identity–password authenticated key agreement scheme based on biometrics. Secur. Commun. Netw. **8**(13), 2350–2360 (2015)
26. Upadhyay, G., Nene, M.J.: One time pad generation using quantum superposition states, pp. 1882–1886. In: IEEE (2016)
27. Gruska, J.: Quantum Computing, vol. 2005. McGraw-Hill, London (1999)
28. Rieffel, E.G., Polak, W.H.: Quantum Computing: A gentle Introduction. MIT Press, Cambridge (2011)
29. Yanofsky, N.S.: An introduction to quantum computing. arXiv preprint arXiv:0708.0261 (2007)
30. Kim, Y.H.: Single-photon two-qubit entangled states: preparation and measurement. Phys. Rev. A **67**(4), 040301 (2003)

Hopfield Network Based Approximation Engine for NP Complete Problems

T. D. Manjunath, S. Samarth[✉], Nesar Prafulla, and Jyothi S. Nayak

Department of Computer Science and Engineering,
B.M.S College of Engineering, Bengaluru, India
{1bm15cs115, 1bm15cs088, 1bm15cs063,
jyothinayak.cse}@bmsce.ac.in

Abstract. NP Complete problems belong to a computational class of problems that have no known polynomial time solutions. Many popular and practically useful problems in the field of optimization and graph theory which have real life application are known to be NP Complete and solving them exactly is intractable. The existing problem with these is that there is no known efficient way to locate a solution in the first place, the most notable characteristic of NP-complete problems is that no fast solution to them is known. However approximate solutions can be obtained in polynomial time. Hopfield networks are one of the ways to obtain approximate solution to the problems in polynomial time. Exploiting the reducibility property and the capability of Hopfield Networks to provide approximate solutions in polynomial time we propose a Hopfield Network based approximation engine to solve these NP complete problems.

Keywords: NP complete problems · Approximate algorithms · Hopfield Networks · Reduction

1 Introduction

P denotes the class of all decision problems that can be solved by a Turing machine in polynomial time, for example sorting and searching. The set of problems for which there exists an algorithm that can verify the given solution in polynomial time is called Non-deterministic Polynomial (NP) time indicating that the problem can be solved in polynomial time on a non-deterministic Turing machines, for example factorization of integers. NP-Hard is the class of problem where no known efficient verification algorithms exist and these are the hardest problems in NP. NP complete problems are the intersection of NP and NP-Hard, meaning that they have efficient verification algorithms and are as hard as any other problems in the NP class problems. The advantage of solving NP-Complete problems is that, they are all different forms of the same problem and can be expressed as others in polynomial time. This concept is known as reduction. Reduction is the process of representing a given problem (here NP) as another problem, say P2 (also in NP), such that the solution to the so obtained P2 can be used to determine the solution to teh original problem at hand. So solving one NP-complete problem means that you can obtain results to other NP-complete

© Springer Nature Switzerland AG 2020
J. S. Raj et al. (Eds.): ICIDCA 2019, LNDECT 46, pp. 319–331, 2020.
https://doi.org/10.1007/978-3-030-38040-3_37

problems using these reductions. Formally reductions are polynomial time algorithms that give a mapping from instance A of problem instance P to instance A' of problem instance P' so that the output of A' w.r.t P' is same as output of P w.r.t to A.

Many real life problems are known to be NP-Complete problems [1]. Some of these problems exist in the domain of graph theory and have been studied for many decades now. Since the time to determine the exact solution to these problems grows exponentially with the size of the input instance, it is often infeasible to get the exact solution to these problems. Also, in many cases obtaining an approximate/near optimal solution to these problems in a relatively very short time is more reasonable [2]. These algorithms, called approximation algorithms, draw inspiration from different techniques and approaches like bio inspired algorithms, neural networks and genetic algorithms. One such approach is the use of Hopfield Networks.

Hopfield Neural Networks are Recurrent neural network with symmetric weights and zero valued self-loop. They have a global energy function due to the symmetry in the weights. There also exists a simple update policy that converges to energy minima (local/global) [3].

2 Related Work

Optimization problems generally have a set of parameters that have to be determined to maximize or minimize an objective function. Along with this there are constraints introduced on the parameters that are to be determined. The way the mapping onto the Hopfield network is done is such that these constraints are modeled as penalty terms and are added to the global energy function. Now the problem of satisfying the constraints has been modeled into the task of minimizing or maximizing the objective function that is reducing the global energy function which is typically applying the update policy repeatedly until we reach a stable state. The introduction of the penalty terms in the Energy function modifies the update rule. The modified update policy also takes into account the constraints introduced.

The general steps taken to map a constrained optimization problem onto a Hopfield Network is succinently given here [4].

- Selection of an appropriate representation of the original problem as a Hopfield Network so that the result obtained from the stable state of the Hopfield Network can be mapped to the original problem's near optimal solution.
- Selection of the energy function such that the optimal solution to the problem coincides with the global minima of the Hopfield Network.
- Appropriate selection of the symmetric weights depending on the given problem instance and Choosing the initial values of the neurons.

One major issue with this approach is the existence of local minimas or spurious minimas. These don't correspond to the actual optimal solution but are just unwanted minima. Techniques have been proposed to overcome this problem. One of them is that of gradient ascent proposed by Chen [5]. Here the idea is, once the Hopfield network settles in a local minima (stable state), the energy surface is modified by varying the weights and the threshold values such that the local minima of the energy surface raises and the Hopfield Network can continue moving in the direction of reducing energy. In this paper they talk about the minimum vertex cover problem as a hard problem to solve even if some constraints are placed on them, like, limiting the degree of the vertices. Light is thrown on the existence of a factor 2 approximate algorithm for the vertex cover problem and also on the existence of faster algorithms on k-vertex cover problems. The gradient ascent technique proposed basically has 2 phases. The first phase is that of the Hopfield network phase where the entire system moves to a lower energy state. The second one being the gradient ascent phase where the parameters of the Hopfield network are modified in such a way to fill up the local minima. Then these 2 steps are performed repeatedly till they reach the global minima. The results they have obtained are promising considering that their algorithm ran faster and gave a better solution than the factor 2 approximate algorithm on the randomly generated data set. Another valuable fact that this paper points out is that of the inherent difficulty in obtaining a data set for NP-Complete/Hard class of problems or even creating a synthetic one. Although there are lot of methods to obtain approximate solutions to NP-Complete problems they mostly focus on a particular problem with a particular method.

Hopfield and Tank proposed their approach of mapping a given TSP problem onto a Hopfield Network [3]. The results obtained showed that the idea of using Hopfield Networks for tackling TSP is an area to be explored further.

3 Proposed System

There are several methods for approximating solutions to NP-Complete problems like greedy algorithms, bio inspired algorithms. Most of the systems giving approximate solutions to the problems are specific to a single problem and a single method. The idea behind the proposed system is to generalize the methods for a given problem and also to increase the different types of problems solved by using the concept of approximation preserving reductions [6]. In the current system the approximation methods are based on hopfield neural networks.

The proposed system can be viewed in 2 parts. The first part deals with the reduction of the input instance graph to a problem that has a corresponding kernel in model. The second part deals with actually solving the input instance or the problem instance obtained after a reduction or a chain of approximation preserving reductions (Fig. 1).

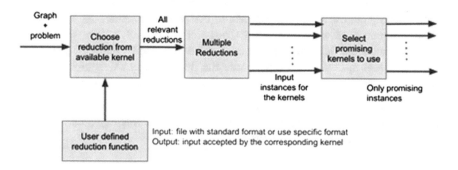

Fig. 1. Part 1 of engine

Part 1: Reducer and Selector Part. This part of the engine deals with the reduction of the given problem with the graph. This consists of-

- A module where a reduction function is chosen based on the problem and the input instance of the graph. If the reduction function is not available from the kernel module, then the user has the option to define his own reduction function. It also adds the function to the existing kernel.
- With all the possible reduction functions that can be applied, there will be multiple reductions that result in multiple input instances for further processing by their corresponding kernels.
- All the resulting input instances go through a module where only instances that seem like they may give a good approximate solution will go through. The way it is decide if something is promising or not is based on certain properties of the input graph that can be determined in polynomial time.

The second part of the engine is where the actual solution to the problem is searched for (Fig. 2).

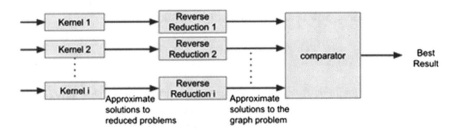

Fig. 2. Part 2 of engine

Part 2: Approximating Core Part. This part deals with finding the approximate solution of the problem with the kernels obtained. This consists of:

- With the promising kernels obtained, they are solved individually to obtain solutions that will go to the inverse of the reducer.
- The inverse of the reduction maps the approximate solution to the problem obtained after reduction to an approximate solution of the original problem. More than one solution may be obtained, one for each of the promising reductions.
- Since we can have multiple solutions, one from each of the different kernels, we need to choose the best one amongst these. Since the quality of the solution depends on the problem, the comparator choose the best solution using problem specific evaluation metric.

Three kernels (TSP, Max Cut and Vertex Cover) were experimented with. One approximation reduction from Vertex Cover to Max Cut was explored. The results obtained from these tests look promising.

Travelling Salesman Problem

The initial paper by Hopfield and Tank proposed the idea of using Hopfield Networks for tackling NP Complete Problems by targeting the Travelling Salesman Problem. In this kernel we map the given problem as follows. If the number of cities is N then we create a NxN matrix. The rows represent the different cities and the columns represent the possible positions of the city in the itenary. Each sell in this matrix is a neuron and the matrix gives the value of the neurons. The energy function is composed of 4 parts, they are:

- The row penalty
- The column penalty
- The global inhibition.

These three are required to ensure that the configuration of the network is meaningful. When there is exactly one occurrence of 1 in each row and each column, this part of the energy function becomes zero.

$$E = A/2 \sum_X \sum_i \sum_{j \neq i} V_{Xi} V_{Xj} + B/2 \sum_i \sum_X \sum_{X \neq Y} V_{Xi} V_{Yi}$$
$$+ C/2 \left(\sum_X \sum_i V_{Xi} - n \right)^2$$

And the 4th term in the penalty is used to push the model to prefer shorter routes over longer routes.

$$1/2D \sum_X \sum_{Y \neq X} \sum_i d_{XY} V_{Xi} \left(V_{Y,i+1} + V_{Y,i-1} \right)$$

The term V followed by the subscript is the term in the matrix obtained by applying the activation function on the matrix that has the values of the neurons. This energy function is differentiated to obtain the equation of motion for the neurons so them move towards a lower energy state.

$$du_{Xi}/dt = -u_{Xi}/\tau - A \sum_{j \neq i} V_{Xj} - B \sum_{Y \neq X} V_{Yi}$$
$$- C \left(\sum_X \sum_j V_{Xj} - n \right)$$
$$- D \sum_Y d_{XY} \left(V_{Y,i+1} + V_{Y,i-1} \right)$$
$$V_{Xi} = g(u_{Xi}) = \frac{1}{2}(1 + tanh(u_{Xi}/u_o)) \quad (for\ all\ X, i)$$

The 'u' term with the subscript represents the value of the neurons. 'n' is the number of cities.

Vertex Cover

Definition: Given a graph G(V,E) minimum vertex cover problem is to find a subset V'
of V such that for every edge (v1,v2) in E either v1 belongs to V' or v2 belongs to V'

Hopfield network based algorithm for finding approximate MVC:

The constraint of the vertex cover is added as a term to energy of hopfield neural
nets.

The size of vertex cover is also added as another term in the energy function

The energy function looks as follows:

v = vector representing vertex cover

u = input potential of the neuron

d = adjacency matrix

$$E = A \sum_i v_i + B \sum_i \sum_j dij \overline{v_i \bigvee v_j}$$

$$= A \sum_i v_i + B \sum_i \sum_j d_{ij} \left(1 - v_i \bigvee v_j \right)$$

$$= A \sum_i v_i + B \sum_i \sum_j d_{ij} \left(1 - v_i - v_j + v_i v_j \right)$$

$$= A \sum_i v_i + B \sum_i \sum_j d_{ij} \left(v_i v_j - v_i - v_j \right) + B \sum_i \sum_j d_{ij}$$

The final energy function looks like this:

$$E = A \sum_i v_i + B \sum_i \sum_j d_{ij} \left(v_i v_j - v_i - v_j \right)$$

After deriving the energy function the motion equation of the hopfield network
corresponding to this problem is:

$$\frac{du_i(t)}{dt} = -u_i(t) - \frac{\partial E}{\partial v_i(t)}$$

After partial differentiation the motion equation for the hopfield net is:

$$\frac{du_i(t)}{dt} = -u_i(t) - A - 2B \sum_j d_{ij} v_j + 2B \sum_j d_{ij}$$

Based on this motion equation the update rule is given by

$$u_i(t+1) = u(t) + \frac{du_i(t)}{dt}$$

The output potential is calculated using sigmoid function as follows

$$v_i(t) = 1 / \left(1 + e^{-u/\theta} \right)$$

The hyper parameters were set like this empirically:
A = 1.0, B = 0, θ = 5, threshold = 0.5. The above motion equations were used from the paper about hopfield neural network by X. Chen et al. [7].

Max-Cut

Definition: Given a graph G(V, E) the objective of max cut is to partition the vertex set into v1 and v2 such that the sum of the weights of edges crossing the partition is maximized.

In max cut the hopfield network technique to minimize the energy which is the size of the cut then the local update rule is applied to get to an energy minimum which corresponds to approximately large cut close to max cut.

A modified version of this algorithm which gives a running time of O(nlog(W)) was used, and a theoretical lower bound was established for this algorithm w(max-cut) >= (1/(2+e))*sum(w(all-edges)).

4 Experimental Results

TSP

The key challenge in building a model for tackling the Travelling Salesman problem was that of using the right set of parameters. The parameters used by Hopfield and Tank seemed to work well. The experiments were performed with city locations (x, y) that were randomly sampled in the range of x $\in$ [0, 1] and y $\in$ [0, 1]. The number of cities were set to 10. The remaining parameters were varied and the observations have been recorded below:

- In the range of 10^{-5} to 10^{-6} there was roughly an inverse proportionality between the delta_t value and the number of iterations the model took to converge to a low energy state. The notable observation was that the quality of the solution wasn't affected by varying the delta_t value. This verified the observation made in the original paper too.
- It was observed that in majority of the cases the number of iterations taken to arrive at a solution was less than 2000 iterations and in many more it took less than 1000 iterations. The conclusion we can draw from this is that if we aren't approaching a solution even after 2000 iterations it is better to restart the convergence with a different initial state (Fig. 3).

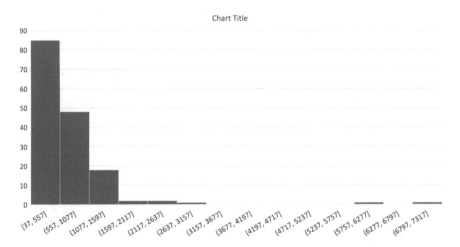

Fig. 3. Number of instances vs the Number of iterations to convergence

- We can see the effect of the distance term on number of iterations it takes to converge and also the quality of the solution. The following plots show the energy of the system with the number of iterations (Fig. 4).

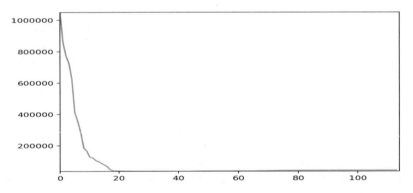

Fig. 4. Plot with the coefficient of the distance term set to 100.

By increasing the D value to 150 the below plot was obtained (Fig. 5).

It is observed that with the increase in the coefficient of the penalty terms the number of iterations taken to reach a valid configuration increases. The corresponding tours are as given below. It is observed that with the increase in the coefficient of the penalty term the quality of the solution improves (Figs. 6 and 7).

It is seen that the Hopfield networks offer a mechanism to trade-off between the running time and the quality of the solution obtained.

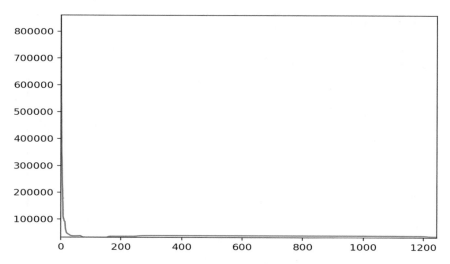

Fig. 5. Plot with the coefficient of the distance term set to 300.

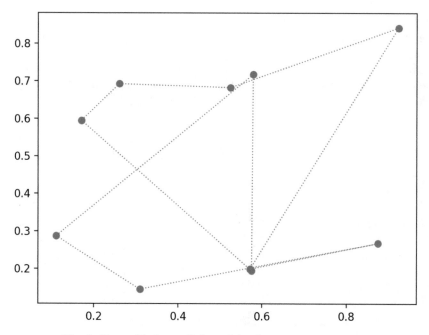

Fig. 6. Tour with the coefficient of the distance term set to 100.

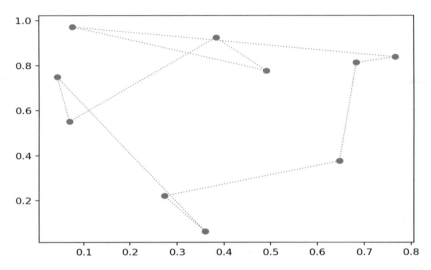

Fig. 7. Plot with the coefficient of the distance term set to 300.

Vertex Cover

The key challenge is to identify the best values for the parameters of the motion equation.

The vertex cover seemed to outperform the greedy algorithm on randomly generated graphs statistically but since the number of possible input instances are large the performance of the algorithm is limited to statistical confidence.

density = n(n − 1)/2 m	vc kernel (hnn)	Greedy ration-2
0.1	876	124
0.2	852	148
0.3	774	226
0.4	811	189
0.5	861	139
0.6	702	298
0.7	736	264
0.8	692	308
0.9	733	267
1	796	204

Random graphs of various densities were generated and the hopfield network based algorithm seemed to outperform the greedy ratio 2 algorithm in many cases.

Max-Cut

The experiments were performed on graphs that were generated randomly, first by fixing the number of vertices and then selecting the edges based on the density parameter. The edges were distributed evenly. The hopfield minimization based local search algorithm discussed in algorithm design by Kleinberg and Tardos [8] were implemented with an alteration of starting with 4 different initial random states that are [n/6] hamming distance away and the best parameter e was found to be e = 0.1.

The running time on intel i3 processor versus the input size of the graph for dense and sparse graph are shown here (Figs. 8, 9 and 10).

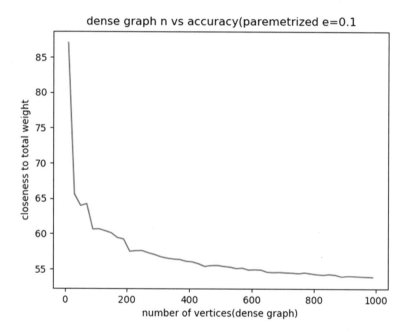

Fig. 8. Closeness to total weight vs number of vertices

It can be observed here that there is no exponential explosion of the time with increase of the number of vertices and that the cut obtained as the solution will always be at least half the size of the original cut.

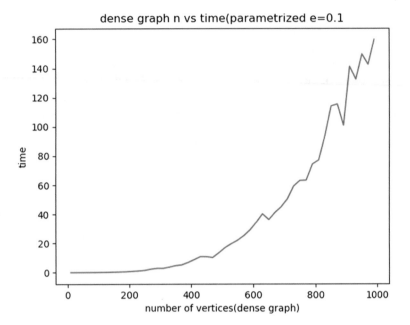

Fig. 9. Time to convergence vs number of vertices

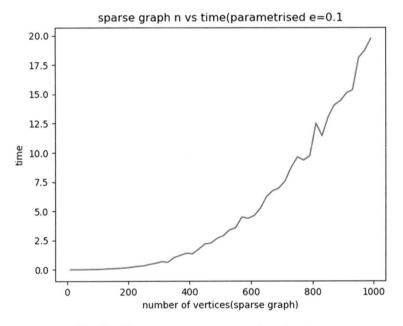

Fig. 10. Time to convergence vs number of vertices

5   Conclusion

The particular system acts as a platform to evaluate the performance of various approximation algorithms for NP-complete problems and compare their performance statistically and to compare the performance of the approximation scheme based on some polynomial time checkable feature. The engine was aimed at providing a platform for more methods and new approximation preserving reductions to be implemented and to enhance the knowledge of the best approximation scheme for the given graph input. Hopfield networks provide a mechanism to tune the quality of the solution. This is usually done by trading off the running time of the algorithm. As shown, approximation preserving reductions can be used in conjunction with Hopfield Networks.

References

1. Papadimitriou, C.H.: NP-completeness: A retrospective. LNCS, pp. 2–6 (1997)
2. Vazirani, V.V.: Approximation Algorithms. Springer Science. Business Media, Berlin (2013)
3. Hopfield, John J.: Neural networks and physical systems with emergent collective computational abilities. Proc. Nat. Acad. Sci. **79**(8), 2554–2558 (1982)
4. Ramanujam, J., Sadayappan, P.: Optimization by neural networks. In: IEEE International Conference on Neural Networks, vol. 2 (1988)
5. Chen, X., et al.: An Hopfield network learning for minimum vertex cover problem. In: SICE 2004 Annual Conference, vol. 2. IEEE (2004)
6. Orponen, P., Mannila, H.: On approximation preserving reductions: complete problems and robust measures (1987)
7. Chen, X., et al.: An algorithm based on Hopfield network learning for minimum vertex cover problem. In: International Symposium on Neural Networks. Springer, Berlin, Heidelberg (2004)
8. Kleinberg, J., Tardos, E.: Algorithm design. Pearson Education India (2006)

The Evolution of Cloud Computing and Its Contribution with Big Data Analytics

D. Nikhil$^{(\boxtimes)}$, B. Dhanalaxmi, and K. Srinivasa Reddy

Department of Information Technology, Institute of Aeronautical Engineering,
Dundigal, Hyderabad 500043, India
Dnikhilkumar05@gmail.com, dinnul8@gmail.com,
kondasreenu@gmail.com

Abstract. In recent years, Big Data Analytics is used for the specification of a large amount of data to uncover or extract the information such as the hidden data. This allows the user or developer to utilize the large chunk or data in order to extract the correlations and patterns which may be present inside the large unstructured chunk of data. The term, Big Data was initially coined in the mid-1990s due to the seemingly increased volumes of data which were often stored in hardware devices, up until we obtained a permanent solution to the excess amount of storage space we know today as Cloud Computing. Cloud Computing, known as the phenomenon which involves mass storage of data and information among a certain entity which hosts no hardware, i.e. stored in an imaginary circumstance termed as 'cloud', has deemed beneficial to the Technology surrounding us and enabled an immense advantage when it comes to artificial storage. Over the years, there have been various technological improvements and studies surrounding this phenomenon in which this Technology is deemed to be present among almost every hardware component in the near future. In this paper we are going to view some of the advancements and achievements this technology has brought in our rapidly growing technological world.

Keywords: Cloud · Computing · Technology · M2M sensor · Predictive analysis · Business data analytics

1 Introduction

The Big Data Analytics plays a key role with regards to Cloud Computing as it initializes in the 5Vs, which are:

- Volume
- Value
- Velocity
- Variety
- Veracity

The Volume indicated the excess amount of data received per second from social media platforms which includes multimedia such as audio, video, photos, files, which may be shared through various devices such as smart phones, laptops, M2M Sensors,

© Springer Nature Switzerland AG 2020
J. S. Raj et al. (Eds.): ICIDCA 2019, LNDECT 46, pp. 332–341, 2020.
https://doi.org/10.1007/978-3-030-38040-3_38

Cars, Credit Cards, etc. Due to the increase in the volume of data obtained from these devices, we find it problematic to store heavy chunk of data in hardware components. Due to the induction of a Cloud, this is made possible and also drastically reduces the risk such as data loss or data theft, due to the secured protocols followed by the internet cloud. The Value indicates the worth of the data being put to good use. As the large unstructured amount of data is often regarded as junk, there are however, certain crucial information being present in the storage which can be deemed useful only if we collect and analyze the data. This correlates with the cloud, as we often neglect large amount of data which can be put into an internet cloud and can be accessed with ease. This is known as the Value of data in Big Data Analytics. The Velocity refers to how fast the data can be processed with lightning speed in a confined amount of time. Every second, huge amount of information are being generated and analyzed such as emails, attachments, photos, videos, audio files, which travel in real time, and received almost instantly. The speed of transmissions of data in real time is essential in order to withstand the sending and receiving of data or information instantly. This is known as the Velocity in Big Data Analytics. The Variety of data is varied over the years in regards to how the data was stored. In the past, the data was often neatly structured such as name, number, date, etc. which enables information to be obtained with ease. The data present today, however, consist of mostly unstructured data. In recent years, most of the data is filled with clustered multimedia files, such as audio, video, etc., and the remaining included semi structured data which adds to the cluster. This is known as the variety of data in Big Data Analytics. The Veracity of data is often known as the quality of data as to how much of it can be genuine. Due to the availability of large amount of data, it is often perceived that the majority of data is not of quality, i.e. cannot be trusted. For example, if we invoke a GPS to a certain location, the chances are that it may recalculate the data and may drift off into an unknown location. This usually happens due to the data which is not genuine and cannot be trusted. This is known as the Veracity or Data in Big Data Analytics.

There are various aspects of Big Data surrounding the induction of a Cloud. Over the years, the placement of large unstructured data within a cloud has been initiated through the process we know today as Cloud Computing. Cloud computing, over the years, has widely grown since its emergence is 2006. Although theorized in the 1960s the technology of the time lacked compatibility. It wasn't until the late 2010s when it was put into good use due to the increased usage and accessibility among the people as well as the industries. Over the years, due to the advancement in technology, the increase in size of data has grown rapidly to such an extent that Terabytes of data had to be stored within a confined hardware component which is liable to the loss of data in many ways such as damage, or contact with water. The utilization of Cloud Computing was the best way to manage this scenario and it has given rise to a massive advantage in which gigabytes and terabytes of data can be store and retrieved from an electronic device, irrespective of the location or the nature of the device. This change has allowed the data to be much more secure and portable, allowing the user to be in full control of the data or personal information regardless of using various appliances such as a computer or a mobile phone, in any part of the world given there is access to the internet. Not only has it benefited the users, but has also enabled various business and software organizations to store and modify their data with must ease, in turn enabling the technological growth and development in the modern technological world. Various fields such as Artificial Intelligence and Embedded Systems heavily rely on the phenomenon we know as Cloud Computing. The growth of the technology in various Multinational Companies has created many job opportunities and opened fields which were irrelevant if it weren't for the artificial store of data. There are companies such as Google, which enable the user to access personal information associated with their respective email ids and allow them to have full access of their data, irrespective of size, hence giving them full control on their privacy, without being displayed on the internet without their consent. Although Cloud Computing is primarily cost inductive, however, various companies such as Apple, Google, Microsoft, etc. have internet storage clouds such as iCloud, Amazon Drive, etc. which allow the users to manage storage up to 5 GB and can go up to 1 TB upon certain cost. This initiation plays a great role in the storage point of view, as it allows various fields such as the film industry, startup companies, etc. to store heavy files and process a secure and risk free way to store, and obtain the data. This is usually known as the primary use of Cloud Computing, i.e. Storing heavy files or data within a physically non existing space known as an internet cloud, or just cloud (Fig. 1).

2   Literature Survey

Fig. 1. Big data analytics

In the above pictorial diagram, we can observe the various types of Big Data analytics. They include big data analysis, predictive analytics, real-time streaming analytics, business intelligence and enterprise data warehousing (Fig. 2).

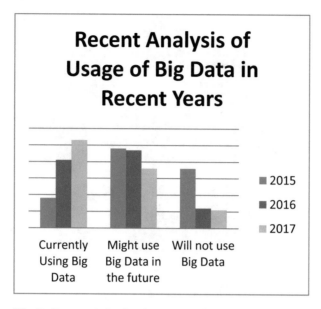

Fig. 2. Bar graph for development of big data in recent years

In the above bar diagram, we can observe how the adoption of big data has changed over a span of 3 years in various companies. The numbers have drastically increased over the years of 2015 to 2017 in the case of using big data. Some companies have gotten over the dilemma of whether to use big data or not. As a result the analysis for this has slightly reduced from 2015 to 2017. Though we can clearly observe hoe drastically the analysis for the companies that have claimed that they will never use big data has substantially decreased, In the year 2015, the amount of companies that claimed that they would never use big data were around were around 38% and this has subsequently decreased to around to 11%. This shows how big data has proved to be a burning topic in the present day and age. Most of the companies have shifted from saying that they will never use big data to depending on big data. We can probably see more companies using big data in the coming years (Fig. 3).

Here, we can observe the various big data analytics tools like No SQL Database which is used by Apache Hbase, Realtime Processing (used by Apache Spark), Software Integration Platform (used by talend), Log Analysis Platform (used by splunk), Data Warehousing (used by HIVE), Messaging System (used by kafka), Data storage and processing Platform (used by hadoop) and finally Data Analysis.

There are various types of Internet Clouds when it comes to Cloud Computing, such as:

- Public Cloud
- Private Cloud
- Community Cloud
- Hybrid Cloud

The above types of Internet Cloud serve different purpose according to the users' or the communities need.

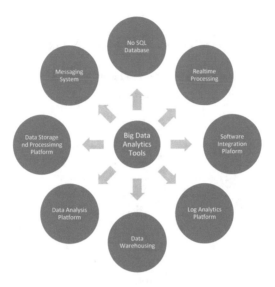

Fig. 3. Tools for big data

A Public Cloud is known as the type of internet cloud in which is accessible to a wide group of people within a circumscribed location which is accessible up to a certain distance [5]. This type of cloud is usually used in order to make resources accessible to the users publicly [7]. Examples of Public Cloud are: Amazon Compute (EC2), Sun Cloud, Google AppEngine, etc. A Private Internet Cloud, or a Private Cloud, is known as the type of Cloud which is accessible to people in which the data or information stored can only be accessed, modified or deleted by a certain individual, and can store personal data which can be accessed via Cloud, known as the Private Internet Cloud [12]. Examples for a Private Cloud are: iCloud, Google Drive, etc. [14]. A Community Cloud is known as the type of internet cloud, in which various organizations obtain access to a specific type of cloud which holds the data or infrastructure of the organizations. Community Clouds are often shared due to their extensive size and its capability to store and access multiple organizations and also allow the sharing of data within the Cloud [16]. Examples of Community Clouds include: Salesforce, IBMSoftLayer, etc. [18]. A Hybrid Cloud is known as the type of cloud in which multiple variations are clouds are combined together or clubbed together thus creating a new cloud which is capable of containing multiple capabilities, such as the merging of a private cloud and a public cloud, Community Cloud and a third party Cloud, etc. [23]. This type of Cloud is usually used in Multinational Software Companies in which the data can be configured from Private to Public, or vice versa, according to the developer or users' need [22]. This type of cloud is known as a Hybrid Cloud [15]. Although Cloud Computing is extensively known for its wide storage or data, however, we must take into consideration on how we are able to access the data. This can be done by the utilization of various devices which are connected to the internet [19]. These devices include smart phones, smart TVs, laptops, desktop computers, databases, binary coded components, app servers, etc. [24] (Fig. 4).

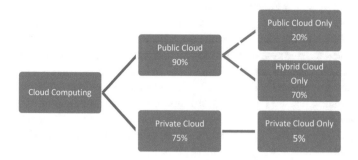

Fig. 4. Usage of cloud computing by market leaders

According to the above figure, the majority of the market leaders are known to use a public cloud rather than a private cloud due to public cloud's ability to share and store large amount data and which is accessible to the vast majority within this firm. This type of cloud is classified further into two sub usage categories, in which the hybrid usage is larger than the latter as it allows the user to modify the cloud according to certain requirements. The Private Cloud however, is primarily used for personal data, and serves minimal use when it comes to the market industry, when compared to the public cloud.

3   Proposed Methodology

As Cloud Computing is known to generate many such usages, there are, however, different categories in which various types of clouds are known to serve a certain purpose when it comes to usage and functionality (Fig. 5).

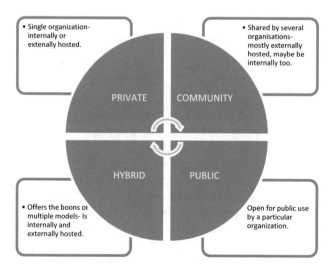

Fig. 5. Categories of cloud computing

As provided by the above diagram, the subcategories such as the private cloud, the hybrid cloud, the community cloud and the public cloud are known to serve different purposes which are utilized by various organizations and individuals. Private cloud is generally used in a single organization and it can be both internally or externally hosted. Whereas, for the case of community cloud, it can be shared by several organizations and it is primarily externally hosted, but it may be internally hosted by one of the organizations. In the case of hybrid cloud, t is the combination of two or more clouds- private, community or public. It may remain as unique elements but they are bound together and offer the merits of various deployment models. Hybrid cloud is both internally and externally hosted. For the case of public cloud, this type of cloud is open to all-the public- by an organization. This organization also hosts the service (Fig. 6).

Fig. 6. Cloud growth by segment and market leaders

From the above diagram, we can notice the cloud growth by segments with the market leaders and the annualized revenue growth of various cloud computing services. We can note that the cloud growth for Infrastructure as a Service (IaaS) in the year 2015 was about 50% for Amazon and Microsoft. For a Private and Hybrid cloud service it was around 45% for the companies of IBM and Amazon. We can also note that for Software as a Service (SaaS), the cloud growth was around 30% for SalesForce and Microsoft. For Unified Communications as a Service (UCaaS), it was 30% for the companies- Cisco and Citrix. Similarly, for public cloud, it was 30% for Cisco and HPE and 18% for Private Cloud for the companies HPE and Cisco. From this we can notice how the cloud infrastructure services, infrastructure hardware and software clouds and how any other cloud services were widely used in well-known companies with their annualized revenue growth in the year 2015 (Fig. 7).

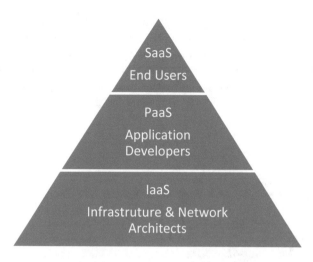

Fig. 7. Pyramid diagram of cloud computing users, developers, and architects.

The above diagram displays the variation in the usage of Cloud Computing as a service. There are three such categories which involve this process, which are:

- Software as a Service (SaaS)
- Platform as a Service (PaaS)
- Infrastructure as a Service (IaaS)

In SaaS, the usage if the operating environment is highly irrelevant, and functional applications are provided, such as CRM, ERP, Email, HIVE, etc.

In PaaS, the operating environment is involved, unlike Saas. In this service, Windows/.NET, Linux/J2EE, and certain choice of applications are deployed.

In IaaS, virtual platforms are used, along with operating environment in which certain platforms are deployed. In this service, storage is included.

4   Conclusion

Big Data has played a crucial role in the way we store the data and has also increased the capacity of more structured and unique data rather than the unstructured data we used to have prior to the introduction of cloud computing. Due to the availability of the internet cloud, we now are able to store and retrieve large amount of data from the cloud, which has also shown a great amount of improvement towards the field of technology. Cloud Computing, has undergone immense development over the years in various fields such as marking, software companies, etc. and are used by various industries such as the hospital, multinational companies, educational institutions, government organizations, and many such departments in which the storage and sharing of data among an entity known as an internet cloud, takes place. Over the years, the development of Cloud Computing has contributed a lot to modern technology. As the technology which was present during the initial start of Cloud Computing is far different from today's technology, it has opened doors to various platforms which showcases and utilizes this process which has highly supported the development of the modern world we know today. As the title suggests, the contribution of cloud computing among big data analytics has proven to be successful in terms of increasing storage space, maintaining multiple infrastructures, increasing the cloud market value, and one of the crucial contributing factors being the portability of the data storage.

References

1. Dhanalaxmi, B., Raju, Ch.S.K, Kameshwar, P., Madhuravani, B.: An enhanced aesculapian management system. Int. J. Pure Appl. Math. **119**(14), 573–579 (2018). (ISSN 1314-3395 Scopus Indexed)
2. Dhanalaxmi, B., Appa Rao Naidu, G., Anuradha, K.: A survey on software inspection improvisation techniques through probabilistic fault prediction method. J. Adv. Res. Dyn. Control Syst. **10**(7), 617–621 (2018). (ISSN 1943-023X, Scopus Indexed)
3. Dhanalaxmi, B., Appa Rao Naidu, G., Anuradha, K.: A review on different defect detection models in software systems. J. Adv. Res. Dyn. Control Syst. **10**(7), 241–243 (2018). (ISSN 1943-023X, Scopus Indexed)
4. Ayyappa, P., Mohana, R.M., Reddy, K.V.S.: Efficient argument checks for different assertions using various packages in R programming. International Journal of Mechanical Engineering and Technology **9**(5), 742–749 (2018). (ISSN (P) 0976-6340 and ISSN (O) 0976-6359, Scopus Indexed)
5. Dhanalaxmi, B., Kumar, I.R., Vishal, Madhuravani, B.: An user-friendly recital teller system. Int. J. Pure Appl. Math. **119**(16), 1801–1805 (2018). (ISSN 1314-3395, Scopus Indexed)
6. Dhanalaxmi, B., Rahul, D., Sowjanya, S.J., Krishna Rao, N.V., Madhuravani, B.: An adaptable architecture for actual-time traffic control using big data analytics. Int. J. Pure Appl. Math. **119**(16), 1543–1548 (2018). (ISSN 1314-3395, Scopus Indexed)
7. Kambatla, K., Kollias, G., Kumar, V., Gram, A.: Trends in big data analytics. J. Parallel Distrib. Comput. **74**(7), 2561–2573 (2014)
8. del Rio, S., Lopez, V., Bentez, J.M., Herrera, F.: On the use of mapreduce for imbalanced big data using random forest. Inf. Sci. **285**, 112–137 (2014)
9. Kuo, M.H., Sahama, T., Kushniruk, A.W., Borycki, E.M., Grunwell, D.K.: Health big data analytics: current perspectives, challenges and potential solutions. Int. J. Big Data Intell. **1**, 114–126 (2014)
10. Nambiar, R., Sethi, A., Bhardwaj, R., Vargheese, R.: A look at challenges and opportunities of big data analytics in healthcare. In: IEEE International Conference on Big Data, pp. 17–22 (2013)
11. Huang, Z.: A fast clustering algorithm to cluster very large categorical data sets in data mining. In: SIGMOD Workshop on Research Issues on Data Mining and Knowledge Discovery (1997). (IJACSA) International Journal of Advanced Computer Science and Applications, vol. 7, no. 2, 2016 517
12. Das, T.K., Kumar, P.M.: Big data analytics: a framework for unstructured data analysis. Int. J. Eng. Technol. **5**(1), 153–156 (2013)
13. Das, T.K., Acharjya, D.P., Patra, M.R.: Opinion mining about a product by analyzing public tweets in Twitter. In: International Conference on Computer Communication and Informatics (2014)
14. Zadeh, L.A.: Fuzzy sets. Inf. Control **8**, 338–353 (1965)

15. Pawlak, Z.: Rough sets. Int. J. Comput. Inf. Sci. **11**, 341–356 (1982)
16. Molodtsov, D.: Soft set theory first results. Comput. Math Appl. **37**(4/5), 19–31 (1999)
17. Karthiban, M.K., Raj, J.S.: Big data analytics for developing secure internet of everything. J. ISMAC **1**(02), 129–136 (2019)
18. Wille, R.: Formal concept analysis as mathematical theory of concept and concept hierarchies. Lect. Notes Artif. Intell. **3626**, 1–33 (2005)
19. Jolliffe, I.T.: Principal Component Analysis. Springer, New York (2002)
20. Al-Jarrah, O.Y., Yoo, P.D., Muhaidat, S., Karagiannidis, G.K., Taha, K.: Efficient machine learning for big data: a review. Big Data Res. **2**(3), 87–93 (2015)
21. Changwon, Y., Ramirez, L., Liuzzi, J.: Big data analysis using modern statistical and machine learning methods in medicine. Int. Neurourol. J. **18**, 50–57 (2014)
22. Singh, P., Suri, B.: Quality assessment of data using statistical and machine learning methods. In: Jain, L.C., Behera, H.S., Mandal, J.K., Mohapatra, D.P. (eds.) Computational Intelligence in Data Mining, vol. 2, pp. 89–97 (2014)
23. Jacobs, A.: The pathologies of big data. Commun. ACM **52**(8), 36–44 (2009)
24. Zhu, H., Xu, Z., Huang, Y.: Research on the security technology of big data information. In: International Conference on Information Technology and Management Innovation, pp. 1041–1044 (2015)

Robust Methods Using Graph and PCA for Detection of Anomalies in Medical Records

K. N. Mohan Kumar[1(✉)], S. Sampath[2], and Mohammed Imran[3]

[1] Department of CSE, SJBIT, Bangalore, India
mohan4183@gmail.com
[2] Department of ISE, AIT, Chikkamagaluru, India
23.sampath@gmail.com
[3] NTT Data, Bangalore, India
emraangi@gmail.com

Abstract. Wellbeing in basic words is normalcy in health of human body and disease is unusual condition that influences typical working of human body with no outer wounds. Health care is about the aversion of maladies by finding and treatment. There are several methods and models developed in health care to predict and classify the chronic diseases, but what if the health care record contains some anomalies. A well designed model also would provide wrong results, if the input data contains anomalies. Any wrong decision in health care management would cost the life of the patient. In this work we have modeled Graph method and PCA method to detect anomalies in health care records by the means of frequency of incidences of disease codes in graph and correlation among the disease codes. We have used CMS medi-claim dataset in which diseases are expressed in terms of International disease code (ICD) and Hierarchical Condition Category (HCC) code to indicate the patient health condition. Game theory approach is used for evaluation of the model. The results of this work have been proved to be promising when compared with existing techniques. Since this work is related life of a patient's, the results should be re looked by the domain experts before taking any decisions.

Keywords: Graph · Game theory · PCA · Anomaly · HCC · ICD

1 Introduction

"All the money in the world can't buy you back good health" [1] this quote aptly summarizes the importance of Health. From the days when Edner Jenner invented first vaccine for Smallpox followed by Louis Pasteur's invention of vaccine for Rabies many vaccines against diseases like Diphtheria, Tetanus, Cholera, Plague etc. have got invented [2]. Healthcare had been vital entity then and now. In recent times understanding the need for a sound body and mind has gained still more impetus and focus. Today the culprit of incorrect diagnosing of health problems is haphazardly recorded patient data. According to [3] Medical records are the documents that provide insight about the patient's history, clinical findings and medication. Also health records play a vital role in legal-battles that get initiated between medicos and patients whenever a mishap occurs.

© Springer Nature Switzerland AG 2020
J. S. Raj et al. (Eds.): ICIDCA 2019, LNDECT 46, pp. 342–352, 2020.
https://doi.org/10.1007/978-3-030-38040-3_39

According to [4], "Chronic illness is a long-term health condition that may not have a cure." Few examples of chronic diseases are Diabetes, Cancer and Psychosomatic conditions. Chronic diseases press upon the need to maintain well documented data of patients. As per World Health Organization 57% of world population will be suffering by chronic diseases by 2020. So, monitoring health records to fight against chronic disease has become more relevant in current context. The mammoth cost of healthcare, particularly for the treatment of chronic disease, is rapidly getting unmanageable. This emergency has spurred the drive towards preventative care for chronicl ailments, where the essential concern is perceiving ailment hazard and making a move at the most punctual signs. Machine learning (ML) techniques like prediction and classification [5] help us to address the preventive healthcare issues. Also we have to agree that 'Patient-Data' goes a long way in treating chronic diseases. But the most unwanted guest in a well documented patient record is Anomalies. As per [6] Anomaly detection implies the issue of discovering patterns in data that don't concur with the normal conduct. Usually anomalies are problems that can occur in poorly planned, un-normalised databases.

Anomaly prevention in patient data is crucial for health care industry to diagnose and treat patients with chronic illness efficiently. According to [7] the increased availability of patient medical records in electronic form helps to replan patient treatment process adopted in a given clinical setting. But at the same time it becomes important to handle any deviations from these patterns to manage and treat patients effectively. Today Anomaly detection has gained importance in wide array of applications like Banking, Insurance or Healthcare, Cyber-security, Fault detection in safety critical systems and military surveillance. This paper is an effort to utilize anomaly detection methods to cleanse Healthcare records so that the life of stakeholders in healthcare industry becomes less strenuous.

Anomaly detection doesn't have a single and straight forward method. Researchers have identified concepts from various domains [6] like statistics, machine learning, data mining, information theory, spectral theory to develop a well conceived technique for solving anomalies occurring in various fields of studies like Banking, Healthcare, Defence, Insurance etc.

The article is structured into different parts. Part 2 details the related work. Part 3 explain the methods and techniques used in this work. Part 4 talks about experimental setup. Part 5 discusses about results. Part 6 concludes by summarizing the research work.

2 Related Work

Investigation of existing methods and techniques would give better perceivability in the context we are dealing with. Anomaly detection in medical records is a challenging task. The existing solution might not fit to all the context and data set. The detailed exploration of this topic is required. The following are some of the related work referred to.

Varun et al. [6] have conducted thorough survey on anomaly detection on health care data. They have discussed on various machine learning techniques contextually applied to different kinds of health records for anomaly detection. They have explored various applications of anomaly detection in life science domain and also have detailed about methods used handle collective anomalies. In their survey they have identified gaps such as handling anomalies in complex systems with multiple components.

Dario et al. [7] have illustrated a model for anomaly detection on clinical data. The medical records are collected from treatment log of local health centers. The records chosen are specific to diabetes, pregnancy and colon cancer. They have used clinical segmentation for clinical records and diagnostic segmentation for diagnostic records for feature extraction. The methods used for anomaly detection are frequent correlation and sequence. Performance evaluation measures are not much discussed. The scope for future work is to include more clinical records to cover different diseases.

Richard et al. [8] have modeled anomaly detection within Medicare specialties using multiple variables of by adopting regression and probabilistic programming. The source of data set is Centers for Medicare and Medicaid Services. They have used automatic feature selection method with masking for feature extraction. The techniques used are bayesian probability method using probability programming and non-parametric regression method. The evaluation techniques such as mean absolute error and root mean square errorn is used for evaluation. It has been recommended to expand the work more number of specialties and also to do detailed comparison this method with other outlier detection techniques.

Luiz et al. [9] have developed solution for detection of anomalies in health care providers using producer consumer model. In their experiment they have used real database of public health system of Brazil. The techniques used are KNN, reverse KNN and Local Outlier Factor for identifying odd health care providers. Evaluation of the model is done manually. The future scope of this work is to work on larger geographical area covering large number of producers and consumers.

Yamini et al. [10] illustrates the mechanism to identify abnormalities social insurance. They have used CMS Medicare-B dataset for their experiment. They have used supervised and unsupervised Machine Learning methods to detect the abnormalities. The future scope is to consider various parameters of social insurance to dig out anomalies.

Jiwon et al. [11] have used a non conventional approach to identify abnormalities social insurance. They have developed a similarity graph on insurance records to compute similarity of occurrences and any dissimilarity could prove the record to be odd. The future scope is to work on the distinguished peculiarities.

The insight into the anomaly detection related work has cleared that the existing techniques are applied on the samples but on the features. The existing techniques are experimented on small set of records. It is also evident that anomaly detection is hardly done on patient records using disease codes. In this work we have implemented the mechanism to detect anomalies in health care records using disease code to indicate the patient condition.

3 Methods and Techniques

In this work we have modeled three algorithms such as Graph method to find anomalies, Principal component analysis (PCA) [12] and Game theory method [13]. The first two methods are for anomaly detection and the third is for evaluation of the outcomes of first two techniques. The input data to the above algorithms is Term Document Matrix (TDM), which discussed in detail in data modelling and pre-processing step.

Graph is a data structure used to represent the complex data by retaining the relationship among the data [13]. Graph contains vertex set 'V' representing the records and edge set 'E' interconnecting the vertices (records), indicating the relationship among the records. Using this graph representation, we will able establish the correlation among the records, with the construction the incidence matrix in which, a row represents the records and column represents the features. By this way, the correlation between the records and features can be established [14].

In Algorithm 1 i.e. Graph Method, the records are represented as incidence matrix A, which is the input to the algorithm. Step 2 HCC code in row 1 is added into HccList, which will be used to map the final results. Step 5 to Step 16 calculates the total column count with respect to each column representing the HCC codes. Step 17 prints the 20 infrequent HCC codes indicating them to be the anomalies. The column count of the incidence matrix would indicate the frequency of occurrences of that feature. Lesser the frequency is lesser is the correlation. The infrequent features can be declared as anomalies, as they will have less correlation the records.

Principal component analysis (PCA) [12] is a subspace or dimensionality reduction algorithm. When we have data samples, which look alike by appearance in first sight but when we get the insight of the samples, records would appear to be composite in nature with varying patterns. The major challenge, is to establish the correlation among the records [15]. PCA converts the correlations among the records into a cluster together. PCA is interpreted by Eigen values used to partition the clusters. Algorithm 2 details the process of finding Eigen values.

The distance between the clusters decide the how much is the correlation between the records. Lesser the correlation, larger would be the distance between the clusters. PCA is one way to make sense of variations in data. The less correlated records can be declared as anomalies upon evaluations [15].

In Algorithm 2 i.e. PCA method, the records are input as a matrix A. Step 2 the first row is added to hccCode list, which will be used map the HCC codes with the resultant Eigen values and it is removed from the matrix a. Step 3 calculates the mean of the matrix A. Step 5 and 6 computes the co variance of matrix A. Step 7 digs out the Eigen values and vectors. Steps 8 maps the HCC codes in hccCode list with the resultant Eigen values. Step 9 sorts the Eigen values in descending order. The PCA Eigen value1 is more imported than the second PCA Eigen value2 and so on.

Algorithm 1: Graph Method to Find Anomalies.

```
1: Input: A [m * n]
2: Initialize hccPair<>="", HccSumList [0..n] =0,
        AnamolySize=20, HccList [0..n] ="",
        AnamolyList [] =""
3: Copy all hcccodes to HccList
4: for i=0 to m
5:      for j=0 to n
6:              HccSumList[j]=HccSumList[j]+a[i][j]
7:      end for
8: end for
9: for i =0 to m
10:     hccPair=hccPair.add (HccList[i]: HccSumList[i])
11: end for
12: for i=0 to m
13:     if Hccpair[i] <anamolySize
14:             AnamolyList.add (Hccpair[i])
15:     end if
16: end for
17: print AnamolyList
```

Algorithm 2: PCA to Identify Anamolies.

```
1: Input: A [n * m]
2: Initialize: hccCode [0..n]="", output[0..n]="",
        EigenValues [0..n] =""
3: Add first row of input to hccCode
4: M = mean (A)
5: C=A-M
6: V=cov(C)
7: values, vectors=eig (V)
8: output hccCode, values
9: sort values in desc order
10: output to csv (anamoly.csv)
```

The game theory method is used for model evaluation [16]. In this method, the top contibuting players are the dominant features contributing to the context. Every player/feature is associated with weight; the value of the weight is domain dependant. Winning weight (WT) is decided based on the domain and context. In our work WT is calculated manually for various chronic diseases. The weight WT is achieved by adding the weights of contributor (features) in different arrangements. Consider that there are 'n' features and 'k' (k < n) top contibutors. The top contibutors are calculated by arranging the 'n' feature in n! ways. The complex part of the process is to find the top 'k' contributors. The weights of the features are added in different arrangements until WT is achieved; the last feature in the sequence achieving WT is declared to be the winner. The winner features in every arrangement is listed in frequency list 'L'. The top 'k' features are selected from frequency list by considering the contributor with more frequency. The Frequency List 'L' is sorted in descending order and the top 'k' features are the top contributors to the context [17].

Algorithm 3 is a Game theory approach to find the disease supporting HCC codes. In step 1 HCC code is paired with its associated weight and the required initializations are done. Step 2 initializes the winningHCC empty array to store the end result and the variable winningweight with X, the value of X is disease dependant. Step 3 invokes the permute function which arranges all the HCC codes in n! Ways where n is the total number of HCC codes. Step 4 invokes the collectOccurancePairs function and the result is stored in winningHCC array. Step 5–24 computes and lists the winningHCC and returns it to the caller of the function. Step 25–30 is swap function which is used as an intermediate operation. Step 31–40 is a function which groups the most contributing HCC code in all the arrangement done in the permute function [16].

```
Algorithm 3: Game Theory to Find Top Contributing HCC Codes

1: Input: hccPair<hcc:count>=0, HccSumList [0...n]=0,
         AnamolySize=20, HccList [0...n]=""
2: Initialize winningHCC [] ="" winningweight=X
3: winners=permute (HccList, 0, n)
4: winningHCC =collectOccurancePairs (winners)
5: func: permute (array, startIndex, endIndex)
6:      if startIndex>=endIndex
7:          sum=0
8:          for j=0 to sizeof (array)
9:          for j=0 to sizeof (array)
10:             sum += HccPair. Get (i)
11:             if sum > winningweight
12:                 winningHCC. add (hcc)
13:                 break;
14:             endif
15:         end for
16:     else
17:         for i=startIndex to endIndex
18:             array = swap (array, startIndex, i)
19:             permute (array, startIndex + 3, endIndex)
20:             array = swap (array, startIndex, i)
21:             end for
22:     endif
23: return winningHCC
24: end func
25: func: swap (array, i, j)
26:     temp=array[i]
27:     array[i]=array[j]
28:     array[j]=temp
29:     return array
30: end func
31: func: collectOccurancePairs (winners)
32:     winningHCC <>=""
33:     for i=0 to size (winners)
34:         if WinningPlayers contains winners[i]
35:             winningHCC. add (winners[i], count+1)
36:         else
36:             winningHCC. add (winners[i], 1);
37:         endif
38:     end for
39:     return winningHCC
40: end func
```

4 Experimental Setup

The experiment started with acquisition of medi-claim data from CMS repository, which is the center for medicare services http://www.cms.gov/ [18]. The acquired data was dumped into big data file system (HDFS). The medi-claim data contains obfuscated records of the patients. The patient medical insurance claims data are classified into beneficiary, inpatient, outpatient, carrier and prescription records. Beneficiary files contain demographic information of the patients, inpatient and outpatient files contain the records of the inpatient and outpatient respectively, carrier file contains the information of the frequent scheduled treatment information and prescription file contains the details of the drugs prescribed to patients. The prescription and beneficiary records are not considered in this work [19]. The health status of the member patient is indicated by the ICD codes in inpatient, outpatient and carrier files [18, 19]. The details of the entire experiment are detailed in Fig.1.

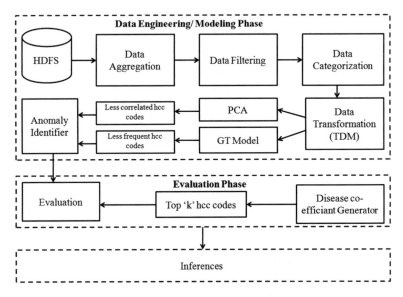

Fig. 1. Anomaly detection model.

4.1   Data Modelling and Pre-processing

Data modelling is a process of careful selection of attributes as per the required context and data pre-processing is a process of transforming the data into the form required by the model. The acquired data contains lots of attributes but the context demands only the attributes containing ICD code [18, 19]. ICD indicates the disease, the patient is suffering. Hive programming is used for querying the data. Hive tables are created for outpatient, inpatient and carrier claims files and records from each file are persisted into the tables. A special table called super data is created into which records from the core tables are loaded. Beneficiary id is the primary key for all the records in every tables. Our main objective is identifying anomalies in chronic diseases such as diabetes, heart, liver, kidney and cancer. Another set of hive tables are created with respect to the selected chronic diseases. The super table is then queried to capture the records in such a way that it includes data after the occurrence of the chronic disease and the records are persisted to respective diseases table.

These tables contain the ICD's and these ICD codes are need to mapped with HCC codes. The resultant HCC codes are stored in separate tables with respect to each chronic disease. The data from these HCC tables are extracted into disease CSV files [18–20]. These steps are included in data aggregation, filtering and categorizing step as shown in Fig 1.

However the data contained in CSV file cannot be processed by ML algorithm, hence the data should be transformed into form required for any ML. The records in the CSV file are transformed into a term document matrix (TDM). TDM is a sparse matrix in which a row represents the patient and the columns represent features (HCC codes). In this matrix, the term '1' represents the occurrence of respective HCC and the term '0' is its absences. This phase of work is show by data transformation block in Fig 1.

5 Results and Discussion

The experiment was conducted on data set ofvarious chronic diseases such as diabetes, heart, liver, kidney and cancer. Graph method is used to detect the anomalies. In this method TDM is represented as a graph. The vertices in the graph represent the patient records and the edges represent columns (HCC codes). Value '1' and '0' in a column represents presence or absence of the respective problem. To identify the anomalies we have calculated the frequency of occurrence of all HCC codes by calculating the column count. Maximum the column count indicates that majority patients suffer from that problem and minimum column values indicates that HCC code is an anomaly [13].

The above method is non conventional and experiment was continued with another approach called PCA [12]. In this approach the patient records were input as matrix and Eigen values were calculated. The results were arranged in ascending order. The first set of HCC codes in the arrangement is the most correlated and the HCC codes with smaller Eigen values are poorly correlated. The poorly correlated HCC codes were listed as anomalies in the records.

The results of graph and PCA are tabulated together as shown in Table 1. The interesting part that was observed is that around 60% to 80% were found to be similar in PCA and Graph method. This motivated us to further extend our research to device a mechanism to list the HCC codes that were representing the respective disease and this could become the evaluation process of methods discussed previously [12].

We have used a game theory method to find the HCC codes supporting the disease, in which HCC codes are used as players [16, 17]. The players are supposed to play the game to achieve the winning weight WT. The algorithm 3 details the process and The results this method for various chronicle diseases is shown in the Table 2.

What seems to be interesting is that, the intersection of HCC codes from the results anomaly detection and game theory is a null set. The inference is that the combination of graph and PCA method is the efficient way to detect anomaly in the health care records [12, 13].

The details of the anomalies with respect to different chronic disease is shown in Table 1. The poor correlation of HCC codes are listed in the PCA column. The less frequent HCC codes obtained from graph method is listed in second and third column. It is noticed that around 11 to 14 HCC codes are common in both PCA and graph approach [12, 13].

Table 1. Anomalies resulted from graph and PCA method

Diabetes			Heart			Liver			Kidney			Cancer		
PCA	CODE	FREQ	PCA	CODE	FREQ	PCA	CODE	FREQ	PCA	CODE	FREQ	PCA	CODE	FREQ
	hcc60	1	hcc110	hcc110	1	hcc157	hcc157	1		hcc110	1	hcc124	hcc124	1
hcc73	hcc73	1		hcc124	1		hcc158	1		hcc29	1		hcc157	1
	hcc76	1	hcc157	hcc157	1	hcc34	hcc34	1	hcc73	hcc73	1	hcc73	hcc73	1
hcc83	hcc83	1	hcc73	hcc73	1		hcc158	2		hcc83	1		hcc83	1
	hcc124	2	hcc158	hcc158	2	hcc74	hcc74	2	hcc124	hcc124	2		hcc34	3
	hcc74	2		hcc29	2	hcc60	hcc60	3	hcc157	hcc157	2		hcc60	3
hcc157	hcc157	3		hcc28	5		hcc70	3		hcc158	2	hcc28	hcc28	4
	hcc28	3		hcc60	5	hcc124	hcc124	4		hcc76	2	hcc71	hcc71	4
	hcc56	3		hcc1	7	hcc162	hcc162	4		hcc60	4		hcc74	4
hcc158	hcc158	4		hcc166	7	hcc71	hcc71	6		hcc28	5		hcc158	6
hcc162	hcc162	6		hcc186	7		hcc75	6	hcc162	hcc162	6		hcc166	6
	hcc1	8		hcc34	7		hcc6	7		hcc166	9		hcc162	9
hcc71	hcc71	9	hcc71	hcc71	7		hcc106	7	hcc71	hcc71	9	hcc75	hcc75	10
	hcc166	10	hcc162	hcc162	10	hcc115	hcc115	8		hcc1	11	hcc1	hcc1	11
	hcc70	10	hcc82	hcc82	10		hcc122	8		hcc34	11		hcc6	11
hcc115	hcc115	11		hcc74	11		hcc17	8		hcc74	11	hcc47	hcc47	12
hcc27	hcc27	11	hcc99	hcc99	11		hcc82	8	hcc99	hcc99	13	hcc99	hcc99	12
hcc99	hcc99	11		hcc70	12		hcc186	9	hcc47	hcc47	14		hcc17	14
hcc82	hcc82	12	hcc115	hcc115	14	hcc99	hcc59	9	hcc6	hcc6	15		hcc186	14
hcc6	hcc6	13		hcc122	14	hcc1	hcc1	10	hcc80	hcc80	17	hcc54	Hcc54	15
hcc80	hcc80	13	hcc80	hcc80	14	hcc80	hcc80	10		hcc106	18	hcc82	Hcc82	16
	hcc122	14	hcc77	hcc77	15		hcc104	11		hcc54	18		Hcc122	17
	hcc75	16	hcc27	hcc27	16	hcc47	hcc47	11	hcc82	hcc82	18	hcc173	Hcc173	17
	hcc22	18		hcc17	17	hcc173	hcc173	12	hcc77	hcc77	19	hcc110		
	hcc47	19	hcc6	hcc6	17		hcc137	15	hcc72			hcc40		
hcc86	hcc86	19		hcc47	19	hcc72	hcc72	15	hcc9			hcc138		
hcc110			hcc55				hcc188	15	hcc173			hcc55		
hcc55			hcc9				hcc22	16	hcc27			hcc77		
hcc72			hcc173				hcc137	17	hcc75			hcc86		
hcc9			hcc86				hcc87	17	hcc107			hcc87		
hcc77			hcc83			hcc82			hcc100			hcc107		
hcc87			hcc87			hcc73			hcc115			hcc80		
hcc173			hcc72			hcc100			hcc87					
hcc189						hcc83								
						hcc110								
						hcc107								

From the results of this work, it is evident that there are some HCC codes in PCA and Graph method which are not common but are acceptable because we have selected only few Eigen values in PCA approach and frequency in graph method is limited to 20 and it can be finetuned further.

The inference is that the combination of graph and PCA method is the efficient way to detect anomaly in the health care records [12, 13]. The game theory is applied as an evaluation technique seems to be the best when there are no patterns in the data to match the anomalies.

Table 2. Supporting HCC codes for different Chronic diseases.

Diabetes	Heart	Cancer	Liver	Kidney
hcc167	hcc96	hcc48	hcc55	hcc122
hcc55	hcc57	hcc86	hcc21	hcc22
hcc58	hcc88	hcc114	hcc107	hcc167
hcc12	hcc169	hcc2	hcc189	hcc88
hcc78	hcc137	hcc57	hcc57	hcc96
hcc87	hcc58	hcc161	hcc111	hcc39
hcc84	hcc55	hcc188	hcc170	hcc170
hcc77	hcc2	hcc78	hcc12	hcc48
hcc23	hcc134	hcc58	hcc136	hcc8
hcc136	hcc78	hcc136	hcc78	hcc78
hcc59	hcc75	hcc77	hcc48	hcc79
hcc40	hcc59	hcc79	hcc79	hcc18
hcc79	hcc23	hcc23	hcc39	hcc58
hcc8	hcc84	hcc55	hcc9	hcc19
hcc85	hcc40	hcc170	hcc59	hcc84
hcc96	hcc18	hcc85	hcc84	hcc85

6 Conclusion

Life science domain is one of the challenging areas for research. The results of the research cannot be compromised at any cost. Any compromise would lead to loss of life of patient; hence anomaly detection in health care records is so important. The anomalies in medical record could be because of human error, misinterpretation of the symptoms expressed by the patients or it could also be intentional to make claims by providing false information. In this work we have demonstrated the Graph method and PCA method to detect anomalies in health records by identifying irrelevant HCC codes representing the diseases. For the evaluation, Game theory approach is used along with manual evaluation is also done to cross verify the model. The results of this work have been proved to be promising when compared with existing techniques. What so ever may be the accuracy of the model, the results should be relooked by the domain experts for confirmation as the decisions made out of the results are made at the cost of patients life. Our work can be extended further with different health care data sets, with the records of all age groups, because the data sets that we have worked on contains only the records of old age patients. Better evaluation techniques can developed as Game theory approach consumes lot of time and demands high configuration computing infrastructure.

References

1. Quote by Reba McEntire
2. https://www.historyofvaccines.org/timeline/all
3. Bali, A., Deepika Bali, D., Iyer, N., Iyer, M.: Management of medical records: facts and figures for surgeons. J. Maxillofac. Oral Surg. **10**(3), 199–202 (2011)
4. https://medlineplus.gov/ency/patientinstructions/000602.html
5. Mohan Kumar, K.N., Sampath, S., Imran, M.: An overview on disease prediction for preventive care of health deterioration. IJEAT **8**(5S), 255–261 (2019)
6. Chandola, V., Banerjee, A., Kumar, V.: Anomaly detection: a survey. ACM Comput. Surv. **41**, 1–72 (2009)
7. Antonelli, D., Bruno, G., Chiusano, S.: Anomaly detection in medical treatment to discover unusual patient management. IIE Trans. Healthc.e Syst. Eng. **3**, 69–77 (2013)
8. Bauder, R.A., Khoshgoftaar, TM.: Multivariate anomaly detection in medicare using model residuals and probabilistic programming. In: Proceedings of Thirtieth International Florida Artificial Intelligence Research Society Conference, pp. 418–422 (2017)
9. Carvalho, L.F.M., Teixeira, C.H.C., Meira, W., Jr., Ester, M., Carvalho,O., Brand, M.H.: Provider-consumer anomaly detection for healthcare systems. In: IEEE International Conference on Healthcare Informatics, pp. 229–238 (2017)
10. Yamini Devi, J., Keerthi, G., Jyotsna Priya, K., Geethasree Lakshmi, B.: Anomaly detection in health care. Int. J. Mech. Eng. Technol. **9**(1), 89–94 (2018)
11. Seo, J., Mendelevitch, O.: Identifying Frauds and Anomalies in Medicare-B Dataset, pp. 3664–3667. IEEE (2017)
12. Bai, Z.-J., Chan, R., Luk, F.: Principal component analysis for distributed data sets with updating. In: Proceedings of International workshop on Advanced Parallel Processing Technologies (APPT) (2005)
13. Sharpnack, J.L., Krishnamurthy, A., Singh, A.: Near-optimal anomaly detection in graphs using lovasz extended scan statistic. In: NIPS Proceeding (2013)
14. Von Luxburg, U., Radl, A., Hein, M.: Hitting and commute times in large graphs are often misleading. ReCALL (2010)
15. Qu, Y., Ostrouchovz, G., Samatovaz, N., Geıst, A.: Principal component analysis for dimension reduction in massive distributed data sets. In: Proceedings of IEEE International Conference on Data Mining (ICDM) (2002)
16. Colman, A.M.: Game Theory and Its Applications in the Social and Biological Sciences, 2nd edn. Routledge, London (1999)
17. Cooper, Russell, John, Andrew: Coordinating coordination failures in keynesain models. Q. J. Econ. **103**(3), 441–463 (1988)
18. https://www.cms.gov/
19. Schapire, R.E.: The strength of weak learnability. Mach. Learn. **5**, 197–227 (1990)
20. Hickey, S.J.: Naive Bayes classification of public health data with greedy feature selection. Commun. IIMA **13**, 87–97 (2013)

Prediction of Cardiac Ailments in Diabetic Patient Using Ensemble Learning Model

Charu Vaibhav Verma[✉] and S. M. Ghosh

Department of CSE, CVRU, Bilaspur, Chhattisgarh, India
kharecharu111@gmail.com, samghosh06@rediffmail.com

Abstract. Due to modern life style there are so many medical issue arises, hence there is increase in risk of disease, here we focus on cardiac ailments in Diabetic Patients. Data driven approach may help in this medical field because data collected by health organization contains useful information by applying Machine learning approach prediction can be done. There are many research has been carried out in this field but less effort drawn in cardiac aliment for diabetic patients. This paper provides an experimental comparison among different machine learning classifiers meant of data forecast to perfectly describe research gap and we have proposed a prediction model named Pretreat-Ensemble, in which we will apply 3-phase preprocessing over input dataset before ensemble machine learning classifier. We have achieved almost 99.7% average accuracy of proposed prediction model.

Keywords: NB · TP · FN · FP · DT · SVM

1 Introduction

In medical domain, the hidden patterns can be explored from datasets with the help of medical data mining. It is essential to extract and integrate the voluminous heterogeneous medical data for exploring hidden data patterns and efficient organization. The metabolic defect in the ability of the human body to convert glucose into energy is termed as diabetes mellitus. Digested food is converted into carbohydrates, protein and fats. The prominent energy source to the body is obtained from carbohydrates that is converted into glucose in blood. A hormone called insulin is required to transfer the generated glucose into the blood cells. This insulin is produced by the organ called pancreas. In case of the organ fails to produce adequate amount of insulin, it leads to diabetes. Type1, Type2 and Gestational diabetes are the three major types of diabetes. Type 1 diabetes occurs mostly in children, type 2 in adults and gestational in pregnant women. It is impossible to completely cure diabetes mellitus. It can instead be regulated and controlled with the help of food and medicines like insulin. Diabetes Mellitus can lead to cancer, kidney diseases and nerve damages. Researches and statistics show high glucose levels causes heart diseases and even cancer and leads to death of several people. Damages to blood vessels, nerves and other human body systems are caused as these effects are uncontrollable and worsens over time.

© Springer Nature Switzerland AG 2020
J. S. Raj et al. (Eds.): ICIDCA 2019, LNDECT 46, pp. 353–361, 2020.
https://doi.org/10.1007/978-3-030-38040-3_40

There are some machine learning classifiers which are used of data prediction.

1. Naïve Bayes (NB)
2. Decision Tree Classifier (DT)

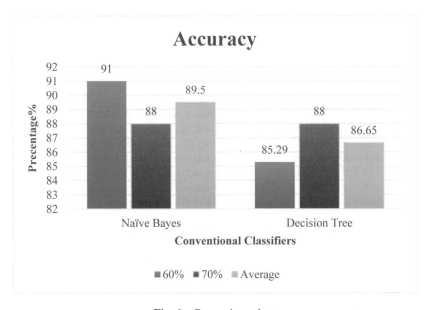

Fig. 1. Comparison chart

Figure 1 shows the 2D plot of accuracy of different machine learning classifiers, implemented using Weka tool, concluded that we need to work upon improving the accuracy. The main objective of our research is to prediction of cardiac ailment in diabetic patient by reducing the complexity of machine learning classifier and improve the accuracy.

Further in next Sect. 2 we will give a brief about research carried out in this field and tabular comparison among some literature, in Sect. 3 we will discuss problem statement, in Sect. 4 we will elaborate out prediction model and dataset used, in Sect. 5 we will describe our experimental result, at last we will conclude our study.

2 Literature Survey

Brisimi et al. focused on chronic heart disease, author used machine learning algorithm random forests, sparse logistic regression, kernelized and sparse Support Vector Machines (SVM), and algorithm validated on Boston Medical Center dataset [1]. Shouman et al. investigated the research gape on heart disease diagnosis and treatment and proposed data mining technique for the same [2]. Amin et al. developed algorithm for Prediction of Heart Disease using neural networks and genetic algorithms [3]. Kaur et al. proposed modified J48 classifier for Prediction of Diabetes [4]. Kalaiselvi et al.

did research which concentrated on chronic diseases like cancer and heart diseases, and proposed Adaptive Neuro Fuzzy Inference System (ANFIS) for prediction [5]. Singh et al. proposed genetic algorithm established feature selection eliminates the unrelated features by reducing dimension and increased accuracy [6]. Pouriyeh et al. made a detailed investigation and comparison for predicting the possibility of heart disease using different classifiers [7]. Perveena et al. had proposed the adaboost and bagging ensemble techniques using J48 for analyzing diabetes [8]. Radha et al. diagnosing heart disease in type2 diabetic patient by cascading the data mining techniques [9]. Parthiban, et al. proposed timely discovery of the susceptibility of a diabetic patient to heart ailment using Naïve bayes and SVM classification algorithm [10].

Shouman et al. reviewed different algorithms for prediction of heart Disease Datasets as follows (Table 1):

Table 1. Accuracy comparison [2]

Author, Year	Classifier used	Accuracy
Tantimongcolwata et al. 2008	Direct kernel self-organizing map	80.4%
	Multilayer perceptron	74.5%
Hara et al. 2008	Automatically defined groups	67.8%
	Immune multi-agent neural network	82.3%
Sitar-Taut et al. 2009	Naïve bayes	62.03%
	Decision trees	60.40%
Rajkumar et al. 2010	Naive bayes	52.33%
	KNN	45.67%
	Decision list	52%

3 Problem Statement

- Accuracy of machine learning algorithm i.e. how much accurate prediction done by machine learning classifiers, as we cited the Shouman et al. [2] table accuracy varies from 50 to 81% which is not permissible in the field of medical diagnosis (Life Care).
- Figure 1 in this paper is the 2D plot of accuracy of some machine learning classification algorithm, from there we can conclude the there is need of improvement in prediction accuracy because medical field need higher accuracy as it related with human being life.
- Entropy of outlier in gathered data is high which may reduce the classification accuracy. Entropy of missing values in gathered data is high, it may also reduce the classification accuracy.
- Most of the researcher in this field may not concentrate on dimension reduction approach.

4 Solution Methodology

As we discussed in Sect. 3, we need to improve the prediction accuracy, we have proposed Pretreat-Ensemble machine learning model, Fig. 2 shows the flow of proposed prediction model, in which we have applied 3 phase data preprocessing which is followed by the proposed algorithm.

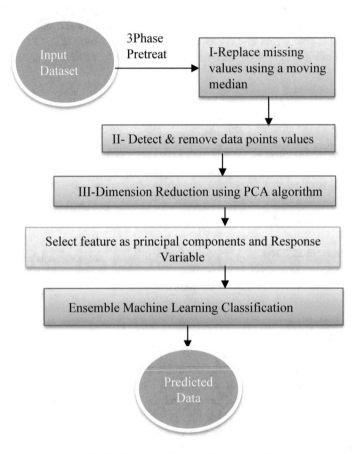

Fig. 2. Proposed data prediction model

Proposed Algorithm

Step-1. //Read Input dataset from csv file.
    a. Check encoding (.csv,.txt etc)
    b. Read the sheet from csv
    c. Set top row as variable name
Step-2. //Convert data D to table array.

$$D_{rc} = \sum_{k=1}^{m} a_{rk}b_{kc}$$

Step-3. //Identification of missing values and replace with NaN.
    a. For each row in D
           if D_{ij}= Null
              D_{ij}=NaN
           end if
Step-4. Fill the missing values(Dij)
Step-5. Detect and remove Outlier(Dij)
Step-6. Normalize the data using feature scaling and then apply PCA
Step-7. Apply Ensemble Boosting algorithm
Step-8. Validate Prediction Model //Predicted data validation
Step-9. Calculate Accuracy, Sensitivity, Specificity, F-Score.

PCA is used for reducing number of variables in the dataset that are correlated. The reduced set of variables are weighted linear combination of the original variable. The principle problem with the data are noise and redundancy. Data generated in the field of medical diagnosis are very large in size and contains noise which require high computation henceforth dimension reduction is needed. PCA uses covariance matrix to quantify redundancy.

Ensemble Machine Learning Model: In this machine learning paradigm number of classification model (Weak Learners) trained to solve the problem. Hypothesis behind ensemble model is when weal learners are correctly integrated then we can attain accurate prediction. There are major three kinds of approaches to combine weak learners are Bagging, Boosting and Stacking. In our proposed prediction model we have used Boosting for combining weak learners. So let us describe how Boosting works. It uses weighted average to make weak learner to strong learner, boosting can be easily described by Fig. 3 followed by algorithm.

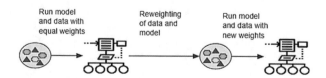

Run model
and data with
equal weights

Reweighting
of data and
model

Run model
and data with
new weights

Fig. 3. Boosting

Ensemble

There are different types of boosting algorithms, opted based on nature of response variable of dataset, if the response variable having binary class then we have to go for following boosting algorithm as: Ada BoostM1, Logit Boost, Gentle Boost, Robust Boost, LP Boost, RUS Boost, Total Boost.

Validation of classifier can done by evaluating following parameters using validation phase generated confusion matrix values.

Dataset Used and Result & Discussion

Dataset: For experimental evaluation we have gathered data from UCI data repository [11], Fig. 4. depicts the snippet of input dataset.

	A	B	C	D	E	F	G	H	I	J	K	L	M	N
1	age	sex	cp	trestbps	chol	fbs	restecg	thalach	exang	oldpeak	slope	ca	thal	num
2	63	1	1	145	233	1	2		0		3	0	6	absent
3	53	1	4	140	203	1	2	155	1	3.1	3	0	7	present
4	56	1	3	130	256	1	2	142	1	0.6	2	1	6	present
5	52	1	3	172	199	1	0	162	0	0.5	1	0	7	absent
6	58	0	1	150	283	1	2	162	0	1	1	0	3	absent
7	60	1	4	117	230	1	0	160	1	1.4	1	2	7	present
8	61	1	3	150	243	1	0	137	1	1	2	0	3	absent
9	59	1	3	150	212	1	0	157	0	1.6	1	0	3	absent
10	65	0	3	140	417	1	2	157	0	0.8	1	1	3	absent

Fig. 4. Input dataset Snippet

For implementation we have used Weka 3.8 tool, where it is used for data mining, which finds valuable information hidden in large volume of data and prediction model validated using percentage split (for splitting the test and train data in percent). Dataset feature which we have used as 14 attributes with different range as age, sex, cp, trestbps, chol, fbs, restecg, thalach, exacng, oldpeak, slope, ca, thal, num.

dm final4-weka.filters.unsupervised.attribute.ReplaceMissingValues													
: sex	3: cp	4: trestbps	5: chol	6: fbs	7: restecg	8: thalach	9: exang	10: oldpeak	11: slope	12: ca	13: thal	14: num
imeric	Numeric	Numeric	Numeric	Numeric	Numeric	Numeric	Numeric	Numeric	Numeric	Numeric	Numeric	Nominal
1.0	1.0	145.0	233.0	1.0	2.0	153.79	0.0	1.10009	3.0	0.0	6.0	absent
1.0	4.0	140.0	203.0	1.0	2.0	155.0	1.0	3.1	3.0	0.0	7.0	pres...
1.0	3.0	130.0	256.0	1.0	2.0	142.0	1.0	0.6	2.0	1.0	6.0	pres...
1.0	3.0	172.0	199.0	1.0	0.0	162.0	0.0	0.5	1.0	0.0	7.0	absent
0.0	1.0	150.0	283.0	1.0	2.0	162.0	0.0	1.0	1.0	0.0	3.0	absent
1.0	4.0	117.0	230.0	1.0	0.0	160.0	1.0	1.4	1.0	2.0	7.0	pres...
1.0	3.0	150.0	243.0	1.0	0.0	137.0	1.0	1.0	2.0	0.0	3.0	absent
1.0	3.0	150.0	212.0	1.0	0.0	157.0	0.0	1.6	1.0	0.0	3.0	absent
0.0	3.0	140.0	417.0	1.0	2.0	157.0	0.0	0.8	1.0	1.0	3.0	absent
1.0	3.0	130.0	197.0	1.0	2.0	152.0	0.0	1.2	3.0	0.0	3.0	absent
0.0	3.0	135.0	304.0	1.0	0.0	170.0	0.0	0.0	1.0	0.0	3.0	absent
1.0	4.0	125.0	257.0	1.0	0.0	163.0	0.0	0.2	2.0	2.0	7.0	pres...
1.0	3.0	180.0	274.0	1.0	2.0	150.0	1.0	1.6	2.0	0.0	7.0	pres...
0.0	3.0	110.0	265.0	1.0	2.0	130.0	0.0	0.0	1.0	1.0	3.0	absent
1.0	4.0	125.0	249.0	1.0	2.0	144.0	1.0	1.2	2.0	1.0	3.0	pres...
0.0	4.0	132.0	341.0	1.0	2.0	136.0	1.0	3.0	2.0	0.0	7.0	pres...
1.0	3.0	140.0	211.0	1.0	2.0	165.0	0.0	0.0	1.0	0.0	3.0	absent
1.0	4.0	130.0	330.0	1.0	2.0	132.0	1.0	1.8	1.0	3.0	7.0	pres...
1.0	4.0	130.0	256.0	1.0	2.0	153.79	1.0	0.0	1.0	2.0	7.0	pres...
1.0	1.0	138.0	282.0	1.0	2.0	174.0	0.0	1.4	2.0	1.0	3.0	pres...

Fig. 5. File snippet of Phase-1

As we have elaborated our proposed "Pretreat Ensemble" prediction model having three preprocessing steps. Figure 5 shows the output of the first phase of the "Pretreat-Ensemble" Prediction Model. We have applied filter as 'ReplaceMissingValues' i.e.. it is used to fill the missing values if present in the input data. In the phase 2 we have detected the outlier by using the filter named 'InterquartileRange'. It is a filter for detecting outliers and extreme values based on interquartile ranges (Fig. 6).

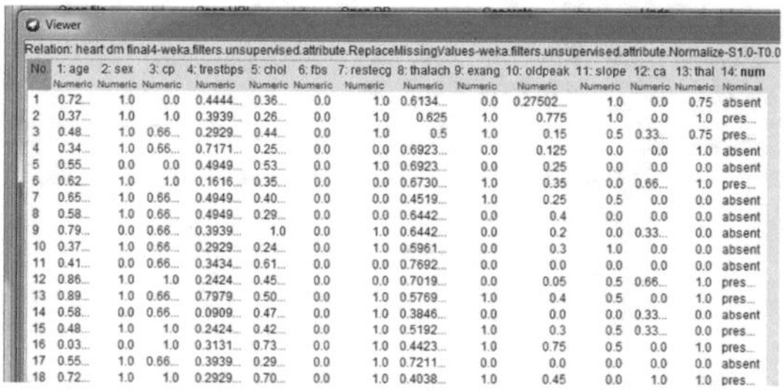

Fig. 6. File snippet of normalized data in Phase-2

Now in the third phase principal components are calculated which we have already discussed. Which is applied by selecting principal component in 'Attribute Evaluator'. Dimensionality reduction is accomplished 0.75 (75%) (Fig. 7).

Fig. 7. Values calculated by principal component in Phase-3

The output of the ensemble model, data passed through 100 conventional machine learning classifier, "LogitBoost" boosting algorithm used in prediction model. Further for validation of proposed prediction we have validate model with different values of percentage split. Figure 8 shows the predicted values of accuracy, error after implementing ensemble machine learning modal for 70% of training data.

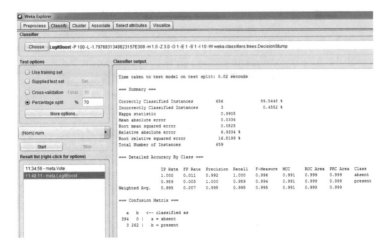

Fig. 8. Output of pretreat ensemble model at 70% of training data

From the output of the implemented ensemble LogitBoost model we have achieved the accuracy at different percentage split values and calculate the average accuracy which is shown in Table 2.

Table 2. Accuracy achieved at different Percentage Split values

Percentage split%	Accuracy
90%	99.539%
80%	99.30%
70%	99.537%
Average accuracy	99.459%

5   Conclusion

Data gathered from medical centers are very large in size contains noise, missing values of attributes, as a result it affects the machine learning algorithm performance, henceforth we have proposed 3phase preprocessing over input data before apply in machine learning classifier, after gone through experimental evaluation we can conclude that proposed "Pretreat-Ensemble" prediction model achieved highest accuracy then other convention machine learning classifiers, with average accuracy 99.459%.

References

1. Brisimi, T.S., Xu, T., Wang, T., Dai, W., Adams, W.G., Paschalidis, I.Ch.: Predicting chronic disease hospitalizations from electronic health records: an interpretable classification approach. Proc. IEEE **106**(4), 690–707 (2018). https://doi.org/10.1109/jproc.2017.2789319

2. Shouman, M., Turner, T., Stocker, R.: Using Data Mining Techniques in Heart Disease Diagnosis and Treatment (2012). 978-1-4673-0483-2 © 2012 IEEE
3. Amin, S.U., Agarwal, K., Beg, R.: Genetic Neural Network Based Data Mining in Prediction of Heart Disease Using Risk Factors (2013). 978-1-4673-5758-6/13/$31.00 © 2013 IEEE
4. Kaur, G., Chhabra, A.: Improved J48 classification algorithm for the prediction of diabetes. Int. J. Comput. Appl. **98**(22), 13–17 (2014). 0975-1887
5. Kalaiselvi, C., Nasira, G.M.: Prediction of heart diseases and cancer in diabetic patients using data mining techniques. Indian J. Sci. Technol. **8**(14), 1 (2015). https://doi.org/10.17485/ijst/2015/v8i14/72688. ISSN (Print) 0974-6846 ISSN (Online) 0974-5645
6. Singh, D.A.A.G.: Dimensionality reduction using genetic algorithm for improving accuracy in medical diagnosis. Int. J. Intell. Syst. Appl. **1**, 67–73 (2016)
7. Pouriyeh, S., Vahid, S., Sannino, G., De Pietro, G., Arabnia, H., Gutierrez, J.: A comprehensive investigation and comparison of machine learning techniques in the domain of heart disease. In: IEEE Symposium on Computers and Communications (ISCC), pp. 204–207 (2017)
8. Perveena, S., Shahbaza, Md., Guergachib, A., Keshavjee, K.: Performance analysis of data mining classification techniques to predict diabetes. In: Symposium on Data Mining Applications, SDMA 2016 (2016). 1877-0509 © 2016 Published by ELSEVIER B.V
9. Radha, P., Srinivasan, B.: Diagnosing heart diseases for type 2 diabetic patients by cascading the data mining techniques. Int. J. Recent Innov. Trends Comput. Commun. **2**(8), 2503–2509 (2014). ISSN 2321-8169
10. Parthiban, G., Rajesh, A., Srivatsa, S.K.: Diagnosing vulnerability of diabetic patients to heart diseases using support vector machines. Int. J. Comput. Appl. **48**(2), 45–48 (2012). 0975–888
11. UCI data repository. https://archive.ics.uci.edu/ml/datasets/Heart+Disease

Emotion Recognition of Facial Expression Using Convolutional Neural Network

Pradip Kumar$^{(\boxtimes)}$, Ankit Kishore, and Raksha Pandey$^{(\boxtimes)}$

Department of CSE, SoS (Engineering & Technology),
Guru Ghasidas Vishwavidyalaya, Bilaspur, CG, India
pradipkumar99688@gmail.com, ankitkishore0@gmail.com,
rakshasharma10@gmail.com

Abstract. Recognition of emotion using facial information is an interesting field for computer science, medicine, and psychology. Various researches are working with automated facial expression recognition system. Convolutional neural network (CNN) for facial emotion recognition method is basically used to recognize different-different human facial landmarks, geometrical poses, and emotions of faces. Human facial expression gives very important information to understand the emotions of a person for an interpersonal relationship. Since it is a classification problem, so the performance of any classifier is dependent on features extracted from the region of interest of the sample. In this paper, we are going to train the machine to recognize different types of emotions through human facial expressions using the Convolutional Neural Network (CNN). We have used sequential forward selection algorithms and softmax activation function.

Keywords: Human facial emotion recognition · Facial expression recognition · Face recognition and Human-Computer interaction (HCI) · CNN

1 Introduction

Humans generally uses different-different forms of communication technology to understand the unspoken words from facial expressions, hand gestures, and emotions. Facial emotions are very much helpful to understand the feelings of man. It is very helpful for those people who are unable to speak about their emotions, then we can easily identify their emotions. The most common way of emotion recognition is facial analysis. Here we used only facial expressions but there is some technique also available by which whole body activity of a human can be recognized. Emotion recognition is among the few methods that have the advantage of both high accuracy as well as low intrusiveness. The approach it chooses has high accuracy with minimal intrusive. For this reason, since the early '70s, Emotion recognition using facial expression has drawn the attention of scholars and researchers in fields from Internet security, image processing, and psychology, to computer vision (Fig. 1).

© Springer Nature Switzerland AG 2020
J. S. Raj et al. (Eds.): ICIDCA 2019, LNDECT 46, pp. 362–369, 2020.
https://doi.org/10.1007/978-3-030-38040-3_41

Fig. 1. Different emotions through facial expressions [2].

Since various algorithms have been proposed by the researchers for Face recognition and Emotion recognition and some are still on the way to release.

This part shows some surveyed research paper:

1. An Introduction to Convolutional Neural Networks [3] - This paper has covered the basic concepts of Convolutional Neural Networks (CNN), by explaining the different layers required to build one and elaborating how to structure the network in most image analysis tasks.

2   Proposed Approach

We are going to propose an approach that assists to recognise the emotion of human faces as Angry, Happy. So this is a classification problem among Angry, Disgust, Fear, Happy, Sad, Surprise, Neutral. In order to do that, we will need dataset first, so we favour dataset of KAGGLE [9]. Then we apply face detection technique so that it can get only facial feature not other at all. For face detection, we have to use Haar algorithms. Since haar detects the whole face that's why it extract all features of faces and it is a supervised classification problem that's why we need their feature and label both. According to their extracted feature, we give these extracted feature and corresponding label to CNN model for training. And After training of the model, we can give input for testing the model.

In order to train the model, we first detect the face from image for getting feature extractions and then use a convolutional neural network to training the model using those extracted features and classification of emotions.

3   Face Detection

Without having the face of an image, we can not recognize the emotion of that image. So in emotion detection, it is very important to detect the face and according to the facial expression of that face, we extract the feature of that image. Face Detection is a computer technology which has many numbers of applications that are widely used to identify the faces in an image. So in order to recognize emotion, we will identify the face of an image first [10]. By seeing eyes and lips, we can easily identify the emotion of a human. So if we give the feature of eyes and lips to the machine for learning, using learnt knowledge, machine can also identify the emotions like a human. Since to recognize the expression of human we need eye feature and lip feature. But we can not separate eye and lip from the face. In short, If we extract feature of face then we automatically get the feature of eye and lip along with face.

Fig. 2. Face, Eye and lip detection from a digital image [6]

We have used the haar cascade algorithm to detect and extract feature of the face, eye and lip. It detects only face, eye and lip, if other things in the image it will crop face location from an image as shown in Fig. 2.

After getting information of face, eye and lip, there is a problem how to classify the emotion among various facial expressions. There are many classification techniques as K-Nearest Neighbours (KNN), Convolutional Neural Network (CNN), Decision tree, Support vector machine (SVM) etc. Since the preprocessing required in CNN is much lesser as compared to other classification algorithms and it is an advancement in computer vision with deep learning has been constructed and perfected with time and primarily better than the other one. So we have to use a convolutional neural network to classify emotions among various expressions.

3.1   Convolutional Neural Network

Convolutional Neural Networks (CNNs) is basically a combination of several types of layers. These are such as convolutional layers, pooling layers, fully-connected layers, output layers, softmax layers. When we put together all of them then these forms CNN. A simplified CNN architecture for MNIST classification [8]. CNN model is shown in Fig. 3.

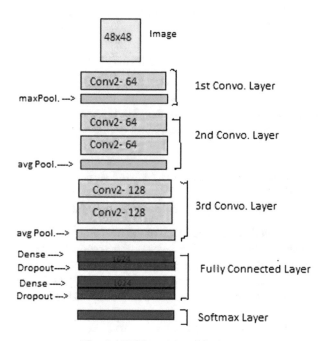

Fig. 3. CNN model and its layers

Convolutional Layer: Its name itself says that in the convolutional layer perform a convolution operation over the input data and filters [Fig. 4]. The resulting output of this layer calculates as [8]. The convolutional layer is the main building block of the convolutional neural network that does most of the computational efforts. Convolution saves the relationship among pixels by learning image information using small squares of input data. Since, the convolution of an image has different - different filters, and by applying filters it can do a different - different task such as edge detection, sharpen and blur. Since the Activation function is the cell of any neural network. So, we have used the activation function ReLu that means Rectified Linear Unit. ReLu activation function is better than the sigmoid activation function and it shares more features than binary units.

$$f(x) = Max(0, x)$$

1. An image matrix of dimension (h*w*3)
2. A filter of dimension (fh*fw*3)
3. Outputs a volume dimension (h-fh+1)*(w-fw+1)*1

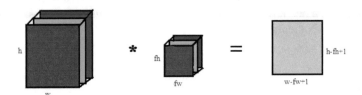

Fig. 4. Image matrix multiplies with filter matrix and gets a new matrix

There is not any exponential function that has to be calculated that's why it has comparatively cheap computation. ReLu function clogs vanishing gradient problem also because the gradient functions are either zero or linear but never to be non-linear function [12].

The purpose of the convolutional layer to extracting the feature of input data. During the feature extraction step, the horizontal and vertical distances between all the landmark pairs are calculated.

It selects feature iteratively which improves the recognition accuracy the most.

Pooling Layer: It is another building block of the convolutional neural network. In the Convolutional neural network, pooling layer is the second layer after convolutional layer. It is very helpful for reducing dimensionality when images are very large. Pooling involves selecting pooling operations which are very much similar to the filter that are used in the feature map. Pooling operation or filter has the size greater than that of the feature map.

Two common functions used in pooling layer. First one is Max Pooling which calculated the maximum value in the feature map by utilisation of maximum function and the other one is Average Pooling which calculated the average value in the feature map.

Fully Connected Layer: Multilayer Perceptron is also known as fully connected layer followers which connect each and every neuron of the previous layer to every neuron of the present layer as shown in Fig. 5. This is analogous to the way that neurons are arranged in traditional forms of ANN. Like a neural network, it takes input in vector form that gets by flattened a matrix into vectors. Flattened output works as an input for a fully connected layer. The output is calculated as [3].

In the Fig. 2, X1, X2, X3… are the vectors produced by the conversion of the feature map matrix. we created a model by combining these features together.

Output Layer: The output layer represents a set of vectors which of all the classes of input images. It has a matrix of all the classes. We can calculate the output of the output layer by a function by Emotion Detection Algorithm [3].

Softmax Layer: Through the softmax layer, the error is back-propagated. Let N be the dimension of the input vector, then the Softmax can be calculated [3].

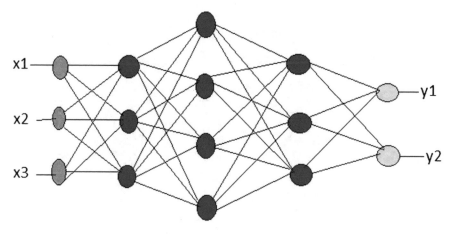

Fig. 5. Flattened as FC layer, after pooling layer

4 Experiment and Result

The performance of emotion detection with facial image model is evaluated. For training this model, We have to use kaggle dataset for this model which have 28,709 grayscale images of 48 × 48 pixel [9]. This data set has seven types of emotions those are Happy, Angry, Fear, Disgust, Sad, Surprise and Neutral. This model uses 25 epochs for training and gave 92.56% validation accuracy. We gave an image as input, is shown in Fig. 6.

Fig. 6. Human face expressing fear [9]

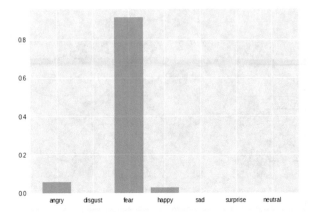

Fig. 7. Bar graph for input image

4.1   Result

After processing, model classified the image very well and gives output as a fear. The model also gives bar graph with corresponding prediction accuracy as shown in Fig. 7.

5   Conclusion

A lot of work has been done in the area of facial recognition. Various models already exist for facial recognition. Our Proposed model uses CNN for Classification. Proposed model gives around 92.56% validation accuracy in near about 25 epochs.

In future we will work in the area of Human emotions along with their activity.

Compliance with Ethical Standards

All author states that there is no conflict of interest.

We used our own data.

References

1. Pillai, V., Upadhyay, A., Pandey, R.: Human Activity Recognition Using Accelerometer Sensor Data, ICEECCMC, 28–29 January 2018
2. Introduction to Emotion Recognition (2018). https://blog.algorithmia.com/introduction-to-emotion-recognition/
3. O'Shea, K., Nash, R.: An Introduction to Convolutional Neural Networks, ArXiv, December 2015
4. Kim, M.H., Joo, Y.H., Park, J.B.: Emotion detection algorithm using frontal face image. In: International Conference of Control, Automation and System, KINTEX, Gyeonggi-Do, Korea (2005)
5. Acharya, D., Huang, Z., Pani Paudel, D., Van Gool, L.: Covariance Pooling for Facial Expression Recognition in ETH Zurich, Switzerland, pp. 480–487. IEEE (2018)

6. Glauner, P.O.: Deep learning for smile recognition. In: FLINS Conference (2017). In Interdisciplinary Centre for Security, Reliability and Trust, University of Luxembourg 2721 Luxembourg, Luxembourg
7. Barsoum, E., Zhang, C., Ferrer, C.C., Zhang, Z.: Training deep networks for facial expression recognition with crowd-sourced label distribution. . In: ICMI Conference (2016). In Microsoft Research One Microsoft Way, Redmond, WA 98052
8. Burkert, P., Trier, F., Afzal, M.Z., Dengel, A., Liwicki, M.: DeXpression: deep convolutional neural network for expression recognition. In: IEEE Conference on Vehicular Electronics and Safety (ICVES) (2015)
9. https://www.kaggle.com/c/challenges-in-representation-learning-facial-expression-recognition-challenge/data
10. Lin, S.-H.: An Introduction to Face Recognition Technology
11. Viola, P., Jones, M.: Rapid object detection using a boosted cascade of simple features. In: Proceedings of the 2001 IEEE Computer Society Conference on Computer Vision and Pattern Recognition, CVPR 2001, vol. 1 (2001)
12. Glorot, X., Bordes, A., Bengio, Y.: Deep sparse rectifier neural networks. In: Gordon, G.J., Dunson, D.B. (eds.) Proceedings of the Fourteenth International Conference on Artificial Intelligence and Statistics (AISTATS-2011), vol. 15 (2011). Journal of Machine Learning Research - Workshop and Conference Proceedings, 2011, pp. 315–323
13. Lucey, P., Cohn, J.F., Kanade, T., Saragih, J., Ambadar, Z., Matthews, I.: The extended Cohn-Kanade dataset (CK+): a complete dataset for action unit and emotion-specified expression. In: Proceedings IEEE Conference on Computer Vision and Pattern Recognition Workshops (CVPRW) (2010)

Information Security Through Encrypted Domain Data Hiding

Vikas Kumar[(✉)], Prateek Muchhal, and V. Thanikasiselvan

Department of Electronics and Communication Engineering,
VIT University, Vellore, Tamil Nadu, India
vikas6094@gmail.com, muchhalpratik@gmail.com,
thanikaiselvan@vit.ac.in

Abstract. In this contemporary world of internet, information security is the most important issue because while transmitting and receiving any information there is a chance of hacking. So to tackle the cyber-attack, this paper proposes a new two-layer information security and data hiding algorithm which involves a block scrambling algorithm for an image encryption and an algorithm for incorporating the message bits into an encrypted image. Firstly, the image encryption has been done by dividing image horizontally into two equal blocks followed by block scrambling algorithm and diffusion scheme. After encryption the data hiding algorithm is used for hiding the message bits into an encrypted image. This technique of hiding secret data into an encrypted image is highly secured and on performance analysis of block scrambling algorithm gives the excellent performance.

Keywords: Image encryption · Block scrambling · Data hiding · NPCR · UACI · LSB

1 Introduction

Security in this contemporary world have been the prime issue with the increase in the developed technologies and internet. Recently hacking and leakage of information is the major threat, hence encryption to secure the data.

Encryption is characterized as the change of plain message into a frame called a figure content that can't be perused by any individuals without decoding the scrambled content. The picture encryption is to transmit the picture safely over the system so that exclusive approved client can decode the picture. Picture encryption, turmoil based encryption have applications in many fields including the web correspondence, transmission, therapeutic imaging, tele-medication and military correspondence, and so on.

In past years there have been many studies and proposed algorithms and techniques in relation to encryption. Logistic map based encryption algorithm was proposed by Matthew in the year 1989. Later many algorithm and methodologies have been introduced like cryptography, watermarking, logistic map, Arnold cat map, etc. The Diffie-Hellman (DH) algorithm, made by White Diffie and Martin Hellman, acquainted open key cryptography with public. Now they can utilize that mutual secret key to encrypt messages ensured with a secret-key cipher, and each can decode what alternate

© Springer Nature Switzerland AG 2020
J. S. Raj et al. (Eds.): ICIDCA 2019, LNDECT 46, pp. 370–379, 2020.
https://doi.org/10.1007/978-3-030-38040-3_42

encodes. Aman and Mohammad Ali Bani Younes present a block construct algorithm called blowfish in which the calculation is based with respect to the fusion of picture change and an outstanding encryption and decoding. Here unique picture is partitioned in pieces, and applying the change calculation it was improved, and after that the calculation is used to encode the transformed picture. By the results it is seen that the expansion in the quantity of blocks is better for encryption as it results in higher entropy.

In 1998, First chaotic image encryption was introduced by Fridrich where under certain initial conditions the pixels are rearranged. The several advantages of the chaotic encryption approach are in the encryption system design there is high flexibility, ergodicity, periodicity, pseudo-randomness. Logistic map is one of the algorithm based on chaotic technique where the plain image is shuffled with hyper–chaos. Some other techniques like S-box, hash tables, optical transform, DNA encoding, etc. are also based on chaotic encryption algorithm.

The paper overcomes weaknesses like unable to resist chosen and known-plain text attacks effectively to achieve the great safety performance in one round. It contains a double layer of information security that includes block scrambling algorithm and diffusion scheme for the image encryption and a new data hiding algorithm for hiding the text message into an encrypted image. Firstly, image encryption has been done by the block scrambling scheme which is done by breaking the image into two equal parts horizontally. If the row number is not even then one random row is added accordingly.

Then the chaotic matrix is used to make X and Y coordinate table and swapping control table. For pixel's swapping in the first half and in the second half Swapping control table is utilized. Processing pixel is located with the help of the X and Y coordinate table. After scrambling, diffusion scheme is applied on the both scrambled image that is followed by data hiding technique.

Data Hiding is a procedure utilized for concealing mystery messages into a cover-media with the end goal that an unapproved individual won't have the capacity to get to the shrouded messages. 8-bit dark scale pictures are taken as a cover pictures in this paper. The pictures are also called as cover images. The secret or the important messages by inserting is hided in cover images which is called stego-images. In the writing the few procedures of information stowing away have been proposed [1–5]. LSB i.e. Least Significant Bit [6] is one of the basic method which depends on controlling the LSB planes by supplanting the LSB's of cover-image with the message bit. Here the encrypted image will act as a cover media because it will be highly secure compared to plain image. Our final output will display only encrypted image which contains the hidden secret message so if someone tries to decrypt the information mostly he will only tries to decrypt the encrypted image and our secret data will be secure.

2   Proposed Methodology

The proposed algorithm is in two phases. First phase is the image encryption by block scrambling algorithm followed by diffusion scheme. Second phase is the hiding of message bits using data hiding algorithm. Block diagram is given in Fig. 1.

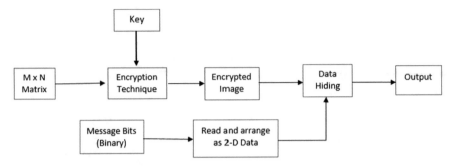

Fig. 1. Flowchart of the proposed algorithm

2.1 Phase I

Consider a source image of size M × N. Let the M and N are even. Now dividing the image horizontally into two equal parts i.e. two image of same size i.e. P × Q where P = M/2 and Q = N. Name it as Img1 and Img2.

2.1.1 Table Construction

Logistic map is the method to produce the chaotic matrix B1 with the help of key key1 (a_0, α_1) that will have the elements ranging from 1 to P. With the Eq. 1 iterate chaotic matrix L' times where L' = P × Q and obtaining the sequence m and m is quantified by Eqs. 2 and 3 is used to transform into B1.

$$tk + 1 = \mu tk(1 - tk)\mu \in (0, 4), tk \in (0, 1) \tag{1}$$

$$a_1 = mod\left(floor\left(m \times 10^{14}\right), P\right) + 1 \tag{2}$$

$$B1 = reshape(a1, P, Q) \tag{3}$$

Similarly, use the secret key key2(b_0, α_2) to produce chaotic matrix B2.

Now construct the X coordinate table, Y coordinate table and swapping control table with the chaotic matrix B1 and B2.

(a) Constructing XT

$$\begin{aligned} &if\ abs(B1(i,j) - i) < (P/4) \\ &\qquad XT(i,j) = mod(B1(i,j) + (P/4), P) + 1 \\ &else \\ &\qquad XT(i,j) = P1(i,j). \end{aligned} \tag{4}$$

(b) Constructing YT

$$if\ abs(B2(i,j) - i) < (Q/4)$$
$$YT(i,j) = mod(B2(i,j) + (Q/4), Q) + 1 \tag{5}$$
$$else$$
$$YT(i,j) = B2(i,j).$$

(c) Constructing CT

$$if\ abs(B1(i,j) - i) < (P/4)\ or\ abs(B2(i,j) - i) < (Q/4)$$
$$CT(i,j) = 1 \tag{5}$$
$$else$$
$$CT(i,j) = 0$$

With the help of XT, YT and CT tables swap the elements of plain image i.e. Fig. 2(a) from left upper corner and in order from top to bottom and left to right.

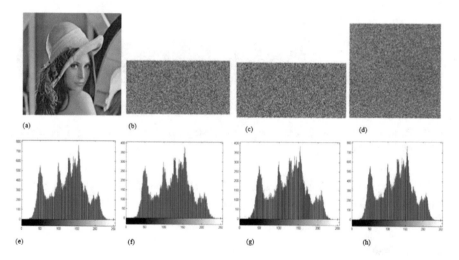

Fig. 2. (a) Lena plain image (b-c) block scramble images of lena and their histograms (d) Scrambled lena image (e) Histogram of plain lena image (f-g) histograms of scramble images of lena (h) Histogram of scrambled image.

(3) Scrambled block images should be combined i.e. Img1 and Img2 to get scrambled image Img3 which has a size of M × N.

Now full scrambled image is of size M × N. Sample image of Lena is taken for the scrambling purpose which shown in Fig. 2(a) and (b) shows the histogram and scrambled images, Img1 and Img2 are shown in Fig. 2(c) and (d) respectively and Fig. 2(e) has shown their respective histograms and Fig. 2(f) Combined scrambled image (i.e. Img3) in Fig. 2(g) and respective histogram is shown in Fig. 2(h).

After observing the respective histograms of Img1 and Img2 is like the histogram of Img3 which indicates that block image information is properly distributed in these scrambled images.

For further analysis four more samples images are analyzed to check and prove that this algorithm has good scrambling performance. The sample images are shown in Fig. 3 (a) and (b) has shown the scrambled images, histograms shown in Fig. 3(c) are of the block scramble images and their scrambled images histograms are shown in Fig. 3(d).

(a)   (c)

(b)   (d)

Fig. 3. Histogram analysis (a) Plain image (b) Histogram of plain image (c) Cipher image (d) Histogram of cipher image

The histogram of sample images is significantly different from each other but after scrambling their histograms are almost similar. This means that this scrambling algorithm is more stable in pixels distribution of both the images.

2.1.2   Diffusion

Step 1: Reshape the above scrambled image into a sequence q of size $L1 = M \times N$. Arrange the pixel values of scrambled image from upwards to downwards, from left to right to make q.

Step 2: Build secret key key3 (c_0, α_3) to make the key stream z_0 where the numbers ranges from 0–255. Iterate Eq. (1) L1 times, and obtain order k1. Then use Eq. (7) to calculate z:

$$z_0 = mod\left(floor\left(z \times 10^{14}\right), 256\right) \tag{7}$$

Step 3: Except the first one, Calculate the summation of the elements in p.

$$sum = \sum_{i=2} q(i) \tag{8}$$

Step 4: Set the value u_0 by Eq. (9). With the help of Eq. (10) encrypt the first element in q which is identified to all plain values i.e. for various plain images.

$$u_0 = mod(sum, 256) \tag{9}$$

$$q(1) = u_0 \oplus q(1) \oplus z_0(1) \tag{10}$$

Step 5: Set i = 2, analyze zt_1 and zt_2 (dynamic indexes), that is used to encrypt the ith element in q.

$$zt_1 = floor(mod(q(i-1) + z_0(i)), 256)/256 \times (i-1)) + 1 \tag{11}$$

$$zt_2 = floor(mod(q(i-1) + z_0(i)), 256)/255 \times (L1 - i - 1)) + i + 1 \tag{12}$$

where $zt_1 \in [1, i-1], zt_2 \in [i+1, L1]$.

Step 6: Encrypt the ith element in q by calculating zt_1 and zt_2.

$$q(i) = q(i) \oplus z_0(i) \oplus q(zt_1) \oplus (z(zt_2)) \tag{13}$$

Step 7: Set i = i + 1, and repeat from step 5 until i reaches L1 − 1.

Step 8: Set i = L and calculate index zt_1 by Eq. (11) to, and by Eq. (14) encrypt the last element

$$q(i) = q(i) \oplus z0(i) \oplus q(zt_1) \tag{14}$$

Step 9: Diffused vector q should be now transferred into the cipher image.

2.1.3 Decryption

Decryption process is the reversal of encryption process. After applying decryption process we can retrieve back the original image from the scrambled image.

The following steps have been taken for the decryption:

Step 1: Produce chaotic matrix B1 and B2 by using secret keys key1(a_0, α_1) and key2(b_0, α_2) respectively.

Step 2: With the help of B1 and B2 construct XT, YT and CT.

Step 3: Produce the key stream z_0 by using the secret key key3(c_0, α_3).

Step 4: Set i = L1, calculate zt_1 by Eq. 11 and last element should be decrypted by Eq. 14.

Step 5: Set i = L1 − 1, calculate zt_1 and zt_2 by Eqs. (11) and (12).

Step 6: Set i = i − 1 and return to step 6 until i = 2.

Step 7: Except the first one, calculate the summation of the elements of q.

Step 8: Calculate u_0 and by Eq. 10 decrypt the first element.

Step 9: Reshape the matrix q into the image M × N. Then partition Img1 and Img2 horizontally in equal parts.

Step 10: Swap the components in two images beginning from lower end right corner from base to top and appropriate to left.

Step 11: Construct an image of size M × N by combining block images Img1 and Img2

Figure 4(c) is the decrypted image which is like the original image and it's histogram is same as original image's histogram.

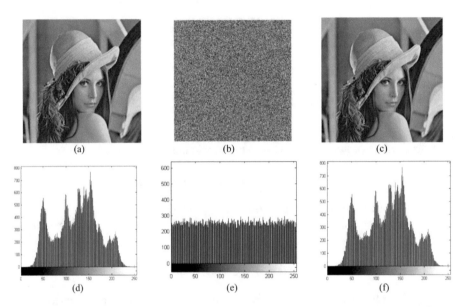

Fig. 4. (a) Base image (b) Encrypted image (c) Decrypted image (d) Histogram of Base image (e) Histogram of encrypted image (f) Histogram of decrypted image

2.2   Phase II: Data Hiding Algorithm

This segment contains the overall operation of hiding data away by basic LSB substitution strategy is depicted. Let C of size M × N be the scrambled picture in which LSB is performed. P be the n bit secret message.

Step 1: Convert P into binary bits (Pb).
Step 2: Take 4 bit at a time for a LSB substitution.
Step 3: Now convert that 4 bit to decimal i.e. z.
Step 4: Now the entire column matrix containing decimal.
Step 5: Reshape that column matrix into a matrix to M × N.
Step 6: Embed the reshaped secret message to get the stego image S by eq.:

$$S(i,j) = pn(i,j) - mod(pn(i,j), 2^\wedge k) + z(i,j)$$

where i = 1:M and j = 1:N.

pn is the image which is used as a cover media, k is the number of bits used for LSB substitution and z is the secret message. The embedding procedure is finished by supplanting the k bits of the pn by z.

Now in the extraction process the embedded messages i.e. P can be easily extracted without hindering the pixels of the cover image. Extract the selected pixels from k LSB's so that secret message bits are reconstructed.

Step 7: Extraction has been done by following equation:

$$R = mod(S, 2^\wedge k)$$

Step 8: Now convert R into binary (Rb).

Hence, extract the original secret message bits that was embedded in the encrypted image. So for the verification of proper extraction, Rb-Pb = 0.

The image of the embedded of the secret message into encrypted image and it's histogram is shown in Fig. 4(e) and (f) respectively. The flat histogram shows that the secret message is properly embedded, and nobody will be able to know that encrypted message contains any hidden information.

3   Performance Analysis

3.1   Histogram Analysis

The histograms of couple, Flower, Horse and Mars are shown in Fig. 3. Flat histogram is seen in Fig. 3(d) which demonstrates that the distribution of pixels of the encrypted image are uniform over the range [0–255] that is entirely not similar from the Fig. 3(b). This flat histogram shows the excellent performance against statistical attack. This block scrambling algorithm shows more uniform distribution of pixels.

3.2   NPCR Analysis and UACI Analysis

NPCR is number of changing pixel rate and UACI is the unified averaged changed intensity are two most used methods to assess the superiority of image encryption algorithms concerning differential assaults (Fig. 5).

$$NPCR = \frac{\sum ij\, H(i,j)}{M \times N} x\, 100\%$$

Let the plain image be P1(i,j) and the encrypted image be P2(i,j) of size M × N (Table 1).

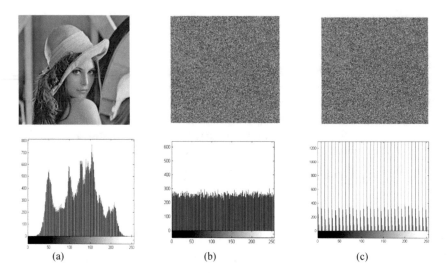

Fig. 5. (a) Base image and their histogram (b) scrambled image and their histogram (c) embedded image and their histogram

Table 1. NPCR & UACI value of sample images

Image	NPCR	UACI
Lena	0.9961	0.3353
Couple	0.9957	0.3737
Flower	0.9959	0.3284
Horse	0.9960	0.2913
Mars	0.9859	0.3250

H(i,j) is defined as

$$H(i,j) = \begin{cases} 0 & P1(i,j) = P2(i,j) \\ 1 & P1(i,j) \neq P2(i,j) \end{cases}$$

$$UACI = \sum_{i,j} \frac{|P1(i,j) - P2(i,j)|}{M \times N \times 255} \, x \, 100\%$$

4   Conclusion

This paper proposed a new dual layer information security and data hiding algorithm which involves a block scrambling algorithm [1] and the diffusion scheme for encrypting an image and data hiding algorithm [6] for hiding the secret message bits into an encrypted image. Firstly the source image is separated into two equivalent block images in horizontal direction and chaotic matrix A1 and A2 is constructed by using secret keys. With the help of A1 and A2, XT (x coordinate table), YT (y coordinate table) and CT (swapping control table) is constructed. Swapping control table is utilized to swap the pixel in the first half or in the second half. Processing pixel is located with the help of the X and Y coordinate table. After scrambling, with two dynamic index-based diffusion algorithms, diffuse the elements of scrambled image. Diffusion scheme is applied on the both scrambled image that is followed by data hiding technique. Data Hiding is a system utilized for concealing secret messages into a cover-media with the end goal that an unapproved individual won't have the capacity to get to the secret messages. Here LSB is used to hide secret message bits into an encrypted image bit. After embedding extraction of the secret message bits has been obtained in such a way that the extracted bits are exactly recovered. This scrambling algorithm uniformly scramble the plain image. The performance analysis and the simulation demonstrates that the block scrambling algorithm can adequately oppose the differential attack with brilliant performance.

References

1. Lu, X., Gou, X., Li, Z., Li, J.: A novel chaotic image encryption algorithm using block scrambling and dynamic index based diffusion. Opt. Lasers Eng. **91**, 41–52 (2017)
2. Zhang, Y.Q., Wang, X.Y.: A new image encryption algorithm based on non-adjacent coupled map lattices. Appl. Softw. Comput. **26**, 10–20 (2015)
3. Zhang, Y.Q., Wang, X.Y.: A symmetric image encryption algorithm based on mixed linear-nonlinear coupled map lattice. Inf. Sci. **273**(20), 329–351 (2014)
4. Liu, W., Sun, K., Zhu, C.: A fast image encryption algorithm based on chaotic map. Opt. Lasers Eng. **84**, 26–36 (2016)
5. Xiang, T., Wong, K.W., Liao, X.: Selective image encryption using a spatiotemporal chaotic system. Chaos **17**(3), 023115 (2007)
6. Chan, C.-K., Cheng, L.M.: Hiding data in images by simple LSB substitution. Pattern Recogn. **37**, 469–474 (2004)
7. Tirkel, A.Z., Van Schyndel, R.G., Osborne, C.F.: A digital watermark. In: Proceedings of ICIP 1994, Austin Convention Center, Austin, Texas, vol. II, pp. 86–90 (1994)
8. Bender, W., Morimoto, N., Lu, A.: Techniques for data hiding. IBM Syst. J. **35**(3/4), 313–336 (1996)
9. Chen, T.S., Chang, C.C., Hwang, M.S.: A virtual image cryptosystem based upon vector quantization. IEEE Trans. Image Process. **7**(10), 1485–1488 (1998)
10. Marvel, L.M., Boncelet, C.G., Retter, C.T.: Spread spectrum image steganography. IEEE Trans. Image Process. **8**(8), 1075–1083 (1999)
11. Qiao, L., Nahrstedt, K.: Ishikawajima-Harima Eng. Rev. **22**, 437 (1998)
12. Servetti, A., Martin, J.C.D.: IEEE Trans. Speech Audio Process. **10**, 637 (2002)
13. Mastronardi, G., Castellano, M., Marino, F.: Steganography effects in various formats of images. A preliminary study. In: International Workshop on 2001 Intelligent Data Acquisition and Advanced Computing Systems: Technology and Applications, pp. 116–119 (2001)
14. Zeng, W., Lei, S.: IEEE Trans. Multimedia **5**, 118 (2003). https://doi.org/10.1109/TMM.2003.808817
15. Liu, J.-L.: Pattern Recogn. **39**, 1509 (2006)

Person Re-identification from Videos Using Facial Features

Ankit Hendre$^{(\boxtimes)}$ and Nadir N. Charniya

Vivekanand Education Society Institute of Technology,
Chembur, Mumbai, India
{2017ankit.hendre, nadir.charniya}@ves.ac.in

Abstract. To precisely re-identify a person is a daunting task due to various conditions such as pose variation, illumination variation, and uncontrolled environment. The methods addressed in related work were insufficient for correctly identifying the targeted person. There has been a lot of exploration in the domain of deep learning, convolutional neural network (CNN) and computer vision for extracting features. In this paper, FaceNet network is used to detect face and extract facial features and these features are used for re-identifying person. Accuracy of FaceNet is compared with Histogram of Oriented Gradients (HOG) method. Euclidean distance is used for checking similarity between faces.

Keywords: Re-identify · Deep learning · Convolutional neural network · Computer vision · Histogram of oriented gradients

1 Introduction

Humans can easily Re-identify a person by various parameters such as facial attributes/features, height, pose, gestures. It is difficult to mimic the same ability of humans to Re-identify a person using machines. As far as security and safety are concerned it is very important to each individual. For surveillance purpose, it is very important to continuously monitor the system, which by human means is not possible. The human monitoring surveillance system can also be biased. Making an intelligent surveillance system is a solution to it. Due to the recent advancement in the technology it possible to relate human intelligence in machines. With the help of machine learning, deep learning, and computer vision techniques we can make machines perform at a similar level like human's do.

Re-Identification (Re-ID) can be formulated to determine the resemblance between images of a person captured from various cameras. In Person Re-ID scenario pose, illumination, and clothing of a person can change over a lapse of time. Considering appearance features, gait or gestures for person Re-ID helps for short term only, but facial features do not change a lot over a course of time. Therefore, using facial features for Re-ID can give better results for long term usage. Due to the availability of high-speed computers, deep learning techniques are widely explored and practiced for computer vision, speech recognition, and medical image analysis [1]. This paper has experimented on HOG and FaceNet techniques for re-identifying persons.

© Springer Nature Switzerland AG 2020
J. S. Raj et al. (Eds.): ICIDCA 2019, LNDECT 46, pp. 380–387, 2020.
https://doi.org/10.1007/978-3-030-38040-3_43

2 Related Work

Initially, the task of identifying a person was carried out by manually extracting the features. A very famous method developed by Viola and Jones [2] which uses Haar features for detecting face and extracting facial features. The algorithm recognizes a face by Haar feature selection, creating an integral image and using AdaBoost and cascade classifier for identifying a face. Another method used was Principal Component Analysis (PCA) [3]. In PCA approach faces are transformed into the Eigenfaces which become initial set as a database. To recognize new a face eigenvalue of that face is calculated and compared with the stored database. Face recognition using Linear Discriminant Analysis (LDA) [4] uses fisher function to discriminate class. Panoramic Appearance Map [5] uses multiple cameras and a triangulation method to locate the person. It extracts features using height and azimuthal angles. Color Histogram based method use RGB color space to extract facial features [6]. Local Binary Pattern (LBP) gives textural information of an image which is proposed by Ojala et al. [7]. It is a simple means for representing local information. The colored image is converted to a grayscale image by comparing a center pixel to its neighboring pixels in a 3x3 matrix. If the magnitude of the pixel to be converted to grayscale is greater or equal to its neighbor's pixel value than it is set 1 or else, it is set to 0. A similar procedure is done for all the pixels in the image to generate a gray image. To obtain LBP, any neighbouring pixel can be considered and direction could be clockwise or anticlockwise. Same direction pattern must be followed for all the pixels. The process of converting to grayscale and obtaining the output decimal value in the LBP array is repeated for each pixel in the image [8]. Scale Invariant Feature Transform (SIFT): David and Lowe [9] proposed SIFT in 2002. The algorithm is divided into multiple steps such as generating scale-space which is done using the Gaussian operator. Then, Difference of Gaussian is found for locating key points and unnecessary key points are discarded i.e., features located at edges are discarded. Orientations are assigned to key points and then a descriptor is computed having 128 elements [10].

2.1 Limitations

Most of the methods cited in the literature didn't consider illumination variation, pose variation and occlusion. Moreover, the accuracy of these techniques is not good and also PCA and LDA methods are not reasonable as they are very slow in processing.

3 Methodology

3.1 Histogram of Oriented Gradients (HOG)

The following steps are used for calculating HOG:

(a) Initially, the image is convolved with the kernels $[-1\ 0\ 1]$ and $[-1\ 0\ 1]^{\mathbf{T}}$ which gives image gradient x and y direction respectively. Gradients magnitude and direction is calculated using Eqs. (1) and (2)

$$g = \sqrt{g_x^2 + g_y^2} \tag{1}$$

$$\theta = arctan\frac{g_y}{g_x} \tag{2}$$

(b) The gradient magnitude and direction calculated in above step are put into a histogram of 9 bins having angles $0°$, $20°$, $40°$....$160°$ i.e. $0-180°$ is divided into 9 parts. The image is divided into cells (for e.g. 4×4 or 8×8) and for each cell, the histogram is calculated.

(c) After generating HOG for each cell, blocks of histograms are normalized to eliminate intensity variation. Finally, all the normalized histograms are connected into one element vector [11].

3.2   Convolutional Neural Network (CNN)

Convolutional Neural Networks (CNN) are highly inspired by the human brain and the concept of the neuron to neuron connectivity. Compared to the descriptor-based method CNN based method is more precise for Person Re-ID. CNN is more preferred for image processing and computer vision applications since its structure can manage large data creating fewer parameters. The simple architecture of CNN is shown in Fig. 1. CNN architecture consists of the following blocks:

(a) Convolutional layer: In convolutional layer the dot product of input image with kernel/filter takes place. Different filter size can be used for getting the best results but odd filter size gives good results.

(b) Non-linearity or Activation Function: It is used to introduce non-linearity since the convolution taking place is linear.

(c) Pooling: There are various pooling methods but usually max pooling is mostly used in image processing application because it reduces the dimension of the input preserving the useful

(d) Fully connected layer (FC): For classifying FC layer is used, as input is transformed into features by convolution and pooling. The matrix obtained after pooling is flattened and fed for classification. For classification softmax or sigmoid function is used depending upon the classification problem i.e., binary or multiple classes.

There can be multiple Convolutional and pooling layers stacked for fine feature extraction. For more accuracy running the model for a different number of epochs and selecting the model which gives the best accuracy. Filter's weights are automatically updated after each epoch. Similarly, there are different activation functions which can be used for classification [12].

Where, Conv: Convolution layer
Pool: Pooling layer
Relu: Activation function
FC: Fully connected layer

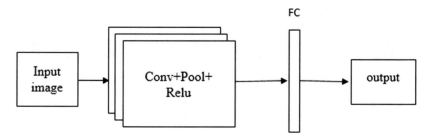

Fig. 1. Simple CNN architecture.

3.2.1 FaceNet Model Structure

In this paper, the deep network used is FaceNet which is proposed by researchers at Google in 2015. The model is experimented on around 200 million images consisting of about 8 million identities. The network was trained for 2000 h on a cluster of CPU. It is a 22-layer network structure. The output dimension of the network is 128 which gives validation accuracy of 87.9%. The smaller number of dimension would give less accuracy and a large number of dimension would require more training. It gives an accuracy of 98.87% for classification [13].

First, from videos frames are extracted and these frames are resized. Faces from frames are detected and their facial features are extracted and then compared with stored facial features. If the detected facial features match with the stored facial features from the database, then the label of the corresponding person will be displayed and if not, unknown will be displayed (Figs. 2 and 3).

3.3 Flowchart

Step 1: Generating facial features of the desired person.

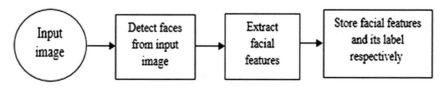

Fig. 2. Generating facial features of desired person.

Step 2: Re-identifying person.

In step 1 facial features of the desired person are extracted using Facenet. These facial features are stored with its respective label. In step 2 image or video is given as input. Faces from input are detected and facial features are extracted using FaceNet. These facial features extracted are compared with the stored facial feature using Euclidean distance. If the Euclidean distance calculated is less than the threshold then person is re-identified or else person is unidentified.

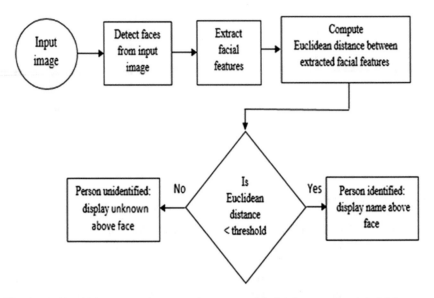

Fig. 3. Re-identifying person by comparing extracted facial features stored facial features.

The spatial distances used for verifying similarity are as follows:

(1) Cosine distance between x and y is given as

$$d_c(x, y) = 1 - \frac{x.y}{||x||_2.||y||_2} \tag{3}$$

$||x||_2$ is second-order norm

(2) Euclidean distance between x and y is as follows

$$d_e(x, y) = \left(|(x - y)|^2 \right)^{\frac{1}{2}} \tag{4}$$

(3) Manhattan distance between x and y is

$$d_m = |x - y| \tag{5}$$

(4) Minkowski distance is given by

$$d_{mi}(x, y) = \left(|(x - y)|^3 \right)^{\frac{1}{3}} \tag{6}$$

(5) Covariance between two variables is

$$Cov(x, y) = \frac{(x - \overline{X})(x - \overline{Y})}{n} \tag{7}$$

$\overline{X}$ *and* $\overline{Y}$ are mean of x and y respectively

(6) Correlation Co-efficient is given by

$$r = \frac{\left(\sum xy\right) - \left(\sum x\right)\left(\sum y\right)}{\sqrt{\left(\sum x^2 - \left(\sum x\right)^2\right)\left(\sum y^2 - \left(\sum y\right)^2\right)}} \tag{8}$$

from Eqs. 1 to 6 x and y are 1-dimensional vectors of two images.

4 Results

Figure 4 shows three faces are detected and identified accurately. The person's faces were detected at a very odd angle.

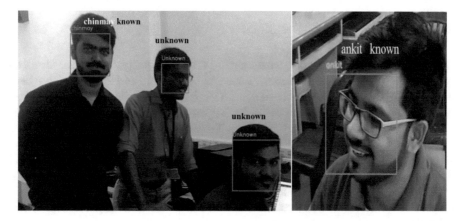

Fig. 4. Three persons detected; known (chinmay and ankit) and unknown identified accurately. The Person's faces were detected and identified correctly at a very odd angle.

Accuracy of HOG and FaceNet is shown in Table 1. It is very evident from the table that FaceNet is more accurate than the HOG method. Applying more strict threshold value to HOG and FaceNet, the accuracy of both methods increase significantly but FaceNet's performance is still better than HOG.

Table 1. Accuracy of HOG and FaceNet at threshold 0.6 and 0.45

Method	Accuracy	
	Threshold 0.6	Threshold 0.45
HOG	73%	81%
FaceNet	91%	95%

Various spatial distances are computed for similar faces and different faces. From Table 2 it is clear that the distance between the faces of the same person is small as compared to the distance between the faces of different persons.

Table 2. Various spatial distances for similarity check

Sr. No.	Distances	Distance between similar faces	Distance between different faces
1	Cosine distance	0.06812	0.1092
2	Euclidean distance	0.50681	0.64247
3	Manhattan distance	4.5835	5.8064
4	Minkowski distance	0.26228	0.33457
5	Covariance	0.000022	0.000037
6	Correlation Co-efficient	0.1305	0.2079

5   Conclusion

Facial features can help in long term scenario for Re-ID people i.e., facial attributes extracted can be stored and used in the future. In this paper, FaceNet network has been used for Re-ID people and an accuracy of 95% is achieved. The network can detect faces at different angles and can easily Re-ID person. Comparing with HOG method, FaceNet provides better accuracy shown in Table 1. The HOG method cannot detect faces at odd angles whereas FaceNet can do it accurately. Various distances are computed between similar faces and different faces shown in Table 2. The spatial distance obtained is small for similar faces and large for different faces. This method is useful in Re-ID of persons from video frames and it can also be used for security purposes to provide authority to the selected person for accessing the confidential information.

References

1. Bedagkar-Gala, A., Shah, S.K.: A survey of approaches and trends in person Re-IDentification. Image Vis. Comput. **2**(4), 270–286 (2014)
2. Viola, P., Jones, M.J.: Robust real-time face detection. Int. J. Comput. Vis. **57**(2), 137–154 (2004). Springer
3. Kaur, R., Himanshi, E.: Face recognition using principal component analysis. In: International Advance Computing Conference, Bangalore, India, pp. 585–589. IEEE, June 2015
4. Lu, J., Plataniotis, K.N., Venetsanopoulos, A.N.: Face recognition using LDA-based algorithms. IEEE Trans. Neural Netw. **14**(1), 195–200 (2003)
5. Gandhi, T., Trivedi, M.M.: Person tracking and Re-IDentification: introducing Panoramic Appearance Map (PAM) for feature representation. Mach. Vis. Appl. Sydney, Aust. **18**, 207–220 (2007). Springer

6. Poongothai, E., Suruliandi, A.: Survey on colour texture and shape features for person Re-IDentification. Indian J. Sci. Technol. 9(29) 2016
7. Ojala, T., Pietikäinen, M., Mäenpää, T.: Multiresolution gray-scale and rotation invariant texture classification with local binary patterns. IEEE Trans. Pattern Anal. Mach. Intell. **24**(7), 971–987 (2002)
8. Ahonen, T., Hadid, A., Pietikainen, M.: Face description with local binary patterns: application to face recognition. IEEE Trans. Pattern Anal. Mach. Intell. **28**(12), 2037–2041 (2006)
9. Lowe, D.G.: Distinctive image features from scale-invariant keypoints. Int. J. Comput. Vis. **60**(2), 91–110 (2004). Springer
10. Geng, C., Jiang, X.: Face recognition using sift features. In: Image Processing International Conference, Cairo, Egypt, IEEE, November 2009
11. Julina, J.K.J., Sharmila, T.S.: Facial recognition using histogram of gradients and support vector machines. In: Computer, Communication and Signal Processing, Chennai, India, pp. 1–5. IEEE, January 2017
12. Chollet, F.: Deep Learning with Python Version 6. Manning Publications, New York (2017)
13. Schroff, F., Kalenichenko, D., Philbin, J.: Facenet: a unified embedding for face recognition and clustering. In: Proceedings of the IEEE Conference on Computer Vision and Pattern Recognition, pp. 815–823. October 2015

Analysing Timer Based Opportunistic Routing Using Transition Cost Matrix

Chinmay Gharat[✉] and Shoba Krishnan

Vivekanand Education Society Institute of Technology,
Chembur, Mumbai, India
chinmay1992.cg@gmail.com, shoba.krishnan@ves.ac.in

Abstract. For some wireless sensor network applications, Energy consumption and Delay are important parameters for selecting routing scheme. To date, there has not been any simple analytical method to calculate these parameters. Analytical method proposed in this paper uses transition cost matrix and existing Markov model with slight modifications to calculate average energy consumption and average end to end delay. This method is developed only for timer based opportunistic routing scheme, this scheme synchronises neighbour nodes using waiting time, where waiting time is calculated using specified parameter. Analytical results obtained from this method are compared with simulation results, exhibiting minimal difference between them. The proposed method gives accurate results and is easier to implement than previously known methods. This model can be used for selecting suitable parameter for timer based opportunistic routing scheme as per the performance requirement of wireless sensor network applications.

Keywords: Wireless sensor network · Markov model · Transition cost matrix · Average energy consumption and average end to end delay

1 Introduction

Routing enables communication in Wireless Sensor Network (WSN). Routing for WSN is categorised into: Best path routing (BPR), Stateless routing (SR) and Opportunistic Routing (OR). BPR uses various metrics to find suitable fixed path between source and destination and works best for networks with stable and reliable wireless links [1]. SR performs random forwarding of packets and is suitable for networks with sparse resources [2]. OR facilitates all neighbour nodes (NN) to act as potential forwarding candidates, best node from them is selected for relaying. If it fails, then the next best node acts as relay and so on. OR is suitable for networks with unreliable links [2].

Based on co-ordination of NN, OR scheme is classified as: ACK based co-ordination (ABC) and Timer based co-ordination (TBC) [3]. ABC co-ordinates the NN using control messages, this prevents redundant transmissions and in process consumes extra bandwidth. TBC co-ordinates the NN using timer which is calculated using priorities assigned to NN. First node to complete the timer countdown will relay the packet. Use of timer for co-ordination nullifies the requirement of control messages, in process, saving

© Springer Nature Switzerland AG 2020
J. S. Raj et al. (Eds.): ICIDCA 2019, LNDECT 46, pp. 388–394, 2020.
https://doi.org/10.1007/978-3-030-38040-3_44

energy and bandwidth consumed, but is susceptible to duplicate packet transmissions. This features of TBC makes it lucrative for applications requiring high throughput and low processing power. Such applications usually occur in monitoring of environment, Air Quality, Agriculture fields, Life Stock intrusion etc. where numerous low cost motes (inheriting low processing speed and battery life) are necessity. Due to nature of such applications, energy consumption and delay becomes important parameters for selection of routing scheme.

Opportunistic routing satisfies Markov property (Memory less property of stochastic process) making it possible to model them using Markov chains. Numerous Discrete Time Markov Models (DTMC) [4] are proposed to mathematically analyze the performance of OR scheme. To best of our knowledge, none of them provides simple and elaborate method to calculate average energy consumption and average end go end delay. In this paper a method using Transition cost matrix is presented to calculate average energy consumption and average end to end delay required to transfer packet form source to destination. Existing Markov model with suitable modifications is used for analyzing these parameters.

2   Related Work

Various methods using DTMC have been proposed to model WSNs and other networks in general. Markov modeling is done for Throughput analysis of slotted CSMA/CA. super-frame structure, acknowledgements and retransmissions are taken into consideration while modelling [5]. DTMC and Green's function is used for modelling various wireless routing scheme, fixed point approximation is used for calculating transmission cost estimation [6]. DTMC model based on probability mass function is used for calculating average relay transmit power and network lifetime [7]. Recently, a model based on connection of various DTMC is proposed to calculate delay incurred by opportunistic routing in wireless cache networks [8].

3   Implementation of Proposed Technique

3.1   System Model

DTMC can be used for modelling Timer Based Opportunistic Routing Scheme (TB-ORS). Model proposed in this paper, shown in Fig. 1 is based on [9]. Proposed model is designed for linear network topology with each node having two NN. States $N_i = \{N_0, N_1, \cdots\cdots, N_n\}$ in the model represents network with n number of nodes. Sink node and failure to select next relay node (Due to packet drop) Are represented in the model as N_n and F, and are modelled as absorbing states.

Transition matrix from above model is called as T and is conveyed in canonical form as shown in Fig. 2. Transition probabilities in T are obtained from simulation of network whose parameters are to be analysed. Transition probabilities $S_{i,j} = \{S_{0,1}, S_{0,2}, \cdots\cdots, S_{n-1,n}\}$ in matrix T represents the probability of selecting next relay as node j when current relay is node i. Transition probabilities $D = \{D_0, D_1, \cdots\cdots, D_{n-1}\}$ in matrix T, represents the probability of failure in selecting next relay node.

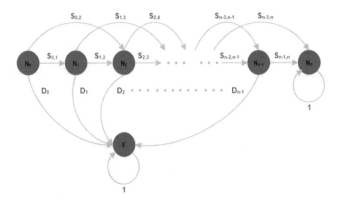

Fig. 1. System model

$$T = \begin{array}{c} \\ N_0 \\ N_1 \\ \vdots \\ N_{n-1} \\ N_n \\ f \end{array} \begin{array}{c} N_0 \quad N_1 \quad N_2 \quad \cdots \quad N_n \quad N_f \\ \left[\begin{array}{cccccc} 0 & S_{0,1} & S_{0,2} & \cdots & S_{0,n} & S_{0,f} \\ 0 & 0 & S_{1,2} & \cdots & S_{1,n} & S_{1,f} \\ \vdots & \vdots & \vdots & \cdots & \vdots & \vdots \\ 0 & 0 & 0 & \cdots & S_{n-1,n} & S_{n-1,f} \\ 0 & 0 & 0 & \cdots & 1 & 0 \\ 0 & 0 & 0 & \cdots & 0 & 1 \end{array} \right] \end{array}$$

Fig. 2. Transition matrix

3.2 Cost Calculation Method

Method to calculate average energy consumption and average end to end delay is not described in previous models. In this paper the above mentioned parameters are calculated using Transition cost matrix. Transition cost matrix is defined in next section and is expressed as C. For calculating the above mentioned parameters, information about initial state distribution is required. Initial state distribution is defined as $I_0 = [100 \cdots 0]$. Values in I_0 states that the probability of Markov model to be in N_0^{th} state is 100%. In the context of OR it means that initially the source node has data packet and is selected for relaying. To calculate average energy consumption and average end to end delay, distribution of state vector at l^{th} step is required. Value of l equals to the number of iterations required to obtain stead state and is given by (1) as

$$I_l = I_0 T^l \tag{1}$$

Expected cost for transmitting from l to $(l+1)^{th}$ state is given by (2) as

$$\mathcal{E}_{cost_l} = I_l H v \tag{2}$$

Where, H is the Hadamard product [10] of Transition matrix and Transition cost matrix and is given by (3) as

$$H = T \odot C \tag{3}$$

v Is a column vector composed of all 1's and is given by (4) as

$$v = [111 \cdots 1]^\tau \tag{4}$$

Cost for transmitting single packet from source to sink is calculated by summing expected cost required for n steps and is given by (5) as

$$\mathcal{A}_{cost} = \sum_{m=0}^{m=l-1} \mathcal{E}_{cost_m} \tag{5}$$

Total cost for transmitting N such packets is given by (6) as

$$\mathcal{T}_{cost_N} = N\mathcal{A}_{cost} \tag{6}$$

Substituting Eqs. (1), (2), (3) and (4) in Eq. (6) we get

$$\mathcal{T}_{cost_N} = N\left(\sum_{m=0}^{m=l-1} (I_0 T^m((T \odot C)v))\right) \tag{7}$$

3.3   Defining Transition Cost Matrix

$$C_\delta = \begin{array}{c} \\ \\ \\ \\ \\ \\ \end{array} \begin{array}{c} N_0 \\ N_1 \\ \vdots \\ N_{n-1} \\ N_n \\ f \end{array} \overset{\begin{array}{cccccc} N_0 & N_1 & N_2 & \cdots & N_n & N_f \end{array}}{\begin{bmatrix} 0 & C_{\delta_{0,1}} & C_{\delta_{0,2}} & \cdots & C_{\delta_{0,n}} & C_{\delta_{0,f}} \\ 0 & 0 & C_{\delta_{1,2}} & \cdots & C_{\delta_{1,n}} & C_{\delta_{1,f}} \\ \vdots & \vdots & \vdots & \cdots & \vdots & \vdots \\ 0 & 0 & 0 & \cdots & C_{\delta_{n-1,n}} & C_{\delta_{n-1,f}} \\ 0 & 0 & 0 & \cdots & 0 & 0 \\ 0 & 0 & 0 & \cdots & 0 & 0 \end{bmatrix}}$$

Fig. 3. Transition cost matrix for delay

In proposed Markov model cost required for transitioning from one state to another state is defined in transition cost matrix. Two transition cost matrix: C_∂ and C_ε are defined, as shown in Figs. 3 and 4 respectively, C_ε represents transition cost matrix for average energy consumption and C_∂ represents transition cost matrix for end to end delay. Cost of energy expended by the network for selecting next relay as node j when current relay is node i is represented in C_ε by elements $C_{\varepsilon_{i,j}} = \{C_{\varepsilon_{0,1}}, C_{\varepsilon_{0,2}}, \cdots\cdots, C_{\varepsilon_{n-1,n}}\}$. Cost of energy expended by network when none of the neighbour node is selected as relay is represented in C_ε by elements $C_{\varepsilon_{i,f}} = \{C_{\varepsilon_{0,f}}, C_{\varepsilon_{1,f}}, \cdots\cdots, C_{\varepsilon_{n-1,f}}\}$. Similarly $C_{\partial_{i,j}}$ and $C_{\partial_{i,f}}$ are delay cost represented in C_∂. Average energy consumption is calculated by substituting C_ε in eq. (7) and is given as,

$$E_{cost_N} = N\left(\sum_{m=0}^{m=l-1} (I_0 T^m((T \odot C_\varepsilon)v))\right) \qquad (8)$$

Similarly, Average end to end delay is given as,

$$D_{cost_N} = N\left(\sum_{m=0}^{m=l-1} (I_0 T^m((T \odot C_\partial)v))\right) \qquad (9)$$

$$
C_\varepsilon =
\begin{array}{c}
\\
N_0 \\
N_1 \\
\vdots \\
N_{n-1} \\
N_n \\
f
\end{array}
\begin{array}{cccccc}
N_0 & N_1 & N_2 & \cdots & N_n & N_f \\
\left[\begin{array}{cccccc}
0 & C_{\varepsilon_{0,1}} & C_{\varepsilon_{0,2}} & \cdots & C_{\varepsilon_{0,n}} & C_{\varepsilon_{0,f}} \\
0 & 0 & C_{\varepsilon_{1,2}} & \cdots & C_{\varepsilon_{1,n}} & C_{\varepsilon_{1,f}} \\
\vdots & \vdots & \vdots & \cdots & \vdots & \vdots \\
0 & 0 & 0 & \cdots & C_{\varepsilon_{n-1,n}} & C_{\varepsilon_{n-1,f}} \\
0 & 0 & 0 & \cdots & 0 & 0 \\
0 & 0 & 0 & \cdots & 0 & 0
\end{array}\right]
\end{array}
$$

Fig. 4. Transition cost matrix for energy

4   Results

Analytical results obtained from proposed cost calculation method is compared with simulation results of three different parameters from Energy saving via opportunistic routing algorithm (ENSOR) [11], Most forward progress algorithm (MFR) [12] and Optimum distance algorithm (OPD) [13] used for calculating waiting time in TB-ORS. OMNeT++ based Castalia network simulator is used for simulation. Up to 100 iterations are performed. Transition matrix and transition cost matrix values vary for every iteration.

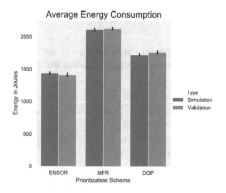

Fig. 5. Energy validation

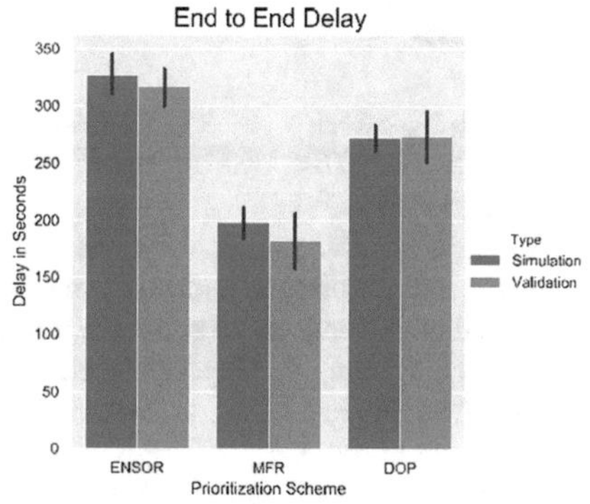

Fig. 6. Delay validation

Average energy consumption and average end to end delay is calculated for all three parameters and is shown in Figs. 5 and 6 respectively. Variance of difference between Analytical and Simulation results is calculated and are plotted as shown in, Figs. 7 and 8. From this results it can be observed that the difference between analytical and simulation results is negligible.

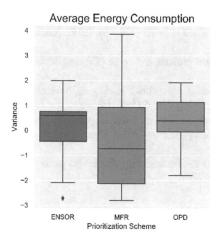

Fig. 7. Energy validation variance

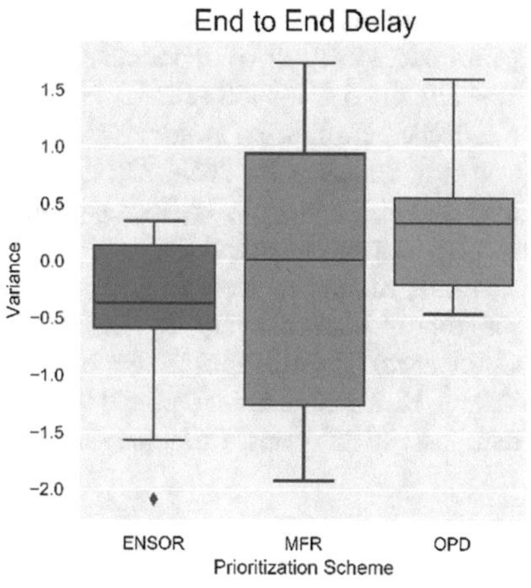

Fig. 8. Delay validation variance

5 Conclusion and Future Scope

In some WSN applications Energy and Delay play important factor in selection of routing scheme. Method developed in this paper precisely calculates average end to end delay and energy consumption. Analytical results obtained from this model can be used for comparing and validating the Simulation results of TB-ORS. This method is limited to analysis of TB-ORS, further modifications will enable it to analyze other types of OR schemes.

References

1. Babaee, E., Zareei, S., Salleh, R.: Best path cluster-based routing protocol for wireless sensor networks. In: 2013 UKSim 15th International Conference on Computer Modelling and Simulation (2013)
2. Shiming, H., Dafang, Z., Kun, X., Hong, Q., Ji, Z.: A simple channel assignment for opportunistic routing in multi-radio multi-channel wireless mesh networks. In: 2011 Seventh International Conference on Mobile Ad-hoc and Sensor Networks (2011)
3. Patel, T., Kamboj, P.: Opportunistic routing in wireless sensor networks: a review. In: 2015 IEEE International Advance Computing Conference (IACC) (2015)
4. Pierro, A.D., Wiklicky, H.: Probabilistic abstract interpretation: from trace semantics to DTMC's and linear regression. Semant. Logics, Calculi Lect. Notes Comput. Sci. **9560**, 111–139 (2015)
5. Jung, C., Hwang, H., Sung, D., Hwang, G.: Enhanced markov chain model and throughput analysis of the slotted CSMA/CA for IEEE 802.15.4 under unsaturated traffic conditions. IEEE Trans. Veh. Technol. **58**, 473–478 (2009)
6. Li, Y., Zhang, Z.-L.: Random walks and greens function on digraphs: a framework for estimating wireless transmission costs. IEEE/ACM Trans. Netw. **21**, 135–148 (2013)
7. Mousavifar, S.A., Leung, C.: Lifetime analysis of a two-hop amplify-and-forward opportunistic wireless relay network. IEEE Trans. Wireless Commun. **12**, 1186–1195 (2013)
8. Herath, J.D., Seetharam, A.: A markovian model for analyzing opportunistic request routing in wireless cache networks. IEEE Trans. Veh. Technol. **68**, 812–821 (2019)
9. Darehshoorzadeh, A., Grande, R.E.D., Boukerche, A.: Toward a comprehensive model for performance analysis of opportunistic routing in wireless mesh networks. IEEE Trans. Veh. Technol. **65**, 5424–5438 (2016)
10. Neudecker, H., Liu, S., Polasek, W.: The hadamard product and some of its applications in statistics. Statistics **26**, 365–373 (1995)
11. Luo, J., Hu, J., Wu, D., Li, R.: Opportunistic routing algorithm for relay node selection in wireless sensor networks. IEEE Trans. Ind. Inform. **11**, 112–121 (2015)
12. Yang, S., Zhong, F., Yeo, C.K., Lee, B.S., Boleng, J.: Position based opportunistic routing for robust data delivery in MANETs. In: 2009 IEEE Global Telecommunications Conference, GLOBECOM 2009 (2009)
13. Takagi, H., Kleinrock, L.: Optimal transmission ranges for randomly distributed packet radio terminals. IEEE Trans. Commun. **32**, 246–257 (1984)

Survey of Load Balancing Algorithms in Cloud Environment Using Advanced Proficiency

Dharavath Champla[1(✉)] and Dhandapani Siva Kumar[2]

[1] Department of CSE, Easwari Engineering College, Anna University,
Chennai, Tamilnadu, India
champla.805@gmail.com
[2] Department of ECE, Easwari Engineering College, Chennai, Tamilnadu, India
dgsivakumar@gmail.com

Abstract. Cloud computing has turned out to be well known because of its appealing highlights. The heap on the cloud is expanding hugely with the improvement of modern applications. Burden Adjusting is a crucial a piece of cloud domain setting that protects that each processor or devices execute equal quantity of labor in same quantity of your duration. Various types' calculations to the burden adjusting in distributed computing with the created intend to create cloud assets open end of the clients easily, intrigue. In this paper, we mean to give an organized, extensive diagram of the examination on burden adjusting algorithms in distributed computing. This paper reviews the cutting-edge burden adjusting apparatuses and systems. We gathering existing methodologies planned for give load balancing in a reasonable way. With this arrangement we give a simple and succinct perspective on the basic model embraced by every approach.

Keywords: Algorithms · Cloud computing · Load balancing · Proficiency

1 Introduction

Cloud computing gives adaptable approach to hold document & information's which includes virtualization, circulated figuring, & services of web. Additionally it has a few components like customer and dispersed servers. The point of distributed computing is to give most extreme administrations least cost whenever. These days, there is in excess of hundred more amounts of PC gadgets with associated to Internet. These gadgets present their solicitation and get the reaction immediately. Associate and approach the information from cloud at some random duration. The primary destinations of distributed to lessen value, upgrade reaction of time, give good execution; consequently, Cloud is additionally a pool of called administrations. Weight has different sorts like, CPU load, organize weight, stored limit problem and so forth. With regards to distributed computing, burden adjusting is share heap of virtual machines over total hubs to increase assets, administration usage, gives huge fulfillment of the clients. Because of burden distribution, each hub proficiently works; information been gotten and share immediately. Dynamic burden adjusting algorithm utilizes framework data while dispersing the heap. A powerful plan is increasingly adaptable and flaw tolerant.

© Springer Nature Switzerland AG 2020
J. S. Raj et al. (Eds.): ICIDCA 2019, LNDECT 46, pp. 395–403, 2020.
https://doi.org/10.1007/978-3-030-38040-3_45

Burden adjusting empowers advance system offices and assets for better reaction and execution. A few calculations are utilized adjust cloud information amid hubs. The entire client burden is taken care of the distributed supplier for easy machinery of administrations. In this way, the scheduled method will be utilized by cloud service provider (CSP).

Burden adjusting as typically connected to more measure, collection of data traffic, computing machine disperse job. Propelled models in distributed are received to accomplish hurry and proficiency. There is a various quality balancing of load, for example, equivalent work of division over the hubs, assistance of in accomplishing client fulfillment, enhance generally execution the framework, decrease reaction time, and give administrations to accomplish total asset usage. Diagram 2 demonstrates of the burden adjusting in distributed environment. For instance, on the off chance that we develop software about distributed several clients relied upon get to at a different point. In this way, reaction time to hundred individuals will be moderate and servers will end up occupied in all respects rapidly, bringing about moderate reaction and inadmissible clients. On the off chance that we proposed on our application burden adjusting, at that point job conveyed different hubs; Show signs of improvement reaction. The current study does not basically examine the accessible apparatuses and strategies in cloud computing that are utilized.

This paper gives in this paper an extensive outline of intuitive load balancing calculations in distributed computing. Every algorithm tends to various issues from various viewpoints and gives various arrangements. A few confinements previous calculations, execution problem, bigger preparing point, deprivation, constrained earth which point burden varieties not many and so on. A decent load balancing algorithm ought to maintain a strategic distance from the over stacking of one hub. In this purpose to assess presentation of distributing system burden adjusting calculations those are created more time. The rest of paper as sorted out, pursues. This part 2, we expect regarding survey numerous loads adjusting calculations. Part 3, said the exhibition assessment as varied distributed system calculations is examined & assessed with guides of various charts. Words, discoveries ar printed, therefore in this paper is finished up in part 4.

Diagram 1. Computing in cloud

2 Corresponding Works

The procedure wherein the heap is isolated among a few hubs of appropriated framework is called burden adjusting in distributed computing. Load adjusting helps algorithms through the distributed computing. Plenty of jobs have been success to regulate the heap therefore on improve execution and keep one's usages from the distance of assets. Completely various burden adjusting calc are talked concerning together with Min-Min, spherical robin (RR), Max-Min then on. Load equalization rule is isolated in 2 principle categories, particularly static and dynamic. Diagram 3 demonstrates the grouping of burden adjusting calculations. During this section, we have a tendency to provide a purpose by purpose exchange on this burden adjusting calculations for cloud.

I. Static Algorithm

These calculations depend on fulfillment time of an assignment. In this static algorithms choice around burden adjusting is set aside a few minutes. These are restricted to the earth where burden varieties are not many. These calculations aren't presents on the dependent state of the framework. In this static burden balance calculation separates servers can traffic as same. It is not utilize the framework knowledge whereas dispersing the heap and is a smaller amount amazing. Specific of the weight allotted to server. This server having most elevated weight gets a lot of associations comparatively. Employment is appointed by the capability of hub. Static calculation, energetic variation at run-time ar often no pondered. Aboard static, these algorithms do not be able to upset burden modifications at some point of run-time. Radojevic planned central load balancing decision model can be a development variety of spherical Robin calculation. This algorithmic rule works suitably during a framework with low type of burden. In CLBDM association job among user and hub is decided. Those algorithmic rules may be problematic as a result of unlooked-for circles. Static calculation moves just fix measure of information. It has no capacity to adaptation to non-critical failure.

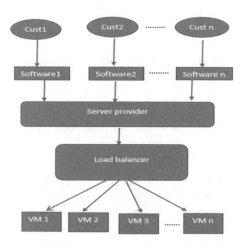

Diagram 2. In cloud load balancing

(a) Max – Min algorithm

This algorithmic program is same as Min-Min. Yet, max-min picks the trip with greatest value, provides for the actual system. When apportion the enterprise, automobile jobs as per refreshes. This allotted undertaking expel from chart. The picked hub and errands orchestrate during a explicit example refreshes concerning the ready time ar given by consolidating the period of time of the activity.

(b) OLB + LBMM

Wang et al. proposed a mix of load balancing min-min (LBMM) and opportunistic load balancing algorithms to improve the presentation of Undertakings. By this calculation, assets can be utilized all the more adequately and it builds the undertaking capability. All assignments are given to the hubs in a particular way. Its outcomes are superior to every other calculation & it's utilized in Load balancing min-min. Load balancing min-min working level three: level primary acts as solicitation administrator. LBMM figures out how long get, dispense of undertaking to support director. At point when the solicitation is gotten by administration chief, the errand is separated into lumps to accelerate the procedure. After that it apportions these lumps to the hub. The doling out of errands depends on accessible hubs, CPU limit and remaining memory. Errand finish, moving work the hub isn't deal with Opportunity LB. That is the reason the undertakings set aside much effort for consummation. Frequently demands are found in holding up rundown, till hubs become allowed to take different errands.

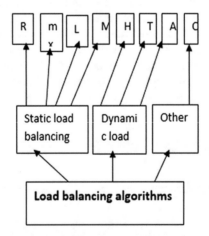

Diagram 3. Various types of Load balancing algorithms

(c) Round Robin Load Balancer

Round robin algorithm activity given by fixed amount of time. Fundamental accentuation to round robin is reasonableness, period restriction. This utilizes the round to be arranging of these gathered undertakings. It uses equivalent time to finish every undertaking. In the event of substantial Burden, RR catches complete to the long haul all the given undertakings. The event of bigger errands, it requires some finished investment. RR burdens ar similarly dispersed all VM. Hardly any impediments those calculation recommend that, accomplish superior, more than one customer associations ought not to begin simultaneously. The RR appears, it jobs in a roundabout example. Every hub is fixed with a period cut and plays out an assignment at assigned time on its turn. It is less intricate. Subsequently, at any minute some hub may have substantial burden and others may have no solicitation. In this manner, it isn't helpful for distributed computing. This issue was handled by weighted round robin where every hub is permitted to get explicit number of solicitations as indicated by the doled-out weight.

(d) Min – Min

Among every one of the assignments least tedious errand is search in the initial step. The undertaking is organizing, as indicated by that littlest time an incentive on machine. The current minute for different undertakings as additionally refreshed. Couples of trends identified with static burden adjusting are as per the following, Expected Time of Computer, Opportunity LB & Min Execution Time. Min-min looks good outcomes there where are little errands huge in number. Deprivation is noteworthy inconvenience. Variety in machine and assignments can't be anticipated through this calculation (Chart 1).

Chart 1. Advantage and Drawbacks of LB algorithms

Scheduling Algorithms	Advantages	Drawbacks
Statistic load balancing	Choice about burden adjusting is set aside a few minutes. Partitions the traffic similarly among the server Less structure	Constrained to the earth where burden varieties are not many Try not to have capacity to deal with burden changes throughput runtime
Max-Min	Prerequisites ar earlier familiar. Better works	Complete the job its take long time
Round robin	Quantum time fixed; Comprehend to easy; Reasonableness works well for small CPU crust Likewise utilized need (landing time and running time)	Takes long time for bigger tasks More occurrence switched context because of small time quantum Job ought to same to be accomplish superior
Min-Min	Littlest fulfillment value of time. Nearness of all the littler assignments, it presence good outcome	Machine Starvation and variation tasks cannot be anticipated
Dynamic load balancing	In run time work distribute; error resilience Only present condition required framework	Want consistent node checking More complicated considered
Honey bee	Response time minimize and increase throughput	Without VM can't work high priority works
Throttled	Performance good, manage the task used to list	Job needs to wait
Ant colony	Quicker data can be gathered by the ants; Minimizes make range; Independent errands; Computationally escalated	Long time to takes searching because of over headed network. Ant's quantity is not clarity
Carton	Reasonableness; Good execution; Equal conveyance of reactions Low correspondence is required	Depends upon lower cast

II. Dynamic Load Balancer

Dynamic rule depends on the varied resources of hubs, as an example, capacities & system information transfer capability. This would like steady checking the hub & square measure commonly onerous to actualize. Dynamic algorithms square measure acceptable in distributed computing condition since they flow into work run time and relegate cheap masses to the server. A lowest loaded server is search in system and favored by this rule. Dynamic algorithms square measure thought of more and more baffled. Ran planned WLC (weight least connections), novel burden adjusting rule for distributed computing. The WLC attach out errands supported range of associations for existing node. In distinctive burden leveling the heap circulates among the hubs in the course of run-time. On the off likelihood that heap balancer discovers high use of electronic equipment the solicitation is send to the subsequent hub. To affect the heap, current condition of the framework is employed. In powerful burden leveling, archives and knowledge is downloaded with at any confinement of specific memory. Its vantage approaches, once any hub is fizzled. In such circumstance, it does not stop the framework; simply its presentation is influenced.

(a) Honey bee algorithm

Dhinesh et al. proposed algorithm completed detail examination of scavenging conduct, honey bees. VM assigns a task when an under stacked, this modified count of need errands & heap of VM to different assignments in holding up rundown. Their VM pick here different procedure to causes this methodology. On the off chance that an errand has high need, at that point it chooses VM keeping least number of need assignments. It does not mull over just burden adjusting yet in addition monitors needs of errands which presently expelled from substantial stacked machines. It builds throughput and limits reaction time.

(b) Throttled LB Algorithm

VM search suitable of theory depends upon this algorithm. The job administrator gives records of VMs. By the list using, relevant machine allotted to requested user. Requested machine suitable to capability and size, at that point the activity is given to that machine. RR alg is not better than this alg.

(c) Ant colony

Various ant colony algorithms are additionally acquainting with parity the heap applying subterranean insect conduct for looking through nourishment. Bigger weight implies that asset has high calculation control. Load balancing ant colony optimization balance the heap as well as limits makes range. All assignments are thought to be commonly free and computationally serious.

(d) Carton

Combination of distributed rate limiting & Load Balancing is carton technology. Along Load Balancing, the servers have taken jobs fairly. While distributed rate limiting surly the same dissemination of assets. Outstanding burden is contemporarily dole out, improve the exhibition & extend the heap comparably to all of the servers. Carton alg can without much of a stretch be executed as low correspondence required.

Chart 2. Load balancing algorithms measures

LB Algorithms	Reasonableness	Overhead	Response time	Throughput	Fault tolerance	Speed	Resource utilization	Performance	Complexity
Static	Good	NA	Fast	High	No	High	Greater	Fast	Low
OLB+ LBMM	Not good	Less	Slow	High	No	Low	Greater	Fast	High
Round Robin	Good	Greater	Fast	High	No	NA	Greater	Fast	Low
Max-Min	Not good	Greater	Fast	High	No	Low	Greater	Fast	Low
Min-Min	Not good	Greater	Fast	High	No	High	Greater	Fast	Low
Dynamic	Not good	Greater	Slow	High	Yes	High	Greater	Slow	High
Honey bee	Not good	Less	Slow	High	No	High	Greater	Slow	Low
Throttled	Not good	Less	Fast	Low	Yes	High	Greater	Fast	Low
Ant colony	Not good	Greater	Slow	High	NA	High	Greater	Slow	Low
Carton	Good	NA	Fast	High	NA	High	Greater	Fast	High

3   Realization Estimate

Diagram 1 depicts the assessment of the examined Load leveling calculations towards varied specifications related decency, throughput, holding up duration. Chart 2, the correlation of those calculations shows advantage and disadvantage outcomes & that we call this term as high and low. As examined permeable varied calculations show varied outcomes. With the tip goal that, Static calculation accept cheap for disperse the heap. Yet, it's less incredible & not blame tolerant. Min-Min algorithmic rule is not cheap, defect tolerant. If there ought to be an occasion of very little assignments, it looks good outcome. The Max-Min, stipulations ar earlier noted. Therefore, it works higher and offers high output. Aboard this, dynamic burden adjusting needs simply current condition of the framework and has all the lot of overhead and adaptation to internal failure. Honey bee has slow response and high throughput. It's low overhead and execution since high want assignments cannot work while not VM machine. Subterranean insect province is easy calculation and fewer complexes. Instrumentality algorithmic rule needs low correspondence and its operating is affordable. Table two offers associate itemized examination {of varied of varied of assorted} algorithms over various parameters like decency, execution, speed, intricacy. we tend to dictate that, spherical Robin is increasingly productive as per following realities, spherical Robin accept cheap for convey the heap, it's high output, nice latency and fewer incredible than totally different calculations. The $64000 little bit of leeway of RR is time constraint and utilize equivalent amount to complete every endeavor.

4   Conclusion

In this paper, we've got introduced correlation of assorted load adjusting algorithms for cloud computing, for instance, round robin (RR), Min-Min, Max-Min, Ant colony, Carton, honey bee (4) then forth. We tend to delineated focal points and an impediment for these algorithms demonstrating brings regarding numerous conditions. The essential piece of this paper is correlation of assorted algorithms considering the attributes like reasonableness, throughput and adaptation to internal failure, overhead, execution, and interval and quality use. The constraint of existing work is that every distributed computing calculation doesn't address the connected problems like reasonableness, high turnout and correspondence. Future work is to moderate the higher than issue, and utilize the half and half ways that to influence accomplish higher execution and secure the framework.

References

1. Katyal, M., Mishra, A.: A comparative study of load balancing algorithms in cloud computing environment. Int. J. Distrib. Cloud Comput. 1(2) (2013)
2. Rajan, R.G., Jeyakrishnan, V.: A survey on load balancing in cloud computing environments. Int. J. Adv. Res. Comput. Commun. Eng. 2(12), 4726–4728 (2013)
3. Sakthivelmurugan, V., Saraswathi, A., Shahana, R.: Enhancedload balancing technique in public cloud. IJREAT Int. J. Res. Eng. Adv. Technol. 2(2), 1–4 (2014)
4. Aslam, S., Shah, M.A.: Load balancing algorithms in cloud computing: a survey of modern techniques. In: 2015 National Software Engineering Conference (NSEC) (2015)
5. Somani, R., Ojha, J.: A hybrid approach for vm load balancing in cloud using cloudsim. Int. J. Sci. Eng. Technol. Res. (IJSETR) 3(6), 1734–1739 (2014)
6. Li, K., Xu, G., Zhao, G., Dong, Y., Wang, D.: Cloud task scheduling based on load balancing ant colony optimization. In: Sixth Annual Chinagrid Conference, pp. 3–9, August 2011
7. Abraham, A.: Genetic algorithm-based schedulers for grid computing systems Javier Carretero, Fatos Xhafa. Int. J. Innov. Comput. Inf. Control 3(6), 1–19 (2007)
8. Haryani, N., Jagli, D.: Dynamic method for load balancing in cloud computing. IOSR J. Comput. Eng. 16(4), 23–28 (2014)
9. Kansal, N.J., Chana, I.: Cloud load balancing techniques: a step towards green computing. IJCSI Int. J. Comput. Sci. Issue 9(1), 238–246 (2012)
10. Chen, H., Wang, F.: User-priority guided min-min scheduling algorithm for load balancing in cloud computing. In: National Conference on IEEE Parallel Computing Technologies (PARCOMPTECH) (2013)
11. Cse, M.T.: Comparison of load balancing algorithms in a Cloud Jaspreet kaur. Int. J. Cloud Comput.: Serv. Arch. (IJCCSA) 2(3), 1169–1173 (2012)
12. Sran, N., Kaur, N.: Comparative analysis of existing load balancing techniques in cloud computing. Int. J. Eng. Sci. Invention 2(1), 60–63 (2013)
13. Subashini, S., Kavitha, V.: A survey on security issues in service delivery models of cloud computing. J. Netw. Comput. Appl. 34(1), 1–11 (2011)
14. Kaur, R., Luthra, P.: Load balancing in cloud computing. In: International Conference on Recent Trends in Information, Telecommunication and Computing, pp. 1–8. ITC (2014)
15. Chaczko, Z., Mahadevan, V., Aslanzadeh, S., Mcdermid, C.: Availability and load balancing in cloud computing. Int. Proc. Comput. Sci. Inf. Technol. 14, 134–140 (2011)
16. Roy, A.: Dynamic load balancing: improve efficiency in cloud computing. Int. J. Emerg. Res. Manage. Technol. 9359(4), 78–82 (2013)
17. Kaur, J., Kinger, S.: A survey on load balancing techniques in cloud computing. Int. J. Sci. Res. (IJSR) 3(6), 2662–2665 (2014)
18. Zhou, M., Zhang, R., Zeng, D., Qian, W.: Services in the cloud computing era: a survey. In: 4th International Universal Communication Symposium (IUCS) (2010)
19. Raghava, N.S., Singh, D.: Comparative study on load balancing techniques in cloud computing. Open J. Mob. Comput. Cloud Comput. 1(1) (2014)

20. Radojevis, B., Žagar, M.: Analysis of issues with load balancing algorithms in hosted (cloud) environments. In: 2011 Proceedings of the 34th International Convention. MIPRO, May 2011
21. Rajeshwari, B.S.: Comprehensive study on load balancing. Int. J. Adv. Comput. Technol. **3** (6), 900–907 (2014)
22. Randles, M., Lamb, D., Taleb-Bendiab, A.: A comparative study into distributed load balancing algorithms for cloud computing. In: 24th International Conference Advanced Information Network Application Workshops, pp. 551–556. IEEE (2010)
23. Kumar, D.: Review on task scheduling in ubiquitous clouds. J. ISMAC **1**(01), 72–80 (2019)
24. Shameem, P.M., Shaji, R.S.: A methodological survey on load balancing techniques in cloud computing. Int. J. Eng. Technol. (IJET) **5**(5), 3801–3812 (2013)
25. Abdullah, M., Othman, M.: Cost-based multi-qos job scheduling using divisible load theory in cloud computing. Procedia Comput. Sci. **18**, 928–935 (2013)

Comparative Study on SVD, DCT and Fuzzy Logic of NOAA Satellite Data to Detect Convective Clouds

B. Ravi Kumar[✉] and B. Anuradha

Department of Electronics and Communication Engineering,
S.V.U.C.E, Tirupati, India
rav.kumar37@gmail.com

Abstract. Information available in the Infrared and visible NOAA (National Oceanic Atmosphere Administration) satellite imagery is in the form of brightness temperature and albedo. Three Techniques are constructed to retrieve the convective cloud portion using Singular Value Decomposition (SVD), 2D-DCT (Discrete Cosine Transform) and fuzzy logic methods provided within MATLAB®, and compared the accuracy of these methods. The results delineates that there is tradeoff between accuracy of effective cloud amount and execution time. Though the fuzzy logic method takes the time to train the data but gives the accurate detection and classification of clouds

Keywords: SVD · 2D-DCT · Fuzzy logic · NOAA data

1 Introduction

Now a day's analysis of weather condition is quite simple due to the fact that the huge data of satellite imageries are generated at receiving stations. Reliable Interpretation of cloud detection and classification has been needed with advanced modeling. Therefore, processing of satellite imagery for cloud classification is to inference the cloud physical properties for climatological applications like rain estimation and cyclone prediction.

Cloud types can be segregated from earth surface based on height they exist or develop their vertical profile. The cumuliform clouds are vertically strong enough, whereas stratus cloud appears horizontally. Cirrostratus, cirrocumulus comes under high altitude clouds, altostratus and altocumulus are found in the middle altitude and one can find cumulus in the low-altitude. Cumulonimbus is spread from low to high altitude.

Generally clouds are determined by a lower temperature values and higher reflectance values than remaining earth surface. So simplest way of detection of clouds in visible and infrared window are threshold approach [1, 10].

Although NOAA infrared and visible images are more acquainted with cloud data, but cloud properties varied with different atmosphere conditions and geographical locations. Earlier, there are many basic approaches to detect the clouds from satellite data which are threshold method, spectral ratio, brightness temperature difference (BTD). BTD is a best method for identifying optical thick clouds from optically thin cirrus clouds. Optically thick cumulus clouds have smaller BTD than thin cirrus clouds [2, 11].

© Springer Nature Switzerland AG 2020
J. S. Raj et al. (Eds.): ICIDCA 2019, LNDECT 46, pp. 404–409, 2020.
https://doi.org/10.1007/978-3-030-38040-3_46

Perhaps threshold approaches are not performed well in cloud classification of multispectral satellite images then several classification approaches namely supervised classification-means clustering maximum likelihood and support vector machine (SVM) are used to classify the clouds on multispectral data [3, 4, 12].

In this paper mainly emphasis on comparative study of three different approaches namely SVD, 2D-DCT and Fuzzy logic for the same satellite data.

2 Methods Used

2.1 Methodology of SVD Applied to AVHRR NOAA Data

SVD is well known decomposition of a matrix A and factorization of matrix A into the product of three matrices $A = USV^T$ where the S is Diagonal square matrix (r x r) with real positive values and columns of U (n × r) and V (r× d) are orthonormal matrices [5].

Properties of SVD
The first property of SVD applied to the NOAA satellite image is shown in Fig. 2a and distribution of Eigen values observed in S matrix. First diagonal value of matrix S for cloud portion is dominant and rest of singular values are diminished to zero, whereas in case of snow, land and ocean first singular value is less compared to cloud value and energy for remaining singular values are spread to last value [6].

2.2 Methodology of DCT Applied to AVHRR NOAA Data

Consider the f(m, n) is 2D image of size M × M, the 2D-DCT of the image f(m, n) is given by Eq. (1) (Fig. 1).

$$F(k,l) = \alpha(k)\alpha(l) \sum_{m,n=1}^{M-1} f(m,n) \left(\cos\left[\frac{(2n+1)\pi k}{2M}\right] \cos\left[\frac{(2n+1)\pi l}{2M}\right] \right) \quad (1)$$

where $\alpha(K) = \sqrt{\frac{1}{M}}$ if $k = 0$; $\sqrt{\frac{2}{M}}$ if $k \neq 0$

Similarly $\alpha(l) = \sqrt{\frac{1}{M}}$ if $l = 0$; $\sqrt{\frac{2}{M}}$ if $l \neq 0$

2.3 Methodology of Fuzzy Logic Applied to AVHRR NOAA Data

In this method fuzzy logic approach is applied to NOAA satellite data with spatial resolution of 1.1 km^2 to detect clouds. This fuzzy logic system follows triangular, Gaussian and trapezoidal membership functions as inputs and outputs of membership function [7–9] (Table 1).

2.4 Accuracy Measurements

Accuracy measurement for convective cloud coverage in the output of above three different methods is estimated from Confusion matrix.

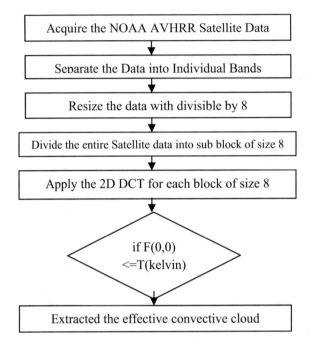

Fig. 1. Flow chart to extracted clouds of NOAA satellite visible and TIR data using 2D-DCT

Table 1. Fuzzy logic membership function for three spectral bands

Class	Visible (K)	NIR (K)	TIR (K)
Convective clouds	(500 700)[a]	(700 850)[a]	(−80 −50)[a]
Middle level clouds	(300 500)[a]	(500 700)[a]	(−50 −20)[a]
Low level clouds	(200 250 300)[b]	(200 300 400)[b]	(−20 −10 0)[b]

[a]Gaussian membership function; [b]Triangular membership function

2.4.1   Confusion Matrix

A confusion matrix is a table in which the performance of a classifier on a set of known data. It is a summary of predicted correct and incorrect values which will makes us more insight errors made by methods [13].

Table 2. Confusion matrix

	Class 1 (predicted)	Class 2 (predicted)
Class 1 (Actual)	True positive (TP)	False negative (FN)
Class 2 (Actual)	False positive (FP)	True negative (TN)

Accuracy of predicted image is calculated from the confusion matrix parameters as shown in Table 2 and is given by

$$Accurcy = \frac{TP + TN}{TP + TN + FP + FN}$$

3 Comparative Results

This paper discussed about convective cloud detection methods using the NOAA AVHRR satellite data available on 13th Dec 2018, at 15:30PM is shown in Fig. 2a. The rain baring cloud output of SVD, 2D-DCT and Fuzzy logic methods as show in Fig. 2(b–c–d) respectively. The ground truth image was taken from ERDAS software provided with temperature and albedo values to NOAA multispectral data. Comparative study of three different cloud detection methods is as shown in Table 3. SVD method is better to find cloud pixels from non-cloud pixels and take less time to process the data but accuracy of cloud convective coverage is less due to varying the top surface of the clouds. 2D-DCT and statistical parameters method improves the accuracy to cover convective thick cloud portion but discrimination of clouds is quite difficult compare to SVD method. Fuzzy logic system introduces the more cloud

Table 3. Comparative study between SVD, 2D-DCT & statistical parameters and Fuzzy logic methods for cloud detection for NOAA AVHRR on 13th Dec 2018; 15:30PM

Method name	Confusion Matrix			Accuracy of Convective Cloud coverage
SVD	N=16,384	Predicted yes	Predicted no	75.4%
	Actual yes	8,250(50.3%)	1,250(7.6%)	
	Actual no	2,780(16.9%)	4,104(25.0%)	
	N=total no .of pixels in selected area			
2D-DCT and statistical parameters(mean, standard deviation, entropy, skewness)	N=16,384	Predicted yes	Predicted no	84.9%
	Actual yes	10,298(62.8%)	1,230(7.5%)	
	Actual no	1,232(7.5%)	3,624(22.1%)	
Fuzzy logic	N=16,384	Predicted yes	Predicted no	91.3%
	Actual yes	11,820(72.1%)	412(2.5%)	
	Actual no	1,002(6.1%)	3,150(19.2%)	

physical properties with triangular and Gaussian membership functions and better to get convective cumulonimbus clouds and accuracy improved compare to SVD and statistical parameters but it takes lot of time process data.

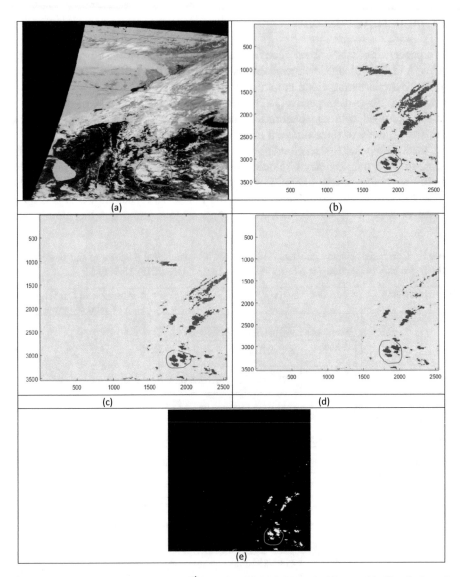

Fig. 2. NOAA Satellite images on 13th Dec 2018 (a) Multispectral image (b) Clouds detection using 2D-DCT Top (c) Cloud detection suing SVD (d) Cloud detection using fuzzy logic (e) ground truth image

4   Conclusion

In this paper, a comparative study is conducted on SVD, 2D-DCT and Fuzzy Logic methods for effective convective cloud detection from the multispectral NOAA satellite imagery. It was found that SVD has taken less execution time (64.89 s) and effective cloud coverage is 75.4%. This SVD method is extremely well to discriminate the cloud pixels from snow, land and ocean pixels but accuracy of convective cloud is less due to fact that similar characteristics of thick cirrus clouds as that of convective clouds. 2D-DCT and statistical parameters method provide effective cloud coverage with accuracy is 84.9% and execution time (110.06 s) greater than SVD. Fuzzy logic provides accuracy of 91.3% which is better than SVD and spatial distribution parameters but elapsed time to execute is more. Discrimination of different clouds can be best achieved from Fuzzy logic than SVD and DCT.

Acknowledgements. The data used in this work was browsed from Center of Excellence (CoE), Department of ECE, S V University college of Engineering, Tirupati. Authors expressed their gratitude to the CoE for data and UGC for providing Research fellowship.

References

1. Ackerman, S., et al.: Discriminating clear-sky from cloud with MODIS algorithm theoretical basis document (MOD35). MODIS cloud mask team, cooperative institute for meteorological satellite studies, University of Wisconsin (2010)
2. Jedlovec, G.J., Haines, S.L., La Fontaine, F.J.: Spatial and temporal varying thresholds for cloud detection in GEOS imagery. IEEE Trans. Geosci. Remote Sens. **46**(6), 1705–1717 (2008)
3. Tso, B., Mather, P.: Classification Methods for Remotely Sensed Data, 2nd edn. Taylor and Francis Group, London (2009). Chapter 2-3
4. Li, P., Dong, L., Xiao, H., Xu, M.: A cloud image detection method based on SVM vector machine. Neuro Comput. **169**, 34–42 (2015). https://doi.org/10.1016/j.neucom.2014.09.102
5. Baker, K.: Singular value decomposition tutorial, January 2013
6. Ravi Kumar, B., Anuradha, B.: Detection of clouds using SVD and spectral properties for NOAA AVHRR imagery. i-Manag. J. Image Process. **4**, 10–15 (2017)
7. Kim, K.B., Woo, Y.W.: Cloud analysis using a fuzzy reasoning method. J. Korea Inst. Marit. Inf. Commun. Syst. **13**(6), 1181–1187 (2009)
8. Ravi Kumar, B., Anuradha, B.: Fuzzy logic based convective cloud detection from the Kalpana data. Int. J. Eng. Technol. **7**, 316–318 (2018)
9. Suseno, D.P.Y., Yamada, T.J.: Two dimensional threshold based cloud type classification using MTSAT data. Remote. Sens. Lett., 1–10 (2012)
10. Roca, R., Viollier, M., Picon, L., Desbois, M.: A multi-satellite analysis of deep convection and its moist environment over the Indian Ocean during the winter monsoon. J. Geophys. Res. Atmos. **107**(D19), INX2-11 (2002)
11. Foga, S., et al.: Cloud detection algorithm comparison and validation for operational Landsat data products. Remote Sens. Environ. **194**, 379–390 (2017)
12. Taravat, A., Del Frate, F., Cornaro, C., Vergari, S.: Neural networks and support vector machine algorithms for automatic cloud classification of whole-sky ground-based images. IEEE Geosci. Remote Sens. Lett. **12**, 666–670 (2015)
13. https://www.geeksforgeeks.org/confusion-matrix-machine-learning

Question Classification for Health Care Domain Using Rule Based Approach

Shubham Agrawal[✉] and Nidhi Mishra

Poornima University, Jaipur, Rajasthan, India
s.agrawal781@gmail.com, Nidhi.mishra@poornima.edu.in

Abstract. Question Classification (QC) System for Health Care domain is a specific component of Health Care Question Answering System (QAS). A Question Classification System implementation is usually a computer program that may find its answer type of the question by input the question. The QC system for Health Care domain is very important role in medical field. The Experiment on 427 health questions collected from many different web sites. We have used 9 type of health questions like what, when, how, how many, who, which, why, how much, where, when i.e. that we have get after question preprocessing. After preprocessing we have create question classifier and the experimental result overall accuracy of our system is achieved 80.79% accuracy. Future work has also been discussed as proposed system can be designed a healthcare based question answering system and as the targeted system mismatched in giving answers type and some fruitful efforts will be made so that this drawback of the proposed system can be resolve.

Keywords: Question Classification · Machine learning · Rule based approach · Question answering system · Answer Extraction algorithms

1 Introduction

Question Classification (QC) System is a specific type of Question Answering System (QAS). Today, in the world of the internet, information stored in the large repositories or in the form of large documents. The big Problem we face is to retrieve specific/needed information from the large text documents or repositories available on the web which cannot be managed with out automatic search. This problem can be solved by using Question answering System but first initial step can be solved Question classification. The main goal of Question classification system is generate answer type of the question.

A Question Classification System implementation is usually a computer program that may find its answer type of the question by using input the question in text.

Question Classification Systems have inherited many techniques from machine learning, information retrieval, and NLP. The QC has valuable role in QAS. Classification is to label a question into a class that represents the answer type. The QC system has represented both syntactic and semantic structure. The QC system is based on rule based method. We have made some rules for QC system.

© Springer Nature Switzerland AG 2020
J. S. Raj et al. (Eds.): ICIDCA 2019, LNDECT 46, pp. 410–419, 2020.
https://doi.org/10.1007/978-3-030-38040-3_47

Questions are two type categorized. It is 'factoid and Non-Factoid'. The simple type question is factoid and the complex question is Non-Factoid. The taxonomy of question type is a predefined categories set. It has consist of six course and set of fine grained classes such as abbreviation, descriptive, entity, human, location and numeric [1].

In the experiment, we have collected 427 question based health care from web sites. We have used 9 type questions like what, when, how, how many, who, which, why, how much, where, when i.e. We create five rules for find the answer type of questions for medical based question. They used QC system doctor and that obtained simply answer that time treatment. The QC system is valuable for Health care domain.

2   Literature Survey

QC is very popular topic of this era. Kim et al., has developed a new Question Answering architecture that used sentences. Question that are automatically generated and in which used MySQL DBMS tool. Gaikwad et al. proposed AGRI-QA System is based Agriculture and Authors analyzed 100 questions and used Lucence tool and for input authors used different website. The System achieved the accuracy is 69% [2, 3].

Topez et al. proposed the QAS based ontology and used Ontological approach. Authors analyzed 10 EGP task and finally achieved the accuracy of 64%. Toti et al. proposed the AQUEOS QAS that gaves best result only for System simple question correct answer. Used Lucence tool. The System achieved the accuracy is 100% [6, 8].

Chien Ta et al. proposed Algorithm for an information extraction system for all types of Academic Library domain. Authors analyzed 2000 questions and for input authors used ACM digital library and used Stanford lexical dependency parser tool. Wang et al. proposed an An automated QAS for Mobile service consulting area domain. Authors analyzed 160000 questions and for input authors used MSC dataset, and achieve the accuracy 82% [7, 9].

Chen et al. designed an FAQ system for Education domain. They used Concept Hierarchy approach for Accuracy of system improved. Authors analyzed 12481 questions. Jamgade et al. proposed IR system and concluded that Semantic search technique. The resulted system was effective for simple queries. They used Stanford parser tool [4, 15].

Agrawal et al. proposed Automated Question Answering system for IPC section and law domain. The Proposed question answering system give suitable solution. They used input different website [26].

Vijeta et al. proposed Medical QAS for medical domain using Preprocessing, ranking. Authors used Factiod type question. The question gained 70% accuracy. They used internet input dataset and They used Standford parser toolkit for experiment works [11].

Jiang et al. designed Conversational agent System for Singapore tourism domain using NLP. Authors used Factiod type question. They Proposed system show good result. They used for input different web site and used Lucence and Apollo toolkit for experimental work [24].

Liu et al. Measure entities semantic relationship using ODP search interface. Authors analyzed 30 questions. They Proposed very effective semantic similar b/w real world entities. They used for input different web site and they used NER tool for experimental results [14].

3 Architecture of Question Classification for Health Care System

In above diagram first we used web site for collected input question than create offline text input question. Than third step in which question pre-processing are performed like tokenization, stop word removal, stemming, POS tagging. After next step select the keyword first and second keyword than next step question classifier used created rule find the answer type of the question (Fig. 1).

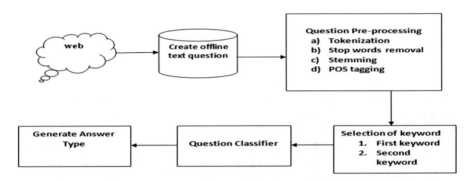

Fig. 1. Architecture of question classification for health care system

4 The Taxonomy of Question Type

The taxonomy of question type is fifty fines and six coarse grained classes that are show as below in Table 1. These sets are predefined.

Table 1. Coarse and fine grained classes

Coarse	Fine
Abbreviation	abbreviation, expression.
Human	group, individual, title, description.
Descriptive	definition, description, reason.
Numeric	code, count, date, distance, money, period, percent, speed, temperature, size.
Location	city, country, mountain, other , state.
Entity	animal, body, word, color, creative, currency, disease.

5 Methodology

In the proposed work, we have taken health-based question in text file using different web sites. Moreover, the each question is tokenized into words then after tokenized removed the stop word then perform stemming. After stemming, the word has tagged. After then question classification select the first keyword and second keyword then perform rule using rule-based approach and generate the answer type given input questions. The rule based approach rules are:

Rule: 1 if question first word tagged with "WP" or "WRB" then question is "descriptive". In which rule are question type category are like what, how, why fulfills. We used tagged word WP is pronoun and WRB is adverb. Example-

(a) What is the covering of heart?
(b) How is a deviated septum treated?
(c) How can I relieve neuropathy pain?
(d) Why do people take iron?
(e) Why do people take turmeric?

The above all questions are generally "descriptive" category.

Rule: 2 if question first word tagged with "WP" and second word tagged "VBD" then question is "human". In which rule are question type category are like 'who' fulfills. We used tagged word WP is pronoun and VBD is verb. Example-

(a) Who discovered penicillin?
(b) Who discovered DNA?
(c) Who discovered cell in 1665?

The above all questions are generally "Human" category.

Rule: 3 if question first word tagged with "WRB" and second word tagged "JJ/NN" then question is "Numeric". In which rule are question type category are like how many, how much, when fulfills. We used tagged word JJ/NN is adjective/noun and WRB is adverb. Example-

(a) How many facial muscles are involved in speaking?
(b) How many ribs in the human thorax?
(c) How much protein should I eat if I have HIV?
(d) How much blood is in the average human body?
(e) When was first robotic surgery done?
(f) When is world cancer day celebrated?

The above all questions are generally "Numeric" category.

Rule: 4 if question first word tagged with "WDT" and second word tagged "JJ/RB" then question is "descriptive". In which rule are question type category are like 'which' fulfills. We used tagged word JJ/NN is adjective/noun and WDT is determiner. Example-

(a) Which acid is secreted from stomach?
(b) Which is a communicable disease?

The above all questions are generally "descriptive" category.

Rule: 5 if question first word tagged with "WRB" and second word tagged "JJ/NN" then question is "Location". In which rule are question type category are like 'where' fulfils. We used tagged word JJ/NN is adjective/noun and WRB is adverb. Example-

(a) Where is the tibia found?
(b) Where on the human body is the mandible located?

The above all questions are generally "Location" category.

6 Flowchart for the Proposed Question Classification for Health Care System

See Fig. 2

Algorithm :
Step 1: Start
Step 2: User input the question.
Step 3: Tokenization the input question on that breaks the question into token.
Step 4: Remove the stop word after tokenization
Step 5: Perform the stemming of the word and convert into root word.
Step 6: Perform the POS tagging.
Step 7: After POS tagging question classification perform that generate the answer type according to the following case.

    Case1: Question type = "Why"
            Return ("Descriptive");
    Case2: Question type = "Who"
            Return ("Human");
    Case3: Question type = "What"
            Return ("Descriptive");
    Case4: Question type = "Where"
            Return ("Location");
    Case5: Question type = "How"
            Return ("Descriptive");
    Case6: Question type = "How many"
            Return ("Numeric");
    Case7: Question type = "How much"
            Return ("Numeric");
    Case8: Question type = "Which"
            Return ("Descriptive");
    Case9: Question type = "When"
            Return ("Numeric");

Step 8: End

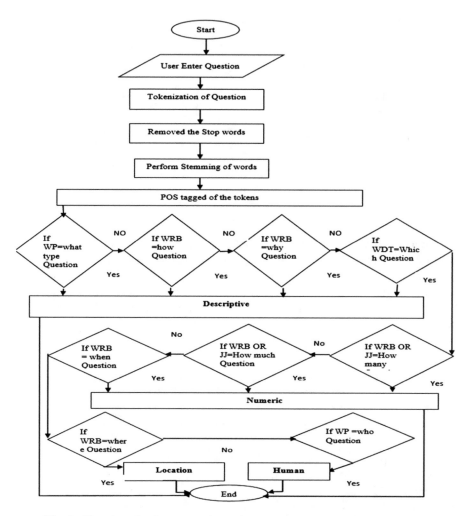

Fig. 2. Flowchart for the proposed question classification for health care system

7 Evaluation and Result

For the performance of our system, we developed input set of 427 questions. We calculated accuracy for each question type. They are show as performance measurement of the system. These are show as below Table 2.

$$\text{Accuracy} = \frac{\text{Question correctly answered type} * 100}{\text{Total number of question}}$$

Table 2. Accuracy performance of health care questions classifier

Question type	Total question	Matched	Mismatched	Accuracy
What	142	106	36	74.64%
How	70	57	13	81.42%
How many	20	20	0	100%
How much	4	4	0	100%
Who	43	36	7	83.72%
When	18	13	5	72.22%
Which	72	53	19	73.61%
Where	19	17	2	89.47%
Why	39	39	0	100%

From the Table 2 Total Question, matched and accuracy rate for each question type were calculated accuracy for health care question classification system. The category to questions for the average performance has accuracy 86.12%. In which highest accuracy is 100% for the classifying Why type, how many and how much of question. The lowest accuracy was when type 72.22% and which type 73.61%. The main difficult question type classifying was what type that accuracy of 74.64%. The chart for the data given in Table 3 can be shown in Fig. 3 given below.

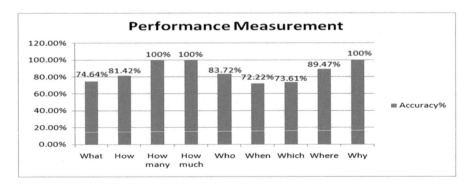

Fig. 3. Performance measurement

Total 427 questions were experimented to prove the overall system performance and experimental result is shown in the Table 3 given below.

Table 3. Overall System performances

S. No.	Total no. of question	Matched	Mismatched
1	427	345	82

Table 3 shows that out of 427 questions 345 questions generated correct answer matched while 82 questions were found mismatched which means that the questions did not fulfill the system requirement. However, user may get fluctuations in the accuracy rate according to the entered question. The chart for the data given in Table 3 could be shown in Fig. 4 given below.

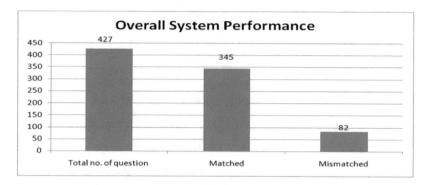

Fig. 4. Overall System performances

Above Fig. 4 the overall system performance based on correct response observed. The system performance is measured accuracy rate are approximately 80.79%. Total correct generate answer type of the system is observed as 80%.

8   Conclusion

Question classification for healthcare domain using rule based approach and found out current challenges and scope of work in the area. They used created rules for find the answer type of question and then second step that simply resolved. It could be found that most of the researchers used Answer Extraction algorithms for extracting the precise and to find the expected answer type. We evaluated our system for take 427 health based question then achieved an accuracy of 80.79%. Future work has also been discussed as proposed system can be designed a health care based QAS and as the targeted system mismatched in giving answers type and some fruitful efforts will be made so that this drawback of the proposed system can be resolve.

References

1. Dodiya, T., Jain, S.: Question classification for medical domain question answering system. In: 2016 Pacific, IEEE International WIE Conference on Electrical and Computer Engineering (WIECON-ECE) AISSMS, Pune, India, 19–21 December 2016
2. Kim, M., Kim, H.: Design of question answering system with automated question generation. In: Fourth International Conference on Networked Computing and Advanced Information Management, Washington, USA, pp. 365–368 (2008)

3. Gaikwad, S., Asodekar, R., Gadia, S.: AGRI-QAS question-answering system for agriculture domain. In: IEEE International Conference on Advances in Computing, Communications and Informatics (ICACCI), pp. 1474–1478 (2015)
4. Jamgade, A.N., Karale, S.J.: Ontology based information retrieval system for academic library. In: 2nd IEEE International Conference on Innovations in Information, Embedded and Communication systems (ICIIECS), pp. 1–6 (2015)
5. Parmar, A., Desai, T.: Survey on different approaches of question answering system. Int. J. Adv. Res. Eng. Sci. Manag. **1**(5) (2015). ISSN: 2394-1766
6. Topez, V.: Ontology-based Inference for Information-seeking in natural language dialog system. In: Proceedings of the 16th IEEE International Symposium on Robot and Human Interactive Communication, pp. 178–181 (2007)
7. Chien Ta, D.C.: Identifying semantic and syntactic relations from text documents. In: The Proceedings of IEEE RIVF International Conference on Computing & Communication Technologies Research, Innovation, and Vision for Future (RIVF), Can Tho, pp. 127–131 (2015)
8. Toti, D.: AQUEOS: a system for question answering over semantic data. In: IEEE International Conference on Intelligent Networking and Collaborative Systems, pp. 716–719 (2014)
9. Wang, D.S.: A domain-specific question answering system based on ontology and question templates. In: The Proceedings of 11th IEEE ACIS International Conference on Software Engineering, Artificial Intelligence, Networking and Parallel/Distributed Computing, pp. 151–156 (2010)
10. Najmi, E., Hashmi, K., Khazalah, F., Malik, Z.: Intelligent semantic question answering system. In: The proceedings of IEEE International Conference on Cybernetics, pp. 255–260 (2013)
11. Vijeta, B.V.: A restricted domain medical question answering system. Int. J. Sci. Res. (IJSR) **3**(5), 2319–7064 (2014)
12. Madabushi, H.T., Lee, M.: High accuracy rule-based question classification using question syntax and semantics. In: Proceedings of COLING 2016, the 26th International Conference on Computational Linguistics: Technical Papers, Osaka, Japan, 11–17 December 2016, pp. 1220–1230 (2016)
13. Supriana, I., Purwarianti, A., Suwarningsi, W.: Estimation question type analyzer for multi close domain indonesian QAS. Int. J. Res. Sci. Manag. **4**(6) (2017). ISSN: 234-5197
14. Liu, J., Birnbaum, L.: Measuring semantic similarity between named entities by searching the web directory. In: IEEE/WIC/ACM International Conference on Web Intelligence, Washington, USA, pp. 461–465 (2007)
15. Chen, L., Shen, R.: FAQ system in specific domain based on concept hierarchy and question type. In: The Proceedings of IEEE International Conference on Computational and Information Sciences, pp. 281–284 (2011)
16. Pizzato, L.A., Molla, D.: Indexing on semantic roles for question answering. In: The Proceedings of the 2nd workshop on Information Retrieval for Question Answering (IR4QA), Manchester, UK, pp. 74–81, August 2008
17. Devi, M., Dua, M.: ADANS: an agriculture domain question answering system using ontologies. In: International Conference on Computing, Communication and Automation (ICCCA2017). IEEE (2017)
18. Mollaei, A., Rahati-Quchani, S., Estaji, A.: Question classification in Persian language based on conditional random fields. In: 2nd International eConference on Computer and Knowledge Engineering (ICCKE) (2012)

19. Lee, M., Cimino, J., Zhu, H.R., Sable, C., Shanker, V., Ely, J., Yu, H.: Information retrieval-medical question answering. In: Symposium Proceedings/AMIA Symposium, February 2006
20. Sarrouti, M., Lachkar, A., El Alaoui Ouatik, S.: Biomedical question types classification using syntactic and rule based approach. In: KDIR 7th International Conference on Knowledge Discovery and Information Retrieval (2017)
21. Van-Tu, N., Anh-Cuong, L.: Improving question classification by feature extraction and selection. Indian J. Sci. Technol. **9**, 1–8 (2016)
22. Biswas, P., Sharan, A., Kumar, R.: Question classification using syntactic and rule based approach. In: International Conference on Advances in Computing (2014)
23. Huang, P., Bu, J.: Learning a flexible question classifier. In: IEEE International Conference on Convergence Information Technology, pp. 1608–1613 (2007)
24. Jiang, R., Banchs, R.E., Kim, S., D'Haro, L.F., Niculescu, A.I., Yeo, K.H.: Design and evaluation of a conversational agent for the touristic domain. Configuration of dialogue agent with multiple knowledge sources. In: Signal and Information Processing Association Annual Summit and Conference (APSIPA), pp. 840–849. Asia-Pacific (2015)
25. Jayalakshmi, S., Sheshasaayee, A.: Question classification: a review of state-of-the-art algorithms and approaches. Indian J. Sci. Technol. (2015)
26. Agrawal, A.J., Kamdi, R.P.: Domain specific question answering system. Int. J. Electr. Electron. Comput. Syst. (IJEECS) **3**(2), 2347–2820 (2015)
27. Gunawardena, T., Pathirana, N.: performance evaluation techniques for an automatic question answering system. Int. J. Mach. Learn. Comput. **5**(4), 294–300 (2015)
28. Tahri, A.: DBPEDIA based factoid question answering system. Int. J. Web Semant. Technol. **4**(3), 23–38 (2013)
29. Heiner, C., Zachary, J.L.: Improving student question classification. Educ. Data Min. (2009)
30. Li, F., Zhang, X., Yuan, J., Zhu, X.: Classifying what-type questions by head noun tagging. In: Proceedings of the 22nd International Conference on Computational Linguistics (Coling 2008), Manchester, pp. 481–488, August 2008

A Conjoint Edifice for QOS and QOE Through Video Transmission at Wireless Multimedia Sensor Networks

S. Ramesh[1]([✉]), C. Yaashuwanth[1], and Prathibanandhi[2]

[1] Department of Information Technology,
Sri Venkateswara College of Engineering (Autonomous), Sriperumbudur, India
swami.itraj@gmail.com, yaashuwanth@gmail.com
[2] Department of Electrical and Electronics Engineering,
Anna University, Chennai, India
prathiraj90@gmail.com

Abstract. The Myth of distributing multimedia content came to reality with the emancipation of Wireless Multimedia Sensor Networks (WMSNs). The proper management is needed to avoid the excessive packet drop during transmission of multimedia data over WMSNs. Existing QoS does not lead to increase in sensor nodes and volume of data in wireless sensor networks. This paper proposes to emphasis the QoS in WMSNs inspite of insubstantial flimsy Error Concealment (EC) scheme. The sustainable quality for retaining the receiving ends is the prime core for the Quality of Experience (QoE). The proposed key objectives of the edifice are to reduce the effects of leaped video packets and to maximize its network. The variable Quantization Parameters (QPs) are used to control the data rate at the multimedia sensor with Scalable High-efficiency Video Coding (SHVC). The real-time video transmission is exploited by the multipath routing. Experimental outcome reveals that the proposed edifice able to proficiently regulate network distortions under large volumes of video data and produce better objective measurements for lost video frames.

Keywords: WMSNs · QoS · EC · QoE · QPs · SHVC

1 Introduction

Researchers' community attention was recently grabbed by dynamic Wireless Multimedia Sensor Networks (WMSNs). Many application layer services were solved by these network supports like observing the health, predicting the weather, managing the transport, systematizing the industry and inspecting the security [1]. In olden days the audios, videos, and images [2] are transmitted and received through multimedia data by the lower power transceiver, small cameras and battery source are used for the multimedia sensor nodes. A lotus mote is the newly evolving superior configuration node which would perform much better than the former which shown in Table 1. The variety of sensor nodes includes Base Station (BS), simple and multimedia sensor nodes in traditional WMSN as depicted in Fig. 1. Upcoming advancement in the domain of

© Springer Nature Switzerland AG 2020
J. S. Raj et al. (Eds.): ICIDCA 2019, LNDECT 46, pp. 420–431, 2020.
https://doi.org/10.1007/978-3-030-38040-3_48

embedded system, IPv6 above Low power Wireless Personal Area Networks (6Low-PAN) and internet of things has facilitated WMSNs to control and perform remote surveillance through the internet [3].

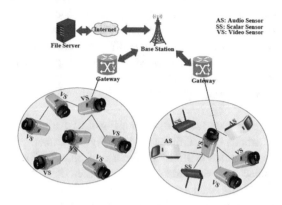

Fig. 1. Conventional wireless multimedia sensor network

A assorted range of multimedia devices and server were used in the traditional multimedia transmission system. A transmission link which is wide spread accelerated the device involving multimedia to seize the data and directly transmit to multimedia server. The renowned codecs HEVC/H.265 aided the multimedia server to store and encode the multimedia data. At the same time, the lightweight application was running on the handler side. Some of the limitations of WMSN are inadequate bandwidth, storage computation, and energy. Many to one are the communication paradigm used by WMSN. The single entity multimedia sensor nodes were up streamed by the data traffic also called sink. The major role of the sink is to perform the computational task. The excessive bandwidth resulted in producing enormous volume of multimedia raw data which increases the density of multimedia sensor nodes.

Table 1. Generations of sensors

Sensor utilities	Specifications
Generation	Old (TelosB) New (Lotus Mote)
Microcontroller	Atmega128/ARM7 Cortex M3
Frequency	8–48 MHz 10–100 MHz
Transceiver	ChipCon CC2420 802.15.4 Radio Antenna
Memory	4–10 KB RAM, 512 KB Flash 64 KB SRAM, 512 KB Flash, 64 MB Serial Flash

Wireless Sen*Transmission systems based Wireless Sensor Networks (WSNs) has chief confrontation in Real-time data delivery. Because of a lack of vitality and bandwidth that is accessible, scalable data distribution with energy-efficiency schemes are needed. The board range of WMSN-based applications like observing the health, predicting the weather, managing the transport, systematizing the industry and inspecting the security requires a seamless delivery of data in factual phase. In the present scenario, many kinds of research are being evolved in dynamic compression of multimedia data and its transmission [4–6]. The pioneering effort was taken into account for dynamic real-time processing and video transmission which is efficient in minimal energy consumption through WMSNs. The minimal available bandwidth and an ensured Quality of Service (QoS) is the confronting challenge present in WMSNs. An adequate QoS data is dispersed in WSNs [7–9] is approached by the number of cross layers. Moreover, it only improves the individual performance instead of the whole. These researches resulting in poor quality of video heavy pocket drop and channel error.

For avoiding video packet loss [10] Concealment of errors (EC) and Resilience in error (ER) are the methods often utilized for tracking impairments in the network. The error propagation was handled skillfully by the ER schemes hick acts as an encoder for data packets associated with the video. These schemes delay the architecture of the encoder; produce immense data streams that would lead to delay in computation. Therefore the limited bandwidth and energy resources are unsuitable for real-time video transmission through transmission networks. However, EC schemes are decoder based techniques. Cumulative data size or higher bandwidth is not demanded this scheme. Instead, they use spatiotemporal statistics from data received from the video for evaluating the missing video packets. This situation will lead to the fall of single video packet may drop the whole video frame. Burst packet-loss frequently occurs in sensor networks and it generates significant quality deprivation in real-time video transmission. Burst whole frame losses both conventional block-based and spatiotemporal EC techniques. Therefore, a high demand for developing real-time energy-efficient is needed to calculate the estimation of lost video frames.

The iconic topics for the modern trend are multi-path packet scheduling, video encoding and error concealment. These domains act separately and never blending would be occurred. Many challenges have to be faced while combining different research domains through implementations and evaluations. The lower resolution video encoded with the H.264 standard was used in the research purposes at the olden period. Latest video coding standard was implanted in upcoming researchers i.e., H.265/High efficiency video coding. The principle of HEVC is complex and it does not provide greater than 50% bit rate compared to MPEG-4 Part 10. It formally structured for encoding HD and Ultra HD (UHD) videos. A large number of data was produced by HD video streaming and it as intolerant towards streaming delays. Visual disturbances can be showed through loss of packets occurring through streaming of video.

This research paper focuses on proficient use for scalable streaming of video based on the standards of HEVC besides EC scheme in WMSNs. There is a failure in finding solution for using WMSNs on aerial scalable video streaming. This paper focuses on existing problems on how to maintain the quality in the transmitted videos and QoS alternatively transmit the video sources in an increasing number. Prevailing studies concentrate on end-to-end delay in gathering data sources and ignoring the load balance at the transmission of video and affecting the pocket drop of transmitted video data. Distinctively our method proposes a multi-hop scalable video transmission scheme to maximally utilize the available resources.

- The researcher intends to establish energy-constrained WMSNs for encoding video scheme. The video coding method is used to generate bit-rates of wide range in order to recompense enormous volume of video data generated by sensor nodes in multimedia.
- The researcher proposes for progress in WMSNs multi-path video packet scheduling algorithm and power efficient was used to produce QoS. Initially the proposed algorithm shifts in uncorrelated paths existing in sensor node of multimedia and its BS and then schedule the video packets based on their need.
- The researcher intends to use an effective scheme for assessing the missing frames in video vanished during transmission. Peak Signal-to-Noise Ratio (PSNR) is the main objective for evaluation metric used in concealing the quality of video frames.

The continuing of this paper is organized as follows. Section 2 deals with a summary of existing approaches related to QoS and QoE in WMSNs. An elaborated version of model for system streaming is presented in Sect. 3. Section 4 describes the anticipated video scalable encoding outline, video streaming and EC technique. The simulation result and experimental arrangements are deliberated in Sect. 5. Finally the conclusion of the paper in analyzed Sect. 6.

2 Related Work

This section deals with the existing studies related to QoS and QoE sustenance for WMSNs. QoS studies mask the cross-layers and simple approaches whereas QoE analyses High Definition (HD) and the concealment of non-HD video streaming.

2.1 Quality of Service

As like TCP/IP architecture, the WMSN has the default layered architecture with the same set of layers. A WMSN comprises of wide range of multimedia sensor nodes and multiple form of multimedia data. The interaction of nodes is needed to share different forms of data for maintaining an assured level of QoS. The different research has already conducted on diverse layers to attain a efficient QoS in a layered architecture [12–14]. In order to improve the performance of QoS in the system, the information was gathered from all the layers [8, 15]. The lower layer was directed by application layer in defining QoS criteria. Routing and MAC has played a vital role in layered as well as cross layered platforms in transmission of the video.

MAC Layer: The standard IE*The conventional IEEE 802:15:4 designs was adopted to created system supporting minimal power and minimal range sensors by 250 kbps [16] data- rate. Another technologies are ZigBee and Ultra Wide Band (UWB) which increases the data-rate with the proposal of 250 kbps to 480 Mbps [17, 18]. Due to short-range of communication WMSNs is not feasible but it is suitable for simple WSNs. IEEE 802:15:4 has gained demand among other technologies for supporting video devices in minimal range wireless ad-hoc networks in security surveillance and monitoring health-care applications. IEEE 802:11 has been providing an ideal platform for provisioning the suitable QoS intended for WMSNs [19, 20] for a low-range communication with high data-rates. A IEEE 802:15:4 is a cross layered platform utilized for improving the energy and consumption of QoS. The conventional IEEE 802:15:4 architecture outperformed by reducing the energy consumption and it is not to be compared by long-range multimedia consumption. For maintaining application-specific QoS services [9] Sched Ex-GA, an alternate cross-layered framework was proposed to predict the network configuration. The higher-end to end reliability was provided by this approach. The network management cost would increase the additional sink sources. The protocol form MAC proposed for cross layered sensor-network for reducing the energy consumption and increase the life of the battery. This scheme is not suitable for bi-directional video communication but it works well on unidirectional communication.

Routing Layer: The proper optimization of routing protocols can maintain the improvement of network performance. The comparison of mobile ad-hoc's geographic routing protocols and sensor networks was presented in certain research work. The shortest path selection between sources and destinations is efficiently showed in this comparison. For delivering the multimedia contents in available bandwidth is required to be optimized for the WMSNsprotocols. The optimization of bandwidth multimedia contents is to be utilized in WMSNs. In multi-path routing, an energy bottleneck issue was proposed to solve the pair-wise directional geographical routing observed in WMSNs. The proposed network increases the lifetime and minimizing the delayed time. The sensor nodes energy has conversed through straight line routing. It ensures good performance but cannot be used for wide range of WMSNs in video transmission. The cluster-based multi-hop WSNs is delivered through data streaming with mobile nodes were proposed. For energy consumption, the cross-cluster handover will maximize the delivery latency path redirection strategies. Increasing range of the sensor network would deteriorate the performance. The addition of multimedia sensor nodes is the finest example.

2.2   Experiencing Quality

Scalable Video Transmission: The technical requirements for video broadcasting technologies were reviewed and the heterogeneous demands for scalable video streaming is combined with the subscription-based multicasting system. Delay in transmission combined with the scalable video coding technique will analyze the prior data on wireless channels through constrained deadline. QoE level is described as a cross-layered transmission improvement is analyzed. A scalable video delivery through wireless networks was optimized by cross-layered transmission. The channel distortion was maximized and increases the achievable data-rate. For wireless scalable video streaming the switching strategy was proposed for the smoothness constraint-based on the dynamic layering. To minimize the service disruption IEEE 802:11 networks schemes are used. Opting video streaming instead of IEEE 802:11 wireless networks QoE-based link was proposed earlier. Dynamic video streaming in wireless networks through switching algorithm was proposed in [35]. The methods depicted in conventional methodologies execute effectively in multicast environments during the release of energy intake required for hardware platforms which are complex.

Error Concealment: The transmitted video quality has been needed to improve the extensive efforts which transmitted video quality beside network impairments. H.265 encoded videos survey of error concealment was conducted. The 2D video streaming error concealment techniques were summarized in certain research work. The moving and static video cameras based on error concealment for proposed video in painting. The optical results were produced by this technique. The multimedia sensor nodes are in need of 20 video frames for concealing the data which is missed. The missing video frames are concealed by the interpolation-based technique in the real-time. The demand for high computing memory and power has concealed the frames. The video frames which are missed are concealed by depth mapping through transmission over WMSNs. The blocks which are mixed of pixels in the video frames as concealed by the proposed spatial error. The efficient objective and subjective results are produced by the proposed algorithms. Besides, it is applicable for multi-view and non-HD videos. H.265 encoded Video streaming over wireless networks was having an impact over network impairments. The results obtained improves the QoE through implementation of efficient concealment error techniques. To improve QoE levels the investigation as made on a real-time scalable video streaming on LTE and 802:11p wireless networks. The paper proposes the use of QoE in a proper error concealment scheme for maintaining acceptable real-time scalable video streaming.

3   System for Streaming

WMSN involves K Multi-media Sensor Nodes (MSNs) and the BS monitors and collects data available in existing nodes for upcoming processing and analysis. The following assumptions are made in the anticipated approach.

1. The prior knowledge of MSNs and BS location has already known.
2. The omnidirectional antenna will cover the geographical range R for MSNs.

In MSNs the density d at given geographical region R with radius r is obtained by using Eq. 1.

$$d = \pi r^2 / RK \tag{1}$$

3.1   Model for Video Encoding

The following drawback arises while capturing every event by more than one MSN at various angles in a dense geographical region.

1. Consumption of Energy is observed for the identical video shot in MSN.
2. In the transmission of the same event, Bandwidth consumption occurs.
3. In BS load occurs in processing, memory consumption and data redundancy is observed.

These drawbacks are overcome by using different Quantization Parameters (QPs) to capture MSN encode videos. The video model is designed for capturing MSNs video from SHVC encoder at different settings in configuration available in mode at default layer. SHVC originates from two layers in this mode i.e., a layer for Enhancement Layer (EL) and a Layer in Base region (BL). A BL comprises of Intra (I) frames and an EL contains Predicted (P) frames. The first group of pictures (GoP) will be encoded in the start. Initially 10 video frames from both BL and EL are sent to BS for response. This step reduces computational load and saves the bandwidth on the BS and MSNs.

3.2 Model for Streaming Video

The high critical factors needed for WSN are reliability, latency and bandwidth. For carrying the video traffic at multiple path for adopting the best routing scheme. The nodes fail to connect directly with BS besides the nodes which are intermediate act as a relay node for carrying video data. There is no fixed path between senders and receivers. The region dedicated for transmission in between the sender nodes which are sending and the BS is partitioned into free paths termed as n, and an interference to obtain best effort delivery as depicted in Fig. 2. The sender node is identified as S, the BS is specified as D, the nearby nodes are termed as relay nodes denoted by RNi and upcoming relays are mentioned as Rj. Initially the paths are manipulated before initiating video streaming in off-line multi-path routing.

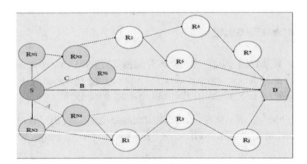

Fig. 2. Multipath forwarding

For measuring the reliability of the routing schemes end-to-end delay and data-rate plays a vital role for underlying transmission medium. Access points deliver the available bandwidth information for MSNs.

3.3 Network Impairment Model

While transmission of the network the video quality is affected by three types of distortions. (i) During the encoding process QP is used (ii) The False decoding and (iii) packet- loss during a transmission. This proposed paper mainly focus on distortions occurs through pocket-loss. For maintaining the quality of the video, threshold level is fixed on the receiver's side. Instead of getting crashed the delayed message is concealed by the decoder through suitable EC technique.

3.4   Problem Description

MSNs value is denoted as S and k is assumed to be the scope of S where $0 \leq k \leq K$. The nearest reliable node to transfer the data $Si \in S$. The data in the set video frames denoted as V. The bit rate is encoded as *bit* at time t. Si followed to reduce two steps in the video frames. The video frames in V bit frames are reduced and it controls the existing status of the relay node. The transfer of data requested is determined by Eq. 2.

$$S_j^{sta} = 0 \leq i \leq k, 0 \leq j \leq K \left(b_{i,t} - b_{i,t+1}\right) + \omega_i - \left(b_{j,t+1}^{in} - b_{j,t+1}\right) \tag{2}$$

$$i \leq T_g \ where \ i \in K \tag{3}$$

$$\left(b_{j,t+1}^{in} + b_{j,t+1}\right) \leq b_{j,t+1} \tag{4}$$

$$q_j \leq q_i \tag{5}$$

$$l_i + l_j \leq T \tag{6}$$

To sustain the persistent quality of video data it should not exceed in Eq. 3. Equation 4 guarantees the bit rate of the incoming data aggravated by the forwarded data and it should not reach beyond the bit rate for maintaining the time. Equation 5 ensured that should not exceed Eq. 6 warranties that and must not surpass a predetermined threshold T. The important role of routing schemes in both data-rate and end-to-end delay play is to measure reliability, latency and bandwidth available in the transmission medium.

4   Streaming of Video and Framework Based on Error Concealment

As depicted in Fig. 3 Streaming of video and Concealment of Error (VSEC) is depicted as a cross-layered framework. The multi-hop scenario is used in SHVC standard videos encoded in the VSEC framework. This standards-based framework helps to generate many number of video frames with flexible bit rates. The feedback message would improve video quality. BS checks whether all data's are received successfully. The lightweight Scalable EC (SEC) scheme is utilized for recovering the frames which are being missed.

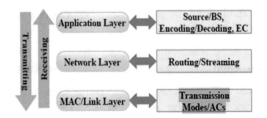

Fig. 3. Cross layer architecture

The process of streaming the video and high concealment framework use different cross-layer architecture to maximize the equality of HD Videos. It is categorized under A. Scalable High-Efficiency Video Coding to adopt the framework of VSEC framework. It is having four sub-frameworks to improve the quality of HD videos. It is categorized as

1. Reducing Frames per Retransmission
2. Controlling Coding Complexity
3. Selecting Quantization Parameter for Intra Frames
4. Obtaining Quantization Parameter for Inter Frames

For efficient video streaming multipath routing, the concept is used in VSEC framework. Under B. Scalable Video Streaming four routing scheme is followed such as

1. Selection of Next Hop:
2. Path Length/Maximum Hop Count
3. Traffic Control
4. Packet Drop

The third part falls under Concealment with Scalable Error which recovers the data which is missed by using lightweight quality-driven SEC scheme. (1) Frame concealment in base layer scheme process the BL video frame under five phases as

1. Block Size
2. Frame segmentation
3. Scanning Technique
4. Motion Vector Extrapolation
5. Frame Generation
6. Denoising

The enhancement Layer Frame Concealment used EL frame to conceal the missing video sequence. This approach modifies the information through EL frames and sends the data to SEC scheme. This computed average video frames will produce the master video frame as shown Fig. 4.

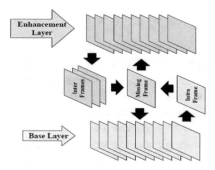

Fig. 4. Master frame referencing

5   Experimental Setup and Evaluating the Performance

The proposed framework is evaluated under three subsections as (a) Processing the video, (b) video streaming and (c) concealment of error. While processing the video, for testing purpose, two video sequences are chosen and they encode the default layer. The obtained video frames are collected as data packets. Initially, the GoP is set as 10 and at the end, it reduced to 1 because of network impairments. In video streaming the problem of processing delay is controlled through proposed routing algorithm testing which fixes the bit rates that are varied. For analyzing the performance of the network simulation is executed thrice.

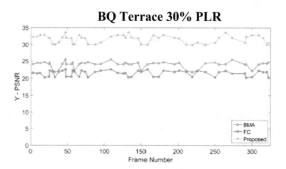

Fig. 5. PSNR comparison

In error concealment, Matlab2016a is used to implement in EC algorithm. In the experimental setting the conventional Block Matching Algorithm (BMA) and Frame Copy (FC) is implemented under the same experimental settings. Under various average Computation time the proposed algorithm related to ECa proximity of 99% minimal computational time when compared to the classic module. In real time processing the computational time and concealed video quality is considered. The processing is simple and fewer Hardware resources are highly on demand to produce fluctuations that are visible and freezing video effects. The anticipated EC technique outperforms equally in BMA and FC technique in the average of PSNRs. This average QP pairs in different PSNR is shown in Fig. 5.

6 Conclusion

The paper has proposed a complete deduction of the problem that occurs in dynamic video streaming over WMSNs and has proposed a design which is cross-layered in video processing, error concealment and streaming through WMSNs. An iconic adoptive style has been adopted for changing the amount of video frames in various network impairments like buffer overloading, bandwidth consumption and packet dropping. In MSNs an innovative acknowledgement based feedback is introduced. The QoS has been conserved through adaptive streaming, fluctuating bit-rate, a hop count, and the response from BS. In the meantime the QoE is maintained by an error concealment algorithm; lightweight quality oriented video processing. The evaluation of proposed streaming is done by bit rates which are constant or varied values, concerns based on single/multi-path. In average computational time and PSNR, EC technique has given better performance. Thereby it has been proved that it shall be applied to WMSNs for maintaining dynamic streaming with good quality. This paper will provide a brief idea for developing a future framework and utilize them in unique lossy wireless networks similar to LTE, WiMAXand5G with an incredible resolution video.

References

1. Akyildiz, I.F., Melodia, T., Chowdhury, K.R.: A survey on wireless multimedia sensor networks. Comput. Netw. **51**(4), 921–960 (2007)
2. Ehsan, S., Hamdaoui, B.: A survey on energy-efficient routing techniques with QoS assurances for wireless multimedia sensor networks. Commun. Surv. Tutorials IEEE **14**(2), 265–278 (2012)
3. Shelby, Z., Bormann, C.: 6LoWPAN: The Wireless Embedded Internet, vol. 43. Wiley, Hoboken (2011)
4. Zou, Z., Bao, Y., Deng, F., Li, H.: An approach of reliable data transmission with random redundancy for wireless sensors in structural health monitoring. IEEE Sens. J. **15**(2), 809–818 (2015)
5. Tapparello, C., Simeone, O., Rossi, M.: Dynamic compression transmission for energy-harvesting multihop networks with correlatedsources. IEEE/ACM Trans. Netw. (TON) **22**(6), 1729–1741 (2014)
6. Wu, X., Xiong, Y., Yang, P., Wan, S., Huang, W.: Sparsest random scheduling for compressive data gathering in wireless sensor networks. IEEE Trans. Wireless Commun. **13**(10), 5867–5877 (2014)
7. Al-Jemeli, M., Hussin, F.A.: An energy efficient cross-layer network operation model for ieee 802.15.4-based mobile wireless sensornetworks. IEEE Sens. J. **15**(2), 684–692 (2015)
8. Kumar, R.P., Smys, S.: Analysis of dynamic topology wireless sensor networks for the internet of things (IOT). Int. J. Innov. Eng. Technol. (IJIET) **8**, 35–41 (2017)
9. Dobslaw, F., Zhang, T., Gidlund, M.: QoS-aware cross-layer configuration for industrial wireless sensor networks. IEEE Trans. Industr. Inf. **12**, 1679–1691 (2016)
10. Schierl, T., Hannuksela, M.M., Wang, Y.-K., Wenger, S.: System layer integration of high efficiency video coding. IEEE Trans. Circuits Syst. Video Technol. **22**(12), 1871–1884 (2012)

11. Lainema, J., Bossen, F., Han, W.-J., Min, J., Ugur, K.: Intra coding of the HEVC standard. IEEE Trans. Circuits Syst. Video Technol. **22**(12), 1792–1801 (2012)
12. Olwal, T.O., Djouani, K., Kurien, A.M.: A survey of resource management towards 5G radio access networks. IEEE Commun. Surv. Tutorials **18**(3), 1656–1686 (2016)
13. Chen, X., Wu, J., Cai, Y., Zhang, H., Chen, T.: Energy-efficiency oriented traffic offloading in wireless networks: a brief survey and a learning approach for heterogeneous cellular networks. IEEE J. Sel. Areas Commun. **33**(4), 627–640 (2015)
14. Ahmad, A., Ahmad, S., Rehmani, M.H., Hassan, N.U.: A survey on radio resource allocation in cognitive radio sensor networks. IEEE Commun. Surv. Tutorials **17**(2), 888–917 (2015)
15. Fu, B., Xiao, Y., Deng, H., Zeng, H.: A survey of cross-layer designs in wireless networks. IEEE Commun. Surv. Tutorials **16**(1), 110–126 (2014)
16. Molisch, A.F., Balakrishnan, K., Cassioli, D., Chong, C.-C., Emami, S., Fort, A., Karedal, J., Kunisch, J., Schantz, H., Schuster, U., et al.: IEEE 802.15.4a channel model-final report. IEEE P802, vol. 15, no. 04, p. 0662 (2004)
17. Alliance, Z., et al.: Zigbee specification (2006)
18. Chehri, A., Fortier, P., Tardif, P.M.: UWB-based sensor networks for localization in mining environments. Ad Hoc Netw. **7**(5), 987–1000 (2009)
19. Khorov, E., Lyakhov, A., Krotov, A., Guschin, A.: A survey on IEEE 802.11 ah: an enabling networking technology for smart cities. Comput. Commun. **58**, 53–69 (2015)
20. Kuo, Y.-W., Liu, K.-J.: Enhanced sensor medium access control protocol for wireless sensor networks in the ns-2 simulator. IEEE Syst. J. **9**(4), 1311–1321 (2015)
21. Usman, M., Yang, N., Jan, M.A., He, X., Xu, M., Lam, K.-M.: A joint framework for QoS and QoE for video transmission over wireless multimedia sensor networks. IEEE Trans. Mob. Comput. **17**(4), 746–759 (2018). IEEE Transactions on Consumer Electronics, vol. 60

Analysis of Temperature Prediction Using Random Forest and Facebook Prophet Algorithms

J. Asha[1(✉)], S. Rishidas[2], S. SanthoshKumar[3], and P. Reena[4]

[1] Department of Electronics and Communication Engineering, Government Engineering College, Affiliated to APJ Abdul Kalam Technological University, Thrissur, Kerala, India
ashaj@gectcr.ac.in
[2] Department of Electronics and Communication Engineering, Government Engineering College, Barton Hill, Kerala, India
rishidas19731999@gmail.com
[3] Department of Electronics and Communication Engineering, College of Engineering, Trivandrum, Kerala, India
santhosh4678@gmail.com
[4] Department of MCA, Government Engineering College, Thrissur, Kerala, India
reenabasant@gmail.com

Abstract. Forecasting temperature daily has been a great challenge that the meteorological department faces today. For Kerala, the year 2019 has one of the warmest summers on record. Increase in temperature has adverse effects on health and agriculture fields. Accurate prediction of daily temperature enables the Government and people to take proper precautionary steps. This paper focuses on analysing two algorithms- Random Forest and Facebook Prophet- for temperature prediction. Their performance-based on five different stations in Kerala, India, are compared based on Accuracy and Mean Absolute Error. Both the models gave comparable results, but Random Forest gave better accuracy and Mean absolute error. The forecasts of Random Forest is more consistent than Facebook Prophet.

Keywords: Random forest · Facebook Prophet · Temperature prediction · Accuracy · Mean absolute error

1 Introduction

The world has been witnessing rapid climate changes, which has adverse effects on the environment. According to a study conducted by NASA, since 1880, the average global temperature on Earth has increased by about 0.8 °C [1]. This one-degree global rise in temperature is significant because it takes a considerable amount of heat to warm up the oceans, land and atmosphere, by that small amount of temperature. This has resulted in the warming up of atmosphere and oceans, melting down of snow and ice, rising of sea level, changes in precipitation levels, expansion of deserts, wildfires, droughts, heatwaves and extinction of ecosystems. It harms the agricultural field as well as health and security.

© Springer Nature Switzerland AG 2020
J. S. Raj et al. (Eds.): ICIDCA 2019, LNDECT 46, pp. 432–439, 2020.
https://doi.org/10.1007/978-3-030-38040-3_49

Kerala state, popularly known as God's own Country owing to its pleasant climate, lies in the south-west corner of India. Although Kerala lies close to the equator, its proximity with sea and the presence of Western Ghats gives a pleasant climate. Kerala reported an increase in maximum temperature by 0.64 °C, over the last 50 years [2]. The state has also been witnessing seasonal extremes in rainfall resulting in floods and droughts, which are indicators of climate change. Kerala experienced its worst flood during the year 2018, affecting one-sixth of the state's population. Increasing temperature and extremes in rainfall have resulted in severe setbacks in significant areas like agriculture, health and even depletion of marine resources. Even sunstrokes and wildfires have become common during the summer season. Thermosensitive cash crops like pepper, tea, cashew, cardamom, tea, coffee, cocoa and other spices constitute a significant portion of Kerala's agriculture. These crops are severely affected, as the difference between the maximum and minimum temperatures increase. Following the global phenomenon, the mean sea level is rising in Kerala, resulting in the inundation of coastal areas. Sunstrokes have been reported in recent years in many places in Kerala. Palakkad and Punalur usually report the highest of temperatures, even reaching up to 41 °C.

Considering all these adverse effects, owing to the increase in temperature, predicting temperature facilitate people and government to take necessary precautions against temperature rise. It is quite challenging to predict the temperature as well as other weather conditions accurately. In this paper, we are predicting the daily maximum temperature of five different stations in Kerala using two algorithms. Random Forest and Facebook Prophet are the selected algorithms, and their performances are compared based on Accuracy and Mean Absolute Error.

2   Related Works

Both machine learning algorithm and statistical models are being used for predicting various weather conditions. Radhika et al. [3] in 2009 used Support Vector Machine to analyse time-series data of daily maximum temperature at a location to predict the maximum temperature of the next day. Singh et al. in 2011 [4] developed a time series based temperature prediction model using integrated Back Propagation with Genetic Algorithms technique. Kadu et al. [5] in 2012, proposed a model of temperature prediction, which used new wireless technology for data gathering and with the combination of statistical software. Vamitha et al. [6] in 2012, proposed a forecasting model based on multivariate Markov chain on categorical sequences for forecasting the daily average temperature of the Taipei, Taiwan. Categorical data sequences were obtained through the fuzzification of temperature data. Noval et al. [7] in 2014 did the analyses of multi-year verification of precipitation forecasts and medium-range maximum temperature forecasts from weather prediction centres and were compared to automated NWP guidance. Naing et al. [8] in 2015 utilised a Random Forest Model for predicting the one-month temperature in Kuala Lumpur, Malaysia. Karevan et al. [9] in 2015 proposed a black-box modelling technique for temperature forecasting. In their work feature selection, was done in two steps with K-nearest Neighbours and Elastic net and Least-squares Support Vector machine regression was applied to generate the forecasting model. Murat et al. [10] in 2018 used methods of the Box-Jenkins and Holt-Winter Seasonal Autoregressive Integrated Moving Average with external regressors in the form of Fourier terms and the time series regression, including trend and seasonality components with R software for predicting daily air temperature and precipitation time series recorded for period of 20 years in four European Sites. Feng et al. [11] in 2019 showed that the Prophet model and Keras stateful LSTM perform better than conventional neural network models in predicting the crimes for the three US cities in 2018.

3　Prediction Algorithm

The prediction problem can be defined as the estimation of temperature at day $(n+1)$ from previously observed maximum temperatures i.e., $x(n), x(n-1), \ldots \ldots \ldots$ $\ldots x(n-L)$, where the evaluation is performed using 'L' previous date's temperature.

$$x(n+1) = f(x(n), x(n-1), \ldots \ldots \ldots \ldots x(n-L+1)) \tag{1}$$

$\hat{x}$ (n + 1) is the predicted temperature. Figure 1 shows the general flow chart for training and testing a prediction model. The prediction model 'f' is obtained by fitting with the known dataset. For training the model input is taken as delayed sequences $x(n-1), x(n-L)$ and output as x(n) for different values of n.

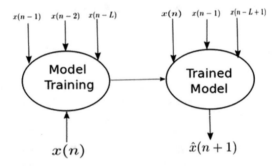

Fig. 1. Flow chart of prediction

The performance metrics of the prediction model is evaluated using two parameters Mean absolute error (MAE) in degree Celsius and accuracy in percentage.

$$MAE = \frac{1}{N} \sum_{n=1}^{N} |x(n) - x(\hat{n})| \tag{2}$$

$$Accuracy = 100 - 100 \times \frac{1}{N} \sum_{n=1}^{N} \frac{|x(n) - x'(n)|}{x(n)} \tag{3}$$

where N denotes the number of samples used for prediction.

4 Methods and Data

4.1 Dataset

The daily maximum temperature, minimum temperature, humidity and daily rainfall data for all the five stations are obtained from the website of Indian Meteorological Department. The data is being taken for 226 days, ranging from 1/1/2019 to 14/08/2019. The Acquired data is tabulated in CSV format.

4.2 Methodology

4.2.1 Prediction Using Random Forests

Random Forest is an ensemble learning method used for classification and regression [12]. Breiman developed it in 2001 [12], whose ideas were influenced by the works of Amit and Geman on feature selection, Ho's random subspace method and Dietterich's random split selection approach. The Random Forest are trained using the bagging method or Bootstrap Aggregating method. The algorithm has two phases.

This work aims to predict the next day maximum temperature of the selected station using past M days temperature data. Here, the problem is a supervised, regression problem with present days temperature as target and previous days temperature as inputs. The proposed methodology uses the python random forest library for testing and the python library, Pandas for data analysis. The temperature data is represented as a data frame, as shown in Table 1 below.

Table 1. Dataframe representation.

Index	Year	Month	Date	Actual Temp(T_0)	T_{-1}	T_{-2}	..	T_{-14}	
1	2019	1	1	32	33	33	...	34	
2	2019	1	2	33	32	33	...	35	
3	2019	1	3	34	33	32	...	34	Training vectors
:	:	:	:	:	:	:	:	:	
n-1	2019	8	29	28	30	31	...	32	
n	2019	8	30	38	28	30	...	20	Test Input

In the data frame T_0 represents maximum temperature of current day. While T_{-1} represents maximum temperature of previous day and T_{-2} represents maximum temperature two days before and so on. To predict the temperature of nth day, the training dataset consisting of n − 1 rows of the data frame, as shown in Table 1, is used.

The forecast object here is a new dataframe that includes a column 'yhat' with the forecast. The training dataset consists of target labels as elements of T_0 and input features, are the corresponding row elements of T_{-1}, T_{-2}, ...T_{-L}. The sci-kit random forest library is used for training and testing the model. In this work, number of decision trees used is 150, and the number of random states is 20. Random Forest Regressor library is used for fitting the training model. Then the predict function in random forest library is applied to the fitted model for predicting n^{th} day temperature. The predict function accepts input features as $T_{-1}(n)$, $T_{-2}(n)$ ..-...$T_{-L}(n)$ and outputs the predicted values as a 1D array. The size of the array is the number of future days to be predicted. In this work, the training model is updated using the actual temperature of the current day to predict next days temperature.

4.2.2 Prediction Using Facebook Prophet

Prophet is a procedure used for the forecasting of time series data, based on an additive model where non-linear trends are fit with yearly, weekly, and daily seasonality [13], including the holiday effects. The Prophet is available in Python, and R. Prophet is an open-source software released by Facebook's Core Data Science team. It is used by many applications on Facebook, for producing forecasts for planning and goal setting. Prophet uses a decomposable time series model [14] with three main model components: trend, seasonality and holidays. Seasonality represents periodic changes like weekly and yearly seasonality. Holidays occur on irregular schedules over one or more days.

They are combined in the following equation [13]:

$$y(t) = g(t) + s(t) + h(t) + \epsilon_t \qquad (4)$$

g(t): piecewise linear or logistic growth curve for modeling non-periodic changes in time series, s(t): periodic changes (e.g. weekly/yearly seasonality), h(t): effects of holidays with irregular schedules and error term ϵ_t. The error term accounts for any unusual changes not accommodated by the model. Seasonal effects of s(t) are represented by the period parameter P and fourier order N. In this work Facebook prophet library fbprophet is used for training and testing the model. The fbprophet requires dataframe in a two column format as y – Target and ds – Date time. In this work the value of y is the actual temperature of the days shown in Table 2.

Table 2. Data frame for Facebook Prophet with predicted value

DS	Y	Yhat
2019-01-01	32	
2019-01-02	33	
:	:	
2019-08-14	35	
2019-08-15	36	35.2

Instantiating of Prophet object is done by passing appropriate parameters like seasonality parameter, period and Fourier order. The model is fitted using the fit function in prophet library by using the data frame as shown above. In this work seasonality parameter used is yearly seasonality and Fourier order N as 3. To predict the temperature make_future_dataframe function of prophet library is used, which accepts the number of future dates for prediction as an argument. Value of 1 is used to forecast the next day's temperature. Once the future data frame is created, predict function in fbprophet library is used for prediction, which produces a future object. The future object contains a column field 'yhat' which contains the predicted values.

5   Results and Discussions

From the website of Indian Meteorological Dept., maximum temperature data of five different stations in the Kerala state of India is obtained in CSV format for the period from 01/01/2019 to 14/08/2019. January and February mark the ending of winter in Kerala, March to May is Summer season with temperatures reaching up to 40 °C, and the southwest monsoon starts by June first week. The performance of both RF and Facebook Prophet are measured using the parameters: Accuracy in percentage and Mean Absolute error in °C. Table 3 shows the comparison of performance.

Table 3. Comparison of performance of daily maximum temperature of selected five stations.

Station	Random Forest		Facebook Prophet	
	Accuracy (%)	MAE (°C)	Accuracy (%)	MAE (°C)
Trivandrum	97.84	0.70	97.06	0.96
Punalur	97.62	0.78	95.82	1.37
Kochi	97.07	0.91	96.39	1.13
Palakkad	96.99	0.94	96.33	1.27
Kannur	96.72	1.01	95.45	1.44
Average	**97.25**	**0.868**	**96.21**	**1.23**

Figures 2 and 3 shows the observed and predicted values of the daily maximum temperature of Palakkad using Random Forest and Facebook Prophet, respectively. Palakkad and Punalur stations experience the hottest climate in Kerala, and the observed temperatures show sharp fluctuations. Random Forest prediction is closer to the observed values, and it tries to follow the changes. However, the forecasts of Facebook Prophet is smoother than Random Forest, implying that it is not capable of capturing the sharp changes in the temperature.

For all the five stations, the accuracy and Mean Absolute Error demonstrated by Facebook Prophet are lower than that of Random Forest. So it can be concluded that Random Forest is a better predictor compared to Facebook Prophet since its performance is consistent. The average value of accuracy is 97.25% for Random Forest and is 96.21% for Facebook Prophet, while MAE is 0.868 °C and 1.23 °C for Random Forest and Facebook Prophet respectively.

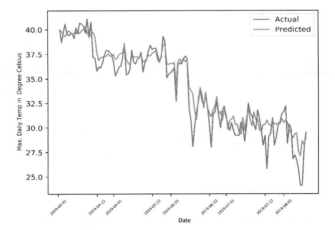

Fig. 2. Observed and predicted daily maximum temperatures from 02/04/2019 to 14/08/2019 for Palakkad station using random forest.

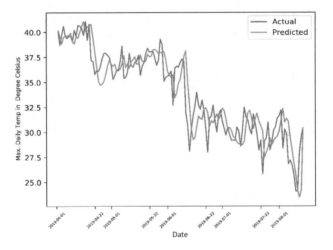

Fig. 3. Observed and predicted daily maximum temperatures from 02/04/2019 to 14/08/2019 for Palakkad station using Facebook Prophet.

References

1. NASA Earth Observatory. https://earthobservatory.nasa.gov
2. Indian Meteorological Department. https://www.imdtvm.gov.in
3. Radhika, Y., Shashi, M.: Atmospheric temperature prediction using support vector machine. Int. J. Comput. Theory Eng. **1**(1), 1793–8201 (2009)
4. Singh, S., Bhambri, P., Gill, J.: Time series based temperature prediction using back propagation with genetic algorithm technique. IJCSI Int. J. Comput. Sci. Issues **8**(5), 28 (2011)

5. Kadu, P.P., Wagh, K.P., Chatur, P.N.: Review on efficient temperature prediction system using back propagation neural network. Int. J. Emerg. Technol. Adv. Eng. **2**(1) (2012). www.ijetae.com. ISSN 2250-2459

6. Vamitha, V., Jeyanthi, M., Rajaram, S., Revathi, T.: Temperature prediction using fuzzy time series and multivariate Markov chain. Int. J. Fuzzy Math. Syst. **2**, 217–230 (2012). ISSN 2248-9940

7. David, R.N., Bailey, C., Rill, F.B., Burke, P., Hogstt, W.A., Ausuch, R., Chichtel, M.S.: Precipitation and Temperature Forecast Performance at the Weather Prediction Center NOAA/NWS/NCEP/Weather Prediction Center, College Park, Maryland (2014)

8. Naing, W.Y.N., Htike, Z.Z.: Forecasting of monthly temperature variations using random forests. ARPN J. Eng. Appl. Sci. **10**(21), 10109–10112 (2015)

9. Karevan, Z., Mehrkanoon, S., Suykens, A.K.: Black-box modeling for temperature prediction in weather forecasting. In: International Joint Conference on Neural Networks (IJCNN) (2015)

10. Murat, M., Malinowska, I., Gos, M., Krzyazczak, J.: Forecasting daily meteorological time series using ARIMA and regression models. Int. Agrophysics **32**(2), 253–264 (2017)

11. Feng, M., Zheng, J., Ren, J., Hussain, A., Li, X., Xi, Y., Liu, Q.: Big data analytics and mining for effective visualization and trends forecasting of crime data. IEEE Access **7**, 106111–106123 (2019)

12. Breiman, L.: Random forests. Mach. Learn. **45**(1), 5–32 (2001)

13. https://github.com/hanhanwu/Hanhan_Data_Science_Practice/blob/master/sequencial_analysis/ReadMe.md

14. Taylor, S.J., Letham, B.: Forecasting at scale. PeerJ Preprints **5**, e3190v2 (2017). https://doi.org/10.7287/peerj.preprints.3190v21

15. Kerala Geography - Kerala Physiography, location, Kerala. https://www.prokerala.com/kerala/geography.html

16. Ho, T.K.: Random decision forests (PDF). In: Proceedings of the 3rd International Conference on Document Analysis and Recognition, Montreal, QC, pp. 278–282 (1995)

Wireless Monitoring and Control of Deep Mining Environment Using Thingspeak and XBee

B. Ramesh[✉] and K. Panduranga Vittal

National Institute of Technology Karnataka, Surathkal, India
rbant96@gmail.com, vittal.nitk@gmail.com

Abstract. The possibility of remotely monitoring and controlling the deep mining environment using Raspberry Pi is studied in this paper. The use of sensor and thingspeak to get the sensor data in the web and to obtain its graph in real-time is explored. Then the controlling of the raspberry pi with the help of XBee communication and remotely controlling by computer is studied. This is done for the moisture level control by using relay and pump as an example. This can be extended to other type of sensors which are of relevance in the deep mining environment and for internet of things applications.

Keywords: Raspberry Pi · XBee · Thingspeak · Control · Mining · Communication

1 Introduction

Mining activities are responsible for environmental damage at a large scale. It may create soil erosion, water pollution or air pollution. The environmental effects are severe in less developed nations, which produce higher percentage of the world's minerals. The mining environment is hazardous in nature with machinery, polluted air, low illumination and in some cases high temperature.

The underground mining has comparatively more hazardous environment, with the mining staff constantly working under extreme situations like high methane and other gases in the mining atmosphere. The gold and coal mines being examples of deep mining, have posed as a threat for miners if not monitored properly. Safety measures in these mines are of paramount importance as regulated by the mining authority.

When mining activity is near to water sources, there is always a threat of water entering the mining area, resulting in flooding at the deep mining area. Due to the sudden inrush of water into mining area, the miners' escape route may get blocked. When the water inrush leads to lock in of miners, the primary reaction is generally to fetch extra pumps to reduce the water in the mining area which helps miners to escape.

The measured parameters can be sensed or detected using industrial grade sensors like industrial grade RTD or thermocouple. The parameters measured can lead to a large number of data sets which needs to be stored in cloud or a nodal server. The cloud storage can be updated in real time or the data can be stored in a server computer and then uploaded to the web. The different platforms like Thingspeak can be used to plot the data variation in real time or offline.

© Springer Nature Switzerland AG 2020
J. S. Raj et al. (Eds.): ICIDCA 2019, LNDECT 46, pp. 440–446, 2020.
https://doi.org/10.1007/978-3-030-38040-3_50

1.1 Safety Procedures

The safety in the deep coal mining industry should be taken very seriously. It is necessary to report annual figures of fatal and non-fatal incidents as per the regulatory board.

Deep mining has issues related to ventilation and mine collapse which are complicated to handle. There is a possibility of a mine collapse in some cases. The safety risk will be there in all types of mining due to the use of heavy machinery. To avoid explosions, methods are necessary to remove methane prior and during excavation.

In one of the works, the images and videos from the mining area are sent to a cloud server. If the server is not accessible, the data could be stored locally in the Raspberry Pi and transmitted when the link is re-established.

By this, the home monitoring is made easier [1]. Raspberry Pi can be the used for smart purposes and client-server communications. The various attributes available in Raspberry Pi are introduced [2]. The possibility of live sensor data transmission from raspberry pi to ubuntu operating system using ZigBee technology is explored [3]. A metering device which senses and sends the reading by use of ZigBee is then processed. It then proposes that the data should be sent for billing purposes. [4]. An autonomous robot which can recognise commands provided by the hand gestures is exhibited with a help of microcontroller and an accelerometer [5]. A powerful system using Raspberry Pi as an intelligent device using which numerous things can be interconnected and can be managed from an extended range was proposed [6].

Suggestions to design a wireless sensor network making use of microcontroller which is able to observe the different parameters like humidity, methane and carbon dioxide in an underground mine was proposed [7]. Various digital and analog sensors could be used along with Raspberry Pi to obtain mine data. Analog to digital converter like MCP3008 is required when using analog sensors because Raspberry Pi does not have inbuilt ADC [8]. It was proposed to build gateways for IoT with Raspberry Pi. Low power wireless packet communication RF modules provides components to build IoT applications quickly and easily [9]. Using wireless communication, sending environmental parameter data from sensors located in mines and transmitted to the control room was proposed. [10]. Environment monitoring using both Raspberry Pi and Aurdino with wireless sensor networks is also proposed [11]. For exterior and offline analysis purpose, the data can be taken from the Thingspeak database as CSV or Jason or PHP format [12].

2 Experimental Analysis

The block diagram of experimental setup with sensor, raspberry pi, XBees, relay and pump is indicated in Fig. 1. Its actual setup is shown in Fig. 2.

2.1 Moisture Sensors

The resistive sensor is made up of two probes. The probe measures the moisture content of the soil by measuring the resistance value. When the moisture content is higher, the conductivity of soil increases leading to change in the value of the resistance probe. When the moisture level is low, the conductivity of soil is low and correspondingly the resistance value will be lower.

Being analog sensors, both resistive and capacitive sensors need to be connected with ADC so that the digital data will be supplied to the Raspberry Pi.

Capacitive measuring eliminates corrosion of the probe and gives an improved measurement of the moisture level of the soil. Non corrosive industrial grade sensors are necessary for the mining environment.

2.2   Thingspeak

Thingspeak is open source which is helpful to collect, store and display data in the cloud. Through Thingspeak it is possible to create public channels. Hence it is advantageous compared to some other similar platforms. The sensor data can be uploaded to the cloud with the help of raspberry pi. Simple visualisations powered by Matlab are possible in this application website. Hence it is visually attractive and is comparatively easier to analyse the collected data.

After creating a thingspeak account, the required number of channels are to be created depending on the number of sensors used for monitoring. The required python application programming interface (API)s and libraries need to be downloaded for uploading the measured data to the cloud. If the data needs to be analysed separately, it can be downloaded from Thingspeak website as CSV or Jason format. The measured percent moisture is indicated by Fig. 3 in Thingspeak.

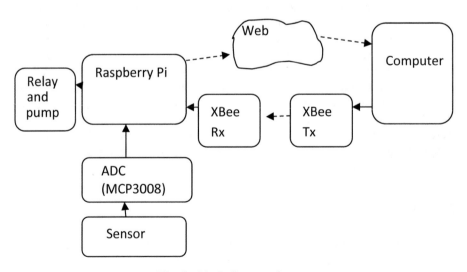

Fig. 1. Block diagram of setup

Fig. 2. Setup with sensor, pump and relay

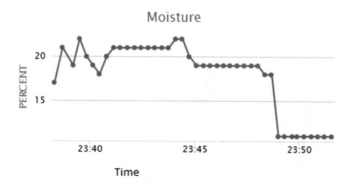

Fig. 3. Real time Graph in the Thingspeak website

2.3   XBee for Data Transfer

Data transfer between the computers and controllers can be achieved by an Xbee device. It consumes low power and uses a serial port. By using a network of intermediate devices, data communication can be achieved over long distances. It becomes simple and less costly when the Xbee is used in the wireless sensor network. A wireless communication is also made possible with Xbee and a Raspberry Pi combined either using a ZigBee Dongle or Tx and Rx pins of raspberry Pi.

IEEE 802.15.4 PHY protocol is used in XBee S2C wireless communication. Wireless communication to terminal devices is provided by XBee in any ZigBee mesh networks. Other units using ZigBee technology are also well-suited to the XBee RF Module. XBee has a defined host interface, which makes programming the device simpler. The host interface API is significant part of XBee. It is interchangeable and can manage diverse types of communication, including ZigBee, 802.15.4, and WiFi.

2.4   X-CTU Software

X-CTU software is used to configure the XBee. One XBee B2C is set for receiving mode and connected to the Raspberry Pi. The exact similar XBee B2C is configured for the transmitter mode.

After installing explorer drivers and setting the communication port number, the window also allows to indicate more specific serial features like data bits and baud rate. Then, the configuration settings of Xbee are displayed on screen.

The two X-bees must be connected with the same baud rate with the software. One XBee is connected to Tx and Rx pins of the Raspberry Pi after installing software and setting it as receiver. Other XBee is configured for transmitter mode and connected to the USB port of a Laptop. The correct com port and the baud rate were selected for the connected device.

The control signal transferred to the raspberry Pi, will initiate the switching ON of the pump with the help of the relay. The controlling can be done from the remote computer. When letter a is pressed in remote computer, the pump becomes ON. When the computer and Raspberry Pi are communicating through XBee, the computer screen will be as indicated in Fig. 4.

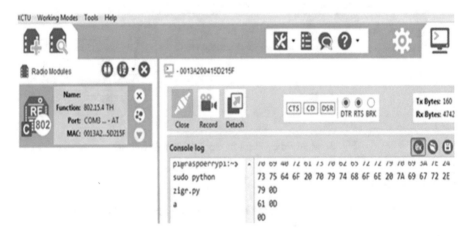

Fig. 4. The wireless communication between computer and Raspberry pi

The overall workflow is indicated by the flowchart indicated in Fig. 5. The website is automatically updated. This information on the situation is used to control the mine situation remotely by Xbee communication between control center and Raspberry Pi which will do the final control action.

3   Result

One method of monitoring and control of the mining environment with wireless communication using X-bee is studied in this paper. The measured parameter was displayed real-time in the thingspeak website. The pump in the field is controlled

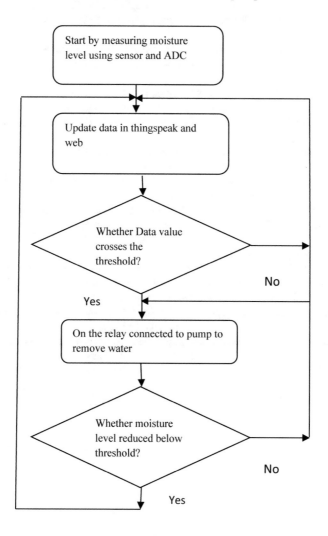

Fig. 5. Flowchart

wirelessly using XBees connected to both Raspberry pi and computer. This method can be extended to monitor the methane gas, temperature and other parameters of the deep mining environment.

References

1. Sugumaran, N., Vijay, G.V., Annadevi, E.: Smart surveillance monitoring system using Raspberry Pi and PIR sensor. Int. J. Innov. Res. Adv. Eng. 4(4), 23–25 (2017)
2. Zhao, C.W., Jegatheesan, J., Loon, S.C.: Exploring IOT application using Raspberry Pi. Int. J. Comput. Netw. Appl. 2(1), 27–34 (2015)
3. Mala, H., Chaithra, R., Ranjitha., L, Sindhu, G., Raghavendra, M.: Zig Bee communication of sensor data between Raspberry Pi and ubuntu system. J. Environ. Nanotechnol. 6(1), 44–50 (2017)
4. Chaudhary, D., Gite, S.V.: Automated electric meter readings and monitoring system using ZigBee integrated Raspberry Pi. Int. J. Adv. Res. Electr. Electron. Instrum. Eng. 2(4), 9082–9087 (2016)
5. Kausthub, N.P., Rounak, S., Bhakthavathsalam, R., Gowranga, K.H., Saqquaf, S.M.: Hand gesture initiated motor operations for remote robot surveillance system using Raspberry Pi processor on a ZigBee communication link. Int. J. Innov. Res. Electron. Commun. 2(5), 55–62 (2015)
6. Vimal, V., Sumalatha, J., Amal, J., Gangadhara: IOT based home automation using Raspberry PI 3B+. Int. J. Eng. Res. Technol. 7(08) (2019). ISSN: 2278-0181
7. Dange, K.M., Patil, R.T.: Design of monitoring system for coal mine safety based on MSP430. Int. J. Eng. Sci. Inven. 2(7), 14–19 (2013)
8. Ramesh, B., Vittal, K.P.: Situation awareness of deep mining environment using Raspberry Pi. J. Comput. Theor. Nanosci. 16, 2604–2608 (2019)
9. Calvo, I., Gil-García, J.M., Recio, I., López, A., Quesada, J.: Building IoT applications with Raspberry Pi and low power IQRF communication modules. Electron. J. 5(54), 1–17 (2016)
10. Ledange, S.M., Mathurkar, S.S.: Robot based wireless monitoring and safety system for underground coal mines using ZigBee. Int. J. Electron. Commun. Eng. 3(10), 24–27 (2016)
11. Ferdoush, S., Li, X.: Wireless sensor network system design using Raspberry Pi and arduino for environmental monitoring application. In: The 9th International Conference on Future Networks and Communications, vol. 34, pp. 103–110 (2014)
12. Jang, R., Soh, W., Jung, S.: Design and implementation of data-report service for IoT data analysis. In: International Conference on Chemical, Material and Food Engineering, pp. 694–697(2015)

Assistive Device for Neurodegenerative Disease Patients Using IoT

Saravanan Chandrasekaran[1(✉)] and Rajkumar Veeran[2]

[1] School of Engineering and Technology, Department of Computer Science
and Engineering, JGI Global campus, Jakkasandra Post, Kanagapura Taluk,
Ramanagara District, Bangalore, Karnataka, India
doctratesaravanan@gmail.com
[2] Department of Computer Science and Engineering, Krishnasamy College
of Engineering and Technology, Cuddalore, India
raj_win7@yahoo.com

Abstract. Today the Neurodegenerative disease is one of the major concerns in the Healthcare industry. There are several types of Neurodegenerative disease and it directly affects the driving force of the body, Brain. As a result the involuntary movement occur such as tremor and this causes disability, which leads the person to depend on others. To support the disabled person, an assistive device for self-feeding is designed using IoT. The main theme of the project is to design a handle, which senses the movement of the hand and accordingly it create a counter action on the quivering action of the hand. The measurement parameters, the deviation of yaw, pitch and tilt of the spoon is measured using MPU-6050 and it is fed into the Arduino Uno Board. The Arduino Uno takes counter measure to make the spoon in stable position using servo motors and this helps the patients to act independently. This project uses the advantage of IoT to serve better for the Neurodegenerative disease affected community and also to monitor the condition of the patient health.

Keywords: IoT · Arduino · Servo motor · Neurodegenerative disease

1 Introduction

Neurodegenerative disease is a acute disease which affects the nervous system of the body. Due to this the proper functioning of the and the routine activities is severally affected. This type of disease takes longer time to cure. It is more prevalent around the age of 60 years. This paper deals with designing a assistive device for people suffering from Neurodegenerative disease, where they can consume their food properly and not wasting their food. Today, this type of disorder is prevalent across the globe [1] and the people are affected physically in large number, and they have to consume their food with the help of spoon. The main problem is that the person with such kind of disorder cannot able to act independently. To resolve this problem, an assistive device is designed which acts in opposite direction, if the involuntary movement of hand is in negative direction and vice versa. This causes the food spillage to zero percentage and the shaking of the hand is reduced. The device designed can be extended to hold the

© Springer Nature Switzerland AG 2020
J. S. Raj et al. (Eds.): ICIDCA 2019, LNDECT 46, pp. 447–452, 2020.
https://doi.org/10.1007/978-3-030-38040-3_51

small objects, which is useful for them. To address the issue, the Arduino board is fed with input as the movement of hand and it provides a pair of servo motor and it equalizes with the negative and opposite value to ensure the spoon remains stable. The research in medical field concludes the disease may be subjected to genetic and issues pertaining to environment [2].

The rapid development of IoT, makes all the systems automated in real-world. IoT extents its research in medical field, secure devices, Agriculture, smart city, and the integration of big data analytics with the industry makes the prediction of future sale in detail. Due to increase in the number of patients related to Neurodegenerative disease, the healthcare industries are in a mandatory situation to investigate the causes of the disease and assist the patients with hassle-free. The survey in North America has predicted that one million peoples are affected with different types on nervous disorders [3]. In healthcare field it can provide huge amount of success. The workflow of the paper is as follows: Sect. 2 details about the Related works. Section 3 details the Hardware used for the prototype, Sect. 4 details the proposed prototype and Sect. 5 discusses the result analysis. At last, Sect. 6 briefly discusses about the conclusion and future work.

2   Related Works

Neurodegenerative diseases is more prevalent in worldwide, among them Alzheimer's disease and Parkinson's disease are more common [11]. In 2016, it is estimated that 5.4 million Americans were affected with this type of disease. It is predicted that in the year 2020, around 930,000 people will be affected in United States. Various Neurodegenerative disorders; models and mechanism have been discussed in the paper [4]. This survey shows many model designed and drugs user to activate the motor function of the body. The paper [5], construct a spoon using PID Controller which controls the change in angle of spoon to maintain it very stable. The early prediction of Neurodegenerative disease is by collecting the speech signal of a person and classifying based on certain characteristics [6]. The work in [7], suggest a self-balancing spoon for tremor patients and it creates a handy prototype for the Parkinson's patient. Non-motor symptom such as Cognitive impairment is more prevalent in Neurodegenerative disease [8]. Several current treatment approaches [9] for cognitive symptoms are studied for Parkinson's disease to assist the affected patients. The interfacing unit is described detailed in [10]. In [11], the machine learning model is proposed to analyze the gait disorder patients. A vibration is generated using the prototype to study the simulation [12] of the tremor patients and the simulation shows the good result when compared to other proposed mechanism. An active game is developed to evaluate the person motor impairment [13]. Authors designed to rehabilitate the upper limb for aged peoples with Neuro disease [14] using motion senor. Although many medical practitioners are taking necessary steps to handle the pain of the patients, there should be an assistive device to handle the day-today activities. This paper comes with the solution of designing an assistive device for the patients to intake the food independently by using the advantage of IoT.

3   Hardware

This Hardware unit describes the components used to assist the patients with tremor. The overall work is based on the Hardware and to design a real-time prototype. The working principle and the advantage of the each hardware is shown below.

3.1   MPU-6050

The chip is made up of gyro sensor and accelerometer sensor and it is available in the market at an affordable price. The MPU-6050 module is shown in Fig. 1. This module consist of 3-axis Gyroscope with MEMS technology. It is used to find the rotational velocity along the X, Y, Z axes as shown in Fig. 2.

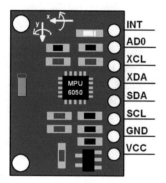

Fig. 1. MPU-6050

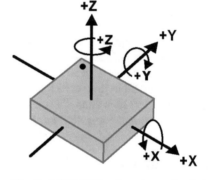

Fig. 2. MPU-6050 orientation and polarity of rotation

The MPU-6050 also consist of 3-axis Accelerometer with MEMs technology. It is used to detect angle of tilt or angle of inclination along the X, Y and Z axes as depicted in below Fig. 3.

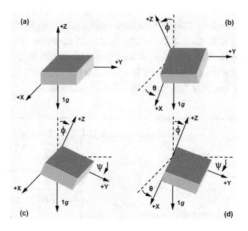

Fig. 3. 3-axis accelerometer

The sensor data points are read from the gyro sensor and the data points provides the crucial values such as roll, pitch, etc. In order to collect the data points from the sensor, the sleep mode should be switched off. The rotation and the tilt of the handler are detected from the sensor values.

3.2 Arduino UNO

The Arduino.cc developed a miniature computer, based on ATmega328P microcontroller. This microcontroller consists of digital and analog input/output (I/O) pins and these pins can be integrated to different circuit boards. This chip has 14 Digital pins, 6 Analog pins and it can be programmed with Arduino Integrated Development Environment. The chip is powered between the range of voltage 7 and 20 volts. The Arduino Uno is designed in such a way it is analogous to Arduino Nano and Leonardo.

3.3 Servo Motors

Servo motor is used to change the angle of rotation at the accurate level. The motor can be powered with AC and DC supply and available in different types of rating. The application of Servo motor is used in handheld devices, Robotic machines, etc. The feeding handle is controlled by the servo based on the data points fetched from MSPU-6050. The basic mechanism behind the operation of the motor is that, it works in opposite direction towards the negative motion.

4 Proposed Prototype

The proposed model Workflow is depicted in Fig. 4. The raw sensor data from the gyro sensor is fed into the Arduino Uno. From the data points, the change in angle is calculated. The motion of tremor hand is tackled by the change in angle and the servo motor is changed in a specific direction to counter the angle of assistive device. The servo motor provides a counter motion based on the output of the Arduino board. For the better results, the gyro sensor data is pre-processed to remove the unwanted noise and to normalize the resultant value. The angle change is calculated based on the sensor data and accordingly servo motor is turned to act in opposite direction to equalize the negative value. This makes the device to be rigid and stable to aid the patients suffering from neurodegenerative disease to take the food without spilling. The sensor data can be stored in online to analyze the patient's health condition.

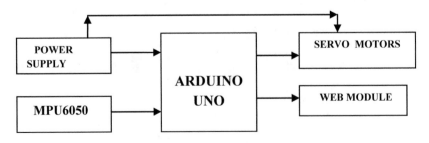

Fig. 4. Workflow of prototype

5   Results and Discussion

This section discusses the analysis of the prototype and the aim of the project is to assist the neurodegenerative disease patient. It is possible to analyze the sensor data online by storing the data points in the cloud. It is possible to create an alert system when the condition of tremor increases in day-to-day life.

The merits of this device are:

- The device helps the patients to eat without spilling the food.
- The device can be replaced by other handler so that any object can be handled by the patient.
- The consumption of power is less to automate the device.

The proposed device can handle the patients with moderate tremor. However, it is difficult to handle the patients with high risk of tremor.

6   Conclusion and Future Work

An assistive device is designed for the neurodegenerative patients, which aids in all the way for day-today activities. The interfacing unit MPU-6050 and Arduino Uno plays a major role and the power consumption used is less for operating. This prototype outbreaks the healthcare industry and the sensor data can be analyzed in detail, to know the seriousness of affected patients. It also to help to track the History of patients and take remedial measures to assist the patient. This system allows the patient to work independently without any help. The Future work is to analyze the data using machine learning algorithm and to predict the degree of seriousness of a patient.

References

1. Van Den Eeden, S.K., Tanner, C.M., Bernstein, A.L., Fross, R.D., Leimpeter, A., Bloch, D. A., et al.: Incidence of Parkinson's disease: variation by age, gender, and race/ethnicity. Am. J. Epidemiol. **157**, 1015–1022 (2003)
2. Lesage, S., Brice, A.: Parkinson's disease: from monogenic forms to genetic susceptibility factors. Hum. Mol. Genet. **18**, R48–R59 (2009)
3. Lang, A.E., Lozano, A.M.: Parkinson's disease. N. Engl. J. Med. **339**, 1044–1053 (1998)
4. Gitler, A.D., Dhillon, P., Shorter, J.: Neurodegenerative disease: models, mechanisms, and a new hope. Dis. Model. Mech. **10**, 499–502 (2017)
5. Baby, C.J., Mazumdar, A., Sood, H., Gupta, Y., Panda, A., Poonkuzhali, R.: Parkinson's disease assist device using machine learning and Internet of Things. In: International Conference on Communication and Signal Processing (ICCSP), Chennai, pp. 0922–0927. IEEE (2018)
6. Froelich, W., Wróbel, K., Porwik, P.: Diagnosing Parkinson's disease using the classification of speech signals. J. Med. Inform. Technol. **23**, 187–194 (2014)
7. Gifty, E.B., Vandana, M.: Parkinson's tremor stabilization spoon. Int. J. Res. Sci. Innov. **5**, 27–28 (2018)

8. Wolter, A.F., Weijer, S.C.F.V.D., Leentjens, A.F.G., Duits, A.A., Jacobs, H.I.L., Kuijf, M. L.: Resting-state fMRI in Parkinson's disease patients with cognitive impairment: a meta-analysis. Park. Relat. Disord. **62**, 16–27 (2019)
9. Rektorova, I.: Current treatment of behavioral and cognitive symptoms of Parkinson's disease. Park. Relat. Disord. **62**, 16–27 (2019)
10. Electronic Wings. https://www.electronicwings.com/arduino/mpu6050-interfacing-with-ard uino-uno
11. Tang, S.Y., Hoang, N.S., Chui, C.K, Lim, J.H, Chua, M.C.H.: Development of wearable gait assistive device using recurrent neural network. In: IEEE/SICE International Symposium on System Integration (SII), pp. 626–631. IEEE (2019)
12. Nandan, S., Zheng, Z.K.: Design and validation of a tremor stabilizing handle for patients with Parkinson disease and essential tremor. In: International Conference on Human-Computer Interaction, pp. 274–283. Springer (2019)
13. Duraipandian, M., Vinothkanna, R.: Cloud based Internet of Things for smart connected objects. J. ISMAC **1**(02), 111–119 (2019)
14. Oana, G., Postolache, O.A., Chiuchisan, I., Prelipceanu, M., Hemanth, D.J.: An intelligent assistive tool using exergaming and response surface methodology for patients with brain disorders. IEEE Access **7**, 21502–21513 (2019)

Ablation of Artificial Neural Networks

Y. Vishnusai[✉], Tejas R. Kulakarni, and K. Sowmya Nag

Department of Electronics and Communication Engineering,
RV College of Engineering, Bengaluru, India
{vishnusai.ecl5, tejasrkulkarni.ecl5,
soumyanagk}@rvce.edu.in

Abstract. Recent research in the field of Artificial Neural Networks (ANNs) has led to a lot of new discoveries in their applications. Many complex problems have been solved using ANNs. But in most cases the question of "how such complex information is stored" or "how such classification or differentiation of input data happens" in the ANNs is still not completely understood. Almost all ANNs developed and introduced are in a way black boxes, in which the way of information processing is not understood at the neuron level considering that single neuron in the network as a whole. It so happens that the networks are trained and the final weights of the networks depend on the initialization and subsequent change of weights during training and there is no proper understanding of what different parts of the network and different neurons contribute to the task as a whole. Ablation is one technique which tries to understand and answer those questions and is a technique as proposed in this paper, which can be used for better training and enabling faster inference of ANNs [1]. And also from the results achieved in this literature it is clear that ablation as a technique can be used to fine tune most ANN architectures and gain more insight over how each individual neurons or parts of the network contribute to the task to be performed.

Keywords: Deep learning · Artificial neural networks · Ablation · Datasets · Dropout · Pruning

1 Introduction

The recent breakthrough in the field of Artificial Intelligence (AI) has resulted in a large number of remarkable applications like object detection and classification in the field of Computer Vision [2–6], speech recognition and segmentation in Natural Language Processing [7–10]. Also the rise of hardware accelerators like GPU's have resulted in the increase of the average size, i.e., the number of weights that could be trained in the network. These networks exhibit cryptic behavior which cannot be fully explained by considering the operations of individual components like the units, their activations and their regularization mechanisms, etc. Despite the development in Deep Learning (DL) research, much of the focus has been towards increasing metrics such as top-1 and top-5 accuracies, speed of computation and competing towards standard benchmark standards. But, there has been no comprehensive study which examines the behavior of these networks internally while the operation of training of the network's weights is

© Springer Nature Switzerland AG 2020
J. S. Raj et al. (Eds.): ICIDCA 2019, LNDECT 46, pp. 453–460, 2020.
https://doi.org/10.1007/978-3-030-38040-3_52

being carried upon. With this motivation, we used a method from medical human neuroscience called as ablation, to analyse the structure of information presented in Deep Neural Networks (DNNs).

Ablation is a neuroscience inspired method wherein specific neural tissues inside the brain are damaged in a controlled manner, while parallely measuring its functionality by its performance against a specific task. While, this won't give an exact measure of the contribution of the damaged neural tissue, it fairly gives an estimate of its functionality and importance of a given neuron against the task performed. The same technique can also be used in a DNN where the function of its individual or a group of neurons is not known. Similar to neural tissues in the brain, the neurons of the deep neural network contain certain information, which cannot be quantised through any methods proposed till now. So ablation can be applied on these neurons to estimate the amount of vital information present in these neurons. This paper presents few novel intuitive ablation techniques that can be applied to analyse the performance of a DNN trained on the standard MNIST dataset. The observed results mentioned in this paper holds good for only the given DNN trained on the MNIST dataset. Generally the ablation results depend on the type of the neural network, the architecture of the neural network, the dataset used for training and the hyper-parameters used. Ablation of neural networks is used to remove neurons which hold no significant or redundant information, thus effectively reducing the computational costs, the training time and the inferencing time.

The rest of the paper is organized as follows. Section 2 deals with the previous works carried out in this area. Section 3 introduces the proposed method for ablation. Section 4 discusses the results and inferences obtained by training a neural network using techniques discussed in Sect. 3. Section 5 concludes the paper and Sect. 6 discusses the future work.

2 Related Work

In this particular field of ablation techniques applied to neural networks, very less work has been done till now. Even though pruning of neural networks are carried out regularly, which essentially increases the speed of training, ablation introduces the idea of indirect visualization of the information distribution in the neural networks. Lillian et al. [10] proposed a method in which ablation was done layer-wise and the results obtained were categorized by computing the principal component analysis (pca). Apart from this, no significant literature could be found on this subject matter.

In a process to further increase research in this field, this paper focuses on two methods. The first method focuses on unit, pair-wise and group wise ablation in a single layer of a network and the second method focuses on layer wise ablation to observe regularization and layer-wise information distribution.

3 Methods for Ablation

This paper focuses on two intuitive techniques that can be considered and applied for ablation. They are:

A. Pair Wise, Group Wise Ablation in a Single Layer

This method performs ablation inside a single layer of a neural network. The types of ablation varies from unit wise, pairwise and group wise. The techniques were applied on a DNN whose architecture is shown in Fig. 1. There are five layers in the neural network. The input layer has 10 neurons, then three hidden layers with 20, 15, 20 neurons respectively. All the layers use Relu activation function. The output layer uses softmax activation for all its 10 neurons.

Initially, all the neurons are initialised with random numbers. Then ablation search is done unit-wise, pairwise and then group-wise. The procedure is described below:

- Train the DNN on the MNIST dataset for a certain number of epochs and observe the overall classification accuracy. This will be the baseline accuracy of the model.
- Perform unit wise ablation on one of the layers by freezing only those neuron's weights to zero. Train the neural network for the same number of epochs chosen earlier, without the neuron's weight(weight effectively disabled) and compare the classification accuracy with the baseline accuracy.
- Similarly perform pair wise and group wise ablation in the layer and compare the overall classification accuracies with the baseline accuracy.
- When these different accuracies are compared with the baseline accuracy, it is clearly noticeable that the ablation of those neurons which caused the least decrease in accuracy suggests that those neurons contribute the least to the final inference.
- Those neurons/pairs/groups are permanently disabled.
- The process can continue until the most optimized model could be obtained, thus reaching the required model with lesser number of trainable weights.

The neurons with the least contribution can form as a basis for pruning or dropout regularization. Instead of randomly dropping neurons, this technique identifies neurons or pair of neurons which could form a basis for dropout regularization.

B. Layer Wise Ablation

In a neural network, the initial layers are considered to hold high level features like edges, contours, etc. and the deeper layers are considered to hold more complex features like a hand, face, etc. This method focuses on analysing the importance of the information held by the initial and the deeper layers, with respect to its contribution to the final accuracy.

In this method, two factors are considered for experimentation. The first one focuses on layer-wise contributions to the total accuracy of the model. The second one focuses on choosing the appropriate layers for regularization, without compromising on information retained by the network

For experimentation, the neural network architecture shown in Fig. 1 is considered.

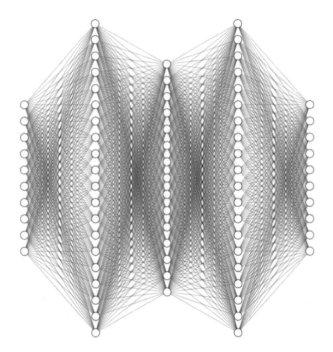

Fig. 1. Architecture of the DNN used

A series of experiments were done on the architecture shown in Fig. 1 to achieve the objectives proposed. They are:

- Train the neural network on the standard MNIST dataset for a specific number of epochs and observe the overall classification accuracy. This will be the baseline accuracy of the model. The number of epochs should be constant after this step.
- Reset all the weights of the neural network by random initialization except for weights in hidden layer 3 of the network. Hidden layer 3 will retain weights from the baseline model. Retrain the neural network and note the overall classification accuracy.
- Follow the same process specified in the previous step, but now freeze the weights in hidden layer 3 of the network and retrain for the specified number of epochs. Observe the classification accuracy.
- Now from the baseline model, retain only the weights of hidden layer 1 and retrain the network. Note down the classification accuracy again.
- Follow the same process again but this time, freeze the weights in hidden layer 1 of the network and observe the classification accuracy.
- After completing all the steps above, now in order to choose the appropriate layers for regularization apply dropout regularization with a specified dropout rate, to hidden layer 1 and 3, separately by retaining weights from the baseline model.

The above mentioned methods provide means to establish and differentiate the importance or the amount of critical information held in the initial layers and final layers, and also gives an intuition while applying dropout regularization in a neural network.

4   Results

In the last section the architecture of the DNN is specified and it is trained on the standard MNIST dataset. This section shows the observations and results obtained by applying the techniques of the previous section. The results are divided into two subsections, one each for the different techniques.

Out of all the images in the dataset, 60,000 images were used for training the model. The batch size was fixed at 32 for all the iterations. The number of epochs was 20 for the first method and 40 for the second method.

A. Experimental Results for Unit Wise, Pair Wise and Group Wise Ablations
The DNN shown in Fig. 1 was initialised with uniformly distributed seeded random numbers and trained for 20 epochs. After noting the baseline accuracy, the ablation techniques mentioned in the previous section were applied. The weights of the neurons in the other layers where ablation was not applied, were retained and in the layer where ablation was applied, the ablated neurons weights were set to zero and frozen and the rest of the neurons in the layer were randomly initialised again. The modified model was trained again for 20 epochs. The results for the various techniques used are shown in Table 1. Only the results having significant change in accuracy are shown.

Table 1. Results of applying group wise ablations to the DNN shown in Fig. 1

Techniques	Accuracy (%)
No ablation (Baseline)	96.2
Ablation applied to first five neurons in layer 1	92.16
Ablation applied to last five neurons in layer 1	92.26
Ablation applied to first 3 neurons in layer 1	95.13
Ablation applied to neurons 4, 5, 6 in layer 1	94.94
Ablation applied to last 3 neurons in layer 1	94.73
Ablation applied to first 10 neurons in hidden layer 3	95.79

From Table 1, the baseline accuracy obtained is 96.2%. When ablation was applied to the first five neurons in Layer 1, and then re-trained, the accuracy dropped to 92.16% and the accuracy observed when the last five neurons were ablated and re-trained the observed accuracy was 92.26%. An intuitive analysis suggests that the first five neurons hold more critical information when compared to the last five neurons, but when ablation was applied to the first three and last three neurons, it was observed that the dip in accuracy was more when the last three neurons were ablated. Also, ablating neurons 4, 5, 6 in Layer 1 showed a drop intermediate to the one discussed before. This might suggest that the pattern of grouping of neurons might have played a role while determining the final overall classification accuracy.

When ablation was applied to individual and pair-wise neurons, there was no significant drop in classification accuracy. The drop was only significant when a minimum of 3 neurons were selected to apply for ablation.

B. Experimental Results for Layer Wise Ablations

For this method, again the DNN in Fig. 1 was initialized with Xavier's initialization and the network was trained for 40 epochs. Then the ablation techniques mentioned in the previous section were implemented. The results for the same are shown in Table 2.

Table 2. Results of applying layer wise ablations to the DNN shown in Fig. 1

Techniques	Accuracy (%)
No ablation (Baseline)	96.06
Retaining the weights of hidden layer 3 only	96.33
Retaining and freezing of weights of hidden layer 3 only	96.05
Retaining of weights of hidden layer 1 only	96.55
Retaining and freezing of weights of hidden layer 1 only	96.45
Dropout applied to hidden layer 3 by retaining weights	88.5
Dropout applied to hidden layer 1 by retaining weights	62.02

When the neural network was trained on for 40 epochs the obtained accuracy was 96.06%. Figure 2 shows the graph of the increasing accuracy, decreasing loss against the number of epochs. This is the baseline accuracy for the model.

From Table 2, it is observed that the initial layers of the DNN holds more important information when compared to the deeper layers, taking into consideration the overall classification accuracy. By retaining the weight of hidden layer 1 of the network, the network's accuracy increased to 95.55% when compared to increase in accuracy of 96.33% obtained by retaining weights of hidden layer 3. Also, when Dropout regularization is applied to hidden layer 1 the accuracy drops down to a low of 62.02% when compared to dropout applied to hidden layer 1 where the accuracy drops to 88.5%. Also holding and freezing of hidden layer 1 gives better accuracy when compared to holding weights to hidden layer 3.

This paper presents ablation applied to hidden layer 1 and 3 of the neural network. It was observed that when the ablation techniques were applied to hidden layer 2 of the neural network, the results were the average of hidden layer 1 and hidden layer 3.

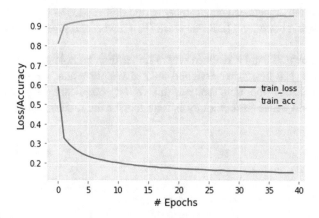

Fig. 2. Epochs vs loss/accuracy

5   Conclusion

From the above studies and results, it is clear that the information processing methods of many of fully connected networks can be understood by stagewise ablation. The parts of the network which contribute to the different tasks, which the network performs can be understood and localised. This knowledge can be mainly used to study that particular network, thus understanding to a better degree as to how the information and activations are localised for a particular output.

This knowledge can also be used to aid in many processes like pruning of the network which will make the network faster, in dropout technique by rather having to drop neurons not performing necessary functions, than randomly removing them and during training where the least contributing neurons can just be removed to have a lighter network at the end of training performing the same task.

6   Future Work

The team feels that the method used above is still in its infancy. The method can be polished more, made to include the study of all the neurons in an even more systematic and finer way. This would lead us closer to the low level understanding of the neural networks. A deeper study and finer ablation would clearly enable understanding of finer details of a network. This ablation method can be applied on other different architectures like Convolutional Neural Networks (CNN's) and Recurrent Neural Networks (RNN's).

The ablation study can be used as a tool to make the network concentrate on what is being required. It can be used to do classwise ablation, so as to perform ablation of neurons that aid to the not required classes/features of the dataset [11]. This would optimise the performance of the network to the maximum and make the network concentrate on required classes if training is continued further.

A clearer contrast between ablation being done for reduction in network complexity and ablation being done for a faster network (lengthwise and layerwise ablation) can be explored.

A further method can be realised where rather than completely disabling a neuron when ablated, it can be further assigned ablated weight where the degree of ablation can be controlled for finer ablation. But this would complicate the process to a larger degree easily and hence further study is required for the methods already being used.

References

1. Lillian, P., Meyes, R., Meisen, T.: Ablation of a robot's brain: neural networks under a knife (2019). arXiv.org. https://arxiv.org/abs/1812.05687
2. Hannun, A.Y., Rajpurkar, P., Haghpanahi, M., Tison, G.H., Bourn, C., Turakhia, M.P., Ng, A.Y.: Cardiologist-level arrhythmia detection and classification in ambulatory electrocardiograms using a deep neural network. Nat. News **25**, 65–69 (2019)
3. Mane, S., Mangale, P.S.: Moving object detection and tracking using convolutional neural networks. In: 2018 Second International Conference on Intelligent Computing and Control Systems (ICICCS), Madurai, India, pp. 1809–1813 (2018)
4. Deshmukh, S., Moh, T.: Fine object detection in automated solar panel layout generation. In: 2018 17th IEEE International Conference on Machine Learning and Applications (ICMLA), Orlando, FL, pp. 1402–1407 (2018)
5. Yu, L., Chen, X., Zhou, S.: Research of image main objects detection algorithm based on deep learning. In: 2018 IEEE 3rd International Conference on Image, Vision and Computing (ICIVC), Chongqing, pp. 70–75 (2018)
6. Shen, W., Wang, W.: Node identification in wireless network based on convolutional neural network. In: 2018 14th International Conference on Computational Intelligence and Security (CIS), Hangzhou, pp. 238–241 (2018)
7. Sintoris, K., Vergidis, K.: Extracting business process models using natural language processing (NLP) techniques. In: 2017 IEEE 19th Conference on Business Informatics (CBI), Thessaloniki, pp. 135–139 (2017)
8. Petridis, S., Li, Z., Pantic, M.: End-to-end visual speech recognition with LSTMS. In: 2017 IEEE International Conference on Acoustics, Speech and Signal Processing (ICASSP), New Orleans, LA, pp. 2592–2596 (2017)
9. Zhao, X., Haihong, E., Song, M.: A joint model based on CNN-LSTMs in dialogue understanding. In: International Conference on Information Systems and Computer Aided Education (ICISCAE), Changchun, China, pp. 471–475 (2018)
10. Luo, Y., Chen, Z., Mesgarani, N.: Speaker-independent speech separation with deep attractor network. IEEE/ACM Trans. Audio Speech Lang. Process. **26**(4), 787–796 (2018)
11. Meyes, R., Lu, M., de Puiseau, C.W., Meise, T.: Ablation studies in artificial neural networks. arXiv:1901.08644v2 [cs.NE]

Precedency with Round Robin Technique for Loadbalancing in Cloud Computing

Aditi Nagar[1(✉)], Neetesh Kumar Gupta[1], and Upendra Singh[2]

[1] Department of CSE,
Technocrats Institute of Technology & Science, Bhopal, India
shanunagar18@gmail.com
[2] Techbeanssolution, Indore, India
Upendrasingh49@gmail.com

Abstract. Load Balancing is a growing field that effort to distribute the dynamic workload as hubs in the cloud. Load Balancing is a crucial test of distributed computing. In Load balancing procedures, execution of the cloud is improved by using all assets ideally. The critical goal of Load balancing is to diminish asset utilization of vitality and to limit carbon emanation; this is the foreboding need of the time. The fundamental objective of this paper is to give a short diagram of different Load Balancing calculations and after that to provide a productive need-based round-robin and Precedency-based algorithm. Load balancing method which organizes different task to virtual machines are based on assets or processor required, the number of clients, time to run, work type, client type, programming utilized, cost and so on and after that passing on them to different accessible has in a round-robin and Precedency-based algorithm design. This methodology improves the capacity of the framework by upgrading different parameters, such as adaptation to non-critical failure, versatility and overhead, and so forth and by limiting asset use and reaction time. This methodology is recreated and tried over Cloud Analyst, which is generally utilized by the device to test cloud-based systems.

Keywords: CloudAnalyst · Round Robin Scheduling · Precedence scheduling · Virtual machine · Resource allocation

1 Introduction

Load Balancing is a procedure to course the workload transversely finished distinctive centres or diverse resources. The essential point is to complete powerful resource utilize, augment in the throughput, reduce the response time, and farthest point the overhead. It moreover achieves the sensibly scattering of the workload.

A DNS server is reserved for making an elucidation of the convey to the particular IP while getting to the specific web benefits through the addresses. This URL understanding is depended to pick a particular point of the centre from the gathering. This relies upon the arranging methodology of the web servers. A period is described to hold each one of the translations to the extent TTL (Time to Live). After a chance to store the understanding is slipped by, the accompanying endeavours coordinated to the

© Springer Nature Switzerland AG 2020
J. S. Raj et al. (Eds.): ICIDCA 2019, LNDECT 46, pp. 461–475, 2020.
https://doi.org/10.1007/978-3-030-38040-3_53

server. Round Robin is the best way to deal with executes the load in a rotating way. Along these lines, the store modifying is proficient in DNS servers [1].

In the framework based approach, programming/gear is acquainted as a front end with modifying the load to the in light of the data found in the traditions including the framework layer tradition or the data layer tradition. The path toward coordinating does this. Apache can, in like manner, be used as an HTTP load balancer which gets the sales from the servers and in the wake of taking care of passes on the requesting to the customers. It is like manner keeps tracks of the sessions that empower a singular customer to deal with the single server [2].

The other trademark is it is versatile, suggests its ability to alter the extended solicitations of the CPU accumulating, exchange speed et cetera. Dispersed registering gives the protected access to the applications as showed up in Fig. 1. Regardless, foreign state security is a test for cloud engineers.

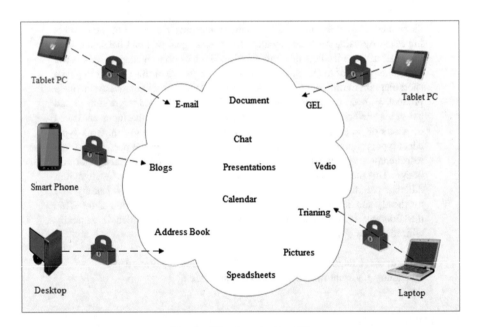

Fig. 1. Cloud computing [3]

The natural figuring resources are versatile yet are not adaptable. More than one occupant can use the advantages of dispersed processing. Similarly, the organizations are metered infers; one can pay only for the benefits it eats up; however, in standard systems, the cost is settled in perspective of the necessities [4–9].

Vishva Joshi et al. Load balancing divide task between available machine using auto-scaling, which is increasing or decreasing machines based on requirements.

M. Kanthimathi et al. Thus, in cloud computing dynamic workload distributed and by the load balancing optimization problem and divide the workload according to the needs. This paper proposed a novel ant colony based algorithm to balance the load by searching under loaded one [11].

Ryuta Mogi et al. Smartphones are having a various sensor, Internet of Things (IoT) and intelligent transport systems. However there are still problems such as cooperation problem, off-load problem and orchestration problem. This paper proposed a load balancing for multi-access edge computing when traffic congestion and guerrilla rainstorms occur [12].

2   Problem and Objective

2.1   Research Problem

- In a base paper apply round-robin algorithm, throttled load balancing algorithm, and Active Load Balancing for cloud load balancing, which is already inbuilt on cloud analyst tool.
- Small task wait time to large.
- The algorithm compares only three parameters, like Data centre processing time, response time, and cost.

2.2   Objectives of Study

- Detailed examination of the load modifying systems and their connection.
- Comparison of (Round Robin Scheduling and Precedency scheduling) and examination of Load Balancing Algorithms.
- Proposal for Efficient Load Balancing Algorithm.
- Execution and Analysis of Results.

3   Research Methodology

3.1   Benefits of Proposed Algorithm

- Cost is less compared to the base paper algorithm for cloud load balancing.
- The resource utilization rate is more for cloud load balancing.
- The scheduling success rate is more for cloud load balancing.
- If come high Precedency task to our algorithm is differentiated.

3.2   Proposed Framework

Figure 2 shows a proposed system structure plan, which is a concentrated framework presented in server ranch and handle other server ranches. There is three-layer first one is justified the application where occupations submitted by a bunch of customers can hold up until the availability of virtual resources. Next layer is store balancer where examines the condor line status, calls or movement, and the last one is a module where operation performed on a virtual machine.

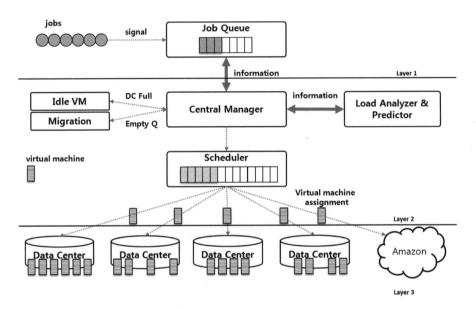

Fig. 2. System architecture

3.3   Proposed Algorithm of Load Balancing Mechanism

In this section Load balancing mechanism proposed which is applied to get authority of proper response for a sole intention. By this mechanism cost, minimization and faithful utilization of cloud resources of a virtual machine are possible. This mechanism explains about heap modifying among virtual resources, enlarge the throughput, and extends the benefits bit rate.

Algorithm: Resource Selection and Monitoring utilizing Precedency-based Allocation.

Algorithm 1 Precedency-based Allocation

1: procedure PRECEDENCY-B_VM_ ALL ()

2: for Dcci; (K ← 1;n) do

3: if DcvipPr == 1 then. Find for Precedency

4: Cloud ← Dcci

5: break

6: end if

7: end for

8: Dcci ← Cloud

9: min ← 0 . Init

10: for DccipHj ; (j ←1; n) do . for all VM

11: if min > DccipHjFV or min == 0 then

12: if DccipHjFV > 0 then

13: Mach id← DccipHj

14: min ← DccipHjFV

15: end if

16: end if

17: end for

18: DccipHj ← Mach id

19: DccipFV ← DccipFV←1

20: DciHjFV← DciHjFV ←1

21: CREATE VM (DccipHj)

22: end procedure

Standard Performance Time (Tek): Standard Performance time of load with possessions is calculated using Eq. (1).

$$Te_k = \sum_{i=1}^{n} \left(t_i(e_k)/n \right) \tag{1}$$

Here n is the number of tasks, t_i is the start time of task and e_k is the execution time of tasks.

Throughput: Throughput (T_t) is scheduled based on the number of loads and Mean performance time of every load.

$$T_t = \text{No. of tasks} \times \text{Average execution time of each tasks } (Te_k) \tag{2}$$

Link Communication and Response Rate: The Link Communication and response rate depend on every request and response from user to sent and received total no. of bytes (kbps) to virtual machines.

Source Consumption Rate (ru_j): Resource utilisation rate is the decrease in the execution time from start to last performance of each task.

$$ru_j = \Sigma t_i \text{ where } t_i \text{ will be executed on } r_j(te_i - ts_i) \qquad (3)$$

Where, te_i is the finishing time and ts_i is the start time of task t_i on resource r_j.

Scheduling Achievement Rate ($SSR_{i,j}$): It is based on useful resource utilisation of resources by the functions.

$$SSR_{i,j} \rightarrow \Sigma t_{i=1}\left(ru_j/m\right) \qquad (4)$$

where, ru_j is the resource utilization rate of resource r_j and m is the number of tasks in each job.

4 Simulation and Result

4.1 Implementations on CloudAnalyst

In our proposition using the CloudAnalyst, Imitated a data centre with two has each with the two PEs. We have made two VMs which require one PE each. These VMs are designated as hosts in light of the no. of PEs open in the host and no. of PEs needed by the VM. The occupations are given to the VMs for execution. The underlying two professions are distributed to the VMs in perspective of First Come First Serve commence. The due date of the accompanying work is checked. If it has a low due date than the underlying two occupations, then it is a high need work. Something unique, the business is a low needs work. The low need work must be executed after any of the occupations that have completed their execution. For the implementation of high need work, any of the occupations in the performance should be suspended. The occupation for suspension is picked upon the lease sort of the business. If no less than two professions are chosen, by then the length of whatever is left of the vocations is checked. Select the work for which the range of extraordinary occupation is generally outrageous. By then, the picked work is suspended. The high need for work is executed in the VM from which work was suspended. Disseminate VM for the suspended work if any of the occupations have completed their execution. A comparable methodology is taken after for all the moving toward rules.

4.1.1 Round Robin and Precedency Based Scheduling Algorithm Configuration

Figure 3 is main configuration like add new user and related region, during execution configure like several users, duration of stimulation, and a number of the data centre. In a data, centre programmer, create a total of six users and 5 data centre regions. If you want to increase the number of user and data centre region the use add new button.

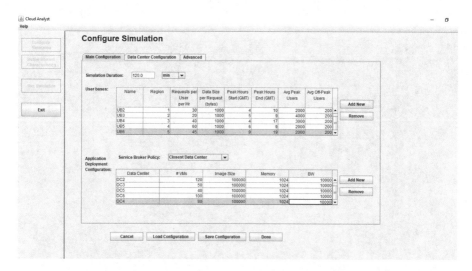

Fig. 3. Main configuration like add new user and related region

Figure 4 is data centre configuration like add new data centre according to user need, during execution, configure who data centre belongs to which region. Means a single data centre has a different area. The programmer creates a total of 5 data centre regions, and each data centre configures physical hardware requirement maintain individual. If you want to increase the number of user and data centre region the use add new button.

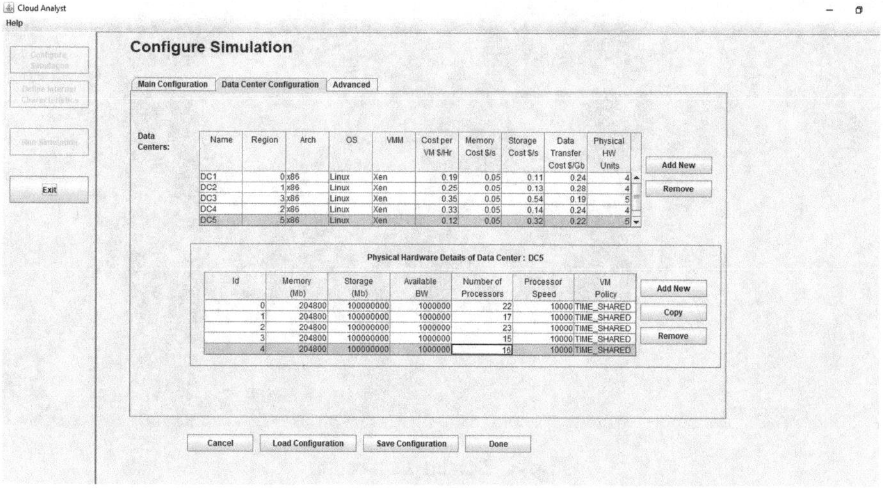

Fig. 4. Data center configuration like add new data center according to user need

Figure 5 is in advanced choose the load balancing policy algorithm (round-robin), programmer finally selects Load Balancing police, and programmer accepts the Round Robin Scheduling algorithm, which belongs to existing work.

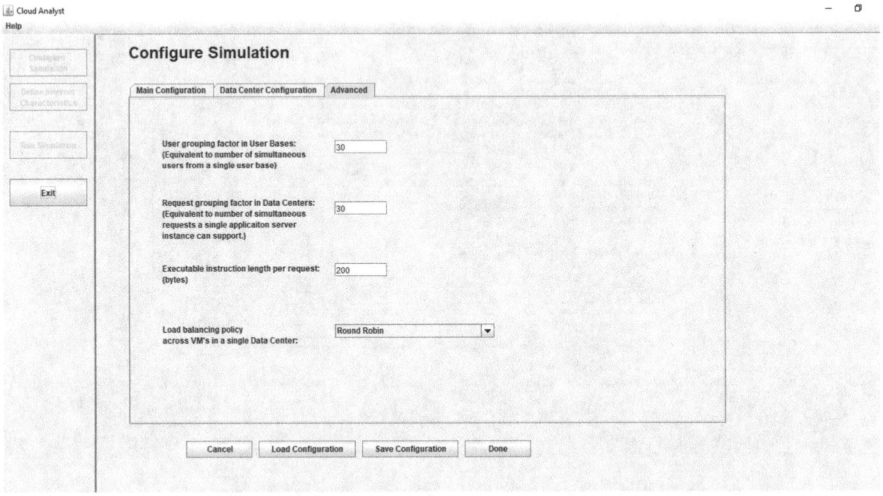

Fig. 5. In advanced choose the load balancing policy algorithm (Round Robin)

Figure 6 eventually, display data centre and user in a glob. In glob divided into five regions, and all areas have a separate data centre, the user chooses appropriate near the data centre.

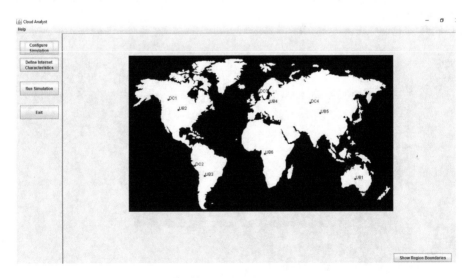

Fig. 6. Display data center

5 Results After Simulation

In this part, the author discusses the existing work and proposed work algorithms result. Proposed work algorithm best compares to existing work algorithm in all parameter describe Sect. 4.3. Suggested work parameters describe below section step wisely.

5.1 Resource CPU Usage

The performance of CPU usage on Graph 7 of VM1, VM2 & VM3. The graph 51 shows when virtual machine VM1 has 55% of CPU usage reach threshold value then-new virtual machine VM2 introduced and when VM2 reached threshold point then-new virtual machine VM3 introduced and incoming requested are redirected to the newly introduced virtual machine.

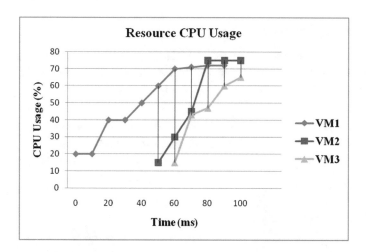

Graph 7. CPU usage

5.2 Resource Memory Usage

Graph 8 indicates the memory utilisation amongst different virtual machines (VM1, VM2 and VM3). In the establishment, each virtual machine has approx the same memory execution. Future primary VM has high memory utilisation and sooner heap adjusted remaining virtual machine at the same level.

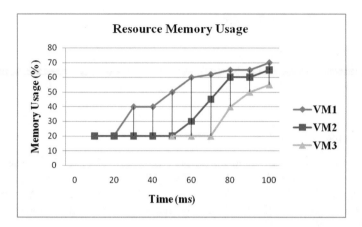

Graph 8. Memory usage

5.3   Standard Performance Time

The Standard Performance Time (Sec.) of each undertaking on comparing VM are taken, and those qualities are designed in the cost network table. Standard Performance Time worth is determined to utilize the Eq. (1).

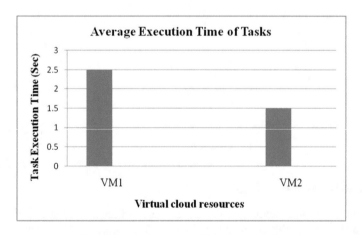

Graph 9. Average execution time of tasks

Graph 9 demonstrates the task execution time for every virtual machine, and it demonstrate task execution time is different for various virtual machines.

5.4   Throughput

The throughput of the virtual machine is determined by the outcome of various undertakings and Standard execution time of every assignment utilising the Eq. (2).

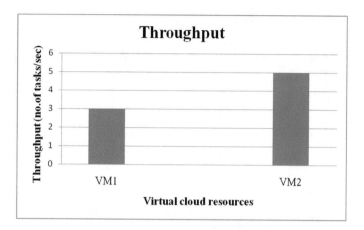

Graph 10. Throughput of virtual resource

Graph 10 demonstrates the throughput estimation of virtual cloud assets. Given the Standard execution time figuring, contrast with VM1, VM2 boosts the throughput esteem.

5.5   Link Communication and Response Rate

The Link Communication rate is the number of byte dispatch from the virtual machine to client. And Reception rate is a number of bytes received to the client.

Graph 11 and Table 1 demonstrate the Communication and Response rate of VM1 is high founded on the number of errands when contrast with VM2 asset.

Table 1. Link communication and response rate

	VM1	VM2
Data sent	170	140
Data received	470	310

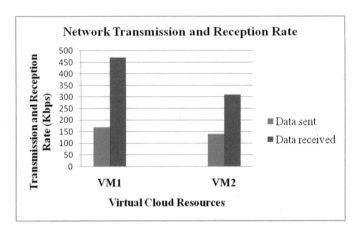

Graph 11. Link communication and response rate

5.6 Source Consumption Rate

Resource utilisation rate is determined by the decreasing in the execution time from start to last performance of each task.

Graph 12 and Table 2 demonstrate the asset use pace of the considerable number of errands on each virtual cloud assets. The diagram shows the Resource utilisation rate is high in VM1.

Table 2. Resource utilization rate

	VM1	VM2
T1	2	0.9
T2	3	1.8
T3	4	2.8
T4	2	2

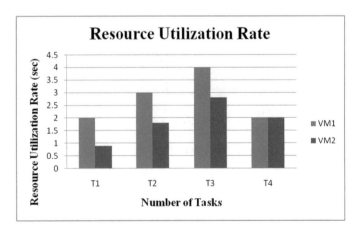

Graph 12. Resource utilization rate

5.7   Scheduling Achievement Rate (SSR)

It is based on effective source consumption of possessions by the tasks. It is determined to utilize the Eq. (4). Graph 13 demonstrates the scheduling success rate is high for VM1 when contrast with VM2.

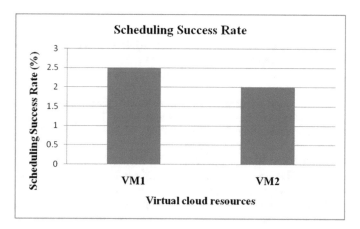

Graph 13. Demonstrates the scheduling success rate is high for VM1 when contrast with VM2

6   Conclusion and Future Scope

6.1   Conclusion

In a distributed computing for Load balancing, base paper apply just round-robin calculation, throttled Load balancing calculation and Active Load Balancing for cloud Load balancing which is already inbuilt on cloud investigator apparatus. In the existing framework, little errand holds up a time to enormous. Existing calculation compare just three parameters like server farm preparing time, reaction time and cost.

Through proposed calculation cost is less compared to the base paper calculation for cloud Load balancing. Same asset usage rate is more for cloud Load balancing. Through offered calculation planning achievement rate is more for cloud Load balancing. Whenever come high need errand to our calculation is separate.

6.2   Future Scope

In the future, the load balancing can be all the more capable by doling out the weights to the parameters logically with a particular real objective to discover the need of the virtual machine. In this proposition, the response time of the computation is improved, yet the effect of doling out different weight on response time can in like manner be evaluated, and the results can be contemplated in the future. So like this, load altering can be more exceptional.

The calculations can be calibrated further to accomplish better predictable outcomes on all the other points of view. Correspondingly, the correlation results ought to be taken for the distinctive activity entry designs on all the three diverse planning and load balancing calculations.

References

1. Chiang, M.-L., Hsieh, H.-C., Tsai, W.-C., Ke, M.-C.: An improved task scheduling and load balancing algorithm under the heterogeneous cloud computing network. In: 8th International Conference on Awareness Science and Technology (iCAST 2017), pp. 290–295. IEEE (2017)
2. Deepa, T., Cheelu, D.: A comparative study of static and dynamic load balancing algorithms in cloud computing. In: International Conference on Energy, Communication, Data Analytics and Soft Computing (ICECDS 2017), pp. 3375–3378. IEEE (2017)
3. Chouhan, P.S., Samvatsar, M., Singh, U.: Energetic source allotment scheme for cloud computing using threshold-based. In: 2017 International Conference of IEEE Electronics, Communication and Aerospace Technology (ICECA), 20–22 April 2017, pp. 1–8 (2017). http://ieeexplore.ieee.org/xpl/mostRecentIssue.jsp?punumber=8168799
4. Geetha, P., Rene Robin, C.R.: A comparative-study of load-cloud balancing algorithms in cloud environments. In: International Conference on Energy, Communication, Data Analytics and Soft Computing (ICECDS 2017), pp. 806–810. IEEE (2017)
5. Gupta, P., Samvatsar, M., Singh, U.: Cloud computing through dynamic resource allocation scheme. In: 2017 International Conference of IEEE Electronics, Communication and Aerospace Technology (ICECA), 20–22 April 2017, pp. 1–5 (2017). http://ieeexplore.ieee.org/xpl/mostRecentIssue.jsp?punumber=8168799
6. Kumar, P., Bundele, M., Somwansi, D.: An adaptive approach for load balancing in cloud computing using MTB load balancing. In: 3rd International Conference and Workshops on Recent Advances and Innovations in Engineering, 22–25 November, pp. 1–6. IEEE (2018)
7. Elrotub, M., Gherbi, A.: Virtual machine classification-based approach to enhanced workload balancing for cloud computing applications. In: 9th International Conference on Ambient Systems, Networks and Technologies, ANT 2018 and the 8th International Conference on Sustainable Energy Information Technology, SEIT 2018, 8–11 May, pp. 683–688 (2018). ScienceDirect
8. Volkova, V.N., Desyatirikova, E.N., Hajali, M., Khodar, A., Osama, A.: Load balancing in cloud computing. In: 2018 IEEE Conference of Russian Young Researchers in Electrical and Electronic Engineering (EIConRus), pp. 387–390. IEEE (2018)
9. Hussain, A., Aleem, M., Islam, M.A., Iqbal, M.A.: A rigorous evaluation of state-of-the-art scheduling algorithms for cloud computing, vol. 4, pp. 1–15. IEEE (2018). IEEE Translations and content mining are permitted for academic research only. Personal use is also permitted, but republication/redistribution requires IEEE permission
10. Joshi, V., Thakkar, U.: A novel approach for real-time scaling in load balancing for effective resource utilization. In: 2018 International Conference on Smart City and Emerging Technology (ICSCET), pp. 1–6. IEEE (2018)
11. Kanthimathi, M., Vijayakumar, D.: An enhanced approach of genetic and ant colony based load balancing in cloud environment. In: International Conference on Soft-computing and Network Security (ICSNS), pp. 1–5. IEEE (2018)

12. Mogi, R., Nakayama, T., Asaka, T.: Load balancing method for IoT sensor system using multi-access edge computing. In: Sixth International Symposium on Computing and Networking Workshops (CANDARW), pp. 75–78. IEEE (2018). https://ieeexplore.ieee.org/xpl/mostRecentIssue.jsp?punumber=8589960

13. Sthapit, S., Thompson, J., Robertson, N.M., Hopgood, J.R.: Computational load balancing on the edge in absence of cloud and fog, pp. 1–14. IEEE (2018). IEEE Translations and content mining are permitted for academic research only. Personal use is also permitted, but republication/redistribution requires

14. Ejaz, S., Iqbal, Z., Shah, P.A., Bukhari, B.H., Ali, A., Aadil, F.: Traffic load balancing using software defined networking (SDN) controller as virtualized network function, pp. 1–13. IEEE (2019). Translations and content mining are permitted for academic research only. Personal use is also permitted, but republication/redistribution requires

15. Singh, A.K., Kumar, J.: Secure and energy aware load balancing framework for cloud data centre networks. Electron. Lett. **55**(9), 540–541 (2019)

Stock Market Prediction
Using Hybrid Approach

Sakshi Jain[1](✉), Neeraj Arya[2], and Shani Pratap Singh[3]

[1] CSE, Acropolis Institute of Technology and Research, Indore, India
j.sakshi1999@gmail.com
[2] Shri Govindram Seksaria Institute of Technology and Science, Indore, India
neerajaryagate2010@gmail.com
[3] ETC, Institute of Engineering and Technology, DAVV, Indore, India
shanipratapsingh018@gmail.com

Abstract. Stock Market is becoming a new trend to make money. It is the fastest-growing system which is changing in every second. It is challenging and complex by nature which can make a drastic change in an investor's life. There are two possibilities either people will gain money, or he will be going to lose his entire savings. So for safe side stock market prediction is required, which is based on historical data. In this paper, we have proposed a hybrid approach for stock market prediction using opinion mining and clustering method. A domain-specific approach has been used for which some stock with maximum capitalization has been taken for experiment. Among all the available approaches our proposed model is different alike existing methods it not only considers general states of mind and sentiments, but it also forms clusters of them using clustering algorithms. As an output of the model, it generates two types of output, one from the analysis of sentiment while another one from clustering-based by taking popular parameters of stock exchange into consideration. The final prediction is based on an examination of both the results. Also, for empirical analysis, we have considered stocks with maximum capitalization from 6 growing sectors of India like banking, oil, IT, pharma, automobile, and FMCG. As a result, we have observed that predicted values from the proposed approach show maximum similarity with the actual values of the stock. The hybrid model returns efficient results in terms of accuracy in comparison with other individual methods of sentiment analysis and clustering.

Keywords: Stock Market · Clustering algorithm · Forecasting techniques · National Stock Exchange · Sentiments · Technical indicators

1 Introduction

Indian stock market [1] has its golden history and plays an important role in the Indian economy. It provides a platform to the investors, which can make them rich or take everything for them. Popular trading in the stock market of India is taking place from these popular stock exchange, namely Bombay Stock exchange (BSE) and National Stock exchange (NSE) [1]. BSE has come into existence in 1875 and has more than 5,000 listed firms till now. While NSE was founded in 1992 and in 1994, it started

© Springer Nature Switzerland AG 2020
J. S. Raj et al. (Eds.): ICIDCA 2019, LNDECT 46, pp. 476–488, 2020.
https://doi.org/10.1007/978-3-030-38040-3_54

trading. It has 1,600 firms listed. Both they follow similar trading mechanism, the process of settlement and hours of trading etc. For trading in both stock exchanges take place by using the book of a limit order in which trading computer is used for order matching, helps in the matching of the value of market orders placed by investors with the best limit order and keeps transparency in the process. In the experiment we have used datasets of the national stock exchange (NSE) [1].

As the trading rate increases, investors need something that will help them in the prediction of the stock market and provides the maximum profit. Many researchers are working on this. There are few tools and techniques available for forecasting which make use of sentiments of users, the general state of minds, historical data, and facts-figures. Some of them generate daily prediction while others generate a monthly prediction. [2] For many years, some traditional techniques have been used like time series, chaos theory and linear regression. But as the uncertainty in the stock market increases with the time, the efficiency of these traditional methods decreases. So, nowadays popular techniques and tools make use of well-known algorithms like Artificial Neural Network, Naïve Bayesian, fuzzy systems, Support Vector Machine and other soft computing techniques which deals with different variables involved in stock market such as GDP, political occasions, market price, face value, earning per share, beta etc. and perform well.

2   Related Work

Billah et al. [3] focused on the improvement of Levenberg Marquardt (LM) algorithm for training of ANN using less memory and taking less time as compare to others. They have taken a dataset of Dhaka Stock Exchange for their research.

Ercan et al. [4] proposed a NARX methodology to predict Baltic Market Value using Artificial Neural Network. As they found that in Baltic countries, ANN is not used to predict financial failures so, their study is based on it. To forecast the value of OMX index, they have been used the nonlinear autoregressive network with exogenous inputs (NARX) in their study. They have taken data in a range of January 1, 2013, to January 1, 2017. As a result, they have observed that NARX methodology predicted Baltic market values successfully.

Vajargah et al. [5] found that in prediction, there are few parameters which are uncertain, and that's why it increases the chances of risk. So, in their study, they used geometric Brownian motion random differential equation and simulation takes place through Monte Carlo and quasi-Monte Carlo methods. They observed that these methods made predictions more exact and better for total stock index and value at risk.

Huynh et al. [6] proposed a model for the prediction that is, BGRU and applied an extended model of RNN such as LSTM and GRU. For the prediction of stock value, it uses both online financial news and historical stock prices data. As a result, they observed that the proposed model is simple and effective, achieves 65% accuracy in individual stock prediction.

Mankar et al. [7] found that the stock market becoming very popular among investors as their side income, and for that, investors are taking help of experts as this field has its risk issues. SO, for the correct prediction of the stock market, they have proposed a system that helps in the prediction of stock price movement for different companies. For building this system, they have used sentiment analysis of the tweets collected from Twitter API and closing value of different stocks. Along with some pros and cons of the proposed model, it returns a better result. However, there is a scope of improvement in future.

Nayak et al. [8] proposed two models, one for daily prediction and another is for monthly prediction. Daily prediction model uses a combination of historical data with sentiments and used supervised learning algorithms for implementation. As a result, 70% accuracy has been found. In the case of monthly prediction, it compares the similarity between two consecutive months. As an evaluation, it has been observed that no two months trends return the same result.

Nivetha et al. [9] proposed a model by analyzing various prediction algorithms as well as presented a comparative study. This model returns next day market price by using monthly prediction and daily prediction. It makes use of correlation of sentiments and the stock value for prediction. Also, a comparative study of three algorithms: Multiple Linear Regression, SVM and ANN are done.

Oncharoen et al. [10] presented a framework to train deep NN for prediction of the stock market. By trading simulation results with risk reward function generates a new loss function. Sharpe ratio and F1 score combination generates a new scoring metric that is, Sharpe-F1 used for model selection. Evaluation of robustness has been checked by using two datasets varying key parameters. As a result, it has been observed that the combination of a risk-reward function and loss function helps to improve financial performance.

Patel et al. [11] performed a prediction and forecasting and accuracy as well as error rate are used as the evaluation standard.

Kato et al. [12] proposed a stock prediction method based on interrelated time series data. They have used data like, other stock data, oil price, foreign exchange and world stock market indices to extract interrelationship between the stock which is predicted and data of various time series. As a result, it has been observed that the proposed method predicts stock direction very well so, it is useful for the manufacturing industry.

Sharma et al. [13] presented a report on popular regression approach for the prediction of the stock market price from data. Also, from their survey report, they have concluded that the result of the multiple regression approach could be improved by using a number of variables.

Somani et al. [14] proposed a predicting method by using Hidden Markov Model (HMM) and a comparison of existing techniques have done. During their research, they have seen that traditional approaches are not able to handle variations in the stock market. However, the neural network and SVM are generating best results for it. Their proposed method improves accuracy as well as performance.

Wang et al. [15] proposed a method which uses mining technology of social media for the quantitative evaluation of the market. As a combination with other information, it has been used to predict the stock price trend in the short term. Results show that the proposed method of combination return more accurate results.

Wang et al. [16] found that ANN is good in the prediction of stock price, but genetic neural network improves the prediction speed and reliability. So, in their proposed method for prediction, they have used a genetic neural network. Also, the basis of market characteristics they expressed the law of stock price by using a genetic neural network.

3 Proposed Approach

This section covers the detailed study of the proposed approach. There is two models proposed in which Model A is based on sentiment analysis while model B is based on clustering technique.

3.1 Model A

This model helps in prediction by using sentiment analysis. For analysis, data has been collected from various sources like micro-blogs (twitter) and new articles. Then data will be processed in multiple levels, firstly it removes stop and makes the data generalized. Then data processed through parsing, tokenization, filter, stemming, Tf-Idf and calculation of score of the post. On the basis of score, the sentiment is analyzed, i.e., negative or positive. Now we will decide the impact of post on the specific sector by matching keywords into sectors specific dictionaries. In the case of keyword matching, for analysis stocks from related sectors are checked. After checking all the stocks using sentiment analysis, a prediction will be made. Finally the nature of the post is determined as negative or positive. On the basis of whole analysis, we predict the impact of post on sector as negative, positive or neutral by using simple moving average (SMA) value for every stock.

$$SMA = \left(SMA \text{ of last } 2^{nd} \text{ month} - SMA \text{ of last month}\right)$$

If it returns positive value that means positive movement will be shown else, it will show negative movement (Fig. 1).

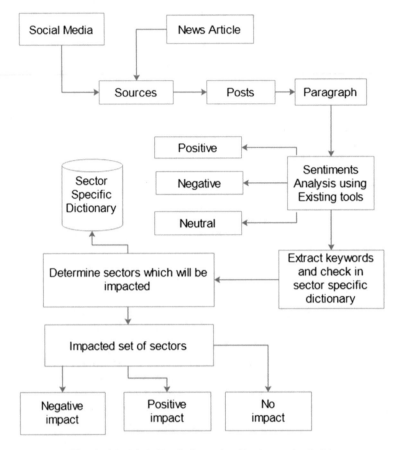

Fig. 1. Model A (Prediction using Sentiment Analysis)

3.2   Model B

Model B helps in prediction using clustering technique known as DENCLUE. It uses technical parameter like SMA. These clusters are positive, negative and neutral. Every set consists of stocks which show the same kind of behaviour like increment (positive set), decrement (negative set) and little bit fluctuation (neutral set) (Fig. 2).

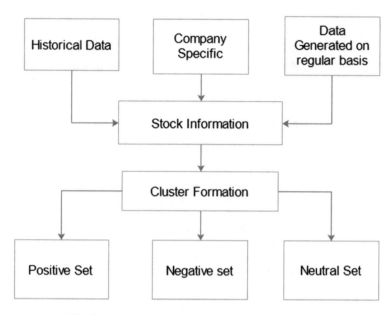

Fig. 2. Model B (Prediction using Cluster Technique)

4   Hybrid Model

The combined output of model A and model B results into Hybrid model. Output of both the models are processed and analyzed, finally provide final prediction. Information from other models will be taken on the basis of which value of technical indicator is computed for input stocks (Fig. 3).

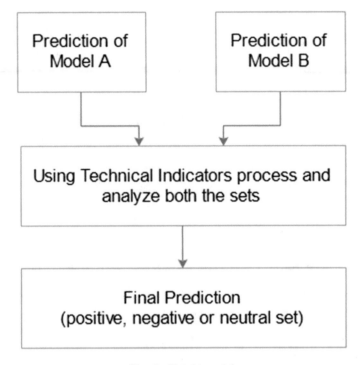

Fig. 3. Hybrid model

5 Dataset

For the experiment, NSE (National Stock Exchange) data has been taken among 21 stock exchanges based on different sectors like banking, IT, Automobile, Oil, pharma and FMCG. NSE is the popular stock exchange of India having more than 1600 companies listed on its platform. 6 popular domains with maximum capitalization have been considering for experiment. From every sector-top companies have been taken for analysis. Market capitalization plays an important role for prediction where,

Market cap = Current price of the share ∗ Total number of shares outstanding.

6 Result

Top 5 companies have been taken from each sector:

The banking sector is an important part of this trading business as it provides investors and market intermediaries to them. There are five companies taken from this sector namely, HDFC, SBI, ICICI, Kotak Mahindra and Axis bank. Table 1 and Fig. 4 shows the actual and predicted value of the stock price.

Table 1. Banking sector

Company name	Actual stock price	Predicted stock price
HDFC Bank	2403.75	2395.05
SBI	363	364.06
ICICI Bank	425.85	427.05
Kotak Mahindra	1479.05	1483.45
Axis Bank	756.45	754.6

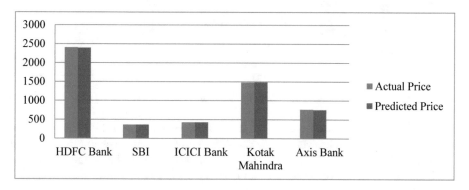

Fig. 4. Banking sector

Information Technology (IT) sector has a market capitalization of approx. 1690317.31Cr. Table 2 shows actual and predicted value of stocks of top-five IT companies like TCS, Infosys, Wipro, HCL and Tech Mahindra. Figure 5 shows a graphical representation of these values.

Table 2. Information technology sector

Company name	Actual stock price	Predicted stock price
TCS	2101.85	2108.9
Infosys	722.4	724.15
Wipro	268	261.7
HCL	1021	1019.55
Tech Mahindra	666.6	665.9

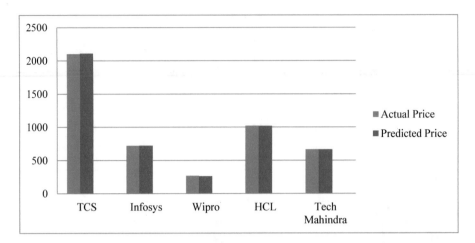

Fig. 5. IT sector

An **oil company** shows a stronger availability of their in NSE. India's biggest refiner is Oil Corp which owns 7.81% on NSE. Top five companies in this sector are ONGC, IOCL, BPCL, GAIL India and Coal India. Table 3 shows the actual and predicted stock values of respective companies, while Fig. 6 shows their graphical representation.

Table 3. Oil sector

Company name	Actual stock price	Predicted stock price
ONGC	153	151.5
IOCL	147.4	146.95
BPCL	350.3	346.65
GAIL India	147.3	147.9
Coal India	234.55	234.35

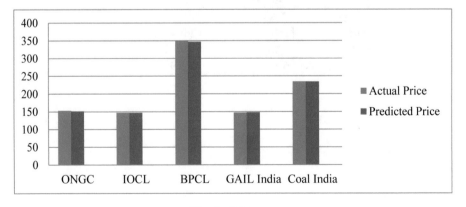

Fig. 6. Oil sector

Pharmaceuticals and health care sector has market capitalization of 713076.54 Cr. Like every other sector it has 5 top companies which shows their impact on NSE like, Sun Pharma India, Lupin, Dr Reddy's Labs, Cipla and Aurobindo pharma. Table 4 shows actual and predicted values for respective companies and Fig. 7 shows a graphical representation of it.

Table 4. Pharmaceutical and health care sector

Company name	Actual stock price	Predicted stock price
Sun Pharma India	399.55	409.25
Lupin	752.05	761.8
Dr Reddy's Labs	2651.25	2636.85
Cipla	552.25	555.6
Aurobindo Pharma	609.6	602

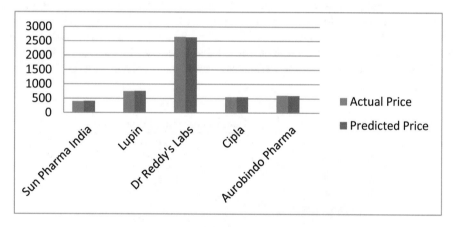

Fig. 7. Pharmaceutical and health care sector

In the recent era, usage of the **automobile** increases day by day. The market capitalization of the automobile sector is 7617.15 Cr. which shows an impact on total capitalization. It includes top five companies like Tata Motors, Maruti Suzuki, M&M, Bajaj Auto and Hero Motocorp. Table 5 shows the actual and predicted value of respective companies and Fig. 8 shows its graphical representation.

Table 5. Automobile sector

Company name	Actual stock price	Predicted stock price
Tata Motors	156.45	160.1
Maruti Suzuki	5955	6029.6
M&M	627.6	632.05
Bajaj Auto	2697.15	2739.95
Hero Motocorp	2490.05	2549

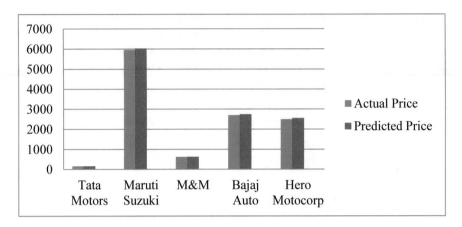

Fig. 8. Automobile sector

Fast Moving Consumer Goods (FMCG) companies are very popular in the market as they sold at low margins but very high volumes and having a high turnover of the product. Top 5 companies are Hindustan Unilever, Nestle India, Dabur India, Godrej Consumer and Britannia Industries. Table 6 shows actual and prediction value of the stock while Fig. 9 shows a graphical representation of the values.

Table 6. FMCG companies

Company name	Actual stock price	Predicted stock price
Hindustan Unilever	1733.35	1721.6
Nestle India	11503	11655.03
Dabur India	409.3	410.2
Godrej Consumer	635.6	633.8
Britannia Industries	2774.85	2773.7

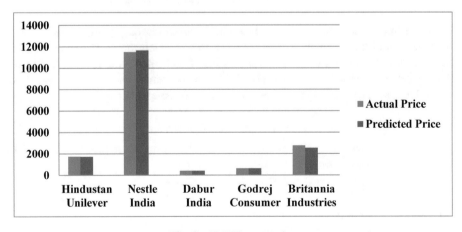

Fig. 9. FMCG companies

7   Conclusions

The stock market is the fastest growing trend which is changing every second. There are two possibilities either people will gain money, or he will be going to lose his entire savings. So for safe side stock market prediction is required, which is based on historical data. In this paper, we have proposed a hybrid approach (opinion mining and clustering method) for prediction in the stock market. The final prediction is based on an examination of both the results. We have done our experiment on data obtained from 5 top companies of different sectors like the banking sector, oil sector, IT sector, automobile sector, pharma sector, and FMCG companies from national stock exchange (NSE) data. As a result, it has been seen that predicted values from the proposed approach show maximum similarity with the actual values of the stock. The hybrid model returns efficient results in terms of accuracy in comparison to other individual methods of sentiment analysis and clustering.

In the future, this study will help to improve more accuracy by integrating more technical integrators and by using different parameters for prediction like values for other stock exchange, including more growing sectors of India. Also, we can use the oldest stock exchange that is, Bombay Stock Exchange for an experiment as it includes approx. 6000 leading companies.

References

1. https://www.investopedia.com/articles/stocks/09/indian-stock-market.asp
2. Kute, S., Tamhankar, S.: A survey on stock market prediction techniques. Int. J. Sci. Res. (IJSR), 1–4 (2015)
3. Billah, M., Waheed, S., Hanifa, A.: Stock market prediction using an improved training algorithm of neural network. In: 2nd International Conference on Electrical, Computer & Telecommunication Engineering (ICECTE), pp. 1–4 (2016)
4. Ercan, H.: Baltic stock market prediction by using NARX. In: 12th International Scientific and Technical Conference on Computer Sciences and Information Technologies (CSIT), pp. 1–4 (2017)
5. Vajargah, K.F., Shoghi, M.: Simulation of stochastic differential equation of geometric Brownian motion by quasi-Monte Carlo method and its application in prediction of total index of stock market and value at risk. Math. Sci. **9**(3), 115–125 (2015)
6. Huynh, H.D., Minh Dang, L., Duong, D.: A new model for stock price movements prediction using deep neural network. In: 8th International Symposium on Information and Communication Technology, pp. 57–62 (2017)
7. Mankar, T., Hotchandani, T., Madhwani, M., Chidrawar, A., Lifna, C.S.: Stock market prediction based on social sentiments using machine learning. In: International Conference on Smart City and Emerging Technology (ICSCET), pp. 1–3 (2018)
8. Nayak, A., Manohara, M.M., Pai, M., Pai, R.M.: Prediction models for Indian stock market. In: Twelfth International Multi-Conference on Information Processing-2016 (IMCIP-2016), pp. 1–9 (2016)
9. Yamini Nivetha, R., Dhaya, C.: Developing a prediction model for stock analysis. In: International Conference on Technical Advancements in Computers and Communications, pp. 1–3 (2017)

10. Oncharoen, P., Vateekul, P.: Deep learning using risk-reward function for stock market prediction. In: 2nd International Conference on Computer Science and Artificial Intelligence, pp. 556–561 (2018)
11. Patel, H., Parikh, S.: Comparative analysis of different statistical and neural network based forecasting tools for prediction of stock data. In: 2nd International Conference on Information and Communication for Competitive Strategies, pp. 1–6 (2016)
12. Kato, R., Nagao, T.: Stock market prediction based on interrelated time series data. In: IEEE Symposium on Computer and Informatics, pp. 1–5 (2012)
13. Sharma, A., Bhuriya, D., Singh, U.: Survey of stock market prediction using machine learning approach. In: International Conference on Electronics, Communication and Aerospace Technology (ICECA), pp. 1–4 (2017)
14. Somani, P., Talele, S., Sawant, S.: Stock market prediction using hidden Markov model. In: Student Conference on Engineering and Systems, pp. 1–4 (2012)
15. Wang, Y., Wang, Y.: Using social media mining technology to assist in price prediction of stock market. In: IEEE International Conference on Big Data Analysis (ICBDA), pp. 1–4 (2016)
16. Wang, H.: A study on the stock market prediction based on genetic neural network. In: International Conference on Information Hiding and Image Processing, pp. 105–108 (2018)

Energy-Efficient Routing Based Distributed Cluster for Wireless Sensor Network

R. Sivaranjani[1(✉)] and A. V. Senthil Kumar[2]

[1] Department of Computer Technology, Hindusthan College of Arts Science,
Coimbatore, India
ranju_rs@yahoo.com
[2] Department of Computer Application, Hindusthan College of Arts & Science,
Coimbatore, India
avsenthilkumar@yahoo.com

Abstract. The wireless sensor network (WSNs) is one of the creating and quickly expanding technologies across the globe. In WSNs, enormous number of sensor hubs inside the cluster has detected the earth and send information to the group head. This paper contain new vitality productive Distributed cluster head planning calculation which is utilized for adequate determination of CH and information social event plot for Wireless sensor systems. Proficient Distributed cluster head booking calculation approach includes three stages: External Cluster Communication (ECC) stage, information group correspondence area and Internal Cluster Communication (ICC) segment. In which these stages are utilized to convey between the group head and portal. Vitality utilization model is keeping up greatest leftover vitality level over the system. In proposed work gave another parcel position which is given to all group part hubs. Planning calculation is displayed to designate availabilities to group the information parcels. The procedure results demonstrate that the proposed plan incredibly adds to greatest system lifetime, high vitality, diminished overhead and most extreme conveyance proportion.

Keywords: Sensor networks · Sensor nodes · Routing protocol · Energy efficient

1 Introduction

Wireless sensor network has gained significant quality in recent years due to the advancement in wireless communication technology and a speedily developing zone for analysis. WSN refers to a system of sensor hubs associated through a wireless medium [1]. Each hub includes of Handling capability (at least one CPU, DSP chips Microcontroller) could contain different types of memory (program, information and flash memories), have a power supply (e.g., batteries and sun-powered cells), and contain completely different sensors and actuators [3]. Energy conservation may be a massive issue in WSN as sensor hubs carry limited non-rechargeable power supply and it's tasking to switch the nodes that make power saving necessary to extend the lifespan of nodes. Energy economical routing protocols are needed to reduce the use of the

© Springer Nature Switzerland AG 2020
J. S. Raj et al. (Eds.): ICIDCA 2019, LNDECT 46, pp. 489–497, 2020.
https://doi.org/10.1007/978-3-030-38040-3_55

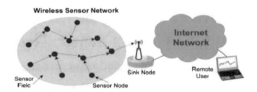

Fig. 1. Shows wirelsss sensor netowrk

facility resources and prolonging the network lifespan path whereas transferring information [4] (Fig. 1).

Remote sensor system increased quality as of late because of the headway in remote correspondence innovation and a rapidly creating zone for investigation. WSN alludes to an arrangement of sensor center points related through a remote medium [1]. Every center point incorporates of Handling capacity (in any event one CPU, DSP chips Microcontroller) could contain various sorts of memory (program, data and glimmer recollections), have a power supply (e.g., batteries and sun-controlled cells), and contain totally various sensors and actuators [3]. Vitality preservation might be a huge issue in WSN as sensor center points convey constrained non-battery-powered power supply and it's entrusting to switch the hubs that make power sparing important to broaden the life expectancy of hubs. Vitality efficient directing conventions are expected to lessen the utilization of the office assets and delaying the system life expectancy way though moving data [4].

2 Literature Review

Ahmad, Latif, Javaid, Khan et al. [1] researched on cluster method that is most all around perceived coordinating methodology in WSNs. due to contrasting need of WSN application beneficial essentialness use in coordinating shows is incredibly still a potential field of investigation. Creators given new vitality efficient coordinating method during this investigation.

Zibouda et al. [3] proposed a circulated vitality prudent versatile grouping convention with data Gathering for WSN lessens the vitality utilization and system life is broadened. The grouping methods are utilized speedily with appropriated cluster heads. The hub's quantitative connection is killed for mounted central measure and rest the executives laws are intended to diminish the cost perform. The situation shows arbitrary preparing of hubs and furthermore the complete recreation time is spoiled exploitation asset reservation. The heap evening out progressed nicely and therefore transmission intensity of the hub is diminish that in this way lessens the vitality utilization.

3 Proposed System

3.1 Proposed Plan: Efficient Distributed Cluster Head Booking Calculation

This area manages system model and calculation for the proposed EDCHS.

Consider a gathering of sensors sent in a field. The accompanying properties were expected about the sensor arrange:

- A homogenous remote sensor system is thought to be the system model any place hubs are arbitrarily spread all through the sensor field.
- Nodes are left unattended once arrangement. Battery energizes and trade is about outlandish for the total activity.
- Base station has no limitation on vitality requirement and is mindful to the topographical areas of the hubs.
- Every sensor hub has comparative highlights (detecting, preparing, and correspondence) and has the capacity of altering their transmission power level progressively dependent on RPR esteem.
- The correspondence among sensor hubs is multihop–symmetric correspondence.

Nodes and base station aren't quality bolstered and thought to be stationary (Fig. 2).

Pseudocode:

Step1: Start

Step2: Sensor Nodes send beacon signals to Base Station and receives RPR values of all the sensor nodes

Step3: Base Station creates Look up table with sensor node ID and its corresponding RPR values (RPR) n, also computes the mean of RPR values (RPR) m

Step4: The condition is IS (RPR) n > (RPR) m. If the condition is lesser than (RPN) m as compared to (RPM) n. It executes Node belongs to primary Tier (T1)

Step5: Compute mean RPR among the primary tier nodes $(RPR)_{T1}$ and Nodes RPR close to mean $(RPR)_{T1}$ are elected as CH and nodes having highest RPR among T1 are elected as Gateway nodes

Step6: If the condition is greater than (RPM) m as compared to (RPM) n. Node belongs to secondary tier (T2)

Step7: Compute mean RPR among the secondary tier nodes $(RPR)_{T2}$ and Nodes RPR close to mean $(RPR)_{T2}$ are elected as CH and nodes having highest RPR among T2 are elected as Gateway nodes

Step8: Either or nor process executes and the process will end.

Fig. 2. Pseudocode proposed method

3.2   The Productive Circulated Group Head Planning Calculation

The proposed EDCHS was created to achieve the accompanying destinations:

- To build up various leveled grouping in WSN.
- To actualize the heap adjusting among the sensor hubs and stay away from vitality gap.
- To recognize a group head that covers the entire field with least correspondence separation
- The group heads ought to be appropriated all through the indicator field in a vitality proficient way.
- To diminish the transmission esteem.

The arranged EDCHS approach separates the system into two levels. The most bit of leeway of the two level plans is that it essentially lessens the correspondence separation between the hubs CH and CN. Thusly this methodology makes the likelihood of lessening the transmitting intensity of CH hubs which would more be able to upgrade the individual hub life expectancy. The EDCHS approach includes three stages: External Cluster Communication (ECC) stage, information group correspondence (DCC) segment and Internal Cluster Communication (ICC) area.

3.3   ExternalCluster Communications Phase

In this stage sensor hubs are speaking with the base station for the development of levels, race of group head and door hubs. At the underlying phase of this part, every hub sends signal message to the base station. From this message the base station makes the look into table with hub id and its RPR (Received Packet Ratio) level for each hub. The BS figures the regular RPR estimations of hubs from the arranging up table by summing up the RPR estimations of the considerable number of hubs and isolating by the whole scope of hubs. As of now the BS isolates the system into essential Tier1 and optional Tier2. The hubs that have the RPR esteems more noteworthy than the run of the mill RSSI worth are characterized by set of hubs, indicated by Node1 and in this manner the rest of the hubs territory unit drawn by Node2.

The BS demonstrates Node1 has a place with essential level and Node2 has a place with optional level. at present the base station chooses three sorts of hubs in Tier1 like cluster head hubs that speak with group hubs, portal hubs GN1 that forward the data between essential level, BS and passage hubs GN2 that go about as hand-off hubs among essential and optional levels. inside the given sensor field the essential group head is set in such how that the CH ought not be far away from the BS and inexact focus area from every one of the hubs in each cluster. To do as such, the hubs that are near the run of the mill RPR worth are frequently chosen as CHs for the Tier1.

The probability of becoming initial CHs $\left(P_{CH(i)}\right)$ was calculated by

$$\left(P_{CH(i)}\right) = \sum_{i=1}^{n} \frac{(RPR)_i}{n}$$

Where n is the quantity of sensor hubs in Tier1 and (RPR)_i is the ith gotten signal quality from a relating hub. Since the vitality level of all the sensor hubs would be same at beginning stage, the vitality level parameter isn't considered at this stage (Fig. 3).

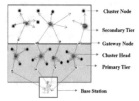

Fig. 3. Shows energy level of sensor data

For discovering first CHs. Thus, the BS figures the normal RPR level all things considered and distinguishes the arrangement of group head hubs in the auxiliary Tier2. Consequently the proposed DCHS plot consistently attempts to limit the transmission separation among the base station, essential level, and optional level.

3.4 Data Cluster Communication Phase

This is the group development stage. The base station communicates the message to all the sensor hubs for giving data about the recently chosen cluster head. The adjacent hubs send Join Request message to CH and CH recognizes the dedication and structures a group. In essential level, each CH chooses one door hub GN1 to guarantee that GN1 is nearer to the base station. Comparable choice of one passage hub GN2 in optional level made so that GN2 is nearer to the limit among essential and auxiliary levels. Presently CH apportions TDMA timetable to every hub in the cluster for transmitting information. During the distributed schedule vacancy group hub will send the information. Presently the cluster head totals and packs the information assembled from group hubs. The CH will advance the compacted information to the passage hubs GN1 if the CH has a place with the essential level or the consequences will be severe, to the entryway hubs GN2.

3.5 InternalCluster Communication Phase

This stage for the most part manages correspondence among the sensor hubs inside two levels and re-appointment of cluster head. The determination of edge estimation of the lingering vitality for cluster head exchanging is tentatively dissected and set as half of the underlying vitality. On the off chance that the lingering vitality of the current CH dips under portion of the underlying vitality, at that point CH will communicate CH_SCHEDULING message to group hubs through multi-throwing strategy. This message speaks to, the current CH having deficient remaining vitality to go about as CH and prompt group head exchanging is required for dragging out the system lifetime. Because of the booking message, the cluster hubs send the parameter Ti which is the summation of its leftover vitality (E_(i)) and RPR level (R_i) to the CH. Presently the cluster head ascertains the limit esteem T_(CH(i)) for distinguishing the following group head. The limit an incentive for getting to be next CH can be communicated by

$$T_{CH(i)} = \sum_{i=1}^{n} \frac{(E_i + R_i)}{n}$$

Where n is the complete number of hubs in the group. Presently the current CH looks at the T_(CH(i)) with T_(i) esteem for every hub. The hub which has ([T] _(i) esteem closer to T_(CH(i)) will be chosen as the following CH. The hint of new CH to group hubs pursues the comparative system as in ICC stage.

The underlying group development in the ICC stage for the proposed EDCHS plot. The underlying CH choice is made based on its closeness to the last mean of RPR estimation of all the sensor hubs. Presently the base station isolates the hubs in essential level or auxiliary level by checking whether the sensor hub's RPR worth is more prominent than the mean RPR of all the sensor hubs. Since it is accepted that underlying vitality level of a sensor hub is same for every one of the hubs, the remaining vitality level parameter isn't considered at the ICC stage. This additionally lessens the calculation overhead to the sensor hubs. After the group design has been framed the base station communicates the directing data to all the sensor hubs.

System life demonstrates anyway long a gadget hub lives with ideal residual vitality level inside the present group. Absolute vitality level is that the general vitality of the cluster that is observed by all CHs. Edge Check Sequence is to spot duplications over the casings transmitted.

At first, the group topology disclosure part is demurely decided to mastermind the WSN as a clustered gadget organizes. Here the gadget hubs are planned as part hubs of a group just bolstered their lingering vitality levels. At that point, with the blending of the vitality utilization model, the vitality spent all through different system errands is measurable. At last, the interim based booking is performed inside the premise of your time Division Multiple Access. The arrangements of sensor hubs that are schedulable for a specific round of correspondence are recognized.

4 Result

The presentation investigation of the proposed DCHS was measurable dependent on four measurements especially all out system vitality expended for different rounds, first hub dead, half of the hub alive and all out assortment of data got at BS. Figure demonstrates the whole system vitality plotted against each circular that will be that the summation of remaining vitality at all detecting component hubs. The proposed framework consistently guarantees that the group head is near the focal point of the cluster by considering a definitive mean of RPR and can't be chosen at the outskirt among the detecting component field which may enormously decrease the transmission separation in the middle of CH, CN, and BS. It additionally can maintain a strategic distance from longer idleness of data parcel. Further, the portal hubs likewise are no selective closer to the BS to reduce the over-burden of CH. so DCHS plan contributes a major amount of vitality sparing and gives higher lifetime to the sensor hubs than group based for the most part steering convention. Sensor nodes than cluster based mostly routing protocol (Figs. 4, 5 and 6).

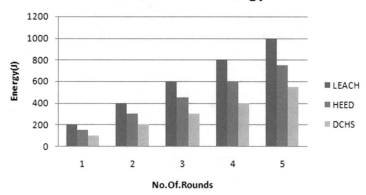

Fig. 4. Show total network energy

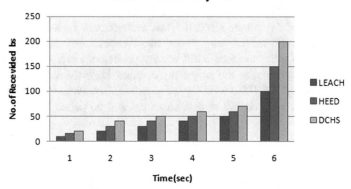

Fig. 5. Show data delivery to based station

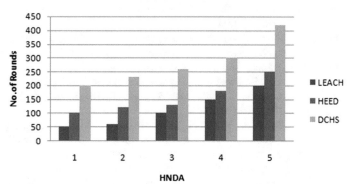

Fig. 6. Shows half of the node alive system

Techniques		Existing System			Proposed System		
		LEACH	HEED	DCHS	LEACH	HEED	DCHS
Total Energy	Network	850	480	325	1000	700	500
Data Delivery BS		75	100	120	100	150	200
Half of the node alive		130	200	395	200	250	425

In this table, we thought about existing framework and proposed arrangement of three calculations like LEACH, HEED and DCHS. In existing framework, all out system vitality, information conveyance BS and half of the hub alive are contrasted and the proposed arrangement of the complete system vitality, information conveyance BS and half of the hub alive is increasingly productive and verified.

5   Conclusion

The proposed DCHS uses booking of CH in each round on the possibility of lingering vitality and RPR level of the gadget hubs and outflanks the favored probabilistic clustering based directing conventions through reasonable dissemination of vitality load among the group arrange. Since the test arrangement includes every ongoing equipment and occasion driven test system, the outcomes got show a great deal of constant conduct. In order to more build vitality proficiency and expand the time of gadget hub, hub level vitality the board topic might be fused with DCHS subject. Thusly, the proposed DCHS might be successfully received for vitality delicate applications in WSN with vitality utilization model and propelled bundle design.

References

1. Ahmad, A., Latif, K., Javaid, N., Khan, A., Qasim, U.: Thickness controlled hole and-rule plot for essentialness viable coordinating in Wireless Sensor Networks. In: 2013 26th Annual IEEE Canadian Conference on Electrical and Computer Engineering (CCECE), pp. 1, 4, 5–8 May 2013
2. Jang, S., Kim, H.-Y., Kim, N.-U., Chung, T.M.: Imperativeness efficient clustering plan with concentric hierarchy. In: 2011 IEEE International RF and Microwave Conference (RFM), pp. 79, 82, 12–14 December 2011
3. Chirihane, G., Zibouda, A.: Passed on essentialness gainful flexible grouping show with data gathering for tremendous scale remote sensor frameworks. In: Twelfth International Conference on Programming and Systems (ISPS), pp. 74–76. IEEE (2015)
4. Roseline, R.A., Sumathi, P.: Local grouping and edge fragile guiding figuring for Wireless Sensor Networks, Devices, Circuits and Systems (ICDCS). In: 2012 International Conference, pp. 12–19. IEEE (2012)
5. Brar, G.S., Rani, S., Song, H., Ahmed, S.H.: Essentialness efficient direction-based PDORP routing protocol for WSN. In: IEEE Special Section on Green Communications and Networking for 5 g Wireless, vol. 4, pp. 3182–3194 (2016)
6. Sinha, J.D., Barman, S.: Imperativeness efficient routing mechanism in wireless sensor network. In: IEEE assembling on Recent Advances in Information Technology, pp. 567, 873 (2012)

7. Pantazis, N.A., Nikolakakos, S.A., Vergados, D.D.: Imperativeness gainful directing shows in remote sensor masterminds: an investigation. IEEE Commun. Surv. Tut. **15**(2), 551–591 (2013)

8. Cheng, B.-C., Yeh, H.-H., Hsu, P.-H.: Schedulability assessment for hard framework lifetime remote sensor frameworks with high imperativeness first grouping. IEEE Trans. Reliab. **60**(3), 675–688 (2011)

9. Awad, F.: Imperativeness efficient and coverage-aware clustering in wireless sensor networks. Wirel. Eng. Technol. **03**(03), 142–151 (2012)

10. Kumar, A.D., Smys, S.: An energy efficient and secure data forwarding scheme for wireless body sensor network. Int. J. Networking Virtual Organ. **21**(2), 163–186 (2019)

11. Awad, F.: Imperativeness efficient and coverage-aware clustering in wireless sensor networks. Wirel. Eng. Technol. **03**(03), 142–151 (2012)

12. Manjeshwar, A., Agrawal, D.P.: Immature: a routing protocol for enhanced efficiency in wireless sensor networks. In: Proceedings of Fifteenth International Parallel and Distributed Processing Symposium, pp. 2009–2015

Autonomous Home-Security System Using Internet of Things and Machine Learning

Aditya Chavan[✉], Sagar Ambilpure, Uzair Chhapra,
and Varnesh Gawde

Rajiv Gandhi Institute of Technology, Versova, India
adityachavan198@gmail.com, sagarambilpure@gmail.com,
uzairchhapra@gmail.com, gawdevarnesh@gmail.com

Abstract. Home Security Systems are mandatory in today's time. But most of the Systems available are highly dependent on Human Beings for working effectively. We need smart Autonomous Home-Security Systems which work effectively with very less or no human efforts. We have tried to solve this problem by using Internet of Things and Machine Learning. Our project solves this problem by making the process of Home security Autonomous.

Keywords: Autonomous · Internet of Things (IoT) · Machine Learning · Raspberry Pi · Django

1 Introduction

Nowadays everything around us is connected to the internet all the time. It is but natural that our house is connected to the internet as well. Using Internet of Things for automating the manual processes seems to be very promising [1]. Traditional Home Security systems available in the market require some kind of Human effort to function effectively whereas the solution developed by us makes this process completely Autonomous using technology. Our solution enables the Homeowner to leave his house without worrying about safety. Another salient feature of our proposed solution is that it is a lot more economical than all the other solutions currently available in the market.

2 Literature Review

All the solutions currently available fundamentally depend upon a human to function effectively [2]. There are some automated systems available which perform partial tasks like CCTV cameras recording video [3], IOT enabled buzzers which lets the home-owner know on their mobile devices whenever a person presses the doorbell on the gate [4], security systems which trigger an alarm whenever motion is detected at unexpected time and place [5]. None of the solutions provide an end-to-end solution to the problem of Home-Security.

© Springer Nature Switzerland AG 2020
J. S. Raj et al. (Eds.): ICIDCA 2019, LNDECT 46, pp. 498–504, 2020.
https://doi.org/10.1007/978-3-030-38040-3_56

3 Proposed System

The device designed in this project can be installed at the main entrance of a house. Our system is divided into two different components that perform different tasks. One of these systems is a higher compute power server and the other is a micro-controller, we are using raspberry pi – 3b as our micro-controller but any raspberry pi computer with support for the camera should work just fine. The system would function as follows. Raspberry pi would receive the signals from all the sensors like pi camera, motion detector, distance sensor, etc. Whenever motion is detected by the ultrasonic sensor the microcontroller is programmed to trigger a get request to the Django server along with the sensor data. Django server is a higher compute power system that has our pre-trained face recognition model and the image database. Once the data is received on the server our pre-trained model compares the image with the database of images to check whether the image contains any pre-approved members. Even if one member is found among the people in the image captured, the door opening protocol is executed else potential intruder protocol is executed. During the entire process, the person standing at the door is intimated about the status by voice and also a LED light.

Components Required

1. Django Server [6]
2. Raspberry Pi 3B [7]
3. Good Wi-Fi Connection and Internet
4. Pi Camera
5. Ultra-Sonic distance sensor HC-SR04 [8]
6. Numpad
7. LED Lights
8. Speaker
9. Android phone to receive a notification on a mobile app
10. Servo motor

4 Proposed Architecture

An autonomous home security system would reduce human efforts significantly as it would operate autonomously without any human efforts. It is an efficient solution that can be implemented using very few resources. This system eliminates the use of a key and a doorkeeper by opening the door automatically without any human intervention. This door opening procedure is emulated by using a servo motor [9] in our proposed system.

The Django server here would be a central server common for all the users using the system. It would act as a central server providing service to all the clients when requested. This would ideally be hosted on a cloud platform like AWS or Digital Ocean. Every home security system would have a unique identification number associated with it. There would be an image database which would contain the image data of the various users. This server would have a pre-trained face recognition model

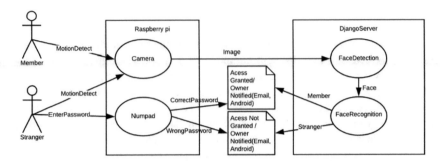

Fig. 1. System architecture

that would be used for face recognition. This server is constantly running waiting for clients to make requests. This is shown in Fig. 1.

Whenever a client (raspberry pi) makes a request along with the image data the following things happen:

1. Image is sent for facial recognition where the algorithm performs the following steps:

    a. **Finding the faces present in the image:** The strategy that we are utilizing for face recognition is HOG (Histogram of Oriented Gradient). This algorithm starts with detecting faces in the input picture for which it converts the image into black and white because color data is not needed for face detection. Due to the conversion into black and white, the processing speed of the model increases making it more efficient. Once the image is converted to black and white it starts Feature description. In feature description, the algorithm looks at every pixel one by one and compares the pixel with its surrounding pixels. It then moves in the direction of an arrow which points where the image is getting darker. This process is repeated for all the pixels at the end of which we get an arrow for every pixel. Saving the gradients for all the pixels gives way too much data which is not useful. So now the algorithm divides the image into small squares of 16 × 16 pixels and then counts the number of arrows in each direction. Now, these squares are replaced by the arrow which is in majority like if in a square of 16 × 16, 20 arrows are pointing upwards then the entire square is replaced by an upward arrow. Now this HOG representation of the original image is compared with the HOG representation which is based on previous training faces [10].

    b. **Posing and Projecting Faces:** We have isolated the faces in our picture. In any case, presently we need to manage the issue that faces turned in different directions appear to be absolutely unique to a computer. To solve this problem, the algorithm tries to twist each image so that the eyes and lips are consistent for future processing. This makes it easier to compare faces in the subsequent stages. The algorithm utilized for this is face milestone estimation. The fundamental thought is to concoct 68 explicit focuses (called tourist spots) that exist on each face — the outside edge of each eye, the top of the chin, the inner edge of each eyebrow, and so on. The algorithm has the ability to locate these 68 explicit focuses on any face. The algorithm does not do any extravagant 3d twists since that would bring mutilations into the picture.

c. **Encoding Faces:** Comparing images directly takes a lot of time and is not efficient. In this step, the algorithm tries to extract 128 measurements for each face in the input image which would be used in the later steps. This algorithm needs an approach to extricate these essential features from each face. For which, it uses features like the size of every ear, the separating between the eyes, the length of the nose, and so on. Deep learning makes a superior showing than people at making sense of which parts of a face are essential to measuring. The algorithm uses a pre-trained convolution neural network which takes input an image and gives out 128 measurements for each face in the image. The network generates nearly the same numbers when looking at two different pictures of the same person. The model used by us is trained by Open Face [11] (Fig. 2).

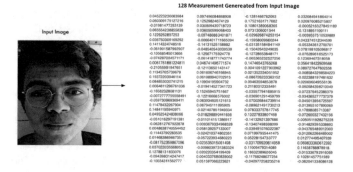

Fig. 2. Encoding for our Test input image [11]

d. **Finding the Person's Name from the Encoding:** In this step, we are trying to find a face in the test image that matches with any of the faces in our image database. For this, the algorithm used is a direct SVM classifier that can find a match for a known individual from the image (Fig. 3).

Fig. 3. Screenshot of how the algorithm works.

2. The result of the facial recognition algorithm is in the form of a Boolean which is returned back to the client through the get request. If the person is a registered user, then returned value is True else the value is false.
3. The result is received at the micro-controller (raspberry pi). Based on the value of the result following steps are executed:
   a. If the value is True: Then a welcome message along with the name is played on the speaker and then the door lock opens up in this case the servo simulates the lock. Also, a notification is sent to the homeowner on the android app and also to his email address along with the name and the picture of the guest at the door. The colour of the status LED also changes to green. This protocol works even if one face matches with the database.
   b. If the value is False: Then the person at the door is intimated by a message from the speaker that he has an option to enter the house by entering the password on the Numpad available. If the password entered by the guest matches with the password set by the homeowner, then the door is opened, and the image is sent to the homeowner on the android app and the registered email address. Notifications on the android app are being sent by using Firebase Cloud Messaging [12].

5   Implementation

In Fig. 4 above we have shown the prototype that has been made by us. The above scenario is before the motion is detected when the red status LED is on. As soon as motion is detected by the ultrasonic sensor the yellow LED lights up which shows the status as motion detected and processing also during this period a message is played on the speaker which lets the person at the door know that motion has been detected and that the person should wait for the processing to be completed.

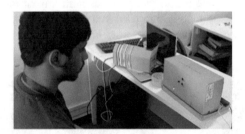

Fig. 4 Working of prototype when motion is detected.

Figures 5 and 6 shows the scenario in which the face matches with a face in the database on the server in which case the person at the door is an authorized user and the user is granted access along with a customized welcome message on the speaker. The owner of the house is also notified of the guest's name and picture on the android app and the registered email address the screenshot for which is above (Fig. 7).

Fig. 5 Access granted protocol executing.   **Fig. 6** Notification on Android app.

Fig. 7. The image above shows the connections of the entire system.

6 Conclusion

The purpose of developing the project was to provide an innovative and yet cost-effective solution for securing people's homes. The working of this project involves taking a picture of the person standing outside the door. This picture would then be mailed to the owner of the house. Meanwhile, facial recognition algorithms would run in the background to try and match the person with its existing set of authorized members. A push notification would then be sent to the owner's Android device notifying him of the name of the authorized member who just reached home, or a stranger if no match was found. This device would open the door only for the authorized members. What this project excels at is the fact that no human intervention is required for it to work and that the hardware used for this project is competent enough for all of its future scope aspects.

7 Future Scope

In the future, there could be an option for live surveillance of the camera outside the house through the android application. We can add an allow-only-once entry feature accessible through our mobile app so that a guest, who is not registered in our database and also unaware of the numeric password, can still enter the house. Another aspect of the project that can be improved is nighttime face detection. IR camera sensors can be added to the normal camera sensor, the synergy of which would improve recognition results significantly.

Compliance with Ethical Standards

All author states that there is no conflict of interest.

We used our own data.

We used author's photo in results section and got approval.

References

1. Hassija, V., Chamola, V., Saxena, V., Jain, D., Goyal, P., Sikdar, B.: A survey on IoT security: application areas, security threats, and solution architectures. IEEE Access **7**, 82721–82743 (2019)
2. Hou, J., Wu, C., Yuan, Z., Tan, J., Wang, Q., Zhou, Y.: Research of intelligent home security surveillance system based on ZigBee. In: 2009 Third International Symposium on Intelligent Information Technology Application Workshops, pp. 554–557 (2008)
3. Abaya, W.F., Basa, J., Sy, M., Abad, A.C., Dadios, E.P.: Low cost smart security camera with night vision capability using Raspberry Pi and OpenCV. In: 2014 International Conference on Humanoid, Nanotechnology, Information Technology, Communication and Control, Environment and Management (HNICEM), Palawan, pp. 1–6 (2014)
4. Quadros, B., Kadam, R., Saxena, K., Shen, W., Kobsa, A.: Dashbell: a low-cost smart doorbell system for home use. ArXiv abs/1706.09269 (2017). https://arxiv.org/pdf/1706.09269.pdf
5. Nosiri, O.C., Akwiwu-Uzoma, C.C., Nmaju, U.A., Elumeziem, C.H.: Motion detector security system for indoor geolocation. Int. J. Eng. Appl. Sci. (IJEAS) **5**(11) (2018). ISSN 2394-3661
6. Suma, V.: Towards sustainable industrialization using big data and internet of things. J. ISMAC **1**(01), 24–37 (2019)
7. Nayyar, A., Puri, V.: Raspberry Pi-a small, powerful, cost effective and efficient form factor computer: a review. Int. J. Adv. Res. Comput. Sci. Soft. Eng. (IJARCSSE) **5**, 720–737 (2015)
8. HCSR04 datasheet. http://www.micropik.com/PDF/HCSR04.pdf
9. Šustek, M., Marcaník, M., Tomasek, P., Úředníček, Z.: DC motors and servo-motors controlled by Raspberry Pi 2B. In: MATEC Web of Conferences, vol. 125, p. 02025 (2017). https://doi.org/10.1051/matecconf/201712502025
10. Dalal, N., Triggs, B.: Histograms of oriented gradients for human detection. In: 2005 IEEE Computer Society Conference on Computer Vision and Pattern Recognition (CVPR 2005), San Diego, CA, USA, vol. 1, pp. 886–893 (2005)
11. Amos, B., Ludwiczuk, B., Satyanarayanan, M.: OpenFace: a general-purpose face recognition library with mobile applications, CMU-CS-16-118, CMU School of Computer Science, Technical report (2016)
12. Khawas, C., Shah, P.: Application of firebase in android app development-a study. Int. J. Comput. Appl. **179**, 49–53 (2018). https://doi.org/10.5120/ijca2018917200

Automatic Skin Disease Detection Using Modified Level Set and Dragonfly Based Neural Network

K. Melbin$^{(\boxtimes)}$ and Y. Jacob Vetha Raj

Department of Computer Science, Nesamony Memorial Christian College
Marthandam Affiliated to Manonmaniam Sundaranar University, Abishekapatti,
Tirunelveli 627012, Tamil Nadu, India
melbinmean@gmail.com, jacobvetharaj@gmail.com

Abstract. Dermatology is an essential fields that are to be analyzed, monitored, and treated to avoid skin disorders. The skin disease is caused because many factors influence humans such as age, sex, and lifestyle. Also, less amount of exposure to sunlight, bacteria, hot weather leads to skin diseases. Hence it is important to detect the skin disease at the earlier stage to avoid the fading condition of skin. In this paper we have developed an efficient automatic detection of skin disease using a two-stage adaptive process. At first stage modified level set approach is used for segmentation of skin images, later using color, shape and texture features are extracted and in the final stage dragonfly optimization-based Neural network is used for classification of types of skin diseases such as normal or abnormal. The proposed dragonfly based NN is evaluated using existing methods such as SVM, ANN for different evaluation metrics such as accuracy, sensitivity, and specificity to show the system efficiency.

Keywords: Skin disease · Neural network · Dragonfly · Level set · Accuracy · Dermatology · Feature extraction · SVM · Segmentation

1 Introduction

These diseases strike without warning and have been one among the primary disease that has life risk for the past ten years [2]. The skin disease detection is one of the critical tasks, and the diagnosis of medical imaging related diseases is the second task to be treated using the latest approaches [3].

Dermatology diseases include common rashes in skin which will spread severe infection, and these infections caused due to heat, allergens, system disorders and medications [4]. Skin disease occurs due to several reasons, it may include carelessness in maintaining the skin, restless, usage of some products that may not be adjusted to the skin or due to some infections. Usually changes in climatic conditions give more impact on the skin leads to some issues. Hence correct maintenance or testing accurately what kind of issues that happened in the skin is essential to be maintained [5].

© Springer Nature Switzerland AG 2020
J. S. Raj et al. (Eds.): ICIDCA 2019, LNDECT 46, pp. 505–515, 2020.
https://doi.org/10.1007/978-3-030-38040-3_57

Using recent development in machine learning approaches the efficiency of skin disease detection has been improved, but the accuracy has not been improved when it comes to classification of skin diseases. Many approaches, such as Neural network, SVM, and many classification algorithms, have been used so far [6].

Artificial Neural Network has some issues while reducing the error and so modification of algorithm has been presented in this work with a combination of dragonfly for weight optimization. The evaluation from experiments shows that ANN-Dragonfly optimization outperforms ANN and SVM in both training and testing. Section 2 is the review of existing approaches, Sect. 3 demonstrate the presented methodology using ANN and modified Level set approach. Section 4 evaluates the results, and Sect. 5 concludes the findings.

2   Literature Review

Bajaj et al. [7] have proffered an automatic skin disease identification approach. The input was skin database images which predicts various types of diseases. The various classification method to identify the disease from plants and human skins. The indications of disease are taken from leaf or stem using innovative approaches using image processing method. The main objective of their system is to detect, quantify and to classify the disease by improving the recognition rate. But the system was not able to recognize disease while using a greater number of diseases.

Maglogiannis et al. [8] have presented an intelligent technique for the segmentation and classification of dermatological images. have presented Support vector Machine for identification of erythematous-squamous skin disease. The solution for arythema disease is not easy because of redness in the blood. Therefore, SVM have been applied for better classification of skin diseases but still fails to achieve better accuracy due to unproper extraction of features extracted during the second stage. Zhang et al. [9] have analysed the effect on skin disease classification. They have mainly classified the early melanoma using deep learning approach. This system is used in order to help human to take a decision easily. The main intention of author is to provide a skin disease detection procedure that is highly accurate. They have used convolutional neural network for train and to classify the disease accurately but the system fails to achieve better accuracy due to unproper extraction of features.

Codella et al. [10] have developed the effect on skin disease classification. They have mainly classified the early melanoma using deep learning approach. This system is used in order to help human to take a decision easily. The main intention of author is to provide a skin disease detection procedure that is highly accurate. They have used convolutional neural network for train and to classify the disease accurately but the system fails to achieve better accuracy due to unproper extraction of features. Codella et al. [11] have proposed a skin disease detection approach using machine learning technique for perfect detection of skin diseases. Initially, they have used pre-processing method, later feature extraction and finally, classification using machine learning approach for skin diseases. The performance metrics shows better output in terms of accuracy, but complexity is high.

3   Proposed Optimal ANN for Skin Disease Detection

Skin diseases can be cured if we detect the disease earlier. Many researchers have found it challenging to classify skin disease efficiently. Recent methodologies have the issue while segmenting the affected region. The recognition rate is less for non-proper segmented images. Figure 1 is the block diagrama of the proffered. ANN-DFO based skin disease detection. At first, the Database images are segmented by the modified level set approach for partitioning the images into different segments. At second stage, color, texture, and shape features are then extracted. Color features are extracted using color histogram modeling, shape features are extracted using edge detection, and the texture of images are extracted using local binary patterns. At third Stage, To identify the disease classes, optimal neural network combined dragonfly optimization algorithm have been used.

3.1   Level Set Segmentation that is Modified

Level set methods [12] is one of the successful method for segmentation of image boundaries perfectly. The proposed method modifies the level set approach using PMD filter for improving the segmented area efficiently. The level set works on the basis of zero-level set of higher dimensional which is said to be set function. The motion of the contour is derived using the set function.

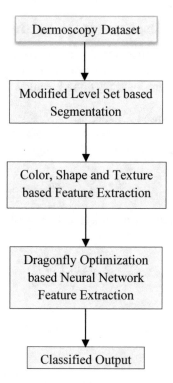

Fig. 1. Proposed structure of skin disease detection method

Stage 1: Find Image Entropy. The Entropy provides the information quantity of an image to facilitate image compression. It estimate the loss of information in the image and can be representes as given below.

$$Entropy = -\sum_{r=0}^{P-1}\sum_{s=0}^{P-1} I(r,s) * \log(I(r,s)) \tag{1}$$

Stage 2: Apply PMD Filter: PMD filter is used instead of Gaussian filtering for smoothing the images.

Stage 3: Calculate image gradient magnitude using

$$M_{norm} = \frac{|\nabla B(p,q)| - \min(|\nabla B(p,q)|)}{\max|\nabla B(p,q)| - \min|\nabla B(p,q)|} \tag{2}$$

Step 4: Compute the modified speed function using

$$sf = \exp\left(-cM_{norm}^2\right) \tag{3}$$

$$\frac{\partial L}{\partial d} = \mu div(x_n(|\nabla L|)|\nabla L|) + \delta G_E(L)div\left(g\frac{\nabla L}{(|\nabla L|)|\nabla L|}\right) + \alpha g G_E(L) \tag{4}$$

Step 5: Repeat

3.2 Feature Extraction Using Color, Shape, and Texture

The extraction has to be done effectively to identify the feature vectors for identifying the disease classes. Now the output of segmentation is applied to feature extraction using color, shape, and texture for extraction of features from the segmented images.

3.2.1 Color Histogram Based Extraction
In this section, a color histogram is used for extraction of color, and the process is as follows. The sum of 256 different arrays of a histogram is calculated using

$$L = \sum_{u=1}^{M} bu \tag{5}$$

Where, M is the total quantity of red, blue, and green mechanisms. The mean of the color histogram is computed using

$$\bar{y} = \frac{\sum_{u=1}^{M}(u+bu)}{N} \tag{6}$$

Now the standard deviation of a color histogram is evaluated and stored using

$$\sigma = \sqrt{\frac{1}{N}\left(\sum_{u=1}^{M} bu * (u - \bar{y})^2\right)} \tag{7}$$

3.2.2 Shape-Based Edge Extraction

Canny shape detection is used for finding the edge of the images. At first Gaussian filter is used for removing the noise from the image using

$$H_{uv} = \frac{1}{2\Pi\sigma^2} \exp\left(\frac{(u - (l+1))^2 + (v - (l+1))^2}{2\sigma^2}\right) \tag{8}$$

Now the intensity gradient and direction of the image has to be calculated using

$$H = \sqrt{H_a^2 + H_b^2} \tag{9}$$

$$\Theta = x\,tan2(H_b, H_a) \tag{10}$$

Select a threshold based on maximum values to minimum values for filtering the pixels.

3.2.3 LBP Based Texture Extraction

The LBP [13] method is used for texture extraction; the size of the window is classified into pixels of 16 * 16 and each pixel is simultaneously compared with other 8 neighbors in a circular from in the clock wise and in its reverse. This increases the pixel value of the median compared to the neighbor means. The histogram is contains feature vectors of 256 dimension and the histogram cells, which to gather produces the window of the whole feature vector.

3.3 Classification Using Artificial Neural Network (ANN)

The training for the network is provided to classify the images that are based on selected features. ANN [14] is used for the classification of facial images. ANN is the computing system inspired by the biological neural networks. It is a framework from many algorithms to process complex data inputs. Each connection like the synapses conveys the signal from neuron to next neuron. The signal connection inbetween neurons is computed as a number with the real values with the output being presented as the non-linear functions with the sum of the input provided. The neuron are connected using the links called the edges. The weights on the edges of each neuron adjusts itself as it learning continues [26]. The procedure of training a network starts with initialization based on three factors: the input layer, hidden layer, and Output Layer. The input layer comprises of several neurons $i_1, i_2, \ldots\ldots\ldots i_n$ that are linked with a hidden layer. The inputs are $W_1, W_2, \ldots\ldots\ldots W_n$. Moreover, each neuron possesses the weight denoted as the ith input layer neuron, which is further associated with

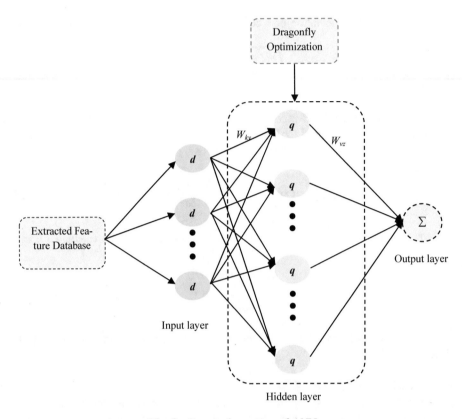

Fig. 2. Proposed structure of ANN

the jth neuron of the hidden layer denoted as $\beta_{12}, \beta_{12}, \ldots \ldots \ldots .\beta_{ij}$. The essential equation function of the input layer is defined as L_l, where L is a Basic function of the hidden neurons; 1 is the number denoting hidden units, β is the weight of the input layer neurons, N denotes the count of data and W is the input value. Hence, the computed basics function is

$$B_f = \sum_{j=1}^{N} W_i \times \beta_{ij} \qquad (11)$$

Where B_f is the basics function, β_{ij} is the input layer weight, and i is the number of input. This layer comprises of several neurons that are named as $h_1, h_2, \ldots \ldots \ldots .h_n$. They are connected to the output layer by using these neurons. Equation α_i denoted the obtained weight whereas L_l is denoted as the total input layers and α_j and β_{ij} as the weight that is obtained from the activation function. Activation function can be categorized as a linear, threshold, and sigmoid function. The sigmoid function is used basically for a hidden layer because it combines all the behaviors, including linear, curvilinear, and also constant behavior depending on the input value. The activation function in ANN modeling process is evaluated by

$$A_f = \sum_{j=1}^{h} \alpha_j * \left(\frac{1}{1 + exp(-\sum_{i=1}^{N} M_i \beta_{ij})} \right) \tag{12}$$

Where, F_i is a fitness function used to weigh the values of α and β, W is the input parameters, i is the number of inputs, j denotes the number of weights, and h is the number of hidden neurons. Figure 2 shows the proposed structure of the ANN.

3.3.1 Network Weight Optimization by Levy Flight-Based Dragon Fly Optimization Algorithm

The trained neural network recognizes the type of disease from the selected features. The weight of the neural network is optimized using levy flight-based dragon fly optimization algorithm. In 2016, Mirjalili proposed Dragonfly optimization algorithm [15], which works based on swarm intelligence and inspired from the natural behavior of dragonflies. Hence this optimization process consists of two phases such as exploitation and exploration. The behavior of dragonflies is static as well as dynamic which entirely depends on the situation like feeding or migration. Here, feeding is demonstrated as static and migration are demonstrated as dynamic. The vital phase of dragonflies is to avoid the enemy while moving near to the food. Hence, these entities avoid the enemy while all the individuals move near to the food sources in the same period. This behavior is demonstrated as follows:

$$U_x = -\sum_{y=1}^{M} A - A_y \tag{13}$$

$$Z_y = \frac{\sum_{y=1}^{M} R_y}{M} \tag{14}$$

$$G_x = \frac{\sum_{y=1}^{M} A_y}{M} - A \tag{15}$$

$$H_x = A^+ - A \tag{16}$$

$$D_x = A^- + A \tag{17}$$

Where A denotes the prompt spot of an individual. A_y denotes the prompt spot of yth individual. M denotes the total quantity of nearby individuals. R_y denotes the speed of nearby individuals. A^- and A^+ denotes the position of food and enemies. Hence to update the location of dragon two equations have been introduced such as step and location vectors as follows:

$$\nabla A_{s+1} = (tT_x + zZ_x + kK_x + lL_x) + f\nabla A_s \tag{18}$$

$$A_{s+1} = A_s + \nabla A_{s+1} \tag{19}$$

$$A_{s+1} = A_s + Levy(q) * A_s \tag{20}$$

$$Levy(q) = 0.01 * \frac{o_1 * \alpha}{|o_2|^{1/\beta}} \tag{21}$$

$$\alpha = \left(\frac{(\varphi(1+\beta) * \sin\left(\frac{\pi\beta}{2}\right)}{\varphi\left(\frac{(1+\beta)}{2}\right) * \beta * 2^{(\beta-1)/2)}} \right)^{1/\beta} \tag{22}$$

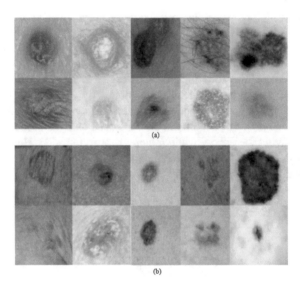

(a)

(b)

Fig. 3. (a) (b) Dataset used for evaluation of proposed skin disease detection using ANN-DFO [12]

4   Results and Discussion

The researches utilize the computer aided design for the dermoscopy to detect the skin disease The frame implementation is done using the mat lab R2017, The dataset of the given skin disease is given in Fig. 3. The Fig. 4 shows GUI of skin disease identification approach.

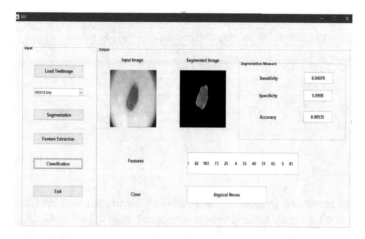

Fig. 4. Matlab GUI of skin disease detection

4.1   Segmentation Performance

In the initial part of the result analysis, performance analysis metric such as the such as Sensitivity, Positive-Predictive-Value (PPV), Specificity, Negative Predictive-Value (NPV), along with the False-Positive Rate (FPR) and False-Negative Rate (FNR) were evaluated and the skin that was segmented was presented.

Table 1. Compare the proffered with the Existing

Segmented Images	Proposed Modified Level set approach						Level set Approach					
	Sensitivity	Specificity	PPV	NPV	FPR	FNR	Sensitivity	Specificity	PPV	NPV	FPR	FNR
A	0.93	0.96	0.65	0.99	0.06	0.07	0.92	0.95	0.61	0.99	0.07	0.09
B	0.97	0.98	0.79	0.99	0.03	0.04	0.96	0.97	0.77	0.99	0.04	0.05
C	0.94	0.92	0.59	0.99	0.01	0.06	0.93	0.91	0.54	0.99	0.1	0.08
D	0.92	0.99	0.94	0.99	0.01	0.09	0.91	0.98	0.91	0.98	0.02	0.1
E	0.90	0.98	0.84	0.99	0.01	0.09	0.89	0.97	0.76	0.99	0.02	0.12
F	0.97	1.00	0.97	0.99	0	0.03	0.96	1.00	0.98	0.99	0	0.05
G	0.92	0.94	0.92	0.95	0.06	0.09	0.91	0.93	0.89	0.91	0.09	0.1
H	0.86	0.91	0.93	0.90	0.06	0.14	0.85	0.90	0.9	0.85	0.09	0.19

Form the Table 1. it is evinced that the segmented images produce 98% of accuracy while classifying the skin database images as normal and abnormal for various metrics.

4.2   Classification Performance

The next part is the classification of the results. Table 2, provides the evaluation of the performance for the various existing models the results obtained clearly states that the DFO with the ANN provides an improved results on the grounds of the of accuracy, sensitivity and specificity.

Table 2. compares the WOA-SVM with the prevailing approaches

Classifiers	Specificity (%)	Sensitivity (%)	Accuracy (%)
k-NN	89	79	90
ANN	91	86	92
SVM-GA	97	73	93
SVM-PSO	94	75	94
ANN-DFO	99	85	98

5   Conclusion

The paper, proposes the Dragonfly based Artificial Neural network for classification of skin diseases as to find the normal or abnormal stage. The main intention is to investigate the proposed automatic detection with existing approaches. The evaluation metrics are estimated for several parameters which are tabulated above and the segmentation accuracy for the presented method is 98%. The segmented images produce 98% of accuracy while classifying the skin database images as normal and abnormal for various metrics.

Compliance with Ethical Standards

All author states that there is no conflict of interest.

We used our own data.

References

1. Güvenir, H.A., Emeksiz, N.: An expert system for the differential diagnosis of erythema-squamous diseases. Expert Syst. Appl. **18**(1), 43–49 (2000)
2. Codella, N.C.F., et al.: Deep learning ensembles for melanoma recognition in dermoscopy images. IBM J. Res. Dev. **61**(4/5), 1–5 (2017)
3. Codella, N., et al.: Deep learning, sparse coding, and SVM for melanoma recognition in dermoscopy images. In: Zhou, L., Wang, L., Wang, Q., Shi, Y. (eds.) International Workshop on Machine Learning in Medical İmaging. Springer, Cham (2015)
4. Olatunji, S.O., Arif, H.: Identification of Erythemato-Squamous skin diseases using extreme learning machine and artificial neural network. ICTACT J. Softw. Comput. **4**(1), 627–632 (2013)
5. Olatunji, S.O., Arif, H.: Identification of erythemato-squamous skin diseases using support vector machines and extreme learning machines: a comparative study towards effective diagnosis. Trans. Mach. Learn. Artif. Intell. **2**(6), 124 (2015)
6. Bajaj, L., et al.: Automated system for prediction of skin disease using image processing and machine learning. Int. J. Comput. Appl. **180**(19), 9–12 (2018)
7. Maglogiannis, I., Zafiropoulos, E., Kyranoudis, C.: Intelligent segmentation and classification of pigmented skin lesions in dermatological images. In: Antoniou, G., Potamias, G., Spyropoulos, C., Plexousakis, D. (eds.) Hellenic Conference on Artificial Intelligence. Springer, Berlin, Heidelberg (2006)

8. Suhil, M., Guru, D.S.: Segmentation and classification of skin lesions for disease diagnosis. arXiv preprint arXiv:1609.03277 (2016)

9. Tasoulis, S.K., Doukas, C.N., Maglogiannis, I., Plagianakos, V.P.: Classification of dermatological images using advanced clustering techniques. In: Engineering in Medicine and Biology Society (EMBC), pp. 6721–6724. IEEE (2014)

10. Zhang, X., et al.: Towards improving diagnosis of skin diseases by combining deep neural network and human knowledge. BMC Med. Inf. Dec. Making **18**(2), 59 (2018)

11. Elgamal, M.: Automatic skin cancer images classification. Int. J. Adv. Comput. Sci. Appl. (IJACSA) **4**(3), 287–294 (2013)

12. Jiang, X., et al.: Image segmentation based on level set method. In: Proceedings of International Conference on Medical Physics and Biomedical Engineering, pp. 840–845 (2012)

13. Liu, L., Lao, S., Fieguth, P.W., Guo, Y., Wang, X., Pietikäinen, M.: Median robust extended local binary pattern for texture classification. IEEE Trans. Image Process. **25**(3), 1368–1381 (2016)

14. Gavrilov, D.A., et al.: Use of neural network-based deep learning techniques for the diagnostics of skin diseases. Biomed. Eng. **52**(5), 348–352 (2019)

15. Kandan, S.R., Sasikala, J.: Multilevel segmentation of fundus images using dragonfly optimization (2017)

Service-Oriented Middleware for Service Handling in Cloud-Based IoT Services

R. Thenmozhi$^{(\boxtimes)}$ and K. Kulothungan

Information Science and Technology,
College of Engineering Guindy, Anna University, Chennai, India
rtmozhi86@gmail.com, kulo@auist.net

Abstract. There is rapid development in technologies of both Cloud Computing and Internet of Things, regarding the field of integrating Cloud-based IoT services. IoT services address the seamless integration of IoT and dynamic interaction with cloud services. In that, most of the services are accessed through the cloud which also responds to the request in a short delay manner. In IoT based cloud service, Load balancing mechanism is one of the key issues where the service request from various services deployed in the cloud system to improve both resource utilization and response time. In this paper, a cloud-Based IoT Service is proposed to improve the efficiency of connectivity between IoT services with less response time based on the load balancing. The Service-Oriented Middleware can be implemented on both Load balancing Service Handling (LBSH) mechanism, which supports deployment services in the cloud and acts as a light complex service processing in IoT. The comparison with existing cloud services the performance of Proposed Load balanced IoT cloud service is evaluated.

Keywords: Internet of things · Service-oriented middleware · Service aggregation · Load balance · QoS

1 Introduction

IoT is the existing escalating technology that utilizes the various existing and new protocols and that protocols which is selected based on the services needed. Many research areas are used the IoT services to provide the smart systems where some examples are a smart home, healthcare and farm industries. And most of the IoT services need the help of the cloud system to enhance data storage and communication as given in [1]. The major advantage of the cloud systems is enabling the devices to access the applications using the existing Internet based protocols. In practical perspectives, the IoT based cloud systems facing different levels of challenges in the data sharing using the software layer [3]. Some protocols create efficient communication in cloud based IoT services to integrate the existing services or to enables the new services. And it is necessary to identify the interface which provides the standard for creating the connectivity even in incompatible between the devices. Some services in middleware for IoT cloud based services have been standard as the system that can provide the necessary communications of services and has become increasingly

© Springer Nature Switzerland AG 2020
J. S. Raj et al. (Eds.): ICIDCA 2019, LNDECT 46, pp. 516–523, 2020.
https://doi.org/10.1007/978-3-030-38040-3_58

important for IoT which adopts the Service-Oriented IoT cloud system are necessary to be reconfigurable [2]. Many existing methods don't fit with the actual need of the consumers, and still the researcher's searching the needed efficient techniques for supporting the dynamic load balancing during service integration of IoT cloud system [8]. The above interpretation of this paper interfered from a Cloud-based IoT service model, taking into consideration, the proposed and developed LBSH algorithms, which quantitatively analyze the less response time during the service process. This paper presents an IoT middleware to handle the aggregation process. This method consists of load balancing, which supports deployment services in the cloud and acts as a light complex event processing in IoT [12]. Combining these two attributes Service-Oriented IoT middleware achieves a great impact on IoT cloud system.

The subsequent part of the paper is structured as follows. Section 2 provides an Overview of the related works trades with cloud-based IoT services using Load balancing methods. Section 3 discusses the IoT cloud system architecture. Section 4 proposed mechanisms. Section 5 concludes the paper with future work.

2 Related Work

Some research articles are explained about the load balancing strategy in the cloud and IoT based services. Most of the existing works concentrate on resource sharing among the same kind of consumers and the load balancing performance is evaluated using several mathematical expressions. The load balancing in cloud systems is carried out by using different tools and techniques which also helps to reduce the time consumption. Because the discussion for load balancing arises only when there is a need for time consumption in the existing system. On other hand, load balancing is needed when the service providers want to increase the scalability of the existing systems [4]. But in our work, load balancing in cloud based IoT services needs to adapt the existing techniques or identifying the new technique or tool which suits the load balancing mechanism in cloud based IoT services. In this section, some of the load balancing mechanism for IoT cloud system is discussed to justify the need for the proposed system.

Lua et al. [5] discuss the load balancing strategy which is applicable for web services. The strategy is named as Join-Idle-Queue algorithm which dispatch the existing services among the users without any delay. For this purpose, this algorithm uses the queue which is not specified with the actual length where it increases dynamically based on the request increases from the users. This will reduce the critical path in the overall process and also reduce the bottleneck problem by maintaining the existing scalability level without any packet loss. James et al. [6] gives the technique for reducing the processing time by modified the existing approach called Active Monitoring Algorithm (AMA). The existing algorithm is modified by assigning weightage to the node to reconsider the algorithm as Weighted AMA (WAMA). In this modified approach, each virtual machine is assigned with the jobs based on the processor capability. The processor capability includes the scalability, availability, power consumption and processor speed. Based on the mentioned parameters, the weightage of the nodes is changed after each service execution and updation takes place in the virtual machine which enables the load balancing in all nodes. This approach provides the optimized and assured load balancing when compared to other techniques. But while you consider or cloud based IoT services, we have to reconstruct the whole process in a different manner. Shobana et al. [7] provide the Central Load Balancer that helps to manage the resource sharing among the virtual machines and avoid the load balancing problems in Cloud-data center. In their work, they suggest data controller and central load balancer where the data centre acts as the centre for handling the request and response to the end user. Central load balancer which connects all virtual machines with the data centre controller to optimize the load balancing. This system uses the keywords busy and available for virtual machines to notify their status to the central load balancer through which the request will be given to the virtual machine. Tang et al. [8, 9], there is another approach discusses the load balancing in the dynamic way which named as dynamical load balanced scheduling (DLBS) approach. This approach uses the time slot for each request and based on the request actual time needed for the task is evaluated by applying set of heuristic scheduling algorithms Chaczko et al. [10]. In most of the cases, Round Robin scheduling is suitable for assigning the time slot because it reduces the average waiting time for an individual task. And also, it reduces the imbalance of waiting time between the request at the data center. Based on the discussions in literature survey, the proposed work suggested the load balancing algorithm through the middleware which reduces the incompatible arises between the service provider and clients. And also when middleware handles the scheduling then the workload on the service provider is greatly reduced.

3   IoT Cloud System Architecture

3.1   Cloud-Based IoT Services

IoT application executed across IoT and clouds whereas the IoT services classified in two ways lightweight and Large-scale analytics based on the data size used. The proposed system initially builds the cloud data center for cloud based IoT systems and then the mechanism for load balancing is suggested based on the requests arise. In addition, the system wants to implement any predictive analysis mechanism then there is a space given for managing the sensors and processing the data from that with the proper storage mechanism. The IoT Service-Oriented Middleware requires handling the Service Process which enhances the system building over the top of the cloud to receive the data form various deployed IoT elements. And also the service provided by the proposed system requires the middleware as a part of the service scheduling mechanism which reduces the scheduling time at the service provider end. This configuration helps to connect the various IoT elements with the cloud services and the gateways. Most of the IoT elements are not sensing the actual data due to the load request form the client side. However, if this action is not associated with the control of cloud services, these services might not react rapidly to deal with a huge amount of incoming services, it might take several minutes to obtain and new computing resources for cloud services. The integration of Cloud-based IoT services provides high-accuracy analytics, from the cloud service, to activate more service requests and increase the service processing time.

3.2   Service-Oriented Middleware

The complex of integrating IoT cloud systems discovers the number of services across the cloud system. Various services should be distributed expertly among the available services. So that the cloud system provides without any impact on service processing based on the service Aggregation process. The Fig. 1 shows the service-Oriented middleware provides in such a way as to reduce the average of Response time by the Load balancing. The service process begins by selecting the set of composite service responsible for collecting at the sink or gateway and moreover sent to the cloud to be stored and used by specific applications, or service processed and then sent to the service process to make intelligent service handling using load balancing. There is enthusiastic service processing to handle the load balancing between the IoT Cloud systems to providing high service availability.

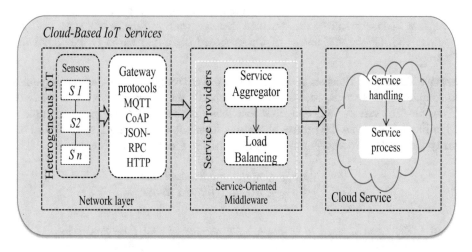

Fig. 1. System architecture for IoT Cloud services

3.3   Service Aggregation Process

Service Aggregation Process consists of solving a sequence of single service requests, each service request concern with the different QoS attributes. Such as Service Response time, reliability. For each service processing, the proposed algorithm consists of finding the service processing time, with respect to the most important of QoS attributes.

4   Load Balancing

The LBSH algorithm utilizing the available services optimally, thereby minimizing the response time. The load balancer determines which services should serve the request. since the existing algorithms had some drawbacks; there is a need to improve the efficiency of existing algorithms. Here, the load balancing mechanism by distributing the services over the cloud. The proposed LBSH mechanism has two major objectives: (1) Response Time in order to ensure high service availability and (2) satisfying the QoS service requirements in terms of integrating IoT services over the cloud. The performance of QoS parameters such as response time and service processing time, based on the average waiting time with balanced service. In the network layer, the service aggregator decides the requested service from the application layer. the service acquirement based on service composition, the aggregation process should arrange the appropriate services or sensor network to execute it. At the integration of cloud service, the load balancing needs to optimize the processing time. the Average waiting time $\left(w(t)_{avg}\right)$ of service processing is defined from the minimized processing time and reduction of the service time allowed by the service requests. If the service processing time, that does not satisfy the obtained average waiting time, are filtered out. To evaluate and analyze the results of the proposed algorithm based on the following Parameters:

Service Time S(t): Consists of service time from the recognition of the service request to completion of service process time. S(t) depends on process capability and service allocation. The service processing time that does not satisfy the Average waiting time w (t) is filtered out from the service allocation. This allows reducing service time. Formally the service processing time solved at $s(p)_t$ is formulated as follows:

$$s(p)_t = \begin{cases} s(p)_{Rt} \leq w(t)_{avg} & Minimum\ time\ value \\ s(p)_{Rt} \geq w(t)_{avg} & filtered\ values \end{cases}$$

Load L(S): Load is used to determine the service of real time processing. Which depends on the request on service arrival time $S(\alpha)$ and service finishing timeS (β). The overall service processing time is formulated as L(S) = $S(\alpha) \times S(\beta)$.

Reliability S(A): Denotes the probability of filtering services(a). Here, we modeled it as a discriminate variable with S(A). if the service processing time, that do not satisfy the obtained average waiting time, are filtered out. the filtering phase allows the reduction of service processing and consequently the service response time.

Response Time: Since the services are integrating into the field of Cloud-Based IoT services. Requested service is serviced; the response time calculated based on the total time taken to respond to the service processing time. hence the response time will improve relatively. The following LBSH algorithm ensures the computation of load balancing with the using of average waiting time (Table 1).

Table 1. List of symbols used in the LBSH Mechanism

$L(S)$	Usage of load service
α	Request arrival time
β	Service Process time
$w(t)_{avg}$	Average waiting time
$S(p)_{RT}$	Service requested from different services
$S(q)_{RT}$	Compute service processing time

Algorithm 1: Load-Balancing in Service Handling [LBSH]

Input: number of services 1 *to n*. Considering the Response time *(RT)* of Services S (p) and S (q) concerning the comparatively finding the response time of $S(p)_{RT}$ and $S(q)_{RT}$.
BEGIN
Step 1:   To calculate the Response Time *(RT)* according to both services p & q.
Step 2:   To calculate the (load) usage of services.
          concerning the Request arrival time α and service process time β.
$L(S) = S(\alpha) \times S(\beta)$
$$S\,(t) = w(t)_{avg} + L(S) \tag{1}$$
Step 3:   To determine the computation of Response time from service requested from different services.

$$S(p)_{RT} = \frac{\sum_{i=1}^{n} S(t)_{RT}}{n(s)} \tag{2}$$

Step 4: *for i =1 to n*
Step 5: Calculate $S(p)$ Using equation 2. Thus equation 2 can be written directly as equation 3.

$$S(p) = S(p)_{RT} \tag{3}$$

Step 6: To find the probability of failure of services (a), we consider using equation 4.

$$S\,(A) = \sum_{1}^{n} \frac{a}{n(s)} \tag{4}$$

Step 7: To compute the response time of service processing time.

$$S(q)_{RT} = \frac{\sum_{1}^{n} L(S)_{RT} S(t)}{n(s)} \tag{5}$$

Step 8: *for j=1 until n*

$$S\,(q) = S(q)_{RT} \tag{6}$$

Step 9: To find the minimum processing time,
        if $(S(p)_{RT} < S(q)_{RT})$ then
        Return(min Response time)
        else if $(S(p)_{RT} > S(q)_{RT})$ then
        Return(filtered vale)
        else
                Return (equation 4)
        End if
END

4.1   Performance Evaluation

The experiments are carried out with the different set of services accessed from different nodes and performance are analyzed by response time. The proposed load balancing algorithm determines the requested services should follow the request from the service processing. Based on that request, service can be offered at different time intervals for the same or different users which processed by the proposed LBSH approach. The performance of the service handling method is evaluated by executing in Java Virtual Machine (JRE 1.6, for Windows 64-bit system). Once the request is received at the service provider end, the current load of the system is evaluated and schedule the request based on the load on each service provider. In Fig. 2, the experiments consider a service request from different services using LBSH mechanism. In this number of Services and Service Processing time is evaluated by the responsibility of different request service and shows that the response time is reduced and improve the load balancing mechanism. And the proposed mechanism is compared with existing HDLB system where LBSH outperforms in the perspective of load balancing as Shown in Fig. 3.

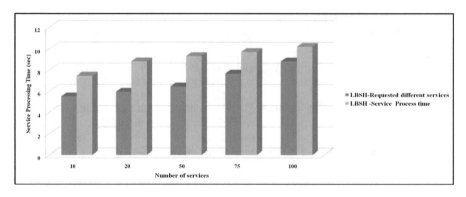

Fig. 2. Response Time for requested services

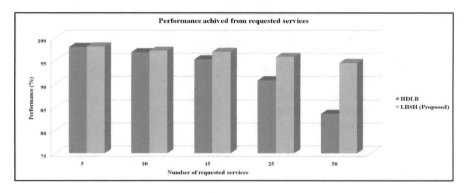

Fig. 3. Service performance compared to HDLB vs. LSHB

5   Conclusions and Future Work

In this paper, a Service-Oriented Middleware approach based on integrating Cloud based IoT services is formulated as LBSH mechanism. The proposed system needs to minimise the service response time through dynamically balancing service flow across the cloud services. While satisfying the QoS requirements. This proposed work shows that this method has a high service availability and ensure that the load is repetitively concerted among all the services. As part of future work, will consider more service requirements, such as IoT service Composition, selection time, composition lifetime, optimize the scheduling performance in the IoT network.

References

1. Ghodake1, S., Sonkamble, S.: Efficient dynamic load-balance flow scheduling in the cloud for big data centers. Int. Res. J. Eng. Technol. (IRJET) **04**(02) (2017). www.irjet.net
2. Li, L., Li, S., Zhao, S.: QoS-aware scheduling of services-oriented Internet of Things. IEEE Trans. Ind. Inform. **10**(2), 1505–1797 (2014)
3. Truong, H.-L., Dustdar, S.: Principles for Engineering IoT Cloud, Vienna University of Technology IEEE cloud computing published by the IEEE computer society 2325-6095/15/$ 31.00 © (2015)
4. Singh, H., Gangwar, R.C.: Comparative study of load balancing algorithms in cloud environment. Int. J. Recent Innov. Trends Comput. Commun. **2**(10), 3195–3199 (2014). ISSN 2321-8169
5. Lua, Y., Xiea, Q., Kliotb, G., Gellerb, A., Larusb, J.R., Greenber, A.: Join Idle-Queue: a novel load balancing algorithm for dynamically scalable web services. Int. J. Perform. Eval. IJPE, 1056–1071 (2011). Amsterdam, The Netherlands, ISSN 0166-5316, https://doi.org/10.1016/j.peva.2011.07.015
6. James, J., Verma, B.: Efficient VM: load balancing algorithm for a cloud computing environment. Int. J. Comput. Sci. Eng. IJCSE **4**, 1658–1663 (2014). ISSN 0975-3397. Accessed Sept 2012
7. Shobana, G., Geetha, M., Suganthe, R.C.: Nature inspired pre-emptive task scheduling for load balancing in cloud datacenter. In: International Conference of Information Communication and Embedded Systems, ICICLES 2014, Chennai, India, pp. 1–6 (2014). ISBN 978-1-4799-3835-3. https://doi.org/10.1109/icices.2014.7033816
8. Tang, F., Yang, L.T., Tang, C., Li, J.: A dynamical and load-balanced flow scheduling approach for big datacenters in clouds. IEEE Trans. Cloud Comput. **6**(4), (2018)
9. Swamy, C.M.: Dynamic load balancing technique in cloud partition computing. Int. J. Eng. Dev. Res. (IJEDR) (2012)
10. Chaczko, Z., Mahadevan, V., Aslanzadeh, S., Mcdermid, C.: Availability and load balancing in cloud computing. In: presented at the international conference on computer and software modeling, Singapore (2011)
11. Kaushik, Y., Bhola, A., Jha, C.K.: A comparison of heuristics algorithm for load balancing in a cloud environment. Int. J. Sci. Eng. Res. **6**(9), 1208–1214 (2015). ISSN 2229-5518
12. Wang, X., Sheng, M.-J., Lou, Y.-Y., Shih, Y.-Y., Chiang, M.: Internet of things session management over LTE—balancing signal load, power, and delay IEEE. Int. Things J. **3**(3), 339 (2016)

Intelligent Educational System for Autistic Children Using Augmented Reality and Machine Learning

Mohammad Ahmed Asif[✉], Firas Al Wadhahi,
Muhammad Hassan Rehman, Ismail Al Kalban, and Geetha Achuthan

College of Engineering, National University Science and Technology,
Muscat, Sultanate of Oman
ahmed_asif96@hotmail.com

Abstract. Autism is a severe disorder affecting 1 in 160 children globally. Autism comprises of several development disabilities such as social, communicational and behavioural challenges. Children being diagnosed by the autism mainly face a hard time studying curriculum in inclusive classrooms based on their IQ level and the autism levels. Although, different strategies and learning teaching tools are available to support autistic children, only few systems aid them in learning efficiently, and are not highly interactive. Thus, the proposed Intelligent Education System primarily focuses on providing interactive learning experience to the autistic children with IQ level >50% and efficient teaching assistance to their tutors using augmented reality and machine learning in both English and Arabic The capability of the education system to perform an action, allows the autistic child to interact with the playable sand and gain interest. In learning stage, once the child scribbles on the sandbox, Kinect 3D camera captures and recognizes the drawn image. After the refinement and recognition of the image using OpenCV and classification model, the stored set of real world object are projected on the canvas. Besides, a webcam captures the facial expression of the child, and emotion detection algorithm determines the reaction of the child. Based on the child's emotion, the current object is projected and pronounced three times to enforce better learning. Once the instructor chooses the language and character to be taught using the developed mobile application, the system displays it over the sandbox and further three objects that starts with the particular character are pronounced and projected. The system is tested rigorously with large set of users, and the results prove the efficiency of the system and happiness of the autistic children in better learning.

Keywords: Autism spectrum disorder · Autistic children · Education · Machine learning · Augmented reality · Image processing

1 Introduction

Autism or autism spectral disorder (ASD) is the most prevalent disability that the children of our generations are facing. According to recent research, autism now affects 1 in 160 children globally [1]. It is a complicated neurobehavioral disorder that

© Springer Nature Switzerland AG 2020
J. S. Raj et al. (Eds.): ICIDCA 2019, LNDECT 46, pp. 524–534, 2020.
https://doi.org/10.1007/978-3-030-38040-3_59

includes impairment in interaction and communication skills combined with rigid, repetitive behaviours and obsessive interests. Autism is found in individual of every race, age, gender, etc. Often children with autism lack empathy and frequently involve in self-abusive behaviour; Biting one's hand or head-banging. There is a growth in the prevalence of the ASD globally. In 2011, Centres for disease control and prevention identified that 12 per 1,000 children are being diagnosed with the stated disorder. Estimated number of autistic disorder cases has increased massively, from 50% to 2000%, i.e. nearly about 67 million people around the globe [2]. The cross-sectional study conducted estimated about 1.4 out of 10,000 children aged 0–14, are found to be autistic in Oman [3]. 74.3% among those are boys. Hiding may be a potential reason for the low prevalence rate in Oman [4]. The families with an autistic child often tend to hide him/her from the outside world, which limits the child from attaining education and medical facilities. As per Dr. Said Al-Lamki, Director of Primary Healthcare, 1 out of 68 children in the sultanate of Oman are being diagnosed with autism [5]. With such high prevalence rate present globally, there is an urgent need to develop a system which can help in better management of kids with autistic symptoms. Several acts were passed to support the children with autism and development disabilities; No Child Left Behind and (2002) and Individuals with Disabilities Improvement Education Act (2004) [4].

Autism children face a hard time studying the curriculum along with other children, nowadays many of the schools have special classroom and a special attention is shown by the teachers for them. Though, the teaching techniques are still not productive and interactive. Several researchers have proposed different educational systems, to facilitate the better learning of autistic children using various methods and procedures. Decristofaro [6] recommends various evidence-based strategies for the tutors to proficiently educate the autistic children in inclusive classrooms, like peer-support, scribing, visual schedules, TRIBES Strategies. Two teachers claimed that adopting these strategies have improved the learning capability of the children in their classes. However, interviewing only two teachers may not be sufficient to gather précised information. A computer based program was developed, based on the daily activities performed during eating and playing, to determine the enhancement in the communication functionalities of the autistic child in a classroom. The system was implemented in a local special education school, by employing five autistic children, who are diagnosed with communication disabilities. Bernard-Opitz, Sriram and Nakhoda-Sapuan [7] idea of using pictures and animations was further enhanced, by developing a Virtual Environment of a café and bus to enhance social understanding among the autistic children. The study was carried out with the purpose of exploring the potential of using virtual environment to be used as an educational tool for autistic individuals.

Artificial Intelligence (AI), is a way of developing intelligent machines or a software, similar to humans. AI continuous to be a major trend in digital transformation in 2018, by affecting every industry and business with rapid advancements in technology [8]. AI has contributed a lot in the field of education, as a report suggests, an increase of AI in education sector will be 47.5% during the year 2017–2021 in United States [9]. A research has been carried out by Smith et al. [10], which discusses about blending human with the artificial intelligence, in order to aid the autistic children through the development of several AI powered tools. The research emphased on the technology named, ECHOES, to enhance social interaction skills of the autistic children. ECHOES was deployed and tested. Children with or without autism were able to interact and perform the activities. However, the results may not be generalised and applied to all autistic individuals due to limitations in each person's highly diverse non-uniform qualities and comorbidity.

As prevalence rate of autism in Oman is increasing from time to time and only a few special need schools and facilities are provided for the education of the autistic children and most of the families are not interested to send their children to those schools. Furthermore, the analysis of the previous literature encouraged the idea of the intelligent education system, as the results of the previously used strategies were impressive. However, the previous system does not allow the child to freely represent their thoughts in an interactive playable environment for the purpose of learning. The Educational System will certainly help the autistic children to learn independently by being at homes, and will support the parents and caretakers of the child to teach the children effectively in both Arabic and English without any distress at low cost.

The remaining part of this research paper is organized as follows. Section 2 discusses about the overall design of the proposed Intelligent Educational System for Autistic Children (IESAC). Section 3 explains the design and methodology adopted in developing the system. Further, a Sect. 4 discusses the implementation details and Sect. 5 covers the result analysis followed by conclusion in Sect. 6.

2 Design

This chapter provides an overview about the design of the proposed system, including flowchart and a 3D model.

2.1 Flowchart

In order to initiate the intelligent autistic learning system, the system must be switched on and the autistic child must be positioned in the right place before carrying any further operations on the system. Once the system is calibrated, all the inputs provided to the intelligent autistic system are captured using the web application. The system then initiated either the teaching or learning process based on the input provided. If the teaching module of the project is initiated then the user must select an object from the drop down list provided on the web application to be displayed on the sand along with the respective audio (Refer to Fig. 1).

Moreover, if the learning module is initiated then the system is as kept on hold, waiting for the autistic child to scribble on the sandbox and raise their hand. Meanwhile, the webcam facing towards the child captures the hand gestures of the child. Once the child raises their hand, the Kinect camera would measure the depth of the play sand within the sandbox, to determine the shape scribbled by the autistic kid. Further processing would be done through image processing, to refine the image and extract the foreground from the background. The processed image is then fed into the machine learning model, through an API call to predict the object.

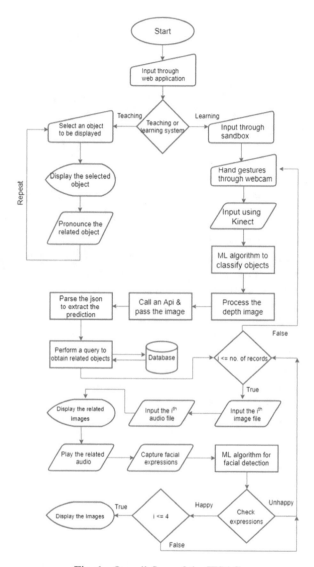

Fig. 1. Overall flow of the IESAC

Once the object is predicted, it is sent as a parameter in a database query to obtain the objects related to it. Each object is projected on the sand box using a projector and its respective audio is played. Further processing is based on the child's reaction on the prediction made by intelligent autistic learning system. A webcam is used to constantly capture and monitor the facial expressions of the autistic child and process it on a machine learning model to determine the emotions. If the projected object is of child's interest, positive emotion would be captured; therefore, the object will be displayed for a longer span of time. In case, if the captured reaction is against the child's preference, then next object from the database will be projected and the process would be repeated.

Once after the entire process has been completed, the system would jump back to the initial phase, to allow the child to scribble the object on the sandbox.

2.2   3D Design

This image demonstrated the 3D design of intelligent autistic learning system (Refer to Fig. 2). It showcases all the components that are used and attached in the intelligent autistic learning system. Starting from the main component of the system being the sandbox, which carries the entire sand. Attached to the sand box is the short throw project on top of the system, which is projecting all the images from the machine learning model onto the sand. Providing the augmented reality on the sandbox. Along with the projector, the Kinect 3D camera is also attached on top of

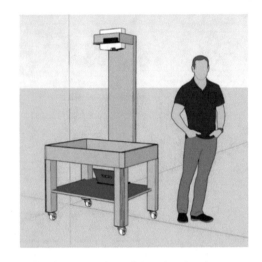

Fig. 2. 3-D design of the educational system

the system, the Kinect 3D camera is used to perform image processing for autistic learning system. A small camera is attached on the centre of the system which is capturing the gestures and expression of the autistic child and makes the decisions accordingly.

3   Methodology

In this chapter, the methodology adopted and the phases of implementation are clearly defined and illustrated, following a structure in which the tasks are undertaken.

3.1   Phase 1

Initially a deep research was done based on machine learning and artificial intelligent in the field of medicine. Many different research papers were reviewed in order to build an efficient and enhanced system without the limitations being faced by the previous systems in the same field. In order to better understand the issues being faced by the autistic child and in order to understand the autism spectrum disorder, different autistic learning centres were visited. Along with the visits, the teachers and students were interviewed in order to understand autism and build the optimum system to aid this disorder.

3.2   Phase 2

Secondly, a process of requirement analysis and gathered all the components was implemented for intelligent autistic learning system. In order to obtain an efficient and optimum solution, a detailed research on each component was done. The outcome of the detailed research concluded that the components that were used in the implementation of this project were the most suitable components for intelligent autistic learning system.

3.3   Phase 3

After gathering all the equipment that are suitable to carry out the implementation process of this project. Implementation phase was started of the project, where each and every component was integrated with the system; this process was a one by one process, in order to test the components integration. Along with the integration of components, the python libraries and code were written to execute machine learning and augmented reality of intelligent autistic learning system. The implementation of this project is done in two languages which was challenging as the objects of both the languages were often similar in nature, which challenged the performance and prediction of the machine learning model. This issue was mitigated by using more datasets of both the languages and often the dataset was captures by drawing the object on the sandbox. It was a long procedure of enhancing the machine learning model in order to make the system intelligent enough for its deployment.

3.4   Phase 4

Upon the completion of integration process of intelligent autistic learning system. The crucial process of testing and debugging was executed where the system was deployed in autistic learning centres and the testing procedures were carried out. This process included the hand on experience of intelligent autistic system with teachers as well as autistic children. This phase enhanced the system by using the approach of trial and error. Different issues were encountered by having real time deployment of the project and the working principles. These issues were then resolved by enhancing the python code and using excessive amount of dataset in order to optimize the machine learning predictions of the system. As, the lack of training data resulted in lower accuracy of the prediction.

The phase of data capturing was repeated several times to gather excessive amount of dataset. Alongside, the scribbles by the autistic child were recoded for the purpose of research and enhancement of the model [11].

4   Implementation

To extract the Kinet Camera frames it was programmed in python with the necessary libraries for the component. Then the kinect captures 3 different streams, namely, color, depth and IR. Through SyncMultiFrameListener the computer receives the frames of the 3 streams. Mainly the depth frame was used to comprehend the diagram on the sand. Kinect camera was operated through Libfreenect2 library. For example, to close/open camera.

The stored frames from the Kinect streams, hence, each frame was processed against an image processing to extract a meaningful image. Set of pre-defined functions in the OpenCV library were utilized to convert the depth frames in a numpy data type, before applying the equalizeHist() function to adjust the colour contrast of the frames. The frame is later masked to set the background pixel values to zero, and further the threshold effect is applied on the frame. Number of frames were extracted on a continuous loop every 3 s and alongside, the sand was scribbled continuously forming different set of images (Refer to Fig. 3). The images were further refined and classified manually before training the model. The classification model was trained employing a third party application, named as Ximilar App. Within the Ximilar app, a new task is created which has a unique ID and Token. Within each task, several categories of images can be defined and the images can be uploaded separately for each. Each category needs be labelled with tags. Once after the categories are defined, other options needs to be configured, to optimize the model. Finally, the model would be deployed, which is later called trough the API calls. After the deployment of trained model, the processed image is passed through API call. An authorization token and taskID is required in API header to pass API. The response of API is in JSON format is parsed to obtain the prediction.

Fig. 3. Processed image

A web application was developed to control the flow of the system by the tutor, such as start/stop the system, display specific objects, etc. Therefore, to design the web application, a readymade bootstrap template was implemented. Moreover, a local database server was configured in the system. Set of related objects were gather and manually stored into the table to perform a query, with the prediction as a where clause. Numbers of records were obtained against each query, which were processed one at a time. The web application has been developed into a fully functional unit by implementing the server-side codes. A python library named as Flask was implemented in this phase, to execute the server side codes during run-time.

Alongside, the Emotion Detection model was obtained from GitHub and was than trained with the dataset. The dataset consist of 35887 greyscale images of size 48 × 48, with total of seven emotions; angry, disgusted, fearful, happy, neutral, sad and surprised. The model was developed using tensorflow, tflearn and keras libraries. Finally, the system was taken for the acceptance testing by the tutors and verification was done by the set of students, 10 + college staff members and the autistic kids.

5   Testing

Throughout the development of the system, it has undergone several types of testing. Several types of testing methodologies and readily available CASE tools are used to test the computer based systems depending on the type and complexity of the system. The Intelligent educational system was divided into two major phases, the functioning of object classification and the functioning of emotion detector model. Therefore, two system tests were performed, each with the gap of 2 months. The interaction of the tutor was required, as no such requirement document was maintained.

To perform the acceptance testing of the intelligent educational system, a visit to an autistic centre was arranged on 29th May, 2019, to allow the autistic kids use the system. Total of 6 participants were involved in the testing of the system. 4 out of 6 participants were male students, within the age group of 5–9. Whereas, the female participants were 4–8 years old. As per the teacher, the kids were not taught English, and only knew the Arabic alphabets and numbers.

The autism level indicated in the Table 1 below, was provided by the tutor and was reported to be obtained using the Childhood Autism Rating Scale (CARS). CARS is an autism assessment tool, to diagnose the autism in children and scale it [12].

Table 1. List of participant along with autism level

Name	Age	Autism level
Participant 1	7	27
Participant 2	9	35
Participant 3	4	30
Participant 4	5	37
Participant 5	7	31
Participant 6	9	30

6   Result Analysis

The improvement in the learning of the autistic kids through the use of educational system is evident. The results achieved by the development of this project bring positivity and motivation to further enhance the project. The intelligent educational system is developed to aid in the learning of the autistic kids, enhance their skill set and build their sense of touch. To ensure the efficient learning of the autistic kids, the system should run smoothly and predicted accurate results, within optimal time duration.

Whereas, in the learning phase, the child should interact with the system. Outcome of the implemented system is successful, as demonstrated in the Figs. 4 and 5. The child scribbled the desired object and the webcam was simultaneously capturing the hand gestures. The emotion detection model was concurrently running in the local machine to capture and store the emotions of the child. However, it was difficult to capture the emotion data, as the child is not expected to be within the frame at all times.

To gain more experience regarding autism and autistic behaviors through interaction with the autistic kids, the autism centre was visited several times.

Fig. 4. Teaching phase with participant 1 in English

The visit to the autistic centre was not only for the purpose of testing, it was to gain more experience regarding autism through the interaction. However, during the testing phase, the emotions of each of the kid were recorded for the purpose of analysis. Participants with less autism level, were the only participants to move to the learning phase within the system, after completion of the teaching phase. They were able to scribble the objects shown on the sand, and was able to name the objects clearly. Moreover, they were able to scribble 2 out of 3 objects dictated by the tutor during the learning phase [13].

Few participants were comparatively not scared to interact initially. They took a little more time to scribble in the sand as compared to others. One of the participants was piling up the sand on the object shown, instead of digging down. Although, the system is designed to work on the depth of the sand, thus this would be a point of further research, to understand the behaviour of the child. Participant with extremely high level of autism were scared of the loud audio, due to which they did not interact with the sand much. Yet, they were excited to see the displayed objects and was able to repeatedly pronounce the words.

Fig. 5. Teaching phase with participant 1 in Arabic

As per the tutor, "This is a good project! There are some children that will be attracted to the project and even without me saying anything; you have seen how the children were happy and interacting with the project. So yeah it will be a new thing and it will not only cover the academic learning only, it will also increase their knowledge and sense of feel (because they play with sand) so it will be a very good project."

On an average, 4 out of 6 participants took 2–3 min to get comfortable with the system, and get involved in scribbling the sand. Fearfulness was observed in 2 out of 6 participants, due to the loud audio or the display of frightful objects such as ants & insects. This required some time for the kids to get used to of the audio; however, 5 out of 6 kids were able to speak out the alphabets by listening to the audio within the first 3–5 min of interaction.

7   Conclusion

In conclusion, the detailed study about autism, the prevalence worldwide and prevalence within Oman is discussed. The problems faced in managing the autistic children were emphasized and the number of strategies and tools available were further discussed in detail. The procedures, findings and limitations were highlighted for each. By keeping in mind the major limitations in the existing education systems for autistic children, an intelligent education system has been designed and developed which aids the interactive teaching and learning of tutors and children.

The developed system was tested by autistic children in a local autism centre, Muscat. The results of the test showed a clear positive impact on the children because the children showed an interest on the project, and they have engaged in learning through the system. However, the centre mentioned that the proposed system will have a higher impact on the autistic children in Oman. Furthermore, the instructors at the centre commented that the proposed system would help the autistic children in the academic learning as well as increase in the sense of feel and overall attention and hence the main aim of the project is successfully accomplished.

Acknowledgment. We would like to thank The Research Council (TRC) of Oman for funding to carry out this research as FURAP project. Our deepest gratitude towards the National University of Science and Technology for providing all the essential assets and facilities. All author states that there is no conflict of interest.

References

1. World Health Organization: Autism spectrum disorders (2017). https://www.who.int/newsroom/fact-sheets/detail/autism-spectrum-disorders. Accessed 10 May 2019
2. Posserud, M.: Autistic features in a total population of 7–9-year-old children assessed by the ASSQ (Autism Spectrum Screening Questionnaire). J. Child Psychol. Psychiatry 47(2), 167–175 (2006)
3. Al-Farsi, Y.M., et al.: Brief report: prevalence of autistic spectrum disorders in the Sultanate of Oman. J. Autism Dev. Disord. 41(6), 821–825 (2011)
4. Al-Farsi, Y.M., et al.: Levels of heavy metals and essential minerals in hair samples of children with autism in Oman: a case–control study. Biol. Trace Elem. Res. 151(2), 181–186 (2013)
5. Ministry of Health: MOH Organizes Autism Workshop for Parents (2017). https://www.moh.gov.om/en/-/—542. Accessed 1 July 2019
6. Decristofaro, A.: Students with Autism in Inclusive Classrooms. University of Toronto (2016). Accessed 25 May 2019
7. Bernard-Opitz, V., Sriram, N., Nakhoda-Sapuan, S.: Enhancing social problem solving in children with autism and normal children through computer-assisted instruction. J. Autism Dev. Disord. 31(4), 377–384 (2001)
8. Shu, L.-Q., Sun, Y.-K., Tan, L.-H., Shu, Q., Chang, A.: Application of artificial intelligence in pediatrics: past, present and future. World J. Pediatr. 15(2), 105–108 (2019)
9. Lynch, M.: 7 Roles for Artificial intelligence in Education (2018). https://www.thetechedvocate.org/7-roles-for-artificial-intelligence-in-education/. Accessed 21 Dec 2018
10. Smith, T.J., et al.: Blending human and artificial intelligence to support autistic children's social. ACM Trans. Comput. Hum. Interact. 25(6) (2018)
11. Tang, T.Y., Xu, J., Winoto, P.: An augmented reality-based word-learning mobile application for children with autism to support learning anywhere and anytime: object recognition based on deep learning. In: International Conference on Human-Computer Interaction, pp. 182–192. Springer, Cham (2019)
12. Special learning: Childhood Autism Rating Scale (2019). https://www.special-learning.com/article/childhood_autism_rating_scale. Accessed 25 Aug 2019
13. The Economic Times: Definition of System Testing | What is System Testing? System Testing Meaning (2019). https://economictimes.indiatimes.com/definition/system-testing. Accessed 25 May 2019

Selective Segmentation of Piecewise Homogeneous Regions

B. R. Kapuriya[1]([⊠]), Debasish Pradhan[2], and Reena Sharma[1]

[1] Centre for AirBorne Systems, Bangalore, India
kapuriyabr@gmail.com, reena@cabs.drdo.in
[2] Department of Applied Mathematics, DIAT, Pune, India
pradhandeb@gmail.com

Abstract. In this article, a novel method for the interactive segmentation of the object and background which has multiple piecewise homogeneous intensities has been introduced. We have proposed a new energy function based on intensity points selected from multiple homogeneous regions in the object and background. To minimize the derived energy function, we use the calculus of variation method and transformed it to PDE. The derived PDE has been solved using an additive operator splitting (AOS) method. Simulation results validate the correctness and accuracy of the proposed method. The performance is assessed by testing the algorithm on synthetic images and results are compared with state of the art methods using Jaccard's similarity index.

Keywords: Active contour · Selective segmentation · Additive operator splitting

1 Introduction

In computer vision applications, image segmentation is one of the very important fields of research. Broadly, it is a task of partitioning images in different regions based on some common characteristics of the image domain. Generally, image segmentation techniques are classified in edge-based and region-based segmentation techniques. In classical segmentation, the full image is segmented. But in many cases, a specific part of the image needs to be segmented which comes under local segmentation. So another way of classifying the image segmentation is global segmentation and local segmentation. In global segmentation, all the objects/edges are segmented while in local segmentation (also called selective segmentation) specific object has to be segmented. One of the segmentation approaches of region merging has been discussed in [28] for SAR images where they merged various subregions with weighted Kuiper's distance. While active contour approach based segmentation technique is one of the widely known technique in global segmentation, which became famous after the work proposed by Kass in [14] and Mumford-Shah in [16]. In these techniques, an initial curve selected near the object which is converged to object boundary with the help of some constraints. In [14], Kass has used intensity gradient information for converging active contour which will converge to the boundary of the object which is classified as edge-based segmentation. Other techniques in the edge-based approach can be seen in [9–11].

© Springer Nature Switzerland AG 2020
J. S. Raj et al. (Eds.): ICIDCA 2019, LNDECT 46, pp. 535–542, 2020.
https://doi.org/10.1007/978-3-030-38040-3_60

In a region information based segmentation model Mumford-Shah (MS) in [16] have proposed one of the first models in this approach. Many methods have been introduced for further refinements in the existing model which can be studied in [21, 25]. One of the effective segmentation technique proposed by Chan and Vese in [6] which is considered a special case of the MS model [16]. In [13], authors have active contour method for segmenting blurred object using pixels blur angle information. The readers can refer [4] for proof of the existence of a global minimum of energy functionals in active contour models.

For local segmentation, many researchers have combined edge and region information based approaches and derived hybrid models which can be studied in [2, 3, 5, 7, 18, 22, 23]. In [24], authors have used a modified model for segmenting inhomogeneous regions. Further research works related to inhomogeneous object segmentation can be studied in [1, 8, 15, 19, 27]. In our work, we have focused on selective segmentation technique for the piecewise homogeneous object. Some related techniques are discussed in the details in Sect. 2.

In this article, a new fidelity term instead of existing terms in active contour-based selective segmentation method has been introduced. The modified terms are derived by selecting initial points from available homogeneous regions in the object and background. The implementation of the proposed model has been done in MATLAB and results are compared with existing methods.

Structure of this paper is as follows: we reviewed some active contour based segmentation models and discuss its shortcomings in Sect. 2. The proposed model is presented in Sect. 3. In Sect. 4, we discussed the segmentation results and comparison with related existing methods. The conclusion is given in Sect. 5.

2   Segmentation Models

Here we have reviewed few active contour based segmentation models which have been compared with the proposed model.

2.1   CV Model [6]

Chan and Vese have used level set for simpler implementation for segmenting two homogeneous regions using the MS model [16]. The model is described as follows:

$$F_{CV}(\Gamma, c_1, c_2) = k_1 length(\Gamma) + k_2 \int_{inside(\Gamma)} (f(z) - c_1)^2 dz$$
$$+ k_3 \int_{outside(\Gamma)} (f(z) - c_2)^2 dz, \tag{1}$$

where c_1 and c_2 are the average intensity values inside and outside the object in the image domain respectively, Γ is the initial curve near the object boundary and $f(z)$ is input image. Parameters k_1, k_2, k_3 are all positive constants and controls weight of each term. For notational convenience, we denote $z = (x, y)$, $d(z) = d(x, y)$, $dz = dxdy$, a double integration for two dimensional images throughout the paper.

2.2 Badshah-Chen Model

In [2], Badshah and Chen have combined an edge-based technique in [6] with a region-based technique in [9] and proposed the following minimization functional to find the desired object.

$$F_{BC}(\Gamma, c_1, c_2) = k_1 \int_{\Gamma} d(z) * g(|\nabla f(z)|) dz + k_2 \int_{inside(\Gamma)} (f(z) - c_1)^2 dz$$
$$+ k_3 \int_{outside(\Gamma)} (f(z) - c_2)^2 dz, \tag{2}$$

where c_1 and c_2 are the average intensity values of the image inside and outside the object repectively. Parameters k_1, k_2, k_3 are all positive constants and controls weight of each term. $d(z)$ and $g(|\nabla f(z)|)$ are distance function and edge detector function defined as in (4) & (5) respectively. In [3], Badshah et al. have modified fidelity terms as $(f(z) - c_1)^2 / c_1^2$ and $(f(z) - c_2)^2 / c_2^2$ for objects having blurry edges.

3 Proposed Model

To segregate object or background having multiple intensity values, the above methods sometimes give an erroneous result. To overcome that problem, we give preference to regions to be segmented by selecting one point from each homogeneous regions of the object or background. So, we introduced a new energy function which is comprised of two new fidelity terms along with one of the terms from the Gout model. The new fidelity terms are derived from selecting points from each homogeneous regions existing in the object regions and background regions. So, the proposed new minimization functional for object segmentation is given by

$$F(\Gamma) = k_1 \int_{\Gamma} d(z) * g(|\nabla f(z)|) dz + k_2 \int_{inside(\Gamma)} \left(\prod_{i=1}^{N_1} (f(z) - b_i')^2 \right) dz$$
$$+ k_3 \int_{outside(\Gamma)} \left(\prod_{j=1}^{N_2} (f(z) - b_j^O)^2 \right) dz \tag{3}$$

Where Γ is the boundary of the object, $d(z)$ is the distance function defined as

$$dist((x, y), A) = d(z) = \prod_{k=1}^{n_1} \left(1 - e^{-\frac{(x-x_k)^2}{2\tau^2} - \frac{(y-y_k)^2}{2\tau^2}} \right) \tag{4}$$

Where A is a set of n_1 geometric points which are selected at the boundary of the object of interest and $g(|\nabla f(z)|)$ is the edge detector function defined as

$$g(|\nabla f(z)|) = \frac{1}{1 + |\nabla G_\sigma(z) * f(z)|^2} \qquad (5)$$

Parameters k_1, k_2, k_3 are all positive constants which controls the weight of the each term, $b_i^I; i = 1, 2, 3\ldots, N_1$ and $b_j^O; j = 1, 2, 3\ldots, N_2$ are selected points from foreground (object) and background regions respectively. The above model can be rewritten in terms of level set [17, 20] formulation for simpler implementation using regularized Heaviside function (6) and Dirac delta function (7).

Level Set formulation: Let $\phi : \Omega \to R$ is a Lipschitz function defined as

$$\Gamma = \{z \in \Omega : \phi(x, y) = 0\}$$

$$\text{inside}(\Gamma) = \{z \in \Omega : \phi(x, y) \geq 0\}$$

$$\text{outside}(\Gamma) = \{z \in \Omega : \phi(x, y) \leq 0\}$$

To differentiate the Eq. (3), the regularised Heaviside function and Dirac delta function are used and they are defined as

$$H_\varepsilon(s) = \begin{cases} 0; s \leq -\varepsilon \\ \frac{1}{2}\left[1 + \frac{s}{\varepsilon} + \frac{1}{\pi}\sin\left(\frac{\pi s}{\varepsilon}\right)\right]; |s| \leq \varepsilon \\ 1; s > \varepsilon \end{cases} \qquad (6)$$

and

$$\delta_\varepsilon(s) = \frac{1}{2\varepsilon}\left[1 + \cos\left(\frac{\pi s}{\varepsilon}\right)\right], \qquad (7)$$

respectively. After introducing level set formulation in energy function (3), we obtain

$$F(\phi) = k_1 \int_\Omega d(z) * g(|\nabla f(z)|)|\nabla H_\varepsilon(\phi(z))|dz$$

$$+ k_2 \int_\Omega \left(\prod_{i=1}^{N_1}(f(z) - b_i^I)^2\right) H_\varepsilon(\phi(z))dz \qquad (8)$$

$$+ k_3 \int_\Omega \left(\prod_{j=1}^{N_2}(f(z) - b_j^O)^2\right)(1 - H_\varepsilon(\phi(z)))dz$$

To minimize the above functional $F(\phi)$ in (8), we use functional derivative

$$\lim_{t \to 0} \frac{F(\phi + t\psi) - F(\phi)}{t} = 0,$$

where function ψ is similar to ϕ function. So, from Green's theorem, (8) can be written as the Euler Lagrange equation.

$$\delta_\varepsilon(\phi) \left[k_1 \nabla . \left(W \frac{\nabla \phi}{|\nabla \phi|} \right) - k_2 \left(\prod_{i=1}^{N_1} \left(f(z) - b_i^I \right)^2 \right) + k_3 \left(\prod_{j=1}^{N_2} \left(f(z) - b_j^O \right)^2 \right) \right]$$
$$= 0$$

(9)

in the image domain Ω with the boundary condition

$$\frac{W \delta_\varepsilon(\phi)}{|\nabla \phi|} \cdot \frac{\partial \phi}{\partial \vec{n}} = 0$$

on $\partial \Omega$. Where $W = d(z) * g(|\nabla f(z)|)$ and $\vec{n}$ is the unit outward normal to the boundary. To solve (9), we have used an additive operator splitting (AOS) method, originally introduced in [25]. This method will help in decomposing derived 2D problem in (9) into two 1D problems which is easy to solve. Final solution is derived by averaging both solutions.

4   Experimental Results

Here we present and discuss results derived after implementation of the proposed model and another related model for comparison. Effectiveness of the proposed method is tested using various user scenarios and results are compared by measuring Jaccard's similarity index [12]. We compared the proposed method with existing methods mentioned as CV [6], BC [2], and CoV [3]. In Fig. 1(a), the synthetic image has three intensities. In the first scenario, our interest is to segment region having gray intensity marked as '1' in Fig. 1(a). All the existing methods give wrong segmentation which can be observed from Fig. 1(b-d), while our method produces true segmentation of the desired region as observed in Fig. 1(e). In scenario 2, an object of interest to be segmented is a black region, denoted as '2' in Fig. 1(f). In this case, CoV and proposed method produce true segmentation as shown in Fig. 1(i) & (j) respectively, while the CV and BC method does not segment the desired region which can be observed in Fig. 1(g) & (h). In scenario 3, where square object denoted as '3' in Fig. 1(k) comprises of gray and black intensities needs to be segmented. In this case, all three methods; CV, BC, and proposed method produces true segmentation as shown in Fig. 1(l), (m) & (o), while CoV method gives produces false segmentation in Fig. 1(n). So the proposed method produces true segmentation in all scenarios. This fact can be observed from Table 1 using Jaccard's similarity index (Here value 1 means exact segmentation and value 0 means completely wrong segmentation).

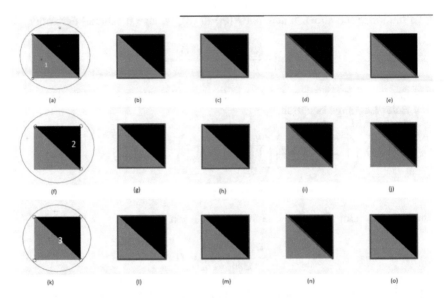

Fig. 1. (a), (f) & (k) Input images with initialization parameters for three scenarios respectively; Geometric constrains points at boundary of the object (marked as 'o' in blue color) for BC, CoV and proposed methods while additional homogeneous region-wise points for proposed methods (marked as '*' for object in red color and '+' for background in blue color). The initial curve is shown as a circle in red color around the object of interest marked as '1','2'&'3' for different scenarios. Constant parameters: $k_{1=0.1}, k_2 = 1, k_3 = 1, \alpha = -0.00151$. Second column: CV method outputs for all scenarios. Third column: BC method outputs. Forth column CoV method outputs. Fifth column: Proposed method outputs. Note: The final boundary is shown in red color.

Table 1. Jaccard's similarity index

Methods	Fig. 1(a)	Fig. 1(f)	Fig. 1(k)
CV [6]	0.4978	0.4924	0.985
BC [2]	0.4978	0.4924	0.985
CoV [3]	0	0.9912	0.492
Proposed	0.9935	0.9911	0.986

5 Conclusion

In this paper, we proposed a novel fidelity terms in the active contour model for segmenting objects/regions which have piecewise nearly homogeneous regions. By observations of experimental results the followings are concluded: (1) The proposed method is flexible and efficient than the other existing methods [2, 3, 6] as initial information related to desired objects/regions helping in merging various non-homogeneous regions. (2) Selecting initial contour near the boundary of the object helps in faster convergence. This method has scope for improvement in continuous nonhomogeneous object domain.

Acknowledgment. Authors thanks to DIAT (DU), Pune and CABS (DRDO), Bangalore for providing necessary support to carry out this research work.

References

1. Ali, H., Rada, L., Badshah, N.: Image segmentation for intensity inhomogeneity in presence of high noise. IEEE Trans. Image Process. **27**(8), 3729–3738 (2018)
2. Badshah, N., Chen, K.: Image selective segmentation under geometrical constraints using an active contour approach. Commun. Comput. Phys. **7**(4), 759–778 (2009)
3. Badshah, N., Chen, K., Ali, H., Murtaza, G.: Coefficient of variation based image selective segmentation model using active contours. East Asian J. Appl. Math. **2**, 150–169 (2012)
4. Bresson, X., Esedoglu, S., Vandergheynst, P., Thiran, J.P., Osher, S.: Fast global minimization of the active contour/snake model. J. Math. Imaging Vis. **28**(2), 151–167 (2007)
5. Caselles, V., Kimmel, R., Sapiro, G.: Geodesic active contours. Int. J. Comput. Vision **22**(1), 61–79 (1997)
6. Chan, T.F., Vese, L.A.: Active contours without edges. IEEE Trans. Image Process. **10**(1), 266–277 (2001)
7. Darolti, C., Mertins, A., Bodensteiner, C., Hofmann, U.G.: Local region descriptors for active contours evolution. IEEE Trans. Image Process. **17**(12), 2275–2288 (2008)
8. Dodo, B.I., Yongmin, L., Tucker, A., Kaba, D., Liu, X.: Retinal OCT segmentation using fuzzy region competition and level set methods. In: 2019 IEEE 32nd International Symposium on Computer-Based Medical Systems (CBMS), pp. 93–98. IEEE (2019)
9. Gout, C., Guyader, C.L., Vese, L.A.: Segmentation under geometrical conditions with geodesic active contour and interpolation using the level set method. Numer. Algorithms **39**, 155–173 (2005)
10. Guyader, C.L., Gout, C.: Geodesic active contour under geometrical conditions theory and 3D applications. Numer. Algorithms **48**, 105–133 (2008)
11. Guyader, C.L., Forcadel, N., Gout, C.: Image segmentation using a generalized fast marching method. Numer. Algorithms **48**, 189–212 (2008)
12. Jaccard, P.: The distribution of the flora in the alpine zone. New Phytol. **11**(2), 37–50 (1912)
13. Kapuriya, B.R., Pradhan, D., Sharma, R.: Detection and restoration of multi-directional motion blurred objects. SIViP **13**(5), 1001–1010 (2019)
14. Kass, M., Witkin, A., Terzopoulos, D.: Snakes: active contour models. Int. J. Comput. Vision **1**, 321–331 (1988)
15. Miao, J., Huang, T.Z., Zhou, X., Wang, Y., Li, J.: Image segmentation based on an active contour model of partial image restoration with local cosine fitting energy. Inf. Sci. **447**, 52–71 (2018)
16. Mumford, D., Shah, J.: Optimal approximation by piecewise smooth functions and associated variational problems. Commun. Pure Appl. Math. **42**, 577–685 (1989)
17. Osher, S., Sethian, J.A.: Fronts propagating with curvature-dependent speed: algorithms based on Hamilton-Jacobi formulations. J. Comput. Phys. **79**(1), 12–49 (1988)
18. Rada, L., Chen, K.: A new variational model with a dual-level set functions for selective segmentation. Commun. Comput. Phys **12**(1), 261–283 (2012)
19. Roberts, M., Chen, K., Irion, K.L.: A convex geodesic selective model for image segmentation. J. Math. Imaging Vis. **61**, 482–503 (2019)

20. Sethian, J.A.: Level Set Methods and Fast Marching Methods: Evolving Interfaces in Computational Geometry, Fluid Mechanics, Computer Vision, and Material Science. Cambridge University Press, Cambridge (1999)
21. Vese, L.A., Chan, T.F.: A multiphase level set framework for image segmentation using the Mumford and Shah model. Int. J. Comput. Vision **50**(3), 271–293 (2002)
22. Wang, X.F., Huang, D.S., Xu, H.: An efficient local Chan-Vese model for image segmentation. Pattern Recognit. Lett. **43**(3), 603–618 (2010)
23. Wang, B., Gao, X., Tao, D., Li, X.: A Nonlinear adaptive level set for image segmentation. IEEE Trans. Cybernet. **44**(3), 418–428 (2014)
24. Weickert, J., Romeny, B.M.T.H., Viergever, M.A.: Efficient and reliable scheme for nonlinear diffusion filtering. IEEE Trans. Image Process. **7**(3), 398–410 (1998)
25. Wenbing, T., Kunquin, L., Kun, S.: SaCoseg: object cosegmentation by shape conformability. IEEE Trans. Image Process. **24**(3), 943–955 (2015)
26. Zhang, K., Song, H., Zhang, L.: Active contours driven by local image fitting energy. Pattern Recogn. Lett. **43**(4), 1199–1206 (2010)
27. Zhang, K., Zhang, L., Lam, K., Zhang, D.: A level set approach to image segmentation with intensity inhomogeneity. IEEE Trans. Cybernet. **46**(2), 546–557 (2016)
28. Zhang, Z., Pan, X., Cheng, L., Zhan, S., Zhou, H., Chen, R., Yang, C., Wang, C., Lin, Y., Lin, J.: SAR image segmentation using hierarchical region merging with oriented edge strength weighted Kuiper's distance. IEEE Access **7**, 84479–84496 (2019)

Blockchain Technology in Healthcare Domain: Applications and Challenges

Chavan Madhuri$^{(\boxtimes)}$, Patil Deepali, and Shingane Priyanka

Ramrao Adik Institute of Technology, Nerul, Navi Mumbai 400706, India
{madhuri.chavan, deepali.patil,
priyanka.shingane}@rait.ac.in

Abstract. Blockchain is a technology which is proposed to address various problems in different domains as like in the banking sector, supply chain. Blockchain is distributed public ledger that can hold permanent records in a secured way. Blockchain ensures that transactions can never be modified. Blockchain also provide data transparency, integrity, security and interoperability. To address the need for patients to manage and control their electronic health records, blockchain technology could be successfully applied in the healthcare sector. Using blockchain decentralized storage of patient information is possible. Health records which are up to date will be accessed by authenticated user only. It is possible for patient to share his records with other doctors for diagnosis of diseases for example chronological disease, thus blockchain provides personal data transaction in health care sector using blockchain. In this paper we reviewed different applications of blockchain in healthcare system and also challenges for deployment of this technology in healthcare industry.

Keywords: Blockchain · Distributed ledger · EHR · Healthcare applications · Security · Interoperability

1 Introduction

Blockchain caused a new uprising in technology which has high influence on the society. The survey done by World Economic Forum report (September 2015) said that in opinion of around 58% of people in survey, 10% of global products will use for storing [1]. Blockchain technology is attracting investors to invest in Blockchain technology. Till 2015 investment amount in Blockchain technology reached almost half a billion dollars and trend is increasing [2].

Blockchain technology is a decentralized system without any third party for transaction control between organizations. Each cryptographically validated transactions and data is transcribed in an immutable ledger in a verifiable, protected, crystalline and permanent way, with a timestamp and other details [3]. Use of this technology is growing continuously and promises applications in every facet of information and communications technology like in financial services [4, 5], in the manufacturing industry to track goods within a supply chain [6], for voting in governments [7], payments and transactions [8] and energy exchange using smart contract [9] and in IOT applications [10].

© Springer Nature Switzerland AG 2020
J. S. Raj et al. (Eds.): ICIDCA 2019, LNDECT 46, pp. 543–550, 2020.
https://doi.org/10.1007/978-3-030-38040-3_61

On paper recording is a traditional method of maintaining medical history of patient. Due to advancement in technology, it is possible to maintain and access records electronically on the internet using Electronic Health Records (EHRs) which provide a convenient storage service of health record. This system allows patients to access and the control of operating and sharing EHRs with individuals and healthcare providers. During course of patient's life, EHRs are scattered in different areas which causes the movement of EHRs from one health care service to another. Due to such scattered nature, sensitive health records can leads to unavailability of information when required and worsening health outcomes. Furthermore, due to increased patient's involvement, there is a on growing need to access and control patient's data. Blockchain is a secure, decentralized online ledger that could be used to manage electronic health records (EHRs), which can address issues such as security, interoperability, integrity and transparency. Blockchain technology assures health stakeholders of authentication, confidentiality, accountability and data sharing while handling sensitive health data. Data recorded in the blockchain cannot be changed or deleted without leaving a trace which is a critical requirement for any health care system.

2   Blockchain Overview and Blockchain Types

2.1   Blockchain Overview

Cryptocurrency Bitcoin is the first implementation of Blockchain introduced by Satoshi Nakamoto in 2008 to perform transactions among two entities using on cryptographic proofs without need of trust of third party [11]. The major application areas of Blockchain technology are [12]

- Blockchain 1.0 – It uses crypto currencies which work by being bundled up in a 'block' connecting the previous block forming a 'chain'. It then gets verified in two ways, either Proof of Work, or Stake like Digital payment transactions [13].
- Blockchain 2.0 – It uses smart contracts which works on Ethereum platform where the developer community can build distributed applications such as finance market for the Blockchain network like shares, future mortgages [14]
- Blockchain 3.0 – New technologies are emerging such as IOTA, NEO, EOS which improves the capabilities and flaws of Bitcoin cryptocurrency and Ethereum networks to some extent like Public Administration, health, etc. [3].

Data modification is impossible in blockchain. Blocks are connected in such a way that data transaction cannot be altered. Blockchain is shared, and hence it allows the system to be transparent and everybody can verify the data present in it. 'Nodes' are the members who participate in network of blockchain and who participate in authorized transactions. These varying types of nodes are permitted to enter the network making use of dedicated software such as Ethereum which performs different functions like mining and storing data. It uses a public key cryptography, so, if anyone tries to change any transaction data, would need to modify the same information in all the respective nodes in the network that is practically impossible. Thus provides secure records of exactly who had done what, when and where [15].

2.2   Blockchain Types

- Public, Private and Consortium are the different types of blockchain which are based on type of application [16].
- A Public blockchain: A public blockchain is decentralized, permissionless blockchain. Here, everyone can participate the consensus process. Examples of Public blockchains are Bitcoin and Litecoin. Many communities are attracted to public blockchain as it is open to all users.
- A Private blockchain: A Private blockchain is fully centralized, permissioned blockchain. Here, only selected and certificated member of an organization can be a part of the blockchain process. Hyperledger Fabric and Corda are examples of private blockchain.
- A Consortium blockchain: A Consortium blockchain are partially- decentralised, permissioned blockchain. A group of approved participants in an organisation can participate in blockchain. Many business applications like R3, EWF (Energy Web Foundation), Quorum are some examples.

3   Blockchain in Healthcare Industry

There are number of potential use case in healthcare industry for the application of blockchain to work in more efficient manner. In post treatment health monitoring system patient is supervised using sensors attached to their body and home which transmits sensitive information to clinical staff and doctors. To apply blockchain set of medical devices will be put in the medical devices blockchain. Thus medical devices are configured for every patient. The Smart Contract can be applied for medical Devices blockchain. The process involves fabric hyperledger configuration. The data validation is done by approval of peer nodes and hence updates a ledger involved in blockchain [17].

3.1   Blockchain in Healthcare Applications

- Clinical data examination: Permissioned Ethereum, which uses smart contract feature in blockchain [16], is analogous to clinical data management systems. The central issue in clinical data is the patient enrolment problem [18]. The related study showed that as compared to bitcoin, Ethereum resulted in quicker transactions, and hence it was concluded that data transparency in clinical trials can be provided by using Ethereum based smart contracts. So blockchain can be used for patient enrolment in clinical research [19, 20].
- Medicinal supply chain management: In medical industry one of the important applications of blockchain is drug supply chain management having a greater concern in healthcare, due to its growing complication. Any threaten to the healthcare supply chain have an impact on the wellness of a patient [21]. Drug supply chains involve moving parts and persons; it is more vulnerable and can be tampered through its phases of supply chain. As the blockchain apply smart contract

so manipulating data is impossible [22]. Hence Blockchains are secure platform to eliminate, detect and prevent fraud by higher data transparency and enhance tracking of product.

- Pharma Industry and Research: The pharmaceutical industry introduces new drugs, develops, and produces it. These are also responsible for marketing of drugs for use as medications to cure patients. It is a leading sector in healthcare domain. Pharma companies face difficulties of tracing their drugs stocks, which can sometimes have severe risks of fraud production, or intrude illegal drugs into the healthcare system. Hence Blockchain technology can be used during the production process of medicinal drugs like for observing, tracking and securing the production processes of medicinal drugs. Recent research foundation, launched a project based on blockchain technology in which Hyperledger [23] is used as a prominent tool, to scrutinize and trace the fake drugs production.
- Neuroscience: Blockchain is also useful in neuroscience where neural devices can decode brain activity patterns and interpret them to direct controlling of external device based brain activity data. It can also detect the current mental state of a person [24].
- Billing Systems: Conventional modes of billing systems are very complicated and also vulnerable to frauds. This process requires maximum resources and also time consuming to receive all the necessary bills. Hence the payment done using blockchain can ease the process of billing as compared with conventional billing systems, where to claim the bills is time consuming. Even to claim the bills health insurance company requires several days to process. Hence blockchain technology can be used in billing systems and the insurance company for the payment, which makes the billing faster, requires less resources, and cost [25, 26].
- Preserve and maintain previous medical records: Blockchain can be used for preserving and maintaining the previous medical records of the patients. Patients may approach different hospitals and due to hospital policies previous records might be discarded and hence it will not be obtainable and it will be inaccessible. These issues can be overcome by using blockchain for maintaining patient's records in each hospital he visits. Even records of medical test can be recorded and maintained so that there are no repeated medical tests for patients to undergo every time [27, 28].

3.2   Advantages of Blockchain in Healthcare Applications [9]

- Transparency: In medical healthcare systems Blockchain can provide transparency, administration costs can be reduced and process faster claims. In some scenarios where a health plan and patient holds contracts, the blockchain can authorize information and guarantees transparency.
- Security: In data exchange systems blockchain enables cryptographically secured data and irreversible. This ensures uninterrupted access to historic and real-time patient data, and removes counterfeit information.
- Integrity: To authenticate the drug supply chain integrity, to maintain the drug development process and improve new drug production blockchain is used.

- Interoperability: Using blockchain technology it preserves and maintains all data and data transactions permanently. Hence multiple parties can use data safely and easily ensuring integrity and privacy.

4   Challenges in Blockchain for Healthcare

Blockchain is a promising technology, which is accepted in various fields for trusted, secure, transparent, irreversible transactions in cost efficient way, however, there are some challenges for deployment in healthcare applications that should be addressed. Following section summarize some of the major challenges.

- Interoperability: Blockchain requires communication between various communicating providers and services like hospitals, laboratory, drug suppliers, insurance company talk to one another impeccably and appropriately. The implementation of healthcare system is difficult as it includes different types and numbers of devices. There should be cooperation between stakeholders and all parties involved in the system for implementation of blockchain in healthcare. This challenge creates obstacle in adoption of this technology as there will not be the effective sharing of data [9].
- Standardization: As healthcare system is related to human life, to deploy blockchain in this system successfully there should be appropriate way of data storage of patients health record and information exchange between different systems involved in blockchain application [9, 29]. For this international standardization authorities should publish well-authenticated and certified standards These standards will provide precautionary safety measures for the shared to avoid misuse of data [28, 30].
- Scalability: Data generated in healthcare system is of large volume. So for blockchain-based healthcare system scalability is a major challenge. Performance of system may cause degradation in performance interims significant latency because of storage of medical data of high-volume on blockchain. For example, the validation mechanism in the current set-up of the Ethereum blockchain platform based applications all the nodes in network take part into validation process [31]. Participation of all nodes causes significant processing delay for major data load. If computing facilities are less as compared to amount of medical transactions in a system then it will limit the scalability of the healthcare system [32].
- Uncertain Development Cost: To build and manage traditional healthcare system for information storage and data exchange requires spending a large amount of time, human resources and money along with this there is overhead of continuous system updation. By implementing blockchain based healthcare system this cost may be reduced. But initial deployment cost is high. Government and health care industry should come with policy that will reduce operational cost and reduce required resources and will specify deployment cost for each stakeholder involved into the system [33].
- Social Challenges Use of blockchain technology in healthcare system will change the traditional healthcare system which is mostly paper based or sometimes online like EHR/EHR [28]. This will bring cultural change. Mostly patients are not

interested to share their data with multiple parties so chances of cultural resistance may occur. To use blockchain in healthcare system some efforts will require changing the behavior of the people so they will accept data sharing in distributed environment [34].

- Security and Privacy Concerns: In blockchain all transactions are verified by entire community instead of one trusted third party so the data send by one node can be accessed by all nodes involved in a system which causes the privacy and security risk to the data in a system. However some features are proposed for security [26] still more solutions to be propose to handle this issue. It is a open research problem to find the way so that allowed entity will access and share the data securely [35–38].

5   Conclusion

In a very short time, blockchain technology is accepted from digital currencies based small applications to vital applications in variety of domains at different levels including health industry. Our study shows that blockchain can be used for variety of applications in healthcare. Blockchain assures safety storage, management and delivery of healthcare data. Use of blockchain in healthcare is making diagnosis and treatment effectively by safe and secure data sharing. Applications using blockchain comes with some challenges like scalability, interoperability, implementation cost, security and privacy that should be addressed. In literature so many solutions proposed to address these challenges. Still additional research is also to explore more solutions needed in addition to address the challenges for deployment of blockchain in healthcare industry.

References

1. Deep shift: technology tipping points and societal impact. Technical report, World Economic Forum, September 2015
2. Coleman, N.: VC Fintech funding sets record in 2015, Fueling Bitcoin and Blockchain Growth, March 2016
3. Krawiec, R.J., et al.: Blockchain: opportunities for health care. In: Proceedings of NIST Workshop Blockchain Healthcare, pp. 1–16 (2016)
4. Poonpakdee, P., Koiwanit, J., Yuangyai, C., Chatwiriya, W.: Applying epidemic algorithm for financial service based on blockchain technology. In: International Conference on Engineering, Applied Sciences, and Technology (ICEAST) (2018)
5. Bosco, F., Croce, V., Raveduto, G.: Blockchain technology for financial services facilitation in RES investments. In: IEEE 4th International Forum on Research and Technology for Society and Industry (RTSI) (2018)
6. Mondragon, A., Coronado, C., Coronado, E.: Investigating the applicability of distributed ledger/blockchain technology in manufacturing and perishable goods supply chains. In: IEEE 6th International Conference on Industrial Engineering and Applications (ICIEA) (2009)
7. Hjálmarsson, F., Hreiðarsson, G., Hamdaqa, M., Hjálmtýsson, G.: Blockchain-based e-voting system. In: IEEE 11th International Conference on Cloud Computing (CLOUD) (2018)

8. Sidhu, J.: A peer-to-peer electronic cash system with blockchain-based services for e-business. In: 26th International Conference on Computer Communication and Networks (ICCCN) (2017)
9. Kang, E., Pee, S., Song, J., Jang, J.: A blockchain-based energy trading platform for smart homes in a microgrid. In: 3rd International Conference on Computer and Communication Systems (ICCCS) (2018)
10. Ferrag, M.A., Maglaras, L., Janicke, H.: Blockchain and its role in the internet of things. In: Strategic Innovative Marketing and Tourism, pp. 1029–1038. Springer, Cham (2019)
11. Satoshi Nakamoto, N.: Bitcoin: a peer-to-peer electronic cash system, March 2009. https://bitcoin.org/bitcoin.pdf
12. Swan, M.: Blockchain: Blueprint of the New Economy, p. 240. O'reily, Sebastopol (2017). 1Olimpbusiness, Moscow
13. Grinberg, R.: Bitcoin: an innovative alternative digital currency. Hast. Sci. Technol. Law J. **4**, 160 (2012)
14. Werbach, K.D., Cornell, N.: Contracts ex machina. SSRN Electron. J. **67**, 313 (2017)
15. Crosby, M., Nachiappan, Pattanayak, P., Verma, S., Kalyanaraman, V.: BlockChain technology beyond bitcoin. Technical Report, Sutardja Center for Entrepreneurship & Technology, 16 October 2015
16. Zheng, Z., Dai, H., Xie, S., Chen, X.: Blockchain challenges and opportunities: a survey. Int. J. Web Grid Serv. **14**, 352–375 (2018)
17. Zyskind, G., Nathan, O., Pentland, A.: Decentralizing privacy: using blockchain to protect personal data. In: IEEE Security and Privacy Workshops, San Jose, pp. 180–184 (2015)
18. Siyal, S., Junejo, A., Zawish, M., Ahmed, K., Khalil, A., Soursou, G.: Applications of blockchain technology in medicine and healthcare: challenges and future perspectives. Cryptogr. **3**, 1 (2019)
19. Nugent, T., Upton, D., Cimpoesu, M.: Improving data transparency in clinical trials using blockchain smart contracts. F1000Res. **5**, 2541 (2016)
20. Buterin, V.: A Next-Generation Smart Contract And Decentralized Application Platform. White Paper; Ethereum Foundation (Stiftung Ethereum), Zug, Switzerland (2014)
21. Wood, G.: Ethereum: A Secure Decentralised Generalised Transaction Ledger. Yellow Paper; Ethereum Foundation (Stiftung Ethereum), Zug, Switzerland (2014)
22. Buterin, V.: A Next-Generation Smart Contract And Decentralized Application Platform; White Paper; Ethereum Foundation (Stiftung Ethereum), Zug, Switzerland (2014)
23. Kevin, A.C., Breeden, E.A., Davidson, C., Mackey, T.K.: Leveraging blockchain technology to enhance supply chain management in healthcare: an exploration of challenges and opportunities in the health supply chain. Blockchain Healthc. Today **1** (2018). https://doi.org/10.30953/bhty.v1.20
24. Mauri, R.: Blockchain for Fraud Prevention: Industry Use Cases, July 2017
25. Taylor, P.: Applying Blockchain Technology to Medicine Traceability, April 2016
26. Swan, M.: Blockchain thinking: the brain as a decentralized autonomous corporation. IEEE Technol. Soc. Mag. **34**, 41–52 (2015)
27. Engelhardt, M.A.: Hitching healthcare to the chain: an introduction to blockchain technology in the healthcare sector. Technol. Innov. Manag. Rev. **7**(10), 2–34 (2017)
28. Boulos, M.N.K., Wilson, J.T., Clauson, K.A.: Geospatial blockchain: promises, challenges, and scenarios in health and healthcare. Int. J. Health Geogr. **17**, 25 (2018)
29. Esposito, C., De Santis, A., Tortora, G., Chang, H., Choo, K.: Blockchain: a panacea for healthcare cloud-based data security and privacy? IEEE Cloud Comput. **5**(1), 31–37 (2018)
30. Abdullah, T., Jones, A.: eHealth: challenges far integrating blockchain within healthcare. In: IEEE 12th International Conference on Global Security, Safety and Sustainability (ICGS3), pp. 1–9 (2019)

31. Azaria, A., Ekblaw, A., Vieira, T., Lippman, A.: MedRec: using blockchain for medical data access and permission management. In: International Conference on Open and Big Data (OBD), Vienna, Austria, pp. 25–30. IEEE (2016)
32. Kumar, T., Liyanage, M., Braeken, A., Ahmad, I., Ylianttila, M.: From gadget to gadget-free hyperconnected world: conceptual analysis of user privacy challenges. In: European Conference on Networks and Communications (EuCNC), Oulu, pp. 1–6, June 2017
33. Stgnaro, C.: White Paper: Innovative Blockchain Uses in Healthcare. Freed Associates, August 2017
34. Yli-Huumo, J., Ko, D., Choi, S., Park, S., Smolander, K.: Where is current research on blockchain technology? - a systematic review. PLoS ONE 11, 1–27 (2016)
35. Linn, L., Koo, M.: Blockchain for health data and its potential use in health IT and health care related research (2016)
36. Puppala, M., He, T., Yu, X., Chen, S., Ogunti, R., Wong, S.T.C.: Data security and privacy management in healthcare applications and clinical data warehouse environment. In: IEEE-EMBS International Conference on Biomedical and Health Informatics (BHI), p. 58, February 2016
37. Omar, A.l., Rahman, M.S., Basu, A., Kiyomoto, S.: MediBchain: a blockchain based privacy preserving platform for healthcare data. In: International Conference on Security, Privacy and Anonymity in Computation, Communication and Storage, December 2017
38. Attia, O., Khoufi, I., Laouiti, A., Adjih, C.: An IoT-blockchain architecture based on hyperledger framework for healthcare monitoring application. In: 10th IFIP International Conference on New Technologies, Mobility and Security (NTMS), pp. 1–5. IEEE (2019)

Comparison on Carrier Frequency Offset Estimation in Multi Band Orthogonal Frequency Division Multiplexing (OFDM) System

C. Rajanandhini$^{(\boxtimes)}$ and S. P. K. Babu

Department of Electronics and Communication Engineering, Periyar Maniammai
Institute of Science and Technology, Thanjavur, Tamilnadu, India
nandhini_selvan@yahoo.com, spkbabu@rediffmail.com

Abstract. The OFDM system is the mainly proficient in addition to well-known variety of modulation for transporting the message with higher speed. The OFDM is essentially the effective method to overcome multiple path fading, resist delay spread, SNR (signal to noise ratio) increments among negligible Inter symbol in addition to Inter carrier Interference. The main problem in the system is offset in carrier signal, that reduces the orthogonality between the signal and reduces amplitude of needed signal. In sequence to overcome this carrier offsets various methods of evaluation with better way suggested.

Keywords: Carrier Frequency Offset · Orthogonal Frequency Division
Multiplexing · SNR · Multipathfading · Orthogonality

1 Introduction

In Modern world through rapid development in addition to adoption of smartphones and mobile data services increases the requirement of high data rate. Orthogonal Frequency Division Multiplexing (OFDM) is a multi-carrier transmission process which sub-divides bandwidth into numerous narrow bands such that multiple datas are transmitted in parallel using FFT along with less effective for frequency selective fading then it have consistent spectral efficiency [1]. Every sub-carrier are allowed to deliver information for different user thats results in simple numerous access. It combines the small data rates bands to form a composite high data rate communications [1]. The signals are Orthogonal till they are independent to every other. Orthogonality property allows multiple signal with data are transmitted into a ideal channel and detected, without any interference [2]. Orthogonality loss would produce distraction within data signals. The spectral overlapping is between the subordinate carriers are permitted due to orthogonality which separates the subordinate carrier at the receiver which provides good spectral efficiency [3]. The OFDM is extensively used, in this sort of broadcasting obtain high data rate its best suitable way for 4G as well as next generation [8]. It has good bandwidth efficiency match up to other methods moreover It exhibits effective frequency selection in wireless channel [1]. The OFDM applicable for wired as well as wireless communication due to its flexible architecture system [2].

J. S. Raj et al. (Eds.): ICIDCA 2019, LNDECT 46, pp. 551–560, 2020.
https://doi.org/10.1007/978-3-030-38040-3_62

In the wireless field transmitting the data at high speed is the major challenges of this field [3]. There is necessity for high frequency bandwidth. The increase in data rate causes distortion in the arriving signal due to multipath channel fading which becomes the major issue in telecommunication field. The OFDM gives high bandwidth efficiency with help of orthogonality principle and to resist multiple path fading and reduces delay [6]. This OFDM is based on spreading datas with high speed is transmitted to low rate signals. In communication field orthogonal means the uncorrelation of the two signals over the intervals, this features of the orthogonality of subordinate carriers wave exhibits the null interference [9].

This system have some drawbacks such as higher peak to average power ratio and Carrier Offset [4]. This OFDM is very reactive for time and frequency synchronization. This synchronization complexity has two important parts: Carrier Frequency offset (CFO) and symbol Time Offset (STO) [16]. The synchronization of OFDM signal is finding the symbol timing and carrier frequency offset.

Different methods are used to calculate approximately the Frequency Offset and overcome the Frequency Offset to enhance the effectiveness of system. A survey carried on the estimation of the CFO.

In Sect. 2 the CFO is defined and effects are discussed. The different estimation algorithms are summarized in Sect. 3 The Results are compared with various algorithm in Sect. 4 furthermore concluded in Sect. 5.

2   Carrier Frequency Offsets

The Carrier Offset is the major disadvantage of the OFDM systems that disturbs the orthoganality of the signals then introduces Inter carrier Interference (ICI) that increase the bit error rate and depicts the SNR [11]. The Offsets in carrier can be caused by two main reasons frequency disparity between sender and receivers that results in left behind CFO. Next is due to Doppler shift that results in relative motion between the transmitting and deliverying end [1].

The OFDM system is being used in Multicarrier modulation where all the sub carriers are orthogonal with each other that improve the efficiency of the bandwidth suitable for high data environment [2]. The Orthogonality of subcarriers depends upon transmitting and receiving end operating at same frequency when the subcarriers are orthogonal, the spectrum is null at center frequency which results is interference free between the carriers. The CFO and Symbol Time offset (STO) occurs due to the frequency mismatch of the signal in-between transmitter and receiver which also reduces the amplitude of the signal [5]. It also causes the Inter Carrier Interference (ICI) which the frequency reference of the receiver end is offset with respect to the transmitter by the frequency error [6]. The CFO occurs by the Doppler shift because of the relative motion across the transmitter and receiver. The Doppler effect is given as

$$f_d = v * f_c/c \tag{1}$$

Where v is the velocity of the receiver, f_d is the doppler frequency and c is the speed of light (Fig. 1).

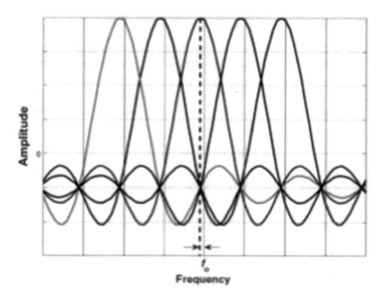

Fig. 1. Carrier frequency offset

2.1 Effects of Carrier Frequency Offset on OFDM

The sub carriers in the OFDM signals will sample at their peak and when there is absence of frequency offset this will occur. However if there is frequency offset the sampling occurs at sampling point not at peak point so the amplitude will be reduced [7]. It also result to raise the Inter Carrier Interference. The frequency error occurs due to the mismatch of frequency between the transmitter and receiver. This can be avoided due to tolerance of the electronic devices.

3 CFO Estimation Techniques

The ICI caused due to CFO reduces the OFDM system performance and its required for the evaluation of the CFO. The CFO estimations is broadly classified as Pilot and non pilot estimation methods. The pilot based method are popular, very fast and reliable estimation methods and it contains well defined pilot symbols for estimation. The non pilot based method utilize the structural and statistical properties of the signals. These Techniques preserves the data rates that leads processing received data multiple times that causes delay while decoding. Some of the algorithms of the CFO is estimated using time and frequency domain methods based on the mean square Error algorithm [9] is explained below.

3.1   CFO Estimation Method Using Cyclic Prefix (CP)

Cyclic prefix (CP) is a part of OFDM signal tends to attract inter symbol interference (ISI) caused by any transmission channel time scattering and it tends to be needed in CFO estimation. Figure 3 indicates OFDM Symbol with CP. CP based estimation technique exploits CP to evaluate the CFO in time area [11]. Considering the channel collision is insignificant and can be ignored, at that point, the lth OFDM image influenced by CFO

$$y_1(n) = x_1(n)e^{\frac{j2\pi\varepsilon n}{N}} \tag{2}$$

Replacing n by (n + N) in Eq. (2) it can be written as a

$$y_1(n+N) = x_1(n)e^{\frac{j2\pi\varepsilon(n+N)}{N}} \tag{3}$$

$$y_1(n+N) = x_1(n)e^{\left(\frac{\sqrt{j2\pi\varepsilon n}}{N} + j2\pi\varepsilon\right)} \tag{4}$$

By comparing the Eqs. 2 and 4 the phase difference between OFDM symbol and Cp is $2\pi\varepsilon$ (Fig. 2).

Then the amount of CFO can be found as

$$\varepsilon = \frac{1}{2\pi}\arg\{y_1^*(n)y_1(n+N)\},$$

$$\text{where } n = -1, -2. \dots - N_g. \tag{5}$$

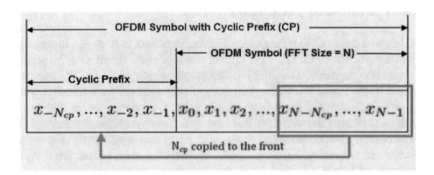

Fig. 2. OFDM symbol with cyclic prefix

To reduce the noise effect in the samples the average of the samples in CP interval as

$$\varepsilon = \frac{1}{2\pi} \arg\left\{ \sum_{n=Ng}^{-1} y1 * (n) y_1 (n+N) \right\}, n = -1, -2\ldots\ldots N_g. \tag{6}$$

This method is suitable for estimation of Fractional CFO. Its doesn't estimate the integer offset [6]. To overcome this problem Training sequence technique is estimated.

3.2 CFO Estimation Method Using Training Sequence

It has been demonstrated that the estimation of CFO method utilizing CP that gauge the CFO just inside the particular range. Since CFO can be vast at the underlying synchronization organize, we may require estimation strategies that can cover a more extensive FO extend. The scope of CFO estimation can be expanded by decreasing the separation between two blocks of the samples for correlation [13]. This is made conceivable by utilizing preparing symbols that are monotonous with some shorter period. Give D a chance to be a number that speaks to the proportion of the OFDM image length to the length of a tedious example as appeared in Fig. 4.

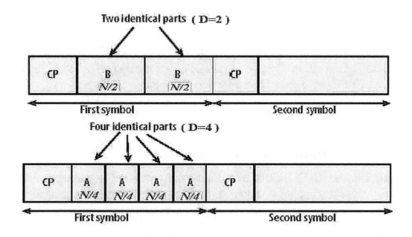

Fig. 3. Training sequence in OFDM symbol

When the transmitter sends the symbols that are trained with the D repetitive patterns in the time domain and it can be generated in frequency domain by the inverse fourier transform and its given as

$$X_1(k) = \left\{ A_m, \text{if } k = D i = 0, 1, 2\ldots, \left(\frac{N}{D} - 1 \right) \right. \tag{7}$$

Or else the $X_1(k) = 0;$

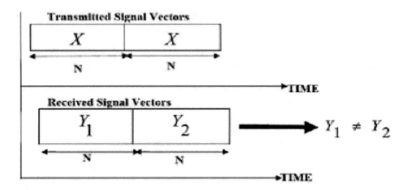

Fig. 4. Schematic of moose technique

Where the A_m represents the mary symbol and $\frac{N}{D}$ is the integer. The $X_1(n)$ and $X_1(n + N\backslash D)$ are identical, then $(y_1 * (n) y_1(n + N/D) = |y_1(n)|^2 e^{j\pi\varepsilon})$. In the receiver the CFO estimation as follows

$$\epsilon = \frac{D}{2\pi} \arg\left\{\sum_{n=0}^{\frac{N}{D}-1} (y1 * (n)y1(n+N/D)\right\} \tag{8}$$

The CFO estimation extend secured by this procedure is which ends up more extensive as D increments. Enlarge in estimation go is acquired at the forfeit of mean square fault (MSE) execution [9]. Consequently, there is an exchange off connection between the MSE execution as well, estimation scope of CFO is unmistakably appeared.

3.3   Frequency-Domain Estimation Methods for CFO

Frequency Domain CFO Estimation systems are connected under presumption that perfect time synchronization is obtained. Besides, the FD strategies depend on transmitting two indistinguishable symbols or pilot tone (pilot inclusion).

3.3.1   CFO Estimation Technique Using Training Symbol Method

Moose considered the OFDM system performance depend on the frequency offsets. The Maximum Likehood based CFO estimation is analysed with two consecutive symbols and identical training symbols. The repetitive data frame and phase value of the carrier symbol are compared and offset is found. CFO estimation strategy dependent on two sequential and indistinguishable preparing symbols [16]. Similar information outline is rehashed and the stage estimation of the every transporter between back to back symbols are looked at as appeared in Fig. 5. The counterbalance is controlled by most extreme probability estimation calculation (MLE).

The receiver signal of the OFDM without noise can be represented as

$$R_n = (1/N)\left[\sum_{k=-k}^{K} X_k H_k e^{2\pi jn(k+\epsilon)/N}\right] \tag{9}$$

Here X_k is the transmitted signal and H_k is the transfer function of the signal then ϵ is the frequency offset. To establish the frequency offset value the consecutive received data should be compared.

The relative frequency offset can be determined by the Maximum Likehood method [9].

$$\tilde{\epsilon} = \frac{1}{2\pi}\tan^{-1}\left[\frac{\left(\sum_{K=-K}^{K} Im\left[Y_{2k}Y_{1k}^*\right]\right)}{\sum_{K=-K}^{K} Re\left[Y_{2k}Y_{1k}^*\right]}\right] \tag{10}$$

Here $\tilde{\epsilon}$ is the relative frequency offset defined as $N\Delta f/B$. The estimated range in moose method is ± 0.5 sub carrier spacing. Moose have increased the range of the shorter training symbols the accuracy has been decreased.

3.3.2   CFO Estimation Method Using Pilot Method

In the pilot tone strategy, we embed some pilot tones in the repetition space and transmitted in each OFDM image, which exploit in CFO estimation at the beneficiary of taking FFT [16]. Figure 5 demonstrates a CFO based structure using pilot tones. Initially, two OFDM symbols, are accessed in the memory after synchronization. At that point of the signals were transformed as means of FFT then in which pilot tones were ejected estimating in frequency domain the carrier offset is compensated by the time domain. Two different modes are utilized for the CFO estimation they are acquisition and tracking modes. In acquisition mode all range of the CFO is estimated but while in tracking mode only the fine CFO is estimated. The integer CFO is given by

$$\epsilon_{ac} = \frac{1}{2\pi Tsub}\max\left\{\left|\sum_{j=0}^{L-1} Y_{1+d}(p(j),\epsilon)Y_1^*(p(j),\epsilon)X_{1+d}^*(p(j))X_1(p(j))\right|\right\} \tag{11}$$

The fine CFO is estimated as

$$E = \frac{1}{2\pi Tsub.D}\max\left\{\left|\sum_{j=0}^{L-1} Y_{1+d}(p(j),\epsilon ac)Y_1^*(p(j),\epsilon ac)X_{1+d}^*(p(j))X_1(p(j))\right|\right\} \tag{12}$$

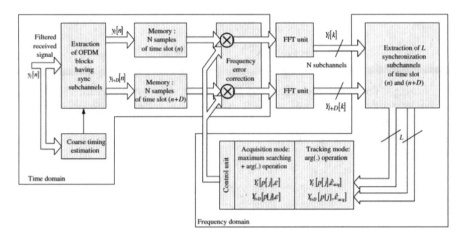

Fig. 5. CFO estimation strategy by using pilot tones.

4   Simulation Results

Here, we demonstrate the correlation between four past CFO assessment systems. Figure 6 demonstrates the MATLAB reproduction after-effects of CFO evaluation utilizing four distinct techniques. The reproduction results demonstrate that, at MSE of 10–4, the pilot tone estimator beats the cyclic prefix estimator utilizing 32 CP-length by

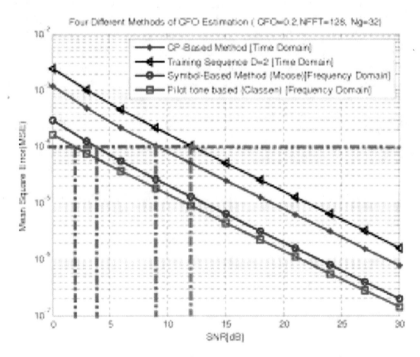

Fig. 6. Comparison between CFO estimation methods in AWGN channel

means of 7 dB as well as the preparation succession estimator (D = 2) by 10 dB. The image based estimator beats the pilot tone estimator by roughly 2 dB. The reproduction results demonstrate the predominance of the pilot tone inference strategy contrasted with other estimation techniques.

5   Conclusions

This survey mainly focused on performance based enhancement on the OFDM interacting model over potentiality of CFO. Since the CFO estimation in frequency domain by some methods are analysed. The Pilot tone method features is better. The evaluation of offset included the Cyclic Prefix (CP) method in time domain. To evaluating the CFO assessment using CP base algorithm and symbol based algorithm. In this OFDM system without decreasing transmission quantity productivity. All the things are measured by which would require as complex execution at the beneficiary side which requires a high handling speed. This model is advanced with the presumption to improve the estimation in OFDM using multiple channel then perform correction of offset frequency in multiples of subcarrier spacing depending on the number of adding extra bits using IFFT with an AWGN channel for transmission. Simulation results characterize the behaviour of Pilot tone based is improved method than other estimations. As a future enhancement analyst could investigate the roads for the enhancement of this model to be utilized in some other corresponding channel.

References

1. Lin, T.-C.: A new cyclic-prefix based algorithm for blind CFO estimation in OFDM systems. IEEE Trans. Wireless Commun. **15**(6), 1 (2016). https://doi.org/10.1109/TWC.2016. 2532325
2. Qiu, S., Xue, L., Wu, P.: Improved interference cancelation channel estimation method in OFDM/OQAM system. Math. Prob. Eng. **2018**, 1–9 (2018)
3. Eslahi, A., Mahmoudi, A., Kaabi, H.: Carrier frequency offset estimation in OFDM systems as a quadratic eigenvalue problem. https://doi.org/10.13164/re.2017.1138
4. Yadav, A., Dixit, A., et al.: A review on carrier frequency offset estimation in OFDM systems. J. Emerg. Technol. Innov. Res. (JETIR) **4**(07), 108 (2017). www.jetir.org, JETIR (ISSN-2349-5162), JETIR1707019
5. Nandi, S., Pathak, N.N.: Performance analysis of cyclic prefix OFDM using adaptive modulation techniques. Int. J. Electron. Electr. Comput. Syst. IJEECS **6**(8), August 2017. ISSN 2348-117X
6. Dugad, P., Khedkar, A.: Performance of OFDM system for AWGN, Rayleigh and Rician channel. Int. J. Innov. Res. Comput. Commun. Eng. **7** (2017). Copyright to IJIRCCE https://doi.org/10.15680/ijircce.2017 0507047 13222
7. Sinha, H., Meshram, M.R.: BER performance analysis of MIMO-OFDM over wireless channel. Int. J. Pure Appl. Math. **118**(5), 195–206 (2018)

8. Tripathi, A., Singh, R.K., Sriwas, S.K., et al.: A comparative performance analysis of O-OFDM-IDMA and O-OFDMA scheme for visible light communication (LOS and NLOS channels) in optical domain. ARPN J. Eng. Appl. Sci. **13**(6), 2229 (2018). www. arpnjournals.com, ISSN 1819-6608 ©2006-2018. Asian Research Publishing Network (ARPN). All rights reserved

9. Nagle, P., Lonbale, N.: DFT based pilot tone approach for carrier frequency offset estimation in OFDM systems. Int. J. Sci. Res. January 2017. ISSN (Online): 2395-566X

10. Rony, S.K., Mou, F.A., Rahman, M.M., et al.: Performance analysis of OFDM signal using BPSK and QPSK modulation techniques. Am. J. Eng. Res. (AJER) **6**(1), 108–117 (2017). e-ISSN: 2320-0847, p-ISSN: 2320-0936

11. Tripathi, A., Singh, S.K.: BER performance of OFDM system in Rayleigh securing on channel using cyclic prefix. Int. J. Adv. Eng. Res. Sci. (IJAERS), **4**(11), November 2017. https://dx.doi.org/10.22161/ijaers.4.11.2, ISSN: 2349-6495(P), 2456-1908(O)

12. Anilbhai, P.B., Rana, S., et al.: BER analysis of digital modulation schemes for OFDM system. Int. Res. J. Eng. Technol. (IRJET), **04**(04), April 2017. www.irjet.net, e-ISSN: 2395-0056. 2017 IEEE 30th Canadian Conference on Electrical and Computer Engineering (CCECE)

13. Han, H., Kim, N., Park, H.: Analysis of CFO estimation for QAM-FBMC systems considering non-orthogonal prototype filters. IEEE Trans. Veh. Technol. **68**(7), 6761–6774 (2019)

14. Khedkar, A., Admane, P.: Estimation and reduction of CFO in OFDM system. In: 2015 International Conference on Information Processing (ICIP) (2015)

15. Kumar, P.S., Sumithra, M.G., Sarumathi, M.: Performance analysis of Rayleigh fading channels in MIMO-OFDM systems using BPSK and QPSK modulation schemes. SIJ Trans. Comput. Netw. Commun. Eng. **04**(02), 01–06 (2016)

16. Tripathi, V., Shukla, S.: Timing and carrier frequency offset estimation using selective extended cyclic prefix for correlation sequence for OFDM systems. Wireless Pers. Commun. **101**(2), 963–977 (2018)

A Reliable Method for Detection of Compromised Controller in Software Defined Networks

Manaswi Parashar$^{(\boxtimes)}$, Amarjeet Poonia, and Kandukuru Satish

Goverment Women Engineering College, Ajmer, India
manaswiparasharengg@gmail.com,
{amar,kandukuru}@gweca.ac.in

Abstract. With the advancement in technology and increased innovations in network managing large networks become easy due to an emerging technology known as SDN software defined networking as accepted by many enterprises. By separation of control plane and data plane, SDN is abstracted by a new layer which is implemented by *controller*. By introducing more practical issues in SDN, today SDN controller is a challenging aspect in terms of security of controller hijacking, compromised controller, malicious controller. The controller is the main nerve of SDN which manages all the actions and functions. Software Defined Networking (SDN) allows network controller to handle and manage the network according to user choice and requirements. SDN Controller is infected by many serious and disastrous attacks and due to centralized behavior of SDN, Controller gets compromised by malwares and attacks. In this paper we introduce framework that completely detect compromised controller in SDN. Our basic idea is to detect percentage of malicious behavior present in network by extracting four open flow parameters from open flow traces and fed them into backup Controller for handling all update information. Finally the information is fed into Random Forest classifier that will test and check the controller is compromised or not. Following this idea and using Random Forest Classifier compromised controller is detected in network with performance analysis of 98% with better feasibility and efficiency than prior work to secure SDN.

Keywords: SDN · Software Defined Network · Denial of service DOS · Distributed denial of services (DDOS) · Network elements · Open flow · Secure SDN Architecture

1 Introduction

Open source proposed a new networking design philosophy "Software Defined Networking (SDN) that can achieve high availability and scalability. It is a paradigm in which management is done by control plane of network. The main characteristic of this SDN is decoupling of control plane and data plane. SDN controller [1] is the heart of system that manages and function overall network. Both the control and data plane are separated. The two planes communicate by using a protocol named as OpenFlow protocol. Open Flow protocol is a protocol in which server tells to network switches

© Springer Nature Switzerland AG 2020
J. S. Raj et al. (Eds.): ICIDCA 2019, LNDECT 46, pp. 561–568, 2020.
https://doi.org/10.1007/978-3-030-38040-3_63

about where to send packets and this protocol is defined in between southbound interface. A communication interface is defined between control and data plane is made up of standard protocol called as OpenFlow protocol. It is a standard API [2] (Fig. 1).

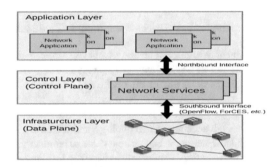

Fig. 1. SDN Architecture

1.1   SDN Controller

SDN controller is main nerve of SDN and it is most attractive target for attacker. It is basically an application that has full control over network. It lies in control plane or in between application end and data plane end. For better network performance controller manages the control flow of network. The communication between applications and other network devices is made by controller itself. In short SDN Controller acts as an operating system in network. The major advantage of controller is its centralized nature. At the same time disadvantage is controller's single point of failure. Once it get failed whole network become bottleneck. So some security policies to defend controller should be created to secure our network. The major threats related to SDN Controllers.

(1) *Packet-In Flooding:* For each unseen flow in Open Flow *Packet_In* messages are generated by attacker which floods the control plane. This flooding of numerous *Packet_In* messages causes a controller to enter into an unreachable state. The vulnerability of controller is identified by using AVANT-GUARD [3] that proposes a solution that limits communication of data plane.

(2) *Flow Table Overflow:* The controller will generate new flow rule for each missed packet and each flow rule has a certain fixed timeout value after which the entries will be return back. Here the attacker dominates and generates many new packets with new flow rules. This overloading of packets cause collapse of SDN. To defend SDN flow table and controller overloading from DDOS attack Kandoi et al. [4].

(3) *Compromised Controller:* When any malicious request arrives on network devices, compromised devices are formed. So, an approach is proposed that make use of Open Flow traces for the detection of compromised devices in control plane.

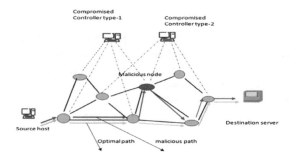

Fig. 2. Type-1 compromised controller issued malicious path

Type-1 Compromised Controller: Controller will create flow rule to forward the packets via switches. But here switch does not form path of shorter length. In Fig. 2, flow rules are generated by malicious controller and packets are forwarded by a switch which do not comes in path of shortest path (shown by red path).

2 Detection Strategy and Architecture

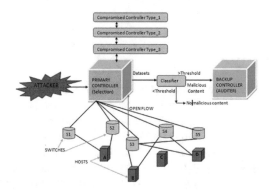

Fig. 3. Architectural model

To detect compromised controller, some parameters that will detect any malicious behavior in network are identified. The percentage analysis of malicious behavior will tell the performance and quality of network. These parameters are analysed as (Fig. 3):

1. Index Participation of switch

For each missed packet flow rule are defined and switch will intimate a PACKET_IN message to controller to install a flow rule. But if network is having compromised controller, then situation becomes bottleneck around switch. So, to defend attacks on switch a parameter knows as switch participation index is defined that will calculate variance of flow rules that are installed in switch port because more the number of flow rules more bottleneck condition at switches.

2. Per flow average switch fraction

To each switch present in path, flow rules are generated. Along with flow rules, a FLOWMOD message is also generated by controller to handle each flow. To identify which flow get affected by flow rule, we can keep track to each FLOWMOD message and analyze their matching conditions. Therefore, to identify, according to FLOW-MOD data, Average switch fraction (ASF) is calculated as the ratio of switches through which flow passes to the overall flows present in network.

3. Ratio of Packet in Packet out

It is the ratio of total number of packet in messages to packet out messages to controller. For each message, controller generates forward rule to forward the packet and send PACKETOUT message to switch to stop corresponding PACKETIN messages. So, PPR PACKETIN PACKETOUT indicates counts of received PACKETIN messages per PACKETOUT messages.

4. Drop Action Variance

When a controller generates flow rules for packets FLOW_MOD messages are generated in response to controller with actions to be performed by switch. Due to presence of malicious behavior in controller DROP ACTIONS are generated to deny services of switches. If action list is empty then by default action is issued as drop action.

3 Simulation SetUp

For learning, classification we generate a very large volume of datasets [5]. We use SDN controllers, applications, Mininet tool Mininet v2.2 simulator [6] to create a SDN environment having switches, hosts, controllers, applications, topology etc. Mininet tool is a network emulator used to create virtual hosts switches topologies of network. One host is always connected to each switch. When simulation set is made, traffic at switches is collected by using *TCPDUMP* [7] utility with simulation are generated on UBUNTU 14.04 machine hosted on APPLE MACBOOK PRO 2015. The switches topology is created by using OPENFLOWv1.3 [8] and SDN controller here we are using is RYU 4.7 [9]. And at last switches and controller are connected via direct link.

In our proposed scheme 7000 different topologies [10] of network used. From these different topologies open flow traces are identified by using existing datasets [5]. These traces are basically repository of data that give complete information about each packet and its fields. By calculating traces, they are fed into machine learning algorithms for classifications and detailed analysis. By classification features like Packet_In, Packed_out, FlowMod, Drop actions are extracted which will identify maliciousness present in controller or not according to threshold value set on controller. Once the parameters are analyzed with malicious content is present in network, we will proceed to second stage known as verification stage. During start of packet capture in network, a backup controller is created with primary controller. This backup controller has all the information regarding network topology, updates to controllers, switches, states of controllers and switches. This backup controller keeps on tracking primary controller. Once malicious content is detected, testing is performed in backup controller by performing matching of response. Based on four parameters, backup controller is capable of performing audit record to verify primary controller is compromised or not or it will inform nodes or operators in network when any compromised device is detected. For each request of packet to primary controller the correct response is tracked in backup controller. So if any deviation in primary controller occurs, matching is performed. If response of primary controller and backup controller is matched then no malicious content is found in network. But if there is a mismatch between backup controller response (the correct response) and primary controller response (hijacked by an attacker), then it is verified as compromised device (Fig. 4).

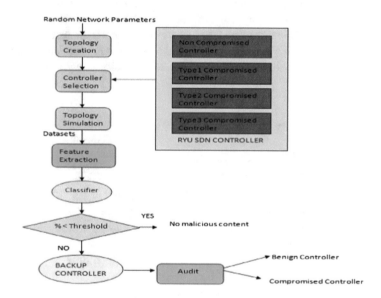

Fig. 4. Flow chart of proposed scheme

According to our strategy, backup controller performs matching test that uses four types of information:

1. Updating request on network Ns.
2. Result of updating request on primary controller Rp
3. Result of updating request on backup controller Rb
4. Network update request on switches Rs
5. State update request at switches Ss

According to this information, backup controller can easily identify and verify whether switch or controller is compromised. Some test cases for stage two are:

a. If Rp = Rb = Rs = Ss, both update request are consistent and both controllers are benign.

b. If Rp = Rs = Ss ≠ Rb, request generated by primary controller is not having same results as backup controller depict primary controller is compromised.
c. If Rp = Rs ≠ Ss, means some improper update is performed at switches and switches are compromised.
d. If Rp ≠ Rs Update request in manipulated or man in middle attack occurs.

By these test cases backup controllers can easily verify controllers are compromised or not. Figure explains the stage two auditing method. It shows message sequence of how network update request is processed. Once four features are known and extracted, they are passed to classifier for detection of compromised controller or any malicious presence. To classify samples, we use 50 fold cross validation method. The total samples of network is divided into 50 parts acc to cross validation method as C1, C2, C3 …C50. After partitioning, we will test the accuracy on part C1 and train the algorithm for parts C2 to C50 by using its samples. For the next step, C2 accuracy is tested while C1, C3 …C50 are trained. In this way all the parts are tested in percentage for their accuracy by using possible combinations and in last calculated accuracy are averaged to get accurate optimized value that will classify compromised of benign controller (Fig. 5).

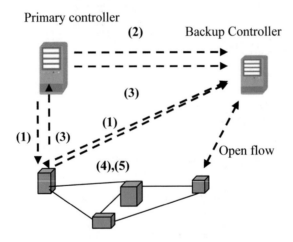

Fig. 5. Architectural framework of proposed scheme

4   Evaluation

This paper reports on detection of compromised controller in SDN and we have present an effective solution to locate compromised controller by calculating four most important parameters of open flow traces to detect malicious behavior in network and analyzed all the test cases with backup controller to verify and detect compromised controller. For that we are using Random Forest Classifier. Random Forest classifier is a group of number of decision tress classifiers. In RF, if performance of any decision tree is poor, then to repay the performance, remaining trees work for betterment.

Each of the decision tree cast vote for decision class and maximum voted class is chosen as Decision of RF. To evaluate our solution scheme we use two parameters. First one is False Alarm Rate that will falsely reject wrong hypothesis for any test case or is expectancy of false positive ratio. Second one is Detection Rate Management that calculates the ratio of malicious controller to benign controller. To calculate:

$$D.A = \frac{TP}{TP + FN} \tag{1}$$

$$F.A = \frac{FP}{TP + FP} \tag{2}$$

where, TP is true positive means controller is truly detected as compromised, FN is false negative, means the percentage of maliciousness we detected is false and controller is benign controller FP is false positive, means benign controller is detected as compromised controller and TN is true positive, means no malicious behavior found in network, controller is compromised. By such evaluation, compromised controller is detected in SDN by using RF classifier with performance analysis of 98%.

5 Conclusion

This paper reports on detection of compromised controller in SDN. We analyze the vulnerabilities of controller. SDN controller is a challenging aspect in terms of security of controller hijacking, compromised controller, malicious controller. As SDN controller is having a single point of failure, so it is prone to major attacks. Every attacker wants to get access to controller so they can get whole SDN system access. The controller is the main nerve of SDN which manages all the actions and functions. So SDN controller security is a major issue. In this paper we proposed a scheme to detect compromised controller by using training samples with four most important open flow parameters and backup controller. We identified: (i) problems to SDN Controller, (ii) Four Open flow Parameters, (iii) Proposed scheme with audit process, (iv) Detection evaluation. This method enables backup controller to audit and verify update information for detecting compromised controller in SDN. We have tested network samples among which some few are in control with primary controller and many samples detects compromised controller. With Random Forest Classifier, the calculated accuracy percentage is 98% with high feasibility to detect malicious behavior present in network. Further, we have tested our proposed framework for detecting compromised controller theoretically and experimentally. Our Future work is to achieve accuracy on percentage detection of maliciousness with some additional features and detailed study of malicious controller and type of attack.

References

1. McCauley, M.: About POX (2015). http://www.noxrepo.org/pox/about-pox/
2. McKeown, N., Anderson, T., Balakrishnan, H., Parulkar, G., Peterson, L., Rexford, J., et al.: OpenFlow: enabling innovation in campus networks. ACM SIGCOMM Comput. Commun. Rev. **38**, 69–74 (2008)
3. Shin, S., Yegneswaran, V., Porras, P., Gu, G.: AVANT-GUARD: scalable and vigilant switch flow management in software-defined networks. In: Proceedings of 20th ACM Conference on Computer Communication Security (CCS), pp. 413–424, November 2013
4. Kandoi, R., Antikainen, M.: Denial-of-service attacks in OpenFlow SDN networks. In: 2015 IFIP/IEEE International Symposium on Integrated Network Management (IM), pp. 1322–1326. IEEE, May 2015
5. Systems and Networks Lab: OpenSDNDataset (2017). https://github.com/iist-sysnet/OpenSDNDataset
6. Lantz, B., Heller, B., McKeown, N.: A network in a laptop: rapid prototyping for software-defined networks. In: Proceedings of the 9th ACM SIGCOMM Workshop on Hot Topics in Networks, Hotnets-IX, pp. 19:1–19:6. ACM, New York (2010). https://doi.org/10.1145/1868447.1868466
7. TCPDUMP/LIBPCAP public repository. http://www.tcpdump.org/. Accessed 26 July 2017
8. Open vSwitch. http://openvswitch.org/. Accessed 25 July 2017
9. Ryu SDN framework. https://osrg.github.io/ryu/. Accessed 27 July 2017
10. Sarath Babu (2018). https://github.com/4sarathbabu

Alphabet Classification of Indian Sign Language with Deep Learning

Kruti J. Dangarwala[1]([⊠]) and Dilendra Hiran[2]

[1] Department of Computer Engineering, Pacific Academy of Higher Education
and Research University, Udaipur, India
`krutidangarwala@gmail.com`
[2] Department of Computer Science, Pacific Academy of Higher Education
and Research University, Udaipur, India
`sigmapawan72@gmail.com`

Abstract. Deaf–mute people interconnect through Indian Sign Language with all communities. Demanding of Sign Language Interpreter is increased day to day for solving gap of communication. Deep CNN plays key role for classification problem. We propose the Deep convolution neural network model for classification of alphabet signs. The model is prepared with addition of six convolutional layers and three fully connected layers. The alphabet signs with 3629 images of A–Z dataset are developed in college laboratory with help of students. Experiment is performed with 1744 sample images for training, 437 sample images for validation and 1448 sample images for testing among 3629 images. Proposed method is achieved 96% model classification accuracy. Training and validation datasets accuracy is achieved with 99% and 94.05% respectively.

Keywords: Pooling layer · Sign language · Training · Classification

1 Introduction

Sign language is popular in deaf-mute community to interact with others [1, 2, 17–19]. Sign language is used to convey messages through visually without any verbal communication. Sign language is divided into manual and non-manual features based on visually communication. Manual features include hand part as communication and Non-manually features include face, lip and other body parts. Indian sign language is mostly work with manual features [3, 4]. Assessment of feature extraction and classification methods like HOG [5–7], SIFT [5, 6], SVM [5, 6], KNN, BAG [6], ANN [3], Fuzzy logic [3, 5], Principal Component Analysis [3, 5] etc. are done by various researchers for sign language recognition. Sign language recognition is based on hand segmentation, tracking, feature extraction and classification [8]. A dynamic skin detector based on the face colour tone is used for hand segmentation [8].

Image classification is foremost study field in various types of applications [9]. Day to Day, study on image classification research is increasing so it is necessary to work more with image classification algorithms [9]. Deep neural network methodologies with sign language classification achieve grand success [1]. Convolution neural network has

© Springer Nature Switzerland AG 2020
J. S. Raj et al. (Eds.): ICIDCA 2019, LNDECT 46, pp. 569–576, 2020.
https://doi.org/10.1007/978-3-030-38040-3_64

acquired grand success in the field of image classification as comparison with other feature extraction [9]. Deep CNN is gaining more successful for image classification [9, 10]. CNN on spatial features and RNN on temporal features are used for real time sign word recognition with American sign language [4]. Researchers use deep neural network to improve accuracy and efficient rate of Sign language recognition system as deaf people require interpretation system for sign language [1]. 3D CNN approach is applied to recognize dynamic signs [11]. The Thai sign language of Alphabet signs recognition with HOG feature extraction method and Back propagation neural network is achieved 85.05% [7]. Earth Observation Classification is performed using dual stage classification method with pretrained model and trainable CNN model [12].

Focus point of this paper starts with collections and preparation of alphabet images of sign language. Alphabet signs A–Z becomes input for our proposed system and 26 different classes of alphabet act as output of our system. Big issue is that to collect image datasets for Indian sign language. We have created 3629 images dataset for alphabet signs. Here image datasets are created by us in our laboratory with different students. Next focusing point is classification of alphabet signs with Deep CNN model. Here performance of proposed model is based on training, validation and testing accuracy & losses. Another measure of performance is done with generation of confusion matrix and classification report through our model.

2   Alphabet Signs Sample Database

Related work specifies that generation of Indian sign Language samples lead to challenging task. Many researchers used various sensors like digital camera, depth camera, kinect etc. for capturing images for inputs [2]. Here we have formed Indian sign language – alphabet datasets using Redmi note 5 camera with consideration of images sizes are 1200 by 1200 RGB pixels. Alphabet datasets made up of image signs which are single handed and double handed both as per visited various deaf institute of Gujarat. Extension of file is considered as JPEG format whenever taking the images. Sizes of Images are changed to 100 by 100 pixels for increasing execution speed. Sample datasets images involve college students with male students as well as female students. These images contains various lighting as background. Here 3629 images datasets are taken for implementation. Following Fig. 1 indicate sample alphabet datasets.

Fig. 1. Alphabet sign samples A to Z [4]

3 Deep Learning Approach

Deep learning extracts features from input database deeply with large no of layers [13]. Convolution Neural network has large number of convolutional, pooling and fully connected layers [1, 10, 14]. Input for convolutional layer is image which has height, width and channels parameters. Image sizes are resized 100 pixels height, 100 pixels width and 3 no. of channels as we consider RGB images. The convolution layer has number of filters with filter size parameters. Filters are used to generate feature map by convolved with original image matrix by using non-linear activation function. The model working of convolution neural network proposed approach is explained in algorithm 1 with 6 convolutional layers, 6 max pooling layers and three fully connected layers. Convolutional layers are used to generate feature matrix which connected with previous layer using filter matrix [14]. The result of convolve operation carry forward through ReLU activation function [10]. First convolution layer use 6 filters with size (2, 2), second layer use 12 filters with (2, 2), third layer use 18 filters with (2, 2), fourth layer use 24 filters with (2, 2), fifth layer use 48 filters with (2, 2) and six layer use 96 filters with (2, 2) filter size. Here pooling layer is applied for reduction of dimension [14]. Each max pooling layer with size (2, 2) is used here. Padding is used to add pixels to edge of images. Here padding argument with value "same" is used on convolution 2D layer as this argument with "same" value do calculation and add padding as per requirement of image by maintaining output and input shape should remain same. Dropout layer is added after first fully connected layer to increase efficiency of model [11]. Softmax classifier is used as more than two classes are here. CNN model algorithm works as follows:

Algorithm 1 . ISL alphabet CNN mdoel Algorithm

Input : Alphabet Sign Images (CSV file)

1. Divide Alphabet samples into training, testing and validation sets.

2 Apply Deep learning convolutional model with six different CNN layers by implementation with python keras library .CNN layer function has no. of filters , filter size and input image size parameters. Convolutional   Layer has learnable parameters so it is calculated using following formula:

Parameters = No. of filters * (previous layers no. of filters *filter size) + 1)   (1)

Convolution 1: $6*(3*(2*2)+1) = 78$
Convolution 2: $12*(6*(2*2)+1) = 300$
Convolution 3: $18*(12*(2*2)+1) = 882$
Convolution 4: $24*(18*(2*2)+1) = 1752$
Convolution 5: $48*(24*(2*2)+1) = 4656$
Convolution 6: $96*(48*(2*2)+1) = 18528$

3. Each convolutional layer is followed by Pooling Layer and activation function. Pooling layer has no learning parameters as it is only find out maximum number based on max pooling function with (2,2). So it   calculates   only particular number. Number of parameters is zero. ReLU   function   is considered as activation function at each layer.

4. We apply Fully Connected Layer (FC) which has learnable parameter as we seen from model summary. Fully connected layer is followed by drop out layer. We find out parameters generated by this   layer   using   following formula:
No. of Parameters = No of classes *(no of previous layer filters +1)   (2)
Dense 1: $100*(96+1) = 9700$
Dense 2: $50*(100+1) = 5050$
Dense 3: $26*(50+1) = 1326$

5. Lastly Softmax activation function is applied for classification of signs as we work with deep neural network[15]. Alphabet signs have 26 classes so it reaches to multiple classification problems. Softmax classifier is suitable for it. Finally, we find out total trainable parameters with summation of all parameters. So we get Total Trainable parameters =4227.

6. Find out training and validation accuracy with epochs size 10.

7 Apply testing datasets to find out confusion metrics and classification report.

Output: Confusion matrix, Classification report, training accuracy, validation accuracy

---

4   Outcome of CNN Model Approach

Experiment is performed with 1744 sample images for training, 437 sample images for validation and 1448 sample images for testing. Image datasets converted into CSV file for our model. Our proposed model shown in Algorithm 1 is applied to our training datasets and validation accuracy is measured. This experiment has taken 0:03:48.853391 times to execute our experiment. We have done this experiment with CPU configuration in python keras library is used here. First here find out training accuracy and validation accuracy. Also find out training loss and validation loss. We have performed this experiment with epoch size 10 and batch size 128. Figure 2 indicate training and validation loss curve with 10 epochs. It indicates that increasing with epochs losses are decreasing. Figure 3 indicate training and validation accuracy curve with 10 epochs. It indicates that increasing with epochs accuracy is increasing. As indicate that we have achieved 94.05% validation accuracy and training accuracy is almost 99%. Confusion matrix is calculated to find out correctly identify alphabet signs so we can calculate individual sign classification accuracy and overall model accuracy [16]. Table 1 indicates Confusion matrix which is generated through implementation. Confusion matrix is two dimensional table which is made up of true label and predicated label. Here we have 26 alphabet classes based on datasets. Confusion matrix diagonal values highlight correctly classify numbers of signs. The system is also generated classification report with performance metrics precision, recall and F1-score value which is shown in Table 2. Classification report indicate that out of 1448 testing images, we have correctly classify 1394 images based on our experiments. Overall model classification accuracy is 96%.

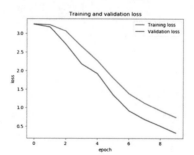

Fig. 2. Samples loss curve with 10 epochs

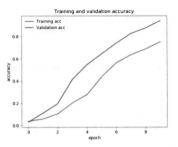

Fig. 3. Samples accuracy curve with 10 epochs

Table 1. Confusion matrix

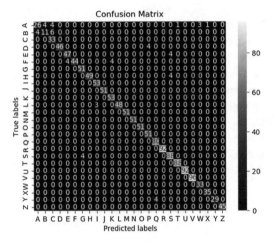

Table 2. Classification report

Signs	Precision	Recall	F1-score	Correctly identify signs	Total signs
A	0.87	0.67	0.75	26	39
B	0.73	0.52	0.61	11	21
C	0.77	1.00	0.87	33	33
D	1.00	0.92	0.96	46	50
E	0.92	0.92	0.92	47	51
F	1.00	0.85	0.92	44	52
G	0.93	1.00	0.96	51	51
H	1.00	0.92	0.96	49	53
I	0.95	1.00	0.97	53	53
J	1.00	1.00	1.00	51	51
K	0.93	1.00	0.96	53	53

(*continued*)

Table 2. (*continued*)

Signs	Precision	Recall	F1-score	Correctly identify signs	Total signs
L	1.00	0.94	0.97	48	51
M	1.00	1.00	1.00	51	51
N	1.00	1.00	1.00	51	51
O	1.00	1.00	1.00	51	51
P	0.93	1.00	0.96	51	51
Q	0.95	1.00	0.98	80	80
R	1.00	1.00	1.00	92	92
S	0.92	0.96	0.94	88	92
T	0.99	1.00	0.99	88	88
U	1.00	1.00	1.00	92	92
V	1.00	1.00	1.00	96	96
W	0.92	1.00	0.96	33	33
X	0.97	1.00	0.99	35	35
Y	1.00	0.88	0.94	29	33
Z	1.00	1.00	1.00	45	45
Average	0.96 (96%)	0.96 (96%)	0.96 (96%)	1394	1448

5 Conclusion

Experiment with proposed Deep Convolution neural network approach with six convolutional layers and three fully connected layers is shown that overall model classification accuracy is 96% with 1448 testing data of alphabet signs. Here almost every alphabet signs is achieved above 90% accuracy. Only problem with alphabet A and B which is shown in Table 2 of classification report indicate that 65% and 52% accuracy respectively. For alphabet signs B, we have required more images so we can get more accuracy like other alphabets. When we work with deep neural network, requirement of datasets should be as large as possible for achievement of better accuracy. In future, we will try same model with daily communication signs videos so it will useful to our deaf and dumb community.

Compliance with Ethical Standards. All author states that there is no conflict of interest. Humans/Animals are not involved in this work. We used our own data. The alphabet signs with 3629 images of A–Z dataset are developed in college laboratory with help of students. We would like to thank Pacific Academy of Higher Education and Research University for providing all the essential assets and facilities.

References

1. Zheng, L., Liang, B., Jiang, A.: Recent advances of deep learning for sign language recognition. In: International Conference on Digital Image Computing: Techniques and Applications (DICTA), pp. 1–7. IEEE (2017)

2. Kumar, P., Gauba, H., Roy, P.P., Dogra, D.P.: A multimodal framework for sensor based sign language recognition. Neurocomputing **259**, 21–38 (2017)
3. Verma, V.K., Srivastava, S., Kumar, N.: A comprehensive review on automation of Indian sign language. In: International Conference on Advances in Computer Engineering and Applications, pp. 138–142. IEEE (2015)
4. Masood, S., Srivastava, A., Thuwal, H.C., Ahmad, M.: Real-time sign language gesture (word) recognition from video sequences using CNN and RNN. In: Intelligent Engineering Informatics, pp. 623–632. Springer, Singapore (2018)
5. Dangarwala, K., Dilendra, H.: A research gap on automatic Indian sign language recognition based on hand gesture datasets and methedologies. Int. J. Comput. Eng. Appl. **12**(3), 46–54 (2018)
6. Dangarwala, K., Dilendra, H.: Deep learning feature extraction using pre-trained Alex Net model for Indian sign language recognition. Int. J. Recent Technol. Eng. (IJRTE) **8**(2), 6326–6333 (2019)
7. Chansri, C., Srinonchat, J.: Hand gesture recognition for Thai sign language in complex background using fusion of depth and color video. In: Proceedings of Computer Science, vol. 86, pp. 257–260 (2016)
8. Ibrahim, N.B., Selim, M.M., Zayed, H.H.: An automatic Arabic sign language recognition system (ARSLRS). J. King Saud Univ. Comput. Inf. Sci. **30**(4), 470–477 (2018)
9. Liu, X., Zhang, R., Meng, Z., Hong, R., Liu, G.: On fusing the latent deep CNN feature for image classification. World Wide Web **22**(2), 423–436 (2019)
10. LeCun, Y., Bengio, Y., Hinton, G.: Deep learning. Nature **521**(7553), 436–444 (2015)
11. Liang, Z.J., Liao, S.B., Hu, B.Z.: 3D convolutional neural networks for dynamic sign language recognition. Comput. J. **61**(11), 1724–1736 (2018)
12. Marmanis, D., Datcu, M., Esch, T., Stilla, U.: Deep learning earth observation classification using ImageNet pretrained networks. IEEE Geosci. Remote Sens. Lett. **13**(1), 105–109 (2015)
13. Chen, Y., Lin, Z., Zhao, X., Wang, G., Gu, Y.: Deep learning-based classification of hyperspectral data. IEEE J. Sel. Top. Appl. Earth Obs. Remote Sens. **7**(6), 2094–2107 (2014)
14. Sajanraj, T.D., Beena, M.: Indian sign language numeral recognition using region of interest convolutional neural network. In: Second International Conference on Inventive Communication and Computational Technologies (ICICCT), pp. 636–640. IEEE (2018)
15. Golovko, V.A.: Deep learning: an overview and main paradigms. Opt. Mem. Neural Netw. **26**(1), 1–17 (2017)
16. Peng, Y., Dharssi, S., Chen, Q., Keenan, T.D., Agrón, E., Wong, W.T., Chew, E.Y., Lu, Z.: DeepSeeNet: a deep learning model for automated classification of patient-based age-related macular degeneration severity from color fundus photographs. Ophthalmology **126**(4), 565–575 (2019)
17. Sinha, S., Singh, S., Rawat, S., Chopra, A.: Real time prediction of American sign language using convolutional neural networks. In: Singh, M., Gupta, P., Tyagi, V., Flusser, J., Ören, T., Kashyap, R. (eds.) Advances in Computing and Data Sciences, vol. 1045, pp. 22–31. Springer, Singapore (2019)
18. Karush, S., Gupta, R.: Continuous sign language recognition from wearable IMUs using deep capsule networks and game theory. Comput. Electr. Eng. **78**, 493–503 (2019)
19. Rumi, R.I., Hossain, S.M., Shahriar, A., Islam, E.: Bengali hand sign language recognition using convolutional neural networks. Ph.D. dissertation. Brac University (2019)

Impact on Security Using Fusion of Algorithms

Dugimpudi Abhishek Reddy, Deepak Yadav, Nishi Yadav$^{(\boxtimes)}$,
and Devendra Kumar Singh

Department of CSE, Guru Ghasidas Vishwavidyalaya,
Bilaspur, Chhattisgarh, India
abhishek.duggimpudi@gmail.com,
deepakyadav28031998@gmail.com, nishidv@gmail.com,
devendra.singh170@gmail.com

Abstract. Cryptography is a technique used in transferring information in a protective and secured manner. Cryptography is usually about building and analysing the set of protocols that prevent the public and the third parties from reading private messages. The importance of cryptography in Network Security has gained a lot of importance and has become a research area for many researchers. Cryptography transforms data into a manner that is unreadable to a person who does not have authorization. This makes the data secured and privacy is maintained between the sender and receiver and the information is limited to themselves. This paper presents the study and examination of two individual encryption-decryption algorithms (T-RSA and Playfair) which are used in cryptography. It also shows how it improves security when combining than the individual ones.

Keywords: Authorization · Cryptography · Decryption · Encryption · Network security

1 Introduction

As days are passing we are going to bound ourselves in the age of innovation and technology. In present day, Computer use requires automated tools to make files and information more secure. Modern cryptography [1] exists at the fields of mathematics, Computer science, and electrical engineering. The security of our data over web is a standout amongst the most fundamental need of technical aspect and hence we require cryptography [1]. Cryptographic applications include passwords, electronic commerce and ATM cards,

The field of cryptography [14] can be classified into several types like:

(i) Symmetric/private key cryptography,
(ii) Asymmetric/public key cryptography.

As it is clear from the Fig. 1 that, in symmetric systems, only one key is used to encrypt and decrypt the plaintext. An asymmetric system uses public key to encrypt the plaintext and private key to decrypt the cipher text.

Examples of Asymmetric systems include RSA and ECC and symmetric systems include AES.

© Springer Nature Switzerland AG 2020
J. S. Raj et al. (Eds.): ICIDCA 2019, LNDECT 46, pp. 577–585, 2020.
https://doi.org/10.1007/978-3-030-38040-3_65

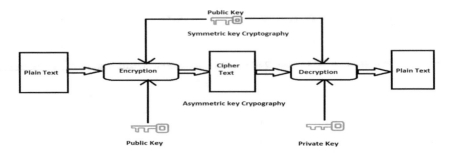

Fig. 1. Cryptography (Both symmetric and asymmetric) [16]

Cryptography [1] in the digital world offers three core areas that protect you and your data from attempt theft or an unauthorized use of your data. Cryptography [1] cover these essential area; authentication, integrity, and confidentiality. Authentication refers to the characteristic that only an authorized person must access the data and integrity is about how you protect your data i.e., data should not be modified whereas confidentiality is how sure that your data is being kept secret throughout the transmission.

1.1 RSA

RSA is one of the earliest public-key cryptosystems and is mainly applied in transmitting data securely. In this type of technique, the encryption key is public which is different from the decryption key that is kept protected. This algorithm was first described in the year 1977 and is named after the first letters of the last names of makers. The encryption and decryption in a simple RSA algorithm was mentioned in [2] and Dual RSA was mentioned in [3]. So this paper mainly focuses on the key generation in RSA with triple modulus i.e., T-RSA [10] and also how encryption and decryption works in T-RSA [10].

1.1.1 Key Generation in T-RSA
Choose six randomly prime numbers of large size say p1, p2 and q1, q2 and r1, r2.

(1) Find the modulus for keys n1, n2, n3.

$$n1 = p1 * p2$$
$$n2 = q1 * q2$$
$$n3 = r1 * r2$$

(2) Find the Euler's totient functions "phi" for each of the modulus keys n1, n2, n3.

$$\text{phi}(n1) = (p1 - 1) * (p2 - 1)$$
$$\text{phi}(n2) = (q1 - 1) * (q2 - 1)$$
$$\text{phi}(n3) = (r1 - 1) * (r2 - 1)$$

(3) Find the values of e1, e2, e3 such that they are co-prime to phi(n1), phi(n2), phi (n3) respectively which will be the public keys.

$$\gcd(e1, \text{phi}(n1)) = 1$$
$$\gcd(e2, \text{phi}(n2)) = 1$$
$$\gcd(e3, \text{phi}(n3)) = 1$$

(4) Compute d1, d2, d3 such that congruence relation does exists which will be the private keys.

$$d1e1 \equiv 1 \,(\text{mod phi}(n1))$$
$$d2e2 \equiv 1 \,(\text{mod phi}(n2))$$
$$d3e3 \equiv 1 \,(\text{mod phi}(n3))$$

(5) **T-RSA Encryption**: Suppose say sender wants to send a message M to Receiver. First Receiver gives its public key (n1, n2, n3, e1, e2, e3) to sender. Private keys d1, d2, d3 will never be shared. First convert M into a number m (m < n) by using a protocol called padding method [15]. And then Cipher-Text c is computed as: c = (((m^e1 mod n1)^e2 mod n2)^e3 mod n3) [4]. Now this c is send to receiver as a cipher text.

(6) **T-RSA Decryption**: To get backplain text from the Cipher-Text, receiver can decrypt c into m by using private keys di {i = 1, 2, 3} as following: m = (((c^d3 mod n3)^d2 mod n2)^d1 mod n1)

1.1.2 Playfair

In the Playfair algorithm a 5 by 5 table is used which contains a key word. Fill the key table with the alphabets of the key word and then complete the table with the remaining alphabets in sequential manner. The entries in the table must be entered in a sequence manner from left to right, top to bottom. The encryption process in the playfair algorithm is done by dividing the message into pairs for example, "Playfair" changes to "PL AY FA IR". These pairs will be placed accordingly in the key table. Sometimes there is an incomplete pair so a different letter is added to that pair say 'X'. The two letters of the pairs are placed opposite corners in the key table. The given four steps must be applied to each pair of letters for performing the substitution:

1. Attach the alphabet "X" after the first alphabet, if both the alphabets are the same (and only one letter is left). Encrypt the new pair of alphabets and continue the process.
2. Replace them with the succeeding right alphabet, if the alphabets are on the same row of the key table.
3. Replace them with the alphabets with the alphabet in the next column if the alphabets are on the same column of the table.
4. If the alphabets are neither in the same row nor in the same column, swap them with the alphabets on the same row respectively but at the opposite pair of corners of the matrix given by the actual pair.

To decrypt, use the opposite of the above three steps, and the first one as it is.

2 Literature Survey

In this paper [1, 2] Devi made a study and provided information regarding the importance of Crytpography in Network Security.

In this paper [3] Gupta, Gupta, Yadav provided the procedure for enhancing security by modifying the conventional RSA algorithm and using blowfish gave a new model B-RSA.

In this paper [4] Manu and AartiGoel used the RSA algorithm using dual encryption and decryption using two public and private keys to provide protection against Brute-Force attack. This paper overcomes the weakness of conventional RSA that is if we can factor modulus into its prime numbers then private key will be in danger.

In this paper [5] Ekka, Kumari proposed a hybrid algorithm of AES and RSA followed by XOR operations. By doing this they concluded that proposed scheme takes lesser time in encryption phase than decryption phase, so security increases.

In this paper [6] Panda and Chattopadhyay proposed a hybrid security algorithm for RSA cryptosystem. Here they computed the public and private keys by using four prime numbers. After that they compared the various process times and speed with conventional RSA.

In this paper [7] Dar, Amit Verma proposed methods for enhancing the security of Playfair cipher by double substitution and transposition techniques.

In this paper [10] Gupta, Gupta, Yadav provided a new model of RSA by making changes to the conventional RSA i.e. T-RSA [10] which is a RSA with tripe modulus.

In this paper [11] Khan had given a detailed study of analysis of playfair ciphers with different sizes and designed some new concepts.

In this paper [12] Hemanth, Raj, Yadav made a mixed algorithm using the RSA and playfair cipher algorithm with 9×6 matrix which includes some of the special characters and obtained good results.

In this paper [13, 14] Abood, Guiguis have made a survey on different types of cryptographic algorithms with repsect to different parameters.

In this paper [15] Peng and Wu provided the Research as well as Implementation part of RSA algorithm using java language. They developed a small software that does the encryption and decryption of text files along with analysing of performance.

3   Proposed Work

This paper proposes a new mixed algorithm of T-RSA [10] and Playfair [11] which ultimately increases the security by increasing decryption time in comparison to both individual algorithms. The T-RSA [10] and Playfair [11] algorithms are mixed with appropriate modifications required and some XOR operations are done wherever needed.

The following are the steps for the mixed algorithm:

(1) Take the plain text 'nt' and key 'sk1' as input for the Playfair matrix.
(2) Make a Playfair matrix of 5 × 5 size and arrange the key 'sk1' in the matrix without duplicating the letters using the steps of the playfair algorithm.
(3) By using plain text 'nt' and the key 'sk1', generate encrypted message 'ent1' using Playfair Cipher algorithm.
(4) Apply XOR operation for the pair (ent1, sk1) and obtain 'ent2'.
(5) Obtain the values of n1, n2, n3 and e1, e2, e3 using RSA algorithm
(6) Apply RSA algorithm on key 'sk1' to convert it into 'sk2'
     sk2 = (((sk1^e1modn1)^e2modn2)^e3 modn3)
(7) Apply XOR operation for the pair (ent2, sk1) to finally get the output 'ent3'.
(8) Then Sender sends the pair (ent3, sk2) to the Receiver.
(9) Then the Receiver decrypts the Cipher text back to plain text.

3.1   Example

Let us take a plain text nt be "This is the key to encrypt" and the key to the Playfair [11] Cipher algorithm sk1 be "mixedcipher" and give this pair (nt, sk1) as input to the Playfair matrix.

Encryption:
The output of this phase is ent1 which is the Cipher Text is
ent1 = QAEOEOQARSRENQMSPAVRQD
Now applying the XOR operation between the pair (ent1, sk1) to obtain ent2
ent2 = < (=*!,81:6 (')(73(&:46
Here T-RSA algorithm is applied on the key sk1 to obtain the encrypted key sk2
sk2 = [B@10dea4e
Again an XOR operation between the pair (ent2, sk2) is applied to get the output ent3
ent3 = gj}H]PS{jgSVI_ot
Finally the pair (ent3, sk2) is sent to the receiver side so that he may decrypt the Cipher text ent3 to obtain the plain text nt.

Decryption:
Firstly the encrypted key sk2 must be decrypted as it helps in decrypting the cipher text. This is done using the T-RSA algorithm.

sk1 = [B@1f96302

Thus with the help of key sk1, the final output is obtained as

TH IS IS TH EK EY TO EN CR YP TX

which is "This is the key to encrypt".

4   Simulation Work

As mentioned throughout this paper that it mainly presents a mixed algorithm of T-RSA [10] and Playfair [11] algorithms. The results presented below prove that this is more secure than the individual ones. Here the comparison is between the T-RSA [10] and the mixed algorithm as we are not including Playfair [11] Cipher algorithm since in T-RSA [10] encryption time is more than that of Playfair. The mixed algorithm had been implemented in java programming language and then the encryption time and decryption times are recorded and presented in a graph for easy viewing and under-standing the results.

From the above Figs. 2 and 3 it is clear that decrypting time is more which ulti-mately draws a conclusion that it is more secure and protected than the T-RSA [10]. As it is also noted that it is taking more than the double time of T-RSA but as we increase the size of the input message time is not that increasing and is secure from all the assaults. The comparison of key sizes [12] as mentioned in the Figs. 4 and 5 of the individual algorithms T-RSA and Playfair [11] and also the mixed algorithm that is mentioned in the paper.

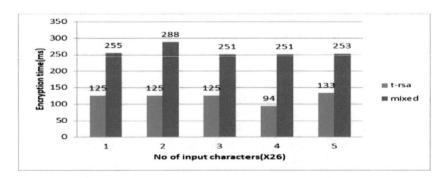

Fig. 2. Encryption time v/s size of input (in characters)

It depicts that as the key size increases, both T-RSA and the mixed algorithm encryption times are not much varying and that concludes that this mixed algorithm is as good as T-RSA and it gives better results if the key size is increased even more.

The proposed algorithm is robust to the security attacks like factoring of public key, brute force attacks. As the factoring of public key was main concern in the conven-tional RSA, large prime numbers are being used to avoid that problem and make the

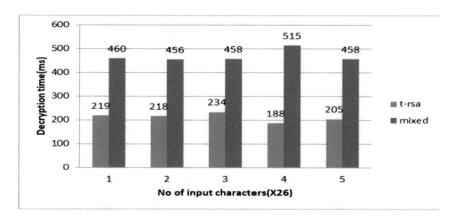

Fig. 3. Decryption time v/s size of input (in characters)

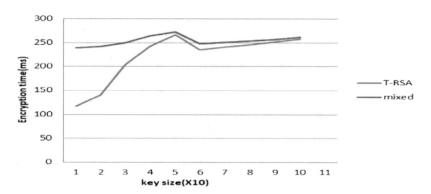

Fig. 4. Encryption time vs key size (in characters)

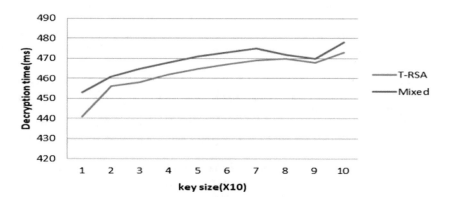

Fig. 5. Decryption time vs key size (in characters)

algorithm even stronger and secure. Many number of brute force method are also done and no one is able to crack the key as it is unpredictable. These all things somehow prove that the algorithm proposed is strong and secure from the third parties.

5 Conclusion

The presented mixed algorithm is assumed to be more secure in terms of decryption time than the other algorithms as it includes all the characteristics of T-RSA [10] and Playfair [11] Cipher Algorithms. It is taking a little longer time than the other mixed but it guarantees that it is so secure and stronger and difficult to break it. This way it satisfies the keen properties of cryptographic algorithms and gives good results along with other stronger encryption-decryption algorithms. As this mixed algorithm uses simple playfair algorithms of 5×5 matrix it can be extended to 9×6 as mentioned in [8] which may increase security and ability of algorithm.

This paper also presents a conclusion that based on the parameter of key size for comparing the strength of encryption-decryption algorithms and gives good results in comparison to T-RSA [10]. As the key size increases the encryption time is almost same with the RSA algorithm with triple modulus which is a good result 0.

References

1. Santoso, P.P., Rilvani, E., Trisnawan, A.B., Adiyarta, K., Napitupulu, D., Sutabri, T., Rahim, R.: Comparison study of symmetric key and asymmetric key algorithm. In: IOP Conference Series: Materials Science and Engineering Published under licence by IOP Publishing Ltd, vol. 420
2. Devi, T.R.: Importance of Cryptography in Network Security. In: 2013 International Conference on Communication System and Information Technology (2013)
3. Gupta, A., Gupta, S., Yadav, N.: Enhancement of security using B-RSA algorithm. In: International Conference on Inventive Computational Technologies (ICICCT 2019). Springer
4. Manu and AartiGoel: ABES Engineering College Ghaziabad, India, 3rd IEEE International Conference on "Computational Intelligence and Communication Technology" (IEEE-CICT 2017)
5. Ekka, D., Kumari, M., Yadav, N.: Enrichment of security using hybrid algorithm. In: International Conference on Computer Networks and Communication Technologies. Springer, Singapore (2019)
6. Panda, P.K., Chattopadhyay, S.: A hybrid security algorithm for RSA cryptosystem. In: International Conference on Advanced Computing and Communication System (ICACCS-2017), Coimbatore, India (2017)
7. Dar, J.A., Sharma, S.: Implementation of one time pad cipher with rail fence and simple columnar transposition cipher, for achieving data security. Int. J. Sci. Res. 3(11) (2014)
8. RSA (Cryptosystem). https://en.wikipedia.org/wiki/RSA_(cryptosystem)
9. Playfair (Cryptosystem). https://en.wikipedia.org/wiki/playfair
10. Gupta, A., Gupta, S., Yadav, N.: Enhancement of Security Using T-RSA Algorithm. Int. J. Sci. Res. Comput. Sci. Appl. Manag. Stud. 7(4) (2018)

11. Khan, S.A.: Design and analysis of playfair ciphers with different matrix sizes. Int. J. Comput. Network Technol. (3) (2015)
12. Hemanth, P.N., Raj N.A., Yadav, N.: Secure data transfer using mixed algorithms. In: IEEE Sponsor 2nd International Conference for Convergence in Technology 2017, Hotel Crown Plaza, Pune Center, Pune, 7–9 April 2017
13. Abood, O.G., Guirguis, S.K.: A Survey on Cryptography Algorithms. Department of Information Technology Institute of Graduate Studies and Researches, Alexandria University, Egypt
14. Vincent, P.M.D.R.: RSA encryption algorithm- a survey on its various forms and its security level. Int. J. Pharm. Technol. (2016). ISSN:0975-766X
15. Peng, J., Wu, Q.: Research and implementation of RSA algorithm in JAVA. In: International Conference on management of e-commerce and e-government (2008)
16. https://hackernoon.com/hn-images

Smart Stick for Blind

B. A. Sujatha Kumari[✉], N. Rachana Shree, C. Radha,
Sharanya Krishnamurthy[✉], and Saaima Sahar

Department of Electronics and Communication,
JSS Science and Technology University, Mysuru, Karnataka, India
sujathakumari@sjce.ac.in, rachanashree4798@gmail.com,
radhachanspur@gmail.com, sharanyakml098@gmail.com,
saharsaaima@gmail.com

Abstract. Independence is the fundamental requirement to achieve dreams, goals, and objectives in life. Despite the fast-paced progress in the world, it is challenging for the visually impaired to carry out their routines independently. Thus, they are always in need of help. For decades, the white cane has become the aid for a blind person's navigation. Continuous efforts have been made to improve the efficacy of the cane. Despite their sharp haptic sensitivity, the blind persons find it difficult to move around alone. The proposed electronic walking stick interfaced with different sensors to detect obstacles in the path helps the blind person by providing more convenient means of life.

Keywords: Blind stick · ATmega328 microcontroller · XBEE · GPS · GSM · LDR · Ultrasonic

1 Introduction

The major problem faced by the blind is to navigate their way to wherever they wish to go. In the paper by Banat et al. [2], they have explained about a navigation stick integrated with components such as GPS, magnetometer, ultrasonic sensor, audio feedback through earphones and a vibrating motor to provide vibratory feedback all of which have been interfaced with an Arduino Mega (ATMEGA) Microcontroller board. The ultrasonic sensor performs the task of alerting the user of an obstacle while the magnetometer +GPS tracks the location and directs the user to the required destination through audio instructs such as Move right/left etc. The alerting mechanism is both by vibration and audio input to the user.

Another noteworthy publication in this regard is the paper by Murali et al. in their work [3] have discussed about an aiding stick for the blind that per- forms the functionalities of obstacle detection, location tracking (GPS), GSM for distress communication, depth or terrain detection, water detection, step sensing and audio feedback to user. The key features of this system include set of recorded audio segments to be played when uneven terrain like elevation, depression and presence of water are detected, encoding in arrays of PCM data, audio amplification and distress message alerting through GSM to the caretakers of the user.

J. S. Raj et al. (Eds.): ICIDCA 2019, LNDECT 46, pp. 586–593, 2020.
https://doi.org/10.1007/978-3-030-38040-3_66

2   Related Work

A paper by Chourasia and Kavitha [4] discusses about a walking stick for the blind which has an IR sensor and two ultrasonic sensors to detect obstacles present at various distances. The IR sensor is placed at the bottom of the stick which is used to detect the objects which are very close to the stick and the two ultrasonic sensors are placed above the IR sensor to alert about objects which are placed at farther distances. The stick also has a GSM module to send messages in case of emergency to the stored contacts. The message is sent upon pressing a switch button which is attached to the handle of the stick.

Another paper published by Agarwal, Kumar and Bhardwaj [5] tells about implementing a stick which consists of GPS, GSM, ultra- sonic sensors and a camera. The stick performs the function of obstacle detection, tracking location of the person and communication via message in case of emergency. Ultrasonic sensors are used to detect obstacles, whereas GPS and GSM are used to send location of the blind person and distress message to the contacts stored on the module. In the areas where the strength of signals is low, the paper proposes alternative method to use camera which is mounted on the helmet on the persons head.

In the paper [6] the authors have discussed about walking stick which consists of ultrasonic sensor to detect object ahead by using ultrasonic waves. It has LDR sensor which is used to detect brightness or darkness in a room and water sensor to detect water level on the ground. Wireless RF based remote is used to identify dislocated stick and all the alerts are given to the blind through buzzer.

Further in their work [7] the authors discuss about a system which includes obstacle detection unit and artificial vision system. The artificial vision system consists of stereo cameras mounted on the helmet and is connected via connectors to a portable computer.

In a paper [8] published in 2015, they have developed a stick which consists of ultrasonic sensors to detect obstacles in the range of 400 cm. The distance between the stick and obstacle is displaced using LCD which is interfaced with the Arduino. Water Sensors connected at the bottom of the stick detects water on the ground. The data collected is passed to the microcontroller which on processing the data activates buzzer.

In the paper [9] the stick has four ultrasonic sensors to detect the obstacles at different sides i.e. on the left, right and front. The detected distance is sent to user headphone via mobile app. Two push buttons are provided at the handle of the stick, one button is used for the emergency text transmission via app and other button is used for tracking location of the person. Application has settings option wherein three emergency contacts are stored, so if the person is lost, he can send the information to his/her caretaker via app.

3   System Description

The block diagram for the proposed work is as shown in Fig. 1. Ultrasonic sensors sense the obstacle and the data is given to the micro controller. Light sensor, LDR checks whether the person is in dark room or light room. GPS and GSM module is used to get the person's location which is then sent to his\her caretaker. All the raw data collected from all the sensors is given to micro controller which processes the data and gives the appropriate buzzing sound. If the person loses his stick, then it can be found by pressing a switch in hand, which transmits the RF signal. These signals are then received by RF receiver and the data is given to micro controller. On processing the data, it activates buzzer on the stick so that the person can easily track the stick.

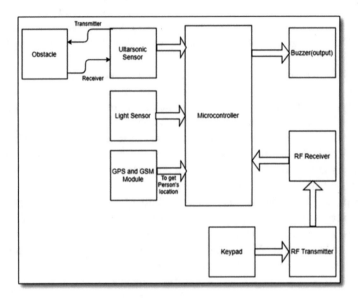

Fig. 1. Block diagram

3.1   Ultrasonic Sensor

The proposed method uses Ultrasonic ranging module HC - SR04 which provides 2 cm–4 m non-contact measurement function, and can detect the obstacle with an accuracy of 3 mm. The module consists of ultrasonic transmitters, receiver and control circuit. It has a working voltage of 5 V and working current of 15 mA. On powering up, transmitter continuously emits ultrasonic waves which are then detected by receiver when they are reflected back after hitting an obstacle.

3.2   LDR Sensor

The features of the LDR sensor chosen include Good reliability, Small volume, High sensitivity, Fast response and Good spectrum characteristics.

3.3   GPS+GSM Module

This component was chosen due to the following reasons: SIM 808 [10] has integrated GPS and GSM modules in a single board with an operating voltage of 3.3–5 V, sleep mode current of 1 mA, burst current of 2A. The module can be put to idle when not in use to save power.

3.4 Xbee S$_2$c

The chosen module has an operating frequency of 2.4–2.5 GHz and a transmission power of 2 mW [11]. In addition, interfacing XBEE module with Arduino is simple and is achieved through S$_2$C protocol.

4 Functional Description

4.1 Ultrasonic System

The ultrasonic sensors activate the buzzer with 1000 Hz and 2000 m/s delay when it detects any objects with in a range of 100 cm (Fig. 2).

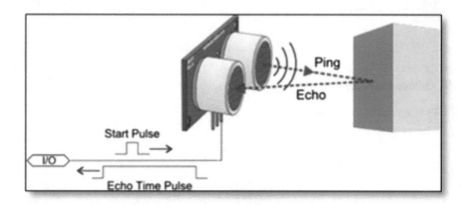

Fig. 2. Ultrasonic sensor

4.2 LDR System

When there is no light, the material resistance increases and buzzer is activated is such situations with Hz having a delay of m/s.

4.3 Gps+Gsm

GSM and GPS module establishes communication between the microcontroller and a GSM-GPS system. The GSM is an architecture used for mobile communication. The GPS tracks and updates the location of the stick continuously and displays it on the serial monitor. A push button on the stick, when pressed, sends the location to the saved contact on the GSM module via the microcontroller.

4.4 Remote System

The XBee module consists of a hardware and a software part (Zigbee protocol). XBee devices (transmitter and receiver) communicate with each other over the air, sending and receiving wireless messages. They use the same radio frequency. The devices only transfer those wireless messages. They communicate via the serial interface. Two XBee's are mounted on the stick and a remote respectively. In case the stick is misplaced, these XBee's communicate wirelessly upon pressing a switch button on the remote and activate a buzzer on the stick (Fig. 3).

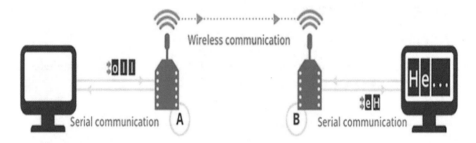

Fig. 3. XBEE transmitter and receiver

4.5 Microcontrollers

The flow chart for the proposed system is shown in Fig. 4.

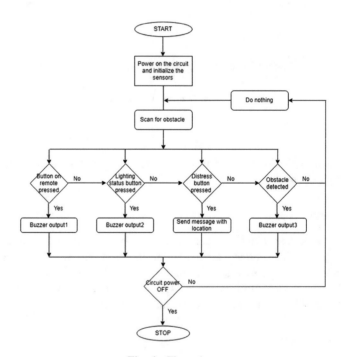

Fig. 4. Flow chart

On powering the micro controller, all the sensors will be activated. The ul- tran-sonic sensors on the stick continuously senses the obstacle and if any present, respective buzzing sound will be heard else if no obstacle is found, it continues sensing. If push button to know lightning condition, is pressed respective buzzing sound will be heard indicating his presence to others in the dark conditions. If the person presses the push button in order to send distress message, GPS location will be tracked and is sent to the contact that is saved on the GSM module. If in case stick is lost, when the person presses respective button on the hand held device, the buzzing sound will be heard from the buzzer that is placed on the stick. Following that sound, the person can track the stick.

5 Result

The prototype of the stick is shown in Fig. 5.

Fig. 5. Working model

The message received by the caretaker displaying the GPS location of the blind person tracked is shown in Fig. 6.

Fig. 6. Received message

6   Conclusion and Discussion

The smart blind stick system can be used to navigate in the straight road and across the bends, avoiding minor obstacles in their path. It also allows the user to send the distress message along with their current location to their contact. The feature of detecting the lost stick is very helpful to the visually impaired as they cannot otherwise locate their stick if misplaced.

All the features listed above have been tested and the results are positive and convincing. Overall, this support system for the blind helps them carry out their daily routine independently without waiting for help from other.

Acknowledgement. The authors of this paper acknowledge and thank their institution for the constant support throughout the duration of this project. The students also thank their guide, professors and classmates from JSS Science and Technology University who provided insight and expertise that greatly assisted the work.

References

1. https://www.who.int/news-room/fact-sheets/detail/blindness-and-visual-impairment.html
2. James, N.B., Harsola, A.: Navigation aiding stick for the visually impaired. In: IEEE International Conference on Green Computing and Internet of Things, Noida, India, October 2015
3. Murali, S., Shrivatsan, R., Sreenivas, V., Vijjappu, S., Joseph, G.S., Rajavel, R.: Smart walking cane for the visually challenged. In: IEEE Region 10 Humanitarian Technology Conference, Agra, India, 21–23 December 2016
4. Chaurasia, S., Kavitha, K.V.N.: An electronic walking stick for blinds. In: International Conference on Information Communication and Embedded Systems, Chennai, India, 27–28 February 2014
5. Agarwal, A., Kumar, D., Bhardwaj, A.: Ultrasonic stick for blind. In: International Journal of Engineering and Computer Science, vol. 4, no. 4, April (2015)
6. Deepika, S., Divya, B.E., Harshitha, K., Komala, B.K., Shruthi, P.: Ultrasonic blind walking stick Int. J. Adv. Electr. Electron. Eng. **5**(6) (2016)

7. Dambhare, S., Sakhare, A.: Smart stick for Blind: Obstacle Detection, Artificial vision and Real-time assistance via GPS. In: 2nd National Conference on Information and Communication Technology, Proceedings published in International Journal of Computer Applications (2011)
8. Gbenga, D.E., Shani, A.I., Adekunle, A.L.: Smart Walking Stick for Visually Impaired People Using Ultrasonic Sensors and Arduino, vol. 5, no. 5, October–November 2017
9. Nowshin, N., Shadman, S., Joy, S., Aninda, S., Minhajul, I.M.: An intelligent walking stick for the visually-impaired people. Int. J. Online Eng. **13**(11) (2017)
10. https://simcom.ee/modules/gsm-gprs-gnss/sim808/
11. https://components101.com/wireless/xbee-s2c-module-pinout-datasheet
12. https://www.electronicsdatasheets.com/manufacturers/arduino/parts/arduino-uno-rev3

A Neural Network Based Approach for Operating System

Gaurav Jariwala$^{(\boxtimes)}$ and Harshit Agarwal$^{(\boxtimes)}$

Sarvajanik College of Engineering and Technology, Surat, India
gjariwala9@gmail.com, 9arshit@gmail.com

Abstract. The operating system is the central element of a computing device. It is the base level on which all the applications run. It allows the user to interact with the hardware with the help of the user interface. Creating a more efficient and capable software means less load on the hardware. Evolving nature of the neural network will help the operating system to learn about the user and will help in creating a better experience for the user. In this paper, we propose the integration of the neural network system at the kernel level of the operating system. Further, we show that the proposed scheme is more efficient and advanced than the current conventional system.

Keywords: Activation function · Backpropagation · Memory management · Neural network · Operating system · Process scheduling · Recurrent neural network

1 Introduction

An advance operating system will help create a better system. Advancements in hardware are at an exponential rate but unfortunately, the software has not been given that importance. Different versions of operating systems are being released, but the kernel has been the same for many years.

In multitasking, context switching is the most crucial part of the process scheduling. Current process scheduling system is quite slow compared to the speed it can achieve. This is where the neural network can be used in the operating system. Knowing what job is to be scheduled by learning the user's behavior and optimally doing that is the most important step. This way the neural network can predict what process the user is going to use next. This will significantly increase the process scheduling speed.

In a multiprogramming system, memory management requires to utilize the memory most efficiently. Memory needs to be allocated to ensure a reasonable supply of ready processes to consume available process time [13]. It is advisable to have as many processes as possible in the main memory for efficient use of the processor. This task can be improved with the use of a neural network in which the processes that are being used frequently will be placed in the main memory.

In the neural network, user's data will be taken as an input and the output will be the pre-arrangement of the processes in the queue along with the necessary memory reservation, so that the next process having the highest priority that the user is going to access will be ready to run. This will work as the neural network will be trained by the

© Springer Nature Switzerland AG 2020
J. S. Raj et al. (Eds.): ICIDCA 2019, LNDECT 46, pp. 594–599, 2020.
https://doi.org/10.1007/978-3-030-38040-3_67

data obtained at the initial stage from the user's usage. While training from this data the neural network will adjust its weights to get the results according to the dataset, so when the input is provided to the trained neural network, it can generate an accurate output.

The remaining paper is organized as follows: Sect. 2 summarizes the related work done in this domain; Sect. 3 describes our method; Sect. 4 summarizes the conclusion and future directions for the concept.

2   Related Work

Kumar and Nirvikar [8] proposed an algorithm (SRT) based on integration of round-robin (RR) and SJF CPU scheduling algorithms. This makes their algorithm have advantages of both, reducing starvation and priority scheduling. This paper has compared the average waiting time of RR with the SRT and drawn the conclusion that SRT takes less average waiting time than RR.

Tani et al. [14] has done a comparison of neural networks algorithms for cloud computing CPU scheduling. They found out that Multi-layer Perceptron ANN was the best among the other algorithms, reducing the average waiting time of the tasks on the execution queue and stimulating the response time.

Ajmani and Sethi [2] proposed a fuzzy logic-based CPU scheduling algorithm for efficient utilization of CPU. They compared priority algorithm with the proposed fuzzy CPU scheduling algorithm (PFCS) and conclude that priority algorithm has more average waiting time and average turnaround time than PFCS.

Miglani et al. [9] dealt with a comparative study of performance measures of CPU burst time fuzzy in nature which is helpful for the designer in selecting the right scheduling algorithm at high abstraction levels which save him from error prone priority assignments at the final stage of system design.

Bektas et al. [3] proposed similarity-based prognostic algorithm that is fed by the use of data normalisation and filtering methods for operational trajectories of complex system. This is combined with a data-driven prognostic technique based on feedforward neural networks with multi-regime normalisation. The proposed algorithm was found to be effective for remaining useful life (RUL) calculation of C-MAPPS dataset.

Samal et al. [12] proposes design of a novel functional link artificial neural network (FLANN) based dynamic branch predictor and compares the performance against a perceptron-based CBP. FLANN is a single layer ANN with low computational complexity. They concluded that the further tuning of the parameters of the implementation will enable the FLANN-based CBP to achieve performance at par with and better than that of perceptron-based CBP implantation.

3   Method

The proposed system will be embedded into the kernel itself and will participate in taking core decision for the system. In our proposed method, several inputs can be considered when an application is being used by the user, but we are considering the inputs which are most significant to operating system and can provide the best possible accuracy and speed. For this system, neural network can be created using Keras [5] and Tensorflow library [1] for backend calculations.

Since the processing will be done in the background without user interaction, chances of human error are virtual reduced to zero. Also, the user will have better experience with the operating system without having to learn anything new.

Table 1. Terminologies used.

Terms	Definitions
Feature scaling	The range of features of data is rescale using this method. For example, min-max normalization to bring data in the range of [0,1]. $x' = \frac{x - \min(x)}{\max(x) - \min(x)}$
Relu activation function	Rectified Linear Unit activation is used to produce non-linearity in output [10]. $y = Max(0, x)$
Adam optimizer function	It used to update the weights of the neural network during training [7]
Mean Squared Error	It is a cost function used to reduce the difference between predicted and actual value. This function guides the optimizer function. $MSE = \frac{1}{n} \sum_{i=1}^{n} (Y_i - Y'_i)^2$
Drop out rate	Rate at which nodes are dropped in each epoch to avoid overfitting

As shown in Fig. 1 The proposed system can run with any system kernel such as Linux or Unix based system or MS-DOS based system, with minimum changes. But the performance may vary from system to system depending on the developer. Integration of the neural network to the core of the operating system will help in overcoming any latency issue that a system may face while processing any request.

Figure 2 shows our proposed method diagram. We are taking background time, usage frequency, usage time of the application (elapsed time) and foreground time of the application to predict scheduling of the processes. Since the range of input values varies, normalization of the data becomes necessary before they are used to train the neural network. This is done using Min-Max normalization.

The output will be received in the form of priority of the application, preloading urgency, memory reserve for the application in both RAM and virtual memory and queueing of the process for CPU usage.

Four hidden layers are being used. Each layer contains 6 nodes where edges are initialized with random weights and have 'Relu activation function'. For the output layer, hyperbolic tangent activation function [11] (tanh activation function) is used which gives output values in range of [−1, 1]. Inverse transformation function can then be applied to obtain the actual value for each of the expected output.

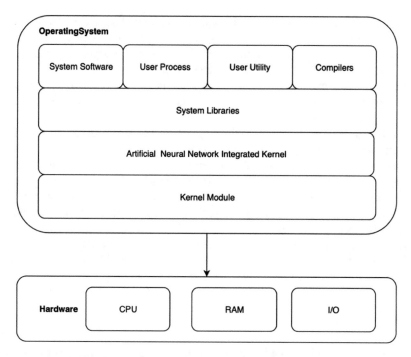

Fig. 1. Architecture of the proposed operating system

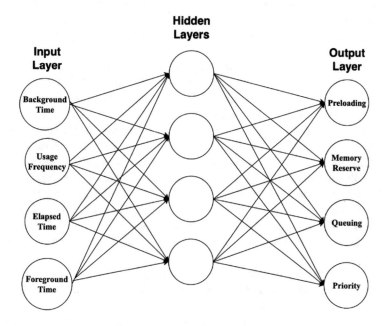

Fig. 2. Proposed System diagram

"Adam" optimizer is used for minimization of loss function. It is better than the rest of the adaptive technique and it rectifies every problem that is faced in other optimization techniques such as vanishing learning rate, slow convergence or high variance in the parameter updates which leads to fluctuating loss function [15].

After studying and analyzing different loss function, "Mean Squared Error" is found to be most appropriate for this system. Dense layer connection will be used with a dropout rate of 15% to avoid overfitting.

The pre-training of the system will be done on the inputs based on generic user behavior such as, training it on application's inputs which are common among the maximum number of users. When the system is being utilized, actual user's usage will be stored in memory block separate from system memory. Using this memory block the neural network will be trained in the background without taking compute power of CPU or GPU from the user. Alternatively, the neural network can be trained when the system is in sleep or idle mode.

The system will over time learn about user habits regarding the usage of an application. When the system is used next time, the kernel will be able to anticipate the needs of the user and plan its internal working while booting.

4   Conclusion

The system will be able to predetermine user needs and configure system preferences. The user will experience a much smoother and faster performance from their machine. Since all the overhead preprocessing such as process scheduling, memory reservation, CPU allocation or virtualization of memory are done ahead of time, user will be able to open an application much faster and significant improvement in performance will be observed.

Results obtained from a different study [4] were also in favor of neural network integrated kernel. An ordinary system was overloaded with two bzip2 processes to test the responsiveness of different kernels. When a generic or conventional kernel was being used, severe drop in the system's performance was observed. The application's loading was slow and for multimedia applications frame rate drop was approximately 96%. When basic neural network-based kernel was used, after few test runs significant improvement in the system was observed. The system was able to recognize which application the user wanted to use and allocated more resources to those processes, thus creating a better system. The frame rate drop was reduced to less than 3%.

However, there are some drawbacks to this proposed system. If multiple people are using the same device (same logging ID) then the neural network will not be able to recognize the pattern in different application behavior of the users, making the system unreliable.

It is possible that due to some error, system might take all the CPU or GPU usage away from the user, for training the neural network thus making the user experience of the device slow and application may lag.

The ANN is not able to deal with subtle changes in user behavior regarding the usage of applications. The system does not consider under which context the applications are being used and hence cannot predict the outcomes precisely. This problem can be solved by using RNN. Usage of Long Term Short Memory [6] will help the system recognize the context under which applications are being used by keeping the immediate previous history of system usage and by implementing changes much faster than ANN.

References

1. Abadi, M., Agarwal, A., Barham, P., Brevdo, E., Chen, Z., Citro, C., Carrado, G.S., Davis, A., Dean, J., Devin, M., Ghemawat, S., Goodfellow, I., Harp, A., Irving, G., Isard, M., Jozefowicz, R., Jia, Y., Kaiser, L., Kudlur, M., Levenberg, J., Mané, D., Schuster, M., Monga, R., Moore, S., Murray, D., Olah, C., Shlens, J., Steiner, B., Sutskever, I., Talwar, K., Tucker, P., Vanhoucke, V., Vasudevan, V., Viégas, F., Vinyals, O., Warden, P., Wattenberg, M., Wicke, M., Yu, Y., Zheng, X.: TensorFlow: Large-scale machine learning on heterogeneous systems (2015). http://tensorflow.org
2. Ajmani, P., Sethi, M.: Proposed fuzzy CPU scheduling algorithm (PFCS) for real time operating systems. BIJIT - BVICAM's Int. J. Inf. Technol. 5(2), 583 (2013)
3. Bektas, O., Jones, J.A., Sankararaman, S., Roychoudhury, I., Goebel, K.: A neural network filtering approach for similarity-based remaining useful life estimation. Int. J. Adv. Manuf. Technol. 101(1–4), 87–103 (2019)
4. Bex, P.: Implementing a Process Scheduler Using Neural Network Technology. Radbound University Nijmegen
5. Fran, C.: Keras (2015). http://keras.io
6. Hochreiter, S., Schmidhuber, J.: Long short-term memory. Neural Comput. 9(8), 1735–1780 (1997)
7. Kingma, D., Ba, J.: Adam: a method for stochastic optimization. In: 3rd International Conference for Learning Representations, San Diego (2015)
8. Kumar, N.: Performance improvement using CPU scheduling algorithm-SRT. Int. J. Emerg. Trends Technol. Comput. Sci. (IJETTCS) 2(2), 110–113 (2013)
9. Miglani, S., Sharma, S., Singh, T.: Modified CPU scheduling for real-time operating system considering performance parameters in fuzzy. Int. J. Appl. Eng. Technol. 2(4), 58–68 (2014)
10. Nair, V., Hinton, G.E.: Rectified linear units improve restricted boltzmann machines. In: ICML 2010 Proceedings of the 27th International Conference on International Conference on Machine Learning, Haifa, Israel, pp. 807–814 (2010)
11. Nwankpa, C., Ijomah, W., Gachagan, A., Marshall, S.: Activation functions: Comparison of trends in practice and research for deep learning. arXiv preprint arXiv:1811.03378 (2018)
12. Samal, A.K., Mallick, P.K., Pramanik, J., Pani, S.K., Jelli, R.: Functional link artificial neural network (FLANN) based design of a conditional branch predictor. In: Cognitive Informatics and Soft Computing, pp. 131–152. Springer, Singapore (2019)
13. Stallings, W.: Operating Systems Internals and Design Principles, 7th edn. Prentice Hall, New Jersey (2012)
14. Tani, H.G., Amrani, C.E., Lotfi, E.: Comparative study of neural networks algorithms for cloud computing CPU scheduling. Int. J. Electr. Comput. Eng. (IJECE) 7(6), 3570–3577 (2017)
15. Wallia, A.S.: Types of optimization algorithms used in neural networks and ways to optimize gradient descent. https://towardsdatascience.com

Randomness Analysis of YUGAM-128 Using Diehard Test Suite

Vaishali Sriram[⊠], M. Srikamakshi, K. J. Jegadish Kumar,
and K. K. Nagarajan

Department of ECE, SSN College of Engineering, Chennai, India
vaishalisriram19@gmail.com, srikam.m30@gmail.com,
{Jegadishkj,nagarajankk}@ssn.edu.in

Abstract. The random key generation for the encryption/decryption process prevails as the significant character of the stream ciphers. Here, we have introduced a novel idea to elucidate the results from the randomness test using the Diehard test suite. The proposed method is used to test the randomness of the YUGAM-128 stream cipher. The randomness tests resulted in p-values. In this paper, Diehard test results have been consolidated, and the p-values obtained for the algorithm are distributed accordingly, such that it indicates improved randomness of the security level for the Safe Arena, while the presence of many p-values in the Failure Arena reports that the produced text is less likely to be random. The outcome of the conducted tests showed that YUGAM-128 resulted in an 81.2% of the p-values distributed over the safe area.

Keywords: Stream cipher · CA · Diehard test · RNG · And the LFSR

1 Introduction

The pseudo random-number generation employed by the cellular automata (CA) has become a prominent arena of research since last 10 years [1]. One of the advantages offered by the CAs in the field of VLSI: cellular automatas are regular simple, modular and a locally interconnected. The CAs serve as an excellent options for the application on board and are often utilized in the encryption decryption components [2] and the BIST [3, 20]. In the various computational fields such as stochastic optimization methods, an significant role is accomplished by the RNG and the pseudorandom-numbers generation in the scientific computation that are parallel [21].

One primary requirement of these domains is that they must satisfy the various statistical tests enumerating the randomly generated numbers quality. Also, the generation of long sequences of these random numbers must be as fast as possible as computational efficiency is of prime import. Thus, one can argue that CAs are an ideal choice in such situations as they can produce rapid random number streams that are of excellent quality.

In the past, there have been comprehensive studies on One-dimensional random number generators [1, 3, 4, 5].

On considering a solution for delay type faults, CA-generated pseudorandom numbers are regarded to be a better choice than any other existing methods such as

© Springer Nature Switzerland AG 2020
J. S. Raj et al. (Eds.): ICIDCA 2019, LNDECT 46, pp. 600–607, 2020.
https://doi.org/10.1007/978-3-030-38040-3_68

linear feedback shift registers based on conclusions reached from the above studies [6]. The various researches carried out in studying the configuration of the progressively generated bit patterns and the threshold that was set by the theoretical results has helped in the development of the CA RNGs.

Stream Ciphers are now extensively used in the various mobile devices/PDAs. This is because they are fast, occupy less space and consume less power for their operations than any other cryptographic algorithms. Usually, the inputs for a stream cipher are in the form of a secret key and a public IV. Each cycle of operation done by the cipher will form a keystream bit. On the encryption side, we produce the ciphertext by XORing the plaintext with the keystream. Decryption involves getting back the plaintext by merely XORing the cipher-text with the keystream.

2 General Theory of the CA

A cellular automaton (CA) is an effective system with the discrete features of the space and time. As stated in a local, identical interaction rule, the cellular automaton comprises an array of cells with the number of finite states [20]. These states are included as time steps that are discrete corresponding to a rule function. The cellular array is d-dimensional, where d = 1, 2, 3 are used in general for cellular automata-based computer applications [20]; however the paper focuses on the d = 1, that cellular array defined in one dimension. Every cell comprises of the identical rule, which is a well-defined Boolean function or transition function that is represented by a rule table as shown in Fig. 1 [20]. In 1-D CAs, each cell state relies on r local neighborhood cell states on either side of the referenced cell, and 'r' is denoted as the radius [20]. Generally, in two-dimensional Cellular automates (2-D CAs), two types of cellular neighborhoods are most commonly practiced in various computing applications. One of the types is Von Neumann neighborhood 2-D CA with five cells; each cell has four adjacent non-diagonal neighbors. The second type is called Moore neighborhood 2-D CA with nine cells, each cell has eight surrounding neighbors. To reduce the large search-space size, the most common choice is 5-neighbor grids whose results make it more responsive to hardware implementation.

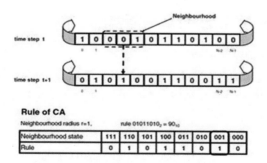

Fig. 1. Cellular Automata

For the finite-size 1-D CA grid, cyclic boundary conditions form a circular lattice of cells for the 1-D and a toroidal one for the 2-D [20], conditions used, in boundary is NULL (that bounds the lattice cells within an outer layer that is fixed to zero) [20]. The design can be easily achieved in hardware.

3 Stream Cipher

The input message fed to the stream cipher has a varying length and can transform plaintext bits at different positions by combining it with a pseudorandom cipher digit stream, i.e., a keystream. A stream cipher contains two functions, one for updating the state and the other for the output. During encryption, the state of a stream cipher is continuously refreshed as the various bits that occupy the consecutive position in the message needs to be encrypted with different states. The encryption or decryption operation is carried out by the output function, which is also responsible for generating the keystream from the state. If the first state of the stream cipher and the key are distinct, then by using a key setup, we generate an initial state. This key setup creates an initial state by using a key that uses different initialization vectors to create keystreams.

The following factors characterize a good stream cipher: should be statistically random with a long period and no repetitions, high linear complexity of LFSR, high correlation immunity (compromising linear complexity), good confusion and diffusion, and using Boolean functions that are non-linear [21] (Fig. 2).

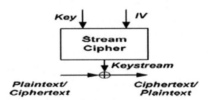

Fig. 2. Block diagram of stream cipher

4 Algorithmic Strategy in Designing

4.1 CA Rule-Based Function

In the book, "A New Kind of Science", the author Stephen Wolfram proposed a new one-dimensional binary CA Class III rule with periodic and chaotic traits known as Rule 30 [20].

Table 1. Rule 30 Neighbourhood State

Current state surroundings	000	001	010	011	100	101	110	111
Emerging state for core cell	0	1	1	1	1	0	0	0

The motive for using Rule 30 is due to its periodic, chaotic nature that provides complex random patterns from simply defined rule functions and also have reversible property [20]. With these exciting facts, Wolfram concluded that Rule 30 based cellular automata is vital for creating complex systems and higher performance. Hence, rule 30 has been most commonly used as a RNG, and it has also been considered to make use of it in cryptography as a potential stream cipher.

Wolfram's proposed basic CA model in which the 1-D array of CA cells with infinite in number and having only two states [0,1] are designed. All of these cells are present in some initial state, and the states are changed in each cell at discrete time intervals that rely on its present state and that of its two neighbors. Table 1 shows the ruleset (for Rule 30), which dictates the subsequent state of the automaton.

The enumerated operation for cellular automata rule 30 is

$$f(x) = \underline{x}_{i-1}x_{i+1} + \underline{x}_{i-1}x_i + x_{i-1}\underline{x}_i\underline{x}_{i+1} \tag{1}$$

The given function is used for the evaluation of Rule 45, set of ruleswhich controls the automaton in the future state.

$$f(x) = \underline{x}_{i-1}\underline{x}_{i+1} + x_{i-1}\underline{x}_ix_{i+1} + \underline{x}_{i-1}x_i \tag{2}$$

For Rule 57, rules the emerging state

$$f(x) = \underline{x}_i\underline{x}_{i+1} + x_{i-1}\underline{x}_i + \underline{x}_{i-1}x_ix_{i+1} \tag{3}$$

The LFSR shown in the Fig. 3 is built using the flip flops that acts as the timed storage elements, and the return path for the feedback. The count of the flip-flops used decides the degree 'm' [20]. The logical response function evaluates the xor-sum of the flipflops of the shift register to feed it as input to the filpflop in the final stage. For every clock pulse the bits in the flip flop are shifted right and the one at output of the final flip flop is considered as the output [20]. The feedback path utilizing the XOR-sum of the some few intermediate bits enumerates the left-most flip-flop state.

This kind of logic circuit is called linear feedback shift registers. Let us consider the initial states to be ($s_2 = 0$, $s_1 = 0$, $s_0 = 1$), the complete sequence of states of the LFSR is given in Table 2.2 [21]. The LFSR operation is given in Eq. (4) to (6). Then the output bits s_i is computed as in (7) [21]:

4.2   Mathematical Portrayal

Figure 3 represents the conventional form of LFSR having a degree m, where m denotes the possible feedback locations and the number of flip-flops. The XOR operation connects these flip-flop feedback locations. p_0, p_1,..., p_{m-1} represents the feedback coefficients and is used to determine the state of the feedback path.

- When the feedback coefficient $pi = 1$, the switch is considered closed, and the feedback is said to be active.

- When the feedback coefficient $pi = 0$, the switch is considered to be open, and the feedback is not active (i.e.) the input does not use the output obtained from the equivalent flip-flop.

The script given below provides us with a feedback paths mathematical portrayal

> **The LFSR of degree m generates a maximum sequence of length 2m-1**

When the coefficient p_i of the flip-flop I is multiplied by its output, the result will give us the state of the feedback (i.e.) if $p_i = 1$, the switch is closed, and the output value provides the result, or if $p_i = 0$ the switch is open, and the result will be equal to zero. The output sequence generated by the LFSR is highly dependent on the values of the feedback coefficients.

Let's consider that the values $s_0, ..., s_{m-1}$ are given as inputs to the LFSR. An XOR sum operation is performed on the product of the feedback coefficients of the flip-flop and their respective outputs to calculate s_m, which is not only the following output of the LFSR serves also as input to the flip-flop in the leftmost [21]. We have

$$sm \equiv sm - 1pm - 1 + \cdots + s1p1 + s0p0 \, mod \, 2 \qquad (4)$$

The output for the subsequent LFSR can be calculated as:

$$sm + 1 \equiv smpm - 1 + \cdots + s2p1 + s1p0 \, mod \, 2 \qquad (5)$$

Hence, the output sequence is given by

$$si + m \equiv \sum_{j=0}^{m-1} pj \cdot sj - 1 mod 2 \qquad (6)$$

The combination of a few of the former output values will precisely determine the output values. The output sequence of an LFSR has a periodic occurrence, and so occasionally, we call it the number of recurring states. Furthermore, depending on the feedback coefficients, an output sequence of varying lengths can be constructed by an LFSR. The LFSR maximum length is computed as a degree from the theorem given below.

It is effortless to prove the validity of the theorem. The minterm register bits are solely responsible for the state of an LFSR. As a result, repetitions start to occur due to LFSR assumption on its preceded state. It is only possible for 2_{m-1} nonzero states to appear in a state vector having m-bits, and 2_{m-1} also gives us the maximum sequence length before repetition. Consider that the LFSR takes up a zero state. In such a case, the LFSR will never leave the state, and it will get "stuck" in it. Hence, it is necessary to ensure that the zero states are omitted. Specific patterns $(p_0, ..., p_{m-1})$ results with LFSR of maximum length. The example shown below illustrates it.

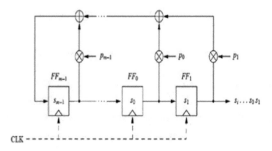

Fig. 3. Block diagram of LFSR with tapping

5   YUGAM-128 Architecture

Proposed YUGAM-128 stream cipher, which is built using a cellular automaton has a straightforward architecture. The Fig. 4 given below represents the simple architecture. The cellular automata (CA) rule is used to convert the input 128-bit key into an unrecognizable form. The shift register with the linear feedback path (LFSR) is fed with the 128 bit that is initialized, and the output of this process is used to generate a keystream per clock cycle by XORing it with CA rule-based update key.

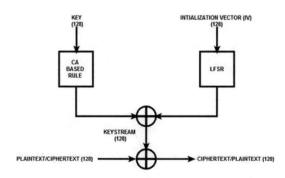

Fig. 4. Architecture of proposed stream cipher

6   Die Hard Test

Essentially to test the quality of random numbers, we need to analyze the keystream generated by the stream cipher by using a diehard test suite. This test suite contains a series of independent tests that obtains an input containing random bits. Each diehard test returns a p-value, which should ideally not be equal to 0 or 1. Also, It is expected that the p-values should have a normal distribution on (0,1). When the p-value is either 0 or 1, we consider the test to be a failure. Based on this, we have performed the analysis on Yugam-128 stream cipher, and the results of which are recorded in the following table. The distribution of the p-values obtained in the above tests has been plotted as a graph and shown in Fig. 5 (Table 2).

Table 2. Die Hard results after evaluation

Name	p-value	Result
Birthday Spacing Test	0.624429	Cleared
The overlapping 5 permutation test	0.460913	Cleared
Binary rank test 6X8 matrix	0.463822	Cleared
Bitstream test	1.000000	Not cleared
Overlapping Pairs Sparse Occupancy test	1.000000	Not cleared
Overlapping Quadruples Sparse Occupancy test	0.056732	Cleared
DNA test	0.155907	Cleared
Count the 1s test (a stream of bytes)	0.930793	Cleared
Count the 1s test (specific bytes)	0.577306	Cleared
Parking lot test	0.571836	Cleared
Minimum distance test	1.000000	Not cleared
3-D spheres test	0.594688	Cleared
Squeeze test	0.994753	Cleared
Overlapping sums test	0.533471	Cleared
Runs test	0.497556	Cleared
Craps test	0.212355	Cleared

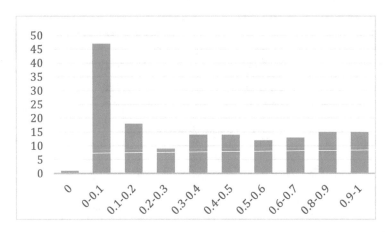

Fig. 5. p-values distribution.

7   Conclusion

From the above table, we find that there are only 3 tests that have failed. Hence, we have a success rate of 81.2%. Although in Fig. 5 there appears to be a large number of p-values within the 0–0.1 range, the rest of the values have a normal distribution. Based on these facts, we conclude that the stream cipher is secure. However, further NIST tests can be conducted to provide substantial support to this statement.

References

1. Chaudhuri, P.P., Chowdhury, D.R., Nandi, S., Chattopadhyay, S.: Additive Cellular Automata: Theory and Applications, vol. 1. IEEE CS Press, Los Alamitos (1997)
2. Nandi, S., Kar, B.K., Chaudhuri, P.P.: Theory and application of cellular automata in cryptography. IEEE Trans. Comput. **43**, 1346–1357 (1994)
3. Bouganim, L., Guo, Y.: Database encryption. In: Encyclopedia of Cryptography and Security, 2nd Edn. Springer, New York (2010)
4. Carlet, C., Dalai, D.K., Gupta, K.C., Maitra, S.: Algebraic immunity for cryptographically significant boolean functions: analysis and constraction. IEEE Trans. Inf. Theory **52**(7), 3105–3121 (2006)
5. Coppersnith, D., Halevi, S., Lutla, C.S.: Cryptanalysis of stream cipher with linear masking. In: Yung, M, (eds.) Advances in Cryptology-Crypto 2002. LNCS 2442, pp. 515–532. Springer, Heidelberg (2002)
6. Pucknell, D.A., Eshraghian, K.: Basic VLSI Design. 3rd Edn, pp. 118–274. Prentice Hall, India (2004)
7. Ekdahl: On LFSR Based Stream Ciphers (Analysis and Design). Ph.D. Thesis, Lund University, November 2003
8. Gammel, B.M., Gottfert, R., Kniffler. O.: An NLFSR-based Stream Cipher. In: ISCAS (2006)
9. Good, T., Benaissa, M.: ASIC Hardware Performance In: New Stream Cipher Designs: The eSTREAM Finalists, LNCS, vol. 4986, pp. 267–293 (2008)
10. Grocholewska-Czurylo: Random generation of boolean function with high degree of correlation immunity. J. Telecommun. Inf. Technol. 14–18 (2006)
11. Kim, J.Y., Song, H.Y.: A nonlinear boolean function with good algebraic immunity. In: IEEE Proceeding Of IWSDA 2007, pp. 94–98 (2007)
12. Kitsos, P., Sklavos, N., Papadomanolakis, K., Koufopavlou, K.: Hardware implementation of bluetooth security. IEEE Pervasive Comput. **2**(1), 21–29 (2003)
13. Maximov: Some Words on Cryptanalysis of Stream Ciphers. Ph.D. dissertation, Lund University, Lund, Sweden (2006)
14. Menezes, A., van Oorschot, P., Vanstone, S.: Handbook of Applied Cryptography, pp. 482–504. CRC Press, Boca Raton (1996)
15. Kitsos, P.: On the hardware implementation of the MICKEY-128 Stream Cipher. In: eSTREAM, ECRYPTStreamCipher Project, Report 2006/059 (2006)
16. Rukhin, A., Soto, J., Nechvatal, J., Smid, M., Barker, E., Leigh, S., Levenson, M., Vangel, M., Banks, D., Heckert, A., Dray, J.: A Statistical Test Suite for Random and Pseudorandom Number Generators for Cryptographic Applications. In: NIST, pp. 1–153. Special Publication 800-22, 15 May 2001
17. Rizomiliotis, P.: On the resistance of boolean functions against algebraic attacks using univariate polynomial representation. IEEE Trans. Inf. Theory **56**(8), 4014–4024 (2010)
18. Rose, G.G., Turing, H.G.: A Fast Stream Cipher. In: Fast Software Encryption FSE 2003, pp. 290–306. Springer, Heidelberg (2003)
19. Hwang, D., Chaney, M., Karanam, S., Ton, N., Gaj, K.: Comparison of FPGA-targeted Hardware Implementations of eSTREAM Stream Cipher Candidates. In: The State of the Art of Stream Ciphers, pp. 151–162 (2008)
20. Kumar, K.J.J., Deepak, G.Y., Sivaji, S.S.: A novel cellular automata based stream cipher: Yugam-128. Int. J. Adv. Innovative Res. (IJAIR) **2**(1), 231–236 (2013)
21. Kumar, K.J.J., Sudharsan, S., Karthick, V.: FPGA implementation of cellular automata based stream cipher: Yugam-128. Int. J. Adv. Res. Electr. Electron. Instrum. Eng. 3(3) (2014)

Detecting Spam Emails/SMS Using Naive Bayes, Support Vector Machine and Random Forest

Vasudha Goswami[✉], Vijay Malviya, and Pratyush Sharma

RGPV Bhopal, Bhopal, India
vasudha.goswami30@gmail.com,
{vijaymalviya,hod_cs}@mitindore.co.in

Abstract. SMS spams are dramatically increasing year by year because of the expansion of movable users round the world. Recent reports have clearly indicated an equivalent. Mobile or SMS spam may be a physical and thriving drawback because of the actual fact that bulk pre-pay SMS packages are handily obtainable recently and SMS is taken into account as a trusty and private service. SMS spam filtering may be a relatively recent trip to deal such a haul. The amount of information traffic moving over the network is increasing exponentially and therefore the devices that are connected thereto are considerably vulnerable. Thus there's a bigger have to be compelled to secure our system from this kind of vulnerability, here network security play a really vital role during this context. In this paper, a SMS spams dataset is taken from UCI Machine Learning repository, and after perform pre-processing and different machine learning techniques such as random forest (RF), Naive Bayes (NB), Support Vector Machine (SVM) are applied to the dataset are applied and compute the performance of these algorithms.

Keywords: Data mining · Review spam · Classification · Comparative analysis · Spam detection · Sentiment analysis

1 Introduction

Short Message Service (SMS) is that the most often and wide used communication medium. The term "SMS" is employed for each the user activity and every one sorts of short text electronic messaging in several components of the planet. it's become a medium of promotion and promotion of product, banking updates, agricultural data, flight updates and net offers. SMS is additionally utilized in direct called SMS marketing. Typically SMS promoting could be a matter of disturbance to users. These types of SMSs area unit referred to as spam SMS. Spam is one or a lot of uninvited messages, that is unwanted to the users, sent or announce as a part of a bigger assortment of messages, all having considerably identical content. The needs of SMS spam area unit promotion and promoting of assorted product, causing political problems, spreading inappropriate adult content and net offers. That's why spam SMS flooding has become a significant downside everywhere the planet. SMS spamming gained quality over alternative spamming approaches like email and twitter, thanks to

© Springer Nature Switzerland AG 2020
J. S. Raj et al. (Eds.): ICIDCA 2019, LNDECT 46, pp. 608–615, 2020.
https://doi.org/10.1007/978-3-030-38040-3_69

the increasing quality of SMS communication. However, gap rates of SMS area unit beyond ninetieth and opened inside quarter-hour of receipt whereas gap rate in email is just 20–25% inside twenty four hours of receipt. Thus, a correct SMS spam detection technique has important necessity. There are many researches on email, twitter, net and social tagging spam detection techniques. However, a awfully few researches are conducted on SMS spam detection.

2 Literature Review

According to [1], conducted a survey of ways and applications for detection and filtering uninvited advertising messages or spam in a very telecommunication network [2]. Building up a classification algorithmic [3] program that channels SMS spam would provide useful equipment for portable suppliers. Since naïve mathematician has been used effectively for email spam detection, it seems to be expected that it may likewise be accustomed build SMS spam classifier. With reference to email spam, SMS spam represents further difficulties for automatic channels [4, 6].

3 Problem Definition

The fact that an email box can be flooded with unsolicited emails makes it possible for the account holder to miss an important message; thereby defeating the purpose of having an email address for effective communication. These junk emails from online marketing campaigns, online fraudsters among others is one of the reasons for this paper. We try to obtain the feature sets that can best represent and distinguish the spams from ham(non-spam). We then follow both supervised and unsupervised methodology to obtain spams from the dataset. We also include sentiment analysis methodology into our spam detection.

4 Proposed Work

A machine learning techniques have been proposed for detecting and classify the reviews through various processing steps which is shown in Fig. 1.

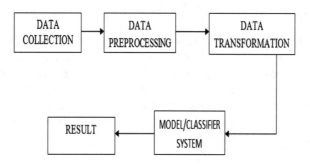

Fig. 1. Proposed flow diagram

Our Steps or Algorithm Steps will follow:

1. Data Collection:- We will use a dataset from the dataset repository of Centre for Machine Learning and Intelligent Systems at the University of California, Irvine!.
2. Data Preprocessing: Data preprocessing is the most important phase in detection models as the data consists of ambiguities, errors, redundancy which needs to be cleaned beforehand.
3. Data Transformation: Data is transformed into lowercase and change the data types according to algorithm needs.
4. Classification System: The attributes are identified for classifying process and system perform feature extraction and then these classification system classify the content into spam or ham.

5   Experimental Analysis

The experimental and result analysis is done by using intel i5-2410M CPU with 2.30 GHz processor along with 4 GB of RAM and the windows operating system is running. For result analysis we use R and R studio for processing the data and then we load the sms dataset which consist a 5574 observation with no missing are present in the dataset. Figure 2 shows the dataset has been loaded.

```
> data_text <- read.delim("SMSSpamCollection", sep="\t", header=F, colClasses="chara
cter", quote="")
>
> str(data_text)
'data.frame':    5574 obs. of  2 variables:
 $ V1: chr  "ham" "ham" "spam" "ham" ...
 $ V2: chr  "Go until jurong point, crazy.. Available only in bugis n great world la
e buffet... Cine there got amore wat..." "Ok lar... Joking wif u oni..." "Free entr
y in 2 a wkly comp to win FA Cup final tkts 21st May 2005. Text FA to 87121 to recei
ve entry question("| _truncated_ "U dun say so early hor... U c already then sa
y..." ...
>
> head(data_text)
    V1
1  ham
2  ham
3 spam
4  ham
5  ham
6 spam
                                                                              V2
1                            Go until jurong point, crazy.. Availab
le only in bugis n great world la e buffet... Cine there got amore wat...
2
                                                   ok lar... Joking wif u oni...
3 Free entry in 2 a wkly comp to win FA Cup final tkts 21st May 2005. Text FA to 871
21 to receive entry question(std txt rate)T&C's apply 08452810075over18's
4
                                            U dun say so early hor... U c already then say...
5                    Nah I don't think he goes to usf, he lives around here though
6      FreeMsg Hey there darling it's been 3 week's now and no word back! I'd like
some fun you up for it still? Tb ok! XXX std chgs to send, £1.50 to rcv
```

Fig. 2. Loading a dataset

After loading, For easy identification of the columns, we rename V1 as Class and V2 as Text. And we have to also convert the Class column from Character strings to factor. Data often come from different sources and most of the time don't come in the right format for the machine to process them. Hence, data cleaning is an important aspect of a data science project. In text mining, we need to put the words in lowercase, remove stops words that do not add any meaning to the model etc. Various data cleaning steps are shown in Fig. 3.

```
            ham      spam
      0.8659849 0.1340151
      > library(tm)
      Loading required package: NLP
      >
      > library(Snowballc)
      > corpus = VCorpus(VectorSource(data_text$Text))
      > as.character(corpus[[1]])
      [1] "Go until jurong point, crazy.. Available only in bugi n great world la e buffet... Cine th
      ere got amore wat..."
      >
      > corpus = tm_map(corpus, content_transformer(tolower))
      > corpus = tm_map(corpus, removeNumbers)
      > corpus = tm_map(corpus, removePunctuation)
      > corpus = tm_map(corpus, removeWords, stopwords("english"))
      > corpus = tm_map(corpus, stemDocument)
      > corpus = tm_map(corpus, stripWhitespace)
      > as.character(corpus[[1]])
      [1] "go jurong point crazi avail bugi n great world la e buffet cine got amor wat"
      >
      >
      > #Creating the Bag of Words for the model
      >
      > dtm = DocumentTermMatrix(corpus)
      > dtm
      <<DocumentTermMatrix (documents: 5574, terms: 6981)>>
      Non-/sparse entries: 43801/38868293
      Sparsity           : 100%
      Maximal term length: 40
      Weighting          : term frequency (tf)
      >
      > dtm = removeSparseTerms(dtm, 0.999)
      > |
```

Fig. 3. Data pre-processing

In text mining, it is important to get a feel of words that describes if a text message will be regarded as spam or ham. What is the frequency of each of these words? Which word appears the most? In other to answer this question; we are creating a DocumentTermMatrix to keep all these words. We want to words that frequently appeared in the dataset. Due to the number of words in the dataset, we are keeping words that appeared more than 60 times. We will like to plot those words that appeared more than 60 times in our dataset. Figure 4 shows the wordcloud of the dataset.

Fig. 4. Presenting the word frequency as a word cloud

Usually in Machine Learning is to split the dataset into both training and test set. While the model is built on the training set; the model is evaluated on the test set which the model has not been exposed to before. We will be building our model on 3 different Machine Learning algorithms which are Random Forest, Naive Bayes and Support Vector Machine for the purpose of deciding which perform the best.

Random Forest

The Random Forest Model is an ensemble method of Machine Learning with which 300 decision trees were used to build this model with the mode of the outcomes of each individual trees taken as the final output. So we can train the random forest model on training dataset and then test the performance of the model on testing dataset which is shown in Fig. 5.

```
>
> # Predicting the Test set results
> rf_pred = predict(rf_classifier, newdata = test_set[-1210])
>
> # Making the Confusion Matrix
> library(caret)
Loading required package: lattice
>
> confusionMatrix(table(rf_pred,test_set$Class))
Confusion Matrix and Statistics


rf_pred  ham spam
   ham  1191   36
   spam    4  154

               Accuracy : 0.9711
                 95% CI : (0.9609, 0.9793)
    No Information Rate : 0.8628
    P-Value [Acc > NIR] : < 2.2e-16

                  Kappa : 0.8687

 Mcnemar's Test P-Value : 9.509e-07

            Sensitivity : 0.9967
            Specificity : 0.8105
         Pos Pred Value : 0.9707
         Neg Pred Value : 0.9747
             Prevalence : 0.8628
         Detection Rate : 0.8599
   Detection Prevalence : 0.8859
      Balanced Accuracy : 0.9036

       'Positive' Class : ham
```

Fig. 5. Performance measure of Random Forest

Naive Bayes

Naive Bayes Classifier is a Machine Learning model that is based upon the assumptions of conditional probability as proposed by Bayes' Theorem. It is fast and easy and the performance outcomes of the model on testing dataset are shown in Fig. 6.

Support Vector Machine

The Support Vector Machine is another algorithm that finds the hyperplane that differentiates the two classes to be predicted, ham and spam in this case; very well. SVM can perform both linear and non-linear classification problems. The figure shows the performance outcomes of the model on the testing dataset (Fig. 7).

Performance Measure

We used accuracy, which are derived using confusion matrix (Tables 1, 2 and Fig. 8).

```
> nb_pred = predict(classifier_nb, type = 'class', newdata = test_set)
>
> confusionMatrix(nb_pred,test_set$Class)
Confusion Matrix and Statistics

          Reference
Prediction ham spam
      ham 1195    7
     spam    0  183

              Accuracy : 0.9949
                95% CI : (0.9896, 0.998)
    No Information Rate : 0.8628
    P-Value [Acc > NIR] : < 2e-16

                 Kappa : 0.9783

 Mcnemar's Test P-Value : 0.02334

           Sensitivity : 1.0000
           Specificity : 0.9632
        Pos Pred Value : 0.9942
        Neg Pred Value : 1.0000
            Prevalence : 0.8628
        Detection Rate : 0.8628
  Detection Prevalence : 0.8679
     Balanced Accuracy : 0.9816

      'Positive' Class : ham
```

Fig. 6. Performance measure of Naive bayes

```
> svm_pred = predict(svm_classifier,test_set)
>
> confusionMatrix(svm_pred,test_set$Class)
Confusion Matrix and Statistics

          Reference
Prediction ham spam
      ham 1195  189
     spam    0    1

              Accuracy : 0.8635
                95% CI : (0.8443, 0.8812)
    No Information Rate : 0.8628
    P-Value [Acc > NIR] : 0.4882

                 Kappa : 0.009

 Mcnemar's Test P-Value : <2e-16

           Sensitivity : 1.000000
           Specificity : 0.005263
        Pos Pred Value : 0.863439
        Neg Pred Value : 1.000000
            Prevalence : 0.862816
        Detection Rate : 0.862816
  Detection Prevalence : 0.999278
     Balanced Accuracy : 0.502632

      'Positive' Class : ham
```

Fig. 7. Performance measure of SVM

Table 1. Confusion matrix

	Classified as normal	Classified as attack
Normal	TP	FP
Attack	FN	TN

Table 2. Accuracy of the models

Model	Accuracy
Random Forest	97.11%
Naive Bayes	99.49%
Support Vector Machine	86.35%

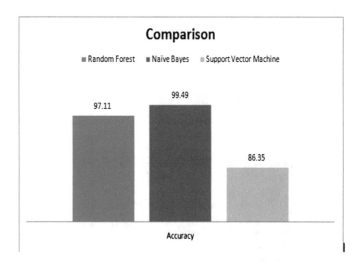

Fig. 8. Accuracy comparison

6   Conclusion

In this paper, we propose a machine learning technique for SMS Spam filtering based on algorithms namely Naïve Bayes, Support vector machine and Random Forest. The dataset that we have used in our work consists of 5574 observations of 2 variables. The first variable is the content of the emails and the second variable the target variable, which is the class to be predicted. In this paper, The Random Forest and Naive Bayes performed exceptionally well as compared to SVM.

References

1. Navaney, P., Dubey, G., Rana, A.: SMS spam filtering using supervised machine learning algorithms. IEEE (2018)
2. Shrivastava, S., Anju, R.: Spam mail detection through data mining techniques. IEEE (2017)
3. Shafi'i, M.A., Latiff, M.S.A., Chiroma, H., Osho, O., Abdul-Salaam, G., Abubakar, A.I., Herawan, T.: A review on mobile SMS spam filtering techniques. IEEE (2017)
4. Jindal, N., Liu, B.: Mining comparative sentences and relations. In: AAAI, vol. 22 (2006)
5. Nitin, J., Liu, B.: Review spam detection. In: Proceedings of the 16th International Conference on World Wide Web, pp. 1189–1190. ACM Press, New York (2007)

6. Nitin, J., Liu, B.: Opinion spam and analysis. In: Proceedings of the 2008 International Conference on Web Search and Data Mining, pp. 219–230. ACM Press, New York (2008)
7. Xie, S., Wang, G., Lin, S., et al.: Review spam detection via temporal pattern discovery. In: Proceedings of the 18th ACM SIGKDD International Conference on Knowledge Discovery And Data Mining, pp. 823–831. ACM Press, New York (2012)
8. Lim, E.-P., Nguyen, V.-A., Jindal, N., et al.: Detecting product review spammers using rating behaviors. In: Proceedings of the 19th ACM International Conference on Information and Knowledge Management, pp. 939–948. ACM Press, New York (2010)
9. Jindal, N., Liu, B., Lim, E.-P., et al.: Finding unusual review patterns using unexpected rules. In: Proceedings of the 19th ACM International Conference on Information and Knowledge Management, pp. 1549–1552. ACM Press, New York (2010)
10. Wang, G., Xie, S., Liu, B., et al.: Identify online store review spammers via social review graph. ACM Trans. Intell. Syst. Technol. 3(4), 61.1–61.21 (2011)
11. Indurkhya, N., Damerau, F.J.: Handbook of Natural Language Processing, 2nd edn. Chapman and Hall/CRC, London (2010)
12. Ohana, B., Tierney, B.: Sentiment classification of reviews using SentiWordNet. In: 9th IT & T Conference, p. 13 (2009)

Feature Based Opinion Mining on Hotel Reviews Using Deep Learning

Kavita Lal$^{(\boxtimes)}$ and Nidhi Mishra

Poornima University, Jaipur, India
kavitaroy2124@gmail.com, nidhi.mishra@poornima.edu.in

Abstract. Social media and networks are being used excessively these days for commenting on any news, product, services etc. We have Facebook, Twitter, LinkedIn for sharing of information with others. The data on these social media sites are in the form of text and everyday many users are commenting on these networks hence we are producing zettabytes of data each day. These data need to be managed properly so that they can be used for the benefit of companies, product manufacturers etc. The analysis of data can be done and find whether people are commenting in favor or against any particular product or service. This is known as mining of opinions. The analysis of priorities of customer's, their needs and their attitude towards any service or product, analyzing and extracting data from reviews of customers is the primary goal of this paper. For targeting the mentioned goal, the research has focused on the approach of deep learning and NN to find the polarity of reviews of customers in the Hotel domain. Research in this dissertation explores new techniques to aggregation, automated analysis, and extraction of opinions and features of customer reviews from text by using data mining and natural language processing techniques. It focuses on aspect-based opinion mining of customer reviews from hotel booking websites. It discusses about customer reviews characteristics and describes different approaches to extract aspects and their corresponding sentiments.

The results of this research show that the proposed model using CNN networks, is capable to find the score as well as polarity of reviews with 98.22% accuracy of combined reviews, 95.345% accuracy for positive reviews and 96.145% accuracy for negative reviews and the output obtained is comparatively more accurate than other given techniques.

Keywords: Deep learning · Machine learning · Neural network · Natural language processing · Opinion mining

1 Introduction

Computer technologies had undergone major advancements in last decades. During last few years analysis of natural languages has become important as Machine Learning, Deep Learning, Artificial Intelligence has arrived. With Machine Learning the machines were made learned and predict the results but again the problem came of same interpretation as humans. But with arrival of Neural Networks and Artificial Intelligence, they enabled the machines to comprehend as humans [7]. Natural Language Processing (NLP) is utilized to evaluate the natural languages and find the

© Springer Nature Switzerland AG 2020
J. S. Raj et al. (Eds.): ICIDCA 2019, LNDECT 46, pp. 616–625, 2020.
https://doi.org/10.1007/978-3-030-38040-3_70

context behind the reviews. The attitude of customers regarding service or product features, sets the base for the businesses to get best information to remove the defects in the services or products provided and generate new products with customers contentment [9]. Hence, it is important to know the opinion of customers as positive, negative or neutral. In this respect, a more particular method should be used as compared to other methods available for analyzing the reviews and finding the polarity. Taking into consideration the points mentioned above, this paper delves into obtaining the polarity of opinions of customers using the approach of deep learning [10]. Section 2 deals with opinion mining and deep learning key terms and basic concepts. Section 3 is related with literature review. Section 4, proposes an explanation based on approach of deep learning. In Sect. 5 dataset of hotel reviews is taken and the proposed model is evaluated and analyzed on it. Last Section provides with the final conclusion of the research.

2   Basic Concepts

2.1   Opinion Mining

Mining of opinions is a modern field of Natural Language Processing that categorizes subjective text into negative or positive or neutral.

Mining of Opinions at various Levels:-

- At Document Level: Opinion holders sentiment helps in classifying a document like reviews, blogs. It is assumed that every document emphasizes on one object and contains opinions of a sole opinion holder. Its main aim is to find the sentiment of the document whether they are positive, negative or neutral [3].
- At Sentence Level: In this level each sentence is considered to be a single unit and it should contain only one opinion [3].
- At Feature/Aspect Level: The objective of opinion mining at the level of features is to develop opinion summary of multiple reviews based on features. It consists of 3 tasks. The first one is identifying & extracting features of objects which has been reviewed by an opinion holder (e.g. "room", "spacious"). The second one is identifying the polarity of opinions on the basis of features that is positive, neutral and negative. Lastly, final task is to group synonyms of features [15].

2.2   Natural Language Processing (NLP)

Natural Language Processing mainly concentrates on the analysis of natural languages in order to find the polarities. The main goal of Natural Language Processing is collecting information and understanding how humans work with language and develop computer programs which are enabled for processing and understanding language in a similar manner [9].

2.3   CNN

It is a neural network that consists of bias and learning weights. Convolutional neural network has proved very effective in area of images recognition, text analytics. This is a fully connected network which uses SVM or softmax function as loss function. It works with multiple different layer which performs uniquely in the network processing. The layers are:-

The Convolutional layer – It is the first layer in CNN network and takes the array of pixel as input.

Pooling layer – This layer is the another mostly used layer in convolutional neural network. This layer is responsible for reduction of spatial size of the representation tin order to reduce the amount of the parameters and computation in the network. Fully connected layer – This is the final layer in the CNN. In this layer every neuron has full connection to all activations of the previous layers [8].

2.4   Deep Learning

Deep Learning is a technique of Machine learning that enables a system to do what a human can do naturally. Driverless cars, systems that can recognize stop signs, voice controlling systems, home assistance devices are the systems which implement deep learning. Deep Learning systems can do classification job by directly learning from text, images and sound [12].

3   Literature Review

Performing literature analysis, it has been found that many researchers had used deep learning approach for determining the polarity of reviews while some had used non deep methods. In research [2] the authors had collected 3000 natural audio streams of 10 users and converted into text. First data collection was done and then part of Speech tagging was done. To access the score from SentiWordNet, PennTreebank tags were used. Then 3-layer LSTM was applied with drop rate of 0.5. As per experimental results it was found that LSTM with unigram gave better results with accuracy of 84.5%. In [4] the authors had taken multilabel datasets. First data preprocessing was done, after that text encoding was done. Then all the learning models like CNN-1, CNN-V, CNN-I, Bi directional LSTM and Bi GRU were implemented. As per experimental results BiGRU gave highest Precision of 0.65, Recall of 0.75 and F1 score of 0.7 and was able to find the toxic sentiments very well. In paper [10] authors had collected data from two domains restaurants and laptop in which restaurants training data was 3844 and testing data was 100 and for laptop training data was 3845 and testing data was 100, then pre-processing was done to prepare the data for deep learning. The length of sentences was kept 300 by padding. Then RNN and LSTM was implemented and F1 measure, Precision and accuracy was found out. As per experimental results. For restaurant data F1 measure was 89.25, precision was 85, Recall was 84.33 and Accuracy was 97.87 and for Laptop data F1 measure was 96.72, Precision was 84.62, Recall was 84.43 and Accuracy was 96.72. It was found that the proposed system gave a suitable platform which improved evaluation measures when compared to other methods. In paper [13] the given system concentrated on collecting the data from Twitter. It consisted of data collected first, Pre-processed data, features were extracted and then classified and finally result was evaluated. 1,048,588 English tweets were used for this work. Accuracy, sensitivity and specificity was found out using various ML algorithms like Naïve Bayes, Random Forest, Decision Tree, RNN-LSTM, CNN etc. RNN classification algorithm provided the accuracy of 83.6% which was highest and sensitivity was 87.1% and specificity was 79.3%.

4 Proposed Approach

With respect to research goal of calculating polarity of customer's opinions in hotel domain based on the approach of deep learning; first task will be introduction of dataset. Then data pre-processing is done, after which features are extracted and dataset is prepared to enter the neural network. The process flow diagram has been explained in Fig. 1.

4.1 Dataset Introduction

The data is collected from online site www.Booking.com which consists of data from various hotels. A web scraper was run and data was collected. Data was stored in CSV file format. The number of positive and negative reviews collected are shown in Table 1. Training dataset and testing dataset are shown in Table 2. Table 3 shows sample of Positive reviews and Table 4 shows sample of negative reviews.

Table 1. Proposed dataset

Positive reviews	Negative reviews	Total reviews
2000	1000	3000

Table 2. Size of dataset used

Domain	Training dataset	Testing dataset	Total
Hotel	2700	300	3000

Table 3. Sample of positive reviews

S. no.	Positive reviews
1	Only the park outside of the hotel was beautiful
2	No real complaints. The hotel was great. Great location surroundings. Room amenities and service were awesome
3	Location was good and staff were ok. It is cute hotel. The breakfast range is nice. Will go back
4	Great location and nice surroundings. The bar and restaurant are nice and have a lovely outdoor area
5	Amazing location and building

Table 4. Sample of negative reviews

S. no.	Negative reviews
1	My room was dirty and I was afraid to walk barefoot on the floor which looked as if it was not cleaned in weeks
2	Backyard of the hotel is total mess. It shouldn't happen in hotel with 4 stars
3	Apart from the price for the breakfast everything was very good
4	Very steep steps in room up to the bed not safe for children. I asked to move rooms and was put in another identical one
5	We did not like the fact that breakfast was not included although you could pay extra for this. The room we stayed in was lacking in space a bit and the bathroom was very badly lit

4.2 Data Pre-processing

Python programming language will be used for pre-processing operation. Before the data preprocessing, the reviews are collected with the help of web scraping and stored in database. The preprocessing of reviews is very essential step for data preparation in opinion mining which includes the basic tasks as:

- Tokenization
- Stop Word Removal
- Lemmatization
- Stemming
- Removing mentions
- Removing one hot sequence.

Here the sentence will be tokenized that is all the tokens will be removed. After which stop words like 'The', 'was', 'were' etc. will be removed. Then Lemmatization and stemming will be done to convert the words into the root words and all the mentioned symbols will be removed like &.@.% etc. Lastly the data will be converted into integers.

4.3 Feature Extraction

In Python Tfidf and CountVectorizer will be used to extract the features. CountVectorizer will count the number of features and Tfidf decides which word has a meaning related to domain (Table 5).

Table 5. Features considered

Hotel	Room	Food	Room service	Bar	Location	Wifi	Bathroom	Staff	AC	Lift	Bed

4.4   Different Cases Handled in Feature Level Opinion Mining

In the proposed work we have used PyCharm where the complete scraped dataset is stored. The complete work has been discussed over here. For considering the reviews Conjunction handling, Negation handling and Intensifier handling has been taken into consideration. Table 6 consists of conjunctive words used. Table 7 gives sample of sentences where conjunctive words are used. Table 8 discusses how negation handing is done and score is calculated. Table 9 gives sample of sentences where intensifier handling is done.

Table 6. List of conjunctive words

Conjunctive words	So, and, nor, but, yet

Table 7. Sample of coordinating conjunction handling in reviews

Input	Output
Example Our entire stay was so relaxing and we will definitely stay there again	Line 1: our entire stay was so relaxing Line 2: we will definitely stay there again
Example: Air conditioner did not work in the room so it was like sleeping in an oven. Location is quite terrible and rooms are not nice	Line 1: air conditioner did not work Line 2: it was like sleeping in an oven Line 3: location is quite terrible Line 4: rooms are not nice

Table 8. Negation handling in reviews

Input		Compound score	Polarity
WiFi is not free		−0.4023	Negative
No lift and narrow staircase. Room is so small	Line 1: No lift and narrow staircase Line 2: Room is so small	−0.396	Negative
There was nothing wrong everything was perfect. Location was nice and staff was always ready to assist	Line 1: There was nothing wrong everything was perfect Line 2: Location was nice Line 3: Staff was always ready to assist	0.8848	Positive

Table 9. Intensifier handling in reviews

Input	Score	Polarity
The food here is tasty	0.49260	Positive
The food here is tasty!!	0.5399	Positive
The rooms were clean	0.4404	Positive
The rooms were extremely clean	0.4927	Positive
Room service was pathetic	−0.6344	Negative
Room service was very pathetic	−0.6932	Negative

4.5   Used Hardware and Software

Table 10 discusses the hardware configuration used.

Table 10. Hardware configuration

Features name	Features detail
Brand	Dell
Model used	Inspiron 5570
Size of RAM	8 GB
Technology of RAM memory	DDR4
Size of hard drive	2 TB
Operating system	Window 10
Processer used	Intel Core i7

Windows 10 is the operating system used. For finding polarity of reviews Anaconda – a collection of Python modules is used. In this research Python version 3 and related libraries like Keras, spacy has been used to create neural networks. PyCharm has been used as a code editor.

4.6   Implementation of Proposed Model

In this research CNN has been implemented to find the polarity of opinions. Figure 1 Shows the process flow diagram after which the model will be evaluated to check for the accuracy and compared with other methods like Naïve Bayes, Random forest, SVM.

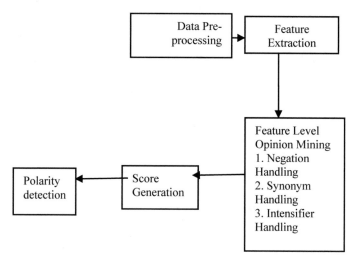

Fig. 1. Process flow diagram

4.7   Evaluation of Proposed Model

In the proposed work we are implementing CNN's three layers. The output received from the 3 CNN layers is joined to a layer called Dense layer which is a completely connected layer and Relu is the activation function applied. Then the output is fed to a Dropout.

Table 11. Parameters defined for polarity calculation

Implementation method	CNN
Epoch done	20
Batch size	512
Inter layers activation function	Relu Softmax
Lost function	categorical_crossentropy
Optimizer function	Rmsprop
Validation set	1000

5   Results

Evaluating the results in line with the hotel domain on the basis of Neural Network given in Table 11, the accuracy has been obtained separately for positive reviews as 95.345%, for negative reviews as 96.145% and for combined reviews as 98.22%. When compared to other models the accuracy of each model obtained is given in Table 12.

Figure 2 shows comparative analysis of different models and proposed model.

Table 12. Comparison in accuracy obtained from different models.

Model	Accuracy of positive reviews	Accuracy of negative reviews	Accuracy of combined reviews
Naïve Bayes	–	–	95.986%
SVM	–	–	96.113%
Random forest	–	–	92.345%
Proposed model	95.345%	96.145%	98.22%

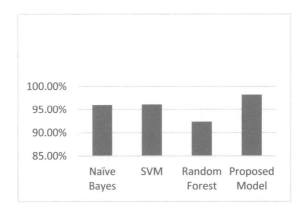

Fig. 2. Comparative analysis of proposed and different models

6 Conclusion

When deep learning is implemented in obtaining the polarity of reviews, it gave better results than other models like Naïve Bayes, SVM and Random Forest. We have collected online hotel reviews from www.booking.com website. The Design and Implementation of the system for Feature Level Opinion Mining on Hotel Reviews has been carried out using different cases and scenarios. The proposed system deals with single reviews and multiple reviews both. Three cases are handled for single feature review as well as multiple feature reviews as Negation Handling, Intensifier Handling and conjunction handling.

The performance analysis has been carried out for Accuracy. Proposed system achieved the performance as 98.22% accuracy. It could be found that the proposed system's performance is better than the existing models performance.

Finally Future work has also been discussed as, to deal with the double intensifiers and to include neutral reviews also.

References

1. Chachra, A., Mehndiratta, P., Gupta, M.: Sentiment analysis of text using deep convolution neural networks. In: 10th International Conference on Contemporary Computing (IC3), Noida, pp. 1–6 (2017)
2. Dash, A.K., Pradhan, R., Rout, J.K., Ray, N.K.: A constructive model for sentiment analysis of speech using deep learning. In: International Conference on Information Technology (ICIT), Bhubaneswar, pp. 1–6 (2018)
3. Hnin, C.C., Naw, N., Win, A.: Aspect level opinion mining for hotel reviews in Myanmar language. In: IEEE International Conference on Agents (ICA), Singapore, pp. 132–135 (2018)
4. Nanda, C., Dua, M., Nanda, G.: Sentiment analysis of movie reviews in Hindi language using machine learning. In: International Conference on Communication and Signal Processing (ICCSP), Chennai, pp. 1069–1072 (2018)
5. Momin, H.G., Kondhawale, C.S., Shaikh, R.E., Gawandhe, K.G.: Feature based evaluation of hotel reviews and ratings for differently abled people. In: International Conference on Energy, Communication, Data Analytics and Soft Computing (ICECDS), Chennai, pp. 2452–2454 (2017)
6. Saeed, H.H., Shahzad, K., Kamiran, F.: Overlapping toxic sentiment classification using deep neural architectures. In: IEEE International Conference on Data Mining Workshops (ICDMW), Singapore, pp. 1361–1366 (2018)
7. Kaur, J., Sidhu, B.K.: Sentiment analysis based on deep learning approaches. In: 2nd International Conference on Intelligent Computing and Control Systems (ICICCS), Madurai, pp. 1496–1500 (2018)
8. Akhtyamova, L., Alexandrov, M., Cardiff, J.: Adverse drug extraction in twitter data using convolutional neural network. In: 28th International Workshop on Database and Expert Systems Applications (DEXA), Lyon, pp. 88–92 (2017)
9. Liu, B.: Sentiment analysis and subjectivity. Handb. Nat. Lang. Process. **2**, 627–666 (2010)
10. Bavakhani, M., Yari, A., Sharifi, A.: A deep learning approach for extracting polarity from customer's reviews. In: 5th International Conference on Web Research (ICWR), Tehran, pp. 276–280 (2019)
11. Bilal, M., Asif, S., Yousuf, S., Afzal, U.: 2018 Pakistan general election: understanding the predictive power of social media. In: 12th International Conference on Mathematics, Actuarial Science, Computer Science and Statistics (MACS), Karachi, pp. 1–6 (2018)
12. Day, M., Lin, Y.: Deep learning for sentiment analysis on Google play consumer review. In: IEEE International Conference on Information Reuse and Integration (IRI), San Diego, pp. 382–388 (2017)
13. Abd El-Jawad, M.H., Hodhod, R., Omar, Y.M.K.: Sentiment analysis of social media networks using machine learning. In: 14th International Computer Engineering Conference (ICENCO), Cairo, pp. 174–176 (2018)
14. Mehra, R., Bedi, M.K., Singh, G., Arora, R., Bala, T., Saxena, S.: Sentimental analysis using fuzzy and naive bayes. In: International Conference on Computing Methodologies and Communication (ICCMC), Erode, pp. 945–950 (2017)
15. Songpan, W.: The analysis and prediction of customer review rating using opinion mining. In: IEEE 15th International Conference on Software Engineering Research, Management and Applications (SERA), London, pp. 71–77 (2017)
16. Quan, W., Chen, Z., Gao, J., Hu, X.T.: Comparative study of CNN and LSTM based attention neural networks for aspect-level opinion mining. In: IEEE International Conference on Big Data (Big Data), Seattle, pp. 2141–2150 (2018)

Using Deep Autoencoders to Improve the Accuracy of Automatic Playlist Generation

Bhumil Jakheliya, Raj Kothari$^{(\boxtimes)}$, Sagar Darji, and Abhijit Joshi

Department of Information Technology Engineering,
Dwarkadas J. Sanghvi College of Engineering,
Mumbai 400056, Maharashtra, India
bhumil2210@gmail.com, rajk3770@gmail.com,
darjisagar7@gmail.com, abhijit.joshi@djsce.ac.in

Abstract. Manual creation of playlists is a trivial and time consuming task due to huge catalogue of songs available online. Automatic Playlist Generation systems are now available on various music platforms such as Spotify, Youtube, etc. Personalized playlists are generated by these systems considering likes and dislikes of users. These systems are developed by following two main approaches: Collaborative Approach and Content-based Approach. Systems based on collaborative approach requires enough user's data to generate accurate results while the accuracy of playlists generated using Content-based approach systems depend highly on features used from the dataset as well as learning algorithm. In this paper, the authors propose a hybrid model which leverages the benefits of above mentioned approaches. The proposed system uses Deep Autoencoders to extract the most important features of songs present in the dataset to form clusters. Data about user's playlist i.e. User-level data is combined with the clusters to build the hybrid model. This approach is validated with the help of prototype implementation and a survey of 100 users.

Keywords: Automatic Playlist Generation · Deep Autoencoders · Deep Learning · Machine Learning · K-Means · Deep neural network · Clustering · Songs · Music · Spotify · Silhoutte score

1 Introduction

Rapid growth in digital distribution of music has had both positive and negative implications on society. Songs are now accessible like never before and emerging artists can showcase their talent just by click of a button. On the other hand, due to the availability of huge catalogue of songs, it has become difficult for users to decide what to listen.

Music Platforms like Spotify and iTunes provide manually compiled playlists, which comprise of tracks that share similar genre or musical qualities. On the other hand, auto-generated radio stations use Machine Learning Clustering Algorithms for musical analysis. A song from user's history is taken as input parameter and a collection of similar songs is provided as output. Auto-generated radio stations suggests tracks according to user's history while manually compiled playlists consists of tracks from service provider's catalogue. This approach of generating playlist is known as

© Springer Nature Switzerland AG 2020
J. S. Raj et al. (Eds.): ICIDCA 2019, LNDECT 46, pp. 626–636, 2020.
https://doi.org/10.1007/978-3-030-38040-3_71

Content-based. Various playlists generated by these platforms are also based on user's data. The reason is, people having similar choices, listen to similar type of music. This approach of generating playlist is known as Collaborative-based [12]. We propose a system which would combine the benefits of both approaches and overcome their limitations.

The rest of the paper is organized as follows: Sect. 2 provides Literature review on existing systems. Section 3 explains the system implemented using Deep Autoencoders and hybrid model in detail. The section ends with experimental results obtained from our model. The paper ends with Conclusion and Future Scope.

2 Literature Review

2.1 Existing Approaches

Major approaches upon which automatic playlist generation systems are built, consists of Most Popular N Songs, Collaborative-based approach and Content-based approach. These approaches are explained in subsequent section.

2.1.1 Most Popular N Songs

It is naive approach for generating automatic playlists. Results are not personalized according to user and generated playlist consists of the N most popular songs with highest ratings/user hits among all songs.

2.1.2 Collaborative Approach

In Collaborative Approach, playlists are generated based on the similarity between different users. This similarity is based on user's preferences, history or activities. Data Collection can be explicit like user ratings or implicit, such as page views, clicks, number of times a song is listened to, etc. [3, 4].

To get a fair idea of Collaborative Approach, consider an example which is discussed shortly. Table 1 denotes a mapping between users and their ratings collected for various songs i.e. User A has rated Song 2 with 4 stars and user B has rated Song 4 as 3 stars, respectively. Question Marks (?) denotes user data, which is not available. It can be observed from Table 1 that Users A and D have similar ratings for songs 1, 2 and 3. This signifies that there is high similarity between likes and dislikes of both users. Models based on this approach can predict how much User A would like to listen to songs 4 and 5 [13].

Table 1. Ratings of various songs as given by a set of users

Users	Song 1	Song 2	Song 3	Song 4	Song 5
A	5	4	5	?	?
B	5	?	5	3	2
C	?	2	1	0	5
D	5	4	5	3	2
E	1	?	1	0	?

2.1.3   Content-Based Approach

This approach does not involve other users as seen in Collaborative Approach. The algorithm simply recommends similar items. A user profile is generated from the data provided by the user, either explicitly (rating) or implicitly (clicking links). The Suggestions are then made to the user based on his/her user profile. For example, for generating playlists, recommendations are provided according to user profile and features of songs present in the dataset. The accuracy of systems based on Content-based approach increases as more and more amount of data is collected from the user [5, 6].

2.2   Observations Based on Existing Approaches

Most Popular N Songs approach is not suitable for current scenario due to huge library of songs available in the market today and unique likes and dislikes of each person.

Accurate results are obtained from Collaborative Approach in many cases. However, Collaborative Approach faces some major drawbacks. High-dimensional Data consists of many inter-dependent attributes and therefore, the generated results can be highly unpredictable as the model does not take into account the content of attributes. It also faces the problem of cold start i.e. when a new data item arrives, the model needs enough information to give equal priority to the newly arrived data item. When a new song is added to the database, ratings from a large number of users has to be collected or when a new user joins the network, enough user history should be available before accurate results can be produced. Hence, collaborative system cannot be used for applications where dataset is changing continuously [4].

There is less diversity in the playlists generated by Content-based Approach. For example, consider a song recommender system, even if a user likes rock songs, but the user will never know this unless he/she decides to give a try on his own because this approach will keep recommending songs similar to user's history. This is one of the major drawbacks of Content based Approach systems. Another drawback is, the accuracy of Content-based filtering systems highly depend on the features selected from the dataset and the learning algorithm used. Thirdly, Content-based filtering systems don't take into account the current trends or current songs, which many users are listening [4, 5]. Now let us see some existing systems.

2.3   Existing Systems

These section describes some of the prominent existing systems such as Last.fm and youtube.

2.3.1   Last.fm

Last.fm, a music website, uses a recommendation system named "Audioscrobbler" in which a detailed user profile is built for every user by recording details of the tracks of the song user listens to, either from user's computer or Internet radio stations (digital audio service transmitted via the Internet). This data is transferred to Last.fm's database and Collaborative-level filtering is used to recommend new playlists [7].

2.3.2   Youtube

Youtube recommendation algorithm uses a two-stage approach. Candidate Generation and Ranking. A user profile is generated for every Youtube user based on videos previously watched by the user and search history. The Candidate Generation stage extracts videos watched by similar users. Ranking is performed on these selected videos considering factors such as how long ago the user watched video on the same topic, number of videos the user watched of same channel, etc. These ranked results are then recommended to users [8].

2.4   Observations on Existing Systems

Last.fm and Youtube both use Collaborative-level filtering to generate playlists. Youtube uses an additional stage known as Ranking to filter among it's huge catalogue of videos. Hence, these systems face drawbacks of Collaborative Approach mentioned above. Purely Content-based systems are not available in the market today due to lack of personalization and various other drawbacks discussed above. The proposed system uses a hybrid model where in the drawbacks of existing systems are addressed. Let us see the proposed system.

3   Proposed System

Collaborative Approach faces issues with High-dimensional data and requires substantial amount of user's data to generate accurate results whereas the Content-based approach provides less diversity in the playlist generated [4].

To address these issues and concerns, we propose a hybrid model which leverages the benefits of both Collaborative and Content-based approaches simultaneously to overcome their drawbacks. The proposed system is divided into three modules: Database Creation, Data Exploration & Preprocessing, and Learning Models. Now let us see the role of these modules one by one.

3.1   Database Creation

The proposed system uses two datasets obtained from Spotify Music Platform [9, 10].

1. Spotify Songs Dataset, that consists of a list of one million songs along with their artist names. In order to collect various important features of songs such as danceability, loudness, energy, and tempo, Spotify API is used [9].
2. Spotify Playlist Dataset, which contains a number of playlists generated by Spotify Users [10].

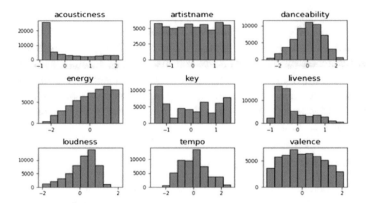

Fig. 1. Histograms of features representing distribution of data

3.2   Data Exploration and Preprocessing

Various Plots such as Histograms (See Fig. 1) and Box Plot (See Figs. 2 and 3) were used to visualize the data distribution. The authors observed from these plots that the data distribution is skewed and in different ranges. Outliers are detected with the help of box plots and categorical data such as artist name is converted to numeric data using one-hot encoding. The preprocessing steps performed to remove outliers, duplicates and normalize various features of dataset are explained below.

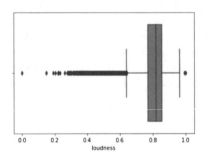

Fig. 2. Box Plot of loudness feature   **Fig. 3.** Box Plot of tempo feature

Removal of Noise/Outliers and Duplicates: Outliers are extreme values that deviate from other percentiles i.e. Q3 and Q1. Data points less than $Q1 - 1.5 * IQR$ and greater than $Q3 + 1.5 * IQR$ were found to be outliers and removed from the dataset. Duplicates present in the dataset are also removed.

Normalization: As seen in Fig. 1, certain features of the dataset such as acousticness, energy and liveness consists of highly skewed data. To obtain high accuracy and precision, data should be evenly distributed. To ensure this, each value of a feature is subtracted by it's mean and then divided by its standard deviation for normalization.

This is known as Z-score Normalization. For example, feature acousticness is normalized as follows:

$$train\_df['acousticness'] = \frac{(train\_df['acousticness'] - train\_df['acousticness'].mean())}{train\_df['acousticness'].std()}$$

One-Hot Encoding: Machine Learning and Deep Learning Algorithms work well with numeric data than categorical data [14]. One-Hot encoding ensures that every artist in the dataset is given equal weightage by the model by converting categorical data to numeric data as each artist is now represented by a unique number.

After removing outliers and duplicates, the dataset is normalized followed by one-hot encoding which leads to a final cleaned dataset of 53,848 rows was. The rows are randomly shuffled to avoid any bias before being used for training and testing.

3.3   Learning Models

First Step is to create Content based Learning Model. Two models are used for experiment purpose:

1. Clustering on features of tracks using K-Means Algorithm
2. Using Deep Autoencoders to extract most relevant features of the tracks and then performing Clustering on them using K-Means Algorithm.

3.3.1   Clustering Model

To generate playlist of similar songs based on their features, clustering is used to create subgroups of dataset. In our system, k-means clustering algorithm is used. K-means is one of the most popular algorithms to be used for clustering. The aim of K-means is to minimize inter-cluster distance, maximizing intra-cluster distance [12]. To perform clustering using K-Means, main challenge is to determine the value of K. Various values of K are experimented and measures such as Inter-cluster similarity, Intra-cluster similarity and Silhouette scores are used to determine the quality of clusters formed.

Inter-cluster similarity is calculated by measuring euclidean distance between features of input song and features of songs belonging to the same cluster. Intra-cluster similarity is calculated by measuring the Euclidean distance between features of Input song and centroids of all other clusters excluding its own cluster.

Various values of K are experimented from 100 to 10000 and the quality of generated clusters is checked using these measures. Most closely related subgroups are obtained for K = 6000.

3.3.2   Deep Learning Model

The architecture of deep learning model is shown below in Fig. 4. The major components of the architecture are Autoencoder and Clustering modules.

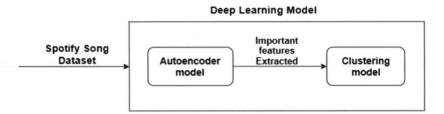

Fig. 4. Deep learning model (content-based)

Deep Learning Model - The function of the Deep learning model is to perform Content-based filtering on Spotify Song Dataset. This is implemented with the help of two sub-models: Autoencoder Model (Sect. 3.3.2.1) and Clustering Model (Sect. 3.3.2.2).

3.3.2.1 Autoencoder Model

Autoencoders are special cases of neural networks with the aim of generating output similar to input. Input is first compressed to a latent-space representation. Latent-space representation describes most important features need to be extracted to reconstruct the input. Deep Autoencoder Networks consist of two main parts: Encoder and Decoder.

The latent representation layer z adapts to useful properties when the autoencoder model is trained to generate output similar to input. These properties depends on various constraints of model such as dimensions of hidden layers, etc. [11].

The Autoencoder model used for this project is a 5-layer model which is trained for 1000 epochs (See Fig. 5). Autoencoder model is applied on dataset containing 9 features from which most important features representation is extracted, which are further used for clustering.

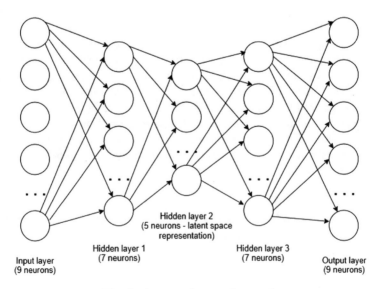

Fig. 5. Autoencoder neural network

Description of layers used in our models are as follows:

Encoding Layers
Layer 1: Dimensionality of Output Space as 9 with activation function Relu.
Layer 2: Dimensionality of Output Space as 7 with activation function Relu.

Latent-Space Representation
Layer 1: Dimensionality of Output Space as 5 with activation function sigmoid.

Decoding Layers
Layer 1: Dimensionality of Output Space as 7 with activation function Relu.
Layer 2: Dimensionality of Output Space as 9 with activation function Relu.

Output of this model has shape (53848, 5) where 53848 represents number of rows of training examples and 5 represents most important features learned by the latent-space representation layer. After training for 1000 epochs, the training set loss is found and it is 0.1994. This output is then given to Clustering Model, which is discussed in the next section.

3.3.2.2 Clustering Model
K-Means Clustering model is used to perform clustering as mentioned in the previous section. Instead of features of song as input, output from autoencoder model is given as input. This leds to the formation of better clusters as only most relevant features extracted by Deep Autoencoder are used [12].

3.3.3   Results Obtained from Learning Models
Results obtained from both learning models are summarized in Table 2:

Inter-Cluster Similarity Percentage, Intra-Cluster Similarity Percentage and Silhouette score for clusters formed using Clustering and Deep learning model are calculated and presented in Table 2.

Table 2. Different measures used to check the performance of models

Model	Inter-cluster similarity %	Intra-cluster similarity %	Silhouette score
Clustering Model	78	53	0.20
Deep Learning Model	86	40	0.37

Inter-Cluster Similarity denotes similarity of data points belonging to the same cluster and is found to be higher in Deep Learning Model. Intra-Cluster Similarity denotes similarity between different clusters and is found to be lower (Refer Table 2) in Deep Learning Model indicating non-overlapping clusters. Higher Silhouette score indicating better clusters are formed by Deep Learning Model.

Few Clusters formed in both the model are also analysed manually to check similarity between tracks by listening to them. More similar tracks are found in clusters formed by Deep Learning Model. Hence, It is observed that Deep Learning Model outperforms Clustering Model and therefore used for Hybrid Implementation.

3.4   Hybrid Implementation

The architecture of the hybrid system is shown in Fig. 6. It consists of three main modules: Deep learning model, Current user's playlist data and Hybrid model. Let us see the working of these models now.

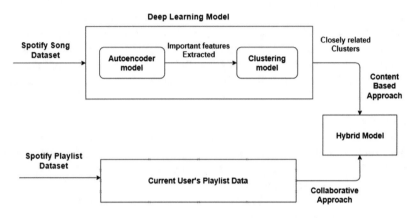

Fig. 6. System architecture

Deep Learning Model - It is explained in detail in the Sect. 3.3.2.

Current User's Playlist Database - Playlist listened by the active users around the world is extracted using Spotify API for collaborative Filtering.

Hybrid Model - It combines the results obtained from Deep Learning model (Content based Approach) and current user's playlist Database (Collaborative Approach) to generate a playlist.

Hybrid Model is a combination of Content-level filtering and Collaborative-level filtering. Content-level filtering is performed by Deep Learning Model, which forms Highly related clusters. To implement Collaborative-level filtering, second dataset mentioned in Database Creation step is used. As stated previously, the database contains a list of user's playlists obtained from Spotify. Playlists, in other words, can be understood as a group of songs which user's listen together i.e. user-level data and hence, can be used to implement Collaborative-level filtering.

To understand the working of our hybrid model, let us consider a song A as an input from user's history. The playlist generated by the hybrid model for song A would be a combination of songs obtained from cluster containing song A as well as from playlist containing song A. Hence, the result would be a combination of songs similar to input song (Content-based filtering) as well as from songs which other users generally listen together (Collaborative-level filtering).

3.5   Experimental Results

Figure 7 depicts experimental results obtained from our models. It takes two inputs from 'user, Track Name and Artist Name (from User's history) and then a playlist is generated by the hybrid model explained in proposed implementation section.

```
In [53]:  ▶  generate_playlist("Waiting For Love","Avicii")

              Generating Playlist...

              Track Name                      Artist
              -------------------------------------------------
              Wake Me Up                      Avicii

              Hey Brother                     Avicii

              Counting Stars                  One Republic

              A Sky Full Of Stars             Coldplay

              Faded                           Alan Walker

              All of Me                       John Legend

              Perfect                         Ed Sheeran

              Photograph                      Ed Sheeran
```

Fig. 7. A playlist generated by the implemented model

The list of tracks generated in the playlist are quite similar to the track given as input. To compare with user-data, these results are also compared with results generated by music platforms like Spotify along with Google and Youtube. It is found that the results are overlapping. To determine the quality of playlists generated, a survey among 100 people is conducted. The results of the survey are tabulated in Table 3.

Table 3. Results of survey

Question	Yes	No	Maybe
Were the playlist generated equal or better than the playlists generated by other music platforms?	39/100	35/100	26/100
Was the playlist generated as expected by you?	44/100	31/100	25/100
Would you like to see this automatic playlist generation system on leading music platforms?	36/100	43/100	21/100

As can be seen in Table 2, a high percentage of users said that the playlist generated was equal or better than that generated by other music platforms and met their expecations. Further, many users asserted that they would like to see this automatic playlist generation system on leading music platforms.

4 Conclusion and Future Scope

As can be seen in the experimental results, the playlist generated is quite accurate. Few tracks of the same artist and other similar tracks of different artists are suggested by the hybrid model. Silhouette score and Euclidean Distance measures indicated Deep Learning Model outperformed Clustering Model. Few playlists generated by the hybrid model are manually checked by listening to songs in the playlist. The results of the survey indicates that the playlists generated by the hybrid model are quite accurate though the accuracy of model may be improved by increasing features of song such as lyrics of song, type of instruments used, etc. Further, performing NLP analysis on the lyrics can help to form better clusters. Algorithms other than K-Means can also be used for Clustering. The number of hidden layers can also be modified to examine its impact on the final result.

Compliance with Ethical Standards

All author states that there is no conflict of interest.

Humans/Animals are not involved in this work.

We used our own data.

References

1. Lin, D., Jayarathna, S.: Automatic playlist generation from personal music libraries. In: 2018 IEEE International Conference on Information Reuse and Integration for Data Science (2018)
2. Chen, Y.-W., Xia, X., Shi, Y.-G.: A collaborative filtering recommendation algorithm based on contents' genome (2012)
3. Towardsdatascience Collaborative Filtering. https://towardsdatascience.com/intro-to-recommender-system-collaborative-filtering-64a238194a26
4. Medium Music Recommendation Systems. https://medium.com/@briansrebrenik/introduction-to-music-recommendation-and-machine-learning-310c4841b01d
5. Recommender-systems. http://recommender-systems.org/content-based-filtering/
6. Medium. https://medium.com/@rabinpoudyal1995/content-based-filtering-in-recommend-ation-systems-8397a52025f0
7. Wikipedia. https://en.wikipedia.org/wiki/Last.fm
8. Quora. https://www.quora.com/How-does-YouTubes-recommendation-algorithm-work
9. Millionsongdataset. http://millionsongdataset.com/
10. Spotify. https://developer.spotify.com/documentation/web-api/
11. Towardsdatascience. https://towardsdatascience.com/deep-inside-autoencoders-7e41f3199-99f
12. Towardsdatascience. https://towardsdatascience.com/k-means-clustering-algorithm-applica-tions-evaluation-methods-and-drawbacks-aa03e644b48a
13. Scikit Learn Documentation. https://scikitlearn.org/stable/modules/generated/sklearn.metrics.silhouette_score.html
14. MachineLearningmastery.com. https://machinelearningmastery.com/why-one-hot-encode-data-in-machine-learning/

Evolutionary Correlation Triclustering for 3D Gene Expression Data

N. Narmadha[✉] and R. Rathipriya

Department of Computer Science, Periyar University, Salem, India
mahanarmadha@gmail.com, rathipriyar@gmail.com

Abstract. Triclustering method is seen as an issue of an optimization problem to discover correlated triclusters with high MCV and high volume. To optimize the tricluster, the Genetic Algorithm (GA) is used. This work dealt with the mining of optimal shifting and scaling patterns from 3D microarray data in the form tricluster. Optimal Tricluster is the tricluster that satisfies the specified objective function. To test the performance, an empirical study of the triclustering algorithm will be attempted at the yeast cell cycle.

Keywords: Genetic algorithm · Yeast cell cycle data · Triclustering · MCV · 3D gene expression data

1 Introduction

The analysis of 3D gene expression data plays a crucial role in identifying the most valuable genes, along with the dimension of gene/sample/time or gene/condition/time dimension (Dede 2013). Analysis of 3D gene expression data represents a computational challenge due to the characteristic of these data. To analyze gene expression, a large number of clustering approaches are proposed (Kapil et al. 2016). Triclustering techniques are the valuable recent simultaneous clustering of genes, samples, and time points (G/S/T) (Gutiérrez-Avilés 2014). Tricluster is a subset of genes that have similar patterns under a subset of a condition along with a subset of time points. This paper discusses the triclustering of 3D gene expression data using a genetic algorithm to find the maximum optimal tricluster. Figure 1 shows the flow chart for the evolutionary triclustering algorithm using a genetic algorithm.

2 Literature Review

The overview of the background study needed for this research work is described in this section. Table 1 explains some of the existing algorithms with GA for clustering, biclustering and triclustering techniques

© Springer Nature Switzerland AG 2020
J. S. Raj et al. (Eds.): ICIDCA 2019, LNDECT 46, pp. 637–646, 2020.
https://doi.org/10.1007/978-3-030-38040-3_72

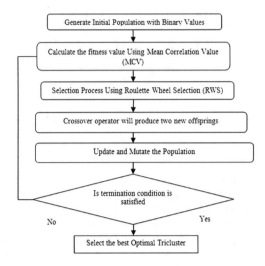

Fig. 1. Flowchart for Evolutionary Triclustering Algorithm Using Genetic Algorithm

2.1 Issues in the Existing Clustering and Biclustering Algorithms for GA

- Most of the algorithms are focused to extract the patterns and to assess the quality of gene expression data using MSR.
- The main limitation of this measure is higher magnitude have higher MSR value. Practically it is difficult to set the MSR threshold to identify the pattern with high correlation.
- Therefore, this MSR based threshold is not able to identify the pattern-based bicluster. Similarly, MSR3D does not have any such range simply minimal is best. Therefore, MSR3D is not a perfect or suitable measure to discover patterns based triclusters

To overcome the issues in the literature, MCV based triclustering is proposed. Mean Correlation Value (MCV) evaluates all kinds of triclusters well, as transformations such as translation and scaling the range of the MCV is from 0 to 1.

Table 1. Some of the Existing Algorithms with GA for Clustering Technique, Biclustering and Triclustering Technique

Clustering techniques			Biclustering techniques			Triclustering techniques		
Algorithm/Methods	Type of data	Description	Algorithm/Methods	Type of data	Description	Algorithm/Methods	Type of data	Description
GA	4 different Dataset Ruspini Data, Randomly Generated Data, Breast Cancer Data, and Iris Plant Data	Find the right clustering from the given dataset (Hruschka and Ebecken 2003)	Two GA with Greedy Search	Yeast Dataset, Lymphoma Dataset	Identify the bicluster with low residue score and high volume (Chakraborty and Maka 2005)	Trigen	Synthetic dataset, Elutriation dataset, Mouse GDS4510 Dataset	Least square line (LSL) measure is used to extract the quality tricluster
Hybrid GA, HGACLUS	Embryonal CNS data	To identify the number of clusters in multi-class gene expression data (Pan et al. 2003)	SEBI, CBEB, CgPGA, DdPGA, PGA	Yeast Dataset	DdPGA is to identify the good quality of bicluster when compared to other algorithms (Mishra and Vipsita 2017)	TriGen	Synthetic data, Saccharomyces cerevisiae, musculus, Homo sapiens	Multi Slope Measure (MSL) Measure the angles of the slopes acquired from the tricluster genes, condition, and time point
K-Means, NClust, GA, Geneclust	3 real dataset Iris, Balance, Breast Cancer	NClust performance better clustering result when compared to other algorithms (Zhang 2018)	FLOC, DBF, CC, Single Objective GA, MOEA, NM, MNM	Yeast Cell cycle data	MNM is extracted the maximum similarity bicluster with low residual and gene variance when compared to other algorithms (Balamurugan 2016)	Genetic algorithm	Yeast cell cycle data, synthetic data	MSR is used to extract the quality of the tricluster
GA	Colorectal Cancer tumor (GDS4382), Small Cell Lung Cancer (GDS4794), DLBCL-FL (GDS4236), Prostate Cancer (GDS4824)	Gene is responsible for the cancer cell. GA is trying to rank the cancer cell based on the gene. (Saha et al. 2018)	CS, GA	• Yeast Dataset • DLBCL • Gasch Yeast • BCLL, PBC • RatStrain	(CS, GA is designed to enhance the quality index and coverage of all bicluster (Yin et al. 2018)	Genetic algorithm	Yeast cell cycle data, synthetic data, inflammation and host response to injury dataset	MSR and LSL evaluation measure is used to identify the tricluster

3   Method and Materials

3.1   Mean Correlation Value for Tricluster

A value close to '1' determines high correlated tricluster otherwise; it is represented as low or null correlated tricluster. Equation 1 shows the formula for MCV, and Table 2 lists the notation used in this paper. Where $\bar{A} = \frac{\sum_m \sum_n (A_{mn})}{m*n}$, $\bar{B} = \frac{\sum_m \sum_n (B_{mn})}{m*n}$

$$\sum_m \sum_n (A_{mn} - \bar{A})(B_{mn} - \bar{B}) \Big/ \sqrt{\left(\sum_m \sum_n (A_{mn} - \bar{A})^2\right)\left(\sum_m \sum_n (B_{mn} - \bar{B})^2\right)} \quad (1)$$

Table 2. Terminologies used in Generation of Initial Population

Notations	Descriptions
nG	No. of genes
nS	No. of samples
nT	No. of time points
kg	Gene cluster
ks	Sample cluster
kt	Time point cluster
nb	Number of bicluster

3.2   Generation of Initial Population

The steps followed for tricluster formation; a two-way K-means clustering algorithm is applied along the two dimensions to generate kg and ks cluster and combined these clusters to get kg*ks initial bicluster. These biclusters are encoded as a binary string of size nb*(nG+nS). Figure 2 represents the encoded bicluster (Rathipriya 2011a, b). Figure 3 shows example for single binary encoded tricluster to get nb*(nG+nS+nT) of nb*nG+nS binary encoded tricluster (Rathipriya 2011a, b).

g1	g2	.....	gnG-1	gnG	s1	s2	.....	snS

Fig. 2. Encoded Bicluster of length nG+nS

g1	g2	.....	gnG-1	gn	s1	s2	.....	snS	t1	t2	......	tnT
8832					9				24			

Fig. 3. Encoded tricluster of length nG+nS+nT

Generate the random binary string of size nb*nT. And then concatenate (nb*(nG +nS) with (nb*nT) binary string of size to get nb*(nG+nS+nT) binary encoded tricluster for further process. Table 3 describes the notations for the Genetic Algorithm.

3.3 Notations Used for Genetic Algorithm (GA)

Table 3. Notations for GA

Crossover probability (cp)	0.7
Mutation probability (mp)	0.01
Population size (pop)	10
Generation (it)	100–1000
MCV (δ)	0.94–0.98

4   Proposed Work

In this work, the initial population is extracted from the given data set using two-way K-means clustering.

Compared to random initialization, this will lead in quicker convergence. Another benefit of initializing with these triclusters is to maintain population diversity.

4.1   Objective Function

Normally, the fitness function is used to convert the objective function value into a relative fitness measure $F(x) = g(f(x))$ where $f(x)$ is the objective function, $g(f(x))$ is transforming the value of the objective function to a non- negative number, and $F(x)$ is the resulting relative fitness to find the maximal volume tricluster with a high correlation.

This implies that a few bits in the binary string can be flipped (Rathipriya 2011a, b). Figure 4 shows the outline of evolutionary triclustering using a genetic algorithm The dataset is used from yeast cell cycle analysis projects is shown in Table 4.

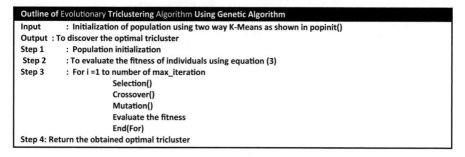

Fig. 4. Outline for Genetic Algorithm

5 Dataset Description

See Table 4.

Table 4. Dataset Description

Dataset	Genes (G)	Sample (S)	Time point (T)
CDC15	8832	9	24
Elutriation	7744	9	14
Pheromone	7744	9	18

6 Results and Discussion

Table 5. Characteristics of Optimal Tricluster using GA for CDC15 Dataset

MCV threshold	nG	nS	nT	Volume	Optimal correlated tricluster for MCV
$\mu = 0.94$	13	9	13	1404	0.9864
$\mu = 0.95$	12	9	12	1296	0.9867
$\mu = 0.96$	11	9	12	1188	0.9868
$\mu = 0.97$	11	9	12	1296	0.9869
$\mu = 0.98$	20	9	12	2160	0.9881

It has been noted that the high volume of tricluster (2160) with MCV threshold $\mu = 0.98$ shows the maximum volume with high MCV (0.9881) in Table 5. The correlation value for optimal tricluster is very high; in most cases, that is nearer to one, which indicates almost perfect homogeneity between the number of genes, number of samples, and number of times points of the tricluster.

Table 6. Characteristics of Initial Population using GA for CDC15 Dataset

Threshold value	Mean volume	Mean MCV
$\mu = 0.94$	670.1250	0.9666
$\mu = 0.95$	900	0.9844
$\mu = 0.96$	724.5000	0.9859
$\mu = 0.97$	945	0.9848
$\mu = 0.98$	1710	0.9855

The Table 6 shows the characteristics of the initial population using GA for CDC15 with threshold value range from 0.94–0.98, the mean volume, and mean MCV. The graphical representation of the mean volume and mean MCV is with different threshold values is shown in Figs. 5 and 6

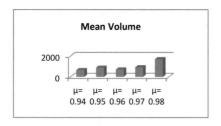

Fig. 5. Mean Volume for CDC15 Dataset using GA

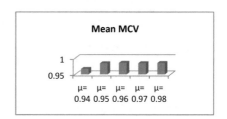

Fig. 6. Mean MCV for CDC15 Dataset using GA

Table 7. Characteristics of Optimal Tricluster using GA for Elutriation Dataset

MCV threshold	nG	nS	nT	Volume	Optimal correlated tricluster for MCV
μ = 0.94	14	9	10	1134	0.9643
μ = 0.95	15	9	10	936	0.9872
μ = 0.96	17	9	10	1080	0.9912
μ = 0.97	23	9	11	2079	0.9832
μ = 0.98	21	9	11	1980	0.9835

Table 7 depicts the Characteristics of Optimal Tricluster using GA for Elutriation Dataset. From the study, it has been observed that MCV threshold μ = 0.96 shows the high MCV value (0.9912). Table 8 shows the characteristics of the initial population using GA for Elutriation with a threshold value range from 0.94-0.98, the mean volume, and mean MCV. The graphical representation of the mean volume and mean MCV is with different threshold values is shown in Figs. 7 and 8. The threshold value of 0.97 shows a high volume.

Table 8. Characteristics of Initial Population using CorTriGA for Elutriation Dataset

Threshold value	Mean volume	Mean MCV
μ = 0.94	776.5714	0.9554
μ = 0.95	712.2857	0.9679
μ = 0.96	962.3571	0.9742
μ = 0.97	1600.71	0.9786
μ = 0.98	1611.64	0.9817

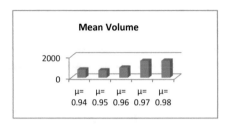

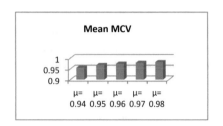

Fig. 7. Mean Volume for Elutriation Dataset using GA

Fig. 8. Mean MCV for Elutriation Dataset using GA

Table 9. Characteristics of optimal tricluster using GA for pheromone dataset

MCV threshold	nG	nS	nT	Volume	Optimal correlated tricluster for MCV
μ = 0.94	15	8	13	1456	0.9797
μ = 0.95	15	8	13	1344	0.9790
μ = 0.96	14	8	13	1456	0.9796
μ = 0.97	13	8	12	1040	0.9795
μ = 0.98	7	8	6	336	0.9864

Form Table 9, it has been observed that the maximum volume tricluster (1456) with high MCV (0,9864). This correlation value is nearer to one, which shows that the perfect homogeneity between the number of genes, the number of samples, and the number of time points. Table 10 depicts the Characteristics of Initial Population using GA for pheromone Dataset, which contains the different threshold value mean volume and mean MCV. The threshold value 0.98 shows the high volume, and high MCV is graphically shown in Figs. 9 and 10.

Table 10. Characteristics of Initial Population using GA for pheromone Dataset

Threshold value	Mean volume	Mean MCV
μ = 0.94	836.4444	0.9538
μ = 0.95	799.1111	0.9789
μ = 0.96	804	0.9732
μ = 0.97	815.5556	0.9761
μ = 0.98	1893.33	0.9845

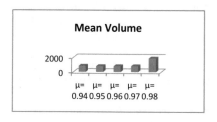

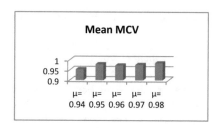

Fig. 9. Mean Volume for Pheromone Dataset using GA

Fig. 10. Mean MCV for Pheromone Dataset using GA

7 Conclusion and Future Enhancement

This paper dealt with the GA based triclustering method that has been developed to extract maximum volume tricluster with high MCV. This MCV value measurement is for the scaling and shifting pattern of the tricluster. The novel proposed algorithm was tested on three different datasets. The results show that GA based triclustering method identifies correlated tricluster for all three datasets. It is approximately equal to one (0.9912). This is significant for the proposed work. In future different hybrid optimization algorithms will be used to enhance the quality of tricluster.

References

Chakraborty, A., Maka, H.: Biclustering of gene expression data using genetic algorithm. In: 2005 IEEE Symposium on Computational Intelligence in Bioinformatics and Computational Biology (2006). Date Added to IEEE Xplore: 21 February

Gutiérrez-Avilés, D., Rubio-Escudero, C., Martínez Álvarez, F., Riquelme, J.C.: TriGen: a genetic algorithm to mine triclusters in temporal gene expression data. Neurocomputing **132**, 42–53 (2014)

Dede, D., Ogul, H.: A three-way clustering approach to cross-species gene regulation analysis. In: 2013 IEEE (2013)

Hruschka, E.R., Ebecken, N.: A genetic algorithm for cluster analysis. Intell. Data Anal. **7**(1), 15–25 (2003)

Pan, H., Zhu, J., Han, D.: Genetic algorithms applied to multi-class clustering for gene expression data. Genomics Proteomics Bioinf. **1**(4), 279–287 (2003). https://doi.org/10.1016/s1672-0229(03)01033-7. Accessed 28 Nov

Zhang, H., Zhou, X.: A novel clustering algorithm combining niche genetic algorithm with canopy and K-means. In: 2018 International Conference on Artificial Intelligence and Big Data (ICAIBD) (2018). IEEE Xplore: 28 June

Yin, L., Qiu, J., Gao, S.: Biclustering of gene expression data using cuckoo search and genetic algorithm. Int. J. Pattern Recogn. Artif. Intell. **32**(11), 1850039 (31 pages) (2018)

Rathipriya, R., Thangavel, K., Bagyamani, J.: Evolutionary biclustering of clickstream data. IJCSI Int. J. Comput. Sci. Issues **8**(3) (2011a)

Rathipriya, R., Thangavel, K., Bagyamani, J.: Binary particle swarm optimization based biclustering of web usage data. Int. J. Comput. Appl. **25**(2) (2011b)

Mishra, S., Vipsita, S.: Biclustering of gene expression microarray data using dynamic deme parallelized genetic algorithm (DdPGA). In: 2017 IEEE Conference on Computational Intelligence in Bioinformatics and Computational Biology (CIBCB) (2017). Date Added to IEEE Xplore: 05 October

Kapil, S., Chawla, M., Ansari, M.D.: On K-means data clustering algorithm with the genetic algorithm. In: 2016 Fourth International Conference on Parallel, Distributed and Grid Computing (PDGC) (2017). IEEE Xplore: 27 April

Saha, S., Das, P., Ghosh, A., Dey, K.N.: Ranking of cancer mediating genes: a novel approach using genetic algorithm in DNA microarray gene expression dataset. In: ICACDS 2018. CCIS, vol. 906, pp. 129–137. Springer (2018)

Balamurugan, R., Natarajan, A.M., Premalath, K.: Biclustering microarray gene expression data using modified Nelder-Mead method. Int. J. Info. Commun. Technol. 9(1) (2016)

A Study on Legal Knowledge Base Creation Using Artificial Intelligence and Ontology

Tanaya Das[✉], Abhishek Roy, and A. K. Majumdar

Adamas University, Kolkata, India
tanayadas.das23@gmail.com, dr.aroy@yahoo.com,
akmajumdar@hotmail.com

Abstract. Artificial Intelligence (AI) as an emerging technology, facilitates the mapping of human intelligence with computerized mechanism. This capability of thinking like human being provides computer with boundless potential for user interaction and prediction of logical solution (i.e. decision) for any particular event using its knowledge base obtained through previous experience. To generate faster solution, Artificial Intelligence (AI) can be utilized in multivariate service sectors, which are accustomed to generate logical solution in slow pace due to various constraints like, manpower, fund, infrastructure, policy paralysis, etc. In a developing country like India, which is striving hard to have a constant growth rate, application of Artificial Intelligence (AI) based tools have wide window open in various service sectors like judiciary, healthcare, business, education, agriculture, banking, etc. Generally in case of judiciary, end users have to wait for a long time to get their desired justice which directly affects their efficiency, contribution and well-being in this society. Using the concept of Artificial Intelligence (AI), Legal knowledge based tools may accelerate the service delivery of legal professionals from typical searching of related case journals to extraction of precise information in a customized manner. This paper focuses on legal representation of legal knowledge base to study the changing trends using Artificial Intelligence (AI).

Keywords: Artificial Intelligence · Legal knowledge base · Ontology

1 Introduction

Human beings are differentiated from other living beings by their enhanced ability of learning and problem solving skills. In psychology, Intelligence is a set of skills consisting of thinking and reasoning capabilities, knowledge gathering, learning from past experience and adopting to environmental changes. The key to design a successful machine is to embed the decision making skill within itself. Knowledge is another terminology which is very much related to intelligence, but applicable for similar kind of situations. Knowledge is all about the facts or information and skills gathered through some experience or understanding about a particular event. There are apparently unsolved problems which can be solved faster using Artificial Intelligence (AI) based approach. Nowadays many subfields of Artificial Intelligence (AI) are widening their domain through various types of software developments. The efficiency

© Springer Nature Switzerland AG 2020
J. S. Raj et al. (Eds.): ICIDCA 2019, LNDECT 46, pp. 647–653, 2020.
https://doi.org/10.1007/978-3-030-38040-3_73

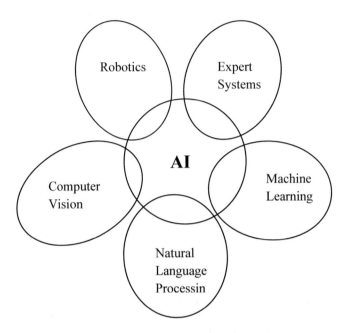

Fig. 1. Interdisciplinary approach of Artificial Intelligence

of new technology is measured by its problem solving capability which determine its sustainability in long run. Figure 1 represents the interdisciplinary approach of Artificial Intelligence which may be applied over multivariate fields to generate faster output.

Knowledge is the building block of any problem solving technique for the humans beings and their creations (i.e. machines). Hence, modeling of knowledge helps to create a knowledge base which is the necessary repository of rules or semantics and facts. For example, to develop any system or products we require sufficient some amount of knowledge which can be transformed into relevant operational models. As Knowledge denotes main working principle of any operation, to launch a new service, the fundamental knowledge base remains the same along with addition of new parameters.

Section 2 states the basic idea of Artificial Intelligence (AI) and studies Artificial Intelligence (AI) based tools in broader perspective. For successful automation of any conventional technique, it is imperative to match Human Intelligence with Machine Intelligence using Artificial Intelligence. Section 3 state the importance of knowledge base as a strong repository of logic and predicate for multivariate service sectors. In order to study legal knowledge base and its ontology, a precise review is mentioned in Sect. 4. Section 5 allude a brief study over application of Artificial Intelligence (AI) in judiciary to deliver prompt services to end users. Finally, the conclusion drawn from this study is mentioned in Sect. 6 which also explores its future scope.

2 Artificial Intelligence (AI) and Its Application Areas

Nowadays, Artificial Intelligence (AI) is becoming a major tool to map human thought process with computational modeling. Artificial Intelligence (AI) can capture, share, develop and transform knowledge as per the customized format of the client. It originates from Greek mythology of artificial beings with thinking ability. McCarthy [8] discussed about the rationalisation of reasoning and common-sense based knowledge representation using Artificial Intelligence (AI). Specific research work focuses on how Artificial Intelligence (AI) is strengthening its root from philosophical thoughts and imagination to some recent issues related to legal domain [3, 16]. Turing [19] had introduced an idea about embedding thinking capability of a machine. The term Artificial Intelligence (AI) went back to a summer conference held in 1956 at Dartmouth College [7] in New Hampshire which was marked as origin of Artificial Intelligence (AI). During this period a popular test was conducted called "Turing Test", which became a landmark in the field of Artificial Intelligence (AI). It is transforming the world by processing enormous amount of data from different government and non governmental source using various techniques like Natural Language Processing (NLP) and Machine Learning (ML), resulting into evolution of Data Science, which can drive multivariate service sectors into faster pace. Many computational models have been proposed using software tools to capture intrinsic knowledge associated with application domains. As a result, Artificial Intelligence (AI) based tools are leading in various service sectors like legal and judiciary, healthcare, business, education, agriculture, banking, etc. For this reason, impact of Artificial Intelligence have been studied over multiple domains like expert system, health care, medical image classification, power system stabilizers using Neural Network, Fuzzy Logic, Network Intrusion Detection System (NIDS) [10]. Intelligent co-pilot system [2], an Artificial Intelligence (AI) based tool in Automobile Industry which can enable safety warning, voice, facial recognition, gesture control emotion control developed by Nvidia, partnered with Volkswagen. Likewise in Culinary practices Chef's Watson [4] help to create a recipe based on the available ingredients. The proposed algorithms will go through the idea of what kind of dishes to be prepared from present ingredients.

Healthcare service sector have equipments like GNS Healthcare [6], a machine learning based tool detects human disease in order to provide right treatment to right patient at right time, Clinical Decision Support System [12] that analyzes medical data to help healthcare specialist to make decisions and improve patient care. It generates alarm for preventive care as well as for dangerous drug detection to alert technicians to take care of their patients.

Entertainment sector is emerging with many applications which are providing services based on personal choices of items like graphics, audio, video frames etc. from an existing movie, and then rank to identify the highest likelihood like Netflix [20]. Authors in above mentioned references have discussed an overall view of Artificial Intelligence (AI) based concepts and its applied areas such as language understanding, problem-solving, learning and adaptive systems, robots, etc. Artificial Intelligence (AI) can be applied to improvise these service sectors, which generally provide logical solution in slow pace due to various constraints like manpower, budget, infrastructure, policy paralysis, etc. To be precise judiciary can be supported using Artificial Intelligence (AI) based mechanisms which are discussed in subsequent sections of this paper.

3 Legal Knowledge Base

Preserving knowledge is like an old story of "Cave Paintings" which were considered as initial form to gain or capture information. Knowledge is meant to build some meaningful information which can be used for storing statements of facts. Any system which consists Knowledge comprises of objects, events, relationship among objects, meta-knowledge, etc. Expert system [15] being the subset of Artificial Intelligence (AI), contains symbolic information, heuristic processing, expert level performance and design of the system. It discusses the vital techniques and concepts of cognitive science to develop and understand how Human Intelligence can transform into computational models. The objective of Knowledge Base is to make the proposed system more cost effective and attain better decision making capability. The knowledge base consists of various rules, facts, proposals that are captured from textual and image documents. In a developing country like India which is striving hard to maintain a constant growth rate, there are several service sectors which have huge scope of improvement using Artificial Intelligence (AI). Particularly in case of judiciary, end users have to wait for a long time to get their desired justice which directly affects their efficiency, contribution and well-being in this society. Artificial Intelligence (AI) can be applied in various fields like contract analysis, case prediction, compliance and document automation using legal tools such as Bloomberg BNA, ROSS Intelligence, Lex Machina, Electronic discovery based on Natural Language Processing (NLP) and Machine Learning (ML) techniques, Lex Predict [9]. As most of the legal documents in the conventional mode are in textual format, Artificial Intelligence (AI) based tools will help both appellant and defendant to establish their views by analysis of those legal texts by introducing a new technology named Semantic Web [13]. Mainly a lawyer's task is to form an opinion or to decide how to set up a defense for a particular case. Over the past few years, lawyers are frequently using search tools that help to identify relevant laws, statutes, rules, judgments associated with different court cases. Recently, Legal Analytics has emerged as an important component of Legal Research. Legal Analytics aims to combine digitized legal data and Machine Intelligence to understand, manage and improve legal services. Artificial Intelligence in legal field requires knowledge representation, automated analysis and Natural Language Processing (NLP). Legal analytics is a concept based on mathematical and statistical tools to extract pattern in various cases and to explore the knowledge using that legal dataset.

4 Ontology in the Legal Domain

The ontology describes domain properties and relationships between the concepts and entities and in a machine-readable format and its precise representation from derived consequences of subject matter. It helps to specify a concept as building blocks of knowledge model using certain relationships and entities. We need to find out the irrelevance in that domain. Even author have addressed approaches based on a case study about building Corporate History Analyzer [17] to manage knowledge-based items. It has been observed that ontology provides precise details than going through a large number of searches generated from a web based search engine. If we go through ontology, we can filter out only the specific parameters suitable for a particular task. This is useful for text document analysis which consists of facts and figures. For example, any newspaper, magazine article, textbooks or any study material contains some amount of information which has inherent semantics. Here in these documents, ontology helps to capture the meaningful information from the available data repository.

Recent research paper [11] focuses on sub domains of Artificial Intelligence like knowledge base, ontology, etc. which supports various service sectors and firms to provide the best services to its clients. Legal knowledge base concept may change the working trends of legal professionals from typical searching of related case journals to more precise information. As law is represented in the form of a specific set of rules, hence it can be easily mapped with a set of well-defined semantics. Representation of legal knowledge in the knowledge base can help lawyers to provide best possible service to their clients within a specific time frame.

5 Literature Survey

Legal knowledge base is different from the legal search engine. It has been observed in most of the legal documents that similar cases are cited using various law journals and statutory laws which are available in digitized form in legal search engine also. As searching and analyzing relevant portion from a document is a time consuming task for legal professionals, application of Artificial Intelligence (AI) based tools can generate faster output. Capturing knowledge from a legal document is about building mean-ingful information structures to store facts and evidence for every case.

Alvarez, Ayuso and Becue [1] address major concepts behind law practices using semantic web technology. This paper is segmented into major sections of legal informatics, methodologies, and information retrieval techniques to handle legal doc-uments. The semantic web is a concept which enables the machine to understand and execute intelligent web based task. It also describes the theory behind the Artificial Intelligence based techniques and law in terms of ontologies for structuring the legal knowledge as a specific set of facts and rules.

A legal system represents the knowledge available in legal documents in the form of arguments and reasoning for any case. Schild and Saban [14] focus on two major approaches in expert legal system i.e. case-based and rule-based problems. This paper mainly focuses on how the Toulmin structures fit within the rule-based concept which is much more convenient to develop legal expert systems.

Machine Learning (ML) and Natural Language Processing (NLP) based tools are working on it to extract minute details, which human brain and human eye may overlook by any chance which is studied by Surden [18].

Legal professionals use some rudimentary filtering, based on search keywords to examine a fairly large subset of related judgments. Specific research paper [5] discusses about changing trends in working principle of legal professionals using Artificial Intelligence and acceptance of legal analytics worldwide.

6 Conclusion

We have studied that how the acquired knowledge of human brain can be modeled into Knowledge Base with the help of Artificial Intelligence based tools and techniques like Machine Learning and Natural Language Processing. With the help of legal ontologies, the semantics of knowledge can be gathered to have the desired knowledge base. With the development of knowledge representation the extracted portion can be modeled using various technologies like knowledge base framework, data mining, and Artificial Intelligence. Reasoning and argumentation are two major issues in any computer-assisted legal system. Machine learning based tools can also be used for contract analysis, case prediction, compliance, and document automation. The legal knowledge base provides specific meaningful information model to legal professionals to deliver prompt and accurate service to its clients. The meaningful information in similar cases represents some model-based approaches like case-based and logic-based techniques. Legal ontologies are acting as supporting tools which help to prepare the structure of the knowledge base for providing effective services.

In India, the working trends of lawyers are to manually search the judgments related to cases which are available through various online search engines. To an Indian law firm, search engines that combine machine learning-based tools will naturally help to extract desired clauses from a large volume of similar legal documents, with greater accuracy. Developing legal knowledge base on civil matters may be possible by observing clauses in every case thereby preparing semantics accordingly. However, as criminal cases are fuzzy in nature, we have to minutely observe the legal documents to provide an efficient output. In this situation, development of legal knowledge based system for criminal cases may be considered as future scope of this study.

References

1. Álvarez, R., Ayuso, M., Bécue, M.: Statistical study of judicial practices. In: Benjamins, V. R., Casanovas, P., Breuker, J., Gangemi, A. (eds.) Law and the Semantic Web, pp. 25–35. Springer, Heidelberg (2005)
2. Engadget. https://www.engadget.com/2018/01/08/nvidia-volkswagen-ai-co-pilot
3. Buchanan, B.G.: A (very) brief history of artificial intelligence. AI Mag. **26**, 53 (2005)
4. IBM Chef's Watson. https://www.ibm.com/blogs/watson/2016/01/chef-watson-has-arrived-and-is-ready-to-help-you-cook/
5. Das, T., Majumdar, A.K.: A survey of changing trends in legal industry from manual search to legal analytics using artificial intelligence. Int. J. Manag. Technol. Eng. **9**, 2404–2410 (2019). http://www.ijamtes.org/gallery/286.%20jan%2019ijmte%20-%20ss.pdf. ISSN: 2249-7455
6. GNS Healthcare: https://www.gnshealthcare.com/
7. Li, D., Du, Y.: Artificial Intelligence with Uncertainty. CRC Press, Boca Raton (2017)

8. McCarthy, J.: Artificial intelligence, logic and formalizing common sense. In: Thomason, R. H. (ed.) Philosophical Logic and Artificial Intelligence, pp. 161–190. Springer, Dordrecht (1989)
9. Mills, M.: Artificial Intelligence in Law: The State of Play 2016. Thomson Reuters Legal executive Institute (2016)
10. O'Leary, D.E.: Using AI in knowledge management: knowledge bases and ontologies. IEEE Intell. Syst. Appl. **13**(3), 34–39 (1998)
11. Pannu, A.: Artificial intelligence and its application in different areas. Int. J. Eng. Innovative Technol. (IJEIT) **4**(10), 79–84 (2015)
12. HealthIT.   https://searchhealthit.techtarget.com/definition/clinical-decision-support-system-CDSS
13. Saias, J., Quaresma, P.: A methodology to create legal ontologies in a logic programming information retrieval system. In: Benjamins, V.R., Casanovas, P., Breuker, J., Gangemi, A. (eds.) Law and the Semantic Web, pp. 185–200. Springer, Heidelberg (2005)
14. Schild, U.J., Saban, Y.: Knowledge representation in legal systems. Syracuse L. Rev. **52**, 1321 (2002)
15. Smith, R.G.: Knowledge-Based Systems: Concepts, Techniques, Examples. Ridgefield, CT USA, Ottawa (1985)
16. Sobowale, J.: How artificial intelligence is transforming the legal profession. ABA J. **1**, 1 (2016)
17. Staab, S., Studer, R., Schnurr, H.P., Sure, Y.: Knowledge processes and ontologies. IEEE Intell. Syst. **16**(1), 26–34 (2001)
18. Surden, H.: Machine learning and law. Wash. L. Rev. **89**, 87 (2014)
19. Turing, A.M.: Can a machine think. Mind **59**(236), 433–460 (1950)
20. How Netflix Uses AI, Data Science, and Machine Learning—From a Product Perspective. https://becominghuman.ai/how-netflix-uses-ai-and-machine-learning

Survey of Progressive Era of Text Summarization for Indian and Foreign Languages Using Natural Language Processing

Apurva D. Dhawale[✉], Sonali B. Kulkarni,
and Vaishali Kumbhakarna

Dept. of CS & IT, Dr. Babasaheb Ambedkar Marathwada University,
Aurangabad, India
addhawale@gmail.com, sonalibkul@gmail.com,
vmk_17@yahoo.co.in

Abstract. The last few years of Data Science definitely show the upward trend in growth of popularity, different industries which are effectively relating with data science & the transformation of world with e-commerce sites, social networking sites, travel aggregators, Google assistants. Here the need of text summarization comes in existence. This text summarization is a conception which actually deals with time saving and giving user the output with minimum text without changing its meaning. This approach is very impressive as the e-contents reading is very important aspect in many fields like academics, industries, hospitals, clinical notes, news papers, crime records, cyber security and many other. This paper shows the advancements which has initiated research for text summarization in many global and local languages. Text mining is coping up with many dimensions like text summarization, question generation, sentiment analysis, and translators. Natural Language Processing is one of the demanding areas now a day, which will definitely improve the human machine interaction.

Keywords: NLP · Text summarization · Extractive · Abstractive text summarization

1 Introduction

Text Summarization is an active area with respect to academia. NLP works with very large dimensions in research, and text summarization is an important area. Also text mining and text summarization techniques are used on input data to cluster & classify it. In recent times various techniques for text summarisations are being developed but still the system is not that much efficient [25]. Compressing the text without changing its meaning is called as text summarization. There is lot of difficulty in getting a higher efficiency in text summarization of Indian regional languages. A lot of work has been found for English language which has better results. So it is needed now to focus on regional languages which are used in various fields for various purposes. Redundancy in text is one of the major problems of web contents and NLP do not have standard

© Springer Nature Switzerland AG 2020
J. S. Raj et al. (Eds.): ICIDCA 2019, LNDECT 46, pp. 654–662, 2020.
https://doi.org/10.1007/978-3-030-38040-3_74

evaluation methods for some problems as mentioned above. So basically, evaluation is a difficult process than the summarization of text (Fig. 1).

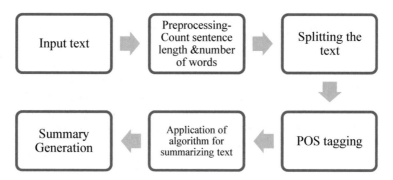

Fig. 1. The process of text summarization

This paper presents introduction, Literature Survey of international and national languages, & then conclusion.

2   Literature Review

There are many parameters for extracting sentences, words, stop words, special symbols. These parameters can be classified on the basis of different methodologies available and techniques to classify and cluster the input data. This classification can be based on single document summarization, Multi document summarization as well as monolingual, bilingual.

2.1   Multidocument and Bilingual Text Summarization

Author	Year	Method	Result
Ravindra, Balakrishnan, and Ramakrishnan [4]	2004	Sentence ranking technique using entropy measures	Results are presented to illustrate the relevance of this method in cases where it is difficult to have language specific dictionaries, translators and document-summary pairs for training
Singh, Kumar, Mangal, Singhal [37]	2016	Deep learning	Features to improve the relevance of sentences in the dataset

There is a lot of emphasis on languages while dealing with text summarization. These are some languages which are mostly used.

2.2   Text Summarizers for Foreign Language

English

Author	Year	Method	Result
Luhn [18]	1958	Word frequency and distribution	Computing relative measures, for each word and sentence, to save readers time
Babar, Patil [32]	2015	Sentence ranking & fuzzy inference system	Text summarization using Latent Semantic Analysis
Jishma, Sunitha, Amal, Dr. Jaya [11]	2016	Abstractive summarization methods	Review on ontology based abstractive summarization methods and its importance in different domains
Liu, Webster, Kit [36]	2009	Extractive text summarization approach	Achieves a state-of-the-art performance
Soumya, Kumar, Naseem, and Mohan [30]	2011	Rule based summarization and keyword based summarization	Extraction of keywords from the document and identifying the sentences containing them
Piskorski and Yangarber [10]	2013	Information extraction algorithm	Facts are structured objects, such as database records
García-Hernández, Ledeneva [15]	2013	Genetic algorithm	Extractive text summariztaion
Nagwani, [22]	2015	Semantic similarity based clustering and topic modeling using Latent Dirichlet Allocation (LDA)	ROUGE and Pyramid score are also measured

Chinese

Author	Year	Method	Result
Hu, He, Ji [23]	2004	Detection of thematic areas using K-medoids clustering	68%
Yu, Lei, Mengge Liu, Fuji Ren, Shingo Kuroiwain [39]	2006	TF-IDF	Average scores are 0.74 and 0.76 at 20% compression ratio

Arabic

Author	Year	Method	Result
Ibrahim, Darwish, Fayek [32]	2006	Trainable Bayesian approach	68.07%
Tarek, El-Ghannam [3]	2012	Extractive summarization approach	66% at 25% compression ratio
Azmi, Al-Thanyyan [1]	2012	Automatic extractive Arabic text summarization system using a two pass algorithm	Used Rouge to evaluate the system generated summaries of various lengths against those done by a (human) news editorial professional
Al Qassem, Wang, Al Mahmoud, Barad, Ahmad, Al-Rubaie, Almoosa [14]	2017	Survey of methodologies & systems for automatic text summarization	The Summarization reviewed systems & its evaluation methodology
Al-Saleh, Menai, [2]	2015	Several research studies of Arabic text summarization	It addresses summarization and evaluation methods, as well as the corpora used in those studies

Turkish

Author	Year	Method	Result
Kutlu, Cığır, Cicekli [21]	2010	Generic text summarization technique	Comparison of output summary with manual summaries of two Turkish data sets

Swedish

Author	Year	Method	Result
Gustavsson, Jonsson [7]	2010	Random Indexing and PageRank	The result shows text types, and other aspects of texts of the same type which has influence on the performance. On government texts, random indexing and PageRank provides the good results

Brazilian Portuguese

Author	Year	Method	Result
Rino, Pardo, Silla, Kaestner, and Pombo, [17]	2004	Five distinct extractive AS systems	Two baseline systems have been considered. An overall performance comparison has been carried out

Persian

Author	Year	Method	Result
Zamanifar and Kashefi [12]	2011	Summarization approach, nicknamed AZOM	Combines statistical and conceptual property of text and in regards of document structure, extracts the summary of text
Shahverdian, Saneifar [8]	2017	Based on semantic and lexical similarities and uses a graph-based summarization	The summaries extracted by the proposed approach reached an average score of 8.75 out of 10, which improves the state-of-the-art summarizer's score about 14%

2.3   Text Summarizers for Indian Language

India has 22 official regional languages. We have considered some of them.

Tamil

Author	Year	Method	Result
Kumar, Ram and Devi [29]	2011	Scoring of sentences, ROUGE evaluation toolkit	Result shows Average Rouge score: 0.4723
Banu, Karthika, Sudarmani, Geetha [20]	2007	Sub graph, Language-Neutral Syntax (LNS), Support Vector Machine (SVM) classifier	Result shows extraction of automatic summaries from the text documents

Kannada

Author	Year	Method	Result
Kallimani, Srinivasa, Eswara [9]	2010	AutoSum	Extracts automatic summaries from the text documents
Sunitha, Jaya, Ganesh, [34]	2016	Survey of various techniques available for abstractive summarization	Currently available work in abstractive summary field of Indian Languages like Hindi, Malyalam Kannada

Punjabi

Author	Year	Method	Result
Gupta, Lehal [5]	2013	Extractive automatic text summarization system	Varies from 81% to 92%
Gupta, Lehal [35]	2011	Preprocessing of punjabi text	Punjabi words boundary identification, stop words elimination, noun stemming, finding Common English Punjabi noun words, finding proper nouns, sentence boundary identification, and identification of Cue phrase in a sentence
Gupta, Lehal, [34]	2012	Text Extractive Summarization System	Text summarization of punjabi language text document

Hindi

Author	Year	Method	Result
Kumar and Yadav [13]	2015	Extractive approach	85%
Sargule, Kagalkar [42]	2016	Bernoulli Model of Randomness	Derive a novel lexical association measure which is useful to generate indexing values for each word
Gupta [6]	2013	Hybrid algorithm for multilingual summarization	New features for summarizing Hindi and Punjabi multilingual text by combining the features

Marathi

Author	Year	Method	Result
Narhari, Shedge [38]	2017	Modified LINGO algorithm & PCA	Used particularly for marathi text documents and results are improved
Patil, Bogiri [39]	2015	LINGO [Label Induction Grouping] algorithm	Efficient for marathi text documents for improving results
Bolaj, Govilkar [41]	2016	Supervised learning methods	Efficient for marathi text documents
Rathod [28]	2018	Domain specific summarization using an algorithm	Works with Summarization of Domain Specific Marathi News
Sethi, Sonawane,	2017		

(continued)

(continued)

Author	Year	Method	Result
Khanwalker, Keskar [42]		Lexical chains and using the WordNet thesaurus, pronoun resolution	Overcome the limitations of the lexical chain approach to generate a good summary
Dangre, Bodke, Date, Rungta, Pathak [43]	2016	Cluster algorithm to collect relevant Marathi news from multiple sources on web	Enables rich exploration of Marathi contents on web

The Human Machine interaction is basic part while dealing with Natural Language Processing; this inspires the users to have a better initiative & to set a milestone in text mining, text summarization of different Languages. Compared with all the above said languages, it is observed that the work done in Indian Regional Languages is not that efficient and have to be improved beyond the simple text mining [19].

3   Conclusion

This paper provides the review of progressive era of text summarization using natural language processing. The text summarization generates significant information content out of input text which helps user to have data without changing its meaning. The work can also be classified into monolingual, bilingual & single document & multi document summarization. In this paper we have taken a review of development in the area of text summarization for foreign & indian languages. In Sect. 2.1 we discussed about Multidocument & Bilingual Text Summarization, in Sect. 2.2 Foreign Language Text Summarization, and in Sect. 2.3 Indian Language Text Summarization.

References

1. Azmia, A.M., Al-Thanyyan, S.: A text summarizer for Arabic. Comput. Speech Lang. **26**(4), 260–273 (2012)
2. AlSaleh, A.B., Menai, M.E.B.: Automatic Arabic text summarization a survey. Artif. Intell. Rev. **45**(2), 1–32 (2015)
3. El-Shishtawy, T., El-Ghannam, F.: Keyphrase based Arabic Summarizer(KPAS). In: International Conference on Informatics & Systems (INFOS). IEEE (2012)
4. Ravindra, G., Balakrishnan, N.: Multi-document automatic text summarization using entropy estimates, pp. 289–300. Springer, Heidelberg (2004)
5. Gupta, V., Lehal, G.S.: Automatic text summarization system for Punjabi language. J. Emerg. Technol. Web Intell. **5**(3), 257–271 (2013)
6. Gupta, V.: Hybrid algorithm for multilingual summarization of Hindi & Punjabi documents, pp. 717–727. Springer, Cham (2013)
7. Gustavsson, P.A.: Text summarization using random indexing & pagerank. In: Proceeding of the 3rd Swedish Language Technology Conference (SLTC-2010). Sweden: SLTC (2010)
8. Shahverdian, H., Saneifar, H.: Text summarization of multi-aspect comments in social networks in Persian Language. (IJACSA) Int. J. Adv. Comput. Sci. Appl. **8**(12), 362–368 (2017)

9. Kallimani, J.S., Srinivasa, K. G.: Information retrieval by text summarization for an Indian regional language. In: International Conference on Natural Language Processing & Knowledge Engineering, pp. 1–4 (2010)
10. Piskorski, J., Roman Y.: Information Extraction: Past, Present & Future, pp. 23–49. Springer, Heidelberg (2013)
11. Mohan, M.J., Sunitha, C.: A study on ontology based abstractive summarization. Procedia Comput. Sci. **87**, 32–37 (2016)
12. Kashefi, A.Z.: AZOM: a Persian structured text summarizer, pp. 234–237. Springer, Heidelberg (2011)
13. Kumar, K.V., Yadav, D.: An improvised extractive approach to Hindi text summarization. In: Informatiove System Design & Intelligence Applications, pp. 291–300. Springer, New Delhi (2015)
14. Al Qassem, L.M., Wang, D.: Automatic Arabic summarization: a survey of methodologies. Procedia Computer Science **117**, 10–18 (2017)
15. García-Hernández, R.A., Ledeneva, Y.: Single extractive text summarization based on a genetic algorithm, pp. 374–383. Springer, Heidelberg (2013)
16. Lehal, V. G.: Preprocessing phase of Punjabi language text summarization, pp. 250–253. Springer, Heidelberg (2011)
17. Rino, L.H.M., Pardo, T. A. S.: A comparison of automatic summarizers, pp. 235–244. Springer, Heidelberg (2004)
18. Luhn, K.P.: The automatic creation of literature abstracts. IBM J. Res. Dev. **2**(2), 159–165 (1958)
19. Hanumanthappa, M., Swamy M.N.: A detailed study on Indian languages text mining. Int. J. Comput. Sci. Mobile Comput. 54–60 (2014)
20. Banu, M., Karthika, C.: Tamil document summarization using semantic graph method. In: Proceeding of International Conference on Computational Intelligence & Multimedia Applications, pp. 128–134 (2007)
21. Kutlu, M., Cığır, C.: Generic text summarization for Turkish. Comput. J. **53**(8), 1315–1323 (2010)
22. Nagwani, N.K.: Summarizing large text collection using topic modeling and clustering based on MapReduce framework. J. Big Data **2**(1), 1–18 (2015)
23. Hu, P., He, T., Ji, D.: Chinese text summarization based on thematic area detection. In: ACL-04 workshop (2004)
24. Poibeau, H. S.: Automatic text summarization: Past, Present & Future, pp. 3–21. Springer, Heidelberg (2013)
25. Sethi, P., Sonawane, S.: Automatic text summarization of news articles. In: International Conference on Big Data, IoT and Data Science, pune, pp. 23–29. IEEE (2017)
26. Shah, P., Desai, N.: A Survey of automatic text summarization techniques for Indian and Foreign languages. In: International Conference on Electrical, Electronics, and Optimization Techniques (ICEEOT), pp. 4598–4601. IEEE (2016)
27. Jayashree, R., Srikanta, K.M.: Document summarization in Kannada using keyword extraction. In: Proceeding of AIAA 2011, CS & IT, pp. 121–127 (2011)
28. Rathod, Y.V.: Extractive text summarization of Marathi news articles. In: IRJET, pp. 1204–1210, July 2018
29. Kumar, S., Ram, V.S.: Text extractionfor an agglutinative language. Language in India, pp. 56–59 (2011)
30. Soumya, S.,Kumar, G. S.: Automatic text summarization, pp. 787–789. Springer, Heidelberg (2011)
31. Babar, S.A., Patil, P.D.: Improving performance of text summarization. Procedia Comput. Sci. **46**, 354–363 (2015)

32. Sobh, I., Darwish, N.: A trainable Arabic Bayesian extractive generic text summarizer. In: Sixth Conference on Language Engineering ESLEC, pp. 49–154. ESLEC (2006)

33. Sunitha, C., Jaya, A., Ganesh, A.: A study on Abstractive text summarization techniques in Indian Languages. Procedia Comput. Sci. **87**, 25–31 (2016)

34. Gupta, V., Lehal, G.S.: Automatic Punjabi text extractive summarization system. In: Proceedings of COLING 2012, Mumbai, COLING, pp. 191–198 (2012)

35. Gupta, V., Lehal, G.S.: Preprocessing phase of Punjabi language text summarization, pp. 250–253. Springer (2011)

36. Liu, X., Webster, J.J.: An extractive text summarizer based on significant words, pp. 168–178. Springer (2009)

37. Singh, S.P., Kumar, A., Mangal, A., Singhal, S.: Bilingual automatic text summarization using unsupervised deep learning. In: International Conference on Electrical, Electronics, and Optimization Techniques (ICEEOT) – IEEE (2016)

38. Narhari, S.A., Shedge, R.: Text categorization of Marathi documents using modified LINGO. IEEE (2017)

39. Patil, J.J., Bogiri, N.: Automatic text categorization-Marathi documents. In: International Conference on Energy Systems and Applications (ICESA 2015), IEEE (2015)

40. Bolaj, P., Govilkar, S.: Text classification for Marathi documents using supervised learning methods. Int. J. Comput. Appl. **155**(8), 0975–8887 (2016)

41. Sargule, S., Kagalkar, R.M.: Strategy for Hindi text summarization using content based indexing approach. Int. J. Comput. Sci. Eng. IJCSE **4**(9), 36 (2016)

42. Sethi, P., Sonawane, S., Khanwalker, S., Keskar, R.B.: Automatic text summarization of news articles. In: International Conference on Big Data, IoT and Data Science (BID) Vishwakarma Institute of Technology, Pune, 20–22 December. IEEE (2017)

43. Dangre, N., Bodke, A., Date, A., Rungta, S., Pathak, S.S.: System for Marathi news clustering. In: 2nd International conference on Intelligent computing,communication & convergence, bhubaneshwar, Elsevier (2016)

Zone Safe Traffic Assist System and Automated Vehicle with Real-Time Tracking and Collision Notification

K. K. Aishwariya$^{(\boxtimes)}$, Sanil K. Daniel, and K. V. Sujeesh

Department of Electronics and Communication Engineering,
Government Engineering College, Idukki, Kerala, India
aishwariyakk@gmail.com, saniiil@rediffmail.com,
kvs.suji89@gmail.com

Abstract. The road accidents and the fatalities are increasing day by day. Negligence in driving and the defiance of traffic rules are causing these accidents. Smarter vehicles embedded with sensors can alert the driver from violating the road traffic regulations and ensure a safe driving. In this paper, we present a smart vehicle that will only allow an undrunk driver to operate the vehicle. The vehicle is also designed to automatically detect the different traffic zones and thereby take appropriate control actions in speed and sounding the horn. An automatic collision notification with a real time tracking is also incorporated in the vehicle. The designed prototype is experimented in different traffic zones and conditions.

Keywords: Sensors · Real–time tracking · Automatic collision notification · Traffic zones · Smart vehicle · Road accidents · IoT

1 Introduction

Every year, along with the tremendous increase in the number of vehicles, thousands of road crashes and its fatalities are being reported all over the globe [1]. The major causes of these road mishaps are negligent driving, the lack of familiarity of the roads to the driver and as well as the defiance of traffic rules. Moreover the death tolls even find an increment due to unnoticed and unattended incidents.

Smart vehicle systems that incorporate road safety features are reported by several research groups. Speed limit alerting and crash detection [2, 3] system that can detect drunken driver [4, 5], vehicle tracking locking and accident alert using GPS and GSM modules [6, 7, 8, 9] are some of them. Systems that can identify its zone of operation and the drunken driving [10] using technologies such as RFID [11, 12] and Zigbee [13, 14] are also reported. The mentioned methodologies used for the smart vehicles have their own advantages and disadvantages. The proposed system, address the causes of accidents due to drunken driving and the violation of red light traffic signal. It also ensures the road discipline with an automated speed control in speed limited areas and also horn control in horn prohibited areas. A notification alert is also incorporated in the case of emergencies.

© Springer Nature Switzerland AG 2020
J. S. Raj et al. (Eds.): ICIDCA 2019, LNDECT 46, pp. 663–669, 2020.
https://doi.org/10.1007/978-3-030-38040-3_75

2   System Design

The proposed method is demonstrated with a system design shown in Fig. 1. The conventional traffic signaling system is modified to include a transmitter module along with the signal boards. And a receiver module is placed on the vehicle. Moreover, an android application provides appropriate alert messages to the driver.

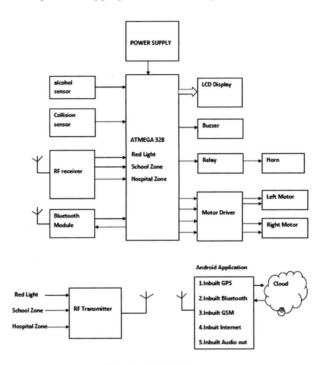

Fig. 1. Block diagram

RF modules that operate at 433 MHz are used in the system. The traffic area is classified into different zones like school zone, traffic signal zone and hospital zones. In addition, these zones are further encoded, and the zone data is transmitted to the receiver module. The receiver module in the vehicle decodes the data and is transferred to the controller unit in the vehicle. The controller module regulates the speed in the speed restricted zones and also provides alert messages to the driver. Similarly, it also deactivates the horn in the horn prohibited areas. It also provides an alert when the red traffic light is ON.

Sensor modules such as alcohol sensor and collision sensor are used in the vehicle to ensure a safe travelling. The alcohol sensor will prevent the drunken driving. If the driver is found drunken then the vehicle will not start and gives the buzzer alert. The bumper switch will act as a collision sensor and if it gets pressed, provides automatic collision notification SMS with the location details of the vehicle and as well as a call to the given emergency numbers which ensures that the injured person gets a faster assistance. The driver mobile phone is interfaced with the controller module with a blue tooth module.

An android application is provided in the driver's phone. An authentication provided from the driver side is necessary for the starting of the vehicle. And this application further allows the driver to get all the alert messages like different zone identification and the measures to be taken in different instances. Another android application in the phone is used to monitor the vehicular system. Real time monitoring of the vehicle is done with IoT enabled android application which can give the real time information about the vehicle to the remote owner to ensure vehicle security and assistance. An LCD module is also provided in the system to show the notifications. Details of hardware and software used are given in Table 1 and Table 2 respectively.

2.1   Hardware Details

Table 1. Hardware Details

Sl. No.	Component	Specification/Details
1	Microcontroller	ATMEGA328P
2	RF Receiver/Transmitter	433 MHz ASK
3	RF encoder	IC- HT12E
4	RF decoder	IC- HT12D
5	Blue tooth module	HC 05

2.2   Software Details

Table 2. Software Details

Sl. No.	Software	Requirements
1	Arduino IDE	For microcontroller
2	MIT app inventor	For application development
3	PCB Wizard	For PCB design

3   Implementation

For the demonstration, a demo vehicle is made as shown in Fig. 2. The receiver module is mounted on the vehicle. The transmitters representing signboard of different zones are placed in such a way to provide required distance between them to avoid RF signal interference. The vehicle operation is initiated by doing an authentication in the monitoring app in the vehicle. The actual starting of the vehicle occurs after the alcohol detection. The live location is updated to the remote owner by the IoT enabled monitoring application that runs in the owner's smart phone. The driver mobile phone is interfaced with the controller module with the help of a blue tooth module. The data obtained from the system is updated in the android application in the driver's phone and the monitoring application in the hands of owner simultaneously with the help of IoT (Internet of Things). Inbuilt GSM and GPS in the mobile phone help in the real time tracking of the vehicle.

Fig. 2. Implemented prototype

4   Results

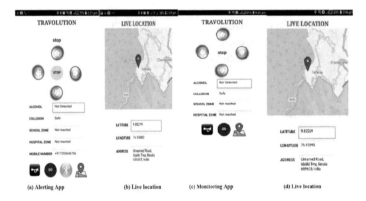

(a) Alerting App   (b) Live location   (c) Monitoring App   (d) Live location

Fig. 3. Real time tracking

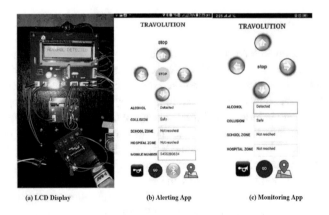

(a) LCD Display   (b) Alerting App   (c) Monitoring App

Fig. 4. Alcohol detection

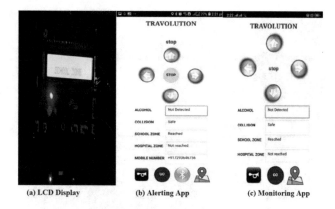

(a) LCD Display   (b) Alerting App   (c) Monitoring App

Fig. 5. School zone identification

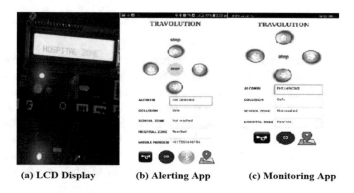

(a) LCD Display   (b) Alerting App   (c) Monitoring App

Fig. 6. Hospital zone identification

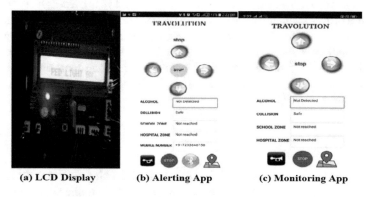

(a) LCD Display   (b) Alerting App   (c) Monitoring App

Fig. 7. Red signal condition

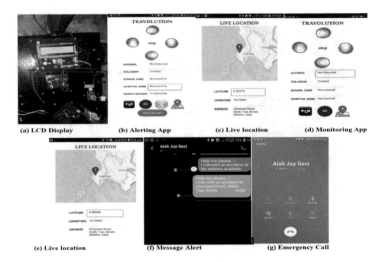

(a) LCD Display　　(b) Alerting App　　(c) Live location　　(d) Monitoring App

(e) Live location　　(f) Message Alert　　(g) Emergency Call

Fig. 8. Accident alert

The IoT enabled android application in the hands of driver and the monitoring application with the owner helps in updating the real time status of the vehicle along with the live location and hence the vehicle can be tracked continuously as shown in Fig. 3.

The alcohol sensor is mounted in the vehicle so that it gets triggered and gives a buzzer sound when exposed to an alcoholic ambience. The application also provides voice alerts along with the text alert as in shown Fig. 4. The response of the vehicle to different speed zones are shown in Figs. 5, 6 and 7. When school zones and hospital zones are detected the controller issues a PWM signal of low average value to the motors, and the vehicle reduces its speed form the normal speed level. And the vehicle gets moved into the normal speed after crossing respective zones. Moreover, in the hospital zones, the horn also gets disabled.

When the vehicle reaches near the traffic signal and if the red light is "ON", the driver gets a buzzer alert and the alerting app shows a red stop signal as given in Fig. 7. On getting this alert the driver can take necessary actions. When the red light turns "OFF" the vehicle starts moving. When any collision happens to the vehicle, a notification is sent in the form of a text message with the current location of the vehicle along with a voice call. The vehicle owner can further watch the collision status remotely as shown in Fig. 8.

5　Conclusion

This paper demonstrates a smart vehicle with real time tracking system that can address the major causes of road accidents like violation of red light traffic signal and drunken driving by an effective alarming system. The different traffic zones are notified by signboards wired with RF transmitter modules. The controller takes the appropriate speed and horn control actions after decoding the traffic zones. As a proof of concept, a cost effective embedded solution is proposed and it is demonstrated with a prototype vehicle. User-friendly android applications are also developed for vehicle owner and drive assistance.

References

1. Tripathy, A.K., Sejalchopra, C.: Travolution-an embedded system in passenger car for road safety. In: ICTSD-Feb 2015
2. Singh, D.N., Teja, R.: Vehicle speed limit alerting and crash detection system at various zones. Int. J. Latest Trends Eng. Technol. (IJLTET) 2(1), 108–113 (2013)
3. Savani, V., Agravat, H., Patel, D.: Alcohol detection and accident prevention ofvehicle. Int. J. Innovative Emerg. Res. Eng. 2(3), 84–91 (2015)
4. Iqbal, J.L.M., HeenaKousa, S.: Automaticvehicle accident detection and reporting with blackbox. In: IJAER, vol. 10, December 2015
5. Malta, L., Miyajima, C., Takeda, K.: A study of driver behaviourunder potential threats in vehicle traffic. IEEE Trans. Intell. Transp. Syst. 10(2), 201–210 (2009)
6. Ramani, R., Valarmathy, S., Vanitha, N.S., Selvaraju, S.: Vehicle tracking and lockingsystem based on GSM and GPS. International Journal of Intelligent Systems and Applications 2 (3), 86–93 (2013)
7. Nishiyama, Y., Hirado, A., Akiyama, A.: The system and the function of positionregulated speed control device. In: Vehicle Navigation and Information Systems Conference, November 1996
8. Gangadhar, S., Shetty, R.N.: An intelligent road traffic control system. In: IEEE Conference Students Technology Symposium (TechSym), April 2010
9. Iyyappan, S., Nandagonal, V.: Automatic accidentdetection and ambulance rescue with intelligenttraffic light system. Int. J. Adv. Res. Electr. Electron. Instrum. Eng. 2(4), 1319–1325 (2013)
10. Kharat, S.B., Sawant, S.D.: Travolution: system for road safety. In: International Conference on Intelligent Computing and Control Systems ICICCS, pp. 1123 – 1125, September 2017
11. Jeevagan, N., Santosh, P., Berlia, R., Kandoi, S.: RFID based vehicle identification during collisions. In: Global Humanitarian TechnologyConference (GHTC), IEEE, pp. 716–720, October 2014
12. Wakure, A.R., Patkar, A.R.: vehicle accident detection and reporting system using GPS and GSM. Int. J. Eng. Res. Dev. 10(4), 25–28 (2014)
13. Tayde, M.S., Phatale, A.P.: Study on intelligent automatic vehicle accident prevention and detection system. In: IJAREEIE, vol. 5, no. 5, May 2016
14. Divya, G., Sabitha, A., SaiSudha, D., Spandana, K., Swapna, N., Hepsiba, J.: Advanced vehicle security systemwith theft control and accident notification using GSMand GPS Module. In: IJIREEICE, vol. 4, no. 3, March 2016

A Survey on Intrusion Detection System Using Machine Learning Algorithms

Shital Gulghane[(✉)], Vishal Shingate, Shivani Bondgulwar,
Gaurav Awari, and Parth Sagar

Department of Computer Engineering,
Rasiklal M. Dhariwal Tech. Campus, Pune, India
shitalgulghane1@gmail.com, vishalshingate26@gmail.com

Abstract. IDS play significant role in the computer network and system. Now a days, research on the intrusion detection that has been use of machine learning applications. This paper proposes novel deep learning technique to empower IDS functioning within current system. The system shows a merging of deep learning and machine learning, capable of accurate analyzing an inclusive range of network traffic. The new approach proposes NDAE for un-supervised feature learning. Moreover, additionally proposes novel deep learning classification display built utilizing stacked autoencoder. Our proposed classifier has been executed in GPU and assessed utilizing the measure using 'KDD' Cup '99' and 'NSL-KDD' datasets. The performance evaluated network intrusion detection analysis datasets, particularly KDD Cup 99 and NSL-KDD dataset.

Keywords: Deep and machine learning · Intrusion detection · Autoencoders · KDD · Novel approach

1 Introduction

The main issues in secure platform are the availability of a strong and adequate NIDS. Regardless of noteworthy progress in NIDS methodology, most of the system still work using techniques based on low skilled companies, disagreeing to usual detection techniques. Current problems are existing techniques that lead to inefficient and inaccurate Attack Detection. There are three main limitations, such as network data volume, detailed monitoring and the granularity required to enhance adequacy and exactness, and finally, the number of protocols and the variety of the information path. The main objective of the NIDS research was the application of machine learning and superficial learning techniques. Initial research into deep learning has shown that his learning of superior characteristics in the strata can improve or at least correspond the result of shallow-learning mechanisms. It is able to facilitate a detailed analysis of network information and a faster examination of any anomaly. In this document, we developed a new model of deep-learning to empower the functioning of NIDS within present network.

© Springer Nature Switzerland AG 2020
J. S. Raj et al. (Eds.): ICIDCA 2019, LNDECT 46, pp. 670–675, 2020.
https://doi.org/10.1007/978-3-030-38040-3_76

2 Related Work

Paper [1] focuses on deep-learning mechanisms that are motivated from the deep framework of the human brain to gain from the characteristic min level to the concept of max levels. Due to multi-level abstraction, the DBN useful to gain the functions assigned from entry to exit. The learning process does not depend on the characteristics created by man. DBN utilizes an un-supervised learning method, a limited RBM for each level. The advantages are in-depth coding is its capacity to transform to modifying contexts related to information, ensuring that method performs an in-depth analysis of the data. It recognizes anomalies in the system which includes the detection of anomalies, the identification of traffic. The disadvantages in this paper are request for faster and more efficient data evaluation.

The objective of the paper [2] is to review the in-depth learning work on monitoring the machines health. The uses of DL in machine checking framework are mainly evaluated based on the multiple characteristics like autoencoder, limited Boltzmann machines including the Deep Belief network. The advantages are DL based MHMS does not need inclusive human job. Uses are: DL models are not limited to specific types of systems. The drawbacks of this paper: The result of DL-based MHMS depends largely on the quality and size of data sets.

It proposes the need of a noise elimination auto-encoder (SDA), It is also a deep-learning methodology, to develop a FDC model for the concurrent extraction and arrangement of features. The SDA model [3] is able to identify the worldwide and an invariant quality in the sensor signals for blemish checking and is powerful against estimation clamor. A LotR consists of a denoising of the automatic encoder which is stacked layers. This multiple layered architecture is able to learn worldwide functionality from difficult information. The advantages are the SdA model is useful in real applications. The SdA model proposes to effectively learn the normal characteristics and relative to the failures of the sensor signals without preprocessing. The disadvantages are the need to study a trained LotR to identify the parameters that have a more significant impact on the classification results.

It proposes a new model of recurrent neural networks (RNN) based on deep learning [4] for the audit provides programmed security of the short messages, which can recognize messages (safe and not unsafe). In this document, word2vec extracts short message function, which acquires information on the order of words and each sentence is assigned to a function vector. In specific, words with a alike meaning are assigned to a alike place in the vector space and then characterize by RNN. The advantages are: the RNNs model obtains an accuracy of 92.7% that has more than SVM. Take advantage of frameworks to integrate different classifier and extraction algorithms of features to increase performance. The disadvantages of this paper it works on only small messages, not on big messages.

The signature-based functionality technique is proposed as a DNN [5] on a cloud for plate localization, text identification and division. The extraction of distinct features ensures that LPRS correctly recognizes the plate in a difficult environment, such as (i) traffic areas there is different plates in picture (ii) representation of the plate to the brightness, (iii) additional data on the plate, (iv) confusion due to corrosion and erosion of images captured in unpleasant weather like blurry pictures. The advantages are: the predominance of the algorithm proposed in the precision of the recognition of LP instead of other traditional LPRS. The disadvantages are: There are some unrecognized images or error detection.

In paper [6], In RBM technique uses an RBM level of a hidden layer to perform function reduction without supervision. The resulting weights of one RBM are transfer to another RBM which produces a network of deep beliefs. The pretrained weights are transfer to a fine regulation level comprising of a Logistic-Regression (LR) classifier. The advantages are it reaches an accuracy of 97.9%. The disadvantages are: the need for improvement of method it required more process of reducing the functions in the deep-learning network and improved the data set.

In paper [7] using sparse autoencoder and soft max regression techniques using that it helps to develop better NIDS. It uses the STL on 'NSL-KDD' - a measure dataset for NIDS. advantages are: Using STL it archived all types of classification adequacy rate is greater than 98%. Disadvantages are in this paper still multiple thing are left like need of real time NIDS, in fetcher learning can be traffic less.

In [8] paper uses multi-core CPU and GPU which enhance the result of the DNN based IDS to manage enormous network data. With accelerated performance, the parallel processing abilities of the neural system put up Deep Neural Network (DNN) to successively monitors through the network traffic. Advantages are: This system are reliable it accurately detects the attacks and classify with the attack classes of input training dataset. Disadvantages are: Need to increase the accuracy in attack identification.

In [9] paper, proposes a methodology for identifying wide scale on network attacks using RNNs for making anomaly detection framework. Labeled data is not required for unsupervised technique. It also adequately detects anomalies on network without assuming that the trained data does not have risk of attack. Advantages are: The proposed technology is able to effectively finds all distinguished DDoS attacks and *SYN Port* scans introduced. Our proposed methodology is flexible when attacks are present. Disadvantages are: Need to improve proposed methodology by using stacked autoencoder DL techniques.

In [10] paper, apply a DL approach for stream-based oddity recognition in a SDN Environment. Advantages are: It mainly choose one type of attack that's why it gives more adequacy. The model picks up the outcome with ampleness of 75.75% which is possible from simply utilizing six fundamental system highlights. Disadvantages are: It is difficult to take a shot at genuine SDN framework.

3 Existing System

Current data on network traffic, which is huge, is a big issue for IDS. These "enormous data" drops the whole recognition procedure and leads to unacceptable categorized adequacy due to problems in calculating computing of such data. ML techniques have usually been used in IDS. However, the long-established machine learning techniques refer to surface learning; they are not able to successfully find the way of dealing the huge problem of intrusion data characterization that found in real world applications.

Additionally, superficial learning is incompatible with tactical intelligence and the predestined requirements of huge-size learning with enormous data.

Disadvantages

PC framework and web is major part of the basic framework. Present network traffic information, which are often tremendous in volume, causes a huge challenge to IDSs. These "enormous data" drops the complete reorganization procedure and which results in unacceptable categorized adequacy because of computational complexities in managing data. Characterizing a huge amount of data usually results into more numerical complexities which then results to greater computational difficulties.

4 System Overview

In this document, propose a new model of deep learning to enable the functioning of NIDS within present networks. The model it proposes is an arrangement of profound and superficial learning, able to properly examine huge areas of network traffic. More particularly, we merge the capacity of stacking our proposed NDAE and the adequacy and rate of RF (surface learning). This document presents NDAE, which is an automatic encoder that has different hidden, non-symmetrical levels. NDAE can be used as a arranged in order of extractor of unsupervised functions that adapts well to large put in data. Find significant features using a imparting system like that of automatic encoder. The NDAE stacking presents an un-supervised representation learning method in layers, which will make our framework to get the different relations between the various functions. It has functionality to extract features, so you can refine the model by giving priority to more descriptive features (Fig. 1).

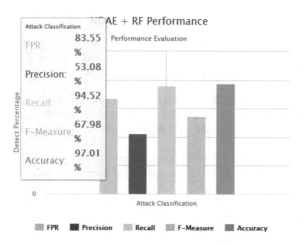

Fig. 1. Performance comparison graph

Advantages:

- Increase accuracy or adequacy of intrusion detection system using deep learning algorithms.
- The framework can be altered dependent on the necessities of customer.
- It gives easy to use interface user which permits easy security board framework.

5 Conclusion

In this survey paper, we have observed the issues appeared by the existing intrusion recognition system. To overcome issues, we have put forward new NDAE system for un-supervised learning. Based on this we preferred novel classification model built from stacked NDAEs and the RF classifier. In this survey paper, it demonstrates that our methodology significant-level precision, adequacy and less training time. The accomplished NIDS system is increased by just 6% of adequacy. Hence there is a requirement for further betterment in precision.

References

1. Dong, B., Wang, X.: Comparison deep leaning method to traditional methods using for network intrusion detection. In: Proceedings of the 8th IEEE International Conference on Communication Software and Networks, Beijing, China, pp. 581–585, June 2016
2. Zhao, R., Yan, R., Chen, Z., Mao, K., Wang, P., Gao, R.X.: Deep learning and its applications to machine health monitoring: a survey. Submitted to IEEE Trans. Neural Netw. Learn. Syst. (2016). http://arxiv.org/abs/1612.07640
3. Lee, H., Kim, Y., Kim, C.O.: A deep learning model for robust wafer fault monitoring with sensor measurement noise. IEEE Trans. Semicond. Manuf. 30(1), 23–31 (2017)
4. You, L., Li, Y., Wang, Y., Zhang, J., Yang, Y.: A deep learning based RNNs model for automatic security audit of short messages. In: Proceedings of the 16th International Symposium on Communications and Information Technologies, Qingdao, China, pp. 225–229, September 2016
5. Polishetty, R., Roopaei, M., Rad, P.: A next-generation secure cloud based deep learning license plate recognition for smart cities. In: Proceedings of the 15th IEEE International Conference on Machine Learning and Applications, Anaheim, CA, USA, pp. 286–293, December 2016
6. Alrawashdeh, K., Purdy, C.: Toward an online anomaly intrusion detection system based on deep learning. In: Proceedings of the 15th IEEE International Conference on Machine Learning and Applications, Anaheim, CA, USA, pp. 195–200, December 2016
7. Javaid, A., Niyaz, Q., Sun, W., Alam, M.: A deep learning approach for network intrusion detection system. In: Proceedings of the 9th EAI International Conference on Bio-inspired Information and Communications Technologies, pp. 21–26 (2016). http://dx.doi.org/10.4108/eai.3-12-2015.2262516
8. Potluri, S., Diedrich, C.: Accelerated deep neural networks for enhanced intrusion detection system. In: Proceedings of the IEEE 21st International Conference on Emerging Technologies and Factory Automation, Berlin, Germany, pp. 1–8, September 2016

9. Garcia Cordero, C., Hauke, S., Muhlhauser, M., Fischer, M.: Analyzing flow-based anomaly intrusion detection using replicator neural networks. In: Proceedings of the 14th Annual Conference on Privacy, Security and Trust, Auckland, New Zealand, pp. 317–324, December 2016

10. Tang, T.A., Mhamdi, L., McLernon, D., Zaidi, S.A.R., Ghogho, M.: Deep learning approach for network intrusion detection in software defined networking. In: Proceedings of the International Conference on Wireless Networks and Mobile Communications, pp. 258–263, October 2016

11. Shone, N., Ngoc, T.N., Phai, V.D., Shi, Q.: A deep learning approach to network intrusion detection. IEEE Trans. Emerg. Top. Comput. Intell. 2(1), 41–50 (2018)

12. Aung, Y.Y., Min, M.M.: An analysis of random forest algorithm-based network intrusion detection system. In: Proceedings of the 18th IEEE/ACIS International Conference on Software Engineering, Artificial Intelligence, Networking and Parallel/Distributed Computing, pp. 127–132, June 2017

Automatic Greenhouse Parameters Monitoring and Controlling Using Arduino and Internet of Things

More Hemlata Shankarrao[✉] and V. R. Pawar

Department of Electronics and Telecommunication,
Bharati Vidyapeeth College of Engineering for Women,
Savitribai Phule Pune University, Dhankawadi, Pune, India
morehemlata123@gmail.com, pawarvr74@gmail.com

Abstract. Greenhouse is provide controlled area environment to grow plants. Growth of the plant always affected by key environmental parameters such as temperature, humidity, light intensity, moisture etc. In the present system, environmental parameters in a greenhouse are monitored and controlled. Sensors are used for data aquisition and are interfaced with microcontroller unit. Android application is developed to display environmental parameters. To control temperature fan is used. To control humidity fogger is used. To control soil moisture water pump is used. To control light intensity artificial light source is used. Wifi module is interfaced with MCU. Greenhouse parameters can be monitored & controlled using IoT.

Keywords: Greenhouse monitoring and controlling · IoT · Arduino · Sensors · Android application

1 Introduction

India is an agriculture country and 90% population of rural India is having agriculture as a main source of livelihood. Nowadays due to natural calamities growth of the plant and production of the plant is affected. Natural calamities are destroying the crops and environmental conditions are greatly affected. In the present system, greenhouse with controlled environment is used for crop production [1, 7]. The crop production and growth of plant depends on environmental conditions maintained inside a greenhouse [3]. Using IoT and android application is possible to develop automatic greenhouse parameters monitoring and controlling system [4, 9]. IoT based greenhouse system is designed using arduino. Environmental data is aquired by using sensors and based on environment conditions water pump, light sources and other actuators are controlled. Android application is developed to monitor the data obtained from arduino after processing [10]. The present system is portable and easily installable. After installing the system farmer can monitor and control environmental parameters inside a greenhouse which may be at a remote location. Any change in the environmental parameters could not lead to financial lose and crop lose to the farmers. So we need immediate controlling actions to prevent this loses.

© Springer Nature Switzerland AG 2020
J. S. Raj et al. (Eds.): ICIDCA 2019, LNDECT 46, pp. 676–683, 2020.
https://doi.org/10.1007/978-3-030-38040-3_77

2 Literature Survey

Many researchers reported the work related to the present domain at national and international level. Cognizance of reported work is taken for the development of present system. Mois et al. [1] has developed a system in which three wireless sensors are used for monitoring environmental parameters. Wifi communication based on User Datagram Protocol (UDP) and communication based on Hypertext Transfer Protocol (HTTP), and a system based on Bluetooth Smart are used. Data can be recorded at remote locations and can visualize with internet connection. Koshy et al. [6] proposed a prototype IoT based system for greenhouse monitoring and control using WSN. The system monitors various environmental parameters such as humidity, soil moisture, temperature etc. The system automization is done if the soil moisture is less than particular value and notification message will be send to the registered mobile number. Vimal et al. [7] has built a system in which monitoring approach is used. Data is aquised through various sensors such as DHT11, moisture sensor, and LDR sensor etc. Message will send through GSM network to control and monitor environmental parameters. Adhao et al. [8] has developed IoT based system for cotton leaf disease detection and disease control. Detected disease name with its remedies send to farmer using mobile app. Also monitor soil parameter and control to achieve optimum plant growth.

In the present work, IoT based approach is used to monitor and control environmental parameters. Android application is used for remote monitoring of greenhouse.

3 System Architecture

System comprises two main units, monitoring unit and controlling unit. The monitoring unit consists of relative humidity and temperature sensor (DHT11), light intensity (LDR), soil moisture sensor. Wifi module is interfaced with arduino and can send aquented data to android phone. Android application is developed to monitor greenhouse parameter information [2]. The controlling unit consists of cooling fan to control temperature. Water pump is used to control soil moisture. Fogger is used to control humidity. Artificial light source is used to control light intensity inside a greenhouse. All the sensors are interfaced with arduino which are the key part of system.

Android application is developed to display the present status of the greenhouse parameters. By using Android app, farmer can control the greenhouse parameters using turning ON/OFF the external devices such as water pump, fan, light to achieve optimum plant growth inside a greenhouse [8].

3.1   Block Diagram of Present System

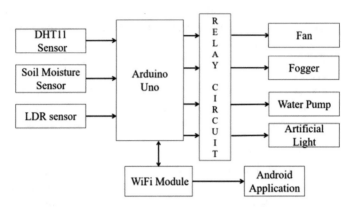

Fig. 1. Block diagram of proposed system

Greenhouse parameters can controlled to ensure optimal growth and productivity [3]. The system designed is as shown in Fig. 1. Arduino Uno is interfaced with temperature and humidity sensor (DHT11), moisture sensor, light dependent resistor sensor (LDR) etc. Arduino Uno is offering necessary firmware essential for physical computing of environmental parameters. Arduino consists ATmega chip which provides various libraries used for sensor control and allows various sensors to be used. Arduino programs are written in the Integrated Development Environment (IDE). Figure 2(a) shows DHT11. Relative humidity and temperature sensor (DHT11) aquires relative humidity and temperature serially over a single wire. DHT11 measure relative humidity (20% RH to 90% RH) and temperature (0 °C to 50 °C) [6]. Soil moisture measures moisture present in soil. Input operating voltage of moisture sensor is 3.3 to 5 V. It has sensitivity adjustable feature [5, 8]. Figure 2(b) shows soil moisture sensor. Light dependent resistor sensor (LDR) measure light intensity present inside greenhouse. LDR sensor has both analog and digital output pin. The cell resistance falls with increasing light intensity. Sensitivity towards light can be adjusted using potentiometer knob. Light dependent resistor sensor consists two leg, one leg is connected to VCC and second to analog pin 0. Resistor of 100 K is connected to the same leg and is grounded. Analog voltage (0 V–5 V) converts into a digital value (0–1023) with the help of inbuilt analog to digital converter (ADC). Figure 2(c) shows LDR sensor. Node MCU ESP8266 wifi module is used to send greenhouse parameters information from sensors to android application which display current environmental parameters.

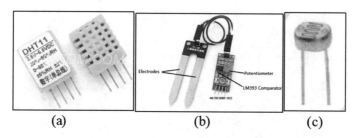

(a)   (b)   (c)

Fig. 2. Different sensors used for greenhouse parameters monitoring (a) DHT11 temperature and humidity sensor (b) Moisture sensor (c) LDR sensor

3.2 System Design

Arduino Uno and various sensors are initialized by providing +5 V power supply. Wifi module requires 3.3 V power supply. Relative humidity, temperature, soil moisture, light intensity values are aquired by using DHT11, soil moisture, LDR sensors and send to arduino uno. Arduino can acquire sensor values with sensor library and analogRead function. The program written in the Integrated Development Environment (IDE) is repeatedly executed and program is uploaded to the Arduino via USB. The threshold value of each parameter is set. For temperature threshold is set to 32 °C. If the temperature is more than threshold value then fan is ON otherwise it remains OFF. For humidity threshold is set to 10% RH. If the humidity is more than threshold value then fogger is ON otherwise it remains OFF. For soil moisture threshold is set to 30%. If the soil moisture is lower than threshold then water pump is ON otherwise it remains OFF. Similarly for light intensity threshold is set to 70. If the light intensity is lower than threshold then light is ON otherwise it remains OFF. Node MCU ESP8266 wifi module is connected with wifi network by updating username and password to wifi access point. Sensors read greenhouse environment parameters data and send to android mobile user through wifi network. This system concerns major role in monitoring environmental parameters.

Android Studio software is used to develop android application for greenhouse automation. Wifi connection is established between Node MCU wifi module along with arduino uno and android mobile application by using socket programming. Greenhouse environment parameters are controlled automatically by simply turning ON_OFF the relay circuit to control actuators like fan, water pump, fogger, light source etc. [8, 12]. Also mobile user can turns ON the actuators by using android app. This system concerns major role in controlling environmental parameters inside a greenhouse. Overall system implementation for monitoring and controlling greenhouse environment parameters is shown in Fig. 3.

Fig. 3. Overall system implementation for monitoring and controlling greenhouse environment parameters.

4 Result and Discussion

The main aim of the primary aim of the proposed system is to monitor and control greenhouse parameters. The secondary aim is to develop an android application for display parameter value and control these using actuators. Prototype of present system is developed. Parameters such as relative humidity, temperature, moisture content in soil, light intensity inside a greenhouse are monitored. Growth of a rose plant observed in various climatic conditions. Quality flowers are obtained with moderate humidity, cool night and absence of strong winds. For proper growth of plant and flower production requires day temperature of 25 °C to 32 °C and night temperature 15 °C to 16 °C. Humidity of 60% RH to 70% RH is requires for proper growth and quality flower production. But development stage is influenced by light intensities. Additional light and temperature is promotes early flowering under protected climate conditions.

The system is having two modes automatic control mode and manual mode. Farmer can turn ON_OFF actuators by converting mode auto to mode manual in android application. In greenhouse environment parameter monitoring and controlling, accuracy is tested along with plant growth in various climatic conditions. Android application displaying greenhouse environment parameters information in different modes in auto mode and in manual mode shown in Fig. 4.

Table 1 reveals that

1. If the temperature exceeds 32 °C then fan is ON otherwise it remains OFF.
2. If the humidity exceeds 10% RH then fogger is ON otherwise it remains OFF.
3. If the soil moisture is lower than 30% then water pump is ON otherwise it remains OFF.
4. If the light intensity is lower than 70 then light is ON otherwise it remains OFF.

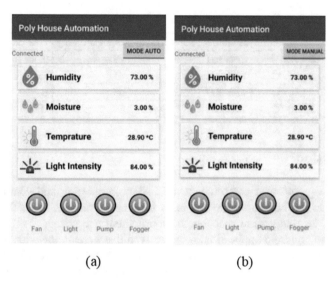

(a)   (b)

Fig. 4. Android app displaying greenhouse environment parameters information in different modes (a) auto mode (b) manual mode

Table 1. Monitoring and controlling of greenhouse parameters

Sr. no	Name of parameter	Time	Monitored values	Controlling actions
1	Temperature	Morning (6 am)	29.90 °C	Fan is OFF
		Afternoon (2 pm)	32.90 °C	Fan is ON
		Evening (7 pm)	32.10 °C	Fan is ON
		Night (11 pm)	29.70 °C	Fan is OFF
2	Humidity	Morning (6 am)	69% RH	Fogger is ON
		Afternoon (2 pm)	73% RH	Fogger is ON
		Evening (7 pm)	71% RH	Fogger is ON
		Night (11 pm)	70% RH	Fogger is ON
3	Soil moisture	Morning (6 am)	34%	Water pump is OFF
		Afternoon (2 pm)	29%	Water pump is ON
		Evening (7 pm)	70%	Water pump is OFF
		Night (11 pm)	30%	Water pump is ON
4	Light	Morning (6 am)	79	Light is OFF
		Afternoon (2 pm)	85	Light is OFF
		Evening (7 pm)	65	Light is ON
		Night (11 pm)	60	Light is ON

Table 2 reveals that plant growth is good when temperature is in range of 25 °C to 32 °C and during night temperature is in range of 15 °C to 16 °C. Humidity is in range of 60% RH to 70% RH then plant growth is good. If humidity exceeds than 70% RH then plant growth is moderate. Additional light and moisture increases growth and production of plant.

Table 2. Plant growth monitored w.r.t. environmental parameters

Sr. no	Temperature	Humidity	Soil moisture	Light intensity	Plant growth
1	29.90 °C	69% RH	34%	79	Good
2	32.90 °C	73% RH	29%	85	Moderate
3	32.10 °C	71% RH	70%	65	Better
4	29.70 °C	70% RH	30%	60	Ok

5   Conclusion

An automatic greenhouse environment parameters monitoring and controlling using arduino and IoT is built on prototype basis which requires less than 50 units power consumption. In this system greenhouse environment parameters are monitored and controlled using android application. It is observed that system operational complexity is low and maintenance cost is less and is easy to use remotely with IoT platform. The proposed system could be extended for more crop related parameters to achieve good crop growth and production.

References

1. Mois, G., Folea, S., Sanislav, T.: Analysis of three IoT-based wireless sensors for environmental monitoring. IEEE Trans. Instrum. Meas. **66**(8), 2056–2064 (2017)
2. Varalakshmi, P., Sivashakthivadhani, B.Y., Sakthiram, B.L.: Automatic plant escalation monitoring system using IoT. In: 2019 3rd International Conference on Computing and Communications Technologies (ICCCT), pp. 212–216. IEEE (2019)
3. Abbasi, M., Yaghmaee, M.H., Rahnama, F.: Internet of Things in agriculture: a survey. In: 2019 3rd International Conference on Internet of Things and Applications (IoT), pp. 1–12. IEEE (2019)
4. Marques, G., Aleixo, D., Pitarma, R.: Enhanced hydroponic agriculture environmental monitoring: an Internet of Things approach. In: International Conference on Computational Science, pp. 658–669. Springer, Cham (2019)
5. Gutierrez, J., Villa-Medina, J.F., Nieto-Garibay, A., Gandara, M.A.P.: Automated irrigation system using a wireless sensor network and GPRS module. IEEE Trans. Instrum. Meas. **63**(1), 166–176 (2014)
6. Koshy, R., Yaseen, M.D., Fayis, K., Shaji, N., Harish, N.J., Ajay, M.: Greenhouse monitoring and control based on IOT using WSN. ITSI Trans. Electr. Electron. Eng. **4**(3), 59–64 (2016)
7. Vimal, P.V., Shivaprakasha, K.S.: IOT based greenhouse environment monitoring and controlling system using Arduino platform. In: 2017 International Conference on Intelligent Computing, Instrumentation and Control Technologies (ICICICT), Kannur, pp. 1514–1519 (2017)
8. Adhao, A., Pawar, V.R.: Automatic cotton leaf diagnosis and controlling using Raspberry Pi and IoT. In: Intelligent Communication and Computational Technologies, pp. 157–167 (2018)
9. Akkaş, M.A., Sokullu, R.: An IoT-based greenhouse monitoring system with Micaz motes. Procedia Comput. Sci. **113**, 603–608 (2017)

10. Patel, U.S., Saiprasad, S., Veerabhadra, K.J.: Green house monitoring and controlling using Android mobile app. Int. J. Comb. Res. Dev. (IJCRD) **5**(5), 710–714 (2016)

11. Nikkam, S.M.G.: Water parameter analysis for industrial application using IoT. In: 2nd International Conference on Applied and Theoretical Computing and Communication Technology (iCATccT), pp. 703–707. IEEE (2016)

12. Khot, S., Gaikwad, M.S.: Green house parameters monitoring system using Raspberry Pi and Web Server. Int. J. Innov. Res. Sci. Eng. Technol. **5**(5), 8424–8429 (2016)

withall: A Shorthand for Nested for Loop + If Statement

Tomsy Paul[1(⌧)] and Sheena Mathew[2]

[1] Rajiv Gandhi Institute of Technology, Kottayam, India
tomsy@rit.ac.in
[2] School of Engineering,
Cochin University of Science and Technology, Kochi, India
sheenamathew@cusat.ac.in

Abstract. The combination of for loop and if statement is very common in array programming. The for loop iterates over the elements in the array, the if statement filters out the required elements and are processed. A shorthand for this combination is developed from the observation that the information required for the loop indexing is often available in the processing statement. We also present a fast and efficient implementation of the loop for Single Dimensional arrays, with a minute increase in time for compilation.

Keywords: Array processing · For loop · If statement · Programming language construct

1 Introduction

A common situation in array programming is the selective processing of elements. However, no programming construct is available in any programming Language in the literature to easily achieve this goal. The conventional technique is to club iteration and selection. A for or similar loop is used to iterate over the entire array. Then the selection of elements are carried out using an if statement followed by the processing of selected elements.

Collective initialization of array elements are supported by many languages. One can use a function like 'fill' function of Julia [1] or 'full' routine of NumPy [2] or use direct assignment of a scalar value to array as in Chapel [3]. Collective processing is also allowed in many languages. However selective processing is available only in very few languages like SAC [4]. Here, 'with' loop [5] is used for processing array elements for a given range of indices. A limitation of 'with' loop is that it can not be used with arbitrary expressions involving array indices.

We present a new programming construct, 'withall', which is a concise representation of the nested for and if combination and is explicitly meant for array processing. The primary goal of this construct is to be more handy from the programmer's view and to be more semantically sound. At the same time we could implement the loop for Single Dimensional arrays that performs equally well in worst cases and far better for other cases, that is, if the selection results in Sparse arrays or if the array size is large. The overhead incurred is only a slight increase in time for compilation.

© Springer Nature Switzerland AG 2020
J. S. Raj et al. (Eds.): ICIDCA 2019, LNDECT 46, pp. 684–691, 2020.
https://doi.org/10.1007/978-3-030-38040-3_78

This paper is outlined as follows. Section 2 gives typical cases of the conventional method. It is followed by the presentation of the proposed loop - The syntax, semantics and illustrations using these cases. Then an implementation of the loop in Chapel, a recent language is given in Sect. 4. In Sect. 5, a number of comparisons of 'withall' and conventional method are given. Then in Sect. 6, the analyses of the results are given, followed by conclusion and future scope in Sect. 7.

2 Background

Consider the initialization of Zero Matrix as shown in pseudo code 1,

$$A[1..M,1..N] : int;$$
$$for\ i:=1\ to\ M\ step\ 1$$
$$for\ j:=1\ to\ N\ step\ 1$$
$$A[i,j]:=0;------->(1)$$

Here, all elements in A are processed. The index variables i and j of the for loops run through the first and second dimensions of A respectively. The left hand side of the assignment statement contains A, which is the name of the array. The for loops need only the ranges of the dimensions of A, that is 1..M and 1..N respectively.

Now consider the initialization of the Anti diagonal elements as shown in pseudo code 2.

$$A[1..N,1..N] : int;$$
$$for\ i:=1\ to\ N\ step\ 1$$
$$for\ j:=1\ to\ N\ step\ 1$$
$$if\ i+j = N+1\ then\ A[i,j]:=1;------->(2)$$

The if statement filters out the diagonal elements. The selection of the elements are decided by the condition, $i+j=N+1$. As in the previous case, the for loops need only the range of the dimensions of A.

In the following pseudo code, an expression involving a logical operator is given. Again, the for loop needs only the range of the dimension of A.

$$A[1..N] : int;$$
$$for\ i:=1\ to\ N\ step\ 1$$
$$if\ i>=N/4\ \&\&\ i<=N/2\ then\ A[i]:=0;------->(3)$$

In pseudo code 4, the condition in if is much more complex and the statement part has reference to three arrays, having the same dimension. However the for loop needs only the range of the dimension.

A[1..N], B[1..N], C[1..N] : int;
for i:=1 to N step 1
if i>=N/4 && i<=N/2 || i>=3*N/4 && i<=N then A[i]:=B[i]+k*C[i];   ------> (4)

Now we present the 'withall' loop.

3 The 'withall' Loop

3.1 Syntax

The syntax of 'withall' loop is,
     withall expression statement
where 'expression' and 'statement' are the condition and the then part of the if statement of the corresponding nested for+if.
    'expression' can also be a list of index variables used in 'statement'.

3.2 Semantics

The 'statement' should contain indexing on at least one dimension of an array and 'expression' should contain at least one of the index variables used in 'statement'. More than one array can be accessed in 'statement'. But all of them should be using exactly the same index variables and also in the same order.

The array index variables in 'statement' are identified. Their occurrence in 'expression' represent the loop indexing for the corresponding dimension of the array. N indices in 'statement' corresponds to N-level nesting of for loops.

If 'expression' is simply a list of index variables, then 'statement' is applied to all elements in the array.

It is also assumed that, as in the case of a 'for' loop, the index variables should not be altered within 'statement'.

3.3 Illustration

The pseudo codes in Sect. 2 can be written respectively as,

(1) withall i,j A[i,j]:=0;
(2) withall i+j=N+1 A[i,j]:=1;
(3) withall i>=N/4 && i<=N/2 A[i]=0;
(4) withall i>=N/4 && i<=N/2 || i>= 3*N/4 && i<=N A[i]=B[i]+k*C[i];

4   Implementation

'withall' could be implemented simply as a wrapper above the nested for+if, which would make it just a syntactic sugar and would perform equally well with List Comprehensions [6] and Boolean (Logical) Indexing [7] used in many functional programming languages. But here a technique is presented that performs remarkably well for Single Dimensional arrays. The idea is to solve expression in Sect. 3.1 for the index variable and use the solutions as array indices to execute the statement. This reduces the number of iterations tremendously. It is more significant when expression results in a Sparse array. Another advantage is the saving of time for evaluation of expression that would happen in every iteration of the corresponding for loop.

Although the solving of expression, in general, is NP Hard and requires much runtime calculations, a simple restriction on the syntax of expression proven to be fruitful.

4.1   Restricted Grammar

```
withall_loop
-> withall withall_expression statement
| withall ID statement
;
withall_expression -> withall_expression && withall_expression
| withall_expression || withall_expression
| (withall_expression)
| !withall_expression
| withall_term
;
withall_term -> ID RELOP expression
;
```

where RELOP can be any of the six operators <, <=, !=, ==, >= and > and ID is the index variable used in statement.

The restriction that LHS of a withall_term consists only of the index variable is in agreement with the commonly used expressions appearing in array processing. It is also easy to represent every withall_term as a set of indices. The indices are stored in the form of a linked list of tuples. Each tuple has the form (low, high) and denotes the range of indices low through high. This representation results in generating the indices on the fly thus making the loop faster.

4.2   Language for Implementation

Chapel language [8] is selected for implementation due to many reasons. They include Free and Open Source nature, excellent support, LR-parsing and the most appealing, the use of modules and iterators. The use of modules [9] allowed much of the coding for 'withall' in Chapel itself. Also iterators [10] are quite suitable for array indexing.

4.3   Implementation Procedure

On reducing a withall_term, a list of tuples (low,high) is created which represents the range of indices corresponding to the withall_term. Assuming the dimension of the array to be 1..N, the term i creates the list {(1,N)}, the term i <= x creates the list {(1, x)} and i != x creates the list {(1,x−1),(x + 1,N)}. The && and || operators are implemented as the set intersection and union of the two lists corresponding to the operands. Since the tuples in the lists are sorted, these operations can be performed in O (m + n) where m and n are the number of tuples in the lists. The expression E can be complemented in O(n) where n is the number of tuples in the list of E.

An issue faced in the implementation is due to the back patching needed to get the array indices from statement to withall_expression which is needed to find the range of dimension. This is essential since the withall_expression precedes statement as per the syntax of 'withall'. It is achieved by using an array of pointers to store dummy values for array names during the reduction of withall_term and updating these during the recognition of array access in statement. Another requirement is regarding the condition that the LHS of every withall_term should be the same as the index variable. An array of strings is used to store the LHS variables during reduction of withall_term and they are compared against the array index at the reduction of withall_loop.

5   Results

The Chapel compiler is modified to accommodate the 'withall' loop. Chapel version 1.14 is used. Sample programs that use both the conventional loop and the equivalent 'withall' loop are created for various array sizes and withall_expressions. The withall_expressions used for testing are summarized in Table 1. The expression i gives the maximum number of iterations (N iterations) whereas i==N/2 gives the minimum (just 1 iteration). The expression !(i>=N/4&&i<=N/2||i>=3*N/4&&i<=N) is the most complex one.

Each program is executed 10 times and the average times are taken. The time for execution of the two loops are measured using the Timer record of the Time module [11] of Chapel. The time for compilation is measured using separate programs for conventional loop and 'withall' loop using GNU 'time' command [12].

The platform used is an Intel Xeon 2 GHz, 6 core processor based system with 8 GB RAM.

Table 1. withall_expressions used for testing

Number	withall_expression
1	i
2	i==N/2
3	i<=N/2
4	!(i>N/2)
5	i<=N/4 || i>=3*N/4
6	i>=N/4 && i<=N/2
7	!((i>=N/4 && i<=N/2) || (i>=3*N/4 && i<=N))

The results are summarized in Tables 2, 3 and 4. Comparisons of execution time, object code size and compilation time are given in the tables. The performance of 'withall' is very effective for array size greater than 10^3. In other cases, the performance is less due to the extra overhead associated with the implementation.

6 Analysis of Results

Three Tables of comparisons between 'withall' and for+if based on Time of Execution, Target code size and Time of Compilation are shown in Tables 2, 3 and 4 respectively. The rows show commonly found expressions in array processing and columns give different array sizes(N).

Table 2. Execution times in seconds, for various withall_expressions and array size

withall_expression		Array size			
		10^3	10^4	10^6	10^8
1	w^a	2.10×10^{-6}	1.93×10^{-5}	1.18×10^{-3}	1.50×10^{-1}
	$f+i^b$	2.50×10^{-6}	1.50×10^{-5}	1.16×10^{-3}	1.61×10^{-1}
2	w	9.00×10^{-7}	1.00×10^{-6}	1.00×10^{-6}	1.80×10^{-6}
	$f+i$	2.50×10^{-6}	1.75×10^{-5}	1.68×10^{-3}	1.71×10^{-1}
3	w	1.80×10^{-6}	1.14×10^{-5}	6.04×10^{-4}	7.52×10^{-2}
	$f+i$	2.90×10^{-6}	1.96×10^{-5}	1.83×10^{-3}	1.84×10^{-1}
4	w	2.50×10^{-6}	1.19×10^{-5}	6.07×10^{-4}	7.44×10^{-2}
	$f+i$	2.70×10^{-6}	1.98×10^{-5}	1.83×10^{-3}	1.84×10^{-1}
5	w	3.40×10^{-6}	1.10×10^{-5}	5.71×10^{-4}	7.23×10^{-2}
	$f+i$	3.30×10^{-6}	2.64×10^{-5}	2.45×10^{-3}	2.45×10^{-1}
6	w	2.00×10^{-6}	7.00×10^{-6}	3.19×10^{-4}	3.75×10^{-2}
	$f+i$	3.30×10^{-6}	2.43×10^{-5}	2.36×10^{-3}	2.34×10^{-1}
7	w	5.30×10^{-6}	1.52×10^{-5}	5.96×10^{-4}	7.00×10^{-2}
	$f+i$	4.20×10^{-6}	3.31×10^{-5}	3.37×10^{-3}	3.13×10^{-1}

[a]w denotes 'withall'
[b]f+i denotes for loop + if statement

It is observed from Table 2 that the speed of execution of 'withall' is better than or equal to that of for+if for large values of N and for expressions where the resultant array has more sparsity. For all other cases, 'withall' has almost the same speed as that of for+if.

In Table 3, the target code size is independent of the value of N. In the case of for +if, the code size is more or less the same for different expressions. However, for 'withall', the complexity of the expression can affect the target code size, especially for logical && and || operators that require set intersection and union respectively. In all cases, the target code size of 'withall' is slightly more than that of for+if. However the increase is quite negligible (nearly 0.1%).

Table 3. Target Code size in Kilobytes, for various withall_expressions and Array Size

withall_expression	Target code size	
	w	f+i
1	3667.89	3667.8
2	3668.04	3667.8
3	3668.04	3667.8
4	3668.08	3667.8
5	3672.37	3667.76
6	3672.27	3667.76
7	3672.40	3667.8

Table 4. Time of compilation in seconds, for various withall_expressions and array size

withall_expression		Array size			
		10^3	10^4	10^6	10^8
1	w	10.19	10.68	10.59	10.26
	f+i	10.52	10.42	10.14	9.98
2	w	10.60	10.38	10.42	10.36
	f+i	10.41	10.19	10.17	10.18
3	w	10.29	10.49	10.40	10.36
	f+i	10.13	10.17	10.21	10.16
4	w	10.48	10.45	10.43	10.39
	f+i	10.43	10.23	10.14	10.13
5	w	10.85	10.96	10.35	10.71
	f+i	10.64	10.09	10.11	10.11
6	w	11.21	10.51	10.43	10.45
	f+i	10.63	10.44	10.10	10.11
7	w	10.57	10.60	10.18	10.56
	f+i	10.50	10.14	10.09	10.02

Table 4 shows that there is a slight increase of 8% in time for compilation and it is attributed to the Chapel compilation process which has around 40 passes and the Chapel flag –fast, used for code optimization.

7 Conclusion and Future Scope

The 'withall' construct is a shorthand for nested for loop+if statement used commonly in array processing. It enhances ease of programming. The implementation is much faster with a slight increase in time for compilation. For all the commonly used expressions, tuple based 'withall' implementation shows better performance compared to for+if. Hence 'withall' can be a better alternative to for loop+if statement.

'withall' can be extended for multidimensional arrays. The authors have already started working for 2-D arrays. The idea of 'list of tuples', is extended to 'list of polygons' and the initial results are promising.

Acknowledgements. The authors express their gratitude to all members of the Chapel development team, especially to Brad Chamberlain and Michael Ferguson for lending a helping hand during the times of dilemma. The time and effort they have spent to prepare lengthy reply mails in the layman's language deserves thanks beyond words.

References

1. Arrays: The Julia Language. https://docs.julialang.org/en/latest/base/arrays/#Base.fill
2. numpy.full—NumPy v1.18.dev0 Manual. https://numpy.org/devdocs/reference/generated/numpy.full.html
3. Arrays—Chapel Documentation 1.20. https://chapel-lang.org/docs/primers/arrays.html
4. index [SaC-Home]. http://www.sac-home.org/doku.php
5. Grelck, C.: Shared memory multiprocessor support for functional array processing in SAC. J. Funct. Program. **15**(3), 353–401 (2005)
6. Data Structures—Python 3.7.5rc1 documentation. https://docs.python.org/3/tutorial/data structures.html
7. Indexing— umPy v1.18.dev0 Manual. https://numpy.org/devdocs/reference/arrays.indexing.html#boolean-array-indexing
8. Chapel: Productive Parallel Programming. https://chapel-lang.org/
9. Modules—Chapel Documentation 1.20. https://chapel-lang.org/docs/primers/modules.html
10. Iterators—Chapel Documentation 1.20. https://chapel-lang.org/docs/primers/iterators.html
11. Time—Chapel Documentation 1.20. https://chapel-lang.org/docs/modules/standard/Time.html
12. GNU Time - GNU Project - Free Software Foundation. https://www.gnu.org/software/time/

Smart Cane-An Aid for the Visually Challenged

Swarnita Venkatraman, Kirtana Subramanian, Chandrima Tolia$^{(\boxtimes)}$,
and Ruchita Shanbhag

Department of Electronics and Telecommunication Engineering,
SVKM's NMIMS Mukesh Patel School of Technology Management
and Engineering, Mumbai, India
swarnital998@gmail.com, kirtana.anu@gmail.com,
chandrimatolia.nmims@gmail.com, rushanbhag@gmail.com

Abstract. This paper deals with the design of a smart cane to cater to the visually impaired in order to make them independent, conscious of their surroundings and able to move with ease. The various features are incorporated on to the cane as the visually impaired have been accustomed to it since many years. The designed cane can detect moving obstacles. It can also detect potholes, depression and elevation even before it makes contact with it. The smart cane is adept with detection of presence of hot object and slippery floor. Additionally, GPS module together with Arduino is used to obtain GPS coordinates of the user which is sent via an email to a loved one making it a great tool in case of an emergency. Moreover, auditory feedback via headphones is provided for every feature. The obtained results prove it to be an empathetic solution for the visually challenged.

Keywords: Smart cane · Visually impaired · Potholes · Elevation · Hot object · Auditory feedback · Ultrasonic · Raspberry Pi · Arduino UNO · GPS · NTC Thermistor

1 Introduction

Leading an independent life is of utmost value for a visually impaired individual. The world, as we know it, is designed with sighted people in mind. This is because vision is the most significant way of receiving information from the surrounding environment [1]. Thus it makes it challenging for the visually impaired to move around places especially when they are unaccustomed to the route. This leads to them travelling with a family member or known sighted person to unfamiliar, new places. Due to this, visually impaired individuals resort to memorizing their surroundings in order to find their way through. The designed smart cane diminishes the absolute need to memorize anything in their surroundings. The designed system gives a visually impaired person the confidence of a sighted individual so that they can carry out their day to day chores independently and autonomously without any third person's help. Even though guide dogs were the initial companion of the blind, later on technologies played a vital role [2]. There are a few mobility aids available in the market as of today but they are restricted by many

© Springer Nature Switzerland AG 2020
J. S. Raj et al. (Eds.): ICIDCA 2019, LNDECT 46, pp. 692–701, 2020.
https://doi.org/10.1007/978-3-030-38040-3_79

factors such as inability to detect potholes or elevation from a distance without actually touching it, detecting obstacles which are more than a meter away or even detecting a moving obstacle amongst a few. While these pose like some of the very common problems that a visually impaired person faces on a daily basis, the designed smart cane exactly focuses on these issues. The proposed design presents a never seen before yet efficient approach with various embedded modules to tackle these issues. The main reason behind choosing a 'cane' to present the technologies on, was due to the traditional use of canes as a mobility aid by the visually impaired and giving them the level of comfort that they are used to.

This paper has been organised in a certain manner such that firstly an overview is given in which what work has been already done in the past and today's scenario has been discussed along with the problem statement, followed by the proposed system which has information about the software and hardware architecture, and lastly the results and discussion. Each section has been divided into various sub-sections which gives a deep insight into each of them for a better interpretation.

2 Overview

This section outlines the problem statement and the issues faced in India along with other countries in the world and certain statistics to support it. It also consists of the related work that other authors and engineers have worked on.

2.1 Problem Statement

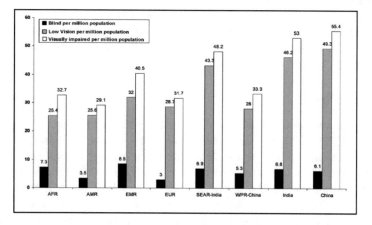

Fig. 1. Global estimates of visual impairment: 2010

As per WHO, based on 'Global data on visual impairment: 2010', globally the number of people visually impaired was estimated to be 285 million, of whom 39 million were blind out of which over 15 million are from India (Fig. 1). With regards to distance vision, 188.5 million people have mild vision impairment, 217 million have moderate to severe vision impairment, and 36 million people are blind. About 90% of the world's visually impaired people live in developing countries [3]. Apart from this, several country specific eye health organizations are also actively working for their own people. India Health Organization, Sightsavers, and many others are dedicated to reducing the number of blind persons in our country also [4]. The white cane is used across the world helping visually impaired, as a matter of fact 15th October is the day marked as "White Cane Safety Day" [5].

The graph shown above gives the per million distribution of blind, low vision and visually impaired people in various countries. This proves to be insightful for considering the gravitas of the issue in today's times as well.

Physical movement is one of the biggest hurdles for blind people, explains World Access for the Blind. A constant effort to make their day to day lives easier, places more accessible and movement more autonomous is being made today. One other common thing observed is that there are hardly any special facilities available in countries like India for the visually impaired with respect to the surroundings, so they have to adapt themselves to a greater extent. The constant thought lingers, when technology is at our disposal it becomes the responsibility of sighted people to exploit it to the optimum and give the visually impaired the easy life that they totally deserve.

In addition to this, especially in a country like India where potholes are found on every other road it becomes extremely difficult for a normal sighted person to get through, let alone the plight of a visually impaired person. Puddles and pits formed due to rain are uncertain, they can appear anywhere at any time hence making it impossible to memorize. Also, it gets difficult for them to detect the presence of any hot object like tea, hot water or even a wet, slippery floor if they do not have anyone around to tell them and this can lead to unforeseen circumstances. The proposed design aims to overcome all these issues and make their daily life simpler.

As the problem statement has been well explained, a gist of the literature survey of various projects on similar lines has been collated to understand the crux of the issue better.

2.2 Related Work

Most of the prototypes found have used Arduino whereas the proposed design uses the Raspberry Pi 3B which gives it immense scope as it is both versatile and protean at the same time. It comes with many inbuilt modules, and is easier to work with in Python and Linux. The use of additional components like a GSM module for emergency alert system has been avoided. The emergency alert system implemented in the proposed design simply sends an email with the coordinates to the loved ones using the SMTP library. A vital feature incorporated in the designed cane is the moving obstacle detection feature which is indicated via auditory feedback. A few projects found have made use of IR sensor for detecting objects in front but that proves to be inefficient as IR sensors cannot detect black objects accurately which is why the designed cane uses ultrasonic sensors for the same.

Design and Implementation of Mobility Aid for Blind People is an analogous project that is an Arduino based jacket with ultrasonic sensors mounted on it solely for obstacle detection. Here the user is notified about obstacles through specific voice commands which are stored in a Micro SD card [6]. This is followed by guidance regarding the best (obstacle free) path via earphones.

Smart Blind Stick is a Raspberry Pi based system that provides obstacle detection via ultrasonic sensors with vibratory feedback. It also provides level detection (holes and steps) via IR sensors with vibratory feedback. A moisture sensor with vibratory feedback is also incorporated. There is an emergency button and upon pressing it, the GPS module integrated in the system informs the user of his or her location via auditory feedback through a Bluetooth device. GSM module is used by the blind person to contact to mobile numbers stored in the microcontroller in case of any emergency [7]. Another work in this domain is *Obstacle Detection for Visually Impaired using Raspberry Pi and Ultrasonic Sensors*. This Raspberry Pi based wearable waist belt is an amalgamation of ultrasonic sensors for obstacle detection in left, right and front directions. Depending on the triggered condition a text output is created. This text output is converted to speech using Text-to-Speech API and as the last step this message is relayed to the visually impaired person via earphone or a speaker [8].

As the overview of the purpose of the proposed design has been discussed, the various modules and sensors used in the proposed design has been highlighted in the next section.

3   Proposed System

This segment of the paper deals with the components and the system hardware/software architecture that has been used to create the project.

3.1   System Architecture

The designed cane is embedded with Raspberry Pi, Arduino Uno, GPS module, and various kind of sensors like moisture detector sensor, ultrasonic sensor and heat sensor that are placed at various parts of the stick making it robust (Fig. 2).

Raspberry Pi 3B. The Raspberry Pi is a cost efficient, credit-card sized 40 pin microcontroller that embodies the core of our project. The Raspberry Pi 3 b model is used as it is highly versatile, powerful, efficient and integrable with various high level programming languages.

Ultrasonic Sensor. The Ultrasonic sensor has been used for elevation and depression detection which is present at the three fourth location of the stick so that presence of staircase and potholes can be detected. The HC-SR04 ultrasonic sensor is used in this system which is a 4 pin (Vcc (Power), Trig (Trigger), Echo (Receive), and GND (Ground)) sensor that comprises of an ultrasonic transmitter, a receiver and a control circuit. On detection it calculates the test distance by using the formula given below.

$$\text{test distance} = (\text{test time} \times \text{velocity of sound (i.e. } 340 \text{ m/s}))/2 \text{ [2]}$$

The test time is the time from sending ultrasonic to returning that sets the Echo pin high. On sensing obstacles, the sensor passes this data to the microcontroller [9].

Water Sensor. The water sensor incorporated with Raspberry Pi is placed above some height. When there is no water detection, the DO (digital out) pin of the sensor is HIGH. When water is detected the pin will change to LOW.

Heat Detection Sensor. The NTC Thermistor based sensor is used in order to detect heat. It follows the principle of Negative Temperature Coefficient wherein there is a decrease in resistance as there is an increase in temperature. A certain threshold is pre-determined and beyond the threshold requisite alarm system for heat detection is utilized in order to alert the user.

GPS Module. The Ublox NEO-6M GPS engine is used. Its module comprises of a power backup battery and EEPROM for storing configuration settings. The antenna is connected to the module via an U.FL cable which provides flexibility in arranging the GPS so that the antenna will always face the sky for best performance. The GPS module is connected to the Arduino which serially communicates with the Raspberry Pi.

3.2 System Software

The following software have been used in the cane:

Linux. Such commands were used to perform the required actions on the program file like; to create files, to save them and run the code as well as view the output on the Terminal. It was used to perform the general coding on the Raspberry Pi 3B.

Python. Many programming languages can be used while coding of the modules on the Raspberry Pi 3B. Python is a powerful multi-purpose programming language created by Guido van Rossum [10]. Python has been used to do the basic coding for each of the modules that has been incorporated in the cane.

Arduino IDE. The Arduino IDE has been used to get the coordinates of the GPS module on the Arduino UNO and then serial communication has been used to transfer these coordinates to the Raspberry Pi to use SMTP for sending an e-mail to the recipient.

3.3 Working

Elevation/Depression and Moving Obstacle Detection Module. A single ultrasonic sensor module has been attached to the Raspberry Pi. This module is used for two purposes.

Detection of Potholes and Elevation. This module works in such a way that it will detect the presence of any potholes or any elevation before the stick can actually touch the surface to detect the presence of any depression or elevation and then inform the user whether there is any of the above present ahead via auditory feedback. This is done so that the visually impaired person can be informed of the presence of any elevation or depression present in front of him/her much in advance, so that the user does not have any accident and has an adequate amount of time to take the necessary measures.

Detection of Moving Obstacle. The ultrasonic module is also used for the detection of any moving obstacle that is approaching the user and continuously give the feedback via audio to the user informing him/her as to how far the obstacle is.

Slippery Floor Detection Module. The water sensor module detects the presence of any water on the floor so that it can notify the user if the floor is slippery. A water sensor is located at the base of the stick to have precaution against the wet surface which it can causing slipping on the floor and thus can hurt [11]. There is an audio feedback for the same.

Hot Object Detection Module. The heat sensor has been attached to the cane so that it can inform user via audio whether there is any hot object present in the surroundings.

Emergency Module. The GPS module has been used along with the Arduino UNO. The coordinates of the GPS module are collected using the Arduino UNO code and then serially transmitted to the Raspberry Pi. After the coordinates are received by the Raspberry Pi, it converts them into a Google Map link and sends the link to the loved one's email ID. This module has been incorporated so that in case the user is lost, he or she can press an emergency button to send an email with their location to a loved one. The email that is sent using this module contains a google map link which has the exact location of the user. This process is carried out by using SMTP (Simple mail transfer protocol) (Fig. 3).

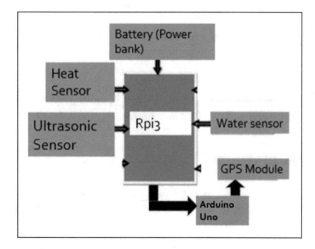

Fig. 2. Block diagram

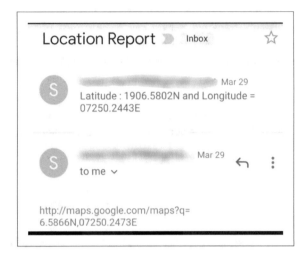

Fig. 3. Screenshot of email sent using SMTP containing GPS location

Fig. 4. Close-up view of the model

3.4 Flowchart and Pictures

The entire working has been discussed briefly in this section, next section throws light upon the results achieved and the final conclusion.

4 Results and Discussion

Certain figures, facts, tables have been discussed below and a few ensuing recommendations have been given which might help to enhance the features of the cane.

4.1 Results

In this project, the modules mentioned above have been successfully incorporated. Each module has been tested to adjust it to a best suited range for the user (Fig. 4). Based on the threshold set for the ultrasonic sensor the moving obstacle is detected and an auditory feedback is given to the person at regular intervals. A certain range of values have been assigned to elevation and likewise depression based on testing. Accordingly, if the value falls in that range respective audio feedback is given via headphones hence alerting the user from a certain distance itself [12]. Similarly, when hot object above 30 °C (threshold set) is detected or wet slippery floor is detected an audio feedback is given for the same to caution the user. In case of an emergency, the GPS coordinates are sent as a converted google maps link via an email to a loved one.

The ultrasonic sensor when used for pothole and elevation detection is set to specific ranges that help it in detection of the same. The range of 130–190 cm is the normal (flat ground) range where the stick will not give any audio feedback. For values lesser than 130 cm and greater than 190 cm please refer to the table below. This wide range had to be selected due to the low accuracy of the sensor and by noting the various distances the beam of the sensor forms with the ground when the angle of the stick with the ground is changed.

For moving obstacle detection, the ultrasonic sensor keeps reading how far the object is starting at 250 cm and gives the audio feedback with the specific distance range.

The heat sensor gives an audio output when the object in its surroundings is more than 30 °C (Table 1).

Table 1. Ranges set based on testing for various values

Sensor	Range	
	Actual	Used (Based on testing)
Ultrasonic (For pothole and elevation detection)	0–400 cm	<130 cm (Depression) 190 cm < (Elevation)
Ultrasonic (For moving obstacle detection)	0–400 cm	Obstacle within: 250 cm, 200 cm, 150 cm, 100 cm
Heat	Upto 80 °C	30 °C<

4.2 Conclusion

The obstacle-detection and alert system for the visually impaired has been made to enhance their mobility. The system reduces dependence on assistance, improves independent movement of the individual and paves the way for an affordable travel aid for the visually challenged particularly in developing countries. It is necessary that these people get access to all the benefits and the luxuries that normal people would otherwise get hands on in order to live their daily life comfortably. In a developing country like India, where people with disabilities are not able to have such functionalities in a single device, having a cane for themselves which does all the work and helps them with their daily life is a blessing. The developed smart cane that is embedded with multiple sensors will help in directing the way while walking and keep alerting the person if any sign of hurdle, obstacle or inconvenience is detected. The developed prototype gives good results in also detecting moving objects spaced at a distance in front of the user. Hot objects and wet floor are also detected well in advance to caution the user. At the same time global positioning system (GPS) is integrated with the system for more accurate results, so that person can know his current position which will be informed to user's loved ones via an email. Overall the smart cane is a blessing in disguise for the people who are visually impaired.

4.3 Future Scope

Image Processing can be used for object detection. An emergency button can be installed which when pressed would send the live GPS location of the cane to loved ones as well as inform the user. Modification can be done in the design so that the same technology could be used by attaching on clothes.

Acknowledgements. The authors would like to thank SVKM's NMIMS Mukesh Patel School of Technology Management and Engineering for providing space and other resources to successfully complete this project. Patent application for the design of the smart cane is in progress.

Compliance with Ethical Standards. All author states that there is no conflict of interest. Humans/Animals are not involved in this work. We used our own data.

References

1. Jain, S., Varsha, S.D., Bhat, V.N., Alamelu, J.V.: Design and implementation of the smart glove to aid the visually impaired. In: 2019 International Conference on Communication and Signal Processing (ICCSP), pp. 0662–0666. IEEE (2019)
2. Megalingam, R.K., Nambissan, A., Thambi, A., Gopinath, A., Nandakumar, M.: Sound and touch based smart cane: better walking experience for visually challenged, p. 4. IEEE (2014)
3. World Health Organization: Blindness and vision impairment, 11 October 2018. https://www. who.int/news-room/fact-sheets/detail/blindness-and-visual-impairment. Accessed 2019
4. Sharma, S., Gupta, M., Kumar, A., Tripathi, M., Gaur, M.S.: Multiple distance sensors based smart stick for the visually impaired, p. 5. IEEE (2017)
5. Daudpota, M.H., Sahito, A.A., Soomro, A.M., Channar, F.S.: Designing and modelling a white cane for the blind people, p. 6. IEEE (2017)
6. Sourab, B.S., Ranganatha Chakravarthy, H.S., D'Souza, S.: Design and implementation of mobility aid for blind people, p. 5. IEEE (2015)
7. Pratik, N.K., Poornesh, V., Shashikant, S.K., Saritha, A.N.: Smart blind stick. Int. J. Latest Trends Eng. Technol. **9**, 273–275 (2018)
8. Wattal, A., Ojha, A., Kumar, M.: Obstacle detection for visually impaired using Raspberry Pi and ultrasonic sensors. In: National Conference on Product Design, Noida (2016)
9. Ashtikar, S., Kalamkar, H., Jain, I., Bhowate, V.: Smart cane for visually impaired people. Int. J. Adv. Res. Ideas Innovations Technol. **4**, 1881–1883 (2018)
10. Programiz: Learn Python programming. https://www.programiz.com/python-programming. Accessed 2019

11. Anwar, A., Aljahdali, S.: A smart stick for assisting blind people. IOSR J. Comput. Eng. (IOSR-JCE) **19**, 86–90 (2017)
12. Zhangaskanov, D., Zhumatay, N., Ali, M.H.: Audio-based smart white cane for visually impaired people. In: 2019 5th International Conference on Control, Automation and Robotics (ICCAR), pp. 889–893. IEEE (2019)

Readmission Prediction Using Hybrid Logistic Regression

V. Diviya Prabha$^{(\boxtimes)}$ and R. Rathipriya

Department of Computer Science, Periyar University, Salem 11, India
diviyaprabha7@gmail.com, rathipriyar@gmail.com

Abstract. Predictive analytics has a prominent role in the field of healthcare. A massive amount of medical data is available such as diagnosing the disease, symptoms of illness, healthcare cost, mortality risk, and so on. Readmission prediction has great significance in improving patient care. This paper represents a Hybrid Logistic Regression (HLR) prediction model for large datasets. This model is the combination of k-means clustering and Logistic Regression in pyspark approach. The patients are clustered based on medical data, and Logistic Regression applied for the prediction approach. Further performance evaluation of the model is calculated and compared with other methods. It achieved better accuracy when compared with the existing feature selection algorithm

Keywords: Machine learning · Logistic Regression · Feature selection · Clustering

1 Introduction

In recent decades, the readmission of the patient to hospitals after discharge is increasing. There is no sufficient model to identify the early readmission prediction of patients. Today upcoming methodology several machine learning approaches are used in prediction of the data model and for other classification problems. Machine learning act as a tool solving many solutions in several healthcare issues [1]. There are several methods used to identify patient similarity [2]. This helps to identify the patients who are at high risk. Identifying high-risk [12] patients will help to reduce medical cost.

Feature selection is a crucial problem in improving the quality of the model. Many studies illustrate us [2] feature selection using machine learning techniques is better when compared with the primary filter and wrapper method. Machine learning can handle different types of attributes simultaneously to identify the relationship between variables on the target feature.

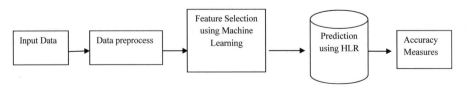

Fig. 1. Flow of the HLR

J. S. Raj et al. (Eds.): ICIDCA 2019, LNDECT 46, pp. 702–709, 2020.
https://doi.org/10.1007/978-3-030-38040-3_80

Recently, many algorithms [13] performs better for feature selection for this readmission prediction Logistic Regression based feature selection performs better out of 55 variables 21 variables are selected as optimal features.

The Fig. 1 describes the flow of the HLR approach input patient data taken from UCI repository https://archive.ics.uci.edu/ml/datasets/diabetes+130-us+hospitals+for +years+1999-2008 is downloaded and given as input to the model. It is of multivariate dataset consist of 1,00,000 patient data and 50 features is preprocessed

The main aim of this paper is to develop a good model and know the significant of patient care. The patient data is learned from history of patients available in dataset and predict the readmission risk. This approach also selects the optimal number of features to improve the accuracy of prediction.

Section 2 explains the methods appropriate to the HLR model. Section 3 briefs the Clustering and Logistic Regression integration, and Sect. 4 discusses the results and measures and, finally Sect. 5 concludes the paper.

2   Related Work

This work compares [3, 4] the clustering and non-clustering methods with a significant importance. Here, multilabel feature selection with clustering improves the accuracy of the model. But in this model overlapping cluster will decrease the features are considered for [5, 6] linear regression with logistic regression is used. Logistic Regression Model [7] with slight indication about cluster effect when it is comparison describes it is much more times greater than the original effect this model suggest that combining cluster with logistic regression improve the performance of the algorithm. The clusters are used to group patient with hypertension [8] the dataset is heterogeneity data it performs better in healthcare data. It categorizes data equally to subgroups so that identifying hypertensive patient is easy. Machine learning can be categorizing it to model based models and the results are evaluated in 5-fold validation to learn the model performance [9].

3   Model and Algorithm

Feature Selection
The features in the data have serious challenges for prediction techniques. There are two methods supervised and unsupervised for which feature selection can be applied. These techniques are based on the labeled and unlabeled data. Labeled data filter and wrapper methods are applied for this data in the paper [15]. So these methods does not give sufficient prediction method. With high dimensional data are applied to machine learning techniques [12] it leads to the better performance with high classification accuracy.

Table 1. Feature selection using machine learning techniques

Algorithms	Features
PCA	$(F_1, F_3, F_4, F_5, F_7, F_{10}, F_{11}, F_{12}, F_{22}, F_{24}, F_{26}, F_{30}, F_{31}, F_{34}, F_{35})$
Random Forests	$(F_1, F_2, F_3, F_4, F_5, F_{10}, F_{11}, F_{12}, F_{22}, F_{24}, F_{26}, F_{30})$
SVM	$(F_1, F_2, F_3, F_7, F_{10}, F_{11}, F_{12}, F_{22}, F_{31}, F_{34}, F_{40})$
Extra Tree Classifier	$(F_1, F_3, F_5, F_7, F_{12}, F_{13}, F_{15}, F_{22}, F_{24}, F_{25}, F_{31}, F_{45})$
MLR	$(F_{14}, F_{15}, F_{21}, F_{24}, F_{30}, F_{45})$
LR	$(F_1, F_2, F_3, F_4, F_5, F_6, F_7, F_8, F_9, F_{10}, F_{11}, F_{12}, F_{13}, F_{14}, F_{15}$ $F_{21}, F_{22}, F_{25}, F_{30}, F_{36}, F_{37})$

The Table 1 designates the feature selection with different machine learning techniques. Principle Component Analysis (PCA) [10] is one of the basic method applied for feature selection for many problems. Similarly other machine learning techniques are applied such as Random forest, Support Vector Machine (SVM), Logistic Regression (LR) and Extra Tree classifier. Multiple Logistic Regression (MLR) is applied for selecting the subset of features [11]. Among the feature selection algorithm LR [14] performs better when compared with other algorithm. Those 21 features are taken for readmission prediction.

HLR Algorithm:
The dataset consist of n patients and labeled features as beta coefficients β to predict the readmission. The function of log and modeling it to express logistic function

$$\text{Logit(y)} = \ln{(\text{odd})} = \log(\frac{p}{1-p})$$
$$= \beta_0 + \beta_1(F_1) + \beta_2(F_2) + \beta_3(F_3) + \beta_4(F_4) + \beta_5(F_5) + \beta_6(F_6) + \beta_7(F_7) + \beta_8(F_8) + \beta_9(F_9) + \beta_{10}(F_{10}) + \beta_{11}(F_{11})$$
$$+ \beta_{12}(F_{12}) + \beta_{13}(F_{13}) + \beta_{14}(F_{14}) + \beta_{15}(F_{15}) + \beta_{21}(F_{21}) + \beta_{22}(F_{22}) + \beta_{25}(F_{25}) + \beta_{30}(F_{30}) + \beta_{36}(F_{36}) + \beta_{37}(F_{37})$$

These are the variables that support the prediction model.
Hybrid Logistic Regression

Input: F = {F1,F2 ....F42} total number of features

X ={X1, X2 ,....} total number of instances

Begin

  Data preprocessing using pyspark
  For F= 1 to 42 do
      R (i) = Select the features using Logistic Regression
  End
  K=2
Initial selected features to the dataset

  Select K from R features

    Calculate the distance from cluster center and object

    Calculate the cluster until new cluster center is identified repeat the steps until criteria is reached

    Calculate Logistic Regression for Readmission Prediction

Stop

The simple logistic regression represents the odd function. This represent a patient is readmitted to hospital readmission. These are the variables that support the prediction model. After preprocessing initially the K-means algorithms used to cluster the positive and negative readmission of patients. K-means clustering row-wise are clustered given as input to the logistic regression model for the readmission prediction.

4   Results and Discussion

The Table 2 describes the features that are selected using LR method are taken for clustering. Cluster 1 represents patients that are readmitted to the hospital and Cluster 2 represents the patients that are not readmitted to the hospital. As the Cluster 1 and Cluster 2 values varies significantly.In cluster1 the values from one feature to other feature increases and decreases subsequently based on the features to the readmission value. Consequently the similar performance are carried out for cluster 2.

Table 2. Two clusters for features

Features	Cluster 1	Cluster 2
F1	2.82	4.88
F2	3.82	4.76
F3	0.08	1.04
F4	0.56	0.33
F5	2.59	1.90
F6	3.35	6.60
F7	5.98	1.91
F8	2.13	3.00
F9	6.33	1.70
F10	1.97	0.23
F11	2.49	3.86
F12	5.47	2.85
F13	0.02	1.34
F14	3.69	0.47
F15	2.77	3.37
F21	3.38	1.14
F22	2.90	5.96
F25	1.56	0.41
F30	3.87	3.50
F36	4.65	4.65
F37	3.54	4.26

The Table 3 describes the co-efficient and p-value of the variable age. The coefficient of positive value suggest that those variables support for readmission prediction and negative value suggest that those values has less support to the readmission of target variable. The patient age 55 and 85 have increase the positive value this suggest that those patient are readmitted when compared with other age patients.

Table 3. Co-efficient and p-value

Age	Co-efficient	p-value
15	−1.55	0.272
25	0.23	0.264
35	−3.61	0.255
45	−2.87	0.245
55	3.13	0.23
65	−3.08	0.22
75	−2.55	0.219
85	2.38	0.211
95	−2.57	0.20

The Table 4 illustrates the accuracy calculation of different machine learning techniques. Among, them Logistic Regression performance better with the high accuracy value of 96% and minimum RMSE value of 0.20.

Table 4. Accuracy calculation of machine learning

Machine learning algorithm	Accuracy	RMSE
Logistic Regression	96%	0.20
Decision Tree	94%	0.26
Random Forest	93%	0.30
SVM	91%	0.37

The Fig. 2 depicts the row clustering of data a sample of first eight rows is mentioned in the figure. R0 to R7 represents the row values after clustering. The value of these varies subsequently based on the features (Fig. 3).

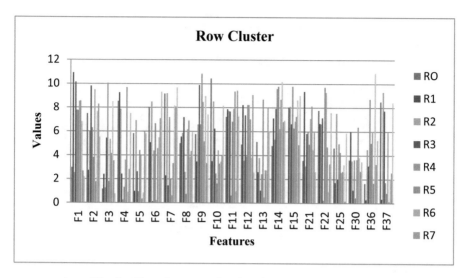

Fig. 2. Clustering row values based on patient readmission

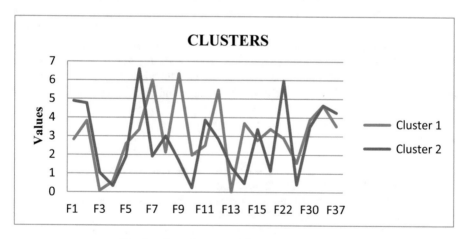

Fig. 3. Clustering based on patient readmission

The Fig. 4 represents the ROC curve for the HLR method the accuracy of the method is 96%. It suggests that the model produce better results than the current one. The model is good because all the predictors in ROC cure value are greater than 0.5.

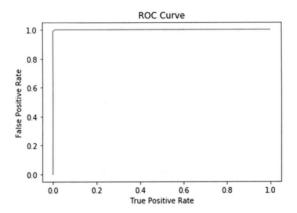

Fig. 4. ROC curve for HLR

5   Conclusion

In this paper, an HLR method has a more considerable significance for readmission prediction. Feature selection, such as classification and regression, are applied to the dataset. Among these methods, the LR method performs better for feature selection. Clustering-based Logistic Regression is performed. It achieves satisfying results of TPR and FPR is high compared with other methods. By using this HLR in the massive amount of data, more accurate results are obtained. The advantage of this method in healthcare helps to analyses patient at the early stage and reduce the readmission risk.

References

1. Li, J., Weng, J., Shao, C., Guo, H.: Cluster-based logistic regression model for holiday travel mode choice. Procedia Eng. **137**, 729–737 (2016)
2. Peng, L., Liu, Y.: Feature selection and overlapping clustering-based multilabel classification model. Math. Probl. Eng. **2018**, 12 (2018)
3. Johnson, P., Vandewater, L.: Genetic algorithm with logistic regression for prediction of progression to Alzheimer's disease. BMC Bioinform. **15**, S11 (2014)
4. Li, X., Yu, L., Hang, L., Tang, X.: The parallel implementation and application of an improved k-means algorithm. J. Univ. Electron. Sci. Technol. China **46**, 61–68 (2017)
5. Sultana, M., Sheikh, N., Mahumud, R.A., Jahir, T., Islam, Z., Sarker, A.R.: Prevalence and associated determinants of malaria parasites among Kenyan children. Trop. Med. Health **45**(1), 25 (2017)
6. Sohail, M.N., Jiadong, R.: A hybrid forecast cost benefit classification of diabetes mellitus prevalence based on epidemiological study on real-life patient's data. Sci. Rep. **9**, 1–10 (2019)
7. Nasierding, G., Li, Y., Sajjanhar, A.: Robustness comparison of clustering - based vs. non-clustering multi-label classifications for image and video annotations. In: Proceedings of the 8th International Congress on Image and Signal Processing, CISP2015, China, pp. 691–696, October 2015

8. Jayatillake, R.V., Sooriyarachchi, M.R., Senarathna, D.L.P.: Adjusting for a cluster effect in the logistic regression model: an illustration of theory and its application. J. Nat. Sci. Found. Sri Lanka **39**(3), 211–218 (2011)
9. Gao, C., Sun, H.: Model-based and model-free machine learning techniques for diagnostic prediction and classification of clinical outcomes in Parkinson's disease. Sci. Rep. **8**, 7129 (2018)
10. Song, F., Guo, Z.: Feature selection using principal component analysis. IEEE (2010)
11. Masoudi-Sobhanzadeh, Y., Motieghader, H., Masoudi-Nejad, A.: FeatureSelect: a software for feature selection based on machine learning approaches. BMC Bioinform. **20**, 170 (2019)
12. Qian, B., Wang, X., Cao, N., Li, H., Jiang, Y.-G.: A relative similarity based method for interactive patient risk prediction. Data Min. Knowl. Discov. **29**(4), 1070–1093 (2015)
13. Xue, B., Zhang, M., Browne, W.N.: A comprehensive comparison on evolutionary feature selection approaches to classification. Int. J. Comput. Intell. Appl. **14**, 1550008 (2015)
14. Diviya Prabha, V., Rathipriya, R.: Prediction of hyperglycemia using binary gravitational logistic regression (BGLR). Int. J. Pure Appl. Math. 105–119 (2018)
15. Yifan, X., Sharma, J.: Diabetes Patient Readmission Prediction Using Big Data Analytic Tools (2016)

Modified Region Growing for MRI Brain Image Classification System Using Deep Learning Convolutional Neural Networks

A. Jayachandran[1(✉)], J. Andrews[1], and L. Arokia Jesu Prabhu[2]

[1] Department of CSE, Presidency University, Bangalore, India
ajayachandran@presidencyuniversity.in,
andrewsj@presidencyuniversity.in
[2] Department of CSE, Sri Sakthi Institute of Engineering and Technology,
Coimbatore, India
arokiajeruprabhu@gmail.com

Abstract. Magnetic Resonance (MR) Imaging is a popular non-invasive modality for the visualization of different abnormalities in the brain due to its good soft-tissue contrast and accessibility of multispectral images. Using information from MR images, CAD systems have been developed to benefit doctors in rapid diagnosis. CAD systems can provide the diagnosis depending upon the specific attributes present in the medical images. The present study proposes a comprehensive method for the diagnosis of the cancerous region in the MRI images. Here, after image noise reduction, optimal image segmentation based on Support Vector Neural Network algorithm is utilized. Afterward, an optimized feature extraction and feature selection based on a modified region growing optimization algorithm are proposed for improving the classification accuracy of brain images. Further, it is also proposed that the input MR brain image be denoised using a non-local Euclidean median in nonsubsampled contourlet space. The classification accuracy of MRG with SVM is 74.24%, MRG with CNN is 82.67% and MRG with ANN is 62.71% and our proposed method MRG with MBCNN is 91.64%.

Keywords: MRI image · CNN · Classification · Denoising · Texture

1 Introduction

The brain is a crucial part of the human body which plays an important role in controlling all parts of the human organ. Sometimes the brain is affected by the unnatural growth of abnormal cells in the tissues. This abnormal growth of cells is called Brain tumor which can be cancerous or noncancerous. Uncontrolled brain tumor leads to death to the patients. So, it is necessary to detect the tumor as early as possible to reduce the chances of death of the patient. Human Brain is made up of numerous number of cells called as neurons. Each cell carries a different function from another. The term growth of a cell includes cell division and cell reproduction. When the cycle of growth of cell fails to control then the cell grows and divides very often irregularly. Then these excess cells make a mass of tissues and termed it as a tumor [1–3]. Thus a

© Springer Nature Switzerland AG 2020
J. S. Raj et al. (Eds.): ICIDCA 2019, LNDECT 46, pp. 710–717, 2020.
https://doi.org/10.1007/978-3-030-38040-3_81

brain tumor, can be defined as a collection of abnormal cells developed inside the brain. Skull of the brain is very hard and has limited space. Any growth inside the brain builds up many problems. Tumors of the brain can be classified as cancerous (malignant) or non-cancerous (benign). As these cell tumors starts to grow they creates a pressure within the skull to increase. Which damages the brain and it is also a life-threatening. There are two types of brain tumors, primary and secondary. A tumor which forms in the brain is termed as primary brain tumor. When the tumor is occurring due to the cancer cells of another organ which is spread to the brain is known as secondary or metastatic brain tumor. Primary brain tumors are more usually occurs in children and older adults. Secondary brain tumor is often found on adults. The MRI sequence images are used here as they represent a particular appearance of tissue which depicted in terms of no. of radio-frequency and gradients [4, 5].

This provides different sequences showing different contrast levels between tissues which helps in diagnosis diseases. This builds a need for classification or separation of MRI sequences into number of different slices namely proton-density (PD) weighted, T1-weighted, T2-weighted, diffusion-weighted, flow sensitive and 'miscellaneous'. In our proposed work 3 main slices for detection of brain tumor used are:

- T1-weighted MRI: Since T1 w allows for an easy annotation of the healthy tissues, it has become the most commonly used sequence images for the brain tumor structure analysis.
- T2-weighted MRI: T2-weighted images reflect the decay speed of magnetic moment in the transverse relaxation.
- PD-Proton Density-weighted MRI: These images reflect the difference between proton density.

Contours usually contain key visual information of an image. In computer vision, contours have been widely used in many practical tasks. Although quite a few contour detection methods have been developed over the past several decades, contour detection is still a challenging problem in the image field. Among the non-learning approaches, many early methods, such as the famous Canny detector, find contours by extracting edges where the brightness or color changes sharply. However, such methods usually employ regular kernels, e.g., Gaussian filter and Gabor filter, to measure the extents of local changes, and thus can hardly deal with textures. To address this problem, many texture suppression methods have been proposed. Examples are the method based on nonclassical receptive field inhibition, the method based on sparseness measures, the method based on surround-modulation, etc. It has been validated that texture suppression can help improve contour detection performance. Nonetheless, these methods still mainly use low-level local features. Moreover, some of them are computationally heavy, which leads to difficulties in practical applications [6–8].

Magnetic Resonance (MR) Imaging is a popular non-invasive modality for the visualization of different abnormalities in the brain due to its good soft-tissue contrast and accessibility of multispectral images. Using information from MR images, CAD systems have been developed to benefit doctors in rapid diagnosis. CAD systems can provide the diagnosis depending upon the specific attributes present in the medical images. Typically, these systems usually employ the steps of preprocessing, attribute extraction, selection, and classification for categorizing normal/ abnormal brain MR images. Numerous methods have been proposed in the literature that employs classical machine learning algorithms for the detection of abnormal brain images. These studies have proposed solutions based on K-nearest neighbour (k-NN), Support vector machine (SVM), Kohonen-Hopfield neural network (KHNN), and Artificial neural networks (ANN), etc. [9, 10]. This paper is organized as follows; Sect. 2 states about the proposed work associated with MRI brain image segmentation. Sections 2.1, 2.2 and 2.3 describes about the proposed framework consists of preprocessing, feature extraction, feature selection, similarity measure and classifier. Section 2.4 contains experimental results. The summarization of the proposed system is given in Sect. 3.

2 Methodology

The proposed system fusses and registers the three MRI scan images, by extracting Gray Level Run Length Matrix (GLRLM) and Centre-Symmetric Local-Binary Patterns (CSLBP) and features are stored in the database, CNN classifier is used to divide the brain images as Benign tumor, Malignant tumor and Normal.

2.1 Image Enhancement and Noise Removal

This section provides some techniques to enhancement of MRI which can perform image enhancement and noise removal techniques that enhance quality of the images for better segmentation accuracy. This primary stage plays a significant role to detect, trace and extract the brain tumor region from hemisphere. Because in this step images are changes to finer, sharper and enhanced. The enhanced image is finer than the original one for the specific application and gives the more accurate segmentation. The main objectives of this step is improve image and quality [11].

Mostly segmentation process depends on the sharp transition of image intensity level. A blur or noisy image is not appropriate for extract information. The average filter has been applied in this method for smoothing the images by reducing the image intensity values variation from one pixel to another. An average filter is a linear smoothing filter that was done by the value of each pixel in an image replaced with the average of the gray levels with a filter mask. In this proposed method, 5×5 filter mask was used for filtering approach that enhanced the image quality and reduced noise. This filter operation had done by the convolution sum of the filter mask with corresponding intensity values in an image [12].

2.2 Segmentation of Brain Tumor Region

As compared with other methods, region growing results with faster and accurate segmentation. Region growing methods can correctly separate the regions that have the same properties. It can provide the original images which have clear edges for the good segmentation results with less time compared to other methods. The approach is more innovative and novel as we can place the seed points inside the edge of the affected region and not on the centre point of the affected region. This method can be carried out manually or automatically. By using region-growing method brain tumor region is segmented [13, 14]. Here the seed point is manually selected and corresponding pixels are grouped by comparing seed pixel with neighboring pixels. Consider if the selected seeds forms n number of regions. $RR1, RR2, RR3, ….. RRi$. For each repetition, one pixel will be added into the regions. Now using the region RRi after m steps and not allotted set of pixel, it is given in Eq. (1)

$$L = \left\{ \bigcup_{i=1}^{n} RRi | I(u) \cap \bigcup_{i=1}^{n} RRi \neq \emptyset \right. \tag{1}$$

Where, $I(u)$ is the adjoining neighbor of pixel u, $u \in L$ means $I(u)$ maps exactly one RRi and $i(u) = \{1, 2, .., n\}$ with satisfying the condition, $I(u) \cap RRi \neq \emptyset$ and $\delta(u)$ finds the difference from the neighboring regions of u. $\delta(u)$ Can be expressed as per Eq. (2):

$$\delta(u) = |g(u) - mean\, y \in RRi(u)g(y)|$$

$$\delta(u) = \min\, u \in L\{\delta(u)\} \tag{2}$$

Where, $g(u)$ is gray-scale value of pixel of existing boundary pixels and $u \in L$ follows that and u belongs to RRi, the above procedure continued till all the pixels are assigned.

2.3   Feature Extraction and Classification

The Feature extraction approach is one of the important methods to classification accuracy. It extracts the relevant information of brain image and it is formed in a feature vector. The obtained feature vector is applied to retrieval process or classification process. In this paper, from the MRI brain image three types of features are extracted on the basis of their shape, margin and their density. The shape features provide the boundaries of the MRI brain image; margin features used to describe the margin characteristics of the MRI brain image and finally the third feature of density feature represents the brightness variation of the MRI brain image. Finally, the obtained three features are formed into a single feature vector for tumor classification. The preprocessing and segmentation results of proposed method is given in Fig. 1.

MRI Brain image is represented in gray levels where density degree is denoted by their brightness variation of MRI image. Density features are obtained through the following steps: First, separate or divide the MRI brain into two regions are inner and outer regions. Inner region minor axis is equal to the half minor axis of the outer region. Second, calculate the average brightness of the inner and outer region. Finally, calculate the density degree for MRI images using the Eq. (3).

$$DensityDegree = \frac{\varphi_{inner}}{\varphi_{outer}} \tag{3}$$

Mostly, the effective classification is subjective to the image features which are associated with an edges and the depth of the network. Hence above mentioned preprocessing techniques are applied to the incoming image sequences to preserve the features edge and the connectivity of each edges for better classification. We boosted

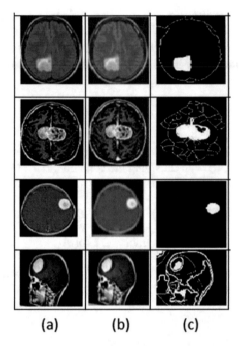

(a)   (b)   (c)

Fig. 1. Segmentation results (a) original image (b) Preprocessed image (c) Segmented image

our network to the maximum in all possible ways and hence named as Maximum Boosted Convolutional Neural Network.

Classification of diseases is a crucial aspect in disease categorization through image processing techniques. The categorization of diseases according to pathogen groups is a significant research domain and potentially a challenging area of work.. Classification and detection are very similar, but in classification primary focus is on the categorization of various diseases and then the classification according to various pathogen groups. It consists of two stages such as training phase and testing phase. Mostly, the effective classification is subjective to the image features which are associated with an edges and the depth of the network. Hence above mentioned pre-processing techniques are applied to the incoming image sequences to preserve the features edge and the connectivity of each edges for better classification. We boosted our network to the maximum in all possible ways and hence named as Maximum Boosted Convolutional Neural Network. CNN structure normally includes convolutional layer, max-pooling layer with an activation function and a fully connected layer. If the input of 2D convolutional layer is $I(x, y)$, and the corresponding feature map $s(x, y)$ will be obtained by convolving the input data with a convolution kernel $w(x, y)$ of size $m \times n$, it is defined in Eq. (4).

$$s(x,y) = \sum_{m=0}^{M-1} \sum_{n=0}^{N-1} I(m,n)w(x - m, y - n) \qquad (4)$$

After each convolution, the input to the hidden layers are changed. Such input distributions limits the layer parameters performance and reduces the learning rate during training. Maxpooling layers are placed after each convolutional layer, which downsamples the input images. With the help of proposed architecture, the features are extracted and forwarded to fully connected layer. The information from fully connected layer are fused and integrated with LSTM layer for classification, Maxpooling and an activation function in each convolutional layer which boost the network to extract the deterministic features.

2.4 Experimental Results

The performance of MRI Brain image classification was evaluated on collected images from National Cancer Institute database (http://cancerimagingarchive.net). Here, the dataset composed comprises of 20 different patients with 200 MRI images. Three orders of MR images has considered for each patient i.e., T1, T2 and FLAIR. However, each volume holds a dissimilar number of slices that is 100–150. In this paper divide the framework into two categories where 100 abnormal and 40 normal images respectively. Classifier performance evaluation in this work is conducted with widely used statistical measures, sensitivity, specificity and accuracy, it is defined in Eq. (5) [15]. The experimental results of sensitivity, specificity and accuracy are given in Fig. 2. Experimental results are also validated using k-fold cross validation method, the overall results are summarised in Table 1.

$$Senstivity = \frac{Tp}{Tp + Fn}$$

$$Specificity = \frac{Tn}{Tn + Fp}$$

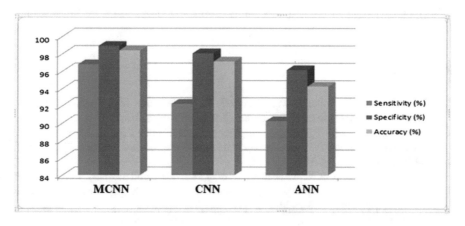

Fig. 2. Experimental results of proposed method

Table 1. Overall summarized result of proposed MRG with different classifiers using k-fold method

	Sensitivity (%)	Specificity (%)	Accuracy (%)
ANN	62.4	63.5	62.7
SVM	74.4	72.5	74.24
CNN	82.65	82.7	82.67
MCNN	91.6	92.2	91.64

$$Accuracy = \frac{Tp + Tn}{Tp + Fp + Tn + Fn} \qquad (5)$$

3   Conclusion

The development of the computer-aided detection systems in recent years turned them into a nondestructive and popular method for the cancer diagnosis in MRI images. The method includes different parts of image processing: in the first step, the original images have been pre-processed by a anisiotropic filter for the noise elimination. Afterward, an optimized image segmentation based on region growing algorithm is used for segmenting the cancer area from the background. Then, several features are extracted for improving the process of classification accuracy. To achieve an optimal feature extraction, an optimal method is used for selecting the useful features and for pruning the remained useless features. After feature extraction, they trained into an SVNN classifier to diagnosis what input image is cancerous or healthy. The image segmentation and the feature selection parts are optimized based on the newly introduced grasshopper optimization algorithm. In future, optimal feature selection method has to be implemented, which should be followed with the development of improved deep learning architecture that may give higher potential in tumor detection and classification.

References

1. Abdel-Maksoud, E., Elmogy, M., Al-Awadi, R.: Brain tumor segmentation based on a hybrid clustering technique. Egypt. Inform. J. **16**(1), 71–81 (2015)
2. Dhanasekaran, R., Jayachandran, A.: Severity analysis of brain tumor in MRI images uses modified multi-texton structure descriptor and kernel-SVM. Arab. J. Sci. Eng. **39**(10), 7073–7086 (2014)
3. Cagney, D.N., Martin, A.M., Catalano, P.J., et al.: Incidence and prognosis of patients with brain metastases at diagnosis of systemic malignancy: a population-based study. NeuroOncology **19**(11), 1511–1521 (2017)
4. Kromer, C., Xu, J., Ostrom, Q.T., et al.: Estimating the annual frequency of synchronous brain metastasis in the United States 2010–2013: a population-based study. J. Neurooncol. **134**(1), 55–64 (2017)

5. Mahiba, C., Jayachandran, A.: Severity analysis of diabetic retinopathy in retinal images using hybrid structure descriptor and modified CNNs. Measurement **135**, 762–767 (2019)
6. Jayachandran, A., Dhanasekaran, R.: Brain tumor detection using fuzzy support vector machine classification based on a texton co-occurrence matrix. J. Imaging Sci. Technol. **57**(1), 10507-1–10507-7(7) (2013)
7. Bar, Y., Diamant, I., Wolf, L., Lieberman, S., Konen, E., Greenspan, H.: Chest pathology detection using deep learning with non-medical training. In: ISBI, pp. 294–297 (2015)
8. Kharmegasundaraj, G., Jayachandran, A.: Abnormality segmentation and classification of multi model brain tumor in MR images using fuzzy based hybrid kernel SVM. Int. J. Fuzzy Syst. **17**(3), 434–443 (2016)
9. Vecht, C.J., Haaxma-Reiche, H., Noordijk, E.M., et al.: Treatment of single brain metastasis: radiotherapy alone or combined with neurosurgery? Ann. Neurol. **33**(6), 583–590 (1993)
10. Jayachandran, A., Dhanasekaran, R.: Multi class brain tumor classification of MRI images using hybrid structure descriptor and fuzzy logic based RBF kernel SVM. Iran. J. Fuzzy Syst. **14**(3), 41–54 (2017)
11. Patchell, R.A., Tibbs, P.A., Walsh, J.W., et al.: A randomized trial of surgery in the treatment of single metastases to the brain. N. Engl. J. Med. **322**(8), 494–500 (1990)
12. Zhang, Y., Wang, S., Dong, Z., Phillip, P., Ji, G., Yang, J.: Pathological brain detection in magnetic resonance imaging scanning by wavelet entropy and hybridization of biogeography-based optimization and particle swarm optimization. Prog. Electromagn. Res. **152**, 41–58 (2015)
13. Jayachandran, A., Dhanashakeran, R., Sugel Anand, O., Ajitha, J.H.M.: Fuzzy information system based digital image segmentation by edge detection. In: 2010 IEEE International Conference on Computational Intelligence and Computing Research, 28–29 December 2010
14. Gudigar, A., Raghavendra, U., San, T.R., Ciaccio, E.J., Acharya, U.R.: Application of multiresolution analysis for automated detection of brain abnormality using MR images: a comparative study. Futur. Gener. Comput. Syst. **90**, 359–367 (2019)
15. Acharya, U.R., Oh, S.L., Hagiwara, Y., Tan, J.H., Adeli, H.: Deep convolutional neural network for the automated detection and diagnosis of seizure using EEG signals. Comput. Biol. Med. **100**, 270–278 (2017)

Effect of Air Substrate on the Performance of Rectangular Microstrip Patch Antenna for the UHF Spaced Antenna Wind Profiler Radar

Jayapal Elluru$^{(\boxtimes)}$ and S. Varadarajan

Department of E.C.E, S.V. University College of Engineering, Tirupati, India
jaijrfscholar@gmail.com

Abstract. The design and analysis of Rectangular microstrip patch antenna is presented with varying patch size and material properties and substrate thickness as parameters at UHF frequencies for the application of an antenna array of UHF Spaced Wind Profiler Radar. The Rectangular microstrip patch antenna has been simulated using High Frequency Structure Simulator (HFSS) software and examined experimentally at 445 MHz. The 50 Ω coaxial probe feed is used for coupling of power to the microstrip patch antenna. As a result of varying airgap substrate thickness, gain, bandwidth, return loss, VSWR, characteristic impedance and of rectangular microstrip patch antenna are improved. Finally, critical needs in the design and development of antenna array of wind profiler Radar are discussed.

Keywords: Wind Profiler Radar · HFSS · Antenna array · Microstrip patch · Substrate · Coaxial probe feed

1 Introduction

Wind Profiler Radars are used to find wind velocity, horizontal and vertical wind directions by detecting backscattered power from turbulence. Wind profiler Radars are giving valuable information for research and operational applications related to climatology and meteorology. In wind profile Radar applications, low profile antennas are required with less weight, reduced size, reasonable cost, good performance and easy installation. To meet these constraints, microstrip patch antennas are used. These patch antennas are simple to design and cost effective to manufacture, conformable to planar and curved surfaces, mechanically robust [1].

Disadvantage of microstrip antennas are low power, low efficiency which are improved by increasing the substrate height and air used as a substrate. The microstrip patch antennas are operate at UHF frequencies, then there is a tradeoff between bandwidth and size [2, 3]. In general, microstrip antennas are also called as patch antennas. The general shapes of radiating patch are geometric square, rectangular shape, circular, elliptical and triangular. Both square and rectangular geometric shapes are used most often due to ease of fabrication, simple analysis, low cross polarization radiation, agreeable radiation characteristics.

© Springer Nature Switzerland AG 2020
J. S. Raj et al. (Eds.): ICIDCA 2019, LNDECT 46, pp. 718–724, 2020.
https://doi.org/10.1007/978-3-030-38040-3_82

Co-axial probe is used preferably to feed the patch antennas as shown in Fig. 1. In co-axial probe feeds, the inner conductor and outer conductor of the coax are connected to the radiating patch and the ground plane respectively. This type of feed is flexible to fabricate and possess spurious radiation [4].

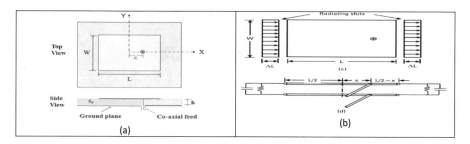

Fig. 1. Rectangular microstrip patch antenna with coaxial probe feed (a) top and side view (b) RMSA two radiating slots

2 Design Considerations of Microstrip Patch Antenna

The length of Rectangular microstrip patch is extended by a distance ΔL due to fringing on each side of the patch as substrate height increases. Extension of *length* (L) depends on the effective dielectric constant ϵ_{reff} and the function of width-to-height ratio (W/h). The practical approximate normalized ΔL is given by

$$\frac{\Delta L}{h} = 0.412 \frac{\left(\epsilon_{reff} + 0.3\right)\left(\frac{W}{h} + 0.264\right)}{\left(\epsilon_{reff} - 0.258\right)\left(\frac{W}{h} + 0.8\right)} \tag{1}$$

$$\epsilon_{reff} = \frac{\epsilon_r + 1}{2} + \frac{\epsilon_r - 1}{2}\left[1 + 12\frac{h}{w}\right]^{-1/2} \tag{2}$$

Since the patch length is expanded by ΔL on each end, the effective length (L_{eff}) of the Rectangular patch is given

$$L_{eff} = L + 2\Delta L \tag{3}$$

Where $L = \lambda/2$ for dominant Transverse Magnetic (TM_{010}) mode without considering fringing effect.

For TM_{010} mode, the resonant frequency of the patch antenna depends on it's length which is formulated as

$$(f_r)_{010} = \frac{1}{2L\sqrt{\epsilon_r}\sqrt{\mu_0\epsilon_0}} = \frac{v_0}{2L\sqrt{\epsilon_r}} \tag{4}$$

Where v_0 is velocity of light in the medium of free space.

Since effective length is given in equation number (3) which does not consider the fringing effect phenomena, modified resonant frequency is formulated with inclusion of fringing effect and written below as

$$(f_{rc})_{010} = \frac{1}{2L_{eff}\sqrt{\epsilon_{reff}}\sqrt{\mu_0\epsilon_0}} = \frac{1}{2(L+\Delta L)\sqrt{\epsilon_{reff}}\sqrt{\mu_0\epsilon_0}} \tag{5}$$

$$= q\frac{1}{2L\sqrt{\epsilon_r}\sqrt{\mu_0\epsilon_0}} = q\frac{v_0}{2L\sqrt{\epsilon_r}}$$

Where

$$q = \frac{(f_{rc})_{010}}{(f_r)_{010}}$$

The quality factor (q) is indicated in terms of *fringe factor*. As the height (h) of dielectric substrate of patch antenna increases, corresponding fringing effect increases which results in enlarge the space between the radiating edges and reduction in operating resonant frequencies [5, 6].

Design

From the simplified equation, design procedure of microstrip antennas has been outlined which realizes practical designs. The procedure assumes few design parameters include the height of dielectric substrate h, the frequency of patch (f_r) and the dielectric constant (ϵ_r).

Design Procedure of Rectangular Patch Antenna Is as Follows

Specify:

    dielectric constant (ϵ_r), resonant frequency (f_r), and height of substrate (h)

Determine:

    width of patch (W), length of patch (L)

Design steps:

1. Width of patch is found by using following equation [7, 8]

$$w = \frac{1}{2f_r\sqrt{\mu_0\epsilon_0}}\sqrt{\frac{2}{\epsilon_r+1}} = \frac{v_0}{2f_r}\sqrt{\frac{2}{\epsilon_r+1}} \tag{6}$$

Where v_0 is velocity of light in the medium of free space.

2. Compute the effective dielectric constant of substrate using Eq. (2).

3. Calculate the extension of the length ΔL with Eq. (1) after finding width W with help of Eq. (6).
4. The physical length of the patch can be find using Eq. (7)

$$L = \frac{1}{2f_r\sqrt{\epsilon_{reff}}\sqrt{\mu_0\epsilon_0}} - 2\Delta L \qquad (7)$$

5. Compute the dimensions of Ground plane using following equations

$$L_g = L + 6h \qquad and \quad W_g = W + 6h$$

6. Coaxial probe feed location from center of the patch

$$(X_f, Y_f) = \left(\frac{L}{4.65}, 0\right) \ in \ cm$$

Table 1. Design parameters of Rectangular microstrip patch antenna.

Air as substrate $(\epsilon_r = 1, \ loss\,tangent\,(\tan\delta = 0))$								
h (in cm)	W (in cm)	Effective dielectric constant (ϵ_{reff})	ΔL (in cm)	L (in cm)	L_{eff}	L_g	W_g	Feed (X_f, Y_f) in cm
1.5	33.707	1	1.057	31.592	33.707	40.592	42.707	(6.79, 0)
2	33.707	1	1.399	30.908	33.707	42.908	45.707	(6.64, 0)
2.5	33.707	1	1.736	30.234	33.707	45.234	48.707	(6.50, 0)
3	33.707	1	2.069	29.568	33.707	47.568	51.707	(6.35, 0)
3.5	33.707	1	2.396	28.914	33.707	28.914	33.707	(6.21, 0)

(a)   (b)

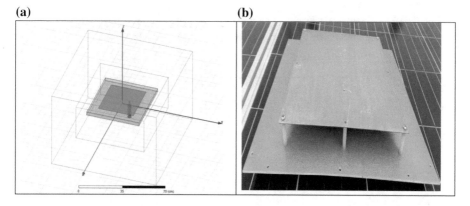

Fig. 2. (a) Rectangular microstrip patch antenna simulated model. (b) Fabricated Rectangular microstrip patch antenna element (Aluminum metallic plate is used as patch and ground).

Design of Rectangular microstrip is simulated using HFSS software and fabricated which is shown in Fig. 2. By using various parameters as shown in Table 1.

3 Results and Discussions

Return losses, VSWR, input impedance, bandwidth, maximum gain and HPBW of microstrip antennas are improved by increasing the air-dielectric substrate thickness (h) from 1.5 cm to 2.5 cm due to more fringing as shown Table 2. But the parameters are decreased after h = 2.5 cm due to the limitation of surface waves. The simulated results are comparable with measured results as shown in the Table 3 [9–12].

Table 2. Simulated results for air as substrate

Antenna parameters	Height of substrate				
	h = 1.5 cm	h = 2 cm	h = 2.5 cm	h = 3 cm	h = 3.5 cm
Return loss (dB)	−16.312	−27.231	**−32.614**	−30.910	−28.213
VSWR	1.360	1.090	**1.001**	1.058	1.081
Input impedance (Ohms)	36.764	45.872	**49.950**	47.256	46.253
Band width (MHz)	10.5	12	**20**	17.5	16.2
Max. gain (dB)	6.7	7.4	**9**	8.1	7.89
HPBW (degrees)	84°	74°	**59°**	66°	71°

Table 3. Comparison of simulated and measured results for the height of air substrate (h) = 2.5 cm.

Antenna parameters	Simulated results	Measured results
Return loss	−32.614 dB	−31.321 dB
VSWR	1.001	1.005
Input impedance	49.950 Ω	49.121 Ω
Band width	20 MHz	19.4 MHz
Max. gain	9 dB	8.7 dB
HPBW	59°	63°

Simulated Results for Air as Substrate at h = 2.5 cm
See Fig. 3.

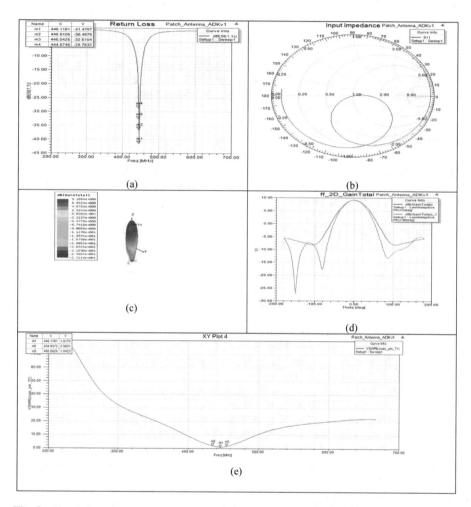

Fig. 3. Simulation of Rectangular micro strip patch antenna (a) Return loss vs frequency plot (b) input impedance (c) 3D Radiation pattern of patch antenna (d) 2D Radiation pattern of patch antenna (e) VSWR vs frequency.

4   Conclusions

In this paper the design of Rectangular microstrip antenna is presented for UHF Spaced antenna Wind profiler Radar with coaxial probe feed. The Return loss, VSWR, Input Impedance, bandwidth, and radiation pattern have been investigated over a range air-dielectric substrate thickness (h) from 1.5 cm to 3.5 cm. As the height of substrate (h) increases, ΔL also increases which give more fringing fields hence Gain increases.

Increasing of height of substrate has a limitations due to the surface waves. Using air as substrate which is having zero loss tangent, improves the Gain. The Rectangular microstrip patch antenna has been fabricated at 445 MHz and air-dielectric substrate height of 2.5 cm. The validity of the finding results is confirmed by comparing simulated results with measured results.

Acknowledgements. Authors would like to express their gratitude towards Dr. P. Srinivasulu, General Manager, Astra Microwave Products Ltd and his team for their abandon support in the design and measurements of antenna for UHF Spaced Antenna Wind Profiler Radar.

References

1. Lindseth, B., Brown, W.O.J., Jordan, J.R., Law, D., Hock, T., Cohn, S.A., Popovic, Z.: A new portable 449 MHz spaced antenna wind profiler radar. IEEE Trans. Geosci. Remote Sens. **50**(9), 3544–3553 (2012)
2. Lindseth, B., Brown, W.O.J., Hock, T., Cohn, S.A., Popović, Z.: Wind profiler radar antenna sidelobe reduction. IEEE Trans. Antennas Propag. **62**(1), 56–63 (2014)
3. Samson, T.K., Kottayil, A., Manoj, M.G., Babu, B., Rakesh, V., Rebello, R., Vasudevan, K., Mohanan, P., Santosh, K.R., Mohankumar, K.: Technical aspects of 205 MHz VHF mini wind profiler radar for tropospheric probing. IEEE Geosci. Remote Sens. Lett. **13**(7), 1027–1031 (2016)
4. Mona, D.F., Sakomura, S., Nascimento, D.C.: Microstrip-to-probe fed microstrip antenna transition. In: IEEE International Symposium on Antennas and Propagation and USNC-URSI Radio Science Meeting, Boston, Massachusetts (2018)
5. Balanis, C.A.: Antenna Theory Analysis and Design. Wiley, Hoboken (2005)
6. Ramesh, G., Prakash, B., Inder, B., Ittipiboon, A.: Microstrip Antenna Design Handbook. Artech House, Norwood (2001)
7. Pozar, D.M., Schaubert, D.H.: Microstrip Antennas: The Analysis and Design of Microstrip Antennas and Arrays. IEEE Press, New York (1995)
8. Elluru, J., Varadarajan, S.: Design of UHF-band microstrip patch antenna for wind profiler radar. Int. J. Adv. Eng. Res. Dev. (IJAERD) **4**(8), 567–572 (2017)
9. Setiadji, H.S., Zulkifli, F.Y., Rahardjo, E.T.: Radiation characteristics of microstrip array antenna for X-band radar application. In: Progress in Electromagnetic Research Symposium (PIERS), Shanghai, China (2016)
10. Liu, Y., Wang, H., Li, K., Gong, S.: RCS reduction of a patch array antenna based on microstrip resonators. IEEE Antennas Wirel. Propag. Lett. **14**, 4–7 (2015)
11. Li, J., Zhang, A., Joines, W.T.: A miniaturized circularly polarized microstrip antenna with bandwidth enhancement. In: IEEE International Symposium on Antennas and Propagation/USNC-URSI National Radio Science Meeting, Fajardo, Puerto Rico (2016)
12. Kottayil, A., Mohanakumar, K., Samson, T., Rebello, R., Manoj, M.G., Varadarajan, R., Santosh, K.R., Mohanan, P., Vasudevan, K.: Validation of 205 MHz wind profiler radar located at Cochin, India, using radiosonde wind measurements. Radio Sci. **51**, 106–117 (2016). https://doi.org/10.1002/2015rs005836

A System-Level Performance of CS/CB Downlink CoMP in Small Cell LTE-A Heterogeneous Networks

Amandeep$^{(\boxtimes)}$ and Sanjeev Kumar

Department of Computer Science and Engineering, Guru Jambheshwar
University of Science & Technology, Hisar, India
am.noliya@gmail.com, sanjukhambra@yahoo.co.in

Abstract. Coordinated Multi-Point (CoMP) MIMO technique is considered to be one of the promising approach to enhance the data rate, spectral efficiency and overall throughput for the cell edge user in OFDMA networks. CoMP can be used with heterogeneous networks to mitigate the inter-cell Interference problem among various sectors and sites in the network. This paper provides a brief description of classification, architecture and identified technical challenges in a CoMP based Heterogeneous network scenario. This article also describes the system model simulation for CoMP and evaluate results on Femto based small cell Heterogeneous cooperative networks. Coordination Scheduling and Coordination Beamforming approach is applied and results were evaluated in terms of edge throughputs, cell throughputs and spectral efficiency.

Keywords: CoMP · CS/CB · JT · MIMO · Clustering · Backhaul

1 Introduction

Multiple-input multiple-output system was designed to meet the requirement of next-generation wireless communication system, and it qualified the requirements of international telecommunication union (ITU) [9]. 3GPP adopted MIMO with OFDM for LTE and LTE-Advanced system. MIMO has distinct feature likes spatial multiplexing and spatial diversity which enhanced the reliability and efficiency of the LTE/LTE-advanced system. Along with this, Deployment of small cell and frequency reuse in the cell edge areas, enhances the efficiently uses of radio resources and effectively improves the cell edge throughputs and network coverage [4, 13]. However, extensive reuse of radio resources at the cell edge may cause strong intercell interference because multiple users of the multiple cells have allocated same resource within that region [14]. Although massive deployment of low power base stations also caused more interference to neighbor cells and users because they share huge information as feedback to their connected users. High power base station also restricts the improvement in the power of small cells, as increased in power generates more interference in the cell [1].

These all obstacle appears when the identical radio resource are reused in different cells in an uncoordinated way. Intercell interference especially degrades (reduce) the

© Springer Nature Switzerland AG 2020
J. S. Raj et al. (Eds.): ICIDCA 2019, LNDECT 46, pp. 725–732, 2020.
https://doi.org/10.1007/978-3-030-38040-3_83

performance of users located at cell edge area. One possible approach to improve this performance is to apply coordinated multipoint transmission reception. Where transmission and reception are coordinated in between multiple geographical antenna in order to improve the overall system performance, particularly for cell-edge users [10, 11]. With the frequency reuse scheme (IFR, FFR, and SFR) in single cell-single user MIMO and multi-user MIMO enabled network, improve the performance of the system but it was not efficient and effective for the improvement in the performance of the cell edge region. Therefore, coordination in MIMO antenna came into existence in 3PP LTE release-11.

2   Classification of Coordination Multipoint (CoMP)

2.1   Coordinated Beamforming/Coordinate Scheduling

Interference alignment also is known as coordinated scheduling/beamforming (CS/CB). In this mode, every eNodeB provides resources to the user which are associated with it. ENodeB operates in such a manner that inter-cell interference can be coordinated and enhanced the system throughputs. UE, the associated with eNodeB be define as serving eNodeB and the other eNodeB will be in a co-operating group. Instead of user data, in CS/CB mode only the channel state information is shared among coordinated eNodeB. So, the user data is available at only one transmission point, an efficient coordinated scheduling and coordinated beamforming designing are required to coordinate between the transmission points. CS/CB require low bandwidth as compare to joint transmission [8].

2.2   Joint Transmission

In joint transmission channel state information, scheduling state information and user data are shared among the coordinated transmission point. Joint transmission requires high backhaul as compare to CS/CB mode. Joint transmission works in two types coherent JT and non-coherent JT. In coherent joint transmission, the same data stream are transmitted simultaneously in between the co-operating eNodeB. It required high synchronization among data transmission. Non-coherent joint transmission is known as soft combining reception of data. In this mode, it doesn't require any precoding and user receive data from several eNodeB. Where data is independently precoded from each transmission point [12].

2.3   Dynamic Point Selection/Joint Point Selection

Dynamic point selection (DPS), in this mode data may be transmitted from one eNodeB. The transmitting eNodeB can be changed from one subframe to another and data would be available simultaneously at multiple nodes. Dynamic point selection approach include dynamic cell selection in which transmission point is chosen according to the best cell channel quality among the coordinated cell. It can be used with the joint transmission [12].

3 Coordination Multipoint Architectures

Coordination Multi-Point (CoMP) is defined as, centralized and distributed, in centralized cooperating a central eNodeB acts as a server and executes all the resources allocation, scheduling and channel state information transmitted to the targeted eNodeB. It uses feedback information from different eNB to reduce intercell interference. Delay, backhaul and signalling overhead are the major issue in this design. In the distributed coordination one of the eNB in the cluster act as a master cell which performs all resources allocation and scheduling and communicates each eNB via x2 interface. All neighbouring eNB share CSI and UE information among them [15].

3GPP proposed four types of CoMP scenarios applied for both homogenous and heterogeneous network. Scenario 1 applied to the homogenous network. There is no backhaul connection required in this scenario. The cooperation among the cell sector is controlled by the identical base station unit. Such types of environments are called intra-site coordination multipoint. Scenario 2 is also applied for the homogenous network, there is coordination between Macrocell and multiple high transmission power node (RRH). Scenario 3 applied to the heterogeneous network, there is cooperation between Macrocell and different small power transmission node. Transmission and reception point generated by small power station have dissimilar cell identity from the macro cell's ID. Scenario 4 applied for the heterogeneous network in which transmission and reception opinions made by Radio Remote Head hold similar cell Identity as the macro cell's identity [15].

4 Challenges in Coordination Multipoint (CoMP) Networks

After reviewing different research papers, the following key challenges are described here

- **Clustering**

  As mentioned in section one that CoMP has the capacity and capability to enhance the spectral efficiency and cell edge throughputs of the system. CoMP required addition signal information from the neighboring cell for coordination. This information may be the interference or useful data for exchanging among the cooperative cells. This extra signal and information are known as an extra overhead signal or overhead symbol. In the field trial, the limited number of cooperative base stations effectively managed the overhead information. Here, the question arises which base stations can be part of the cooperation. This problem was solved by clustering the base stations. In CoMP, clustering is classified as static and dynamic clustering. In static clustering [11] cooperative nodes are used to share information in a time-invariant manner in a static geographical region of the base station. Such type of clustering is useful for serving all UE by the best set of the cell. Static clustering is classified again in overlapping and non-overlapping clustering. While, in dynamic clustering, the system continuously exchanges information according to the feedback of users and other parameters like radio frequency condition, user equipment information, etc. [12]. Commonly used dynamic clustering types are Self-organizing network (SON) and adaptive clustering [3].

- **Synchronization**
  Synchronization in time and frequency among the cooperating nodes is another major challenge in CoMP. Synchronization is a process of maintaining the common notation of time among two or more cooperative devices. Each system whether it is a wireless or wired network used clock cycle for counting cycle by their local oscillator (LO). These clocks are said to be synchronized if they agree exactly on the duration of an interval between two events [11]. Or two clock is said to be synchronized if they agree exactly on the time of occurrence of events at an arbitrary time. Synchronization is classified into different categories:

  (a) Network synchronization
      (i)   The Network Time Protocol Based
      (ii) The precision Time Protocol Based
  (b) Satellite-Based Synchronization
  (c) Endogenous Distributed Wireless Carrier Synchronization

- **Channel Estimation**
  Accurate channel estimation is the base of any CoMP scheme. Channel estimation defines the highest upper bound, attainable in a channel. An efficient estimated channel can feedback to the transmitter. According to this feedback, a transmitter or group of transmitter can allocate effective resources to the users by applying modulation and coding schemes. In 3GPP, channel estimation is classified into implicit and explicit channel estimation. In explicit estimation direct feedback of the channel components. On the other, implicit feedback is based on pre or post-coders codebook. Basically, two types of precoder schemes are used i.e. (CSI RS) channel state information reference signal and coordinated beamforming [5, 6].

- **Backhaul**
  Every Base station involved in the cooperation scheme is required information exchange among the nodes. This information is interchanged by backhaul infrastructure. Backhaul capacity and latency are two major factors that impact the overall throughputs of the cooperative system. The main challenge to backhauling in CoMP is to control load limitation and queuing delay [12].

5 System Model and Simulation Procedure for Coordination Multipoint (CoMP)

In the downlink system model to understand how the coordinated multipoint approach works, a CoMP traditional cellular network where L cooperative eNodeB transmit power to their associated users y received signal is described as:

$$\mathbf{y} = H_{ji}^{[k]} W_{ij}^{[k]} \mathbf{x_i} + \sum_{l \neq i}^{L} H_{li}^{[k]} W_{il}^{[k]} \mathbf{x}_l + \mathbf{n}$$

where $\mathbf{H} \in C^{\text{Nuser}*\text{NcellNunit}}$ and $\mathbf{W} \in C^{\text{Ncell}*\text{Nunit}*\text{Nuser}}$ are the Coordination multipoint processing matrices, which related to cooperative configuration, $x = (x_1, x_2, \ldots, x_{\text{Nuser}})^T \in C^{\text{Nuser}*1}$ is a transmitted signal, $y = (y_1, y_2, \ldots, y_{\text{Nuser}})^T \in C^{\text{Nuser}*1}$ is a N_{user}

signal received by N_{user} UEs, and $\mathbf{n} \in C^{Nuser*1}$ is the additive white gaussian noise of the Nth user. The ij^{th} value $\{H_{ji}^{[k]}\}$ of matrix $\mathbf{H}$ represents channel gain between the ith User and the jth eNodeB unit belonging to the kth cell. The jth element $\{w_{ij}^{[k]}\}$ of the SVD MIMO aggregate precoding matrix $\mathbf{W}$ is the transmission power applied at the jth eNodeB, transmitting toward the ith User, belonging to the kth cell. The first term refers to the received signal for the ith User and the second term defines the interference from other cells.

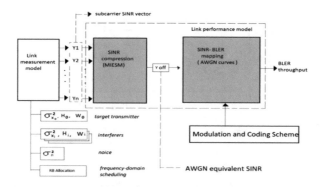

Fig. 1. Physical layer estimation of CoMP

For the performance analysis of heterogeneous network with coordination multipoint with radio remote head and femtocell we use MATLAB based Physical Layer simulator [simulator]. Basically, it is a combination of the LTE physical layer and LTE system layer shown in Fig. 1. LTE-Advanced physical layer simulator basically focused on the performance of different mobile equipments and base station. This is generally providing input in the form of evaluated performance of modulation/demodulation and coding & decoding. The simulation scenario for LTE-advanced system-level consists of different key performance indicators of the application layer. The system-level simulator is a combination of higher layers i.e. radio link control, MAC and radio resource control.

6   Result and Discussion

Results are evaluated by using the heterogeneous networks where different low power base station are deployed with high power macro base station. In the simulation a hexagonal trisector antenna configuration with three cooperative cells used. Only coordinated scheduling and coordinated beamforming approach applied for coordination in the cells. Every cell has deployed femto access point and range of femto access point are deployed in an increasing order from 10 to 50.

Edge throughputs of CSG/OSG cooperative scheduling and cooperative beamforming are shown in Fig. 2, as the number of femto users has increased in cooperation, both CS/CB scheduler increased their edge throughputs for upto 40 femto user (FUE) after that it showed the downward trend. OSG CB mode act as the best performer for the edge throughputs. CSG CS mode outperform the CSG CB mode at 40 FUE.

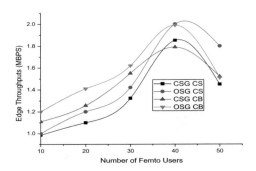

Fig. 2. Edge throughputs of CS/CB

Cell edge throughputs of CSG/OSG coordination configuration are shown in Fig. 3, as the number of FUE increased upto 30 both configuration performs in increasing order after that delined slightly. CSG CB again outperform among all modes.

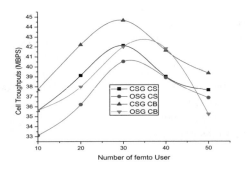

Fig. 3. Cell throughputs of CS/CB

Figure 4 is a description of spectral efficiency of CS/CB cooperative modes. Each modes increase the spectral efficiency at some point, attains higher values than gradually decreases as the number of femto users increased.

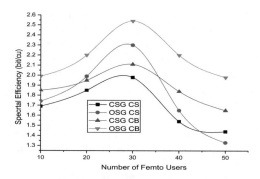

Fig. 4. Spectral efficiency of CS/CB

7   Conclusion

In this paper, CoMP's Coordination scheduling and coordination beamforming techniques for small-cell networks analyzed to achieve the optimal cell edge throughput and to mitigate the problem of intercell interference. These CoMP methods are analyzed in term of spectral efficiency, edge throughputs and cell throughputs. The simulation results revealed the effectiveness in the cell edge throughput, fairness and overall spectral efficiency of Coordination multi-point scheduling.

References

1. Jin, S., Sun, Q., Gao, X., Na, C., Harada, A.: A coordinated scheduling method for ICI mitigation using statistical CSI. In: Wireless Communications & Signal Processing (WCSP), pp. 1–5 (2013)
2. Zhu, J., She, X., Yun, X., Chen, L., Otsuka, H.: A practical design of downlink coordinated multi-point transmission for LTE-Advanced. In: IEEE 71st Vehicular Technology Conference (VTC 2010-Spring), pp. 1–6 (2010)
3. Beylerian, A., Ohtsuki, T.: Service-aware user-centric clustering and scheduling for cloud-RAN with coordinated multi-point transmission. In: Asia-Pacific Signal and Information Processing Association Annual Summit and Conference (APSIPA), pp. 252–257 (2015)
4. Wang, X., Mondal, B., Visotsky, E., Ghosh, A.: Coordinated scheduling and network architecture for LTE macro and small cell deployments. In: 2014 IEEE International Conference on Communications Workshops (ICC), pp. 604–609 (2014)
5. Chung, J., Hwang, C.-S., Kim, K., Kim, Y.K.: A random beamforming technique in MIMO systems exploiting multiuser diversity. IEEE J. Sel. Areas Commun. **21**, 848–855 (2003)
6. Gao, W., Cui, Q.: Adaptive coordinated scheduling/beamforming scheme for downlink LTE-advanced system with non-ideal backhaul. In: International Symposium on Wireless Personal Multimedia Communications (WPMC), pp. 356–361 (2014)
7. Mundarath, J.C., Ramanathan, P., Van Veen, B.D.: A distributed downlink scheduling method for multi-user communication with zero-forcing beamforming. IEEE Trans. Wireless Commun. **7**, 4508–4521 (2008)

8. Sun, H., Fang, W., Liu, J., Meng, Y.: Performance evaluation of CS/CB for coordinated multipoint transmission in LTE-A downlink. In: IEEE 23rd International Symposium on Personal, Indoor and Mobile Radio Communications-(PIMRC) 2012, pp. 1061–1065 (2012)
9. Ahmadi, S.: LTE-Advanced A Practical Systems Approach to Understanding 3GPP LTE Release 10 and 11 Eadio Access Technologies. Acadmic Press Publisher, Cambridge (2014)
10. Sawahashi, M., Kishiyama, Y., Morimoto, A., Nishikawa, D., Tanno, M.: Coordinated multipoint transmission/reception techniques for LTE-advanced. IEEE Wirel. Commun. **17**, 26–34 (2010)
11. Irmer, R., Droste, H., Marsch, P., Grieger, M., Fettweis, G., Brueck, S., Mayer, H.-P., Thiele, L., Jungnickel, V.: Coordination multipoint: concepts, performance, and field trail results. IEEE Commun. Mag. **49**, 102–111 (2011)
12. Qamar, F., Bin Dimyati, K., Nour Hindia, M.H.D., Ariffin Bin Noordin, K., Al-Samman, A. M.: A comprehensive review on coordination multi-point operation for LTE-A. Comput. Netw. **123**, 19–37 (2017)
13. Bai, Y., Juejia, Z., Liu, L., Lan, C., Otsuka, H.: Resource coordination and interference mitigation between macrocell and femtocell. In: IEEE 20th International Symposium on Personal, Indoor and Mobile Radio Communications, pp. 1401–1405 (2009)
14. Li, J., Li, Y., Lu, I.-T.: UE centric coordinated beamforming in multi-cell MU-MIMO systems. In: 34th IEEE Sarnoff Symposium, pp. 1–6 (2011)
15. Amandeep, Kumar, S.: A comprehensive review on resource allocation techniques in LTE-advanced small cell heterogeneous networks. J. Adv. Res. Dyn. Control Syst. **10**(12), 502–516 (2018)

Machine Learning: An Aid in Detection of Neurodegenerative Disease Parkinson

Jignesh Sisodia[⊠] and Dhananjay Kalbande

Sarder Patel Institute of Technology, Mumbai, India
{jsisodia, drkalbande}@spit.ac.in

Abstract. Parkinson is the most common neurodegenerative disease among the elderly people older than 65 years. An appropriate computer-assisted decision support system is the need in diagnosis and evaluation of the progression of Parkinson (PD). It is vital to early diagnose the disease. Machine Learning can assist to classify a normal and patient with Parkinson disease and also in optimising the treatment of the disease. It can be used to determine the various stages of the disease by categorising the symptoms including motor and non-motor symptoms. In this study, we assess the potential of Machine Learning for determining the progression of Parkinson disease and detecting it.

Keywords: Parkinson disease/diagnosis · Machine learning · Performance metrics

1 Introduction

Today the population of elderly people is increasing with the advancement of medical technologies in the health care. Parkinson is the most common neurodegenerative disease among the elderly people older than 65 years. The PD patients are prone to high-risk incidents, which may lead to serious injuries.

1.1 Approach to Parkinson's Disease Examination

The four primary elements of the exams are Bradykinesia, Rigidity, Tremor and Gait & Balance. Bradykinesia is the most important element because you have to have bradykinesia in order to give diagnosis of Parkinson. The other three elements Rigidity, Tremor and Gait and balance problems are also very important but any one of them along with bradykinesia is use to make the diagnosis of Parkinson.

Apart from the four main elements of the examination another important element to diagnose Parkinson is Observation which includes Spontaneous movements, Eye blinks, Fidgeting, Crossing/Uncrossing legs, Hands to gesture and Resting Tremor.

To determine bradykinesia, patients need to be test by rapid alternating movements, with the movements to be big and fast. It is observed that movements of Parkinson patients get slower over time and they get smaller over time in few repetitions. It can be tested with various activities like finger tapping, fist open and close, Pronation/ Supination and Toe tapping where amplitude and speed of the movements need to be observed getting smaller and slower over time.

© Springer Nature Switzerland AG 2020
J. S. Raj et al. (Eds.): ICIDCA 2019, LNDECT 46, pp. 733–741, 2020.
https://doi.org/10.1007/978-3-030-38040-3_84

The next element, Rigidity depends on muscle tone. Rigidity in Parkinson patients is not velocity dependent and not directionally dependent. It is based on resistance to the movements. Tremor is another element to exam the Parkinson disease. Tremor can be observed in patients at rest by small movements in thumb or fingers. Another manoeuvre to detect tremor can be observed using Postural Tremor and action or Kinetic Tremor.

Patients having difficulty in walking or have balance problems are more prone to fall. Thus Gait and balance is important element to examine patients for fall prevention perspective. The various activities to exam gait and balance includes Standing from Chair, walking and Pull test for balance. Step length of Parkinson patients are small and lose balance while taking turns. Also they have reduced arm swing and shuffling gait.

1.2   Machine Learning for Parkinson

Machine learning is use to produce insights both to improve the discovery of new therapy in Health Care and to improve the prevailing ones making it further effective. Machine learning incorporates statistical methods in which machine recognize patterns or correlations in data by inputting huge data sets.

Machine learning improves performance of the approach and also improve over time as they use new data. It also revises the approach as needed without human intervention. "Figure 1" describes the Role of Machine Learning in Parkinson.

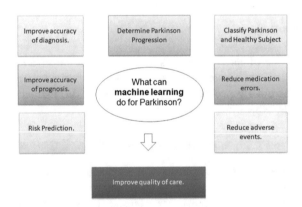

Fig. 1. The role of machine learning in Parkinson detection.

This paper describes the Machine Learning Algorithms used at various research and the data description used to detect Parkinson. It describes the use of performance metrics to determine the accuracy in detecting Parkinson. The accuracy achieved in prediction of Parkinson with different approaches is compared.

2 General Framework to Detect Parkinson

The general framework involves collecting input data from the dataset of the subjects. Several features are then extracted from the data and machine learning approaches are used to create a model. The model is then used by the classification component to classify Heathy Subject and the Parkinson Patient. "Figure 2" describes general framework to detect healthy subject and Parkinson Patient.

The framework can provide measurement of overall symptoms of Parkinson disease to correlate with Unified Parkinson's Disease Rating Scale (UPDRS). UPDRS is the measure of clinical rating scale to assess PD. Numerous research has been done to improvise the UPDRS.

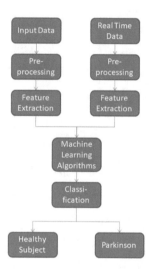

Fig. 2. General framework to detect Parkinson.

3 Data Description

Analytical method in Machine Learning depends on the quality of data to conclude results. Improper data may not yield appropriate results. Parkinson progressive marker initiative (PPMI) is a study to identify the biomarkers in progression of Parkinson. UCI Machine Learning Repository is another huge source of dataset of machine learning which can be used in detection and determining the progression of Parkinson. Another source of dataset for Parkinson is Physionet. Physionet produce dataset for research in clinical study. Various study is carried on Clinical dataset Table 1 describes the data description used in study of Detection of Parkinson. It describes the number of patients and the type of the data used in the dataset by various study. Table 2 describes the machine learning algorithms used at various research.

Table 1. Data description at various research

Authors	Example	Data type	No. of patients/subjects	Data set
Shamir et al. [2]	–	Clinical dataset	10 PD patients	–
Singh et al. [3]	Magnetic resonance images (MRIs)	de novo PD	–	Parkinson Progressive Marker Initiative (PPMI)
Kotsavasi-logloua et al. [4]	–	Clinical dataset	59 Subject	–
Nilashi et al. [5]	UPDRS scale	–	42 Patients	UCI Machine Learning Repository
Singh et al. [6]	Magnetic resonance images (MRIs)	de novo PD	–	Parkinson Progressive Marker Initiative (PPMI)
Grover et al. [7]	Parkinsons Telemonitoring Voice Data Set	Biomedical voice measurements	42 Patients	UCI Machine Learning Repository
Abdulhayet al. [8]	Vertical ground reaction force (VGRF)	–	93 Patients and 73 healthy controls	Physionet
Chen et al. [9]	Parkinsons Telemonitoring Voice Data Set	Biomedical voice measurements	31 People, 23 with PD	UCI Machine Learning Repository
Peng et al. [10]	Magnetic resonance images (MRIs)	–	69 PD Patients and 103 Healthy subject	Parkinson Progressive Marker Initiative (PPMI)

Table 2. Data description at various research

Authors	Machine learning algorithms				
Armananzas et al. [1]	Naïve Bayes (NB)	K-nearest neighbors (k-NN)	Linear Discriminant Analysis (LDA),	C4.5 Decision trees (C4.5)	Artificial Neural Networks (ANN)
Kotsavasilogloua et al. [4]	Naïve Bayes (NB)	AdaBoost (J48)	Log. Regression	Support Vector Machine (SVM)	Random Forest (RF)
Hariharan et al. [15]	Neural network (NN)	EM: Expectation Maximization (EM)	Principal Component Analysis (PCA)	Linear Discriminant Analysis (LDA)	
Shamir et al. [2]	Support Vector Machine (SVM)	Naïve Bayes (NB)	Random Forest (RF)		
Nilashi et al. [5]	Principal component analysis (PCA)	Fisher discriminant ratio (FDR)	Support Vector Machine (SVM)		
Jain et al. [17]	K-Nearest Neighbors (k-NN)	Neural network (NN)	Association Rule (AR)		
Behroozi et al. [18]	Support Vector Machine (SVM)	K-Nearest Neighbors (K-NN)	Naïve Bayes (NB)		
Grover et al. [7]	Medium Tree	Medium Gaussian SVM			
Abdulhay et al. [8]	Extreme Learning Machine (ELM)	Kernel ELM (KELM)			
Chen et al. [12]	Support Vector Machine (SVM)	K-Nearest Neighbors (k-NN)			
Singh et al. [3]	Kohonen Self Organizing Map (KSOM)				
Singh et al. [6]	Deep Neural Network				
Babu et al. [13]	Neural network (NN)				

4   Performance Metrics to Classify PD Patients

Performance metrics are used to evaluate classification problems in Machine Learning. The Confusion matrix metrics is used for finding the correctness and accuracy of the model to classify a PD patient and a healthy subject. Classification performance metrics such as Accuracy, Specificity, Sensitivity are used to predict PD patients and healthy subject. "Figure 3" shows the confusion Matrix. True positives (TP) in a confusion matrix is when a subject is actual PD patient and the model classify him as PD patient whereas True Negatives (TN) is when a subject is not PD patient and the model classify him not as PD patient. False Positives (FP) is when a healthy subject is classified as a PD Patients whereas False Negatives (FN) is when A PD Patients is classified as a healthy subject.

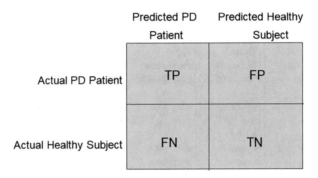

Fig. 3. Confusion matrix.

"Figure 4" shows the results generated using Performance metrics at various research. It specifies Accuracy, Specificity and Sensitivity achieved to detect Parkinson and signifies the substantially high rate, more than 85% in classifying healthy subject and Parkinson patients. "Figure 5" shows the overall accuracy achieved at various research to detect Parkinson.

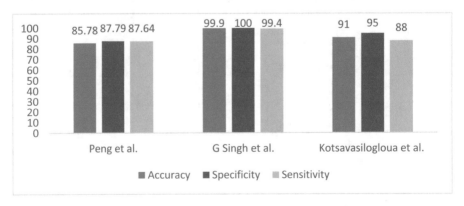

Fig. 4. Performance metrics used to classify PD patients

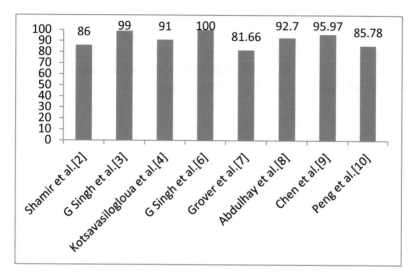

Fig. 5. Overall accuracy to detect PD patients

5   Implications and Potential Challenges

Machine learning is prominently considered in mainstream healthcare research. Machine learning depends on the quality of data available to produce results and will not yield appropriate insights. The input data need to be representative to cover the overall patient population leading to generalizability and correct results. Numerous healthcare data is generated during current practice but information that could result into appropriate insights remains uncollected, or otherwise inaccessible to researchers. Machine learning algorithms are referred as black box as it is intricate to interpret the way it generates insights which makes difficult to determine the working of the algorithms. The conclusions inferred by machine learning algorithms are credible and allied with the clinical consensus.

ML approaches can have a key impact on clinical assessment. Inappropriate results of these algorithms can, therefore, have unfavorable consequences. Performance of ML algorithms is dynamic and depends on the data and clinical practice and should be timely monitored by the clinician.

Machine learning is considered as the solution of all analytic challenges. Thus it would be immensely important to recognize and in understanding when and how to use these approaches.

6   Conclusion

Machine learning demonstrates an exemplary performance in determining the progression and detection of Parkinson. Machine learning approaches proves to assist medical practitioners and patients to judge the condition of the disease. In this study, we identify different dataset used at various research leading to accurate results to identify Parkinson patients. The promising and high accurate results demonstrated at various research using machine learning can provide exhaustive insight that is useful for recognition and determining the progression of Parkinson. Further studies can investigate the obstruction and deployment of various Machine learning approaches in parallel with the clinical care approaches.

References

1. Armananzas, R., Bielza, C., Chaudhuri, K.R., Martinez-Martin, P., Larranaga, P.: Unveiling relevant non-motor Parkinson's disease severity symptoms using a machine learning approach. Artif. Intell. Med. **58**, 195–202 (2013)
2. Shamir, R.R., Dolber, T., Noecker, A.M., Walter, B.L., McIntyre, C.C.: Machine learning approach to optimizing combined stimulation and medication therapies for Parkinson's disease. Brain Stimul. **8**, 1025–1032 (2015)
3. Singh, G., Samavedhama, L.: Unsupervised learning based feature extraction for differential diagnosis of neurodegenerative diseases: a case study on early-stage diagnosis of Parkinson disease. J. Neurosci. Methods **256**, 30–40 (2015)
4. Kotsavasiloglou, C., Kostikis, N., Hristu-Varsakelis, D., Arnaoutoglou, M.: Machine learning-based classification of simple drawing movements in Parkinson's disease. Biomed. Signal Process. Control **31**, 174–180 (2017)
5. Nilashi, M., Ibrahim, O., Ahmadi, H., Shahmoradi, L., Farahmand, M.: A hybrid intelligent system for the prediction of Parkinson's disease progression using machine learning techniques. Biocybern. Biomed. **38**, 1–15 (2017)
6. Singh, G., Vadera, M., Samavedham, L., Lim, E.C.-H.: Multi-class diagnosis of neurodegenerative diseases: a study on parkinson's disease. IFAC-PapersOnLine **49**, 990–995 (2016)
7. Grover, S., Bhartia, S., Akshama, Yadav, A., Seeja, K.R.: Predicting severity of Parkinson's disease using deep learning. Procedia Comput. Sci. **132**, 1788–1794 (2018)
8. Abdulhaya, E., Arunkumar, N., Narasimhan, K., Vellaiappan, E., Venkatraman, V.: Gait and tremor investigation using machine learning techniques for the diagnosis of Parkinson disease. Futur. Gener. Comput. Syst. **83**, 366–373 (2018)
9. Chen, H.-L., Wang, G., Ma, C., Cai, Z.-N., Liu, W.-B., Wang, S.-J.: An efficient hybrid kernel extreme learning machine approach for early diagnosis of Parkinson's disease. Neurocomputing **184**, 131–144 (2016)
10. Peng, B., Wang, S., Zhou, Z., Liu, Y., Tong, B., Zhang, T., Dai, Y.: A multilevel-ROI-features-based machine learning method for detection of morphometric biomarkers in Parkinson's disease. Neurosci. Lett. **651**, 88–94 (2017)
11. Senders, J.T., Staples, P.C., Karhade, A.V., Zaki, M.M., Gormley, W.B., Broekman, M.L. D., Smith, T.R., Arnaout, O.: Machine learning and neurosurgical outcome prediction: a systematic review. World Neurosugery **109**, 476–486.e1 (2017)
12. Chen, H.L., Huang, C.C., Yu, X.G., Xu, X., Sun, X., Wang, G., et al.: An efficient diagnosis system for detection of Parkinson's disease using fuzzy k-nearest neighbor approach. Expert Syst. Appl. **40**, 263–271 (2013)
13. Babu, G.S., Suresh, S.: Parkinson's disease prediction using gene expression – a projection based learning meta-cognitive neural classifier approach. Expert Syst. Appl. **40**, 1519–1529 (2013)

14. Peterek, T., Dohnalek, P., Gajdos, P., Smondrk, M.: Performance evaluation of Random Forest regression model in tracking Parkinson's disease progress. In: 13th International Conference on Hybrid Intelligent Systems (HIS), pp. 83–87 (2013)

15. Hariharan, M., Polat, K., Sindhu, R.: A new hybrid intelligent system for accurate detection of Parkinson's disease. Comput. Methods Programs Biomed. **113**, 904–913 (2013)

16. Froelich, W., Wrobel, K., Porwik, P.: Diagnosis of Parkinson's disease using speech samples and threshold-based classification. J. Med. Imaging Health Inform. **5**, 1358–1363 (2015)

17. Jain, S., Shetty, S.: Improving accuracy in noninvasive telemonitoring of progression of Parkinson'S disease using two-step predictive model. In: Third International Conference on Electrical, Electronics, Computer Engineering and their Applications (EECEA), pp. 104–109 (2016)

18. Behroozi, M., Sami, A.: A multiple-classifier framework for Parkinson's disease detection based on various vocal tests. Int. J. Telemed. Appl. (2016)

19. Pereira, H.R., Ferreira, H.A.: Classification of patients with Parkinson's disease using medical imaging and artificial intelligence algorithms. In: Mediterranean Conference on Medical and Biological Engineering and Computing, pp. 2043–2056. Springer, Cham (2019)

20. Ali, L., Zhu, C., Golilarz, N.A., Javeed, A., Zhou, M., Liu, Y.: Reliable Parkinson's disease detection by analyzing handwritten drawings: construction of an unbiased cascaded learning system based on feature selection and adaptive boosting model. IEEE Access **7**, 116480–116489 (2019)

21. Senatore, R., Della Cioppa, A., Marcelli, A.: Automatic diagnosis of Parkinson disease through handwriting analysis: a cartesian genetic programming approach. In: IEEE 32nd International Symposium on Computer-Based Medical Systems (CBMS), pp. 312–317 (2019)

Elliptical Curve Cryptography Based Access Control Solution for IoT Based WSN

Renuka Pawar[✉] and D. R. Kalbande

Sardar Patel Institute of Technology, Andheri, Mumbai, India
{renuka_pawar, drkalbande}@spit.ac.in

Abstract. With IoT, many apps and services in fields such as police job, health care, safety, etc. are increasing. The services provided are often accessed by the customer from anywhere, anytime and anywhere via sensitive device apps. This makes IoT essential for safety and privacy. Wireless network detector (WSN), a kind of correspondence framework, is regularly conveyed in the unattended condition any place the supposed client gains admittance to the system. The nodes of the detector gather understanding from the environment. If the information is important and private, it requires safety steps to protect them from unlawful access. According to the IoT framework's novel taxonomy, entirely distinct analytical problems are presented, essential alternatives and analytical operations are uncloaked, and instructions are expected for attention-grabbing assessment. Additionally, current access management solutions based on elliptical curve cryptography are surveyed and mentioned to make sure the protection of IoT parts and applications.

Keywords: IoT security · Access control · Wireless sensor networks

1 Introduction

WSNs "Fig. 1" [17] comprise of countless economical sensors that have very constrained resources (e.g., low processing units, low bandwidth, limited battery power, and low memory). Sensors are little in size, and are incorporated with a detecting unit and remote correspondence abilities. These nodes are being conveyed in a wide territory to play out their expected undertakings productively [18]. However, with the increasing ubiquity of WSNs in real applications (e.g., hospitals, military, wildlife monitoring), WSNs data will be available almost everywhere, anytime. Inevitably, much of WSNs information is highly sensitive and critical, and thus it is possible that an adversary may introduce malicious nodes in a network to leak the sensors information (and/or insert false reports) without the consent of the network owner. In addition, an adversary can purposefully interrupt the network smooth functionality by deploying the malicious nodes into the network. Therefore, to protect such a information leakage from the global adversaries and malicious nodes, access control mechanisms are to be enforced to real WSNs from the beginning of a WSN deployment [10, 11].

By using the Internet of Things, the sensor node data collected in WSNs could be given to customers who are allowed access to sensor nodes. As the sensor data is sent

© Springer Nature Switzerland AG 2020
J. S. Raj et al. (Eds.): ICIDCA 2019, LNDECT 46, pp. 742–749, 2020.
https://doi.org/10.1007/978-3-030-38040-3_85

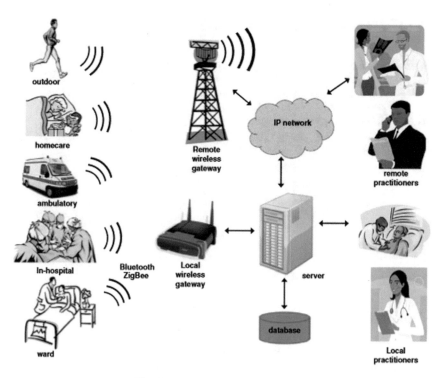

Fig. 1. Typical architecture of WSN

through the open system and the sensor node has a small battery, priority is given to safety and efficiency in WSNs.

1.1   Gartner Prediction

Gartner predicts spending on IoT Endpoint Security arrangements will increment from $240M in 2016 to $631M in 2021, achieving a CAGR of 21.38%. Overall IoT security spends will increment from $912M in 2016, taking off to $3.1B in 2021, achieving a 27.87% CAGR in the estimate time frame. Source: Gartner, Forecast: IoT Security, Worldwide, 2018, February 28.

1.2   Bain and Company Prediction

Bain and Company predicts the Internet of Things market will dramatically increase to $520 billion by 2021. Protection, IT/OT mix and ROI that is indistinct will the very best obstructions in IoT selection today. Bain and business's most recent breakdown of their undertaking clients mirrors the more market that is extensive need that is high verifying and integrating IoT systems. Interoperability, information convey ability, merchant hazard, and system requirements keep on heightening with customers given that the most review that is recent in 2016. August source: Bain and Company, Unlocking Opportunities on the net of Things, 9, 2018 (Fig. 2).

What are the most significant barriers limiting your adoption of
IoT/analytics solutions?

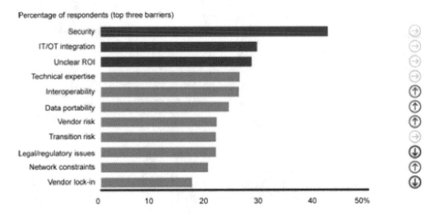

Fig. 2. Bain IoT customer survey

2 Literature Review

Access control solutions are based on different ways; some of it are showed in the
Fig. 3. UPPAC-users' privacy preserving access control, RBAC-Role/Rule based
access control; CBAC-cryptography based access control, Attribute based, and ECC-
Elliptic Curve Cryptography.

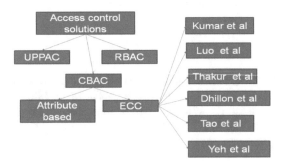

Fig. 3. Taxonomy of access control scheme

Kumar et al. [1] in real WSN, sensors send/receive request to/from the base-station.
In such two-way wireless networks, however, compromising identity privacy (of the
nodes) can inadvertently leak event privacy and it can give away the event occurrence
without the consent of the WSN owner. This paper has proposed an access control
scheme for WSN using ECC, hash function, and cryptosystem. The proposed scheme
achieves the access control of a node and provides robust security.

Samanta et al. [3] Analyzed Dynamic on-body topology, Propagation delay, Network management cost and Packet delivery ratio along with Link failure rate and QoS. Also they have done Dynamic topological disengagements and varieties in connection characteristics expand the vitality utilization pace of sharp WBANs. The Lagrangian enhancement strategy is utilized to get the ideal vitality utilization rate, in view of the diverse restorative needs of WBANs.

Thakur et al. [4] Elliptic Curve Discrete Log Problem (ECDLP) with WSN stable double trap door hash. The base station produces a secret key during the network communication process and then calculates the chameleon's node value and updates as a chamaleon hash value on the base station.

Dhillon et al. [5] the protocol uses IoT system portal node-based architecture with a distant user authentication system for lightweight multi-factor authentication. The suggested protocol is lightweight because it only utilizes computationally less costly one-way hash, perceptual hash functions and XOR activities.

Srinivas et al. [6] proposed an enhancement to Amin-Biswas's scheme. Here is the user authentication procedure and main WSN contract using bio-hashing. Along with the extra characteristics of the dynamic node and user friendly shift of password.

Tao et al. [7] To ensure the Quality of Service (QoS) of dynamic sessions and give satisfactory administrations to new access demands – including both relocated sessions and recently started access solicitations activated by the EVs (electric vehicles) – they have proposed a limit based dynamic access confirmation control plan to acquire the perfect number of acceptable access demands during the flow time frame.

The adjusted CPWABE is likewise used to build salvage adequacy by securely enrolling close by medicinal work force that have the required claims to fame (Tables 1, 2 and 3).

Table 1. Table access control model in WSN.

Access control models	Network architecture and component	Key management	Encryption and decryption	Policy specification and decision making process	Attacks	Tools
Kumar et al. [1]	Base station, coordinator node, sensor node	SHA-1	AES	Node authentication and identity privacy	A1, A2, A3, A4, A18, A19	Test-bed
Luo et al. [2]	Internet user, key generation centre, private key generator, gateway	KGC and PKG	SC- signcryption algorithm And USC- un signcryption algorithm	Certificate less cryptography	A6	–

(*continued*)

Table 1. (*continued*)

Access control models	Network architecture and component	Key management	Encryption and decryption	Policy specification and decision making process	Attacks	Tools
Thakur et al. [4]	Base station, sensor node	Chameleon hash function	Elliptic Curve Discrete Log Problem	Session key security	A1, A2, A3, A7	–
Dhillon et al. [5]	Sensor node, RFID tags	Key agreement protocol	One way irreversible Hash Operation, perceptual hashing operation	Access through anywhere, anytime and anyplace	A1, A7– A15	AVISPA
Srinivas et al. [6]	Home gateway node, smartcard	Bio-hashing	AES 128 SHA-1	Enhancement to Amin and Biswas scheme	A1, A2, A7– A10, A13, A16– A19	AVISPA
Tao et al. [7]	Electric vehicles (EVs), local aggregators (LAGs), certification authorities (CAs) and a control center (CC)	Public-private key	Group-based signature scheme	Capacity-based active access admission control scheme	A1, A7, A8, A16	MATLA

A1-Reply attack, A2-node masquerade attack, A3-forgery attack, A4-Sybil attack, A6-chosen-message attack, chosen cipher text attack, A7-mitm attack, A8 impersonation attack, A9-stolen smart device attack, A10 offline guessing attack, A11- dos, A12 parallel session attack or session key attack, A13 password change, A14 node bypassing, A15 user anonymity, A16 privileged insider attack, A17 eavesdropping

Table 2. Based on features in WSN.

Access control models	Data privacy	User privacy	Flexibility	Support for emergency access
Kumar et al. [1]	✓	–	✓	–
Luo et al. [2]	✓	✓	✓	–
Thakur et al. [4]	✓	✓	–	–
Dhillon et al. [5]	✓	✓	✓	✓
Srinivas et al. [6]	✓	✓	✓	✓
Tao et al. [7]	✓	✓	✓	✓

Table 3. Based on performance evaluation

Performance criterion	Kumar et al. [1]	Luo et al. [2]	Thakur et al. [4]	Dhillon et al. [5]	Srinivas et al. [6]	Tao et al. [7]
Energy consumption (μJ, mj, j) Computation cost or communication cost	ECC-14931 μJ AES 21.06 μJ SHA-1-196.02 μJ	234mj	4*Tem* + 1*T*	8Th + 6Th + 8Th	13Th + 6Th + 10Th + 29Th And 16Th + 5Th + 14Th + 35Th	n × 13.81
	Tpm-1Tpm Th-3Th Tc-3Tc	–	–	–	–	–
Memory usage	ROM-ECC-13.6 AES-10.1 SHA-1-9.7	ROM 1.03mj	–	–	–	–
	RAM-ECC-1.25 AES-0.76 SHA-1-1.9	–	–	–	–	–

Tpm - the time for executing point-multiplication; Thc - the time for executing hash-chain; Th - the time for executing one-way hash function; TC - the time for performing cryptosystem (i.e., encryption and decryption).

3 Proposed Architecture

While providing the solution in IoT WBAN initially we need to do the deployment of the sensors and different nodes according to the scenario for which we are trying to provide the solution. After deployment we need to do distributed network management, so that all the sensors and nodes will communicate efficiently. The major concern in WBAN is Cost minimization, energy minimization. After operating on cost minimization we need to focus on how we can develop the secure communication algorithm using ECC based access control approach, we need to design out own algorithm to do data transmission (Fig. 4).

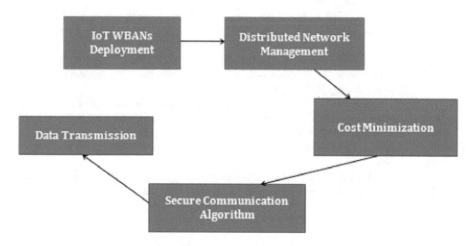

Fig. 4. Proposed architecture

4 Conclusion

This paper presents the security vulnerabilities, safety specifications, and a literature assessment of current ECC-based access control designs, as well as the performance evaluation of the suggested designs in WSNs. In spite of the fact that review has been done on cryptography, execution, highlights and various models in WSNs, few issues still need to be settled, such as selecting reasonable cryptographic methods. Additional constraints such as power, processing capacity and storage must be addressed with the further creation of WSN access control plans. Access control is a fundamental safety management system in sensor systems which is essential to ensure that network services are available to only legal customers on WSNs. The nature of existing access control systems shows that a lot of work still needs to be done. There is a need to develop a strong and secure access control protocol by considering the proposed model.

References

1. Kumar, P., Gurtov, A., Iinatti, J., Sain, M., Ha, P.H.: Access control protocol with node privacy in wireless sensor networks. IEEE Sens. J. **16**(22), 8142–8150 (2016)
2. Luo, M., Luo, Y., Wan, Y., Wang, Z.: Secure and efficient access control scheme for wireless sensor networks in the cross-domain context of the IoT. Hindawi, Security and Communication Networks (2018)
3. Samanta, A., Misra, S.: Energy-efficient and distributed network management cost minimization in opportunistic wireless body area networks. IEEE Trans. Mob. Comput. **17**, 376–389 (2017)
4. Thakur, T.: An access control protocol for wireless sensor network using double trapdoor chameleon hash function. J. Sens. **2016**, 6 (2016)
5. Dhillon, P.K., Kalra, S.: A lightweight biometrics based remote user authentication scheme for IoT services. J. Inf. Sec. Appl. **34**, 255–270 (2017)
6. Srinivas, J., Mukhopadhyay, S., Mishra, D.: Secure and efficient user authentication scheme for multi-gateway wireless sensor networks. Ad Hoc Netw. **54**(1), 147–169 (2017). https://doi.org/10.1016/j.adhoc.2016.11.002
7. Tao, M., Ota, K., Dong, M., Qian, Z.: AccessAuth: capacity-aware security access authentication in federated IoT-enabled V2G networks. J. Parallel Distrib. Comput. **118**, 107–117 (2018)
8. Wazid, M., Das, A.K., Odelu, V., Kumar, N., Conti, M., Jo, M.: Design of secure user authenticated key management protocol for generic IoT networks. IEEE Internet Things J. **5**, 1 (2017). https://doi.org/10.1109/jiot.2017.2780232
9. Yeh, L.Y., Tsaur, W.J., Huang, H.H.: Secure IoT-based, incentive-aware emergency personnel dispatching scheme with weighted fine-grained access control. ACM Trans. Intell. Syst. Technol. **9**(1), 23 (2017). https://doi.org/10.1145/3063716
10. Zhou, Y., Zhang, Y., Fang, Y.: Access control in wireless sensor networks. Ad Hoc Netw. **5**(1), 3–13 (2007)
11. Lee, H., Shin, K., Lee, D.H.: PACPs: practical access control protocols for wireless sensor networks. IEEE Trans. Consum. Electron. **58**(2), 491–499 (2012)

12. Debnath, A., Singaravelu, P., Verma, S.: Privacy in wireless sensor networks using ring signature. J. King Saud Univ. Comput. Inf. Sci. **26**(2), 228–236 (2014)
13. Li, Y., Ren, J., Wu, J.: Quantitative measurement and design of source-location privacy schemes for wireless sensor networks. IEEE Trans. Parallel Distrib. Syst. **23**(7), 1302–1311 (2012)
14. Sfar, A.R., Natalizio, E., Challal, Y., Chtourou, Z.: A roadmap for security challenges in the Internet of Things. Digit. Commun. Netw. **4**(2), 118–137 (2018)
15. Kouicem, D.E., Bouabdallah, A., Lakhlef, H.: Internet of things security: a top-down survey, Comput. Netw. 141, 199–221 (2018). https://doi.org/10.1016/j.comnet.2018.03.012. http://www.sciencedirect.com/science/article/pii/S1389128618301208. ISSN 1389-1286
16. Kumar, P., Lee, H.-J.: Security issues in healthcare applications using wireless medical sensor networks: a survey. Sensors **12**, 55–91 (2012)
17. Al Ameen, M., Liu, J., Kwak, K.: Security and privacy issues in wireless sensor networks for healthcare applications. J. Med. Syst. **36**, 93 (2012). https://doi.org/10.1007/s10916-010-9449-4
18. Kumar, A.D., Smys, S.: An energy efficient and secure data forwarding scheme for wireless body sensor network. Int. J. Netw. Virtual Organ. **21**(2), 163–186 (2019)

Novel Exon Predictors Using Variable Step Size Adaptive Algorithms

Srinivasareddy Putluri[✉] and Md. Zia Ur Rahman

Department of ECE, Koneru Lakshmaiah Education Foundation,
Green Fields, Vaddeswaram, Guntur 522002, India
srinivasphdklu@gmail.com, mdzr55@gmail.com

Abstract. Regions which code for protein in deoxyribonucleic acid (DNA) sequence are crucial to determine in bioinformatics field. Exon region study is significant in drug design also disease detection. Three basic periodicity (TBP) have been noted in exons. Adaptive techniques of signal processing were likely with unique capability to alter genome sequence dependent weight coefficients. From these, we present a new adaptive exon predictor (AEP) using variable step size least mean square (VSLMS) algorithm along with its signed versions that includes SRVSLMS, SVSLMS also SSVSLMS algorithms to decrease computational complexity and compared to LMS. SRVSLMS based AEP was shown to be better based on metrics like Precision 0.6751, Sensitivity 0.6749, also Specificity 0.6625 at threshold value 0.8. Thus, numerous AEPs are able to predict exon positions with distinct real genomic sequences of Homo sapiens as of National Centre for Biotechnology Information (NCBI) gene databank.

Keywords: Adaptive exon predictor · Bioinformatics · Computational complexity · Three base periodicity

1 Introduction

It is an enormous area of Genomics to identify areas which code for proteins. This is because exon areas have an significance in the assessment of diseases and the design of medicines. The DNA sequence comprise of intergenic and genic segments [1]. The research of the structure of the prime protein segments supports the secondary as well as tertiary structure of exon sections. Once total protein segments structure has been assessed, all abnormalities are likely to be detected, heal diseases, and design medications [2, 3]. Whole living beings remain alienated as prokaryotes as well as eukaryotes. Segments involve in coding of proteins part of eukaryotes remain termed as exons, while introns are recognized as non-protein coding segments. The coding areas are only 3% of the sequence in human eukaryotes also non-coding parts comprise of remaining 97%. It is therefore a significant job to identify coding sections in a DNA sequence [4, 5]. The literature presents many exon recognition methods based on different signal processing techniques [6–10]. In several iterations, adaptive algorithms can process very lengthy sequences. A new adaptive exon predictor (AEP) with adaptive algorithms remain presented in our current work. VSLMS technique with its

© Springer Nature Switzerland AG 2020
J. S. Raj et al. (Eds.): ICIDCA 2019, LNDECT 46, pp. 750–759, 2020.
https://doi.org/10.1007/978-3-030-38040-3_86

signed versions is considered to enhance AEP's efficiency over LMS technique. VSLMS algorithm overcomes LMS disadvantages and enhances speed of convergence as well as tracking capacity. In process of exon identification, the excess medium square error (EMSE) is also reduced [11].

Algorithms based on sign use the function of sign also amount of multiplication computations are reduced [12]. The data-independent fixed step-size algorithms results in many errors also not satisfy monitoring needs [13]. Best convergence rate needs lower EMSE as well as larger size of step. The use of VSLMS based techniques results in overcoming this drawback. The error acquired in iteration method shows instances of Variable Step Size adaptive algorithms in [14–18] where the step size is forbidden. These methods perform better than counterparts of Least Mean Squares (LMS) technique. We merge VSLMS algorithm with sign algorithms to decrease computational complexity. Hybrid versions of proposed AEPs include variable step size least mean square (VSLMS), sign regressor variable step size least mean square (SRVSLMS), sign variable step size least mean square (SVSLMS), as well as sign variable step size least mean square (SSVSLMS) techniques. Proposed AEPs are evaluated in terms of performance using actual standard genomic data sets obtained from the gene database of NCBI [19]. Performance measurements used for assessing the effectiveness of numerous AEPs are computational difficulty, convergence features, Precision (Pr), Specificity (Sp), also Sensitivity (Sn). Various techniques for exon identification are discussed in the literature [20–22]. In resulting sections, the theory of adaptive methods and findings of AEPs are also discussed, which also discuss efficiency of distinct AEPs.

2 Adaptive Algorithms for Exon Prediction

Principal stage is to analyze gene sequence taken as of NCBI sequence database based on dimer nucleotide densities also this sequence is then transformed as numerical notation in suggested AEP. This is a major task in the area of genomic processing since signal processing can only be used on signals of discrete or digital nature. Here, DNA sequence is converted to digital information that denotes DNA with binary mapping for instance four digital signals. Digital conversion is a crucial task to process gene sequences since techniques of signal processing can only be used on signals of digital otherwise discrete nature. Therefore, 1 indicates the presence of a nucleotide and 0 is its absence in four digital sequences obtained with binary mapping. The resulting sequence is now appropriate for the adaptive algorithm as an input. An AEP which is created using adaptive methods of signal processing is deliberated. Let, M(n) as digital sequence that is mapped, x(n) as DNA sequence, d(n) as gene sequence that obeys TBP, y(n) indicates result acquired by use of adaptive technique also e(n) represent signal of feedback for altering weight coefficients generated within feedback loop. Length in LMS technique is regarded as 'T'. In accordance with current weight coefficient step size parameter 'S', the next weight coefficient can be estimated in this algorithm with present weight coefficient as p(n), also input binary mapped sequence indicated as M(n) at the moment. The LMS algorithm is mathematically expressed and analyzed in [12]. As shown in Fig. 1 shows block diagram for proposed AEP.

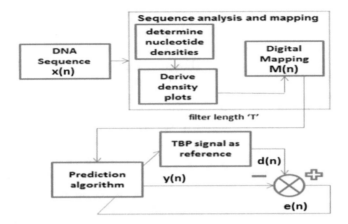

Fig. 1. Block schematic diagram of proposed AEP

The weight expression of LMS adaptive technique is written as

$$p(n+1) = p(n) + S\ x(n)\ e(n) \tag{1}$$

Adaptive algorithms must have minimum computational difficulty in applications pertaining to exon recognition to attract Nano based bioinformatics applications. Applying clipping to gene input information, otherwise a feedback signal or both is feasible by this decreased value. This is discussed in [18] with techniques for this purpose. Three signed versions are covered in these methods.

Signum notation is expressed as

$$V\{x(n)\} = \left\{ \begin{array}{l} 1 : x(n) > 0 \\ 0 : x(n) = 0 \\ -1 : x(n) < 0 \end{array} \right\} \tag{2}$$

These variants prefer to reduce the complexity of LMS computation. LMS has greater computational difficulty compared to these variants. The algorithm of Sign Regressor LMS (SRLMS) is obtained as LMS recursion thru altering the input vector of tap. In this case, the mean values of vector V[x(n)] remains substituted for x(n), while sign function V is used for x(n) on basis of element by element.

The mass renovate expressions for SRLMS technique is denoted as

$$p(n+1) = p(n) + S\ V\{x(n)\}\ e(n) \tag{3}$$

The weight relation for SLMS technique remains obtained thru changing e(n) by signed form as

$$p(n+1) = p(n) + S\ x(n)\ V\{e(n)\} \tag{4}$$

Also, weight relation for SSLMS is resulting thru substitution of x(n), e(n) with use of signed forms as

$$p(n+1) = p(n) + S \ V\{x(n)\} \ V\{e(n)\} \tag{5}$$

For a big input data vector, LMS technique is affected by gradient noise amplification limitation. Effectiveness of LMS can be enhanced easily by varying its size of step at time of its convergence initially. Because of simple hybrid techniques and their performance, our AEP implementations use a variable step strategy to achieve fast convergence rate and tiny stable MSE. By using small step size values near to its steady state, this results in VSLMS algorithm. When the instantaneous error is positive or negative, the step size is increased or decreased by small value in VSLMS algorithms. The mathematical modeling of VSLMS algorithm is clearly depicted in the Fig. 2. In (8) from Fig. 2 of VSLMS, $0 < \mu_{min} < \mu_{max}$ is selected to offer minimum tracking capability and to guarantee algorithm stability. The β parameter remains as positive numeral that offers stability in algorithm design also $\sigma^2 e^{(n)}$ remains as instantaneous error power. These changes are caused by the noisy gradient vectors used to correct p (n). In addition, the sign function is used to decrease the complexity. With use of sign function to VSLMS, three simplified signed versions SRVSLMS, SVSLMS, and SSVSLMS algorithms are derived.

Mathematical Modeling of VSLMS Algorithm

Parameters: T = number of taps (i.e. filter length), S = step size parameter
Let the tap input be x(n) and filter length T is moderate to large
Initialization: Set w(0) = 0 as initial condition
Data: Given x(n) = T-by-1 tap input vector to filter n2 at time n
$$= [x(n), x(n-1) \ldots \ldots x(n-T+1)]^T$$
d(n) = desired response at time n, $\sigma^2 e^{(n)}$ is power of instantaneous error, β is a positive integer, σ^2 is variance of desired response d(n), $(.)^T$ is the transpose of (.)
To be computed:
p(n+1) = estimate of tap-weight vector at time n+1
Computation:
The Output of FIR filter is given by
$$y(n) = x^T(n) p(n) = p^T(n) x(n)$$
The expression for estimation error e(n) is given by
$$e(n) = d(n) - y(n) = d(n) - (p^T(n).x(n))$$
Relation for mean square error or cost function J(n) is $J(n) = E[|d(n) - y(n)|^2] = E[|e(n)|^2]$
The new recursive relation for $VJ(n)$ is written as
$$VJ(n) = E\{V|e(n)|^2\} = VE\{|e(n)|^2\} = E\{e(n) \, Ve^*(n)\} \text{ and}$$
$$Ve^*(n) = -x^*(n)$$
Thus, the resultant expression for gradient vector is given by
$$VJ(n) = -E\{e(n) \, x^*(n)\} \tag{4}$$
Based on the method of steepest descent, the weight update recursion with variable step size S(n) is
$$p(n+1) = p(n) - S(n)VJ(n) \tag{5}$$
The unknown expectation $E[e(n) x^*(n)]$ is replaced with an estimate such as sample mean given by
$$\hat{E}\{e(n) x^*(n)\} = \frac{1}{T}\sum_{l=0}^{T-1} e(n-l) x^*(n-l) \tag{6}$$
Therefore, from (4), the weight update relation of VSLMS algorithm with S(n) is given by
$$p(n+1) = p(n) + S(n) x(n) e(n) \tag{7}$$
Thus, the variable step size parameter S(n) of VSLMS algorithm is calculated as
$$S(n) = S_{max} + (S_{min} - S_{max})e^{-\beta \sigma^2 e^{(n)}} \tag{8}$$

Fig. 2. Mathematical modeling of VSLMS Algorithm

The weight relations for SRVSLMS, SVSLMS, and SSVSLMS techniques are

$$p(n+1) = p(n) + S(n)V[x(n)]e(n) \tag{6}$$

$$p(n+1) = p(n) + S\ x(n)V[e(n)] \tag{7}$$

$$p(n+1) = p(n) + S\ V[x(n)]V[e(n)] \tag{8}$$

We finally developed four AEPs with these algorithms and contrasted their results with AEP using LMS. Performance assessment using Sensitivity, Specificity and Precision shows that SRVSLMS is just below its non-sign regressor variant. Therefore, SRVSLMS is better than its signed versions, referring to metrics for performance also computing difficulty, among various chosen techniques.

3   Computational Complexities and Convergence Issues

The amount of multiplications needed is determined as a metric to estimate and compare algorithm complexity in particular. The focus is not on precise analysis for complexity to perform computations but on the evaluation of distinct VSLMS-based adaptive methods. In addition, those methods based on sign function are without multiply computations which are required for exon identification applications. For instance, LMS needs T+1 multiplications and one addition, when mass update equation is computed as in Fig. 3(i).

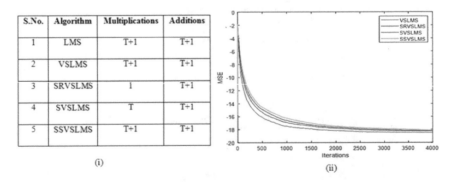

S.No.	Algorithm	Multiplications	Additions
1	LMS	T+1	T+1
2	VSLMS	T+1	T+1
3	SRVSLMS	1	T+1
4	SVSLMS	T	T+1
5	SSVSLMS	T+1	T+1

(i)

(ii)

Fig. 3. (i) Computational complexities of LMS and VSLMS based variants (ii) Convergence Curves of VSLMS Algorithm and its signed variants

While to compute 'S. e(n)', only one multiplication is required for SRVSLMS variant. In case of two other signed VSLMS algorithms, T and T+1 multiply computations remain needed. Figure 3(i) shows the computational complexities of LMS and VSLMS-based variants. The proposed AEPs dependent on VSLMS provide offers less complexity to perform computations than LMS and can be utilized in nano devices [5]. Convergence features of the suggested VSLMS and their signed variants are depicted in Fig. 3(ii). All suggested VSLMS-based adjustable algorithms are obviously more rapidly converging than AEP based on LMS.

4   Results and Discussion

The debate concerns the performance comparison of various AEPs. The AEP outline is illustrated in Fig. 1. Multiple AEPs remain derived from the VSLMS-based algorithms with sign variants. We created an LMS-based AEP comparative study.

Table 1. Gene datasets from NCBI databank

Seq. No.	Accession No.	Sequence definition
1	E15270.1	Human gene for osteoclastogenesis inhibitory factor (OCIF) gene
2	X77471.1	Homo sapiens human tyrosine aminotransferasef (TAT) gene
3	AB035346.2	Homo sapiens T-cell leukemia/Iymphoma 6(TCL6) gene
4	AJ225085.1	Homo sapiens Fanconi anemia group A(FAA) gene
5	AF009962	Homo sapiens CC-chemokine receptor (CCR-5) gene
6	X59065.1	Homo sapiens human acidic fibroblast growth factor (FGF) gene
7	AJ223321.1	Homo sapiens transcriptional repressor (RP58) gene
8	X92412.1	Homo sapiens titin (TTN) gene
9	U01317.1	Human beta globin sequence on chromosome 11
10	X51502.1	Homo sapiens gene for prolactin-inducible protein (GPIPI)

Ten genomic datasets from NCBI databank are used for the purpose of evaluation [19] as shown in Table 1. Determining the output performance thru taking into consideration of parameters Sn, Sp and Pr in [13], presents their expressions and the theory. The resulting output is determined by the results of different algorithms considered in Table 2. Amount of nucleotides during process of locating exon segments as introns remains described to be True Negative (TN), while correctly recognized exons are considered True Positive (TP). In addition, complete number of exon sections situated like nucleotides of intron is indicated for instance False Negative (FN), compared to amount of introns really anticipated to be exon nucleotides calculated to be False Positive (FP). Aimed at coherence of the outcomes, ten DNA sequences are regarded from NCBI as our information gene datasets to assess the efficiency of different algorithms. Gene sequence accessions remain X59065.1, X77471.1, E15270.1, AJ223321.1, AF009962, AB035346.2, AJ225085.1, X92412.1, U01317.1 and X51502.1 respectively as shown in Table 1.

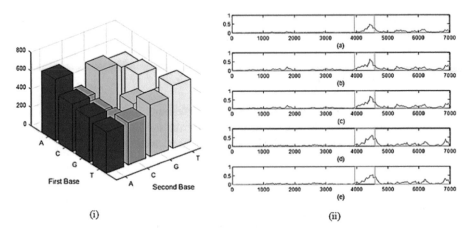

Fig. 4. (i) Nucleotide densities of dimers for gene sequence 5 (ii) PSD plots for location of exon (3934–4581) forecast of sequence using accession AF009962 by numerous AEPs, (a) AEP with LMS, (b) AEP with VSLMS, (c) AEP with SRVSLMS, (d) AEP with SVSLMS, (e) AEP with SSVSLMS. (*Relative Base Location on x-axis and Power Spectrum on y-axis*)

For performance metrics, expressions are

$$Sp = TP/(TP + FP)$$
$$Pr = (TP + TN)/(TP + FP + TN + FN)$$
$$Sn = TP/(TP + FN)$$

The number of exons actually found part of exon segments referred to be Specificity (Sp), while exons quantity that adequately forecast is calculated as Sensitivity (Sn). Exon identification results of gene sequence 5 with use of techniques depending on VSLMS can be found in Fig. 4(ii).

Values of threshold are chosen from 0.4 to 0.9 at an interval of 0.05. The exon forecast is accurate at 0.8. Sections with large percentage of nucleotides for A+T of DNA typically show components that are intergenic; whilst small A+T also greater nucleotides of G+C show potential genes. In bar illustration using MATLAB software, dimer distribution in a gene sequence 5 is portrayed in Fig. 4 which consists of 680 dimers of T–T base pairs. Also, this gene sequence includes 527 A–T and 70 G–C dimers. It is shown that content of G+C is smaller to content of dimers for A+T indicates this comprise lesser genes.

Adaptive exon forecast steps remain explained below:

(a) We analyzed DNA data sets of input of NCBI using density plots in Fig. 4(i) for genes to evaluate presence of gene locations.

(b) The resulting digital signal is now provided as an input to the proposed AEP following analysis. As a reference signal to the suggested AEPs, signal with TBP is provided and signal of feedback e(n) derived was used from Fig. 1 to update filter coefficients. When signal becomes minimal, exons are accurately located with plot derived for PSD.

(c) Plots of PSD are shown for the specified coding areas. In addition, Sn, Pr and Sp are also obtained and compared. The metrics sequence 5 with Accession AF009962 for all sign based VSLMS variants measured using MATLAB. From Table 2, performance measures of VSLMS based AEP are just inferior to SRVSLMS based AEP with less iterations due to low complexity and exon locating ability by locating precisely at 3934–4581 with great intensity and a sharp peak in PSD plot is observed. Hence, this AEP is better compared to LMS with Sn 0.6749 (67.49%), Sp, 0.6625 (66.25%), also Pr 0.6751 (67.51%). These PSD plots for VSLMS and its signed variants are shown in Fig. 4(ii) (b), (c), and (d) respectively.

Table 2. Metrics for performance with numerous VSLMS also its signed AEPs based upon Sp, Sn, and Pr values

Algorithm	Metric	Gene Sequence Serial Number									
		1	2	3	4	5	6	7	8	9	10
LMS	Sn	0.6286	0.6384	0.6457	0.6273	0.6481	0.6162	0.6193	0.6241	0.6268	0.6202
	Sp	0.6435	0.6628	0.6587	0.6405	0.6518	0.6324	0.6529	0.6289	0.6452	0.5965
	Pr	0.5922	0.5894	0.5934	0.5858	0.5904	0.5786	0.5896	0.5856	0.5814	0.5761
VSLMS	Sn	0.6886	0.6797	0.6805	0.6796	0.6865	0.6781	0.6846	0.6845	0.6882	0.6818
	Sp	0.6792	0.6835	0.6887	0.6812	0.6757	0.6832	0.6789	0.6764	0.6802	0.6892
	Pr	0.6878	0.6793	0.6812	0.6788	0.6806	0.6790	0.6814	0.6875	0.6836	0.6884
SRVSLMS	Sn	0.6795	0.6723	0.6749	0.6682	0.6749	0.6672	0.6732	0.6782	0.6806	0.6692
	Sp	0.6642	0.6768	0.6685	0.6671	0.6625	0.6775	0.6642	0.6695	0.6633	0.6761
	Pr	0.6795	0.6656	0.6675	0.6643	0.6751	0.6694	0.6731	0.6727	0.6765	0.6793
SVSLMS	Sn	0.6652	0.6672	0.6647	0.6568	0.6632	0.6567	0.6673	0.6668	0.6756	0.6564
	Sp	0.6581	0.6546	0.6563	0.6475	0.6587	0.6632	0.6574	0.6569	0.6587	0.6653
	Pr	0.6692	0.6515	0.6503	0.6426	0.6654	0.6569	0.6623	0.6596	0.6676	0.6592
SSVSLMS	Sn	0.6424	0.6447	0.6572	0.6538	0.6569	0.6523	0.6607	0.6537	0.6644	0.6508
	Sp	0.6458	0.6435	0.6443	0.6317	0.6432	0.6503	0.6516	0.6504	0.6451	0.6549
	Pr	0.6469	0.6382	0.6451	0.6384	0.6513	0.6435	0.6563	0.6457	0.6575	0.6354

5 Conclusion

The issue based on detection of the exon position in a gene sequence is shown in this paper. A novel methodology is presented here for adaptive exon identification. VSLMS-based adaptive algorithms are used to address the above issue. Measures for exon locations are evident in Table 2 and PSD plots are presented in Fig. 4(ii). The presented AEPs accurately located exon position at 3934–4581 with high intensity in PSD plot. SRVSLMS provides better performance based on complexity to perform computations, performance metrics attained using gene sequence 5 having accession AF009962 at threshold of 0.8 remain just below VSLMS based AEP values. However, it's a better choice due to its reduced computing difficulty also better convergence efficiency to accurately locate exons. The SRVSLMS based AEP can therefore be used in SOC and LOC based nano-bioinformatics applications.

References

1. Ning, L.W., Lin Ding, H., Huang, J., Rao, N., Guo, F.B.: Predicting bacterial essential genes using only sequence composition information. Genet. Mol. Res. **13**, 4564–4572 (2014)
2. Richters, M.M., Xia, H., Campbell, K.M., Gillanders, W.E., Griffith, O.L., Griffith, M.: Best practices for bioinformatic characterization of neoantigens for clinical utility. Genome Med. **56**, 1–21 (2019)
3. Inbamalar, T.M., Sivakumar, R.: Study of DNA sequence analysis using DSP techniques. J. Autom. Control Eng. **1**, 336–342 (2013)
4. Maji, S., Garg, D.: Progress in gene prediction: principles and challenges. Curr. Bioinform. **8**, 226–243 (2013)
5. Putluri, S., Zia Ur Rahman, M.: New adaptive exon predictors for identifying protein coding regions in DNA sequence. ARPN J. Theor. Appl. Sci. **11**, 13540–13549 (2016)
6. Saberkari, H., Shamsi, M., Hamed, H., Sedaaghi, M.H.: A novel fast algorithm for exon prediction in eukaryotes genes using linear predictive coding model and goertzel algorithm based on the Z-curve. Int. J. Comput. Appl. **67**, 25–38 (2013)
7. Wazim Ismail, M., Yuzhen, Y., Haixu, T.: Gene finding in metatranscriptomic sequences. BMC Bioinform. **15**, 01–08 (2014)
8. Ghorbani, M., Hamed, K.: Progress in gene prediction: principles and challenges. Bioinform. Approaches Gene Find. **4**, 12–15 (2015)
9. Putluri, S., Zia Ur Rahman, Md.: Efficient adaptive exon prediction for DNA study using proportionate LMS variants. Int. J. Eng. Tech. **7**, 116–123 (2018)
10. Azuma, Y., Onami, S.: Automatic cell identification in the unique system of invariant embryogenesis in caenorhabditis elegans. Biomed. Eng. Lett. **4**, 328–337 (2014)
11. Liu, G., Luan, Y.: Identification of protein coding regions in the eukaryotic DNA sequences based on Marple algorithm and wavelet packets transform. Abstr. Appl. Anal. **2014**, 1–14 (2014)
12. Simon Haykin, O.: Adaptive Filter Theory, 5th edn, pp. 320–380. Pearson Education Ltd., London (2014)
13. Saberkari, H., Shamsi, M., Hamed, H., Sedaaghi, M.H.: A fast algorithm for exonic regions prediction in DNA sequences. J. Med. Signals Sens. **3**, 139–149 (2013)
14. Nagesh, M., Prasad, S.V.A.V., Rahman, M.Z.: Efficient cardiac signal enhancement techniques based on variable step size and data normalized hybrid signed adaptive algorithms. Int. Rev. Comput. Softw. **11**, 1–13 (2016)
15. Kuang, J., Li, Y.P.: Variable step size LMS algorithm with a gradient based weighted average. IEEE Signal Process. Lett. **16**, 1043–1046 (2009)
16. Kwong, R.H., Edward Johnston, W.: A variable step size LMS algorithm. IEEE Trans. Signal Process. **40**, 1633–1642 (1992)
17. Shin, H.C., Sayed, A.H., Song, W.J.: Variable step size- NLMS and affine projection algorithms. IEEE Signal Process. Lett. **11**, 132–135 (2004)
18. Paula Diniz, S.R.: Adaptive Filtering, Algorithms and Practical Implementation, 4th edn. Springer, New York (2013)
19. National Center for Biotechnology Information. www.ncbi.nlm.nih.gov/. Accessed 25 Jan 2019

20. Pulturi, S., Zia Ur Rahman, M., Amara, C.S., Pulturi, N.: New exon prediction techniques using adaptive signal processing algorithms for genomic analysis. IEEE Access **7**, 80800–80812 (2019)
21. Putluri, S.R., Zia Ur Rahman, M.: Identification of protein coding region in DNA sequence using novel adaptive exon predictor. J. Sci. Ind. Res. **77**, 1–5 (2018)
22. Putluri, S., Zia Ur Rahman, M., Fathima, S.Y.: Cloud based adaptive exon prediction for DNA analysis. IET Healthc. Technol. **5**(1), 1–6 (2018)

Tweets Analysis for Disaster Management: Preparedness, Emergency Response, Impact, and Recovery

Archana Gopnarayan$^{(\boxtimes)}$ and Sachin Deshpande

Computer Engineering, Vidyalankar Institute of Technology,
Wadala, Mumbai, India
archana.gopnarayan@vpt.edu.in,
Sachin.Deshpande@vit.edu.in

Abstract. Social media is very important source for identifying and analyzing various disaster events. The News channels broadcast their headlines of news on twitter as short messages. In case of any disaster situation the news channel immediately broadcast the news. Social media data is categorized into the various phases of disaster management. Disaster researchers and emergency managers use social media data as reference for their analysis. To figure out the disaster management and the social media under various shifts and numerous phases and decide with the an efficient solution "real time" is required. We will be using and comparing Support Vector Machine (SVM), K-nearest neighbors (KNN) and Logistic regression, data mining algorithms for classification of tweets. The most accurate algorithm is applied on real time tweets of news channel for handling disaster situation. The analysis is beneficial for disaster management team to detect the change of state between various stages of disaster management.

Keywords: Disaster · Logistic regression · KNN · SVM

1 Introduction

Social media plays an important role in people communication and information sharing. For Disaster researchers and emergency managers tweeter has become a major channel for communication during any disaster situation. Users and news channels share the recent news, photos, observations and current situation during the case of any disaster.

Social media data is categorized into the various phases of disaster management. Tweeter is the main source for having social media data. Here we can do classification of tweets first, then clustering of tweets and at the end extraction of the tweets. In the classification phase we are using three algorithms (LR, SVM and KNN). In the clustering filters are used to group the similar types tweets of. The extraction phase the tokens can be extracted that have particular information about the disaster.

Disaster management is typically categorized into four phases: Mitigation, Preparedness, Response and Recovery. The traditional data collection methods for handling the different disaster phases of disaster management have many drawbacks.

© Springer Nature Switzerland AG 2020
J. S. Raj et al. (Eds.): ICIDCA 2019, LNDECT 46, pp. 760–764, 2020.
https://doi.org/10.1007/978-3-030-38040-3_87

Social media is having main role while handling disaster situation. Different methodologies like phone calls, direct interview of personal, direct observation were used by disaster responders to handle the positional awareness and for investigation. The disaster management phases cannot be segregated separately or in particular order. Different phases of disaster may co-occurrence with each other and the duration of every stage generally relies over the strictness of the calamity [1].

The investigation identifies the tweets nature produced in every instant of the disaster. Here the tweets are categorized as different classes from best of our knowledge.

In the case of any disasters the fastest way of communication is social media. Through the social media we can get the immediate reply as compared to traditional sources of communication. The affected people can send the feedback in very faster way by using social media.

2 Literature Survey

During different events twitter is well known source for speedy diffusion of views and for getting information of the events. It can be mainly used during different disaster to help calls, damage assessment and to communicate evacuation plans. The author describes the analyses of the tweets during the management of the disaster under various perspectives of the geographical data that is updated in the tweets such as the credibility and the granularity utilizing the naive Bayes in the classification of the tweets [2].

Data generated by social media like twitter for flooding disaster, different assessing techniques can be used in real time so that emergency services can be provided quickly. Artificial Intelligence plays the major role in disaster Management, is a platform for classifying tweets during disaster. Naive Bayes classification technique is used [3].

The author utilizes the mining tools of the twitter to report the responses done on the disaster management [4].

The author put forward the utilization of the API in the analysing the tweets and excerpting the information of the disaster [5].

The main motive of a calamity release and reaction method remains basically passionate about appropriate and correct info concerning the standing of the disaster, the encircling surroundings, and therefore the affected folks. In spite of the availability of various tools for analysing the calamity and provide with the automated response an efficient decision making is still a challenge, so the author in his survey of the technologies for the calamity relief presented the that is intelligent techniques available for the disaster management along with the research direction for the future and the issues in it [8].

3 Problem Statements

The main aim of the project is to classify tweets into various phases of disaster. Apply the data mining algorithms for the classification of tweets and identify the most accurate algorithm.

As working on a live data any miss classification will lead to some hazards situation so proper selection of classification algorithm is required for twitter messages. As there is no mechanism to predict that certain classification algorithm will work better on specific data. It can only be identified by testing and comparing some data mining algorithm on the data we want to classify, performance and reliability of algorithm is highly dependent on type and size of data, taking all this things into consideration in the project we are comparing LR, KNN and SVM data mining techniques for classifying twitter messages and will find the best algorithm among them.

Developed the system that will extract the news as the news is post on the twitter site and will classify into different groups.

4   Proposed System

In proposed system we are comparing SVM, Logistic Regression and KNN data mining techniques for classifying twitter messages and will find the best algorithm for classification of disaster phases (Fig. 1).

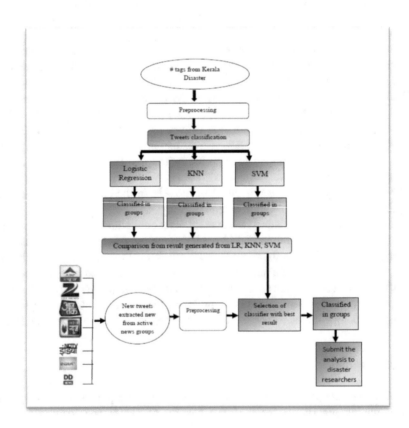

Fig. 1. Proposed system

4.1 Modules

Data Gathering: The tweets from Kerala disaster 2018 are used for training the network. Twitter short messages are required to be collected. Twitter API provides short messages using hash tags.

Pre-processing: Once the tweet has been collected, the features are need to be take out from the short messages.

Data Classification: The tweets are then passed as an input to SVM, Logistic Regression and KNN classifier where tweets will be classify into multiple phases of disaster.

Comparison of Classifier: Classified data is compared in order to find the best algorithm among SVM, Logistic Regression and KNN that will best suited for classifying tweets. The classification will be done based on their processing time and how many tweets they are classifying correctly and incorrectly. Classifier with best processing time and good percentage of correct classification will consider as classifier for twitter and that classifier will be chosen to work on live data.

Current Tweets and Pre-processing: The news channels are fastest source for obtaining disaster news. So tweets of various news channels are used for live data.

Selection of Classifier and Classification: After comparison step we will be able to identify which is the best classifier that particular classifier will be chosen to do the classification, pre-processed data will be given to the classifier as an input and we will get news classified into multiple phases of disaster. The complete analysis can be used as reference for disaster researchers and disaster management team for disaster management.

5   Conclusion

Social media is rich source of data. Huge amount of data we can captured from social media tools which can be used as source of information for handling the disaster situation. The different data mining techniques can be used to classify the collected tweets for any emergency situation. Twitter real time data is more usable during the emergency services and to extract what information can be extracted from the tweets. The emergency managers can the complete analysis to detect the changeover of various stages of disaster management.

References

1. Huang, Q., Xiao, Y.: Geographic situational awareness: mining tweets for disaster preparedness, emergency response, impact, and recovery. ISPRS Int. J. Geo-Inf. **4**(3), 1549–1568 (2015)
2. Zahra, K., Purves, R.: Analysing tweets describing during natural disasters in Europe and Asia. In: AGILE 2017, Wageningen, 9–12 May 2017 (2017)
3. Spielhofer, T., Greenlaw, R., Markham, D., Hahne, A.: Data mining twitter during UK floods (2016)

4. Ashktorab, Z., Brown, C., Nandi, M., Culotta, A.: Tweedr: mining twitter to inform disaster response, May 2014
5. Funayama, T., Yamamoto, Y., Tomita, M., Uchida, O., Kajita, Y.: Disaster mitigation support system using twitter and GIS 978-1-4799-8027-7/14/$31.00©2014 IEEE
6. To, H., Agrawal, S., Kim, S.H., Shahabi, C.: On identifying disaster-related tweets: matching-based or learning-based (2017)
7. Nazer, T.H., Xue, G., Ji, Y., Liu, H.: Intelligent disaster response via social media analysis a survey. ACM SIGKDD Explor. Newslett. **19**, 46–59 (2017)
8. Gupta, A., Lamba, H., Kumaraguru, P.: $1.00 per RT #BostonMarathon #PrayForBoston: analyzing fake content on twitter. In: eCrime Researchers Summit (eCRS), pp. 1–12. IEEE (2013)

CFRF: Cloud Forensic Readiness Framework – A Dependable Framework for Forensic Readiness in Cloud Computing Environment

Sugandh Bhatia[1(✉)] and Jyoteesh Malhotra[2]

[1] Computer Science, Punjab School of Economics, Guru Nanak Dev University,
Amritsar 143005, India
sugandhcs.rsh@gndu.ac.in
[2] Faculty of Engineering and Technology, Guru Nanak Dev University,
Amritsar 143005, India

Abstract. Cloud computing is an important paradigm of information technology that entirely depends upon virtualization technique. Multi-tenancy is one of the most important characteristic and it is applicable in all deployment models. Another significant feature of cloud is its distributed nature which includes jurisdictions of various geographical areas. Due to these two characteristics, it is a complex and tedious task to collect data, proof and evidences in cloud whenever compared with traditional digital forensic. New mechanisms and frameworks are required to apply digital forensic techniques in cloud computing environment. In this research paper, an effort has been made to expound prominent techniques that can be applied to perform digital forensic analysis and investigation in cloud computing environment.

1 Introduction

Cloud computing technology has witnessed exponential growth in the last couple of years and recent studies have revealed that public cloud revenue will be $331B in 2022 [1]. In 2018, 3.6 billion internet users were availing the services of cloud computing. As the number of cloud users is increasing day by day, it reveals that cloud computing is gaining more popularity and acceptance despite the privacy and security issues that still prevails. When any type of security lapse or breach transpired, digital forensic analysis and investigation requires to be implemented. Most of the times, it is noticed that the cases of security breach in cloud computing remain unsolved and the major reason is the lack of cloud forensic tools and techniques. Many academicians and researchers have tried to solve these issues but these endeavors are still not up to the mark. Moreover, cloud security [2] concerns are not related with the cloud users, service providers and other third party stakeholder. In recent times, there have been cases like LabCorp [3], it's a world's leading healthcare diagnostic company in United States of America. LabCorp reported that the crucial data of 7.7 million customers had exposed during the period August 2018 to March 2019. Another famous example of cyber-attack is of Cathay Pacific [4] in 2018 and it is known as largest data breach of aviation industry. More than 9.4 million passengers of Cathay Pacific were affected due

© Springer Nature Switzerland AG 2020
J. S. Raj et al. (Eds.): ICIDCA 2019, LNDECT 46, pp. 765–775, 2020.
https://doi.org/10.1007/978-3-030-38040-3_88

to this pilferage of data. The most interesting part of this data theft is that the objective of this data leakage is still not revealed by the investigative agencies. In all cases of security, privacy and data breach, when an incident happens, digital forensic tools and techniques would be demanded when the cases are presented in the court of law. Therefore, this piece of research tries to furnish a solution and dispense digital forensic techniques from the point of view of cloud computing applications since more important and sensitive data is being transferred and stored in the cloud nowadays.

2 Objectives of the Article

The primary objective of this research article is to design a framework to perform systematic and economically feasible digital forensic analysis and investigation in cloud computing environment. The objectives of this research article can be explained as follows.

- The first and most significant objective of this article is to propose a cloud forensic readiness framework that is dependent upon existing digital forensic tools and it should be economically favorable for the organization to deploy this framework.
- Various digital forensic frameworks will be discussed in the related work section of this article. Pros and cons of every discussed framework are taken into consideration during the literature survey. An effort has been made to remove the gaps from the existing tools while applying these tools for the designing of Cloud Forensic Readiness Framework (CFRF).
- To exhibit the impact of CFRF in the practical world by executing the prototype based on proposed framework.
- Implementing digital forensic triage in the analysis and investigation and to propose a cloud forensic readiness framework [5] that is based on digital forensic service model.

3 Related Work

Lin et al. [6] described an investigation model in case of wireless environments. This model is based upon three phases: preparation, operation and report. Overall forensic part is covered in the second phase of the model and various sub-phases of the second phase are collection of evidence and proof, analysis and forensic.

Shin [7] contemplated a forensic process which involves several phases and the sequence of phases are investigation, classification, priority setup, investigation, criminal profiling and analysis, evidence collection, presenting profiling and report writing. Main advantage of this forensic model is that the operation or task of each phase is explained in detail manner.

Umar et al. [8] described and applied mobile forensic techniques as per the recommendations of National Institute of Standards and Technology (NIST). The mechanism is applied to conduct the analysis of digital artifacts and evidences in case of emails. Collection, testing, analysis and reporting are the four important phases of this mechanism. Wireshark and NetworkMiner tools are used for analyzing the packets. The findings of the comparative study of the Wireshark and NetworkMiner revealed that NetworkMiner is capable to procure digital evidences like delivery time stamp, receiver timestamp, protocols of sender and receiver along with source and destination IP addresses and artifacts from emails.

Kent et al. [9] explained the whole digital forensic process in four steps and these are collection, examination, analysis, and reporting. The most important characteristic of this model is the emphasis on two additional functions. These functions are preservation of the evidences and proper documentation.

Fahdi et al. [10] discussed the criteria to perform digital forensic on the basis of survey or data collected from digital forensic analysts and investigators. Six significant limitations of the digital forensic process were highlighted by the authors. These limitations are size of data, legal matters, time period, tool ability, visualization and forensic training. Moreover, it was specified that cloud computing is such a technology in which digital forensic services and techniques are indispensable.

Rogers et al. [11] proposed a digital forensic model which is known as "Cyber Forensic Field Triage Process Model" (CFFTPM). The primary objective of this model is to perform time bound investigations. It defines a sequence of operations such as identification, investigation or analysis and interpretation of digital artifacts and evidences.

Simou et al. [12] concentrated on the recognition of technical and scientific solutions available these days that can be applied to solve the cloud forensic problems. However, an attempt has made to recapitulate the techniques or methodology and advocate the solutions to solve the problems. Finally, it elucidated the major problems in cloud forensic domain.

Garfinkel [13] adduced in his research paper various directions in the field of digital forensic. Author advised the digital forensic fraternity to apply standardized and transposable techniques for handling and representing the data. This research article also explains succinctly the benefit and importance of scalability and validation of existing digital forensic tools or techniques.

4 Cloud Forensic Readiness Framework

Digital forensic techniques and tools can be applied to collect digital artifacts and evidences from the cloud. Digital forensic mechanism that conforms to these techniques is needed. The steps followed in this article are based on ISO/IEC 27050 [14]. At every level of mechanism such as identification, preservation, collection, processing, review, analysis and production, a perspicuous procedure was specified. ISO/IEC 27050 was selected to implement in this framework as its principal objective is to stimulate fair practice methods and techniques for the collection of digital artifacts and analysis of digital evidence. For the suggested standardized methods to be implemented, an interface is needed in which all these methods can be executed for analyst. To achieve this objective, a cloud forensic readiness framework is proposed. CFRF provides constituents through which the methods can be implemented in a chronological order.

CFRF assists the investigator and performs all the required functions in the investigation process. The proposed framework CFRF is depicted in Fig. 1.

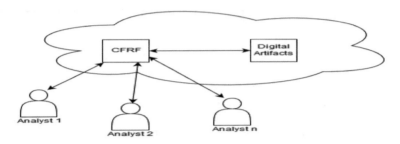

Fig. 1. Proposed framework of CFRF

The ISO standard 27050 divides digital forensic process into seven phases. In this article, digital artifact acquisition procedure is elaborated. The rationale for selecting the digital artifact acquisition procedure as the primary task is that artifact or evidence acquisition is a tedious and complex task. Due to the nature of cloud computing, many times cloud data is not approachable. A ten phase cloud forensic readiness framework is proposed in Fig. 2.

The phases of CFRF are incident detection, connection establishment, strategy and policy making, ready for execution, identification of digital artifacts, artifacts collection and acquisition, organizing artifacts, investigation and analysis, outcome and report writing and closure and preservation.

Incident Detection: To implement CFRF, total ten phases are to deploy and incident detection is the first phase of this framework. This phase is based upon the manual operations performed by the administrators. The phase comprises three sub-phases. These are first information report, case registration and case explanation (Fig. 3).

Connection Establishment: The connection establishment in CFRF is designed to hold incident scene. The overall functioning of cloud computing revolves around the physical hosts, virtual machines, internet connectivity, users and cloud service providers. The first step to be taken in the connection establishment phase is to initiate a secure channel with remote host. The most important function is to establish a secure connection with the incident scene to control the intrusion during analysis and investigation. After establishing the secure channel, the next function is to initiate a remote connection. The function of deploy servers is designed to execute and involve software servers, tools, administrators and investigators. The activate secure logging function initializes logging operations of the incident scene by analysts, offenders and cloud service users (Fig. 4).

Strategy and Policy Making: Proper strategy and policy planning is required to implement an impressive and effective cloud forensic analysis and investigation. This phase revealed in Fig. 5 starts with appointing a research and investigative team to conduct the analysis. There are two possibilities to find out digital artifacts [15]. First

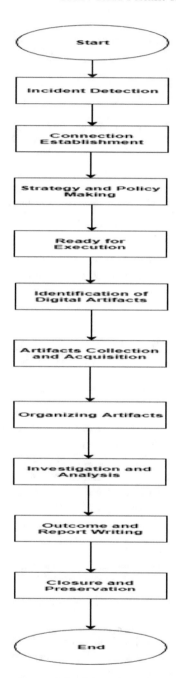

Fig. 2. Ten phases of cloud forensic readiness framework

possibility is that artifacts may be retrieved from the physical network and the second is to fetch the artifacts from virtual network. In case of cloud computing higher

Fig. 3. Process of incident detection

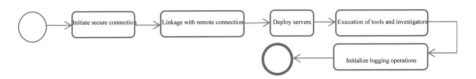

Fig. 4. Connection establishment procedure

probability is to find out the digital artifacts from the virtual network rather that the physical network due to the functioning of cloud computing. The selection of suitable tools and techniques play an important role in acquiring the evidence resources for investigation.

Fig. 5. Policy planning

Ready for Execution: The ready for execution process illustrated in Fig. 6 depicts four main functions namely function allocation, digital artifacts collection, artifacts validation and shortlisting & selection of forensic tools. The sub process of function allocation involves in allotment of duties to the experts of cloud forensic group. Actually, in the phase of ready for execution, a blueprint regarding the collection, analysis of artifacts and selection of digital forensic tool is prepared.

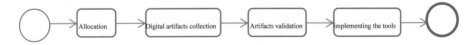

Fig. 6. Ready for execution phase

Identification of Digital Artifacts: The phase of identification of digital artifacts presented in Fig. 7. There are five sub phases included in this phase. These are exploring paging file, RAM categorization, pointing out the corrupted data, artifacts classification, ascertaining the host attached with incident scene. A file saved on the hard drive can be utilized as an expanded component of available RAM. The sub phase of RAM categorization includes the classification of RAM on the basis of incident scene and it retrieves various digital artifacts from the RAM and BIOS (basic input output system). Details procured like capacity and category of RAM, clock rate, frequency and bandwidth. However, these details are used only in the case of Infrastructure as a Service model. The pointing out corrupt data sub phase is associated with the identification of files that are changed recently. Files of any kind with unwanted change will be termed as corrupted. Different type of artifacts can be fetched from these files with the help of logging data [16]. The next sub phase of this phase is the classification of artifacts and evidences and the last step is to find out the host which is directly connected with incident scene.

Fig. 7. Digital artifacts identification

Artifacts Collection and Acquisition: The collection and acquisition of digital artifacts procedure is expressed in Fig. 8. It starts with the initialization of an agent. It is termed as acquisition agent and its major role is to procure the details regarding created files, modified files and corrupted files in a specific timeframe. Another important record which is to be collected by agent is called as network trace [17]. The RAM details recovered in the previous phase will be used for performing the analysis and preservation of all the artifacts or evidences must be ensured at this level for further analysis and investigation.

Fig. 8. Collection and acquisition of artifacts

Organizing Artifacts: To organize the artifacts, movement of evidence and artifacts is required from the point of origin (incident scene). There are two ways to move the artifacts, first is with the help of portable removable media and second by using the secure network channel. In both cases, transportation of artifacts requires to be under the supervision of technical and investigating personnel.

Investigation and Analysis: The most important and prominent phase of any forensic model is investigation and analysis. The same condition applies here. The investigation and analysis of digital artifacts can be classified into two categories: domain based analysis [18] and server based analysis [19]. The working of domain based analysis is discussed here and figure is placed as Fig. 9 on the next page. First of all, examine the connected network of the incident host. For this purpose, MAC addresses and tables of kernel cache will be used. All the TCP and UDP connections (present and previous) must be scrutinized. The connection of attacker can be traced by using all the above said TCP and UDP connections. Interpretation of network protocols such as HTTP, TCP and UDP to be performed. Interpretation makes feasible to fetch the relevant packet for purpose of examination. It is mandatory in the domain based analysis that the fetched packets must be transformed into the database file. Various classification and visualization tools [20] can be applied on this database with the purpose of analysis. The analysis provides all the information regarding operations, functions and movement of data across the network. With the completion of visualization and classification task, a clear network flow will be represented and it highlights distrustful packets. The investigating team defines the scope of operation and it is economically favorable for the organization. Further, the important process is to be accomplished for reconstituting the files. This process includes login files, spooled files, media files and temp files. The reconstituted files may incorporate scripts that are associated with incident host. During the analysis of reconstituted files, the analyst may execute or examine the recovered scripts. Outcome of the domain based analysis would be stored in a data file along with the result received from the host based analysis. In case of server based analysis, operations are entirely dependent upon attributes of the server. By the term attribute, we mean the information associated with the server machine such as server operating system type, version, server name or id, system uptime, file system and hardware details. The information collected at this level plays an important role in process planning and decision making. The first step is to retrieve the details of system of incident scene. Role of kernel module is prominent at this level. If corrupted modules are found, every module is examined rigorously. This examination is performed by specifying the number and details of processes in the running phase and associated with running host. The malicious processes can't be the part of operating system and this way a malicious process can be identified. Moreover, memory assigned to every malicious process is investigated. Scripts can be seized for analysis from the assigned memory. Most of the times, it has been found that attacker initiates the attack with the help of script. It is possible to detect the script from the storage, if the script is still in the execution phase. There is a dire need to apply a search mechanism on the basis of characters on random access memory. Characters or strings can be retrieved from live memory. Thereafter, another function is to find out any malicious software such as rootkit [21]. "A rootkit is a collection of computer software, typically malicious, designed to enable access to a computer or an area of its software that is not otherwise allowed and often masks its existence or the existence of other software". In case, rootkits are found, these are made inactive and procedure is started again to unmask another evidences. If no rootkit is detected, the procedure goes through as it is. Highly skilled investigators and analysts are required for the detection and dismantle of rootkits. It is required that analysts must be given privilege to access remote server which is connected with the incident scene.

Outcome and Report Writing: Results of both domain based analysis and server based analysis are stored, compared and analyzed meticulously. As per the standard of ISO/IEC 27050, after the analysis the next significant step is production which includes various other sub-tasks such as storage of outcome of the analysis on storage media. After the completion of sub-phase of report writing, all the procured evidences are presented in the court. "It is important to note that [ISO/IEC 27050] is not intended to contradict or supersede local jurisdictional laws and regulations. Electronic discovery often serves as a driver for investigations as well as evidence acquisition and handling

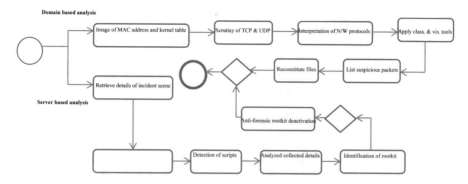

Fig. 9. Investigation and analysis process

activities. In addition, the sensitivity and criticality of the data sometimes necessitate protections like storage security to guard against data breaches."

Closure and Preservation: The last phase of CFRF is to close the case once it is submitted in the court of law and final judgment is given. But an important sub-phase is connected with closure and it is to preserve all the evidences [22] and records for future use. There is a possibility that in future case can be reopened for the investigation. Therefore, it is strictly demanded that whole manuscript must be preserved and secured in a proper manner. As in cloud computing technology, most of the times there is absence of physical storage media which is associated with the incident scene. Therefore, communication with the incident scene must be recorded and interaction should be performed with the automated mechanism.

5   Conclusion

The nature of cloud computing is idiosyncratic whenever compared with another computing technologies. Traditional techniques of digital forensic can't be applied as it is in case of cloud computing. Dedicated methods are needed for cloud computing systems. Primarily, focus of this research article is on the formation of cloud forensic readiness dependable framework. The suggested framework comprises various techniques and tools of digital forensic. These techniques are based upon different phases which were mentioned in ISO/IEC 27050 digital forensic standard. In cloud computing forensic readiness system, the analysis must not be limited to individual incident scene. During the exploration of an incident scene, there is a possibility that area of investigation may be widened if any association is found between two instances.

References

1. Columbus, L.: Public Cloud Soaring to $331B By 2022 According to Gartner. Forbes. https://www.forbes.com/sites/louiscolumbus/2019/04/07/public-cloud-soaring-to-331b-by-2022-according-to-gartner/#323195ec5739. Accessed 7 Apr 2019

2. Alenezi, A., et al.: Experts reviews of a cloud forensic readiness framework for organizations. J. Cloud Comput. **8**(1) (2019)
3. Siegel, R.: LabCorp discloses data breach affecting 7.7 million customers. The Washington Post, Washington, D.C., 5 June 2019
4. Cathay Pacific Cyber Attack Is World's Biggest Airline Data Breach. Insurance Journal. https://www.insurancejournal.com/news/international/2018/10/26/505699.htm. Accessed 26 Oct 2018
5. Alsmadi, I.: Cyber Operational Planning. The NICE Cyber Security Framework, pp. 135–179 (2019)
6. Lin, I.-L., Yen, Y.-S., Chang, A.: A study on digital forensics standard operation procedure for wireless cybercrime. In: 2011 Fifth International Conference on Innovative Mobile and Internet Services in Ubiquitous Computing (2011). https://doi.org/10.1109/imis.2011.58
7. Shin, Y.-D.: New digital forensics investigation procedure model. In: 2008 Fourth International Conference on Networked Computing and Advanced Information Management (2008). https://doi.org/10.1109/ncm.2008.116
8. Umar, R., Riadi, I., Muthohirin, B.: Live forensics of tools on android devices for email forensics. Indones. J. Electr. Eng. **17**(4), 1803–1809 (2019)
9. Kent, K., Chevalier, S., Grance, T., Dang, H.: Guide to integrating forensic techniques into incident response (2006). https://doi.org/10.6028/nist.sp.800-86
10. Al Fahdi, M., Clarke, N.L., Furnell, S.M.: Challenges to digital forensics: a survey of researchers & practitioners attitudes and opinions. In: 2013 Information Security for South Africa (2013). https://doi.org/10.1109/issa.2013.6641058
11. Rogers, M., Goldman, J., Mislan, R., Wedge, T., Debrota, S.: Computer forensics field triage process model. J. Digit. Forensics Secur. Law (2006). https://doi.org/10.15394/jdfsl.2006.1004
12. Simou, S., Kalloniatis, C., Kavakli, E., Gritzalis, S.: Cloud forensics solutions: a review. In: Lecture Notes in Business Information Processing, pp. 299–309 (2014). https://doi.org/10.1007/978-3-319-07869-4_28
13. Garfinkel, S.L.: Digital forensics research: the next 10 years. Digit. Investig. **7**, S64–S73 (2010). https://doi.org/10.1016/j.diin.2010.05.009
14. ISO/IEC 27050 EDiscovery. ISO27k Infosec Management Standards. https://www.iso27001security.com/html/27050.html. Accessed 17 Sept 2019
15. Choo, K.-K.R., Dehghantanha, A.: Contemporary digital forensics investigations of cloud and mobile applications. In: Contemporary Digital Forensic Investigations of Cloud and Mobile Applications, pp. 1–6 (2017). https://doi.org/10.1016/b978-0-12-805303-4.00001-0
16. Kundu, S., Garg, L.: Web log analyzer tools: a comparative study to analyze user behavior. In: 2017 7th International Conference on Cloud Computing, Data Science and Engineering - Confluence (2017). https://doi.org/10.1109/confluence.2017.7943117
17. Mivule, K., Anderson, B.: A study of usability-aware network trace anonymization. In: 2015 Science and Information Conference (SAI) (2015). https://doi.org/10.1109/sai.2015.7237310
18. Analysis cloud - running sensor data analysis programs on a cloud computing infrastructure. In: Proceedings of the 3rd International Conference on Cloud Computing and Services Science (2013). https://doi.org/10.5220/0004371503580365
19. Singh, S.K.: Cloud computing: comparative study own server vs cloud server. In: Recent Advances in Mathematics, Statistics and Computer Science (2016). https://doi.org/10.1142/9789814704830_0056

20. Chen, M.: Information Theory Tools for Visualization (2016). https://doi.org/10.1201/9781315369228
21. Carvey, H.: Rootkits and rootkit detection. In: Windows Forensic Analysis, pp. 307–331 (2007). https://doi.org/10.1016/b978-159749156-3/50011-3
22. Rubsamen, T., Pulls, T., Reich, C.: Security and privacy preservation of evidence in cloud accountability audits. In: Communications in Computer and Information Science, pp. 95–114 (2016). https://doi.org/10.1007/978-3-319-29582-4_6

Face Recognition Based on Interleaved Neighbour Binary Pattern

A. Geetha[(⊠)] and Y. JacobVetha Raj

Department of Computer Science, Nesamony Memorial Christian College,
Marthandam Affiliated to Manonmaniam Sundaranar University, Abishekapatti,
Tirunelveli 627 012, Tamil Nadu, India
geethavijayaragavan1999@gmail.com,
jacobvetharaj@gmail.com

Abstract. Face recognition plays a key role in human face identification. This research work proposes a novel method, Interleaved Neighbour Binary Pattern (INBP) and the similarity measure histogram intersection for face recognition. The INBP is used to extract the facial features and improves the recognition rate under minor variations in expression, illumination and orientation. The proposed approach, INBP compares the interleaved neighbours rather than comparing the adjacent neighbours as in the traditional LBP. Extensive experimental results on three widely used face datasets JAFFE, ORL, YALE and OWN DATABASE show that INBP achieves better performance than traditional LBP for face identification under various conditions. The performance of the proposed approach is measured in terms of recognition rate, precision rate, accuracy, recall rate, f-factor and error rate.

Keywords: LBP · INBP · Histogram intersection · Accuracy · Precision rate · Recognition rate · Recall rate · F-factor · Error rate

1 Introduction

The role of face recognition in still images is identifying the people in a set of test images and compare with a system that has been previously trained with a collection of face images marked with each person's identity. Hence face recognition identifies an unknown input face and matches it against the faces of different known individual's database. For security reason, governments just as, private require solid techniques to precisely distinguish people. Beginning strategies regarded faces as focuses in high dimensional space and after that the Euclidean distance between them is determined. Afterward, some Dimensionality decrease systems including Principal Component Analysis (PCA) [1, 2] have now been effectively applied to the issue, along these lines lessening the multifaceted nature of the recognition procedure and increment the precision.

These days various examples are utilized for feature extraction. LBP, for the most part, marks the pixels of a picture by thresholding the 3×3 sub-cluster. The area of every pixel with the estimation of the focal pixel is compared and concatenated and the bits from the upper left [3, 4]. Because of the simplicity and robustness, it has been widely used in face recognition [5]. Later the LBP operator was reached out to utilize

© Springer Nature Switzerland AG 2020
J. S. Raj et al. (Eds.): ICIDCA 2019, LNDECT 46, pp. 776–784, 2020.
https://doi.org/10.1007/978-3-030-38040-3_89

the areas of various sizes. Sizes are chosen by the span from the focal point of the pixel. Face recognition utilizing Local Binary Pattern (LBP) to be implemented in a Smartphone with Android operating system acquired up to 90% of recognition [13].

This paper proposes a new Interleaved Neighbour Binary Pattern (INBP). The proposed INBP creates a micro pattern which can also be modelled by histogram.

2 Related Works

[6–8] LBP technique gives generally excellent outcomes, both as far as speed and perspicacity execution. In light of the manner in which the texture and shape of images of pictures is portrayed, the technique looks to be very vigorous against face images with various outward expressions.

N_1	N_2	N_3
N_8	N_0	N_4
N_7	N_6	N_5

Fig. 1. Eight – neighborhood around N_0 in LBP

Where N_i, $i = 1, 2,\ldots 8$ is an eight neighbourhood point around N_0 as shown in Fig. 1. The thresholding function for $f(\cdot,\cdot)$ for the basic LBP can be represented by Eq. 1.

$$f(I(N_o), I(N_i)) = \begin{cases} 0, & \textit{if } I(N_i) - I(N_0) \leq \text{threshold} \\ 1, & \textit{if } I(N_i) - I(N_0) > \textit{threshold} \end{cases} \tag{1}$$

Concatenation of the binary gradient direction is called a micro pattern [9]. Figure 2 shows the micro pattern 11010001, when the threshold is set to zero.

Subsequently the LBP operator was extended to use neighborhoods of different sizes [10].

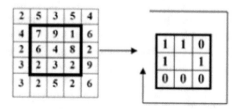

Binary number obtained is 11010001

Fig. 2. Micro pattern obtained by LBP

3 Proposed Method: Interleaved Neighbour Binary Pattern (INBP)

Consequently, the LBP operator was stretched out to utilize neighbourhoods of various sizes [10]. For the most part, the local binary pattern (LBP) example looks at the connection deliver between the referenced pixel and its encompassing 8-neighbours and after that computes the 8-bit pattern. INBP is likewise finished by isolating an image into a few littler 3 × 3 locales from which the features are extracted. These highlights contain binary patterns examples that de-recorder the encompassing neighbouring pixels in the areas has appeared in Fig. 3. This methodology isn't to encode all the eight neighbours, just discover the distinction between the neighbours in a steady progression.

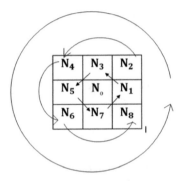

Fig. 3. Eight – neighborhood around N_0 in INBP

In INBP, the comparison starts from N_1 and ends with N_1. It is performed in anti-clockwise. The function for $f(\cdot,\cdot)$ for the INBP can be represented by Eq. (2) with a construct of 7 bits.

$$f\left(I(N_i), I\left(N_{(((i+1)mod8)+1)}\right)\right) = \begin{cases} 0, & if\ I(N_i) - I\left(N_{((i+1)mod8)+1}\right) < 0 \\ 1, & otherwise \end{cases} \quad (2)$$

Where N_i, i = 1, 2,…7. In INBP, Eq. 2 compares all the neighbours around N_0 and constructs a seven bit micro pattern instead of eight bit micro pattern constructed by the traditional LBP. But the eighth bit is computed by Eqs. 3 and 4. The 8 Neighbourhood point around N_0 is shown in Fig. 3. Concatenation of the binary gradient direction is called a micro pattern. Figure 4 shows the INBP micro pattern.

$$(I(N_o), I(N)) = \begin{cases} 1, & if (I(N_0) - I(N) \geq 0 \\ 0, & if (I(N_0) - I(N) < 0 \end{cases} \quad (3)$$

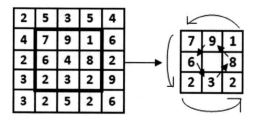

Binary number 0011110

Fig. 4. Micro pattern (7-bit) obtained from the 3×3 array

Where $I(N)$ is calculated by Eq. 4.

$$I(N) = \sum_{i=1}^{8} I(N_i)/8 \qquad (4)$$

Figure 4 shows the 7-bit micro pattern 0011110 constructed by Eq. 2. Eighth bit is constructed after the application of Eqs. 3 and 4. Equation 4 computes the average 5 for the eight neighbours 8, 1, 9, 7, 6, 2, 3, and 2. Equation 3 compares the average 5 with the centre value 4 and produces the eighth bit 0. Finally this approach produces the 8-bit micro pattern 00111100. The decimal equivalent of this micro pattern is 60.

Steps of proposed approach INBP:

1. For INBP Divide the given image into 3×3 sub images as showed Fig. 3.
2. For each 3×3 sub image repeat the steps 3 to 8.
3. Construct the first seven bits by comparing the referenced pixel's neighbours by using Eq. 2.
4. Find the average of 8-neighbours of the referenced pixel by using Eq. 4
5. Subtract average from the centre pixel
   N_0 to obtain the eighth bit.
6. The eighth bit is computed by Eqs. 3 and 4.
7. Compute the decimal equivalent of the eight bits.
8. Assign this decimal value to the referenced centre pixel.

3.1 Histogram Intersection

Before computing the histogram intersection point micro pattern developed by INBP is isolated into rectangular locales regions to by R1, R2... Rn, from which spatial histograms are extricated by Eq. 5.

$$H_{INBP}(i) = \{H_{INBP}(R_i)|i = 1, 2, \ldots n\} \qquad (5)$$

Where H_{INBP} (R_i) is the INBP histogram of features extracted from the rectangular local region R_i. The regions might be of any shape and size [11]. For example, circular regions with different radii can also be used for histogram. Many similarity measures have been proposed for histogram matching. For histogram matching, this paper uses histogram intersection to measure the similarity between two histograms.

$$S(H_{tr}, H_{tst}) \sum_{i=1}^{n} \min(\text{Htrn}_i, \text{Htst}_i) \qquad (6)$$

Where n is the number of regions. Similarity measure finds the minimum histogram value among the testing (Htrn) and training (Htst) data set [12]. The performance of INBP is higher than the LBP is shown in Fig. 5

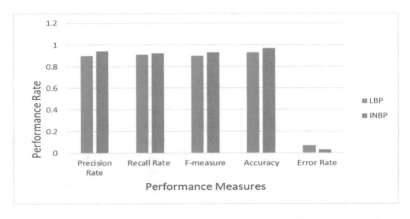

Fig. 5. Performance measures of LBP and proposed INBP using JAFFE

3.2   Performance Metrics

Accuracy, Precision Rate, Recall Rate, F-Measure and error rate can evaluate the performance of the proposed work INBP.

Accuracy: The accuracy of a test is its ability to differentiate the match and mismatch correctly. Equation (7) is used to evaluate the accuracy of the proposed work

$$Accuracy = TP + TN/TP + FP + TN + FN \qquad (7)$$

Precision Rate: The *Precision rate* is the fraction of retrieved instances that are relevant to the find. Precision rate of the proposed work is computed by Eq. (8).

$$Precision = TP/(TP + FP) \qquad (8)$$

Recall Rate: The *Recall rate* is the fraction of relevant instances that are retrieved per query. Recall rate of the proposed work is computed by Eq. (9).

$$Recall = TP/(TP + FN) \tag{9}$$

F-measure: F-measure is the quantitative relation of product exactitude and the sum of exactitude and recall. The F-measure of the proposed work is computed by Eq. (10)

$$F\text{-}measure = 2 * \frac{(Prate \times Rrate)}{(Prate + Rrate)} \tag{10}$$

Where, *Prate* denotes the Precision rate and *Rrate* denotes the Recall rate.

Error Rate: Calculating error rate is used to compare an estimate to an exact value. The *Percentage Error* calculates the distinction between the approximate and actual values as a proportion of the precise value. Error rate of the proposed work is computed by Eq. (11)

$$Error\,rate = (FP + FN)/(TP + TN + FP + FN) \tag{11}$$

The performance measures of LBP and Proposed INBP using JAFFE, ORL, YALE and OWN databases are shown in Figs. 5, 6, 7, and 8. For the JAFFE database, the accuracy of the proposed INBP is higher than the LBP. The accuracy using INBP for the ORL database is 100%.

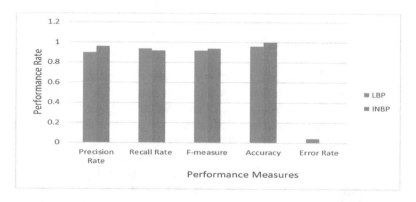

Fig. 6. Performance measures of LBP and proposed INBP using ORL

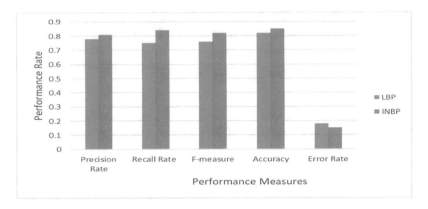

Fig. 7. Performance measures of LBP and Proposed INBP using YALE

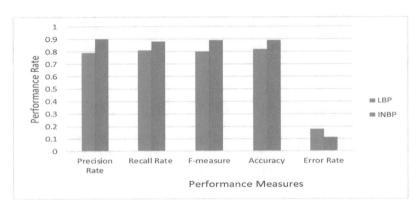

Fig. 8. Performance measures of LBP and Proposed INBP using OWN databases

4   Experimental Results

The performance of the proposed system is computed by using the test images in the database. This work uses four databases for testing purpose. The first database is the standard database JAFFE database with 13 faces for 20 categories.

Second one is a standard database ORL with 40 categories, which contains 10 faces per category. Third database YALE is a standard database with 40 categories. Each category consists of 11 faces with different expressions. Last data base, OWN DATABASE consists are 200 faces in twenty categories. Figure 9 expresses the sample test images that are used in this paper for the categories JAFFE, ORL, YALE and OWN DATABASE.

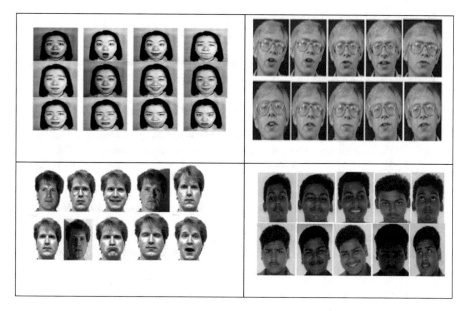

Fig. 9. Sample images of JAFFE, ORL, YALE and OWN data set

5   Conclusion

This paper presents the accuracy, error rate and recognition rate of using INBP for face recognition. To model the distribution of micro patterns, the histogram intersection is used as similarity measurement [11, 12]. The experiments conducted on the database JAFFE, ORL, YALE and OWN-DATASET demonstrate that the proposed approach INBP achieves better recognition and less error rate than the traditional LBP. The average recognition rate of LBP is 89.76% and the INBP is 97.68%. Hence the proposed approach improved by 7.92% of the recognition rate than the traditional LBP. The proposed method focuses on the higher accuracy rather than the traditional LBP. In future, experiments may be carried out to analyze the effect of INBP in a Smartphone with Android operating system.

Acknowledgment. The authors would like to thank Dr. Michael J. Lyons for the use of the JAFFE database, AT&T Laboratories Cambridge for ORL database. The authors gratefully acknowledge the contribution of the reviewer's comments.

References

1. Patil, A.M., Kolhe S.R., Patil, P.M.: 2D face recognition techniques: a survey. Int. J. Mach. Intell. 2(1), 74–83 (2010). ISSN: 0975-2927
2. Datta, A.K., Datta, M., Banerjee, P.K.: Face Detection and Recognition: Theory and Practice. CRC Press, Boca Raton (2015)

3. Zhao, G., Pietikäinen, M.: Dynamic texture recognition using local binary patterns with an application to facial expressions. IEEE Trans. Pattern Anal. Mach. Intel. **29**(6), 915–928 (2007)

4. Zhang, B., Gao, Y., Zhao, S., Liu, J.: Local derivative pattern versus local binary pattern: Face recognition with high-order local pattern descriptor. IEEE Trans. Image Process. **19**(2), 533–544 (2010)

5. Yi, D., Lei, Z., Li, S.Z.: Towards pose robust face recognition. In: 2013 IEEE Conference on Computer Vision and Pattern Recognition (CVPR), pp. 3539–3545 (2013)

6. Xiong, X., De la Torre, F.: Supervised descent method and its applications to face alignment. In: 2013 IEEE Conference on Computer Vision and Pattern Recognition (CVPR)

7. Zou, J., Ji, Q., Nagy, G.: A comparative study of local matching approach for face recognition. IEEE Trans. Image Process. **16**(10), 2617–2628 (2007)

8. Yang, B., Chen, S.: A comparative study on local binary pattern (LBP) based face recognition: LBP histogram versus LBP image. Neurocomputing **120**, 365–379 (2013)

9. Rivera, A.R., Castillo, J.R., Chae, O.: Local directional number pattern for face analysis: face and expression recognition. IEEE Trans. Image Process. **22**(5), 1740–1752 (2013)

10. Li, Q., Ye, J., Kambhamettu, C.: Proceedings of IEEE Computer Society Conference on Computer Vision and Pattern Recognition (CVPR04) (2004)

11. Chai, Z., Sun, Z., Mendez-Vazquez, H., He, R., Tan, T.: Gabor ordinal, measures for face recognition. IEEE Trans. Inf. Forensic Secur. **9**(1), 1–26 (2014)

12. Bhattacharyya, S.K., Rahul, K.S.: Face recognition by linear discriminant analysis. Int. J. Commun. Netw. Secur. **2**(2), 31–35 (2013). ISSN: 2231 – 1882

13. Olivares-Mercado, J., Toscano-Medina, K., Sanchez-Perez, G., Perez-Meana, H., Nakano-Miyatake, M.: Face recognition system for smartphone based on LBP. In: IEEE Xplore, INSPEC Accession Number: 16917437, 29 May 2017

Secure Transmission of Human Vital Signs Using Fog Computing

A. Sonya, G. Kavitha, and A. Paramasivam$^{(\boxtimes)}$

B.S. Abdur Rahman Crescent Institute of Science and Technology,
Chennai 600048, India
sonya.yasmin152@gmail.com, gkavitha.78@gmail.com,
parama.ice@gmail.com

Abstract. Recent years, wearable devices gain more prominence in health care and military applications to monitor the monitor human physiological signs such as heart beat rate, breathing etc. It is essential to secure the human vital signs since it gives complete information human physiological system. In this work, an On-Board Computer based hardware system is developed to read the human vital signs such as heart beat rate, body temperature and Electrocardiogram (ECG) signals. Further, the acquired human vital signs and ECG signals are encrypted using Advanced Encryption Standard (AES) 256-bit encryption algorithm at the node point resulting in secure transmission of human physiological information to a monitoring places or hospitals. Also, the acquired signals are stored in a Comma Separated Value (.CSV) file format and it is encrypted and decrypted with passwords provides high security. Results demonstrate that the encryption and decryption of human physiological information is possible and it is proved that the human vital signs can be transmitted over longer distances with high security.

Keywords: Advanced Encryption Standard · Fog computing · Vital signs · Security

1 Introduction

Recently, there have been many advancements in cloud computing, Big Data analytics etc. [1]. Generally, the cloud refers to accessing computer, software applications etc. over network connectivity. Further, the cloud computing in the field of health care becomes increasingly popular nowadays [2]. Using such technology, the physicians can monitor and proceed further treatment procedures to the patients even at different locations. Also, the cloud computing reduces the size of the wearable devices since it requires internet source with some server for storage.

All the essential data's or information from different sources such as sensors, imaging devices or cameras, patient input electronic files (E-files) etc. can be stored and computed at cloud node results in many advantages. Also, wearable devices are very small and is capable of transmitting human physiological sensor data to the remote cloud servers. It is mandatory to provide security for data's or information which are transmitted from wearable devices to cloud servers. Recent years, many encryption/

© Springer Nature Switzerland AG 2020
J. S. Raj et al. (Eds.): ICIDCA 2019, LNDECT 46, pp. 785–792, 2020.
https://doi.org/10.1007/978-3-030-38040-3_90

decryption algorithms are proposed several researchers for secure data transmission between any two nodes [4–7]. One such data secure algorithm utilized in this work is Advanced Encryption Standard (AES) 256-bit encryption algorithm. Lo'ai et al. have discussed about the networked healthcare, significance of cloud computing and big data analysis. Also, the authors have discussed briefly about the tools and techniques used for big data analytics [2].

Al Hamid et al. have developed a model for preserving healthcare private data using fog computing techniques. Further, the authors have generated session key based on the bilinear paring cryptography and successfully communicated the data with respective session securely. Also, the authors have concluded that the decoy technique is highly useful to access and store private healthcare data in a secure way [3].

Qi and Tao have proposed a hierarchy reference architecture based on edge, fog and cloud computing for smart manufacturing service system [8]. Subahi has presented a architecture, conceptual design and recommendation for edge-based Internet of Things (IoT) medical record system [9]. Gu et al. have discussed about the cost-efficient fog computing supported medical cyber physical systems [10].

Hani et al. have discussed about the design of private cloud storage and prototype development [11]. Rathnayake et al. have discussed briefly about the Internet of Medical Things and its key challenges. Also, the authors have proposed an adaptive model to deal the identified challenges [12].

Shirazi et al. have discussed in detail about the mobile edge computing and fog computing. Also, the authors have discussed about the impact and influence of mobile computing devices namely edge and fog computing on conventional communication and networking service models [13].

The objective of this work is to develop a fog computing device for transmission of human vital information to the cloud server with high data security.

2 Methodology

In this work, a Raspberry PI based On-board Computer system is utilized for fog computing techniques. The proposed system consists of two different sensors namely MAX30205 and heart beat sensor for measurement of human body temperature and heart beat rate respectively. Also, an instrumentation amplifier circuit is designed with the help of Integrated Chip (IC) AD624 for measurement of ECG signals. Figure 1 shows the block diagram for proposed system.

The heart beat rate sensor gives output voltage in a pulsated form and an instrumentation amplifier output is an analog voltage. Since, there is no inbuilt Analog to Digital Converter (ADC) inside Raspberry PI, an external 8-bit ADC converter called as PCF 8591 is utilized in this work to convert applied analog voltage into a digital voltage. In general, the PCF 8591 ADC has 4 different channels in which four different analog input can be applied. In this work, the two different channels are utilized.

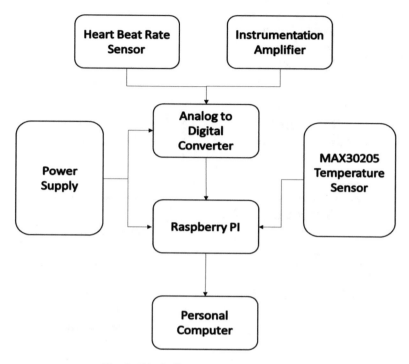

Fig. 1. Block diagram for proposed system

Further, the PCF 8591 converts the analog voltages from the two different sources namely an instrumentation amplifier and heart beat sensor into a digital voltage and it can be transmitted to the Raspberry PI through Inter-Integrated Circuit (I2C) protocol (i.e. two wire communication). Also, the output pin of the MAX30205 temperature sensor module can be directly connected to the another I2C pin of Raspberry PI and the output of the MAX30205 temperature sensor can be read directly.

In this work, a Raspberry PI 3 Model B+ is utilized as a fog computing device. Further, the Debian Linux operating system is loaded inside the Raspberry PI hardware. Also, the Python software with necessary library packages are installed to the Raspberry PI. The human vital signs and ECG signals are acquired with the help of the proposed system. Further, these signals or data's are stored inside the Raspberry PI in .CSV file. Also, the Advanced Encryption Standard (AES) 256-bit encryption algorithm is coded inside the Raspberry PI using Python software [14]. Once the data's from all the sensors and instrumentation amplifier are stored as a .CSV file, the file is further encrypted with the help of AES algorithm. Also, the encrypted file is further stored in a specified location which can be transmitted to any cloud server but cannot be decrypted without appropriate password. For the decryption, the appropriate password has to be given along with the python command. Figure 2 shows overall proposed work carried out.

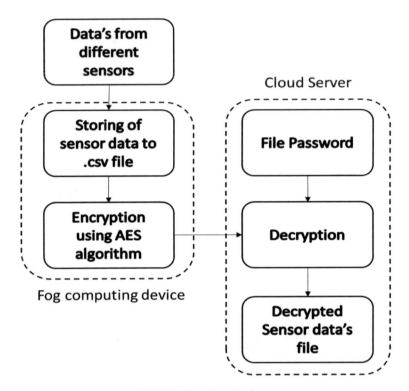

Fig. 2. Overall work flow

3   Results and Discussion

For experimental purpose, the human vital signs namely body temperature, heart beat rate and ECG signals are obtained from a healthy volunteer non-invasively using the developed system. Also, the procedures were explained and the measurement were done with the proper informed consent with the healthy volunteer. Figure 3 shows typical electrocardiogram signal obtained from the healthy volunteer. Further, it is clearly seen that the obtained electrocardiogram signal has necessary P-Q-R-S-T information.

Figure 4 shows the human body temperature acquired from healthy volunteer in degree Celsius. Figure 5 shows the heart beat rate measured from a healthy volunteer. Further, it is observed that the heart beat rate is measured for a period of 5 min.

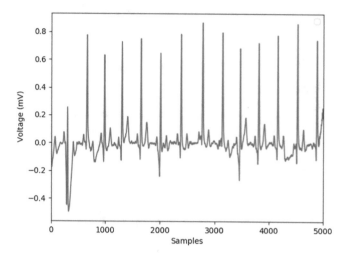

Fig. 3. Typical cardio-signal (ECG) acquired from the proposed system

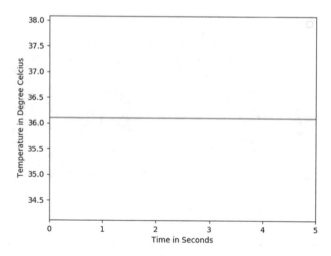

Fig. 4. Human body temperature measured using proposed system

The Mobaxterm software is used as a terminal for Raspberry PI and the AES algorithm is coded using the python software [15]. All the three different sensor data's are stored inside a single .csv file named as Sensor_Data.csv. Figure 6 shows the encryption of Sensor_Data.csv file and it is converted into Sensor_Data.ssb with the given secure password. The encryption can be done by executing the following python command line

python encrypt.py -i /home/pi/Crypt/AES/Sensor_Data.csv -o /home/pi/Desktop/ -p iteee

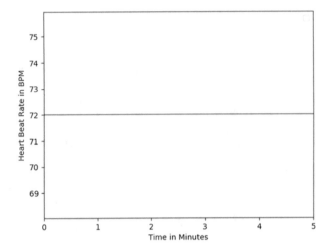

Fig. 5. Heart beat rate measured using proposed system

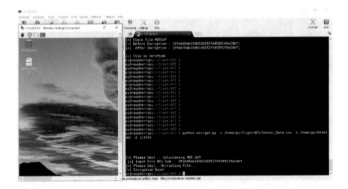

Fig. 6. Encryption using AES algorithm

In this work, the "iteee" is given as an encryption file password and at the time of decryption the mentioned password should be given correctly. Figure 7 shows the decryption using AES algorithm. The original Sensor_Data.csv file can be decrypted by giving encrypted file and its appropriate password along with the following command line.

python decrypt.py -i /home/pi/Desktop/Sensor_Data.ssb -p iteee

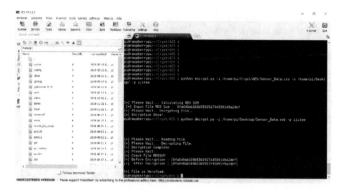

Fig. 7. Decryption using AES algorithm

4 Conclusion

In this work, human vital signs namely body temperature, heart beat rate and ECG signals were acquired using developed hardware system. Further, the acquired data's or signals were stored in a single.csv file. Also, the Raspberry PI based On-Board Computer system was utilized as a fog computing device in which an AES encryption algorithm is utilized to encrypt the sensor data file with a secured password. The experimental result shows that the proposed system is capable of securing the human vital signs from any intruders or strangers when it is transmitted over a public network. Also, it is clear that the secure transmission of human physiological information over any public network is possible with the proposed system.

Acknowledgement. All author states that there is no conflict of interest. We would like to thank the students for acted as healthy volunteer and management of B.S. Abdur Rahman Crescent Institute of Science and Technology for providing space and other resources to successfully complete this research work.

References

1. Farooqui, N.A., Mehra, R.: Design of a data warehouse for medical information system using data mining techniques. In: 2018 Fifth International Conference on Parallel, Distributed and Grid Computing (PDGC), pp. 199–203. IEEE (2018)
2. Lo'ai, A.T., Mehmood, R., Benkhlifa, E., Song, H.: Mobile cloud computing model and big data analysis for healthcare applications. IEEE Access **4**, 6171–6180 (2016)
3. Al Hamid, H.A., Rahman, S.M.M., Hossain, M.S., Almogren, A., Alamri, A.: A security model for preserving the privacy of medical big data in a healthcare cloud using a fog computing facility with pairing-based cryptography. IEEE Access **5**, 22313–22328 (2017)
4. Zheng, D., Wu, A., Zhang, Y., Zhao, Q.: Efficient and privacy-preserving medical data sharing in internet of things with limited computing power. IEEE Access **6**, 28019–28027 (2018)

5. Wang, X., Wang, L., Li, Y., Gai, K.: Privacy-aware efficient fine-grained data access control in internet of medical things based fog computing. IEEE Access **6**, 47657–47665 (2018)
6. Wei, J., Wang, X., Li, N., Yang, G., Mu, Y.: A privacy-preserving fog computing framework for vehicular crowdsensing networks. IEEE Access **6**, 43776–43784 (2018)
7. Jin, H., Luo, Y., Li, P., Mathew, J.: A review of secure and privacy-preserving medical data sharing. IEEE Access **7**, 61656–61669 (2019)
8. Qi, Q., Tao, F.: A smart manufacturing service system based on edge computing, fog computing, and cloud computing. IEEE Access **7**, 86769–86777 (2019)
9. Subahi, A.F.: Edge-based IoT medical record system: requirements, recommendations and conceptual design. IEEE Access **7**, 94150–94159 (2019)
10. Gu, L., Zeng, D., Guo, S., Barnawi, A., Xiang, Y.: Cost efficient resource management in fog computing supported medical cyber-physical system. IEEE Trans. Emerg. Top. Comput. **5**(1), 108–119 (2015)
11. Hani, A.F.M., Paputungan, I.V., Hassan, M.F., Asirvadam, V.S., Daharus, M.: Development of private cloud storage for medical image research data. In: 2014 International Conference on Computer and Information Sciences (ICCOINS), pp. 1–6 (2014)
12. Rathnayake, R.M., Karunarathne, S., Nafi, N.S., Gregory, M.A.: Cloud enabled solution for privacy concerns in internet of medical things. In: 2018 28th International Telecommunication Networks and Applications Conference (ITNAC), pp. 1–4 (2018)
13. Shirazi, S.N., Gouglidis, A., Farshad, A., Hutchison, D.: The extended cloud: review and analysis of mobile edge computing and fog from a security and resilience perspective. IEEE J. Sel. Areas Commun. **35**(11), 2586–2595 (2017)
14. Rijmen, V., Daemen, J.: Advanced encryption standard. In: Proceedings of Federal Information Processing Standards Publications, National Institute of Standards and Technology, pp. 19–22 (2001)
15. https://hackworldwithssb.blogspot.in

Internet of Things Based Smart Baby Cradle

Vedanta Prusty[✉], Abhisek Rath, Pradyut Kumar Biswal,
and Kshirod Kumar Rout

IIIT Bhubaneswar, Bhubaneswar, Odisha, India
vedanta.prusty5@gmail.com, abhisekrath1996@gmail.com,
{pradyut,kshirod}@iiit-bh.ac.in

Abstract. Smart Cradle is an idea of transforming the Traditional Cradle (which requires no power to operate and has no extra features for the security & efficient child care facilities) into a Smart System with the efficient use of technology. By using the concepts of Internet of Things, Embedded Systems & Cloud Technology, we aim to build a smart system that can be productively used for efficient child care and management. In this paper we have focused on the child's security and hygiene issues so as to raise the child in a good and healthy environment. The Cradle will have a transcendental impact which we hope will be having a huge potential to free the stress of today's busy parents.

Keywords: Internet of Things · Embedded Systems · Smart Baby Cradle · Cloud Technology · Security · Sensors · Remote access

1 Introduction

Technology has always kept the world astonished by the various transformations made out of its power. It helps in simplifying normal human lives by embedding and integrating all complexities within it and providing the user with a simplified interface to enjoy the features of it. Technology is present in almost all aspects of our lives and it is irrefutable to imagine life without technology.

In Present scenario, the conventional cradles that are available in the market are manual and are specifically used to make the baby sleep with manual swinging facility to uplift the mood of the baby. Moreover, conventional cradles need the attention of a person to take care of the baby's security, hygiene, etc. In today's busy schedule it has become an onerous job on the part of the parents to take care of the baby in all aspects.

Several automated cradles have been developed out of research and study that have been discussed in [1, 2] and [3]. Features such as collecting sensor data like temperature has been proposed in [4], automatic swing on cry detection has been proposed in [5]. Safety features of smart cradles has been cited in [6]. Similarly analyzing the acoustics of the cry has been introduced in [7]. Automatic cradle movement techniques have been discussed in [8] and [9]. The systems proposed above do not give a flexibility to the parents to control the automatic swing of the baby with fixed intervals on cry detection as well as to swing the cradle manually through wireless communication. The sensor data is collected but the parents have not been able to access the data from any device across the globe by connecting to the internet. Moreover, the baby's

© Springer Nature Switzerland AG 2020
J. S. Raj et al. (Eds.): ICIDCA 2019, LNDECT 46, pp. 793–799, 2020.
https://doi.org/10.1007/978-3-030-38040-3_91

security and safety have not been properly addressed. This encouraged us to combine the aspects of Embedded Systems, Internet of Things and Cloud Technology to design a smart cradle that would curtail the limitations and increase the efficacy of the pre-existing automated cradles.

Its high time to think of making a new, intelligent and feasible product which would be smart enough to help today's busy population and cater to their needs with the efficient use of technology in the real time scenario.

Smart Cradle is a concept of making the Conventional Baby Cradle smarter, secure and more efficient with the collaboration of concepts such as Embedded Systems, Internet of Things and Cloud Technology. The concept behind the cradle has been illustrated in Fig. 1. Our area of focus in this project has been to develop and design a system which will not only keep the baby rapturous and cheerful but also constantly monitor the baby to ensure its security and safety. It will provide an option to the parents of the baby to keep an eye on the baby remotely and will also notify them in case of any emergency situations.

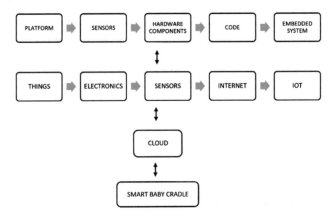

Fig. 1. Concept behind smart baby cradle

The Internet of Things (IOT) is the integration of things used in real life such as physical devices like home appliances, outside usable products, etc. embedded with electronics, software, sensors, actuators, and connectivity (internet) which helps in enabling all these devices to communicate and share work with each other in the real time, thereby successfully co-ordinating all the devices end to end. Wireless communication through bluetooth has been proposed in [10] and sound intensity measuring techniques has been proposed in [11]. The application of servo motors has been discussed in [12]. Integration of all such sensors and components with an IOT network enables them to seamlessly communicate with each other. Developments in IOT wireless sensor networks have been discussed in [13] and enhancements of IOT with fog network has been proposed in [14]. It would not be wrong to say that Internet of Things is the superset of Embedded Systems or Embedded Systems when integrated with internet or connectivity can be termed as Internet of Things.

Cloud Technology is a technology which offers massive storage platforms often termed as cloud which are designed to store/collect and process massive amount of data from various sources such as sensors, IOT Devices, etc. in the real time scenario and this uniquely built platform also provides the features to manipulate, analyse and retrieve the data as per requirement at any place, anywhere just with the requirement of simple internet connectivity in the real time.

2 Problem Statement

With the changing time, the work load on the parents is increasing day by day. Today's parents really face a huge challenge when their baby is just born as they need to take utmost care of the baby and also have to keep parallel pace with their respective work to earn bread and butter. So, with this above constraint many a times the small baby is neglected which brings in a lot of problems to the baby and also to the parents. Lack of time, lack of human resource to take care of the baby brings trouble to the whole family. This results in loss of focus on work, sometimes quarrels between husband and wife, etc. So, with all these things prevailing, the one getting affected is the baby who actually needs to be taken care of at that point of time. This is the real time scenario in almost all families which needs to be addressed soon and a solution to this problem must be generated with the efficient use of technology.

In hospitals there are a lot of issues of baby theft, exchange, nurses not taking proper care, etc. These all issues can be dealt with the power of technology and it is a high time to act upon this issue.

3 Smart Baby Cradle

The workflow of the cradle has been elucidated in Fig. 2 and the block diagram comprising of all the sensors and components has been illustrated in Fig. 3.

3.1 Intruder Detection Unit

What it Does.   Detects the intruder with the help of Ultrasonic Sensor. Whenever intruder comes in the vicinity of the sensor at certain specified distance, sensor detects the intruder.

Alert Email to Parents.   As soon as the Intruder is detected by the sensor, intruder's photo is captured by a webcam and is attached with the mail to send an alert mail to the parents informing them about the approach of an intruder.

Posting the Image of the Intruder on Twitter for Faster Security Purposes.
Today's world of Social Media where everyone is active, makes it one of the best places to post the image of the approached intruder so that immediate help can be generated in case of emergency situation.

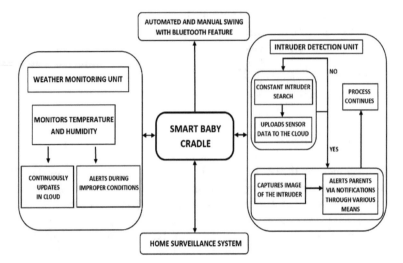

Fig. 2. Workflow of smart baby cradle

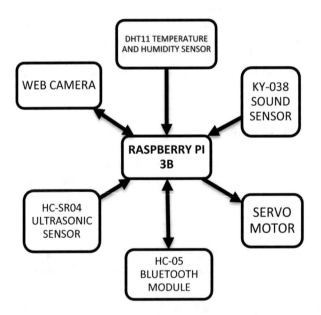

Fig. 3. Block diagram of smart baby cradle

Uploading Ultrasonic Sensor Data to Cloud. With the help of Cloud Technology, it has become easier now-a-days to monitor data sitting far away from the site in the real time scenario. So, the ultrasonic sensor data will be regularly uploaded to the cloud enabling the parents or relatives sitting anywhere in the world with the added advantage of accessing the exact data feed of the ultrasonic sensor and know the real time distance of the intruder from the baby and act accordingly.

3.2   Remote Home Surveillance System

Monitoring the baby by sitting at any corner of the house by just getting connected with the home internet network (to which the smart system is connected) and accessing a website in which regularly the latest intruder image as well as the sensor data are available and are continuously updated.

This is possible by converting our used platform (i.e. Raspberry Pi 3) into a web server in which a local website is being hosted where all the sensor data as well as the latest intruder image is being regularly updated. This web page can only be accessed within the coverage of the home network to which the smart baby cradle is connected. Hence the parents can enjoy accessing the smart cradle remotely sitting at any place and any corner within the house to keep an eye on the baby.

3.3   Weather Monitoring System

Constantly monitors the temperature and the humidity in the room using DHT11 temperature and humidity sensor where the cradle is kept & uploads the data of the sensor regularly to the cloud. The parents can access the cloud data and know the exact scenario being anywhere in the world.

Notifications will be made available to the parents in case of any disturbance in humidity or temperature (which is not suitable for the baby based on set parameters) to take necessary actions. Parents will also have the facility of accessing the sensor data by sitting remotely at any place within the house and accessing the smart cradle through the inbuilt home surveillance system.

3.4   Cradle Swing

This Smart Baby Cradle consists of 2 types of swinging features as follows:

Automatic Swing.   Automated Swing feature allows the parents to set the Smart Cradle in the automatic mode which helps the parents to give control to the smart cradle system to take care of the baby as and when needed.

The Automatic Mode continuously checks if the baby is crying or not with the inbuilt cry detection algorithm running constantly in the backend and when the baby cries it automatically swings the baby for an adjustable fixed interval of time without the need of manual care that helps in making the baby feel good. This may stop the baby cry and make it have a pleasant sleep. This reduces the constant attention of a care taker near the child thereby allowing parents to concentrate on the other important works at that point of time.

Manual Swing.   Manual Mode feature allows the parents to set the smart baby cradle in the manual mode when they may be available at home or free to play with their child. This mode is also automated giving the parents or care takers with the flexibility to swing the cradle without physical presence. They can take the benefit and enjoy the feature of controlling and swinging the cradle with the help of Bluetooth connectivity by directly using the application in their smart phones and relaxing at some other place in the range of Bluetooth connectivity.

With this the child can stay buoyant and happy by constantly offering parents to control the swinging of the cradle on their own or leave it to the system itself to take care of the baby as and when needed.

4   Conclusion

The concept of transforming the Conventional Baby Cradle to Smart Baby Cradle if implemented in today's market has true potential to bring a transformation on the child care sector and can surely benefit its users by its special features which are included in it using the concepts of IOT and Embedded Systems. The working prototype of the cradle has been illustrated in Fig. 4.

Fig. 4. Prototype of smart baby cradle

Technology is improving each and every sector of our lives and now it is time for the child care sector to get transformed with the right and the effective use of it. This transformation needs to be implemented which truly has high potential to help and make today's and upcoming generation's parents stress free and truly help in improving the life of the infants.

References

1. Phillips, R.S.: Baby cradle-like carrier. US Patent 2973889A (1966)
2. Blea, M., Harper, M.: Automatically rocking baby cradle. US Patent 3769641A (1973)
3. George, Y.: Baby cradle rocked by electricity. US Patent 2478445A (1949)
4. Nawaz, A.: Development of an intelligent baby cradle for home and hospital use. B.Tech thesis, NIT Rourkela (2015)
5. Palaskar, R.: An automatic monitoring and swing the baby cradle for infant care. Int. J. Adv. Res. Comput. Commun. Eng. 4(12), 187–189 (2015)
6. Arora, R., Shah, H., Arora, R.: Smart cradle gear to ensure the safety of baby in the cradle. Int. J. Inf. Futur. Res. 4(7), 6768–6777 (2017)

7. Myakala, P.R., Nalumachu, R., Mittal, V.K.: A low cost intelligent smart system for real time infant monitoring and cry detection. In: IEEE Region 10 Conference, Penang, 5–8 November 2017, pp. 2795–2800 (2017)
8. Wong, G.: Automatic baby crib rocker. US Patent 3952343 (1976)
9. Ebenezer, A.: Automatic cradle movement for infant care. Undergrad. Acad. Res. J. **1**(1), 65–66 (2012)
10. Cotta, A., Devidas, N.T., Ekoskar, V.K.N.: Wireless communication using HC-05 Bluetooth module interfaced with Arduino. Int. J. Sci. Eng. Technol. Res. **5**(4), 869–872 (2016)
11. Nurjannah, I., Harijanto, A., Supriadi, B.: Sound intensity measuring instrument based on Arduino board and data logger system. Int. J. Adv. Eng. Res. Sci. **4**(9), 27–35 (2017)
12. Patic, P.C., Ardeleanu, M., Popa, F.: Micro-robots used in control of automatic drilling operations. WSEAS Trans. Circuits Syst. **9**(6), 420–429 (2010). ISSN 1109-2734
13. Raj, J.S.: QoS optimization of energy efficient routing in IoT wireless sensor networks. J. ISMAC **1**(01), 12–23 (2019)
14. Kumar, T.S.: Efficient resource allocation and QoS enhancements of IoT with fog network. J. ISMAC **1**(02), 101–110 (2019)

My Friend – A Personal Assistant for Chauffeur

P. H. Jayanth[1,2(✉)], R. Gowda Nishanth[1,2], K. Gunashekar[1,2],
and Kezia Sera Ninan[1,2]

[1] #47/a MS Layout, Jaraganahalli, JP Nagar Post, Banglore 560062, India
jayanthprash.jp@gmail.com,
nishanthgowdar510@gmail.com
[2] #14, Hunasemaranahalli, International Airport Road, Bangalore 562157, India

Abstract. There has been substantial advancement in car safety field over the last few decades. As a result, car transport seemed to be becoming safer and safer as the time passes. But due to emergence of Smart phones and all the distractions that come along with it, the number of accidents is on the rise once again, thereby putting life at risk. One of key objectives of this project to develop this application is that the digital society is at immense risk to road safety, and there is a need to fix this, to ensure driver and others safety, during the use of Smart phone. In order to perform various task on phone like attending calls, SMS, Whatsapp messages, email and Navigation. we focus at getting accurate routing instructions, during navigation, receiving customized information about present live traffic, which in turn assists drivers to steer to their destination.

Keywords: Speech recognizer · IR Sensor · Ultrasonic sensor · HC-05 bluetooth component · NLP algorithm

1 Introduction

Steering in a Metro environment is chaotic and sometimes dangerous. A major problem associated with the rapid growth in automobile industry is an increase in traffic congestion and accidents. Smart phones are the single factor that has caused the highest number of traffic accidents in recent years. Although cars and their safety measures concepts are developing for few decades, mobile networking and the internet are an integral part of our modern connected lives. This is why 10-in-2 smartphone users tap into social media while driving, 48% surf the net and 20% video chat. Even though everyone knows the danger and prohibition by law. Still lot of people do text messaging, set the navigation or slide through documents quickly on the way to the office while behind the wheels. Back in 70 s, people thought it was acceptable to transport infants without safety. But today it seems not just negligent but fatal. The increasing number of traffic accident victims have caused a massive rethinking process across the globe.

These innovations from decades are totally indispensable in today's world. The car driver sits behind the wheel and fastens his seatbelt as if it were second nature.

J. S. Raj et al. (Eds.): ICIDCA 2019, LNDECT 46, pp. 800–808, 2020.
https://doi.org/10.1007/978-3-030-38040-3_92

But unconsciously the driver quickly glances at his gadget phone as well a huge new safety risk. Smart phone usage are the single factor that has caused the highest number of traffic accidents in recent years.

But they would never think about driving their car without eyes open, would they? Well, that is what millions of people do every day. Some facts, 5 s is the average estimated time that your eyes are off the road while texting. When accelerating at 50 mph, that's enough time to cover the length of a rugby field, you drive blindly on the highway for more than 180 m, if you are distracted by Internet or SMS.

So new safety concepts must be established that are useful and still not a threat to the joys of driving or using the smart phone. The goal is to create a future in which we look back and are just as horrified about today's driving habits as we are today in the absence of the safety belt in the 70's. And, of course: to make the car safe again without turning off the awesome gadgets of toady's digital age.

2   Proposed System

We are focused at obtaining precise routing guidelines, receiving personalized information that assists individual drivers to drive to his destination. To receive/make calls, read/write SMS and whatsapp messages, play music from Music player or youtube. And reach their required destination will significantly decrease the stress of driving, saves fuel and reduces life risk and saves time.

In this paper we propose a system that merges driver diagnostics to notably increases driving comfort, mobility and safety in a Metro city and smart navigation with intelligent assistant.

2.1   Advantages

Life Safety: It avoid road accidents to a larger extent due to use of smart phone. And save Drivers and others life.

Saves Money and Time: The technology can substitute human beings to answer repeatedly asked queries or to give frequently requested information – such as accurate directions, duration of operation, etc. – that doesn't demand analytical thinking skills.

Greater Consumer/Customer Satisfaction. The technology eradicates waiting time by replying to a caller instantly.

Time Irrelevant Service. The technology can function without any interruptions and is available to provide information to callers whenever they need it.

3   Methodology

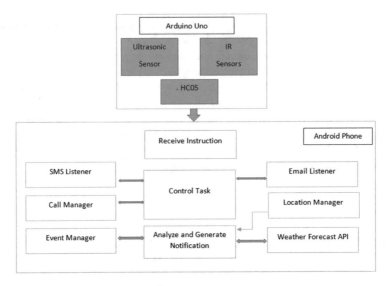

Fig. 1. System architecture

The above Fig. 1 represents the system architecture. The methods used as mentioned below.

Speech Recognizer

Typically, voice input command My Friend, enables the voice recognizing system and this is signaled to the driver by an audio alert. Whenever an audio is prompted, the system contains a "listening window" during which it may accept a voice input for recognition.

Simple audio commands are used to initiate phone calls, send messages via whatsapp, e-mail, SMS or play music from youtube or music-loaded flash drive. My Friend offer natural-language speech recognizing technology in place of a fixed group of commands, allowing the driver to use common phrases and full sentences. With such type of systems the user doesn't need to memorize a fixed set of commands.

Speech recognizing technology is the set of inter-disciplinary technologies that enables the recognition and translation of spoken language into text by systems and sub-fields of computational linguistics that develops methodologies. It is also known as "speech to text" (STT), "automatic speech recognition" (ASR), "computer speech recognition". It researches in linguistics and incorporates knowledge, electrical and computer science engineering fields.

4   Technology Used

4.1   IR Sensor

IR Sensor consists of an emitter and receiver in which it is used to control the music play (Fig. 2).

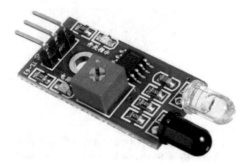

Fig. 2. IR Sensor

4.2   Ultrasonic Sensor

Ultrasonic sensor is used to measure distance using ultrasonic waves. In our system it is used to increase or decrease volume (Fig. 3).

Fig. 3. Ultrasonic sensor

4.3   Bluetooth Component

HC-05 a Bluetooth module which connects the hardware and software (Fig. 4).

Fig. 4. HC-05

4.4 Arduino UNO

It is an open-source platform, means the boards and software are readily made available and anyone can alter and stabilize the boards for better operability and powered by a USB cable or by an external 9 V battery, though it accepts voltages between 7 and 20 V (Fig. 5).

Fig. 5. Arduino uno

5   Results

The Hardware Architecture is shown below in Fig. 6.

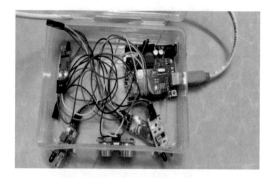

Fig. 6. Hardware architecture

The below Fig. 7 shows you the connectivity of the Bluetooth terminal.

Fig. 7. Bluetooth connectivity

The proposed system alerts you with a drowsy alter buzzer both in day and night mode as shown in Fig. 8.

Fig. 8. Drowsy alert

Our application will send SMS to the specified users by using SMS Manager tools, the contents for SMS will be taken from the voice by speech recognizer with handler for some amount of time and also it will receive and read the incoming SMS, email messages from the different users and we can call to anyone from the contact through our speech (Fig. 9).

Fig. 9. Speech to text

6 Conclusion

We have presented that how our application My Friend a personal assistant for Chauffeur works and how the communication between the driver and the smart gadget through Infra-red sensors and Bluetooth terminal has been carried out. In this system driver can listen to various incoming texts such as SMS, EMAIL and he can also respond back and send the message. The driver can activate the music player by giving commands through voice and also he can get the navigation properly using google server as well as he can also attend the calls. By this application the driver can concentrate only in the driving with no disturbance from messages and calls so that he can reach to the destination safely on time i.e. drivers can avoid time delay and the chances of accidents because the usage of smart gadget can be reduced.

7 Future Scope

The future enhancement can be done by the use of ultrasound sensor with the application can be connected so that in case of accident the sensor will trigger the message to the family member or friends, from that there would be chance of saving lives.

References

1. Ioannou, P., Zhang, Y.: Intelligent driver assist system for urban driving. In: Digital Media Industry & Academic Forum (DMIAF). IEEE (2016)
2. Mohan, P., Padmanabhan, V.N., Ramjee, R.: Nericell: rich monitoring of road and traffic conditions using mobile smartphones. In: Proceedings of the 6th ACM Conference on Embedded Network Sensor Systems. ACM (2013)
3. Shoba, G., et al.: AN interactive email for visually impaired. Int. J. Adv. Res. Comput. Commun. Eng. (IJARCCE), 5089–5092 (2014)
4. Park, J., et al.: Intelligent vehicle power control based on prediction of road type and traffic congestions. In: IEEE 68th Vehicular Technology Conference, VTC 2008-Fall. IEEE (2008)
5. Goregaonkar, R.K., Bhosale, S.: Assistance to driver and monitoring the accidents on road by using three axis accelerometer and GPS system. In: Proceedings of the 1st International Conference, vol. 10 (2014)
6. Ummuhanysifa, U., Nizar Banu, P.K.: Voice based search engine and web page reader. Int. J. Comput. Eng. Res. (IJCER), 1–5 (2013)
7. Betsworth, L., et al.: Audvert: using spatial audio to gain a sense of place. In: IFIP Conference on Human-Computer Interaction. Springer, Heidelberg (2013)
8. Joshi, J., et al.: DASITS: driver assistance system in intelligent transport system. In: 2016 30th International Conference on Advanced Information Networking and Applications Workshops (WAINA). IEEE (2016)
9. Thirupuranthakam, S.K., Negi, A.: An effective and intelligent driver assistance and warning system for sustained safety. In: 2014 IEEE International Conference on Vehicular Electronics and Safety (ICVES). IEEE (2014)
10. Joseph, S.L., et al.: Semantic indoor navigation with a blind-user oriented augmented reality. In: 2013 IEEE International Conference on Systems, Man, and Cybernetics (SMC). IEEE (2013)

11. McGookin, D.K., Brewster, S.A., Christov, G.: Studying digital graffiti as a location-based social network. In: Proceedings of the 32nd Annual ACM Conference on Human Factors in Computing Systems. ACM (2014)

12. Kannan, S., Thangavelu, A., Kalivaradhan, R.: An intelligent driver assistance system (i-das) for vehicle safety modelling using ontology approach. Int. J. UbiComp **1**(3), 15–29 (2010)

13. Meseguer Anastasio, J.E., Tavares De Araujo Cesariny Calafate, C.M., Cano, J.C., Manzoni, P.: Characterizing the driving style behavior using artificial intelligence techniques (2013)

14. Joseph, S.L., Xiao, J., Zhang, X., Chawda, B., Narang, K., Rajput, N., Mehta, S., Venkata Subramaniam, L.: Being aware of the world: toward using social media to support the blind with navigation. IEEE Trans. Hum. Mach. Syst. **45**(3), 399–405 (2015)

15. Zhao, L., et al.: Driver drowsiness detection using facial dynamic fusion information and a DBN. IET Intell. Transp. Syst. **12**(2), 127–133 (2017)

16. Choi, M., et al.: Wearable device-based system to monitor a driver's stress, fatigue, and drowsiness. IEEE Trans. Instrum. Measur. **67**(3), 634–645 (2018)

17. Chowdhury, A., et al.: Sensor applications and physiological features in drivers' drowsiness detection: a review. IEEE Sens. J. **18**(8), 3055–3067 (2018)

Prediction of Type 2 Diabetes
Using Hybrid Algorithm

Aman Deep Singh[1], B. Valarmathi[2(✉)], and N. Srinivasa Gupta[3]

[1] Department of Information Technology, School of Information Technology
and Engineering, Vellore Institute of Technology, Vellore, Tamilnadu, India
aman999999aman@gmail.com
[2] Department of Software and Systems Engineering, School of Information
Technology and Engineering, Vellore Institute of Technology, Vellore,
Tamilnadu, India
valargovindan@gmail.com
[3] Department of Manufacturing Engineering, School of Mechanical
Engineering, Vellore Institute of Technology, Vellore, Tamilnadu, India
guptamalai@gmail.com

Abstract. Diabetes is a major metabolic disease which affects the entire body
system. It becomes a lifelong disease if not handled properly in the early stages
of diagnosis. People produce a hormone called insulin. It is a hormone which is
used to convert glucose to energy. People suffering from diabetes type 2 produce
this hormone, but according to doctors are not able to use it as well as they
should. Data Mining tools can be used to mine the famous Indian Pima diabetes
for more accurate prediction of diabetes. Many attempts have been made by
researchers to improve the efficiency of various models.

Keywords: Data mining · Indian Pima · Random under sampling · Random
over sampling · AdaBoost classifier · Bayesian network · ANN · Random
forest · DTNB

1 Introduction

Diabetes is a very common disease and to cure it, early diagnosis of the disease is
required; Data Mining tools can be used to mine Indian Pima data so as to build good
models to predict results. WEKA is a data mining tool written in JAVA, developed at
the University of Waikato New Zealand. WEKA has a collection of visualization tools
and algorithms which are supported by graphical user interface for easy use. WEKA is
easy to download and install on Windows as well as Linux based systems. WEKA
offers a very user-friendly interface with step by step options for choosing the data set
to predicting the results. Every machine learning project involves the basic steps to
carry out the whole project in a smooth manner. Using the same approach, the project
will be divided into different phases. The first phase of the process is to get the raw
data. Here raw data refers to acquiring of the data set. We will use two data sets for
comparison. The first data set will correspond to the Indian Pima diabetes data set
containing 768 rows and the second is Lab data comprising 15000 rows of similar data

J. S. Raj et al. (Eds.): ICIDCA 2019, LNDECT 46, pp. 809–823, 2020.
https://doi.org/10.1007/978-3-030-38040-3_93

set of diabetic patients, which contains exactly the same types of attributes as that of the Indian Pima data set. The different size of the data set will help us analyse how number of training rows affects the result and answer. The second phase will include training these models directly without any pre-processing and the results are noted for both of the data sets. The third phase will include balancing the data set and then again training them on the suggested models and see the difference in the performance of the models. The first data set after balancing initially has 268 positive samples and 500 negative samples. After balancing it will contain 500 positive and negative samples each. The second data set initially has 5000 positive samples and 10000 positive samples. After balancing the it will contain 10000 positive as well negative samples. The fourth phase will be to select the useful attributes (i.e.) perform attribute selection on the data set and then apply the given model on the data set composed of only selected after attribute selection. The results will be compared to both the previous results obtained. This result will give us the difference obtained after balancing the data set as well as after performing attribute selection on both data sets. Further, plotting graphs and tables will help us analyze the results.

2 Literature Survey

Harleen and Vinita [1] used five supervised machine learning algorithms for classification of the Indian Pima. Some of which are SVM, KNN, ANN and MDR algorithms. They have used R as the programming language and in their models after training highest accuracy is given by linear kernel SVM.

Kannadasan et al. [2] proposed deep learning based solution towards the classification of Indian Pima using stacked auto encoders. They will use these encoders to take out relevant features from the data set and then use the softmax layer for classification. They also try the algorithm after fine tuning it by applying back propagation methods.

Deepti and Dilip [3] proposed that after preprocessing the data Naïve Bayes, SVM and Decision tree can be used to classify the Indian Pima for getting good results. They propose the future work to extend and automate the diabetes analysis using better machine learning algorithms.

Edla and Cheruku [4] proposed a system comprised of three layers which used feed forward architecture called as RBFNN.

Mehrbakhsh et al. [5] proposed a system in which they use SOM clustering along with PCA for attribute selection and Neural networks for classification. Using this method they achieve good results, but they propose a need of larger data sets for evaluation and better classification of the models.

Maniruzzaman et al. [6] proposed classification models for predicting diabetes such as linear discrimant analysis, quadratic discrimant analysis, Naïve Bayes, Gaussian process. The paper makes comparison of these models based on accuracies.

Hayashi and Yukita [7] proposed recursive rule extraction mechanism combined with the J48graft model for predicting diabetes. The J48graft algorithm supported the recursive rule extraction algorithm.

Marcano-Cedeño et al. [8] proposed artificial Metaplasticity neural network model for predicting diabetes.

Patil et al. [9] proposed a model in which they first extract rules using k means clustering and then using these rules they construct a decision tree. Then this tree is used to classify the Kaur and Chhabra [10] proposed improvement of the J48 decision tree classifier for prediction of diabetes. The paper uses pruning for better performance.

Jayalakshmi and Santhakumaran [11] proposed static normalization and back propagation method for classification of diabetes. Various normalization procedures are proposed to increase the accuracy such as z score normalization, min-max normalization, median normalization, sigmoid normalization and statistical column normalization.

Kaur and Wasan [12] proposed applications of data mining in the field of healthcare. Decision tree and ANN are proposed.

Fatima and Pasha [13] proposed disease diagnosis using machine learning algorithms. Weka toolkit was used to predict diabetes using various algorithms.

Jonathan and Harold [14] proposed data mining analysis for predicting diabetes type 2.

Sudharsan et al. [15] proposed machine learning techniques to predict the condition of Hypoglycemia of patients of suffering diabetes type 2.

Marinov et al. [16] proposed a review on data mining technologies used in the prediction of diabetes type 2.

Razavian et al. [17] proposed a model to predict at population level diabetes type 2 and also the analysis of the risks associated with it.

Rajesh and Sangeetha [18] proposed data mining methods and techniques in the prediction of diabetes 2. Decision tree algorithm and it's analysis are used in the paper to predict diabetes.

3 Dataset Description

All data mining process requires a with data for the model to train and then use the trained model to predict the result. A can be described as a matrix where each row represents a unique record of data, the columns represent a unique attribute.

For our study, we will be using the original Indian Pima and the extended version of the Indian PIMA diabetes data set. The attributes include Id of the Patient, number of pregnancies, plasma glucose level, diastolic blood pressure, thickness of the triceps, insulin, BMI, diabetes pedigree, age, diabetic or not. The original data set consists of 768 records of Indian women data, while the extended version contains 15000 rows of patient's data with exactly the same attributes as that of the Indian Pima data set. The class label has two values 'p' and 'n' which represent whether the corresponding patient has diabetes or not respectively. The Indian data set is originally owned by the National institute of diabetes and digestive and kidney diseases. The extended has been taken from Kaggle website. The data set can be found at the respective link https://www.kaggle.com/fmendes/diabetes-from-dat263x-lab01/home. This contains 15000 rows of unique patient data (Tables 1, 2 and 3).

For the reference, in the paper we will be using the terms '1' and '2'.

Table 1. Attributes in the dataset

Attribute name	Type
Number of Pregnancies	Integer
Plasma Glucose Level	Real number
Diastolic Blood pressure	Real number
Triceps	Real number
Insulin serum	Real number
BMI	Real number
Diabetes Pedigree	Real number
Age	Integer
Diabetic	Boolean

Table 2. The Indian Pima (Data set 1)

Name	The Indian Pima Data set
Total number of rows	768
Number of rows with positive class label	268
Number of rows with negative class label	500

Table 3. Extended Indian Pima (Data set 2)

Name	The Extended Data set
Total number of rows	15,000
Number of rows with positive class label	5,000
Number of rows with negative class label	10,000

4 Proposed Work

In the previous stated research papers there was a need to test the data on larger data set. The algorithms tend to work with different accuracies for different sizes of data. The well known Indian Pima dataset comprises of just 768 rows of patient data which is very less for efficient testing and even training of algorithms. This paper aims at working on a larger dataset and tends to display the differences in different approaches and algorithms i.e how the algorithms work on smaller and larger datasets. Therefore for the research work we tend to use a lab data acquired from Kaggle. The larger data set has 15000 rows of patient data having exactly same attributes which will be very helpful is analyzing how algorithms perform with larger data set. All the previous research papers have concentrated on Indian Pima Dataset while this paper contributes to the research by showing the differences in performance of algorithms on Indian Pima dataset (768 rows) and the larger dataset (15000 rows).

We will carry out the work in 4 major steps and find out the difference between the various implementations. Each step in the work will be comparing results based on both the s i.e. the Indian Pima with 768 rows of data and the extended diabetes with

15000 rows of data set. The Fig. 2 shows the Flow chart for the proposed work. First, the raw is directly trained and first pair of results is computed (for both the data sets). Second, the data is balanced, then a second pair of results are computed. Third, we perform attribute selection on the data and the third pair of results are computed. All these results will be compared and analyzed using graphs and tables.

The DTNB hybrid algorithm will also be used. The main steps of the DTNB algorithm are given below:

(1) Divide dataset into two disjoint datasets, first for decision table and the second for naïve bayes.
(2) Model all attributes initially by using decision table.
(3) Use forward search mechanism at each step to select attributes by naïve bayes and decision table.
(4) For each attribute see whether or not the attribute is important at each step and drop the attribute entirely if it is not important.

The model can be represented in steps as the below Fig. 1:

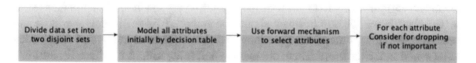

Fig. 1. DTNB hybrid model steps

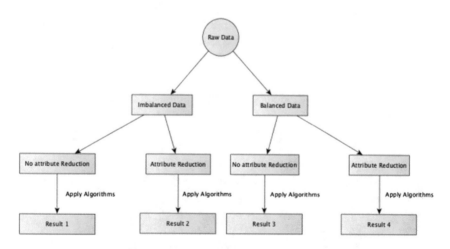

Fig. 2. Flow chart for the proposed work

4.1 Balancing the Dataset

The data are mostly available in the form of unbalanced data. Unbalanced here refers to the difference in the number of rows each class label. Considering our first data set we have 268 rows of positive class labels and 500 negative class labels. Thus, there is a need to balance the data and have an equal number of positive as well as negative samples.

Some dataset balancing algorithms are given below.

4.1.1   Random Under Sampling

In this algorithm the class labels having higher frequency are made equal to another class label by eliminating rows of the higher frequency class label randomly.

4.1.2   Random Over Sampling

In this algorithm the classes with a deficient number of rows are added rows so as to have equal number of rows for each class label. This algorithm results in no loss of data. This algorithm will be used in the proposed work to balance the data sets.

4.2   Attribute Selection

Also known as feature selection. It's not necessary that the data set we choose will have all the attributes contributing significantly to the result. The data set may have redundant attributes which do not contribute or have very low impact on the final result of the class label. These attributes should be removed from the data set, so as to decrease the training time of the algorithms and also to decrease the size of data needed to be stored in the database. There are many efficient algorithms available for attribute selection, some of which are given below:

- Univariate Selection: discussed below in Sect. 4.2.1.
- Feature importance: Some models, mostly tree models have a feature importance property associated with them. Using this property, we can calculate the feature importance of each feature and select the k features having the highest feature score, as these k features will contribute highest to the result of the class variable.

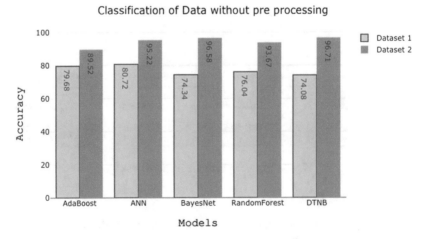

Fig. 3. Comparison of Accuracy between models

– Principal Component Analysis: The algorithm behind the working of Principal
  Component Analysis is that it is used to linearly map high dimensional to lower
  dimensional by maximizing the variance of lower dimensional.

For our work, we will be using the Univariate Selection method for attribute
selection.

4.2.1   Univariate Selection

Using statistical methods, the attributes which have the highest relationship score with
the output variable can be selected. There are many libraries available to implement this
method. It used the k best features technique, and we can select the k number of
features out of n features of our data set.

Steps of the algorithm are given below:

(1) Select the data
(2) Select the statistical methods to use
(3) Apply the statistical method to find the relation between attributes and the class
     label
(4) Select the k best features and use them for learning

4.3   Classification Algorithms

Classification algorithms can be referred as the algorithms which are used in the
situations where we have a set of attributes which have direct or indirect relationship
with the given class label. The classification algorithms can help us find out these
relationships between the given attributes and class labels. After knowing these relations, we can use these relations to predict the unknown class labels.

4.3.1   AdaBoost Classifier

Commonly known as adaptive boosting classifier. The classifier uses other classifiers
for in addition to improve the performance. Sum of other classifiers is used to compute
the effective performance of adaboost model. The model is capable of itself, tuning the
parameters of the original algorithms trained on. It improves their accuracy by
changing the value of parameters used in the algorithm. The algorithm will be used
here with boosting the decision tree classifier. The SAMME algorithm will be used
with n-estimators as 300. The SAMME algorithm refers to an advanced version of
AdaBoost classifier which is basically a multi class adaboost classifier. The SAMME is
powered by the extra Log(k-1) term in the formula and it is shown in the Eq. (1).

$$\alpha(n) = \log((1 - err(n))/err(n)) + \log(K - 1) \tag{1}$$

While the normal adaboost classifier can be represented by the formula and it is
given in the Eq. 2.

$$\alpha(n) = \log((1 - err(n))/err(n)) \tag{2}$$

4.3.2   Bayesian Network

Also known as decision network is a classification model which shows the dependencies between variables using a graph. The graph has directed edges and is acyclic. This model is very good for situations where class labels have direct relationships with attributes. As for our data set the attributes directly contribute to the final class decision and therefore it gives good results as compared to other models. The model makes use of probabilities to calculate the result. The model uses the chaining probability rule and it is shown in the Eq. 3.

$$P(A1, A2, A3, A4, A5) = P(A1|A2, A3, A4, A5) * P(A2|A3, A4, A5)$$
$$* P(A3|A4, A5) * P(A4|A5) * P(A5) \tag{3}$$

4.3.3   Random Forest

Random forest is the machine learning models based on usage of many decision trees together and the final result can be the mean of the result or the mode of the result. The algorithm uses the concept of bagging, bagging refers to the algorithm to increase the stability as well as the accuracy and error of the model by using a statistical approach. Statistics are used to decrease the variance of the data and also it helps in preventing to over fit the data. Feature bagging is applied on all the decision trees used in the random forest algorithm. As the name suggests, all these decision trees combine their result to give a random forest model.

4.3.4   Artificial Neural Network

Similar to the working of our brain this algorithm also uses neurons to train a dense network for classification. It can be single or even multilayer. It is also referred to as the single layer perceptron and multilayer perceptron respectively. The single layer consists of just an input layer and an output layer. On the other hand, multilayer as the name suggests, consists of multiple layers composed of one input layer, followed by hidden layers and at the last an output layer. Each layer has some neurons, which have weights attached to them. The model builds these weights for us. The weights are decided by iterating and back propagating after receiving an error rate. This model can also predict patterns which it is not trained on.

4.3.5   Decision Table + Naive Bayes Hybrid Model (DTNB)

It is a hybrid model available in weka tool for classification. It extends the decision table class in weka to make a hybrid model composed of both Decision table and Naive Bayes. At every iteration of the search part of the algorithm, the algorithm divides the available attributes into two disjoint sets for Decision table and Naive Bayes each. The algorithm also composed of forward search step where some attributes are modelled by Decision table and other by Naive Bayes. Initially, all of the attributes are already modelled by the Decision table. The model at each step can also drop an attribute completely from the data set.

5 Results

The Proposed model after training gives us different accuracies and other metric values for each model. The results are displayed below for each below:

5.1 Direct Training of Algorithms

The two datasets without any balancing or attribute reduction were applied to the classification algorithms produced the following accuracies given below in Table 4.

Table 4. Accuracy for direct training of imbalanced data.

	AdaBoost	ANN	BayesNet	Random Forest	DTNB
1	79.68	80.72	74.34	76.04	74.08
2	89.52	95.22	96.58	93.67	96.71

The table depicts the comparison of accuracies of the five models used i.e Ada-Boost, ANN, Bayes Net, Random Forest and DTNB classifiers.

5.2 Training After Balancing the Data Sets

The dataset was random over sampled to balance the dataset. This was applied to both the datasets, and the classification models were again applied to the datasets. The new accuracies of the data sets with each algorithm are given below in Table 5:

Table 5. Accuracy for training balanced data sets

	AdaBoost	ANN	BayesNet	Random Forest	DTNB
1	85.6	77.2	74.20	88.90	74.80
2	96.06	88.18	96.61	96.90	96.63

The Fig. 4 shows that on balancing the data the accuracies of the models increase as compared to Fig. 3 where the models were trained on imbalanced data.

5.3 Training After Attribute Selection

After performing attribute selection and selecting the top 5 attributes contributing to the data set we have the following 5 attributes: Age, PlasmaGlucose, SerumInsulin, BMI, Pregnancies. The classification algorithms were trained on the new reduced dataset and the following accuracies are obtained in Table 6 (Fig. 5):

Classification of Balanced Data

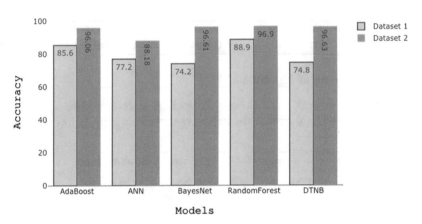

Fig. 4. Comparison of Accuracy between models

Table 6. Accuracy for training after attribute selection

	AdaBoost	ANN	BayesNet	Random Forest	DTNB
1	80.20	79.68	75.65	73.69	73.82
2	94.90	87.44	94.89	93.23	94.99

Classification of Data After Attribute Selection

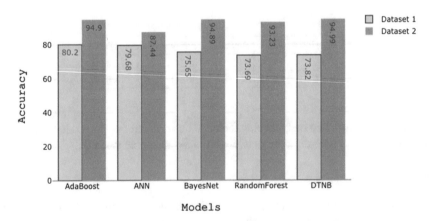

Fig. 5. Comparison of Accuracy between models

5.4   Training After Attribute Selection and Balanced Dataset

The attribute selection of balanced data gives us data set that is balanced along with attributes reduced to five. This is done for both the datasets. The datasets now on applying classification algorithms produce the following accuracies as shown below in Table 7 (Fig. 6):

Table 7. Accuracy of training balanced data with attribute reduction

	AdaBoost	ANN	BayesNet	Random Forest	DTNB
1	82.80	80.00	74.40	87.40	74.40
2	94.82	86.52	94.95	96.41	94.95

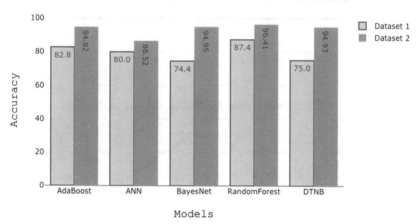

Fig. 6. Comparison of Accuracy between models

5.5   Comparison Between Balanced and Imbalanced Data

The results plotted on a single graph to show the difference between the accuracy of the same algorithm before and after balancing.

Dataset 1 trend chart (Fig. 7):

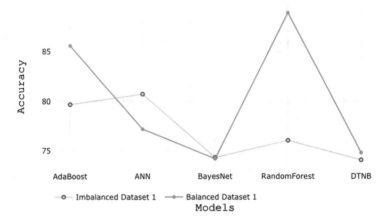

Fig. 7. Comparison between balanced and imbalanced dataset (data set 1)

Data set 2 trend chart (Fig. 8):

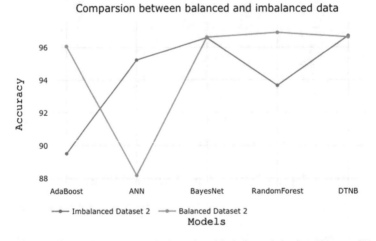

Fig. 8. Comparison between balanced and imbalanced dataset (data set 2)

Almost all algorithms' accuracy increase after performing balancing of data. The best performance is given by DTNB hybrid model and random forest with balanced data sets. Random forest out performs in balanced 1 whereas DTNB has high accuracies in balanced and imbalanced data in data set 2.

5.6   Final Comparison of the Models Using Trend Chart

The models when seen comparatively based on all the methods applied for pre-processing give an interesting conclusion. The Bayes Network and the hybrid decision table + Naive Bayes model gave the highest accuracies overall as compared to other models when seen as a whole. The Random Forest model gave high accuracies in 1 as compared to the DTNB model. The following can be shown using the graph given below in Figs. 9 and 10.

Dataset 1 trend chart:

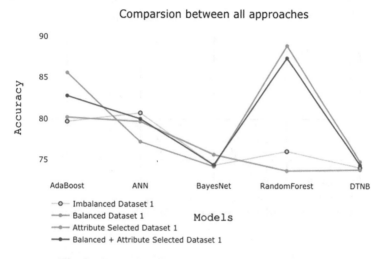

Fig. 9. Comparison between all approaches for dataset 1

The comparison between all approaches for Dataset 2 is shown in Fig. 10.

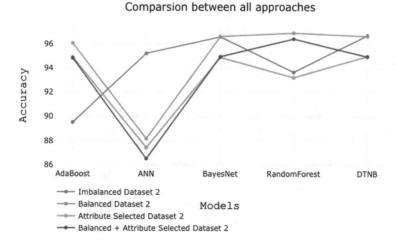

Fig. 10. Comparison between all approaches for data set 2

6   Conclusion

Every algorithm after undergoing the work flow gave different results and different accuracies. Each time we observed an increase in accuracy after balancing the data. Also, after attribute selection 5 attributes were selected to give the highest accuracy. It was also found that attribute selection, along with balancing data does not give the maximum accuracy in models. It was also concluded that the smaller data set (data set 1) i.e. the Indian Pima Data set containing 768 rows of data, gives very low accuracy as compared to the larger data set (data set 2) i.e. the lab data set containing 15,000 rows of data. Therefore, to predict diabetes type 2 using a classification approach more data is needed, which is proved by the larger data set having high accuracy.

References

1. Kaur, H., Kumari, V.: Predictive modelling and analytics for diabetes using a machine learning approach. Appl. Comput. Inf. (2018)
2. Kannadasan, K., Edla, D.R., Kuppili, V.: Type 2 diabetes data classification using stacked auto encoders in deep neural networks. Clin. Epidemiol. Global Health (2018)
3. Sisodia, D., Sisodia, D.S.: Prediction of diabetes using classification algorithms. Procedia Comput. Sci. **132**, 1578–1585 (2018)
4. Edla, D.R., Cheruku, R.: Diabetes-finder: a bat optimized classification system for type-2 diabetes. Procedia Comput. Sci. **115**, 235–242 (2017)
5. Nilashi, M., Ibrahim, O., Dalvi, M., Ahmadi, H., Shahmoradi, L.: Accuracy improvement for diabetes disease classification: a case on a public medical. Fuzzy Inf. Eng. **9**(3), 345–357 (2017)
6. Maniruzzaman, M., Kumar, N., Abedin, M.M., Islam, M.S., Suri, H.S., El-Baz, A.S., Suri, J.S.: Comparative approaches for classification of diabetes mellitus data: machine learning paradigm. Comput. Methods Programs Biomed. **152**, 23–34 (2017)
7. Hayashi, Y., Yukita, S.: Rule extraction using Recursive-Rule extraction algorithm with J48graft combined with sampling selection techniques for the diagnosis of type 2 diabetes mellitus in the Pima Indian. Inform. Med. Unlocked **2**, 92–104 (2016)
8. Marcano-Cedeño, A., Torres, J., Andina, D.: A prediction model to diabetes using artificial Metaplasticity. In: International Work-Conference on the Interplay Between Natural and Artificial Computation, pp. 418–425. Springer, Heidelberg (2011)
9. Patil, B.M., Joshi, R.C., Toshniwal, D.: Hybrid prediction model for type-2 diabetic patients. Expert Syst. Appl. **37**(12), 8102–8108 (2010)
10. Kaur, G., Chhabra, A.: Improved J48 classification algorithm for the prediction of diabetes. Int. J. Comput. Appl. **98**(22), 13–17 (2014)
11. Jayalakshmi, T., Santhakumaran, A.: Statistical normalization and back propagation for classification. Int. J. Comput. Theory Eng. **3**(1), 1793–8201 (2011)
12. Kaur, H., Wasan, S.K.: Empirical study on applications of data mining techniques in healthcare. J. Comput. Sci. **2**(2), 194–200 (2006)
13. Fatima, M., Pasha, M.: Survey of machine learning algorithms for disease diagnostic. J. Intell. Learn. Syst. Appl. **9**, 1–16 (2017)
14. Jonathan, D.W., Harold, R.G.: Data-mining analysis suggests an epigenetic pathogenesis for type 2 diabetes. J. Biomed. Biotechnol. **2005**(2), 104–112 (2005)

15. Sudharsan, B., Peeples, M., Shomali, M.: Hypoglycaemia prediction using machine learning models for patients with type 2 diabetes. J. Diabetes Sci. Technol. **9**(1), 86–90 (2014)
16. Marinov, M., Mosa, A.S.M., Yoo, I., Boren, S.A.: Data-mining technologies for diabetes: a systematic review. J. Diabetes Sci. Technol. **5**(6), 1549–1556 (2011)
17. Razavian, N., Blecker, S., Schmidt, A.M., Smith-McLallen, A., Nigam, S., Sontag, D.: Population-level prediction of type 2 diabetes from claims data and analysis of risk factors. Big Data **3**(4), 277–287 (2015)
18. Rajesh, K., Sangeetha, V.: Application of data mining methods and techniques for diabetes diagnosis. Int. J. Eng. Innov. Technol. (IJEIT) **2**(3), 224–229 (2012)

Discovering Patterns Using Feature Selection Techniques and Correlation

Mausumi Goswami[1,2(✉)] and B. S. Purkayastha[2]

[1] CHRIST (Deemed to be University), Bangalore 560074, India
mausumi.goswami@christuniversity.in
[2] Assam University, Silchar, Assam, India

Abstract. Term Frequency and inverse document frequency is reported to have a significant contribution for various text categorization, document clustering and many other text mining related tasks. A collection of the applications and the enhancements of the Term Frequency and Inverse Document Frequency based document representation technique is examined in this work. The document representation algorithm is essential in the field of text - script mining. In this algorithm, unstructured data is converted into a vector space model where each related document is considered as a point in the vector space. Related documents come in proximity to the other related documents while the documents that are very far away from being coherent remain different from each other. In this paper, four feature selection techniques are implemented to discover the patterns from a repository of unstructured data by using correlation similarity measure. Analysis and comparison with other existing technique is also included. The validation of the patterns formed is performed by using silhouette values. Experiments are conducted to compare performance. Results indicate that TDM_{p1} performance is poor compared to others.

Keywords: Correlation similarity · TF-IDF · Augmented frequency · Applications · Feature selection

1 Introduction

Term Frequency and inverse document frequency is reported to have a significant contribution for various text categorization, document clustering and many other text mining related tasks. In this work an attempt is made to have a cumulative collection of the applications and the enhancements of the Term Frequency and Inverse Document Frequency based document representation technique.

Organization of the paper is as laid in the next section. Section 2 describes few applications of feature selection techniques. Recent trends of the algorithm is inspected to find relevant applications in industry and research related to different aspects of society. Section 3 discusses few important feature selection models and proximity measures. Section 4 discusses a methodology for implementation. In Sect. 5 the Experiments and Results are documented.

© Springer Nature Switzerland AG 2020
J. S. Raj et al. (Eds.): ICIDCA 2019, LNDECT 46, pp. 824–831, 2020.
https://doi.org/10.1007/978-3-030-38040-3_94

2   Applications of Feature Selection in Text Mining

Text mining [1–8, 10–20] is mainly used for responding various engineering and business oriented inquiries and thereby enhancing the day-to-day working adeptness. Feature Selection is a key task in the area of text mining and many problems solving using machine learning. Unsupervised learning is the task of bunching or gathering the set of unlabeled text documents. During the task of bunching, one has to ensure that the text in the same group should be similar to each other. In case of supervised learning, one has to examine the text thoroughly before grouping them into certain categories.

Most of the hospitals in and around India have specific question forums, where people can raise questions that are pertained to certain medical condition that the patients are facing. But the system can seem extremely monotonous when the issue raised is the same and the reply still remains the same, hence making the entire system extremely redundant and pale. Hence, it would be better if we can extract the keywords from the mentioned condition and the give the response accordingly. In literature [29] cosine similarity over the TF-IDF (Terms frequency and inverse document frequency) and perform the classification (categorization). The authors have used a dataset of 95 text documents. This has a collection of unique words counted as 1454. These statistics refers to the step before stop words removal. After stops words removal the count of unique words are 1171. The Model VSM has 1171 features.

With the evolution of the big internet the availability and generation of data has been on an all time high and hence the categorization of the same is extremely important to structure the entire block of data. Categorization of data can help us in designing search engines, spam filters, information filtering etc. Text categorization is the structuring based on the statistical approach and thereby structuring it into a vector form. Once this is performed it is then passed through the TF-IDF [30] along with wither the cosine similarity or even the Euclidean similarity to find the similarity measure between the groups. The dataset consists of 4053 tech news, 3533 non-tech news and 462 tech data which is 3-Dimensional data.

Usage of TF-IDF is reported in analyzing the captions of images [31]. The images are that of most popular stars. Top 20 are considered based on popularity or most followed. Their captions are analyzed and then the keywords are ranked based on the weights obtained from the TF-IDF weights which is a direct characterization of the frequency of the words in the caption. The study was first open to all forms of media which were then narrowed down to only text data as it can easily be analyzed unlike pictures and other forms of data. These are the analyzed with the TF-IDF model and then their ranks are found out which in turn categorizes the words and then helps us in finding out the keywords. ON research the words found out were 'weekend', 'hashtag' etc. The dataset was taken from the captions of the posts of the top 20 most followed stars on Instagram and then trained accordingly.

A major application is to categorize the words in the document into certain bags of words and then give the weight accordingly to the words. The block words or the possible conjunctions that are part of the structure are removed so that they do not have a binding effect on the occurrence of the main words. Following this the words are weighted based on the frequency of the words. The dataset considered in [32] was the FIRE dataset which is basically the corpus of the set of many newspapers.

In the present world that is filled with data the appropriate usage of data for data acquisition, spam filtering and human – emotions analysis is quite essential. The pivotal point among these are the identification of the characteristic of the text. The customers feedback or comments were analyzed in [33] and then clustered into spam, emotional words etc. that are bagged into certain groups with relevant data. Hence, this helps in creating a smarter set that could help the system to understand the emotions of the customers. The data retrieved was from the 4288 buyers on the Taboo shopping site and 23007 words were obtained from the site and then analyzed.

Few other applications reported are to catch the matches between tiny posts (Hashtags) used in social networking, a more practical update on the news cast words [20], damp-heat syndrome categorization applying TF-IDF [21], Ways and means of detection of patterns [22], reduction of feature vector [23], probe for multi word text [24], techniques for resolution of the cons connected with TF-IDF oriented methods [25, 26], Ranking of best Instagram user's image caption [27], usage on farmer's call center dataset. A fuzzy based approach is also reported in [35]. In next section, few models used for document representation will be discussed.

3   Few Models of Feature Selection and Correlation Similarity Measures in Document Clustering

Few feature selection models used in this work are term frequency based TDM, Term frequency and inverse document frequency, Logarithmic frquency, Augmented frequency. The equations are given below:

Term frequency–inverse document frequency, logarithmic and augmented models are depicted in the equations given below:

$$\mathrm{tf}(t, d, D) = \text{frequency of term } t \text{ in document } D \tag{0}$$

$$\mathrm{Tfidf}(t, d, D) = \mathrm{tf}(t, d) \cdot \mathrm{idf}(t, D) \tag{1}$$

$$\mathrm{idf}(d, t) = \log[(1 + n) / (1 + \mathrm{df}(d, t))] + 1. \tag{2}$$

$$\mathrm{tf}(t, d) = \log(1 + \mathrm{ft,d}) \tag{3}$$

$$f(x) = 0.5 + \frac{f(t, d)}{\max\{f(t', d : t' \in d\}} \tag{4}$$

Text Clustering [5, 7–9, 11, 14, 16] also requires a similarity measure. In this work Correlation is being used to find the similarity between two text documents. Correlation value lies between +1 and −1. Maximum similarity is shown by +1 and minimum by −1. This may be converted to distance mesure by subtracting from 1. Also, distance pearson lies between 0 and 2. Mean is subtracted to make it centered also scaling is performed by dividing by the standard deviation.

4   Proposed Methodology

In this section the steps followed to apply feature selection model on real life data set is described.

Step 1: Preprocessing the data after reading it.
Step 2: Construction of Term Document Matrix using four different frequency
Step 2.1 Use Eq. (0) tf = frequency(t,d) to compute TDM_{p1}.
Step 2.2 Use Eq. (1) to compute $TF\text{-}IDF_{p1}$
Step 2.3 Use Eq. (3) to compute $LogF_{p1}$
Step 2.4 Use Eq. (4) to compute $AugF_{p1}$
Step 3: Apply correlation between each pair of documents.
Step 4: Apply Kmeans clustering Algorithm
Step 5: Validation of clustering results
Step 6: Analysis and tabulation of results

The steps mentioned above summarizes the approach. The algorithm used for clustering is shown in the above flow chart.

Table 1. Evelution outcomes obtained using silhouette with four frequencies and correlation

Correlation	TDM_{p1}.	$TF\text{-}IDF_{p1}$	$LogF_{p1}$	$AugF_{p1}$
min	−0.0184	0	0	0
max	0.665	1	1	1
mean	0.2008	0.4556	0.4516	0.4163
median	0.1182	0.3869	0.4287	0.413

5   Experiments and Results

Experiments are conducted using intel i3 processor 8 GB RAM machine with 1 TB hard disk. Python 2.7 is used for implementation. Kmeans clustering algorithm is used. Table 1 summarizes the usage of various feature selection techniques applied. The silhouette values obtained after clustering are tabulated. Each value is positive except for TDM_{p1}. Hence, TDM_{p1} is said to exhibit worst performance (Fig. 1).

Figure 2 is used to visualize the performance using a column chart (Fig. 3).

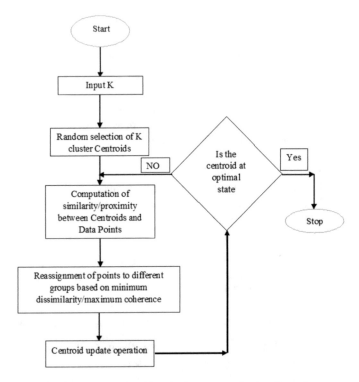

Fig. 1. Kmeans algorithm is demonstrated using flow chart

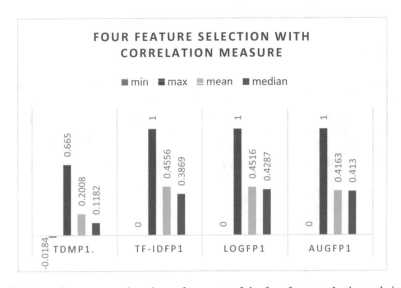

Fig. 2. Above figure summarizes the performances of the four feature selection techniques.

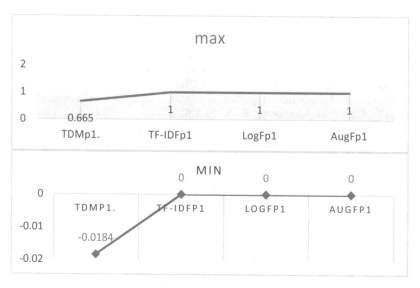

Fig. 3. This figure summarizes the minimum and maximum values obtained after validation of the clusters using silhouette coefficient.

6   Conclusion

Term Frequency and inverse document frequency is reported to have a significant contribution for various text categorization, document clustering and many other text mining related tasks. In this work an attempt is made to have a cumulative collection of the applications and the enhancements of the Term Frequency and Inverse Document Frequency based document representation techniqu. In this paper four different feature selection techniques are empirically evaluated using correlation proximity measure. Clustering is performed using Kmeans. The number of clusters are also varied to understand how cluster quality varies with respect to the choice of a feature selection technique as mentioned in the paper. In this empirical evaluation, TF-IDFp1, LogFp1, AugFp1 have shown consistant behaviour for correlation similarity measure. 1st raw in Table 1 indicates TDM Features selection performs worse than others. Minimum Silhouette Value of TDM is less than 0, the minimum of all others. Maximum Silhouette Value of TDM is less than 1, the maximum of all others. This reflect very poor cohesion and high coupling. In case of correlation similarity measure, Mean of silhouette of TDM is less than mean Mean of silhouette of all other measures. Future work is going to focus on three more similarity measures to realize the relative merits and demerits. An attempt will be made to propose a novel approach for representing documents. Also, work has been carried out across different similarity measures with an objective to identify best similarity measure which gel very well with various frequencies. Future work will include few such comparisons as well.

References

1. Wajeed, M.A., Adilakshmi, T.: Different similarity measures for text classification using KNN. In: 2011 2nd International Conference on Computer and Communication Technology (ICCCT-2011), pp. 41–45. IEEE, September 2011
2. Rousseeuw, P.J.: Silhouettes: a graphical aid to the interpretation and validation of cluster analysis. J. Comput. Appl. Math. **20**, 53–65 (1987)
3. Patidar, A.K., Agrawal, J., Mishra, N.: Analysis of different similarity measure functions and their impacts on shared nearest neighbor clustering approach. Int. J. Comput. Appl. **40**(16), 1–5 (2012)
4. Singh, P., Sharma, M.: Text Document Clustering and Similarity Measures. Dept. of Computer Science & Engg., November 2013
5. Sindhiya, B., Tajunisha, N.: Concept and term based similarity measure for text classification and clustering. Int. J. Eng. Res. Sci. Tech. **2014**(3) (2014)
6. Goswami, M., Babu, A., Purkayastha, B.S.: A comparative analysis of similarity measures to find coherent documents. Appl. Sci. Manag. (2018)
7. Gopala Rao, V., Bhanu Prasad, S.: Space and cosine similarity measures for text document clustering. Int. J. Eng. Res. Technol. **2**(2), 2278–0181 (2013)
8. de Amorim, R.C., Hennig, C.: Recovering the number of clusters in data sets with noise features using feature rescaling factors. Inf. Sci. **324**, 126–145 (2015)
9. Chim, H., Deng, X.: Efficient phrase-based document similarity for clustering. IEEE Trans. Knowl. Data Eng. **20**(9), 1217–1229 (2008)
10. Ester, M., Kriegel, H.P., Sander, J., Xu, X.: A density-based algorithm for discovering clusters in large spatial databases with noise. In: KDD, vol. 96, no. 34, pp. 226–231, August 1996
11. Kalaivendhan, K., Sumathi, P.: An efficient clustering method to find similarity between the documents. Int. J. Innov. Res. Comput. Commun. Eng. **1** (2014)
12. Han, J., Kamber, M., Pei, J.: Data Mining: Concepts and Techniques. Morgan Kauffman, San Francisco (2006)
13. Kaufman, L., Rousseeuw, P.J.: Finding Groups in Data: An Introduction to Cluster Analysis, vol. 344. John Wiley, New York (2009)
14. Lin, Y.S., Jiang, J.Y., Lee, S.J.: A similarity measure for text classification and clustering. IEEE Trans. Knowl. Data Eng. **26**(7), 1575–1590 (2013)
15. Sebastiani, F.: Machine learning in automated text categorization. ACM Comput. Surv. (CSUR) **34**(1), 1–47 (2002)
16. Sruthi, K., Venkateshwar Reddy, B.: Document clustering on various similarity measures. Int. J. Adv. Res. Comput. Sci. Software Eng. **3**, 1269–1273 (2013)
17. Steinbach, M., Karypis, G., Kumar, V.: A Comparison of Document Clustering Techniques. Dept. of Computer Science & Engg. (2000)
18. Tan, P.N., Steinbach, M., Kumar, V.: Introduction to Data Mining. Addision-Wesley, Boston (2006)
19. Tajbakhsh, M.S., Bagherzadeh, J.: Microblogging hash tag recommendation system based on semantic TF-IDF: Twitter use case. In: 2016 IEEE 4th International Conference on Future Internet of Things and Cloud Workshops (FiCloudW), pp. 252–257. IEEE, August 2016
20. Khusna, A.N., Agustina, I.: Implementation of information retrieval using TF-IDF weighting method on detik.Com's website. In: 2018 12th International Conference on Telecommunication Systems, Services, and Applications (TSSA), pp. 1–4. IEEE, October 2018
21. Zhu, W., Zhang, W., Li, G.Z., He, C., Zhang, L.: A study of damp-heat syndrome classification using Word2vec and TF-IDF. In: 2016 IEEE International Conference on Bioinformatics and Biomedicine (BIBM), pp. 1415–1420. IEEE, December 2016

22. Yu, N.: A visualized pattern discovery model for text mining based on TF-IDF weight method. In: 2018 10th International Conference on Intelligent Human-Machine Systems and Cybernetics (IHMSC), vol. 2, pp. 183–186. IEEE, August 2018
23. Elhadad, M.K., Badran, K.M., Salama, G.I.: A novel approach for ontology-based dimensionality reduction for web text document classification. Int. J. Software Innov. (IJSI) 5(4), 44–58 (2017)
24. Darwiyanto, E., Pratama, G.A., Widowati, S.: Multi words quran and hadith searching based on news using TF-IDF. In: 2016 4th International Conference on Information and Communication Technology (ICoICT), pp. 1–6. IEEE, May 2016
25. Li, Y., Shen, B.: Research on sentiment analysis of microblogging based on LSA and TF-IDF. In: 2017 3rd IEEE International Conference on Computer and Communications (ICCC), pp. 2584–2588. IEEE, December 2017
26. Liu, Q., Wang, J., Zhang, D., Yang, Y., Wang, N.: Text features extraction based on TF-IDF associating semantic. In: 2018 IEEE 4th International Conference on Computer and Communications (ICCC), pp. 2338–2343. IEEE, December 2018
27. Kuncoro, B.A., Iswanto, B.H.: TF-IDF method in ranking keywords of Instagram users' image captions. In: 2015 International Conference on Information Technology Systems and Innovation (ICITSI), pp. 1–5. IEEE, November 2015
28. Mohapatra, S.K., Upadhyay, A.: Using TF-IDF on Kisan call centre dataset for obtaining query answers. In: 2018 International Conference on Communication, Computing and Internet of Things (IC3IoT), pp. 479–482. IEEE, February 2018
29. Alodadi, M., Janeja, V.P.: Similarity in patient support forums using TF-IDF and cosine similarity metrics. In: 2015 International Conference on Healthcare Informatics, pp. 521–522. IEEE, October 2015
30. Pan, L., Tang, H., Zhou, L., Wang, L., Zhu, Q.: An identification method of news scientific intelligence based on TF-IDF. In: 2015 14th International Symposium on Distributed Computing and Applications for Business Engineering and Science (DCABES), pp. 501–504. IEEE, August 2015
31. Mishra, A., Vishwakarma, S.: Analysis of TF-IDF model and its variant for document retrieval. In: 2015 International Conference on Computational Intelligence and Communication Networks (CICN), pp. 772–776. IEEE, December 2015
32. Yang, Y.: Research and realization of internet public opinion analysis based on improved TF-IDF algorithm. In: 2017 16th International Symposium on Distributed Computing and Applications to Business, Engineering and Science (DCABES), pp. 80–83. IEEE, October 2017
33. Roul, R.K., Sahoo, J.K., Arora, K.: Modified TF-IDF term weighting strategies for text categorization. In: 2017 14th IEEE India Council International Conference (INDICON), pp. 1–6. IEEE, December 2017
34. Xu, R.: POS weighted TF-IDF algorithm and its application for an MOOC search engine. In: 2014 International Conference on Audio, Language and Image Processing, pp. 868–873. IEEE, July 2014
35. Husaini, M.A.S., Hussain, S.A.: Fuzzy based steganalysis pattern discovery for high accuracy. In: Bhatia, S., Tiwari, S., Mishra, K., Trivedi, M. (eds.) Advances in Computer Communication and Computational Sciences, pp. 59–71. Springer, Singapore (2019)
36. Cummins, M.R.: Nonhypothesis-driven research: data mining and knowledge discovery. In: Richesson, R., Andrews, J. (eds.) Clinical Research Informatics, pp. 341–356. Springer, Cham (2019)

Author Index

A

Achuthan, Geetha, 524
Agarwal, Harshit, 594
Agrawal, Shubham, 410
Ahuja, Komal R., 204
Aisha Banu, W., 45
Aishwariya, K. K., 663
Al Wadhahi, Firas, 524
Alli, P., 148
Amandeep, 725
Ambilpure, Sagar, 498
Andrews, J., 710
Anuradha, B., 404
Arokia Jesu Prabhu, L., 710
Arulanthu, Pramila, 188
Arya, Neeraj, 476
Asha, J., 432
Asif, Mohammad Ahmed, 524
Awari, Gaurav, 670

B

Babu, G. Satish, 229
Babu, S. P. K., 551
Belavadi, Bhaskar, 284
Bhat, Jnanesh, 67
Bhatia, Jitendra, 12
Bhatia, Sugandh, 765
Biswal, Pradyut Kumar, 793
Bondgulwar, Shivani, 670

C

Champla, Dharavath, 395
Chandeliya, Nalin, 94
Chandra, Balina Surya, 276
Chandrasekaran, Saravanan, 447

Chari, Prashanth, 94
Charniya, Nadir N., 204, 380
Chattopadhyay, Manisha, 27
Chavan, Aditya, 498
Chhapra, Uzair, 498

D

Dangarwala, Kruti J., 569
Daniel, Sanil K., 663
Darji, Sagar, 626
Das, Tanaya, 647
Deepali, Patil, 543
Deshpande, Sachin, 760
Deshprabhu, Sumedh, 57
Devadkar, Kailas, 67
Devaraju, Gantakora, 276
Devasia, Aleena, 27
Devendar Rao, B., 297
Dhamodaran, S., 306
Dhanalaxmi, B., 332
Dhawale, Apurva D., 654
Diviya Prabha, V., 702

E

Elluru, Jayapal, 718

F

Felix Enigo, V. S., 109

G

Gawde, Varnesh, 498
Geetha, A., 776
Gharat, Chinmay, 388
Ghosh, S. M., 353
Gondhi, Naveen, 266

© Springer Nature Switzerland AG 2020
J. S. Raj et al. (Eds.): ICIDCA 2019, LNDECT 46, pp. 833–836, 2020.
https://doi.org/10.1007/978-3-030-38040-3

Gopan, Neethu Radha, 156
Gopnarayan, Archana, 760
Goswami, Mausumi, 824
Goswami, Vasudha, 608
Gowda Nishanth, R., 800
Gulghane, Shital, 670
Gunashekar, K., 800
Gupta, Daya Sagar, 1
Gupta, Neetesh Kumar, 461

H
Hafizul Islam, S K, 1
Hansdah, Baha, 87
Hendre, Ankit, 380
Hiran, Dilendra, 569
Hiremath, Shivarajkumar, 237

I
Imran, Mohammed, 342
Indraneel, S., 165

J
Jacob Vetha Raj, Y., 505
JacobVetha Raj, Y., 776
Jain, Sakshi, 476
Jakheliya, Bhumil, 626
Jariwala, Gaurav, 594
Jayachandran, A., 710
Jayanth, P. H., 800
Jegadish Kumar, K. J., 600
Joshi, Abhijit, 626

K
Kalban, Ismail Al, 524
Kalbande, D. R., 742
Kalbande, Dhananjay, 733
Kamoji, Supriya, 194
Kapuriya, B. R., 535
Karia, Deepak C., 57, 94
Karpe, Sameeran, 94
Karthikeyan, C., 297
Karthikeyan, M., 101, 118
Kavitha, G., 785
Khadar Basha, S. K., 37
Khot, Shubhangi, 222
Kiran Kumar, B., 165
Kishore, Ankit, 362
Koshti, Dipali, 194
Kothari, Raj, 626
Kour, Herleen, 266

Krishna, C. V., 134
Krishna Chaitanya Varma, Ch., 306
Krishnamurthy, Sharanya, 586
Krishnan, Shoba, 388
Kulakarni, Tejas R., 453
Kulkarni, Ramesh K., 213
Kulkarni, Sonali B., 654
Kulkarni, Tanay, 67
Kulothungan, K., 516
Kumar, Pradip, 362
Kumar, Sanjeev, 725
Kumar, Vikas, 370
Kumbhakarna, Vaishali, 654

L
Lal, Kavita, 616
Lokesh, Yangalasetty, 276

M
Madhuri, Chavan, 543
Mahadevan, G., 37
Mahendra Prashanth, K. V., 284
Majumdar, A. K., 647
Malhotra, Jyoteesh, 765
Malviya, Vijay, 608
Manikandan, J., 297
Manjrekar, Amrita A., 222
Manjunath, T. D., 319
Mathew, Sheena, 684
Melbin, K., 505
Menon, Shalini, 257
Mishra, Debasis, 87
Mishra, Nidhi, 410, 616
Misra, Sanjana, 87
Mohan Kumar, K. N., 342
Mohan, Bhaskaruni Gopesh Krishna, 276
Mokadam, Purnima, 67
Muchhal, Prateek, 370

N
Nagar, Aditi, 461
Nagarajan, K. K., 600
Narain, Pranay, 140
Narmadha, N., 637
Nayak, Jyothi S., 319
Nene, Manisha J., 312
Nene, Rajas, 140
Nikhil, D., 332
Ninan, Kezia Sera, 800
Nishant, Potnuru Sai, 276

O

Obaidat, Mohammad S., 1
Obiadat, Mohammad S., 12

P

Panda, Ribhu Abhusan, 87
Pandey, Raksha, 362
Panduranga Vittal, K., 440
Paramasivam, A., 785
Parashar, Manaswi, 561
Parikh, Satyen M., 76
Paul, Tomsy, 684
Pavan, S., 37
Pawar, Renuka, 742
Pawar, V. R., 676
Perumal, Eswaran, 188
Peter, David, 127
Peter, Ruth, 194
Poojashree, L. R., 284
Poonia, Amarjeet, 561
Pradhan, Debasish, 535
Prafulla, Nesar, 319
Prathibanandhi, 420
Preetha, K. G., 181
Priyanka, Shingane, 543
Prusty, Vedanta, 793
Purkayastha, B. S., 824
Putluri, Srinivasareddy, 750

R

Rachana Shree, N., 586
Radha, C., 586
Raghu, M. S., 284
Rajanandhini, C., 551
Rajendra Prasad, D., 165
Raju, Divya, 181
Ramanujam, E., 148
Ramesh, B., 440
Ramesh, S., 420
Ramkumar, J., 297
Rasikannan, L., 148
Rath, Abhisek, 793
Rathipriya, R., 637, 702
Ravi Kumar, B., 404
Reddy, Chittepu Dwarakanath, 306
Reddy, Dugimpudi Abhishek, 577
Reddy, K. Srinivasa, 332
Reena, P., 432
Rehman, Muhammad Hassan, 524
Revanth, Madamala, 276
Rishidas, S., 432
Rohit, H. R., 134

Rohra, Anjum, 213
Roja, M. Mani, 140
Rout, Kshirod Kumar, 793
Roy, Abhishek, 647
Ruby Annette, J., 45

S

Sagar, Parth, 670
Sahar, Saaima, 586
Saheb, Shaik Himam, 229
Sajja Priti, S., 247
Samarth, S., 319
Sampath, S., 342
Sanjeev Kunte, R., 237
SanthoshKumar, S., 432
Sathiamoorthy, S., 101, 118
Sathye, Rohit, 57
Satish, Kandukuru, 561
Sebastian, Anu Maria, 127
Senthil Kumar, A. V., 489
Shah Gargi, B., 247
Shah, Mitali K., 76
Shah, Vyom, 12
Shanbhag, Ruchita, 692
Shankarrao, More Hemlata, 676
Shanthi, P., 134
Sharma, Mohit Kumar, 312
Sharma, Pratyush, 608
Sharma, Reena, 535
Shingate, Vishal, 670
Shirly Edward, A., 172
Shobha, B. N., 37
Singh, Aman Deep, 809
Singh, Devendra Kumar, 577
Singh, Rohit Kumar, 87
Singh, Shani Pratap, 476
Singh, Upendra, 461
Sisodia, Jignesh, 733
Siva Kumar, Dhandapani, 395
Sivaranjani, R., 489
Sojitra, Raj, 12
Somalwar, Medha, 140
Sonya, A., 785
Sowmya Nag, K., 453
Sridhar, C. S., 37
Srikamakshi, M., 600
Srinivasa Gupta, N., 809
Sriram, Vaishali, 600
Subash Chandran, P., 45
Subramanian, Kirtana, 692
Sujatha Kumari, B. A., 586
Sujeesh, K. V., 663

Sumithra Sofia, D., 172
Suresh, M., 257
Surve, Mandar, 57

T
Tamgave, Pranoti Annaso, 222
Tanwar, Sudeep, 12
Thakkar, Arkesh, 12
Thanikasiselvan, V., 370
Thenmozhi, R., 516
Tolia, Chandrima, 692

V
Vachhani, Hrishikesh, 12
Valarmathi, B., 809
Varadarajan, S., 718

Varghese, Nimisha Raichel, 156
Vasudevan, M., 101, 118
Veeran, Rajkumar, 447
Velu, Karthika, 188
Venkatraman, Swarnita, 692
Verma, Charu Vaibhav, 353
Vishnusai, Y., 453

Y
Yaashuwanth, C., 420
Yadav, Deepak, 577
Yadav, Nishi, 577

Z
Zia Ur Rahman, Md., 750

Printed in the United States
By Bookmasters